Vital Statistics
of the United States
Births, Life Expectancy, Deaths, and Selected Health Data

First Edition, 2004

Vital Statistics
of the United States
Births, Life Expectancy, Deaths, and Selected Health Data

First Edition, 2004

Editors
Helmut F. Wendel
Christopher S. Wendel

Associate Editors
Katherine A. DeBrandt
Mary Meghan Ryan
Mark Siegal

BERNAN PRESS

ISBN: 0-89059-740-5

ISSN: 1549-8603

Printed by Automated Graphic Systems, Inc., White Plains, MD, on acid-free paper that meets the American National Standards Institute Z39-48 standard.

2005 2004 4 3 2 1

BERNAN PRESS
4611-F Assembly Drive
Lanham, MD 20706
800-274-4447
email: info@bernan.com
www.bernanpress.com

CONTENTS

ABOUT THE EDITORS

Helmut F. Wendel, Ph.D. was the lead editor of the second edition of *State Profiles: The Population and Economy of Each U.S. State*, also published by Bernan Press. Previously, he worked for many years as a monetary and macroeconomist at the Board of Governors of the Federal Reserve System, with one year on detail to the Congressional Budget Office. Dr. Wendel concluded his Federal Reserve career as Deputy Associate Director of Research. Since then he has taught economics at Howard University in Washington, D.C. and at Mohyla Academy in Kiev, Ukraine. He also served as a U.S. Treasury advisor at the Bulgarian National Bank.

Christopher S. Wendel, M.S. is an epidemiologist and biostatistician in the Health Services Research Center at the Southern Arizona VA Health Care System. His past experience includes research and statistical work at the National Research Council and the Arizona Cancer Center. He specializes in design and analysis of studies of chronic diseases that are of particular relevance to veterans, such as diabetes, cancer, substance abuse, and Post-Traumatic Stress Disorder.

ACKNOWLEDGEMENTS

Katherine DeBrandt performed expertly in guiding the collection of materials for this volume, in designing the tables and charts, and in coordinating the flow of the work. She was ably assisted by **Mary Meghan Ryan**, who specialized on the Health chapter, and **Mark Siegal**, who developed the final version of the Birth chapter. Bernan Press' editorial and production departments, under the direction of **Tamera Wells-Lee**, did the copy editing, layout, and graphics preparation. **Kara Prezocki** prepared the layout and graphics, assisted by **Christopher Jorgenson**. **Jacalyn Houston** served as the copy editor. Jacalyn, Kara, and Christopher capably handled all editorial and production aspects of this edition.

The editors are very grateful to each of these colleagues for their contributions, and for the friendly personal relations in carrying out this project.

The editors have great respect for the staff of the National Center for Health Statistics, a component of the CDC (Centers for Disease Control and Prevention). They have performed at a high standard in interpreting and classifying U.S. vital statistics, in putting them into downloadable publications, and explaining the highlights.

INTRODUCTION

For many years ending in 1993, the federal government published *Vital Statistics of the United States* in several thick bound volumes. These offered comprehensive data on births, deaths, and marriage and divorce. The volumes can be found in many libraries, but there is nothing comparable after 1993. However, the Centers for Disease Control and Prevention and its National Center for Health Statistics has continued compiling much of this information and has made it available electronically, in pamphlets, and in news releases. A summary of these data has been provided in the *Statistical Abstract of the United States*, compiled by the U.S. Department of Commerce. These circumstances gave Bernan Press the motivation to bring together a comprehensive collection of tables on births, deaths, and marriage and divorce and to publish them in a single volume on *Vital Statistics of the United States, 2004 Edition*. To supplement this information, numerous tables were added on what comes between birth and death, that is, health and disease. The Health chapter became the longest chapter in this book.

Each statistical table in this volume has descriptive or explanatory text and a number of tables are illustrated by charts. Sources and references are cited at the end of the book in a section called Notes and Definitions. Many of the tables provide historical background, but the latest information currently available is for 2002, and some of the tables do not go beyond 2001 or even earlier.

Current information on marriage and divorce has been condensed to a few tables. Beginning in 1996, the National Center for Health Statistics sharply reduced its data gathering on marriage and divorce, so that only rudimentary current statistics are now available. However, more extensive information on marital status based on large surveys from the mid-1990s has been included in this book.

In light of long-term trends of improved standard of living, disease risk factors, and medical care, we have learned to expect favorable news from our vital statistics. Indeed in 2002 life expectancy at birth rose 0.2 years to an all-time high of 77.4 years; age-adjusted death rates from heart disease declined 3.0 percent and those from cancer declined 1.0 percent. However, the age-adjusted fatality rate from motor vehicle accidents rose slightly.

It is of interest to view some of the U.S. vital statistics in the context of international experience. Table A-1, compiled by the U.S. Census Bureau, compares selected vital statistics among the regions and countries of the world. In particular, data are provided on life expectancy, birth and death rates in 2002, and the natural rate of population increase. Life expectancy for the United States exceeds that of most other countries, but it is somewhat shorter than the average for Western Europe, and distinctly shorter than Japan, which ranks first in longevity. The 2002 natural rate of population increase in the United States was 0.6 percent, as compared to 1.2 percent for Asia as a whole and -0.4 percent for Russia.

Chapter I, BIRTHS, provides information from final data for 2002 and earlier years on many aspects of natality, in particular, birth and fertility rates, the number of children, maternal health and other characteristics of mothers, medical services to pregnant women, and infant health status. **Chapter II, MORTALITY**, focuses on deaths and death rates according to a number of medical, demographic, and social characteristics. Especially detailed are the tables on causes of death, which are shown by age, sex, Hispanic origin, race, and (in briefer form) state of residence. Death rates are also given by marital status and educational attainment. Special tables are provided on infant mortality. Most of the tables were obtained from the National Center for Health Statistics' *Deaths: Final Data for 2001*, but nine tables came from a more recent, but less extensive, issue: *Deaths: Preliminary Data for 2002* (February 2004).

Chapter III, HEALTH, offers a large collection of statistics on health and disease in the United States. While health is not a traditional component of vital statistics, there are, of course, connections between data on health and death statistics. For example, a person looking at death rates from cancer may also be interested in a table on cancer incidence. The Health chapter has a wide range with several sections: Measures of Health; Selected Chronic Diseases; Disabilities,;Use of Addictive Substances; Ambulatory Care; and Inpatient Care. The data were obtained from several different surveys that are presented by the National Center for Health Statistics.

Chapter IV, MARRIAGE AND DIVORCE, presents the relatively limited data that are currently reported by states to the federal government, as well as an interesting collection of historical statistics and survey results.

NOTES AND DEFINITIONS, at the end of the book, provide information on source materials.

Table A-1. Population, Vital Events, and Rates, by Region and Country, 2002

(Population and events in thousands, except as noted. Figures may not add to totals because of rounding.)

Region and country or area	Population	Births	Deaths	Life expectancy at birth (years)		Births per 1,000 population	Deaths per 1,000 population	Annual rate of natural increase (percent)
				Male	Female			
WORLD	6 228 394	128 578	54 997	62	66	21	9	1.2
Less Developed Countries	5 029 539	115 120	42 794	61	64	23	9	1.4
More Developed Countries	1 198 856	13 458	12 203	73	80	11	10	0.1
AFRICA	838 720	30 133	12 097	49	51	36	14	2.2
Sub-Saharan Africa	686 522	26 538	11 285	46	48	39	16	2.2
Angola	10 554	486	272	36	38	46	26	2.0
Benin	6 835	299	93	50	52	44	14	3.0
Botswana	1 579	41	45	33	34	26	29	-0.2
Burkina Faso	12 887	581	242	43	46	45	19	2.6
Burundi	5 965	238	107	42	44	40	18	2.2
Cameroon	15 428	554	235	47	49	36	15	2.1
Cape Verde	409	11	3	66	73	28	7	2.1
Central African Republic	3 623	132	71	41	44	36	19	1.7
Chad	8 971	427	147	47	51	48	16	3.1
Comoros	614	24	6	59	63	39	9	3.0
Congo (Brazzaville)	2 908	88	41	50	51	30	14	1.6
Congo (Kinshasa)	55 042	2 509	832	47	51	46	15	3.0
Côte d'Ivoire	16 598	670	305	40	45	40	18	2.2
Djibouti	447	18	9	42	45	41	19	2.2
Equatorial Guinea	498	19	6	52	57	37	13	2.5
Eritrea	4 306	172	57	52	56	40	13	2.7
Ethiopia	65 254	2 634	1 304	41	42	40	20	2.0
Gabon	1 288	47	14	56	59	37	11	2.6
Gambia, The	1 456	60	18	52	56	41	13	2.9
Ghana	20 163	540	210	56	58	27	10	1.6
Guinea	8 816	377	140	48	51	43	16	2.7
Guinea-Bissau	1 333	52	22	45	49	39	17	2.2
Kenya	31 223	929	491	45	46	30	16	1.4
Lesotho	1 858	51	45	37	37	28	24	0.3
Liberia	3 262	150	58	47	50	46	18	2.8
Madagascar	16 473	699	200	53	58	42	12	3.0
Malawi	11 393	513	254	38	39	45	22	2.3
Mali	11 300	545	218	45	46	48	19	2.9
Mauritania	2 829	120	38	49	54	43	13	2.9
Mauritius	1 200	20	8	68	76	16	7	1.0
Mayotte	171	7	1	58	62	44	9	3.5
Mozambique	17 324	676	507	32	32	39	29	1.0
Namibia	1 897	66	33	46	44	35	18	1.7
Niger	10 760	540	236	42	42	50	22	2.8
Nigeria	130 500	5 120	1 771	51	52	39	14	2.6
Reunion	744	15	4	70	77	21	6	1.5
Rwanda	7 668	308	166	39	40	40	22	1.9
Saint Helena	7	*	*	74	80	13	6	0.7
Sao Tome and Principe	170	7	1	64	67	42	7	3.5
Senegal	10 311	379	114	55	58	37	11	2.6
Seychelles	80	1	1	65	77	17	7	1.1
Sierra Leone	5 565	247	115	40	46	44	21	2.4
Somalia	7 753	363	139	45	49	47	18	2.9
South Africa	42 716	827	708	49	49	19	17	0.3
Sudan	37 090	1 380	364	56	59	37	10	2.7
Swaziland	1 150	35	22	43	40	30	19	1.1
Tanzania	35 302	1 411	612	43	46	40	17	2.3
Togo	5 299	191	60	52	56	36	11	2.5
Uganda	24 889	1 167	431	43	46	47	17	3.0
Zambia	10 149	407	246	35	35	40	24	1.6
Zimbabwe	12 463	382	260	42	39	31	21	1.0
North Africa	152 199	3 595	813	68	73	24	5	1.8
Algeria	32 278	721	166	69	72	22	5	1.7
Egypt	73 313	1 823	396	68	73	25	5	1.9
Libya	5 369	148	19	74	78	28	3	2.4
Morocco	31 168	738	183	67	72	24	6	1.8
Tunisia	9 816	165	49	73	76	17	5	1.2
Western Sahara	256	11	4	. . .	. . .	44	15	2.9
NEAR EAST	178 574	4 905	1 071	67	71	27	6	2.1
Bahrain	656	13	3	71	76	20	4	1.6
Cyprus	767	10	6	75	80	13	8	0.5
Gaza Strip	1 226	51	5	70	73	42	4	3.8
Iraq	24 002	821	144	66	69	34	6	2.8
Israel	6 030	114	37	77	81	19	6	1.3
Jordan	5 307	130	14	75	80	25	3	2.2
Kuwait	2 112	46	5	76	77	22	2	1.9
Lebanon	3 678	73	23	69	74	20	6	1.4
Oman	2 713	102	11	70	75	38	4	3.4
Qatar	793	13	3	70	75	16	4	1.1
Saudi Arabia	23 513	876	138	67	70	37	6	3.1
Syria	17 156	517	88	68	70	30	5	2.5
Turkey	67 309	1 208	400	69	74	18	6	1.2
United Arab Emirates	2 446	45	10	72	77	18	4	1.4
West Bank	2 164	76	9	71	74	35	4	3.1
Yemen	18 701	810	174	59	62	43	9	3.4

* = Between -500 and +500 for events and between -0.05 percent and +0.05 percent for rates.
. . . = Not available.

Table A-1. Population, Vital Events, and Rates, by Region and Country, 2002—*Continued*

(Population and events in thousands, except as noted. Figures may not add to totals because of rounding.)

Region and country or area	Population	Births	Deaths	Life expectancy at birth (years)		Births per 1,000 population	Deaths per 1,000 population	Annual rate of natural increase (percent)
				Male	Female			
ASIA	3 517 862	68 104	26 671	65	68	19	8	1.2
Afghanistan	27 756	1 139	484	47	46	41	17	2.4
Bangladesh	135 657	4 025	1 186	61	61	30	9	2.1
Bhutan	2 094	74	29	54	53	35	14	2.2
Brunei	351	7	1	72	77	20	3	1.7
Burma	42 282	830	515	54	57	20	12	0.7
Cambodia	12 890	354	122	55	60	27	9	1.8
China	1 309 380	17 155	8 741	70	74	13	7	0.6
China excl. Taiwan, Hong Kong S.A.R., and Macau S.A.R.	1 279 161	16 783	8 558	70	74	13	7	0.6
Hong Kong S.A.R.	7 303	80	45	77	83	11	6	0.5
Macau S.A.R.	462	6	2	79	85	12	4	0.8
Taiwan	22 454	287	137	74	80	13	6	0.7
East Timor	953	27	6	63	67	28	7	2.2
India	1 034 173	24 582	8 904	63	64	24	9	1.5
Indonesia	231 326	5 059	1 453	66	71	22	6	1.6
Iran	67 538	1 177	376	68	70	17	6	1.2
Japan	127 066	1 225	1 062	78	84	10	8	0.1
Korea, North	22 215	415	153	68	73	19	7	1.2
Korea, South	47 963	620	285	72	79	13	6	0.7
Laos	5 778	216	73	52	56	37	13	2.5
Malaysia	22 662	549	117	69	74	24	5	1.9
Maldives	320	12	3	62	64	37	8	3.0
Mongolia	2 674	57	19	61	66	21	7	1.4
Nepal	25 874	852	260	59	58	33	10	2.3
Pakistan	147 663	4 489	1 332	61	63	30	9	2.1
Philippines	82 995	2 223	471	66	72	27	6	2.1
Singapore	4 453	57	19	77	83	13	4	0.8
Sri Lanka	19 577	320	126	70	75	16	6	1.0
Thailand	63 645	1 062	432	69	73	17	7	1.0
Vietnam	80 577	1 579	501	67	72	20	6	1.3
LATIN AMERICA AND THE CARIBBEAN	538 680	11 282	3 256	68	74	21	6	1.5
Anguilla	12	*	*	74	80	15	6	0.9
Antigua and Barbuda	67	1	*	69	73	19	6	1.3
Argentina	38 331	679	291	71	79	18	8	1.0
Aruba	70	1	*	75	82	12	6	0.6
Bahamas, The	295	6	3	62	69	19	9	1.0
Barbados	276	4	2	70	74	13	9	0.4
Belize	260	8	2	65	69	31	6	2.5
Bolivia	8 445	223	68	62	67	26	8	1.8
Brazil	179 914	3 247	1 101	67	75	18	6	1.2
Cayman Islands	41	1	*	77	82	14	5	0.9
Chile	15 499	255	87	73	80	16	6	1.1
Colombia	41 008	902	232	67	75	22	6	1.6
Costa Rica	3 835	76	17	74	79	20	4	1.6
Cuba	11 224	136	82	74	79	12	7	0.5
Dominica	70	1	*	71	77	17	7	1.0
Dominican Republic	8 596	209	57	67	70	24	7	1.8
Ecuador	13 447	343	72	69	75	25	5	2.0
El Salvador	6 354	180	39	67	74	28	6	2.2
French Guiana	182	4	1	73	80	22	5	1.7
Grenada	89	2	1	63	66	23	8	1.5
Guadeloupe	436	7	3	74	81	17	6	1.0
Guatemala	13 542	481	92	64	66	36	7	2.9
Guyana	700	13	6	61	67	18	9	0.9
Haiti	7 405	255	100	50	53	34	14	2.1
Honduras	6 514	210	41	66	69	32	6	2.6
Jamaica	2 680	48	15	74	78	18	5	1.2
Martinique	422	6	3	79	78	15	6	0.9
Mexico	103 400	2 312	516	69	75	22	5	1.7
Montserrat	8	*	*	76	80	18	7	1.0
Netherlands Antilles	214	3	1	73	77	16	6	1.0
Nicaragua	5 024	136	24	67	71	27	5	2.2
Panama	2 920	62	18	70	75	21	6	1.5
Paraguay	5 884	179	28	72	77	31	5	2.6
Peru	27 950	653	160	68	73	23	6	1.8
Puerto Rico	3 863	59	29	73	81	15	8	0.8
Saint Kitts and Nevis	39	1	*	68	74	19	9	1.0
Saint Lucia	160	3	1	69	77	21	5	1.6
Saint Vincent and the Grenadines	116	2	1	71	75	18	6	1.1
Suriname	434	9	3	67	72	20	7	1.3
Trinidad and Tobago	1 112	14	9	67	73	13	8	0.4
Turks and Caicos Islands	19	*	*	72	76	24	4	2.0
Uruguay	3 387	59	30	72	79	17	9	0.8
Venezuela	24 288	491	119	71	77	20	5	1.5
Virgin Islands	123	2	1	75	83	16	6	1.0
Virgin Islands, British	21	*	*	75	77	15	4	1.1
EUROPE AND THE NEW INDEPENDENT STATES	803 255	9 186	8 974	69	76	11	11	*
Western Europe	392 237	4 096	3 843	76	82	10	10	0.1
Andorra	68	1	*	81	87	10	6	0.4
Austria	8 170	78	79	75	81	10	10	*
Belgium	10 275	109	104	75	82	11	10	*
Denmark	5 369	63	58	74	80	12	11	0.1
Faroe Islands	46	1	*	75	82	14	9	0.5
Finland	5 184	55	51	74	82	11	10	0.1
France	59 925	763	542	75	83	13	9	0.4
Germany	82 351	723	844	75	81	9	10	-0.1
Gibraltar	28	*	*	76	82	11	9	0.2
Greece	10 645	105	104	76	81	10	10	*

* = Between -500 and +500 for events and between -0.05 percent and +0.05 percent for rates.

Table A-1. Population, Vital Events, and Rates, by Region and Country, 2002—*Continued*

(Population and events in thousands, except as noted. Figures may not add to totals because of rounding.)

Region and country or area	Population	Births	Deaths	Life expectancy at birth (years)		Births per 1,000 population	Deaths per 1,000 population	Annual rate of natural increase (percent)
				Male	Female			
Western Europe—*Continued*								
Guernsey	65	1	1	77	83	10	10	*
Iceland	279	4	2	77	82	14	7	0.7
Ireland	3 883	57	31	74	80	15	8	0.7
Italy	57 927	538	581	76	82	9	10	-0.1
Jersey	90	1	1	76	81	11	9	0.2
Liechtenstein	33	*	*	75	83	11	7	0.4
Luxembourg	449	5	4	74	81	12	9	0.3
Malta	397	5	3	76	81	13	8	0.5
Man, Isle of	74	1	1	74	81	11	12	*
Monaco	32	*	*	75	83	10	13	-0.3
Netherlands	16 068	186	139	76	82	12	9	0.3
Norway	4 525	56	44	76	82	12	10	0.3
Portugal	10 084	116	103	73	80	12	10	0.1
San Marino	28	*	*	78	85	11	8	0.3
Spain	40 153	403	378	76	83	10	9	0.1
Sweden	8 877	87	94	77	83	10	11	-0.1
Switzerland	7 302	72	64	77	83	10	9	0.1
United Kingdom	59 912	666	612	76	81	11	10	0.1
Eastern Europe	120 864	1 296	1 314	69	77	11	11	*
Albania	3 545	66	23	69	75	19	6	1.2
Bosnia and Herzegovina	3 964	51	32	69	75	13	8	0.5
Bulgaria	7 621	61	110	68	75	8	14	-0.6
Croatia	4 391	56	50	71	78	13	11	0.1
Czech Republic	10 257	93	110	71	79	9	11	-0.2
Hungary	10 075	94	132	68	77	9	13	-0.4
Macedonia, The Former Yugoslav Republic of	2 055	27	16	72	77	13	8	0.6
Poland	38 625	397	385	70	78	10	10	*
Romania	22 318	241	274	67	74	11	12	-0.1
Slovakia	5 422	55	50	70	78	10	9	0.1
Slovenia	1 933	18	19	71	79	9	10	-0.1
Yugoslavia	10 658	136	113	71	77	13	11	0.2
New Independent States	290 154	3 794	3 818	61	71	13	13	*
Baltics	7 383	69	100	64	76	9	14	-0.4
Estonia	1 416	13	19	64	76	9	13	-0.4
Latvia	2 367	20	35	63	75	8	15	-0.6
Lithuania	3 601	37	46	64	76	10	13	-0.3
Commonwealth of Independent States	282 770	3 724	3 717	61	70	13	13	*
Armenia	3 330	40	33	62	71	12	10	0.2
Azerbaijan	7 798	147	75	59	68	19	10	0.9
Belarus	10 335	102	145	62	75	10	14	-0.4
Georgia	4 961	57	72	61	68	11	15	-0.3
Kazakhstan	16 742	299	179	58	69	18	11	0.7
Kyrgyzstan	4 822	126	44	59	68	26	9	1.7
Moldova	4 435	61	56	60	69	14	13	0.1
Russia	144 979	1 408	2 017	62	73	10	14	-0.4
Tajikistan	6 720	222	57	61	67	33	9	2.4
Turkmenistan	4 689	133	42	58	65	28	9	1.9
Ukraine	48 396	464	794	61	72	10	16	-0.7
Uzbekistan	25 563	667	204	60	68	26	8	1.8
NORTH AMERICA	319 705	4 432	2 701	74	80	14	8	0.5
Bermuda	64	1	*	75	79	12	7	0.5
Canada	31 902	354	241	76	83	11	8	0.4
Greenland	56	1	*	65	72	16	8	0.9
Saint Pierre and Miquelon	7	*	*	76	80	15	7	0.8
United States	287 676	4 076	2 460	74	80	14	9	0.6
OCEANIA	31 598	536	227	71	76	17	7	1.0
American Samoa	69	2	*	71	80	24	4	2.0
Australia	19 547	248	142	77	83	13	7	0.5
Cook Islands	21	*	*	*	*	22	5	1.7
Fiji	856	20	5	66	71	23	6	1.7
French Polynesia	258	5	1	73	78	18	4	1.4
Guam	161	4	1	76	81	24	4	2.0
Kiribati	96	3	1	58	64	32	9	2.3
Marshall Islands	55	2	*	67	71	34	5	2.9
Micronesia, Federated States of	136	4	1	. . .	. . .	27	6	2.1
Nauru	12	*	*	58	65	27	7	2.0
New Caledonia	208	4	1	70	76	20	6	1.4
New Zealand	3 908	56	30	75	81	14	8	0.7
Northern Mariana Islands	77	2	*	73	79	20	2	1.8
Palau	19	*	*	66	73	19	7	1.2
Papua New Guinea	5 172	163	40	62	66	32	8	2.4
Samoa	179	3	1	67	73	16	6	0.9
Solomon Islands	495	16	2	69	74	33	4	2.9
Tonga	106	3	1	66	71	24	6	1.8
Tuvalu	11	*	*	65	69	21	7	1.4
Vanuatu	196	5	2	60	63	25	8	1.7
Wallis and Futuna	16	*	*	. . .	. . .	20	4	1.6

Source: U.S.Census Bureau, International Population Reports WP/02, Global Population Profile: 2002.

* = Between -500 and +500 for events and between -0.05 percent and +0.05 percent for rates.

. . . = Not available

CHAPTER I: BIRTHS

This chapter presents 2002 data on U.S. births that were derived by the National Center for Health Statistics (NCHS) from an examination of all birth certificates. Many aspects of pregnancy and birth are covered and some will be listed here. Starting with the number of births in 2002 and previous history, the chapter branches out to fertility rates and birth rates by age, by race, by ethnicity and in some cases by country of origin. Several characteristics of mothers are shown, such as smoking status, educational attainment, marital status, and prenatal care.

Detail is presented on birth weight, the duration of gestation, the accompanying weight gain, and on medical complications. Recent and historical data are shown on cesarean deliveries as well as vaginal births with previous cesareans. Recent characteristics of the birth of twins and of other multiple births are also shown.

TABLE I-1. LIVE BIRTHS, BIRTH RATES, AND FERTILITY RATES, BY RACE, SELECTED YEARS, 1940–2002

Table 1 provides data for every 5 years from 1940 to 1960 and every year thereafter. Prior to 1980 some data on race is incomplete and prior to 1985 births in certain states were estimated from 50 percent samples. Beginning in 1980, births were tabulated by race of the mother, rather than race of the child (as determined by race of parents on birth certificate). This change provided a more uniform approach in response to the increasing incidence of interracial parentage and the large proportion of records with father's race not reported.

Births totaled 4,021,726 in 2002, down by approximately 4,000 from 2001. The number of births increased 5 percent between 1997 and 2000, but has declined slightly since. Whites, including White Hispanics, accounted for 79 percent of total births. Blacks (including Black Hispanics) accounted for nearly 15 percent. Asians and Pacific Islanders contributed about 5 percent of births, and American Indians added 1 percent. Births to Asians and Pacific Islanders and American Indians rose between 2001 and 2002, by 5 percent and 1 percent, respectively. Births to Blacks fell by 2 percent, while births to Whites remained essentially unchanged.

The crude birth rate was 13.9 per 1,000 of the total population, men as well as women; this is the lowest birth rate since national data became available in 1909. Blacks experienced the largest drop in the crude birth rate at 4 percent between 2001 and 2002. The crude birth rate has fallen 17 percent since 1990.

The general fertility rate was 64.8, representing a 1 percent drop from 2001 and a 9 percent drop from 1990. This rate measures the number of births per 1,000 women of

Figure I-1. Live Births and Fertility Rates, 1930–2002

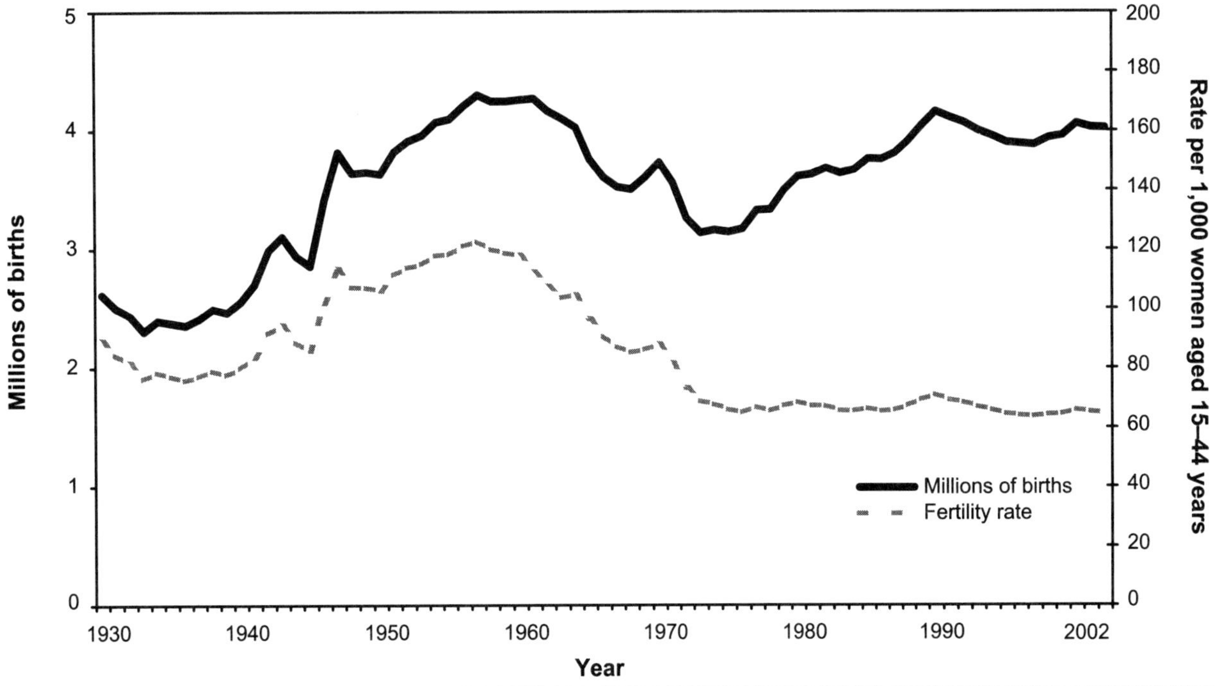

childbearing age, here defined as those aged from 15 to 44 years. Fertility rates for Whites and for Asians were close to the U.S. average. Fertility rates for Blacks have shown a pronounced decline in the last two decades, but they are still somewhat higher than the national average. The American Indian fertility rate is below average and has fallen by over 30 percent since the early 1980s.

Table I-1. Live Births, Birth Rates, and Fertility Rates, by Race, Selected Years, 1940–2002

(Birth rates are live births per 1,000 population in specified group.)

Year and series definition	Number					Birth rate					Fertility rate [1]				
	All races [2]	White	Black	American Indian [3]	Asian or Pacific Islander	All races [2]	White	Black	American Indian [3]	Asian or Pacific Islander	All races [2]	White	Black	American Indian [3]	Asian or Pacific Islander
Registered Births, Race of Mother															
2002	4 021 726	3 174 760	593 691	42 368	210 907	13.9	13.5	15.7	13.8	16.5	64.8	64.8	65.8	58.0	64.1
2001	4 025 933	3 177 626	606 156	41 872	200 279	14.1	13.7	16.3	13.7	16.4	65.3	65.0	67.6	58.1	64.2
2000	4 058 814	3 194 005	622 598	41 668	200 543	14.4	13.9	17.0	14.0	17.1	65.9	65.3	70.0	58.7	65.8
1999	3 959 417	3 132 501	605 970	40 170	180 776	14.2	13.7	16.8	14.2	15.9	64.4	64.0	68.5	59.0	60.9
1998	3 941 553	3 118 727	609 902	40 272	172 652	14.3	13.8	17.1	14.8	15.9	64.3	63.6	69.4	61.3	60.1
1997	3 880 894	3 072 640	599 913	38 572	169 769	14.2	13.7	17.1	14.7	16.2	63.6	62.8	69.0	60.8	61.3
1996	3 891 494	3 093 057	594 781	37 880	165 776	14.4	13.9	17.3	14.9	16.5	64.1	63.3	69.2	61.8	62.3
1995	3 899 589	3 098 885	603 139	37 278	160 287	14.6	14.1	17.8	15.3	16.7	64.6	63.6	71.0	63.0	62.6
1994	3 952 767	3 121 004	636 391	37 740	157 632	15.0	14.3	19.1	16.0	17.1	65.9	64.2	75.9	65.8	63.9
1993	4 000 240	3 149 833	658 875	38 732	152 800	15.4	14.6	20.2	17.0	17.3	67.0	64.9	79.6	69.7	64.3
1992	4 065 014	3 201 678	673 633	39 453	150 250	15.8	15.0	21.1	17.9	17.9	68.4	66.1	82.4	73.1	66.1
1991	4 110 907	3 241 273	682 602	38 841	145 372	16.2	15.3	21.8	18.3	18.3	69.3	66.7	84.8	73.9	67.1
1990	4 158 212	3 290 273	684 336	39 051	141 635	16.7	15.8	22.4	18.9	19.0	70.9	68.3	86.8	76.2	69.6
1989	4 040 958	3 192 355	673 124	39 478	133 075	16.4	15.4	22.3	19.7	18.7	69.2	66.4	86.2	79.0	68.2
1988	3 909 510	3 102 083	638 562	37 088	129 035	16.0	15.0	21.5	19.3	19.2	67.3	64.5	82.6	76.8	70.2
1987	3 809 394	3 043 828	611 173	35 322	116 560	15.7	14.9	20.8	19.1	18.4	65.8	63.3	80.1	75.6	67.1
1986	3 756 547	3 019 175	592 910	34 169	107 797	15.6	14.8	20.5	19.2	18.0	65.4	63.1	78.9	75.9	66.0
1985	3 760 561	3 037 913	581 824	34 037	104 606	15.8	15.0	20.4	19.8	18.7	66.3	64.1	78.8	78.6	68.4
1984 [4]	3 669 141	2 967 100	568 138	33 256	98 926	15.6	14.8	20.1	20.1	18.8	65.5	63.2	78.2	79.8	69.2
1983 [4]	3 638 933	2 946 468	562 624	32 881	95 713	15.6	14.8	20.2	20.6	19.5	65.7	63.4	78.7	81.8	71.7
1982 [4]	3 680 537	2 984 817	568 506	32 436	93 193	15.9	15.1	20.7	21.1	20.3	67.3	64.8	80.9	83.6	74.8
1981 [4]	3 629 238	2 947 679	564 955	29 688	84 553	15.8	15.0	20.8	20.0	20.1	67.3	64.8	82.0	79.6	73.7
1980 [4]	3 612 258	2 936 351	568 080	29 389	74 355	15.9	15.1	21.3	20.7	19.9	68.4	65.6	84.7	82.7	73.2
Registered Births, Race of Child															
1980 [4]	3 612 258	2 898 732	589 616	36 797	. . .	15.9	14.9	22.1	. . .	. . .	68.4	64.7	88.1	. . .	. . .
1979 [4]	3 494 398	2 808 420	577 855	34 269	. . .	15.6	14.5	22.0	. . .	. . .	67.2	63.4	88.3	. . .	. . .
1978 [4]	3 333 279	2 681 116	551 540	33 160	. . .	15.0	14.0	21.3	. . .	. . .	65.5	61.7	86.7	. . .	. . .
1977 [4]	3 326 632	2 691 070	544 221	30 500	. . .	15.1	14.1	21.4	. . .	. . .	66.8	63.2	88.1	. . .	. . .
1976 [4]	3 167 788	2 567 614	514 479	29 009	. . .	14.6	13.6	20.5	. . .	. . .	65.0	61.5	85.8	. . .	. . .
1975 [4]	3 144 198	2 551 996	511 581	27 546	. . .	14.6	13.6	20.7	. . .	. . .	66.0	62.5	87.9	. . .	. . .
1974 [4]	3 159 958	2 575 792	507 162	26 631	. . .	14.8	13.9	20.8	. . .	. . .	67.8	64.2	89.7	. . .	. . .
1973 [4]	3 136 965	2 551 030	512 597	26 464	. . .	14.8	13.8	21.4	. . .	. . .	68.8	64.9	93.6	. . .	. . .
1972 [4]	3 258 411	2 655 558	531 329	27 368	. . .	15.6	14.5	22.5	. . .	. . .	73.1	68.9	99.9	. . .	. . .
1971 [5]	3 555 970	2 919 746	564 960	27 148	. . .	17.2	16.1	24.4	. . .	. . .	81.6	77.3	109.7	. . .	. . .
1970 [5]	3 731 386	3 091 264	572 362	25 864	. . .	18.4	17.4	25.3	. . .	. . .	87.9	84.1	115.4	. . .	. . .
1969 [5]	3 600 206	2 993 614	543 132	24 008	. . .	17.9	16.9	24.4	. . .	. . .	86.1	82.2	112.1	. . .	. . .
1968 [5]	3 501 564	2 912 224	531 152	24 156	. . .	17.6	16.6	24.2	. . .	. . .	85.2	81.3	112.7	. . .	. . .
1967 [6]	3 520 959	2 922 502	543 976	22 665	. . .	17.8	16.8	25.1	. . .	. . .	87.2	82.8	118.5	. . .	. . .
1966 [5]	3 606 274	2 993 230	558 244	23 014	. . .	18.4	17.4	26.2	. . .	. . .	90.8	86.2	124.7	. . .	. . .
1965 [5]	3 760 358	3 123 860	581 126	24 066	. . .	19.4	18.3	27.7	. . .	. . .	96.3	91.3	133.2	. . .	. . .
1964 [5]	4 027 490	3 369 160	607 556	24 382	. . .	21.1	20.0	29.5	. . .	. . .	104.7	99.8	142.6	. . .	. . .
1963 [5,7]	4 098 020	3 326 344	580 658	22 358	. . .	21.7	20.7	108.3	. . .	. . .	103.6	. . .	. . .	. . .	. . .
1962 [5,7]	4 167 362	3 394 068	584 610	21 968	. . .	22.4	21.4	112.0	. . .	. . .	107.5	. . .	. . .	. . .	. . .
1961 [5]	4 268 326	3 600 864	611 072	21 464	. . .	23.3	22.2	117.1	. . .	. . .	112.3	. . .	. . .	. . .	. . .
1960 [5]	4 257 850	3 600 744	602 264	21 114	. . .	23.7	22.7	31.9	. . .	. . .	118.0	113.2	153.5	. . .	. . .
Births Adjusted for Underregistration, Race of Child															
1955	4 097 000	3 485 000	. . .	. . .	. . .	25.0	23.8	. . .	. . .	. . .	118.3	113.7	. . .	. . .	. . .
1950	3 632 000	3 108 000	. . .	. . .	. . .	24.1	23.0	. . .	. . .	. . .	106.2	102.3	. . .	. . .	. . .
1945	2 858 000	2 471 000	. . .	. . .	. . .	20.4	19.7	. . .	. . .	. . .	85.9	83.4	. . .	. . .	. . .
1940	2 559 000	2 199 000	. . .	. . .	. . .	19.4	18.6	. . .	. . .	. . .	79.9	77.1	. . .	. . .	. . .

[1]Fertility rates are live births per 1,000 women aged 15–44 years in specified group.
[2]For 1960–1991, includes births to races not shown separately. For 1992 and later years, unknown race of mother is imputed.
[3]Includes births to Aleuts and Eskimos.
[4]Based on 100 percent of births in selected states and on a 50-percent sample of births in all other states.
[5]Based on a 50-percent sample of births.
[6]Based on a 20- to 50-percent sample of births.
[7]Figures by race exclude New Jersey.
. . . = Not available.

TABLE I-2. LIVE BIRTHS BY AGE OF MOTHER, LIVE-BIRTH ORDER, AND RACE OF MOTHER, 2002

This table classifies the number of births in 2002 by age group of mother. In addition it shows the birth order, that is, whether the newborn was the mother's first child, second, or third child, etc. As expected, the percentages of first borns tend to decrease as the age of the mother increases, while the percentages of higher birth orders increase. At age 35–39, for example, second children were the most frequent, followed by third children and first children. However, above age 45, the percentage of first born becomes prominent again. There were 263 births to women aged 50–54 years in 2002. Although the number of births to this group has increased 13 percent annually since 1997, it remains too small to compute age-specific birth rates in Table 3.

Table I-2. Live Births by Age of Mother, Live-Birth Order, and Race of Mother, 2002

Live-birth order and race of mother	All ages	Age of mother													
		Under 15 years	15–19 years						20–24 years	25–29 years	30–34 years	35–39 years	40–44 years	45–49 years	50–54 years
			Total	15 years	16 years	17 years	18 years	19 years							
All Races	4 021 726	7 315	425 493	18 703	43 147	76 881	118 651	168 111	1 022 106	1 060 391	951 219	453 927	95 788	5 224	263
1st child	1 594 921	7 149	336 061	17 909	39 736	66 132	93 489	118 795	472 976	378 647	276 110	102 180	20 433	1 302	63
2nd child	1 306 786	145	74 552	717	3 100	9 611	21 518	39 606	349 692	362 440	342 782	148 922	26 902	1 284	67
3rd child	675 270	2	12 118	40	184	881	3 015	7 998	142 240	199 106	195 416	105 468	19 991	884	45
4th child	264 263	1	1 470	2	13	56	289	1 110	41 554	76 926	80 687	51 192	11 818	584	31
5th child	95 922	-	183	-	1	9	36	137	10 126	26 541	30 295	22 007	6 410	333	27
6th child	38 759	-	24	1	-	2	4	17	2 283	9 264	12 909	10 468	3 593	211	7
7th child	17 481	-	5	-	-	-	1	4	506	3 230	5 817	5 469	2 268	178	8
8th child and over	18 102	-	2	-	-	-	-	2	203	1 661	4 897	6 906	4 002	418	13
Not stated	10 222	18	1 078	34	113	190	299	442	2 526	2 576	2 306	1 315	371	30	2
White	3 174 760	3 884	305 988	11 665	29 501	54 698	85 957	124 167	783 000	851 142	779 535	369 833	76 928	4 235	215
1st child	1 258 492	3 784	245 314	11 219	27 365	47 722	69 011	89 997	373 535	308 459	225 965	83 646	16 638	1 097	54
2nd child	1 049 590	87	51 458	401	1 944	6 258	14 731	28 124	271 643	298 090	284 502	120 964	21 736	1 053	57
3rd child	536 533	2	7 502	22	105	532	1 805	5 038	102 307	159 468	162 924	87 595	15 984	709	42
4th child	202 693	-	816	2	11	34	165	604	26 693	57 619	65 415	42 174	9 484	464	28
5th child	69 349	-	98	-	1	5	23	69	5 563	17 705	23 117	17 508	5 080	259	19
6th child	26 680	-	7	-	-	2	-	5	1 105	5 432	9 128	8 041	2 797	166	4
7th child	11 700	-	4	-	-	-	1	3	215	1 678	3 861	4 039	1 759	141	3
8th child and over	12 028	-	1	-	-	-	-	1	104	759	2 834	4 837	3 162	324	7
Not stated	7 695	11	788	21	75	145	221	326	1 835	1 932	1 789	1 029	288	22	1
Black	593 691	3 188	103 795	6 336	12 142	19 411	28 266	37 640	194 704	136 591	95 006	48 388	11 443	562	14
1st child	222 832	3 129	78 374	6 016	10 978	16 046	21 011	24 323	75 810	33 608	20 854	8 973	1 987	96	1
2nd child	173 136	53	20 347	291	1 057	2 990	5 975	10 034	64 682	42 511	28 804	13 877	2 731	130	1
3rd child	105 565	-	4 155	17	71	322	1 083	2 662	34 835	31 262	21 719	11 005	2 486	103	-
4th child	49 306	1	585	-	2	19	114	450	13 305	15 911	11 400	6 396	1 631	75	2
5th child	21 689	-	76	-	-	2	11	63	4 104	7 481	5 630	3 386	961	45	6
6th child	9 910	-	17	1	-	-	4	12	1 072	3 283	3 021	1 877	606	33	1
7th child	4 696	-	1	-	-	-	-	1	267	1 335	1 588	1 103	375	26	1
8th child and over	4 768	-	1	-	-	-	-	1	88	773	1 650	1 589	614	51	2
Not stated	1 789	5	239	11	34	32	68	94	541	427	340	182	52	3	-
American Indian [1]	42 368	133	7 707	406	805	1 452	2 163	2 881	14 343	10 139	6 338	2 976	701	30	1
1st child	14 837	127	5 921	390	745	1 226	1 652	1 908	5 289	2 069	1 005	350	72	4	-
2nd child	11 784	4	1 482	16	53	208	436	769	5 216	2 887	1 518	563	113	1	-
3rd child	7 568	-	256	-	5	13	66	172	2 622	2 505	1 429	619	132	4	1
4th child	4 087	-	29	-	-	1	4	24	892	1 507	1 049	495	109	6	-
5th child	2 006	-	-	-	-	-	-	-	227	694	607	379	94	5	-
6th child	980	-	-	-	-	-	-	-	49	290	361	212	66	2	-
7th child	482	-	-	-	-	-	-	-	9	107	172	154	38	2	-
8th child and over	495	-	-	-	-	-	-	-	3	53	176	188	69	6	-
Not stated	129	2	19	-	2	4	5	8	36	27	21	16	8	-	-
Asian or Pacific Islander	210 907	110	8 003	296	699	1 320	2 265	3 423	30 059	62 519	70 340	32 730	6 716	397	33
1st child	98 760	109	6 452	284	648	1 138	1 815	2 567	18 342	34 511	28 286	9 211	1 736	105	8
2nd child	72 276	1	1 265	9	46	155	376	679	8 151	18 952	27 958	13 518	2 322	100	9
3rd child	25 604	-	205	1	3	14	61	126	2 476	5 871	9 344	6 249	1 389	68	2
4th child	8 177	-	40	-	-	2	6	32	664	1 889	2 823	2 127	594	39	1
5th child	2 878	-	9	-	-	2	2	5	232	661	941	734	275	24	2
6th child	1 189	-	-	-	-	-	-	-	57	259	399	338	124	10	2
7th child	603	-	-	-	-	-	-	-	15	110	196	173	96	9	4
8th child and over	811	-	-	-	-	-	-	-	8	76	237	292	157	37	4
Not stated	609	-	32	2	2	9	5	14	114	190	156	88	23	5	1

[1]Includes births of Aleuts and Eskimos.
- = Quantity zero.

TABLE I-3. FERTILITY RATES AND BIRTH RATES BY AGE OF MOTHER, LIVE-BIRTH ORDER, AND RACE OF MOTHER, 2002

This table shows the information from Table 2 in terms of general fertility rates (the first column only) and birth rates for mothers in specific age groups. These birth rates are computed by relating the number of births by mothers in an age bracket per 1,000 women in that age bracket.

Focusing on births by teenage mothers, the birth rates for Asians and Pacific Islanders are much below those of other races. Teenage birth rates of Blacks, on the other hand, exceed those of Whites by a large margin. (See Table 8 for comparable statistics on Hispanics.) At a later age, beginning with age 25, Black birth rates are lower than those of Whites. Birth rates of Asian and Pacific Islander mothers from age 30 and up are above those of Whites.

The highest birth rate of 113.6, for all races combined, occurred between 25 and 29 years. For all age groups combined, Blacks had slightly higher fertility rates than Whites (when Hispanic Whites are included). Previously, for example, in the 1960s, Black fertility rates were much above those of Whites, and the recent data represent a convergence of rates.

Table I-3. Fertility Rates and Birth Rates by Age of Mother, Live-Birth Order, and Race of Mother, 2002

(Rates are live births per 1,000 women in specified age and racial group. Figures for live-birth order not stated are distributed.)

Live-birth order and race of mother	Fertility rate [1]	Age of mother									
		10–14 years	15–19 years			20–24 years	25–29 years	30–34 years	35–39 years	40–44 years	45–49 years [2]
			Total	15–17 years	18–19 years						
All Races	64.8	0.7	43.0	23.2	72.8	103.6	113.6	91.5	41.4	8.3	0.5
1st child	25.8	0.7	34.0	20.8	54.1	48.1	40.7	26.6	9.3	1.8	0.1
2nd child	21.1	0.0	7.5	2.3	15.6	35.5	38.9	33.1	13.6	2.3	0.1
3rd child	10.9	*	1.2	0.2	2.8	14.5	21.4	18.8	9.6	1.7	0.1
4th child	4.3	*	0.1	0.0	0.4	4.2	8.3	7.8	4.7	1.0	0.1
5th child	1.5	*	0.0	*	0.0	1.0	2.9	2.9	2.0	0.6	0.0
6th and 7th child	0.9	*	0.0	*	0.0	0.3	1.3	1.8	1.5	0.5	0.0
8th child and over	0.3	*	*	*	*	0.0	0.2	0.5	0.6	0.3	0.0
White	64.8	0.5	39.4	20.5	68.0	101.6	117.4	95.5	42.4	8.2	0.5
1st child	25.7	0.5	31.7	18.5	51.6	48.6	42.7	27.7	9.6	1.8	0.1
2nd child	21.5	0.0	6.6	1.8	13.9	35.3	41.2	34.9	13.9	2.3	0.1
3rd child	11.0	*	1.0	0.1	2.2	13.3	22.1	20.0	10.1	1.7	0.1
4th child	4.1	*	0.1	0.0	0.2	3.5	8.0	8.0	4.8	1.0	0.1
5th child	1.4	*	0.0	*	0.0	0.7	2.4	2.8	2.0	0.5	0.0
6th and 7th child	0.8	*	*	*	*	0.2	1.0	1.6	1.4	0.5	0.0
8th child and over	0.2	*	*	*	*	0.0	0.1	0.3	0.6	0.3	0.0
Black	65.8	1.8	66.6	40.0	107.6	127.1	99.0	64.4	31.5	7.4	0.4
1st child	24.8	1.8	50.4	35.0	74.2	49.6	24.4	14.2	5.9	1.3	0.1
2nd child	19.2	0.0	13.1	4.6	26.2	42.3	30.9	19.6	9.1	1.8	0.1
3rd child	11.7	*	2.7	0.4	6.1	22.8	22.7	14.8	7.2	1.6	0.1
4th child	5.5	*	0.4	0.0	0.9	8.7	11.6	7.8	4.2	1.1	0.1
5th child	2.4	*	0.0	*	0.1	2.7	5.4	3.8	2.2	0.6	0.0
6th and 7th child	1.6	*	*	*	*	0.9	3.4	3.1	1.9	0.6	0.0
8th child and over	0.5	*	*	*	*	0.1	0.6	1.1	1.0	0.4	0.0
American Indian [3]	58.0	0.9	53.8	30.7	89.2	112.6	91.8	56.4	25.4	5.8	0.3
1st child	20.4	0.9	41.4	27.3	63.1	41.6	18.8	9.0	3.0	0.6	*
2nd child	16.2	*	10.4	3.2	21.4	41.0	26.2	13.5	4.8	1.0	*
3rd child	10.4	*	1.8	*	4.2	20.6	22.7	12.8	5.3	1.1	*
4th child	5.6	*	0.2	*	0.5	7.0	13.7	9.4	4.2	0.9	*
5th child	2.8	*	*	*	*	1.8	6.3	5.4	3.2	0.8	*
6th and 7th child	2.0	*	*	*	*	0.5	3.6	4.8	3.1	0.9	*
8th child and over	0.7	*	*	*	*	*	0.5	1.6	1.6	0.6	*
Asian or Pacific Islander	64.1	0.3	18.3	9.0	31.5	60.4	105.4	109.6	56.5	12.5	0.9
1st child	30.1	0.3	14.8	8.1	24.4	37.0	58.3	44.2	15.9	3.2	0.2
2nd child	22.0	*	2.9	0.8	5.9	16.4	32.0	43.6	23.4	4.3	0.2
3rd child	7.8	*	0.5	*	1.0	5.0	9.9	14.6	10.8	2.6	0.1
4th child	2.5	*	0.1	*	0.2	1.3	3.2	4.4	3.7	1.1	0.1
5th child	0.9	*	*	*	*	0.5	1.1	1.5	1.3	0.5	0.1
6th and 7th child	0.5	*	*	*	*	0.1	0.6	0.9	0.9	0.4	0.1
8th child and over	0.2	*	*	*	*	*	0.1	0.4	0.5	0.3	0.1

[1]Fertility rate computed by relating total births, regardless of age of mother, to women aged 15–44 years.
[2]Birth rates computed by relating births to women aged 45–54 years to women aged 45–49 years.
[3]Includes births to Aleuts and Eskimos.
0.0 = Quantity more than zero but less than 0.05.
* = Figure does not meet standards of reliability or precision; based on fewer than 20 births in numerator.

TABLE I-4. TOTAL FERTILITY RATES AND BIRTH RATES BY AGE AND RACE OF MOTHER, SELECTED YEARS, 1970–2002

This table introduces a different definition of fertility, called total fertility. The objective of this rate is to show how many children would be born during the lifetime of a representative 1,000 women, if the fertility rates in each age bracket remained as currently prevailing. The rate thus shows the average number of children that would be born per 1,000 women. For 2002, total fertility is measured to be 2,013.0. A rate of 2,100 is considered to be the replacement rate, at which a given generation can exactly replace itself. The 2002 rate suggests that there were 2.01 births during the lifetime of a representative woman. To calculate total fertility, fertility rates are added for the 5-year age groups between 10 and 49 years, and the sum is multiplied by 5 since each woman is assumed to pass through each year of these age groups.

Historically the total fertility rate dropped between 1970 and 1978, and then moved intermittently higher to 2,056 in the year 2000. Since then it has dropped 2 percent. The fertility rate for Blacks has been higher than that for Whites, but this difference has narrowed in recent years. Since 1980, the fertility rate for Whites has tended upward, but the fertility rates for other races have been recently lower than in 1980. The influx of Hispanics (most of whom are classified as White) has strengthened the fertility rates of Whites.

The teenage birth rate has fallen in every year since 1991. It fell by 37 percent since 1970. The rate for teenagers, aged 15–17 years, fell by 40 percent since 1970, and it declined 6 percent between 2001 and 2002. Factors explaining these declines include a steady reduction in the proportions of teenagers who are sexually experienced and who have had multiple partners, as well as increases in condom use among sexually active teenagers [See: Brener N. et al. Trends in Sexual Risk Behavior among High School Students, 1991–2001, available at http://www.cdc.gov/mmwr/preview/mmwrhtml/mm5138a2.htm]. Teenage birth rates by race and ethnicity continue to differ considerably (see also Tables 8 and 9 for tabulation by ethnicity). For ages 15–19 the rate per 1,000 females were 18.3 in Asian and Pacific Islanders, 39.4 in Whites, 53.8 in American Indians, and 66.6 in Blacks.

Women in their 20s historically account for the largest share of births. However, the proportion of births to these women has declined in recent years, dropping from two-thirds of all births in 1980 to about half in 2002. Birth rates for women in their 20s dropped steadily in the first half of the 1990s. In recent years, the rates for ages

Figure I-2. First Birth Rates by Age of Mother, 1990 and 2002

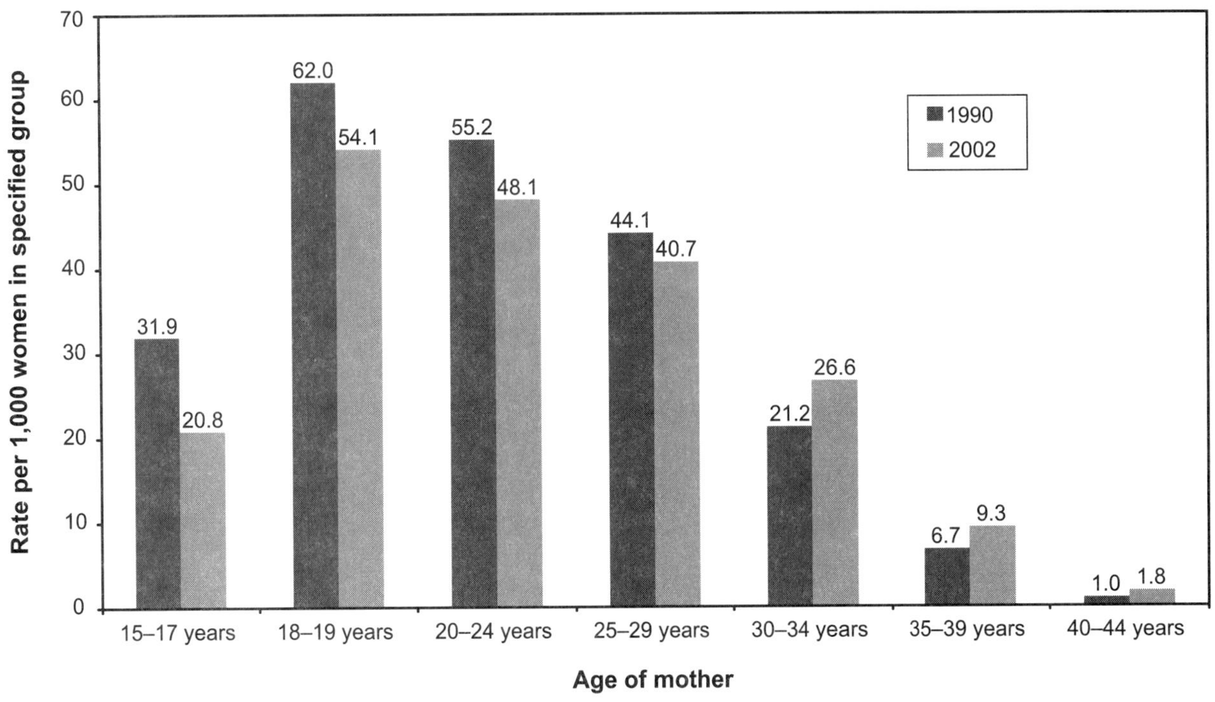

Table I-4. Total Fertility Rates and Birth Rates by Age and Race of Mother, Selected Years, 1970–2002

(Birth rates are live births per 1,000 women in specified group.)

Year and race	Total fertility rate [1]	Age of mother									
		10–14 years	15–19 years			20–24 years	25–29 years	30–34 years	35–39 years	40–44 years	45–49 years [2]
			Total	15–17 years	18–19 years						
All Races [3]											
2002	2 013.0	0.7	43.0	23.2	72.8	103.6	113.6	91.5	41.4	8.3	0.5
2001	2 034.0	0.8	45.3	24.7	76.1	106.2	113.4	91.9	40.6	8.1	0.5
2000	2 056.0	0.9	47.7	26.9	78.1	109.7	113.5	91.2	39.7	8.0	0.5
1999	2 007.5	0.9	48.8	28.2	79.1	107.9	111.2	87.1	37.8	7.4	0.4
1998	1 999.0	1.0	50.3	29.9	80.9	108.4	110.2	85.2	36.9	7.4	0.4
1997	1 971.0	1.1	51.3	31.4	82.1	107.3	108.3	83.0	35.7	7.1	0.4
1996	1 976.0	1.2	53.5	33.3	84.7	107.8	108.6	82.1	34.9	6.8	0.3
1995	1 978.0	1.3	56.0	35.5	87.7	107.5	108.8	81.1	34.0	6.6	0.3
1994	2 001.5	1.4	58.2	37.2	90.2	109.2	111.0	80.4	33.4	6.4	0.3
1993	2 019.5	1.4	59.0	37.5	91.1	111.3	113.2	79.9	32.7	6.1	0.3
1992	2 046.0	1.4	60.3	37.6	93.6	113.7	115.7	79.6	32.3	5.9	0.3
1991	2 062.5	1.4	61.8	38.6	94.0	115.3	117.2	79.2	31.9	5.5	0.2
1990	2 081.0	1.4	59.9	37.5	88.6	116.5	120.2	80.8	31.7	5.5	0.2
1989	2 014.0	1.4	57.3	36.4	84.2	113.8	117.6	77.4	29.9	5.2	0.2
1988	1 934.0	1.3	53.0	33.6	79.9	110.2	114.4	74.8	28.1	4.8	0.2
1987	1 872.0	1.3	50.6	31.7	78.5	107.9	111.6	72.1	26.3	4.4	0.2
1986	1 837.5	1.3	50.2	30.5	79.6	107.4	109.8	70.1	24.4	4.1	0.2
1985	1 844.0	1.2	51.0	31.0	79.6	108.3	111.0	69.1	24.0	4.0	0.2
1984 [4]	1 806.5	1.2	50.6	31.0	77.4	106.8	108.7	67.0	22.9	3.9	0.2
1983 [4]	1 799.0	1.1	51.4	31.8	77.4	107.8	108.5	64.9	22.0	3.9	0.2
1982 [4]	1 827.5	1.1	52.4	32.3	79.4	111.6	111.0	64.1	21.2	3.9	0.2
1981 [4]	1 812.0	1.1	52.2	32.0	80.0	112.2	111.5	61.4	20.0	3.8	0.2
1980 [4]	1 839.5	1.1	53.0	32.5	82.1	115.1	112.9	61.9	19.8	3.9	0.2
1979 [4]	1 808.0	1.2	52.3	32.3	81.3	112.8	111.4	60.3	19.5	3.9	0.2
1978 [4]	1 760.0	1.2	51.5	32.2	79.8	109.9	108.5	57.8	19.0	3.9	0.2
1977 [4]	1 789.5	1.2	52.8	33.9	80.9	112.9	111.0	56.4	19.2	4.2	0.2
1976 [4]	1 738.0	1.2	52.8	34.1	80.5	110.3	106.2	53.6	19.0	4.3	0.2
1975 [4]	1 774.0	1.3	55.6	36.1	85.0	113.0	108.2	52.3	19.5	4.6	0.3
1974 [4]	1 835.0	1.2	57.5	37.3	88.7	117.7	111.5	53.8	20.2	4.8	0.3
1973 [4]	1 879.0	1.2	59.3	38.5	91.2	119.7	112.2	55.6	22.1	5.4	0.3
1972 [4]	2 010.0	1.2	61.7	39.0	96.9	130.2	117.7	59.8	24.8	6.2	0.4
1971 [5]	2 266.5	1.1	64.5	38.2	105.3	150.1	134.1	67.3	28.7	7.1	0.4
1970 [5]	2 480.0	1.2	68.3	38.8	114.7	167.8	145.1	73.3	31.7	8.1	0.5
White											
2002	2 027.5	0.5	39.4	20.5	68.0	101.6	117.4	95.5	42.4	8.2	0.5
2001	2 040.0	0.5	41.2	21.4	70.8	103.7	117.0	95.8	41.3	8.0	0.5
2000	2 051.0	0.6	43.2	23.3	72.3	106.6	116.7	94.6	40.2	7.9	0.4
1999	2 007.5	0.6	44.0	24.5	72.8	105.0	114.9	90.7	38.5	7.4	0.4
1998	1 991.0	0.6	44.9	25.6	73.9	105.4	113.6	88.5	37.5	7.3	0.4
1997	1 955.0	0.7	45.5	26.6	74.8	104.5	111.3	85.7	36.1	6.9	0.3
1996	1 960.5	0.7	47.5	28.0	77.4	105.3	111.7	84.6	35.3	6.7	0.3
1995	1 954.5	0.8	49.5	29.7	80.0	104.7	111.7	83.3	34.2	6.4	0.3
1994	1 957.5	0.8	50.5	30.4	81.0	105.0	113.0	82.2	33.5	6.2	0.3
1993	1 961.5	0.8	50.6	30.1	81.3	106.1	114.7	81.3	32.6	5.9	0.3
1992	1 978.0	0.8	51.4	29.9	83.1	107.7	116.9	80.8	32.1	5.7	0.2
1991	1 988.0	0.8	52.6	30.6	83.2	108.8	118.0	80.2	31.8	5.2	0.2
1990	2 003.0	0.7	50.8	29.5	78.0	109.8	120.7	81.7	31.5	5.2	0.2
1989	1 931.0	0.7	47.9	28.1	72.9	106.9	117.8	78.1	29.7	4.9	0.2
1988	1 856.5	0.6	44.4	26.0	69.6	103.7	114.8	75.4	27.7	4.5	0.2
1987	1 804.5	0.6	42.5	24.6	68.9	102.3	112.3	73.0	25.9	4.1	0.2
1986	1 776.0	0.6	42.3	23.8	70.1	102.7	110.8	70.9	23.9	3.8	0.2
1985	1 787.0	0.6	43.3	24.4	70.4	104.1	112.3	69.9	23.3	3.7	0.2
1984 [4]	1 748.5	0.6	42.9	24.3	68.4	102.7	109.8	67.7	22.2	3.6	0.2
1983 [4]	1 740.5	0.6	43.9	25.0	68.8	103.8	109.4	65.3	21.3	3.6	0.2
1982 [4]	1 767.0	0.6	45.0	25.5	70.8	107.7	111.9	64.0	20.4	3.6	0.2
1981 [4]	1 748.0	0.5	44.9	25.4	71.5	108.3	112.3	61.0	19.0	3.4	0.2
1980 [4]	1 773.0	0.6	45.4	25.5	73.2	111.1	113.8	61.2	18.8	3.5	0.2
Black											
2002	1 991.0	1.8	66.6	40.0	107.6	127.1	99.0	64.4	31.5	7.4	0.4
2001	2 051.0	2.0	71.8	43.9	114.0	133.2	99.2	64.8	31.6	7.2	0.4
2000	2 129.0	2.3	77.4	49.0	118.8	141.3	100.3	65.4	31.5	7.2	0.4
1999	2 082.5	2.5	79.1	50.7	120.1	137.9	97.3	62.7	30.2	6.5	0.3
1998	2 111.5	2.8	83.5	55.5	124.3	138.4	97.5	63.2	30.0	6.6	0.3
1997	2 091.5	3.1	86.3	59.4	127.4	135.2	95.0	62.6	29.3	6.5	0.3
1996	2 088.5	3.5	89.6	63.4	130.1	133.2	94.3	62.0	28.7	6.1	0.3
1995	2 127.5	4.1	94.4	68.6	134.6	133.7	95.6	63.0	28.4	6.0	0.3
1994	2 258.5	4.5	102.9	75.2	145.8	142.9	101.5	65.0	28.7	5.9	0.3
1993	2 351.0	4.5	107.3	78.9	149.9	150.2	106.4	66.6	29.0	5.9	0.3
1992	2 416.0	4.6	111.3	80.6	156.2	156.2	109.7	67.0	28.6	5.6	0.2
1991	2 462.0	4.7	114.8	83.6	157.6	159.7	112.0	67.3	28.2	5.5	0.2
1990	2 480.0	4.9	112.8	82.3	152.9	160.2	115.5	68.7	28.1	5.5	0.3
1989	2 432.5	5.1	111.5	81.9	151.9	156.8	114.4	66.3	26.7	5.4	0.3
1988	2 298.0	4.9	102.7	75.7	142.7	149.7	108.2	63.1	25.6	5.1	0.3
1987	2 198.0	4.8	97.6	72.1	135.8	142.7	104.3	60.6	24.6	4.8	0.2
1986	2 135.5	4.7	95.8	69.3	135.1	137.3	101.1	59.3	23.8	4.8	0.3
1985	2 109.0	4.5	95.4	69.3	132.4	135.0	100.2	57.9	23.9	4.6	0.3
1984 [4]	2 070.5	4.4	94.1	69.2	128.1	132.2	98.4	56.7	23.3	4.8	0.2
1983 [4]	2 066.0	4.1	93.9	69.6	127.1	131.9	98.4	56.2	23.3	5.1	0.3
1982 [4]	2 106.5	4.0	94.3	69.7	128.9	135.4	101.3	57.5	23.3	5.1	0.4
1981 [4]	2 117.5	4.0	94.5	69.3	131.0	136.5	102.3	57.4	23.1	5.4	0.3
1980 [4]	2 176.5	4.3	97.8	72.5	135.1	140.0	103.9	59.9	23.5	5.6	0.3

[1] Total fertility rates are sums of birth rates for 5-year age groups multiplied by 5.
[2] Beginning in 1997, rates computed by relating births to women aged 45–54 years to women aged 45–49 years.
[3] For 1970–1991, includes births to races not shown separately. For 1992 and later years, unknown race of mother is imputed.
[4] Based on 100 percent of births in selected states and on a 50-percent sample of births in all other states.
[5] Based on a 50-percent sample of births.

20–24 have continued to fall, while those for ages 25–29 have stabilized. Birth rates for women in their 30s rose steadily throughout the 1990s—1 percent per year for those 30–34 and 3 percent per year in those 35–39.

The birth rate for women in their 40s has risen in most years since the mid-1980s. From 1990 to 2002, births by women aged 40–44 years nearly doubled from 48,607 to 95,788. The birth rate for women aged 45–49 years has been unchanged at 0.5 since 2000, but it rose considerably in the 1990s. Because the number of births to women 50 years and older are too small to compute age-specific rates, these births are folded into the age 45–49 bracket.

Table I-4. Total Fertility Rates and Birth Rates by Age and Race of Mother, Selected Years, 1970–2002 *—Continued*

(Birth rates are live births per 1,000 women in specified group.)

Year and race	Total fertility rate [1]	Age of mother									
		10–14 years	15–19 years			20–24 years	25–29 years	30–34 years	35–39 years	40–44 years	45–49 years [2]
			Total	15–17 years	18–19 years						
American Indian [6]											
2002	1 735.0	0.9	53.8	30.7	89.2	112.6	91.8	56.4	25.4	5.8	0.3
2001	1 746.5	1.0	56.3	31.4	94.8	115.0	90.4	55.9	24.7	5.7	0.3
2000	1 772.5	1.1	58.3	34.1	97.1	117.2	91.8	55.5	24.6	5.7	0.3
1999	1 783.5	1.4	59.9	36.5	97.9	120.7	90.6	53.8	24.3	5.7	0.3
1998	1 851.0	1.5	64.7	39.8	106.5	125.1	92.0	56.8	24.6	5.3	*
1997	1 834.5	1.5	65.2	41.1	106.8	122.5	91.6	56.0	24.4	5.4	0.3
1996	1 855.0	1.6	68.2	42.7	112.9	123.5	91.1	56.5	24.4	5.5	*
1995	1 878.5	1.6	72.9	44.7	121.8	123.1	91.6	56.5	24.3	5.5	*
1994	1 950.0	1.8	76.4	48.6	122.9	126.5	98.2	56.6	24.8	5.4	0.3
1993	2 048.5	1.4	79.8	51.7	125.5	134.2	103.5	59.5	25.5	5.6	*
1992	2 135.5	1.6	82.4	52.6	129.3	142.3	107.0	61.0	26.7	5.9	*
1991	2 142.5	1.6	84.1	52.2	132.9	143.8	105.6	60.8	26.4	5.8	0.4
1990	2 184.5	1.6	81.1	48.5	129.3	148.7	110.3	61.5	27.5	5.9	*
1989	2 248.5	1.5	82.7	51.6	128.9	152.4	114.2	64.8	27.4	6.4	*
1988	2 155.0	1.7	77.5	49.7	121.1	145.2	110.9	64.5	25.6	5.3	*
1987	2 100.5	1.7	77.2	48.8	122.2	140.0	107.9	63.0	24.4	5.6	*
1986	2 083.0	1.8	78.1	48.7	125.3	138.8	107.9	60.7	23.8	5.3	*
1985	2 129.5	1.7	79.2	47.7	124.1	139.1	109.6	62.6	27.4	6.0	*
1984 [4]	2 137.5	1.7	81.5	50.7	124.7	142.4	109.2	60.5	26.3	5.6	*
1983 [4]	2 182.0	1.9	84.2	55.2	121.4	145.5	113.7	58.9	25.5	6.4	*
1982 [4]	2 215.0	1.4	83.5	52.6	127.6	148.1	115.8	60.9	26.9	6.0	*
1981 [4]	2 092.5	2.1	78.4	49.7	121.5	141.2	105.6	58.9	25.2	6.6	*
1980 [4]	2 165.0	1.9	82.2	51.5	129.5	143.7	106.6	61.8	28.1	8.2	*
Asian or Pacific Islander											
2002	1 819.5	0.3	18.3	9.0	31.5	60.4	105.4	109.6	56.5	12.5	0.9
2001	1 840.0	0.2	19.8	10.3	32.8	59.1	106.4	112.6	56.7	12.3	0.9
2000	1 892.0	0.3	20.5	11.6	32.6	60.3	108.4	116.5	59.0	12.6	0.8
1999	1 754.5	0.4	21.4	11.8	36.5	58.9	100.8	104.3	52.9	11.3	0.9
1998	1 731.5	0.5	22.2	13.2	36.9	59.2	98.7	101.6	51.4	11.8	0.9
1997	1 757.5	0.5	22.3	13.5	37.0	61.2	101.6	102.5	51.0	11.5	0.9
1996	1 787.0	0.6	23.5	14.3	38.6	63.5	102.8	104.1	50.2	11.9	0.8
1995	1 795.5	0.7	25.5	15.1	42.2	64.2	103.7	102.3	50.1	11.8	0.8
1994	1 834.0	0.7	26.6	15.8	43.2	66.4	108.0	102.2	50.4	11.5	1.0
1993	1 841.5	0.7	26.5	15.8	42.5	68.1	110.3	101.2	49.4	11.2	0.9
1992	1 894.5	0.7	26.5	15.1	42.9	71.7	114.6	102.7	50.7	11.1	0.9
1991	1 928.0	0.8	27.3	16.1	43.0	73.8	118.9	103.3	49.2	11.2	1.1
1990	2 002.5	0.7	26.4	16.0	40.2	79.2	126.3	106.5	49.6	10.7	1.1
1989	1 947.5	0.6	25.6	15.0	40.4	78.8	124.0	102.3	47.0	10.2	1.0
1988	1 983.5	0.6	24.2	13.6	39.6	80.7	128.0	104.4	47.5	10.3	1.0
1987	1 886.0	0.6	22.4	12.6	37.0	79.7	122.7	97.0	44.2	9.5	1.1
1986	1 836.0	0.5	22.8	12.1	38.8	79.2	119.9	92.6	41.9	9.3	1.0
1985	1 885.0	0.4	23.8	12.5	40.8	83.6	123.0	93.6	42.7	8.7	1.2
1984 [4]	1 892.0	0.5	24.2	12.6	40.7	86.7	124.3	92.4	40.6	8.7	1.0
1983 [4]	1 943.5	0.5	26.1	12.9	44.5	94.0	126.2	93.3	39.4	8.2	1.0
1982 [4]	2 015.5	0.4	29.4	14.0	50.8	98.9	130.9	94.4	39.2	8.8	1.1
1981 [4]	1 976.0	0.3	28.5	13.4	49.5	96.4	129.1	93.4	38.0	8.6	0.9
1980 [4]	1 953.5	0.3	26.2	12.0	46.2	93.3	127.4	96.0	38.3	8.5	0.7

[1]Total fertility rates are sums of birth rates for 5-year age groups multiplied by 5.
[2]Beginning in 1997, rates computed by relating births to women aged 45–54 years to women aged 45–49 years.
[4]Based on 100 percent of births in selected states and on a 50-percent sample of births in all other states.
[6]Includes births to Aleuts and Eskimos.
* = Figure does not meet standards of reliability or precision; based on fewer than 20 births in numerator.

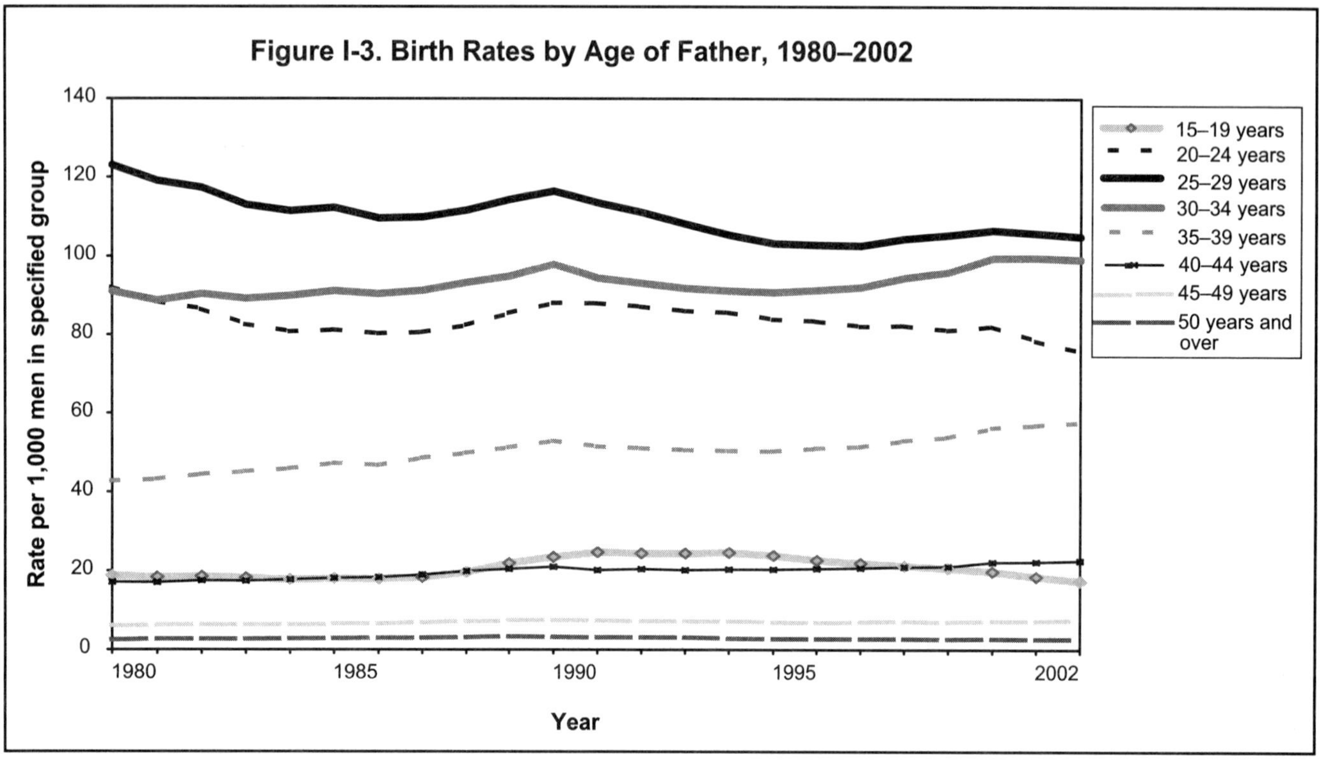
Figure I-3. Birth Rates by Age of Father, 1980–2002
Rate per 1,000 men in specified group
140
120
100
80
60
40
20
0
1980
1985
1990
1995
2002
Year
15–19 years
20–24 years
25–29 years
30–34 years
35–39 years
40–44 years
45–49 years
50 years and over

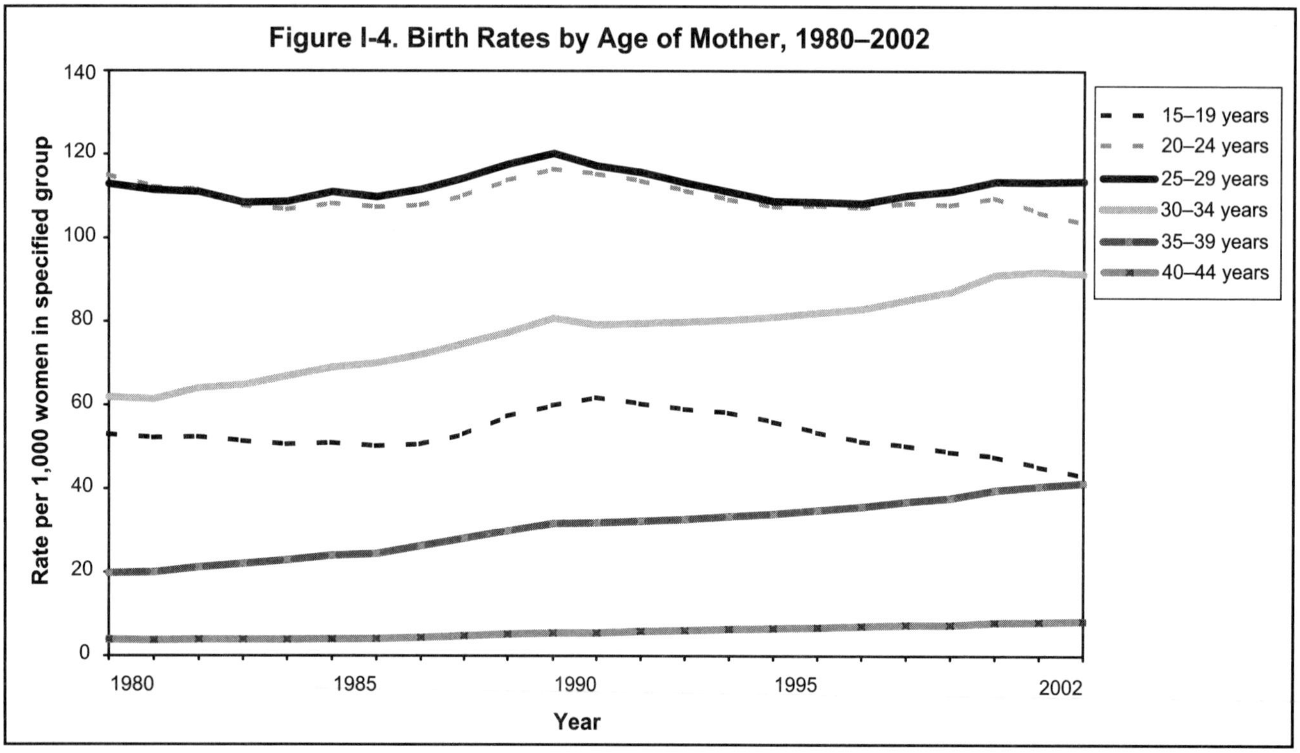
Figure I-4. Birth Rates by Age of Mother, 1980–2002
Rate per 1,000 women in specified group
140
120
100
80
60
40
20
0
1980
1985
1990
1995
2002
Year
15–19 years
20–24 years
25–29 years
30–34 years
35–39 years
40–44 years

Figure I-5. Mean Age of Mother at First Birth by Race and Hispanic Origin, 2002

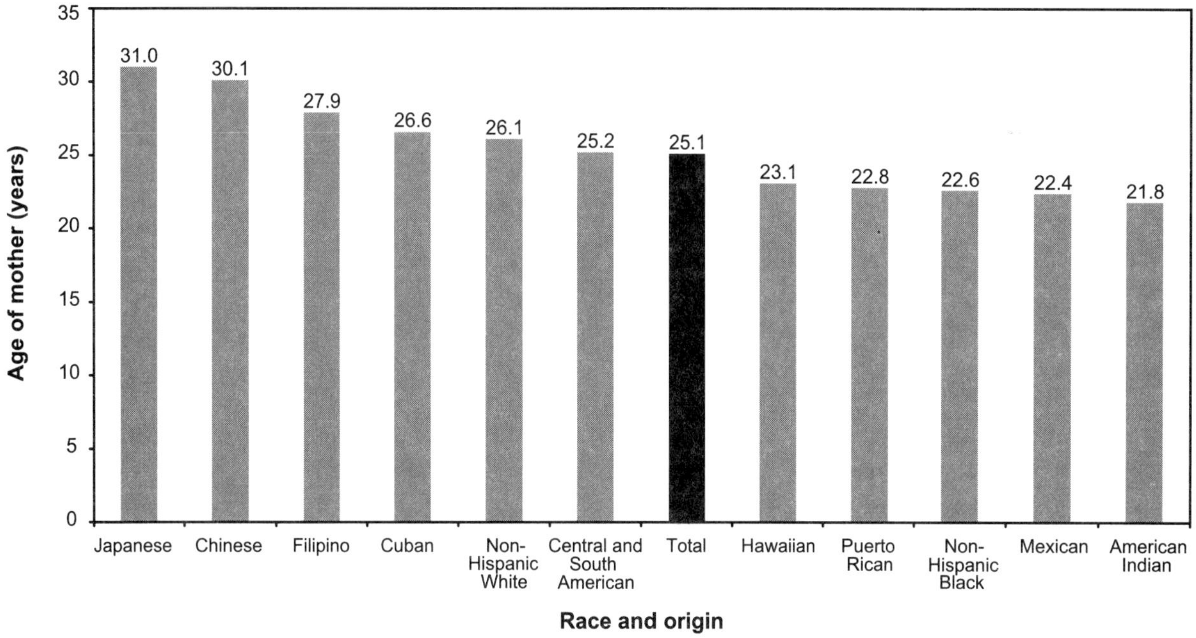

TABLE I-4A. BIRTH RATES FOR WOMEN AGED 15–19 YEARS, BY AGE, RACE, AND HISPANIC ORIGIN, SELECTED YEARS, 1991–2002

Teenage birth rates by race and ethnicity differ considerably. All groups have experienced marked declines in teenage birth rates between 1991 and 2002, with Hispanics showing the smallest declines.

Table I-4A. Birth Rates for Women Aged 15–19 Years, by Age, Race, and Hispanic Origin, Selected Years, 1991–2002

(Rate per 1,000 women in specified group, percent change.)

Age, race, and Hispanic origin of mother	2002	2001	1991	Percent change, 1991–2002
10–14 Years				
All races [1]	0.7	0.8	1.4	-50
Non-Hispanic White	0.2	0.3	0.5	-60
Non-Hispanic Black	1.9	2.1	4.9	-61
American Indian total [2]	0.9	1.0	1.6	-44
Asian or Pacific Islander total [2]	0.3	0.2	0.8	-63
Hispanic [3]	1.4	1.6	2.4	-42
15–19 Years				
All races [1]	43.0	45.3	61.8	-30
Non-Hispanic White	28.5	30.3	43.4	-42
Non-Hispanic Black	68.3	73.5	118.2	-34
American Indian total [2]	53.8	56.3	84.1	-36
Asian or Pacific Islander total [2]	18.3	19.8	27.3	-33
Hispanic [3]	83.4	86.4	104.6	-20
15–17 Years				
All races [1]	23.2	24.7	38.6	-40
Non-Hispanic White	13.1	14.0	23.7	-45
Non-Hispanic Black	41.0	44.9	86.2	-52
American Indian total [2]	30.7	31.4	52.2	-41
Asian or Pacific Islander total [2]	9.0	10.3	16.1	-44
Hispanic [3]	50.7	52.8	69.3	-27
18–19 Years				
All races [1]	72.8	76.1	94.0	-23
Non-Hispanic White	51.9	54.8	70.6	-26
Non-Hispanic Black	110.3	116.7	162.0	-32
Asian or Pacific Islander total [2]	89.2	94.8	43.0	-33
American Indian total [2]	31.5	32.8	132.9	-27
Hispanic [3]	133.0	135.5	155.1	-14

[1]Includes races other than White and Black.
[2]Race and Hispanic origin are reported separately on the birth certificate. Data for persons of Hispanic origin are included in the data for each race group according to the mother's reported race.
[3]Includes person of Hispanic origin of any race.
. . . = Not available.

Figure I-6. Birth Rates for Unmarried Women, by Age of Mother, 1980–2002

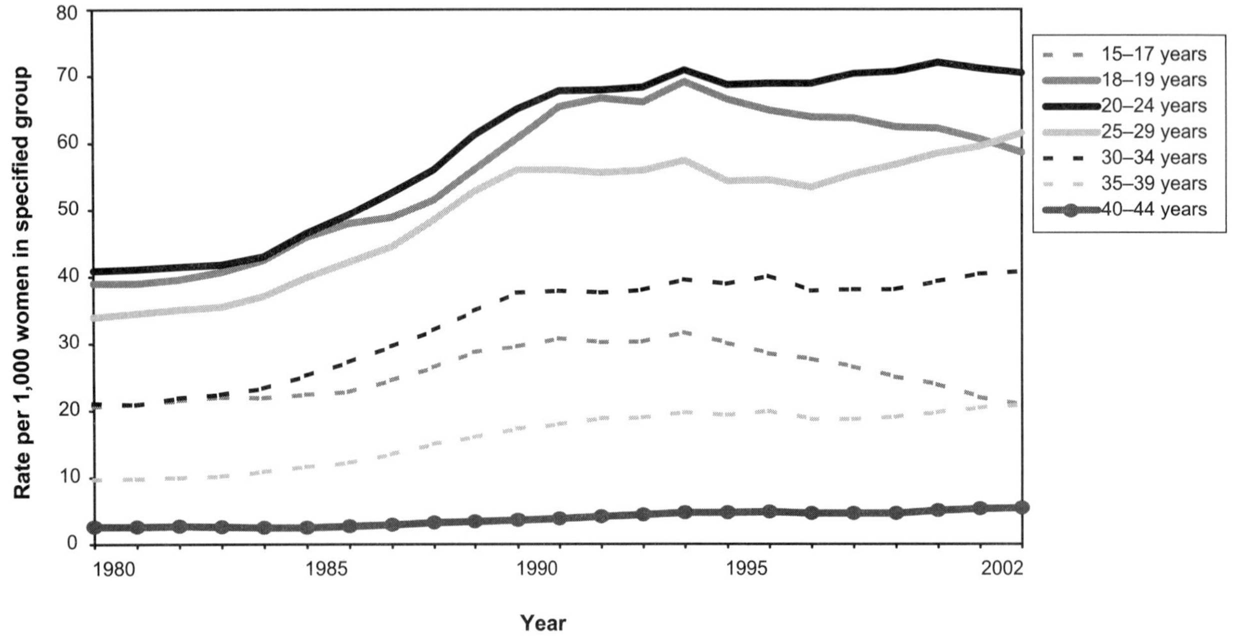

Figure I-7. Birth Rates for Teenagers by Race and/or Hispanic Origin for 1991 and 2002

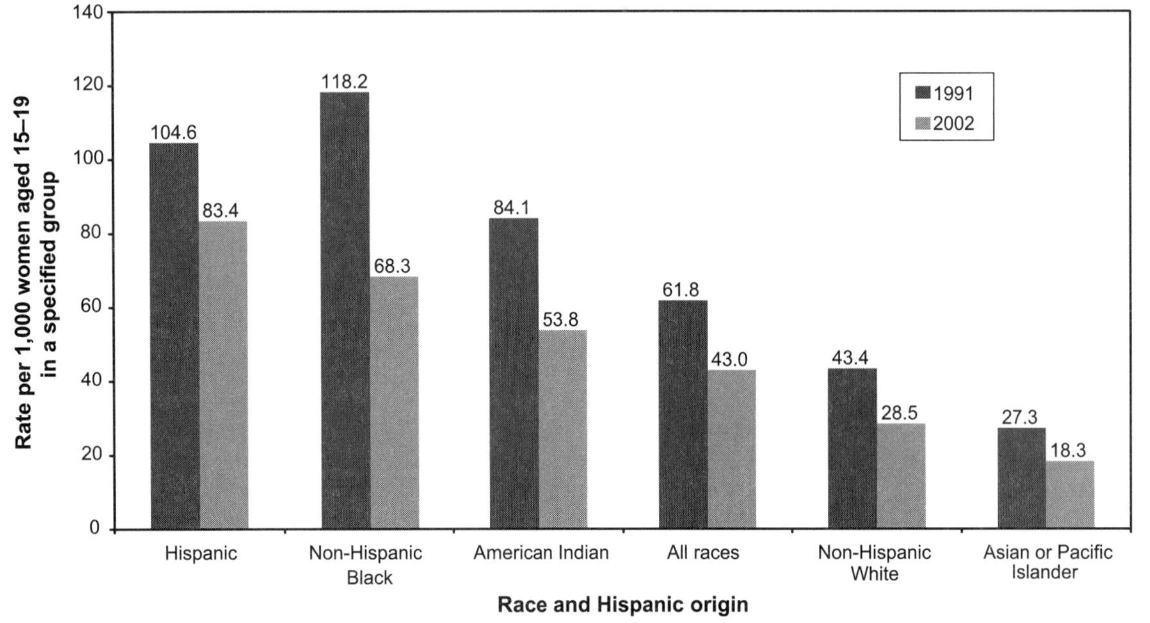

TABLE I-5. FERTILITY RATES AND BIRTH RATES BY LIVE-BIRTH ORDER AND RACE OF MOTHER, 1980–2002

This table gives a history, since 1980, of general fertility rates as well as birth rates by live-birth order; it includes detail by race (White and Black only). The declines in fertility rates for Blacks and their convergence toward White rates are clearly shown. Fertility rates of Whites are fairly stable over the two decades since 1980, but there is a moderate tendency for the frequency of first-borns to decrease, while larger families—third and fourth children—slightly increase in frequency.

Table I-5. Fertility Rates and Birth Rates by Live-Birth Order and Race of Mother, 1980–2002

(Rates are live births per 1,000 women aged 15–44 years. Figures for live-birth order not stated are distributed.)

Year and race	Fertility rate	Live-birth order						
		1	2	3	4	5	6 and 7	8 and over
All Races[1]								
2002	64.8	25.8	21.1	10.9	4.3	1.5	0.9	0.3
2001	65.3	26.0	21.3	11.0	4.3	1.6	0.9	0.3
2000	65.9	26.5	21.4	11.0	4.2	1.6	0.9	0.3
1999	64.4	26.0	21.0	10.7	4.1	1.5	0.9	0.3
1998	64.3	25.9	21.0	10.6	4.1	1.5	0.9	0.3
1997	63.6	25.9	20.7	10.4	4.0	1.5	0.9	0.3
1996	64.1	26.3	20.7	10.4	4.0	1.5	0.9	0.3
1995	64.6	26.9	20.7	10.3	4.0	1.5	0.9	0.3
1994	65.9	27.1	21.2	10.6	4.1	1.6	0.9	0.3
1993	67.0	27.3	21.7	10.9	4.3	1.6	1.0	0.3
1992	68.4	27.6	22.2	11.2	4.4	1.7	1.0	0.3
1991	69.3	28.2	22.3	11.4	4.4	1.7	1.0	0.3
1990	70.9	29.0	22.8	11.7	4.5	1.7	1.0	0.3
1989	69.2	28.4	22.4	11.3	4.3	1.6	0.9	0.3
1988	67.3	27.6	22.0	10.9	4.1	1.5	0.9	0.3
1987	65.8	27.2	21.6	10.5	3.9	1.4	0.8	0.3
1986	65.4	27.2	21.6	10.3	3.8	1.4	0.8	0.3
1985	66.3	27.6	22.0	10.4	3.8	1.4	0.8	0.3
1984[2]	65.5	27.4	21.7	10.1	3.7	1.4	0.9	0.3
1983[2]	65.7	27.8	21.5	10.1	3.7	1.4	0.9	0.3
1982[2]	67.3	28.6	22.0	10.2	3.8	1.4	0.9	0.3
1981[2]	67.3	29.0	21.6	10.1	3.8	1.5	0.9	0.4
1980[2]	68.4	29.5	21.8	10.3	3.9	1.5	1.0	0.4
White								
2002	64.8	25.7	21.5	11.0	4.1	1.4	0.8	0.2
2001	65.0	25.9	21.6	11.0	4.1	1.4	0.8	0.2
2000	65.3	26.3	21.5	11.0	4.0	1.4	0.8	0.2
1999	64.0	25.9	21.2	10.6	3.9	1.3	0.7	0.2
1998	63.6	25.7	21.2	10.5	3.9	1.3	0.7	0.2
1997	62.8	25.7	20.8	10.2	3.8	1.3	0.7	0.2
1996	63.3	26.2	20.9	10.2	3.8	1.3	0.7	0.2
1995	63.6	26.6	20.9	10.2	3.7	1.3	0.7	0.2
1994	64.2	26.7	21.2	10.3	3.8	1.3	0.7	0.2
1993	64.9	26.8	21.5	10.4	3.9	1.4	0.8	0.2
1992	66.1	27.1	21.9	10.7	4.0	1.4	0.8	0.2
1991	66.7	27.7	21.9	10.8	4.0	1.4	0.8	0.2
1990	68.3	28.4	22.4	11.1	4.0	1.4	0.8	0.2
1989	66.4	27.6	21.9	10.7	3.8	1.3	0.7	0.2
1988	64.5	26.8	21.6	10.4	3.6	1.2	0.7	0.2
1987	63.3	26.5	21.3	10.0	3.5	1.2	0.7	0.2
1986	63.1	26.6	21.3	9.8	3.4	1.2	0.7	0.2
1985	64.1	27.0	21.8	9.9	3.4	1.2	0.7	0.2
1984[2]	63.2	26.8	21.4	9.6	3.3	1.2	0.7	0.2
1983[2]	63.4	27.2	21.2	9.5	3.3	1.2	0.7	0.2
1982[2]	64.8	28.0	21.6	9.6	3.4	1.2	0.7	0.3
1981[2]	64.8	28.4	21.1	9.5	3.4	1.2	0.8	0.3
1980[2]	65.6	28.8	21.3	9.6	3.4	1.3	0.8	0.3
Black								
2002	65.8	24.8	19.2	11.7	5.5	2.4	1.6	0.5
2001	67.6	25.4	19.9	12.1	5.6	2.5	1.6	0.5
2000	70.0	26.2	20.8	12.5	5.8	2.5	1.7	0.6
1999	68.5	25.9	20.4	12.1	5.5	2.4	1.6	0.5
1998	69.4	26.5	20.6	12.0	5.6	2.5	1.7	0.6
1997	69.0	26.7	20.2	11.8	5.5	2.5	1.7	0.6
1996	69.2	27.0	20.1	11.7	5.5	2.5	1.7	0.6
1995	71.0	28.2	20.4	11.8	5.6	2.6	1.8	0.6
1994	75.9	29.4	21.9	12.9	6.2	2.9	2.0	0.6
1993	79.6	29.8	23.1	14.0	6.8	3.1	2.1	0.6
1992	82.4	30.3	24.1	14.8	7.1	3.3	2.1	0.6
1991	84.8	31.3	24.9	15.3	7.3	3.3	2.1	0.6
1990	86.8	32.4	25.6	15.6	7.4	3.2	2.0	0.6
1989	86.2	32.9	25.4	15.3	7.1	3.0	1.9	0.6
1988	82.6	31.8	24.6	14.4	6.6	2.8	1.8	0.5
1987	80.1	31.2	23.8	13.9	6.3	2.7	1.7	0.5
1986	78.9	31.0	23.4	13.5	6.1	2.6	1.7	0.5
1985	78.8	31.0	23.4	13.4	6.1	2.6	1.7	0.5
1984[2]	78.1	30.9	23.0	13.2	6.0	2.6	1.7	0.6
1983[2]	78.7	31.1	23.1	13.2	6.1	2.7	1.8	0.6
1982[2]	80.9	31.7	23.9	13.8	6.3	2.7	1.8	0.7
1981[2]	82.0	32.3	24.2	13.7	6.3	2.8	1.9	0.8
1980[2]	84.9	33.7	24.7	14.0	6.5	2.9	2.1	0.9

[1]Includes races other than White and Black.
[2]Based on 100 percent of births in selected states and on a 50-percent sample of births in all other states.

TABLE I-6. LIVE BIRTHS, BIRTH RATES, AND FERTILITY RATES BY SPECIFIED HISPANIC ORIGIN AND RACE OF MOTHER, 1989–2002

In this table, births for Hispanic mothers are subdivided by country (or region) of origin, such as Mexico and Cuba. Birth rates are especially high among mothers from Mexico and from Central and South America. Cuban-American fertility rates are fairly close to those of non-Hispanic Whites and were relatively stable since 1989, whereas Puerto Ricans in the United States had falling fertility rates since 1989 that remain somewhat higher than those of Whites. The Black non-Hispanic birth rate is below the Hispanic average, but it is well above the rate for Whites—after the Hispanic component is removed from Whites.

Table I-6. Live Births, Birth Rates, and Fertility Rates by Specified Hispanic Origin and Race of Mother, 1989–2002

(Birth rates are live births per 1,000 population in specified group. Fertility rates are live births per 1,000 women aged 15–44 years in specified group.)

Year and race	All origins [1]	Hispanic						Non-Hispanic		
		Total	Mexican	Puerto Rican	Cuban	Central and South American	Other and unknown Hispanic	Total [2]	White	Black
Number										
2002	4 021 726	876 642	627 505	57 465	14 232	125 981	51 459	3 119 944	2 298 156	578 335
2001	4 025 933	851 851	611 000	57 568	14 017	121 365	47 901	3 149 572	2 326 578	589 917
2000	4 058 814	815 868	581 915	58 124	13 429	113 344	49 056	3 199 994	2 362 968	604 346
1999	3 959 417	764 339	540 674	57 138	13 088	103 307	50 132	3 147 580	2 346 450	588 981
1998	3 941 553	734 661	516 011	57 349	13 226	98 226	49 849	3 158 975	2 361 462	593 127
1997	3 880 894	709 767	499 024	55 450	12 887	97 405	45 001	3 115 174	2 333 363	581 431
1996	3 891 494	701 339	489 666	54 863	12 613	97 888	46 309	3 133 484	2 358 989	578 099
1995	3 899 589	679 768	469 615	54 824	12 473	94 996	47 860	3 160 495	2 382 638	587 781
1994	3 952 767	665 026	454 536	57 240	11 889	93 485	47 876	3 245 115	2 438 855	619 198
1993	4 000 240	654 418	443 733	58 102	11 916	92 371	48 296	3 295 345	2 472 031	641 273
1992 [3]	4 049 024	643 271	432 047	59 569	11 472	89 031	51 152	3 365 862	2 527 207	657 450
1991 [3]	4 094 566	623 085	411 233	59 833	11 058	86 908	54 053	3 434 464	2 589 878	666 758
1990 [4]	4 092 994	595 073	385 640	58 807	11 311	83 008	56 307	3 457 417	2 626 500	661 701
1989 [5]	3 903 012	532 249	327 233	56 229	10 842	72 443	65 502	3 297 493	2 526 367	611 269
Birth Rate										
2002 [6]	13.9	22.6	24.2	16.5	10.0	22.4	(6)	12.6	11.7	16.1
2001 [6]	14.1	23.0	24.8	17.8	10.3	21.8	(6)	12.8	11.8	16.6
2000 [6]	14.4	23.1	25.0	18.1	9.7	21.8	(6)	13.2	12.2	17.3
1999 [6]	14.2	22.5	24.2	18.0	9.4	21.7	(6)	13.0	12.1	17.1
1998 [6]	14.3	22.7	24.6	17.9	9.7	21.7	(6)	13.2	12.2	17.5
1997 [6]	14.2	23.0	25.3	17.2	10.0	21.3	(6)	13.1	12.2	17.4
1996 [6]	14.4	23.8	26.2	17.2	10.6	22.5	(6)	13.3	12.3	17.6
1995 [6]	14.6	24.1	25.8	18.9	10.7	24.2	(6)	13.5	12.5	18.2
1994 [6]	15.0	24.7	26.1	20.8	10.7	24.9	(6)	13.9	12.8	19.5
1993 [6]	15.4	25.4	26.8	21.5	10.5	26.3	(6)	14.3	13.1	20.7
1992 [6,7]	15.8	26.1	27.4	22.9	10.1	27.5	(6)	14.8	13.4	21.6
1991 [6,7]	16.2	26.5	27.6	23.3	9.8	28.3	(6)	15.2	13.9	22.4
1990 [4,6]	16.7	26.7	28.7	21.6	10.9	27.5	(6)	15.7	14.4	23.0
1989 [5,6]	16.3	26.2	25.7	23.7	10.0	28.3	(6)	15.4	14.2	22.8
Fertility Rate										
2002 [6]	64.8	94.4	102.8	65.4	59.0	86.1	(6)	59.6	57.4	67.4
2001 [6]	65.3	96.0	105.7	72.2	56.7	82.7	(6)	60.1	57.7	69.1
2000 [6]	65.9	95.9	105.1	73.5	49.3	85.1	(6)	61.1	58.5	71.4
1999 [6]	64.4	93.0	101.5	71.1	47.0	84.8	(6)	60.0	57.7	69.9
1998 [6]	64.3	93.2	103.2	69.7	46.5	83.5	(6)	60.0	57.6	70.9
1997 [6]	63.6	94.2	106.6	65.8	53.1	80.6	(6)	59.3	56.8	70.3
1996 [6]	64.1	97.5	110.7	66.5	55.1	84.2	(6)	59.6	57.1	70.7
1995 [6]	64.6	98.8	109.9	71.3	52.2	89.1	(6)	60.2	57.5	72.8
1994 [6]	65.9	100.7	109.9	78.2	53.6	93.2	(6)	61.6	58.2	77.5
1993 [6]	67.0	103.3	110.9	79.8	53.9	101.5	(6)	62.7	58.9	81.5
1992 [6,7]	68.4	106.1	113.3	87.9	49.3	104.7	(6)	64.2	60.0	84.5
1991 [6,7]	69.3	106.9	114.9	87.9	47.6	105.5	(6)	65.2	60.9	87.0
1990 [4,6]	70.9	107.7	118.9	82.9	52.6	102.7	(6)	67.1	62.8	89.0
1989 [5,6]	69.2	104.9	106.6	86.6	49.8	95.8	(6)	65.7	60.5	84.8

Note: Race and Hispanic origin are reported separately on birth certificates. Persons of Hispanic origin may be of any race. In this table, Hispanic women are classified only by place of origin; non-Hispanic women are classified by race.

[1] Includes origin not stated.
[2] Includes races other than White and Black.
[3] Excludes data for New Hampshire, which did not report Hispanic origin.
[4] Excludes data for New Hampshire and Oklahoma, which did not report Hispanic origin.
[5] Excludes data for Louisiana, New Hampshire, and Oklahoma, which did not report Hispanic origin.
[6] Rates for the Central and South American population includes other and unknown Hispanic.
[7] Rates are estimated for the United States based on birth data for 49 states and the District of Columbia. Births for New Hampshire, which did not report Hispanic origin, are included in the rates for non-Hispanic women.

TABLE I-7. LIVE BIRTHS BY AGE, LIVE-BIRTH ORDER, SPECIFIED HISPANIC ORIGIN, AND RACE OF MOTHER, 2002

This table shows the number of births in the various age and birth order categories. Detail is provided for countries of origin of Hispanics and data are shown for non-Hispanic Whites as well as non-Hispanic Blacks. The organization of the data is similar to that in Table 2. The ethnic and racial data are along the lines of Table 6.

Table I-7. Live Births by Age, Live-Birth Order, Specified Hispanic Origin, and Race of Mother, 2002

(Live-birth order refers to number of children born alive to mother. Includes births with stated origin of mother only.)

Live-birth order and origin of mother	All ages	Age of mother													
		Under 15 years	15–19 years						20–24 years	25–29 years	30–34 years	35–39 years	40–44 years	45–49 years	50–54 years
			Total	15 years	16 years	17 years	18 years	19 years							
HISPANIC															
Total	876 642	2 421	127 900	6 647	15 321	24 772	35 123	46 037	265 235	236 143	157 887	71 480	14 809	751	16
1st child	320 582	2 354	97 956	6 318	13 917	20 666	26 388	30 667	114 879	62 031	30 560	10 696	1 991	110	5
2nd child	268 910	61	24 913	299	1 274	3 658	7 440	12 242	95 358	81 706	46 440	17 420	2 890	121	1
3rd child	166 128	-	4 131	16	85	350	1 083	2 597	39 493	57 209	43 082	18 842	3 218	151	2
4th child	72 829	-	511	2	10	25	105	369	11 491	23 431	22 402	12 255	2 624	111	4
5th child	27 473	-	55	-	1	4	15	35	2 684	7 668	9 039	6 289	1 662	74	2
6th child	10 634	-	4	-	-	2	-	2	584	2 415	3 573	2 993	1 005	59	1
7th child	4 504	-	2	-	-	-	1	1	116	772	1 482	1 459	624	49	-
8th child and over	3 636	-	-	-	-	-	-	-	52	375	1 001	1 365	767	75	1
Not stated	1 946	6	328	12	34	67	91	124	578	536	308	161	28	1	-
Mexican	627 505	1 849	97 744	5 126	11 838	19 098	26 811	34 871	196 864	170 146	106 177	45 129	9 137	448	11
1st child	221 757	1 798	74 314	4 862	10 731	15 821	19 941	22 959	82 578	40 046	16 831	5 195	935	56	4
2nd child	189 759	45	19 473	238	1 002	2 914	5 841	9 478	71 987	58 810	28 733	9 246	1 408	56	1
3rd child	122 873	-	3 264	14	67	283	869	2 031	30 264	43 912	31 304	12 200	1 851	76	2
4th child	55 841	-	404	2	9	22	85	286	8 911	18 292	17 346	8 978	1 843	66	1
5th child	21 249	-	46	-	1	3	11	31	2 081	5 963	7 063	4 821	1 226	48	1
6th child	8 319	-	4	-	-	2	-	2	468	1 878	2 793	2 356	773	46	1
7th child	3 541	-	2	-	-	-	1	1	92	602	1 155	1 165	487	38	-
8th child and over	2 809	-	-	-	-	-	-	-	42	286	755	1 066	597	62	1
Not stated	1 357	6	237	10	28	53	63	83	441	357	197	102	17	-	-
Puerto Rican	57 465	205	10 006	524	1 254	1 992	2 721	3 515	18 724	13 841	9 415	4 385	836	53	-
1st child	22 369	200	7 684	513	1 133	1 694	2 068	2 276	7 565	3 602	2 248	883	181	6	-
2nd child	17 742	5	1 955	11	113	274	568	989	6 589	4 607	3 039	1 318	216	13	-
3rd child	10 268	-	302	-	6	20	68	208	3 179	3 298	2 223	1 070	183	13	-
4th child	4 145	-	44	-	1	3	9	31	1 026	1 403	987	565	111	9	-
5th child	1 673	-	3	-	-	-	1	2	269	572	493	267	64	5	-
6th child	625	-	-	-	-	-	-	-	50	205	209	128	31	2	-
7th child	275	-	-	-	-	-	-	-	12	81	103	61	15	3	-
8th child and over	266	-	-	-	-	-	-	-	2	42	97	88	35	2	-
Not stated	102	-	18	-	1	1	7	9	32	31	16	5	-	-	-
Cuban	14 232	13	1 146	37	108	222	349	430	2 410	4 025	3 881	2 283	442	30	2
1st child	6 554	13	967	36	99	198	298	336	1 490	1 967	1 402	591	115	9	-
2nd child	5 103	-	156	1	8	20	46	81	690	1 499	1 700	907	148	3	-
3rd child	1 866	-	21	-	1	4	5	11	183	419	578	544	110	11	-
4th child	486	-	1	-	-	-	-	1	39	90	138	171	42	4	1
5th child	118	-	-	-	-	-	-	-	4	27	36	39	10	1	1
6th child	57	-	-	-	-	-	-	-	3	9	18	20	6	1	-
7th child	16	-	-	-	-	-	-	-	-	3	6	4	3	-	-
8th child and over	18	-	-	-	-	-	-	-	-	3	2	6	6	1	-
Not stated	14	-	1	-	-	-	-	1	1	8	1	1	2	-	-
Central and South American	125 981	170	10 580	473	1 070	1 830	2 923	4 284	31 547	35 429	29 222	15 366	3 481	185	1
1st child	49 915	165	8 642	449	1 002	1 613	2 365	3 213	16 721	12 748	7 806	3 207	594	32	-
2nd child	40 241	5	1 625	23	62	189	479	872	10 383	12 513	10 111	4 680	881	43	-
3rd child	21 981	-	263	-	4	22	69	168	3 427	6 677	6 743	3 953	874	44	-
4th child	8 619	-	22	-	-	-	4	18	770	2 423	2 887	1 976	510	30	1
5th child	3 084	-	2	-	-	1	1	-	140	725	1 036	881	286	14	-
6th child	1 098	-	-	-	-	-	-	-	34	192	373	346	147	6	-
7th child	457	-	-	-	-	-	-	-	7	52	139	164	88	7	-
8th child and over	338	-	-	-	-	-	-	-	3	21	76	134	95	9	-
Not stated	248	-	26	1	2	5	5	13	62	78	51	25	6	-	-
Other and Unknown Hispanic	51 459	184	8 424	487	1 051	1 630	2 319	2 937	15 690	12 702	9 192	4 317	913	35	2
1st child	19 987	178	6 349	458	952	1 340	1 716	1 883	6 525	3 668	2 273	820	166	7	1
2nd child	16 065	6	1 704	26	89	261	506	822	5 709	4 277	2 857	1 269	237	6	-
3rd child	9 140	-	281	2	7	21	72	179	2 440	2 903	2 234	1 075	200	7	-
4th child	3 738	-	40	-	-	-	7	33	745	1 223	1 044	565	118	2	1
5th child	1 349	-	4	-	-	-	2	2	190	381	411	281	76	6	-
6th child	535	-	-	-	-	-	-	-	29	131	180	143	48	4	-
7th child	215	-	-	-	-	-	-	-	5	34	79	65	31	1	-
8th child and over	205	-	-	-	-	-	-	-	5	23	71	71	34	1	-
Not stated	225	-	46	1	3	8	16	18	42	62	43	28	3	1	-

Note: Race and Hispanic origin are reported separately on birth certificates. Persons of Hispanic origin may be of any race. In this table Hispanic women are classified only by place of origin; non-Hispanic women are classified by race.

- = Quantity zero.

Table I-7. Live Births by Age, Live-Birth Order, Specified Hispanic Origin, and Race of Mother, 2002 —*Continued*

(Live-birth order refers to number of children born alive to mother. Includes births with stated origin of mother only.)

Live-birth order and origin of mother	All ages	Age of mother													
		Under 15 years	15–19 years						20–24 years	25–29 years	30–34 years	35–39 years	40–44 years	45–49 years	50–54 years
			Total	15 years	16 years	17 years	18 years	19 years							
NON-HISPANIC [1]															
Total [2]	3 119 944	4 840	295 235	11 949	27 607	51 688	82 883	121 108	750 956	817 969	787 077	379 113	80 128	4 396	230
1st child	1 264 626	4 744	236 288	11 488	25 622	45 104	66 613	87 461	355 455	314 379	243 571	90 703	18 253	1 175	58
2nd child	1 030 613	83	49 287	415	1 813	5 915	13 976	27 168	252 494	278 831	294 380	130 517	23 806	1 152	63
3rd child	505 259	2	7 920	24	99	528	1 916	5 353	101 967	140 808	151 222	85 945	16 633	721	41
4th child	189 824	1	951	-	3	31	181	736	29 829	53 056	57 791	38 618	9 095	462	21
5th child	67 804	-	127	-	-	5	21	101	7 380	18 695	21 071	15 568	4 691	251	21
6th child	27 840	-	20	1	-	-	4	15	1 686	6 782	9 244	7 397	2 558	147	6
7th child	12 826	-	3	-	-	-	-	3	382	2 428	4 279	3 970	1 628	128	8
8th child and over	14 264	-	2	-	-	-	-	2	149	1 272	3 843	5 463	3 188	335	12
Not stated	6 888	10	637	21	70	105	172	269	1 614	1 718	1 676	932	276	25	-
White	2 298 156	1 493	179 511	5 130	14 421	30 205	51 196	78 559	519 153	614 909	620 173	297 436	61 853	3 444	184
1st child	938 374	1 461	148 559	5 002	13 677	27 293	42 929	59 658	259 314	246 081	194 681	72 685	14 568	975	50
2nd child	780 784	27	26 820	114	685	2 653	7 372	15 996	176 900	216 472	237 547	103 246	18 790	929	53
3rd child	370 714	2	3 395	6	24	181	724	2 460	63 068	102 448	119 838	68 616	12 757	552	38
4th child	130 046	-	310	-	1	9	59	241	15 294	34 330	43 007	29 912	6 826	349	18
5th child	41 940	-	42	-	-	1	8	33	2 899	10 063	14 113	11 222	3 408	180	13
6th child	16 074	-	3	-	-	-	-	3	527	3 047	5 557	5 039	1 794	104	3
7th child	7 178	-	2	-	-	-	-	2	95	907	2 367	2 579	1 133	92	3
8th child and over	8 354	-	1	-	-	-	-	1	50	391	1 840	3 444	2 376	246	6
Not stated	4 692	3	379	8	34	68	104	165	1 006	1 170	1 223	693	201	17	-
Black	578 335	3 132	101 494	6 191	11 846	18 980	27 656	36 821	190 241	132 824	92 155	46 831	11 097	547	14
1st child	216 525	3 073	76 481	5 880	10 699	15 676	20 513	23 713	73 705	32 466	20 152	8 626	1 925	96	1
2nd child	168 579	53	20 006	282	1 043	2 931	5 877	9 873	63 217	41 220	27 903	13 420	2 633	126	1
3rd child	102 961	-	4 109	17	69	321	1 076	2 626	34 229	30 487	20 995	10 643	2 399	99	-
4th child	48 263	1	577	-	2	19	113	443	13 128	15 578	11 122	6 191	1 592	72	2
5th child	21 302	-	76	-	-	2	11	63	4 057	7 370	5 507	3 303	939	44	6
6th child	9 742	-	17	1	-	-	4	12	1 059	3 230	2 970	1 840	592	33	1
7th child	4 630	-	1	-	-	-	-	1	263	1 321	1 569	1 082	367	26	1
8th child and over	4 693	-	1	-	-	-	-	1	88	765	1 627	1 563	599	48	2
Not stated	1 640	5	226	11	33	31	62	89	495	387	310	163	51	3	-

Note: Race and Hispanic origin are reported separately on birth certificates. Persons of Hispanic origin may be of any race. In this table Hispanic women are classified only by place of origin; non-Hispanic women are classified by race.

[1]Includes origin not stated, not shown separately.
[2]Includes races other than White and Black.
- = Quantity zero.

TABLE I-8. FERTILITY RATES AND BIRTH RATES BY AGE, LIVE-BIRTH ORDER, SPECIFIED HISPANIC ORIGIN, AND RACE OF MOTHER, 2002

This table converts the numerical birth data of Table 7 into general fertility rates and birth rates. It highlights the high fertility rate of Mexican-origin mothers, which is 102.8. This compares with a rate of 57.4 for non-Hispanic White mothers. The fertility rate of all Hispanics averages 94.4 and thus well exceeds that of non-Hispanic Whites as well as non-Hispanic Blacks. Hispanic birth rates are especially high in the age groups for mothers under 25 years. The birth rates for non-Hispanic Blacks prior to age 25 are below the ones for Hispanics, but they are high when compared with non-Hispanic Whites. Birth rates per 1,000 females aged 15–19 were 94.5 for Mexicans, 28.5 for non-Hispanic Whites, 61.4 Puerto Ricans, 63.0 "other" Hispanics, and 68.3 non-Hispanic Blacks. For age groups 25 and older, the birth rates of non-Hispanic White mothers are somewhat higher than those of non-Hispanic Blacks.

Table I-8. Fertility Rates and Birth Rates by Age, Live-Birth Order, Specified Hispanic Origin, and Race of Mother, 2002

(Birth rates are live births per 1,000 women in specified age and racial group. Figures for live-birth order not stated are distributed.)

Live-birth order and race of mother	Fertility rate [1]	Age of mother									
		10–14 years	15–19 years			20–24 years	25–29 years	30–34 years	35–39 years	40–44 years	45–49 years [2]
			Total	15–17 years	18–19 years						
HISPANIC											
Total	94.4	1.4	83.4	50.7	133.0	164.3	139.4	95.1	47.8	11.5	0.7
1st child	34.6	1.4	64.1	44.5	93.7	71.3	36.7	18.5	7.2	1.6	0.1
2nd child	29.0	0.0	16.3	5.7	32.3	59.2	48.3	28.0	11.7	2.3	0.1
3rd child	17.9	*	2.7	0.5	6.0	24.5	33.8	26.0	12.6	2.5	0.1
4th child	7.9	*	0.3	0.0	0.8	7.1	13.9	13.5	8.2	2.0	0.1
5th child	3.0	*	0.0	*	0.1	1.7	4.5	5.5	4.2	1.3	0.1
6th and 7th child	1.6	*	*	*	*	0.4	1.9	3.1	3.0	1.3	0.1
8th child and over	0.4	*	*	*	*	0.0	0.2	0.6	0.9	0.6	0.1
Mexican	102.8	1.5	94.5	58.6	147.5	176.9	144.5	97.9	47.5	12.3	0.8
1st child	36.4	1.5	72.1	51.1	102.8	74.4	34.1	15.5	5.5	1.3	0.1
2nd child	31.2	0.0	18.9	6.8	36.7	64.8	50.1	26.5	9.8	1.9	0.1
3rd child	20.2	*	3.2	0.6	7.0	27.3	37.4	28.9	12.9	2.5	0.1
4th child	9.2	*	0.4	0.1	0.9	8.0	15.6	16.0	9.5	2.5	0.1
5th child	3.5	*	0.0	*	0.1	1.9	5.1	6.5	5.1	1.7	0.1
6th and 7th child	1.9	*	*	*	*	0.5	2.1	3.6	3.7	1.7	0.1
8th child and over	0.5	*	*	*	*	0.0	0.2	0.7	1.1	0.8	0.1
Puerto Rican	65.4	1.4	61.4	39.7	*	136.5	90.6	61.5	31.3	6.3	0.5
1st child	25.5	1.3	47.3	35.2	*	55.2	23.6	14.7	6.3	1.4	*
2nd child	20.2	*	12.0	4.2	*	48.1	30.2	19.9	9.4	1.6	*
3rd child	11.7	*	1.9	0.3	*	23.2	21.6	14.6	7.6	1.4	*
4th child	4.7	*	0.3	*	*	7.5	9.2	6.5	4.0	0.8	*
5th child	1.9	*	*	*	*	2.0	3.8	3.2	1.9	0.5	*
6th and 7th child	1.0	*	*	*	*	0.5	1.9	2.0	1.3	0.3	*
8th child and over	0.3	*	*	*	*	*	0.3	0.6	0.6	0.3	*
Cuban	59.0	*	*	*	*	*	*	*	*	*	*
1st child	27.2	*	*	*	*	*	*	*	*	*	*
2nd child	21.2	*	*	*	*	*	*	*	*	*	*
3rd child	7.7	*	*	*	*	*	*	*	*	*	*
4th child	2.0	*	*	*	*	*	*	*	*	*	*
5th child	0.5	*	*	*	*	*	*	*	*	*	*
6th and 7th child	0.3	*	*	*	*	*	*	*	*	*	*
8th child and over	*	*	*	*	*	*	*	*	*	*	*
Other Hispanic [3]	86.1	1.1	63.0	34.7	110.3	143.3	147.2	98.4	56.1	12.2	0.8
1st child	34.0	1.1	49.9	30.9	81.5	70.7	50.3	25.9	11.5	2.1	0.1
2nd child	27.4	*	11.1	3.5	23.8	48.9	51.5	33.3	17.0	3.1	0.2
3rd child	15.1	*	1.8	0.3	4.3	17.8	29.4	23.0	14.4	3.0	0.2
4th child	6.0	*	0.2	*	0.5	4.6	11.2	10.1	7.3	1.7	0.1
5th child	2.2	*	*	*	*	1.0	3.4	3.7	3.3	1.0	0.1
6th and 7th child	1.1	*	*	*	*	0.2	1.3	2.0	2.1	0.9	*
8th child and over	0.3	*	*	*	*	*	0.1	0.4	0.6	0.4	*
NON-HISPANIC [4]											
Total [5]	59.6	0.6	35.5	18.2	61.8	91.8	107.9	90.8	40.4	7.9	0.5
1st child	24.2	0.6	28.5	16.5	46.8	43.5	41.6	28.2	9.7	1.8	0.1
2nd child	19.7	0.0	5.9	1.6	12.5	30.9	36.8	34.0	13.9	2.3	0.1
3rd child	9.7	*	1.0	0.1	2.2	12.5	18.6	17.5	9.2	1.6	0.1
4th child	3.6	*	0.1	0.0	0.3	3.7	7.0	6.7	4.1	0.9	0.1
5th child	1.3	*	0.0	*	0.0	0.9	2.5	2.4	1.7	0.5	0.0
6th and 7th child	0.8	*	0.0	*	0.0	0.3	1.2	1.6	1.2	0.4	0.0
8th child and over	0.3	*	*	*	*	0.0	0.2	0.4	0.6	0.3	0.0
White	57.4	0.2	28.5	13.1	51.9	84.3	109.3	94.4	40.9	7.6	0.5
1st child	23.5	0.2	23.7	12.2	41.1	42.2	43.8	29.7	10.0	1.8	0.1
2nd child	19.5	0.0	4.3	0.9	9.4	28.8	38.5	36.2	14.2	2.3	0.1
3rd child	9.3	*	0.5	0.1	1.3	10.3	18.3	18.3	9.5	1.6	0.1
4th child	3.3	*	0.0	*	0.1	2.5	6.1	6.6	4.1	0.8	0.0
5th child	1.1	*	0.0	*	0.0	0.5	1.8	2.2	1.5	0.4	0.0
6th and 7th child	0.6	*	*	*	*	0.1	0.7	1.2	1.1	0.4	0.0
8th child and over	0.2	*	*	*	*	0.0	0.1	0.3	0.5	0.3	0.0
Black	67.4	1.9	68.3	41.0	110.3	131.0	102.1	66.1	32.1	7.5	0.4
1st child	25.3	1.9	51.6	35.8	75.9	50.9	25.0	14.5	5.9	1.3	0.1
2nd child	19.7	0.0	13.5	4.7	27.0	43.6	31.8	20.1	9.2	1.8	0.1
3rd child	12.0	*	2.8	0.5	6.3	23.6	23.5	15.1	7.3	1.6	0.1
4th child	5.6	*	0.4	0.0	1.0	9.1	12.0	8.0	4.3	1.1	0.1
5th child	2.5	*	0.1	*	0.1	2.8	5.7	4.0	2.3	0.6	0.0
6th and 7th child	1.7	*	*	*	*	0.9	3.5	3.3	2.0	0.6	0.0
8th child and over	0.5	*	*	*	*	0.1	0.6	1.2	1.1	0.4	0.0

Note: Race and Hispanic origin are reported separately on birth certificates. Persons of Hispanic origin may be of any race. In this table Hispanic women are classified only by place of origin; non-Hispanic women are classified by race.

[1] Fertility rates computed by relating total births, regardless of age of mother, to women aged 15–44 years.
[2] Birth rates computed by relating births to women aged 45–54 years to women aged 45–49 years.
[3] Includes Central and South American and other and unknown Hispanic.
[4] Includes origin not stated.
[5] Includes races other than White and Black.
0.0 = Quantity more than zero but less than 0.05.
* = Figure does not meet standards of reliability or precision; based on fewer than 20 births in the numerator or, for the Hispanic subgroups, fewer than 75,000 women in the denominator.

TABLE I-9. TOTAL FERTILITY RATES, FERTILITY RATES, AND BIRTH RATES BY AGE, SPECIFIED HISPANIC ORIGIN, AND RACE OF MOTHER, 1989–2002

This table is organized similarly to Table 4: it shows the total fertility rate (for definition, see Table 4 comment), the general fertility rate, and birth rates by age of mother. Time series starting in 1989 are presented—in contrast to a starting date of 1970 in Table 4. The focus in Table 9 is on Hispanics (and subgroups of Hispanics), and these are compared with non-Hispanics of different races. As would be expected, the total fertility rate of Hispanics in 2002 is high, at 2,718 per 1,000 women during the course of their childbearing age. This compares with an average of 2,047 for non-Hispanic Blacks and 1,828.5 for non-Hispanic Whites. Although the total

Table I-9. Total Fertility Rates, Fertility Rates, and Birth Rates by Age, Specified Hispanic Origin, and Race of Mother, 1989–2002

(Birth rates are live births per 1,000 women in specified age and racial group.)

Year and origin/race of mother	Total fertility rate [1]	Fertility rate [2]	Age of mother									
			10–14 years	15–19 years			20–24 years	25–29 years	30–34 years	35–39 years	40–44 years	45–49 years [3]
				Total	15–17 years	18–19 years						
ALL ORIGINS												
2002	2 013.0	64.8	0.7	43.0	23.2	72.8	103.6	113.6	91.5	41.4	8.3	0.5
2001	2 034.0	65.3	0.8	45.3	24.7	76.1	106.2	113.4	91.9	40.6	8.1	0.5
2000	2 056.0	65.9	0.9	47.7	26.9	78.1	109.7	113.5	91.2	39.7	8.0	0.5
1999	2 007.5	64.4	0.9	48.8	28.2	79.1	107.9	111.2	87.1	37.8	7.4	0.4
1998	1 999.0	64.3	1.0	50.3	29.9	80.9	108.4	110.2	85.2	36.9	7.4	0.4
1997	1 971.0	63.6	1.1	51.3	31.4	82.1	107.3	108.3	83.0	35.7	7.1	0.4
1996	1 976.0	64.1	1.2	53.5	33.3	84.7	107.8	108.6	82.1	34.9	6.8	0.3
1995	1 978.0	64.6	1.3	56.0	35.5	87.7	107.5	108.8	81.1	34.0	6.6	0.3
1994	2 001.5	65.9	1.4	58.2	37.2	90.2	109.2	111.0	80.4	33.4	6.4	0.3
1993	2 019.5	67.0	1.4	59.0	37.5	91.1	111.3	113.2	79.9	32.7	6.1	0.3
1992	2 046.0	68.4	1.4	60.3	37.6	93.6	113.7	115.7	79.6	32.3	5.9	0.3
1991	2 062.5	69.3	1.4	61.8	38.6	94.0	115.3	117.2	79.2	31.9	5.5	0.2
1990	2 081.0	70.9	1.4	59.9	37.5	88.6	116.5	120.2	80.8	31.7	5.5	0.2
1989	2 014.0	69.2	1.4	57.3	36.4	84.2	113.8	117.6	77.4	29.9	5.2	0.2
HISPANIC												
Total												
2002	2 718.0	94.4	1.4	83.4	50.7	133.0	164.3	139.4	95.1	47.8	11.5	0.7
2001	2 748.5	96.0	1.6	86.4	52.8	135.5	163.5	140.4	97.6	47.9	11.6	0.7
2000	2 730.0	95.9	1.7	87.3	55.5	132.6	161.3	139.9	97.1	46.6	11.5	0.6
1999	2 649.0	93.0	1.9	86.8	56.9	129.8	157.3	135.8	92.3	44.5	10.6	0.6
1998	2 652.5	93.2	1.9	87.9	58.4	131.7	159.3	136.1	90.5	43.4	10.8	0.6
1997	2 680.5	94.2	2.1	89.6	61.0	132.9	162.6	137.5	89.6	43.4	10.7	0.6
1996	2 772.0	97.5	2.4	94.6	64.1	140.5	170.2	140.7	91.3	43.9	10.7	0.6
1995	2 798.5	98.8	2.6	99.3	67.9	146.7	171.9	140.4	90.5	43.7	10.7	0.6
1994	2 839.0	100.7	2.6	101.3	69.6	148.3	175.7	142.4	91.1	43.4	10.7	0.6
1993	2 894.5	103.3	2.6	101.8	68.4	151.5	180.0	146.0	93.2	44.1	10.6	0.6
1992 [4]	2 957.5	106.1	2.5	103.3	68.9	153.9	185.2	148.8	94.8	45.3	11.0	0.6
1991 [4]	2 963.5	106.9	2.4	104.6	69.3	155.1	184.6	150.0	95.1	44.7	10.7	0.6
1990 [5]	2 959.5	107.7	2.4	100.3	65.9	147.7	181.0	153.0	98.3	45.3	10.9	0.7
1989 [6]	2 903.5	104.9	2.3	100.8	. . .	. . .	184.4	146.6	92.1	43.5	10.4	0.6
Mexican												
2002	2 879.5	102.8	1.5	94.5	58.6	147.5	176.9	144.5	97.9	47.5	12.3	0.8
2001	2 928.5	105.7	1.7	95.4	59.3	147.0	177.0	146.4	101.9	50.0	12.6	0.7
2000	2 906.5	105.1	1.9	95.4	60.6	146.7	174.9	144.7	102.3	49.2	12.2	0.7
1999	2 823.0	101.5	2.1	94.3	60.7	146.0	170.8	141.4	97.4	47.2	10.7	0.7
1998	2 878.0	103.2	2.1	96.4	62.8	149.6	176.5	147.4	94.9	46.9	10.8	0.6
1997	2 957.0	106.6	2.3	103.4	71.1	152.1	180.9	150.0	95.3	47.4	11.5	0.6
1996	3 052.0	110.7	2.6	112.2	77.5	162.1	185.3	154.7	96.5	46.4	12.0	0.7
1995	3 034.0	109.9	2.8	115.9	78.7	172.2	190.4	146.6	93.0	45.5	11.9	0.7
1994	3 024.0	109.9	2.7	109.2	73.4	164.2	189.1	153.6	92.5	45.3	11.7	0.7
1993	3 041.5	110.9	2.5	103.6	68.3	157.0	187.9	159.5	97.2	45.5	11.3	0.8
1992 [4]	3 107.0	113.3	2.4	105.1	. . .	. . .	196.6	160.2	97.1	47.4	11.8	0.8
1991 [4]	3 103.5	114.9	2.5	108.3	70.2	164.3	192.4	156.1	99.7	49.1	11.9	0.7
1990 [5]	3 214.0	118.9	2.5	108.0	69.7	162.2	200.3	165.3	104.4	49.1	12.4	0.8
1989 [6]	2 916.5	106.6	2.0	94.5	. . .	. . .	184.3	153.7	96.1	41.0	11.1	0.6

Note: Race and Hispanic origin are reported separately on birth certificates. Persons of Hispanic origin may be of any race. In this table Hispanic women are classified only by place of origin; non-Hispanic women are classified by race.

[1]Total fertility rates are sums of birth rates for 5-year age groups multiplied by 5.
[2]Fertility rates computed by relating total births, regardless of age of mother, to women 15–44 years.
[3]Beginning in 1997, rates computed by relating births to women aged 45–54 years to women aged 45–49 years.
[4]Excludes data for New Hampshire, which did not report Hispanic origin.
[5]Excludes data for New Hampshire and Oklahoma, which did not report Hispanic origin.
[6]Excludes data for Louisiana, New Hampshire, and Oklahoma, which did not report Hispanic origin.
. . . = Not available.

fertility rate for most groups was below the "replacement rate," it was above "replacement" (2,100 births per 1,000 women) for Mexican and "other" Hispanics.

Beginning in 1993, all states reported data for the subcategories of Table 9. Total fertility for Hispanics has declined some 6 percent since 1993, while that for non-Hispanic Blacks declined 15 percent during this period. The rate for non-Hispanic Whites edged higher by 2 percent in the same period.

The drop in teenage birth rates, which was noted in Table 4, occurred widely among races and ethnic groups. For data since 1989, the drop was especially pronounced among non-Hispanic teenagers. Although Hispanics overall also participated in this decline, their teenage birth

Table I-9. Total Fertility Rates, Fertility Rates, and Birth Rates by Age, Specified Hispanic Origin, and Race of Mother, 1989–2002—*Continued*

(Birth rates are live births per 1,000 women in specified age and racial group.)

Year and origin/race of mother	Total fertility rate [1]	Fertility rate [2]	Age of mother									
			10–14 years	15–19 years			20–24 years	25–29 years	30–34 years	35–39 years	40–44 years	45–49 years [3]
				Total	15–17 years	18–19 years						
Puerto Rican												
2002	1 947.5	65.4	1.4	61.4	39.7	*	136.5	90.6	61.5	31.3	6.3	0.5
2001	2 165.0	72.2	1.7	82.2	*	*	147.2	93.6	70.5	30.7	6.7	0.4
2000	2 178.5	73.5	1.7	82.9	54.7	120.4	149.5	101.6	61.1	32.0	6.6	0.3
1999	2 104.5	71.1	1.6	74.0	49.3	*	146.0	106.5	58.0	27.3	7.2	0.3
1998	2 043.5	69.7	1.8	76.2	51.7	*	146.7	88.7	61.9	25.8	7.2	0.4
1997	1 931.5	65.8	1.7	68.9	44.9	*	136.0	92.9	54.1	26.1	6.2	0.4
1996	1 965.0	66.5	1.9	76.5	48.5	*	133.7	95.6	54.3	25.2	5.6	*
1995	2 076.5	71.3	2.6	82.8	57.0	*	138.1	97.9	61.2	26.9	5.5	0.3
1994	2 341.5	78.2	3.1	99.6	68.5	*	169.0	103.8	59.5	27.5	5.6	0.2
1993	2 416.0	79.8	3.1	104.9	70.0	*	184.6	102.8	54.4	26.7	6.2	*
1992 [4]	2 569.0	87.9	3.4	106.5	. . .	. . .	199.1	102.7	65.3	29.9	6.6	*
1991 [4]	2 573.5	87.9	2.7	111.0	*	*	193.3	108.9	68.1	23.9	6.5	*
1990 [5]	2 301.0	82.9	2.9	101.6	71.6	141.6	150.1	109.9	62.8	26.2	6.2	0.5
1989 [6]	2 421.0	86.6	3.8	112.7	. . .	. . .	171.0	98.0	65.2	26.9	6.3	*
Cuban												
2002	1 940.5	59.0	*	*	*	*	*	*	*	*	*	*
2001	1 792.5	56.7	*	*	*	*	*	*	*	*	*	*
2000	1 528.0	49.3	*	23.5	14.2	43.4	64.2	104.0	68.1	37.3	7.9	*
1999	1 388.5	47.0	*	*	*	*	*	*	*	*	*	*
1998	1 402.5	46.5	*	*	*	*	*	*	*	*	*	*
1997	1 619.5	53.1	*	*	*	*	*	*	*	*	*	*
1996	1 617.0	55.1	*	*	*	*	*	*	*	*	*	*
1995	1 583.5	52.2	*	*	*	*	*	*	*	*	*	*
1994	1 587.0	53.6	*	*	*	*	*	*	*	*	*	*
1993	1 570.0	53.9	*	*	*	*	*	*	*	*	*	*
1992 [4]	1 447.5	49.3	*	*	. . .	. . .	*	*	*	*	*	*
1991 [4]	1 352.5	47.6	*	*	*	*	*	*	*	*	*	*
1990 [5]	1 459.5	52.6	*	30.3	18.2	46.1	64.6	95.4	67.6	28.2	4.9	*
1989 [6]	1 479.0	49.8	*	*	. . .	. . .	*	*	*	*	*	*
Other Hispanic [7]												
2002	2 610.5	86.1	1.1	63.0	34.7	110.3	143.3	147.2	98.4	56.1	12.2	0.8
2001	2 519.5	82.7	1.1	65.3	35.6	115.2	136.0	143.3	95.4	50.3	11.6	0.9
2000	2 563.5	85.1	1.2	69.9	44.4	102.0	133.2	143.9	103.6	47.7	12.5	0.7
1999	2 517.0	84.8	1.5	75.5	53.0	100.7	130.2	138.4	98.3	46.5	12.3	0.7
1998	2 448.5	83.5	1.8	75.0	53.2	100.5	122.7	133.6	97.8	45.4	12.8	0.6
1997	2 376.5	80.6	1.8	66.4	44.4	98.4	129.3	125.8	95.6	43.9	11.8	0.7
1996	2 516.5	84.2	2.2	64.8	43.3	95.9	149.6	127.9	98.0	49.1	11.0	0.7
1995	2 628.5	89.1	2.1	72.1	51.0	100.2	144.3	147.7	97.9	49.4	11.6	0.6
1994	2 693.0	93.2	2.5	82.6	62.5	105.6	151.2	137.0	104.4	48.4	11.9	0.6
1993	2 914.5	101.5	2.6	102.0	74.6	135.0	167.5	139.4	106.7	51.7	12.5	0.5
1992 [4]	2 990.0	104.7	2.4	108.2	. . .	. . .	168.0	152.0	104.4	50.0	12.5	0.5
1991 [4]	3 064.5	105.5	2.2	100.7	67.4	145.3	184.1	164.5	100.2	49.2	11.4	0.6
1990 [5]	2 877.0	102.7	2.1	86.0	57.2	123.8	162.9	155.8	106.9	49.4	11.6	0.7
1989 [6]	2 683.0	95.8	1.7	66.4	. . .	. . .	159.2	150.4	85.1	60.3	12.7	0.8

Note: Race and Hispanic origin are reported separately on birth certificates. Persons of Hispanic origin may be of any race. In this table Hispanic women are classified only by place of origin; non-Hispanic women are classified by race.

[1]Total fertility rates are sums of birth rates for 5-year age groups multiplied by 5.
[2]Fertility rates computed by relating total births, regardless of age of mother, to women 15–44 years.
[3]Beginning in 1997, rates computed by relating births to women aged 45–54 years to women aged 45–49 years.
[4]Excludes data for New Hampshire, which did not report Hispanic origin.
[5]Excludes data for New Hampshire and Oklahoma, which did not report Hispanic origin.
[6]Excludes data for Louisiana, New Hampshire, and Oklahoma, which did not report Hispanic origin.
[7]Includes Central and South American and other and unknown Hispanic.
. . . = Not available.
* = Figure does not meet standards of reliability or precision; based on fewer than 20 births in the numerator or, for the Hispanic subgroups, fewer than 50 women for census years and 75,000 women for noncensus years in the denominator.

rate is considerably higher than that for other groups. This is driven by very high teenage birth rates among Mexicans and moderately high teenage birth rates among Puerto Ricans and "other" Hispanics (excluding Cubans). The cumulative declines in the birth rate for mothers 15–17 years of age is shown for all ethnic and racial groups. This decline is particularly large for non-Hispanic Blacks.

As indicated in Table 4, the birth rate for women in their 30s and 40s has been increasing since 1989, while birth rates for women in their 20s have declined. Blacks, Whites, and Hispanics all participated in these trends.

Table I-9. Total Fertility Rates, Fertility Rates, and Birth Rates by Age, Specified Hispanic Origin, and Race of Mother, 1989–2002—*Continued*

(Birth rates are live births per 1,000 women in specified age and racial group.)

Year and origin/race of mother	Total fertility rate [1]	Fertility rate [2]	Age of mother									
			10–14 years	15–19 years			20–24 years	25–29 years	30–34 years	35–39 years	40–44 years	45–49 years [3]
				Total	15–17 years	18–19 years						
NON-HISPANIC [8]												
Total [9]												
2002	1 877.0	59.6	0.6	35.5	18.2	61.8	91.8	107.9	90.8	40.4	7.9	0.5
2001	1 898.5	60.1	0.6	37.9	19.6	65.2	94.9	107.7	90.9	39.5	7.7	0.5
2000	1 931.5	61.1	0.7	40.7	21.9	68.2	99.5	108.4	90.2	38.8	7.6	0.4
1999	1 894.0	60.0	0.8	42.2	23.2	70.2	98.4	106.7	86.2	37.0	7.1	0.4
1998	1 887.5	60.0	0.8	44.0	25.2	72.4	98.9	105.8	84.4	36.2	7.0	0.4
1997	1 853.0	59.3	0.9	45.0	26.7	73.7	97.4	103.5	82.0	34.8	6.7	0.3
1996	1 852.0	59.6	1.0	47.0	28.5	75.7	97.3	103.6	80.8	33.9	6.5	0.3
1995	1 856.5	60.2	1.1	49.3	30.6	78.4	97.4	104.1	79.9	33.0	6.2	0.3
1994	1 883.5	61.6	1.2	51.7	32.4	81.3	99.5	106.5	79.1	32.4	6.0	0.3
1993	1 901.5	62.7	1.2	52.7	32.9	82.2	101.7	108.7	78.4	31.6	5.7	0.3
1992 [4]	1 929.0	64.2	1.2	54.3	33.2	85.2	104.3	111.4	77.9	31.1	5.4	0.2
1991 [4]	1 953.0	65.2	1.3	56.1	34.4	86.1	106.5	113.1	77.5	30.8	5.1	0.2
1990 [5]	1 979.5	67.1	1.3	54.8	33.8	81.4	108.1	116.5	79.2	30.7	5.1	0.2
1989 [6]	1 921.0	65.7	1.3	53.4	. . .	. . .	107.8	113.4	74.7	28.6	4.8	0.2
White												
2002	1 828.5	57.4	0.2	28.5	13.1	51.9	84.3	109.3	94.4	40.9	7.6	0.5
2001	1 843.0	57.7	0.3	30.3	14.0	54.8	87.1	108.9	94.3	39.8	7.5	0.4
2000	1 866.0	58.5	0.3	32.6	15.8	57.5	91.2	109.4	93.2	38.8	7.3	0.4
1999	1 838.5	57.7	0.3	34.1	17.1	59.3	90.6	108.6	89.5	37.3	6.9	0.4
1998	1 825.0	57.6	0.3	35.3	18.4	60.8	91.2	107.4	87.2	36.4	6.8	0.4
1997	1 785.5	56.8	0.4	36.0	19.4	61.9	90.0	104.8	84.3	34.8	6.5	0.3
1996	1 781.0	57.1	0.4	37.6	20.6	63.8	90.1	104.9	82.8	33.9	6.2	0.3
1995	1 777.5	57.5	0.4	39.3	22.0	65.9	90.2	105.1	81.5	32.8	5.9	0.3
1994	1 782.5	58.2	0.5	40.4	22.8	67.3	90.9	106.6	80.2	32.0	5.7	0.2
1993	1 786.0	58.9	0.5	40.7	22.7	67.5	92.2	108.2	79.0	31.0	5.4	0.2
1992 [4]	1 803.5	60.0	0.5	41.7	22.7	69.7	93.9	110.6	78.3	30.4	5.1	0.2
1991 [4]	1 822.5	60.9	0.5	43.4	23.7	70.6	95.7	112.1	77.7	30.2	4.7	0.2
1990 [5]	1 850.5	62.8	0.5	42.5	23.2	66.6	97.5	115.3	79.4	30.0	4.7	0.2
1989 [6]	1 770.0	60.5	0.4	39.9	. . .	. . .	94.7	111.7	75.0	27.8	4.3	0.2
Black												
2002	2 047.0	67.4	1.9	68.3	41.0	110.3	131.0	102.1	66.1	32.1	7.5	0.4
2001	2 104.5	69.1	2.1	73.5	44.9	116.7	137.2	102.1	66.2	32.1	7.3	0.4
2000	2 178.5	71.4	2.4	79.2	50.1	121.9	145.4	102.8	66.5	31.8	7.2	0.4
1999	2 134.0	69.9	2.6	81.0	52.0	123.1	142.1	99.8	63.9	30.6	6.5	0.3
1998	2 164.0	70.9	2.9	85.7	57.0	127.4	142.5	99.9	64.4	30.4	6.7	0.3
1997	2 137.5	70.3	3.2	88.3	60.9	130.4	138.8	97.2	63.6	29.6	6.5	0.3
1996	2 140.0	70.7	3.6	91.9	65.0	133.4	137.0	96.7	63.2	29.1	6.2	0.3
1995	2 186.5	72.8	4.2	97.2	70.6	138.5	137.8	98.5	64.4	28.8	6.1	0.3
1994	2 314.5	77.5	4.6	105.7	77.2	149.8	146.8	104.1	66.3	29.1	6.0	0.3
1993	2 412.5	81.5	4.6	110.5	81.3	154.2	154.5	109.2	68.1	29.4	5.9	0.3
1992 [4]	2 482.5	84.5	4.8	114.7	83.0	160.8	160.8	112.8	68.4	29.1	5.7	0.2
1991 [4]	2 532.0	87.0	4.9	118.2	86.2	162.0	164.8	115.1	68.9	28.7	5.6	0.2
1990 [5]	2 547.5	89.0	5.0	116.2	84.9	157.5	165.1	118.4	70.2	28.7	5.6	0.3
1989 [6]	2 424.0	84.8	5.2	111.9	. . .	. . .	156.3	113.8	65.7	26.3	5.3	0.3

Note: Race and Hispanic origin are reported separately on birth certificates. Persons of Hispanic origin may be of any race. In this table Hispanic women are classified only by place of origin; non-Hispanic women are classified by race.

[1]Total fertility rates are sums of birth rates for 5-year age groups multiplied by 5.
[2]Fertility rates computed by relating total births, regardless of age of mother, to women 15–44 years.
[3]Beginning in 1997, rates computed by relating births to women aged 45–54 years to women aged 45–49 years.
[4]Excludes data for New Hampshire, which did not report Hispanic origin.
[5]Excludes data for New Hampshire and Oklahoma, which did not report Hispanic origin.
[6]Excludes data for Louisiana, New Hampshire, and Oklahoma, which did not report Hispanic origin.
[8]Includes origin not stated.
[9]Includes races other than White and Black.
. . . = Not available.

TABLE I-10. NUMBER OF BIRTHS, BIRTH RATES, FERTILITY RATES, TOTAL FERTILITY RATES, AND BIRTH RATES FOR TEENAGERS 15–19 YEARS BY AGE OF MOTHER FOR EACH STATE AND TERRITORY, 2002

The largest number of births, over one-half million, took place in California, but the highest birth rate is found in Utah, with Texas second and Arizona third. The lowest birth rates were recorded in Maine, Vermont, New Hampshire, West Virginia, and Pennsylvania. There has been a marked westward shift in the distribution of births, which is attributable to a similar shift in the overall population, as well as relatively high birth rates in states with substantial proportions of births to Hispanic women. Fertility rates, which are a more specific indicator of child-bearing, ranged from a low of 49 births per 1,000 women age 15 to 44 years in Vermont to a high of 91 in Utah. Fertility rates tended to be below 60 per 1,000 on the East Coast, and above 60 per 1,000 in the South, Midwest, and West. Total fertility rates, which

Table I-10. Number of Births, Birth Rates, Fertility Rates, Total Fertility Rates, and Birth Rates for Teenagers 15–19 Years by Age of Mother for each State and Territory, 2002

State	Number of births	Birth rate [1]	Fertility rate [2]	Total fertility rate [3]	Teenage birth rate [4]		
					15–19 years		
					Total	15–17 years	18–19 years
United States [5]	4 021 726	13.9	64.8	2 013.0	43.0	23.2	72.8
Alabama	58 967	13.1	61.2	1 882.0	54.5	31.5	88.7
Alaska	9 938	15.4	73.5	2 550.5	39.5	18.9	73.7
Arizona	87 837	16.1	77.8	2 362.0	61.2	35.0	102.5
Arkansas	37 437	13.8	66.6	2 040.0	59.9	31.6	101.7
California	529 357	15.1	68.3	2 074.0	41.1	22.6	69.1
Colorado	68 418	15.2	69.3	2 150.5	47.0	26.2	79.1
Connecticut	42 001	12.1	58.8	1 881.5	25.8	14.1	45.1
Delaware	11 090	13.7	62.2	1 938.5	46.3	24.7	77.8
District of Columbia	7 498	13.1	52.9	1 582.5	69.1	44.8	101.5
Florida	205 579	12.3	62.5	1 997.0	44.5	23.2	78.4
Georgia	133 300	15.6	68.4	2 109.0	55.7	31.4	92.8
Hawaii	17 477	14.0	68.6	2 137.0	38.2	17.7	66.4
Idaho	20 970	15.6	73.8	2 273.0	39.1	18.4	69.1
Illinois	180 622	14.3	66.1	2 035.5	42.2	23.4	70.5
Indiana	85 081	13.8	64.8	2 015.5	44.6	22.6	78.5
Iowa	37 559	12.8	61.7	1 948.0	32.5	16.4	55.4
Kansas	39 412	14.5	68.7	2 146.0	43.0	21.4	74.2
Kentucky	54 233	13.3	60.5	1 881.0	51.0	26.5	84.8
Louisiana	64 872	14.5	65.4	1 990.5	58.1	31.7	96.1
Maine	13 559	10.5	49.8	1 661.5	25.4	11.9	45.2
Maryland	73 323	13.4	60.6	1 922.0	35.4	20.0	59.6
Massachusetts	80 645	12.5	56.7	1 711.0	23.3	12.5	39.6
Michigan	129 967	12.9	60.7	1 915.0	34.8	18.0	60.8
Minnesota	68 025	13.6	62.0	1 972.0	27.5	14.2	47.3
Mississippi	41 518	14.5	65.7	1 975.5	64.7	37.6	103.3
Missouri	75 251	13.3	62.1	1 942.5	44.1	22.2	76.6
Montana	11 049	12.1	60.3	1 973.5	36.4	17.8	63.3
Nebraska	25 383	14.7	69.5	2 166.0	37.0	18.3	64.2
Nevada	32 571	15.0	72.5	2 245.5	53.9	28.0	96.7
New Hampshire	14 442	11.3	52.4	1 733.0	20.0	8.1	39.0
New Jersey	114 751	13.4	63.5	2 011.5	26.8	14.7	46.1
New Mexico	27 753	15.0	70.7	2 207.5	62.4	37.8	99.5
New York	251 415	13.1	59.8	1 829.5	29.5	15.7	50.1
North Carolina	117 335	14.1	65.4	2 015.5	52.2	28.6	89.3
North Dakota	7 757	12.2	58.7	1 858.5	27.2	11.7	48.7
Ohio	148 720	13.0	61.7	1 943.0	39.5	20.1	69.4
Oklahoma	50 387	14.4	68.8	2 096.5	58.0	30.1	97.6
Oregon	45 192	12.8	61.9	1 909.5	36.8	18.2	64.8
Pennsylvania	142 850	11.6	56.4	1 803.5	31.6	17.2	53.7
Rhode Island	12 894	12.1	54.6	1 703.5	35.6	19.6	59.0
South Carolina	54 570	13.3	60.7	1 868.0	53.0	29.2	87.2
South Dakota	10 698	14.1	68.3	2 186.0	38.0	17.3	67.8
Tennessee	77 482	13.4	62.2	1 940.0	54.3	28.2	94.2
Texas	372 450	17.1	77.1	2 354.0	64.4	38.2	104.3
Utah	49 182	21.2	90.6	2 544.0	36.8	17.8	62.4
Vermont	6 387	10.4	48.9	1 604.0	24.2	10.4	44.4
Virginia	99 672	13.7	61.9	1 927.5	37.6	19.0	66.0
Washington	79 028	13.0	60.2	1 886.0	33.0	16.8	57.6
West Virginia	20 712	11.5	57.0	1 780.0	45.5	21.5	80.7
Wisconsin	68 560	12.6	59.0	1 893.5	32.3	15.9	57.1
Wyoming	6 550	13.1	63.6	2 030.5	39.9	17.7	72.1
Puerto Rico	52 747	13.7	61.6	1 822.5	62.2	42.8	91.1
Virgin Islands	1 634	15.0	71.1	2 286.5	56.8	26.1	121.5
Guam	3 212	19.9	88.3	2 634.5	64.7	41.7	102.1
American Samoa	1 627	28.2	126.7	3 858.0	46.2	20.5	93.6
Northern Marianas	1 290	17.4	45.1	1 266.5	42.3	27.7	60.1

[1]Birth rates are live births per 1,000 estimated total population in each area.
[2]Fertility rates are live births per 1,000 women aged 15–44 years estimated in each area.
[3]Total fertility rates are sums of birth rates for 5-year age groups multiplied by 5.
[4]Live births per 1,000 women in specified group.
[5]Excludes data for the territories.

provide a cross-sectional summary of lifetime fertility, was highest in Alaska (2.6 births per woman), Utah (2.5), and Arizona (2.4) and lowest in the District of Columbia (1.6), Vermont (1.6), and Maine (1.7).

The teenage birth rate was highest in the District of Columbia, with Mississippi and Texas nearly tied for second, followed by New Mexico. It is interesting to note that the high teenage birth rate in the District of Columbia does not lead to high lifetime fertility. Birth rates above 60 per 1,000 female teenagers also occurred in Guam and Puerto Rico. Low teenage birth rates, between 20.0 and 25.8 were reported for the New England states of New Hampshire, Vermont, Massachusetts, Maine, and Connecticut.

Figure I-8. Birth Rates for Teenagers 15–19 Years, by State, 2002

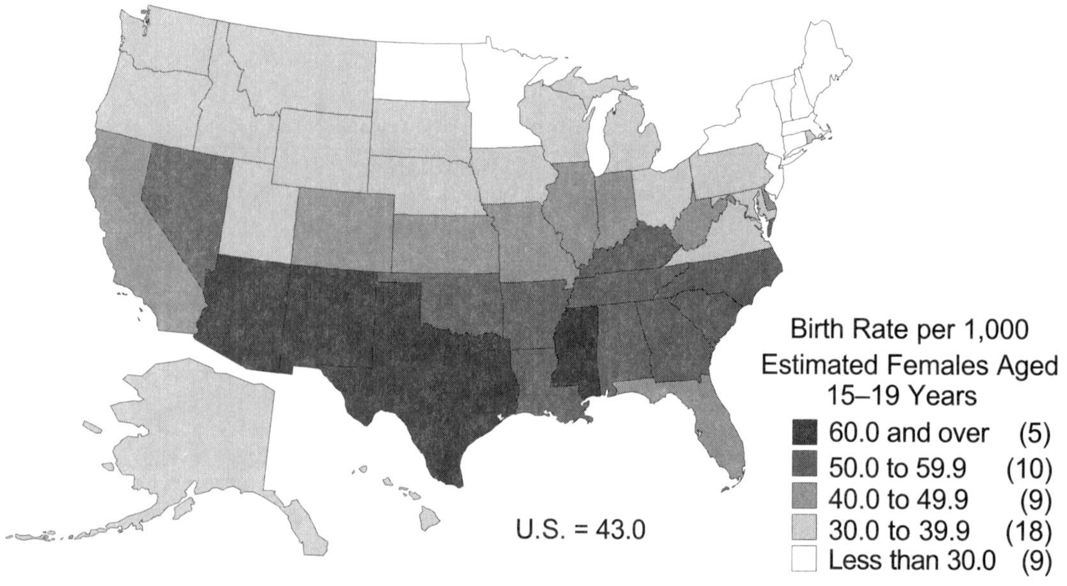

TABLE I-10A. BIRTH RATES FOR TEENAGERS 15–19 YEARS, FOR EACH STATE AND TERRITORY, 1991 AND 2002

This table shows the drop in teenage birth rate by state over a little more than a decade. California, Michigan, New Hampshire, and Alaska had declines by 40 percent or more. Nebraska, Texas, Rhode Island, and South Dakota had the least amount of change. The highest absolute rates in 2002 were the District of Columbia (69 per 1,000 women age 15–19), Mississippi (64.7), and Texas (64.4).

Table I-10A. Birth Rates for Teenagers 15–19 Years, for each State and Territory, 1991 and 2002

(Rate per 1,000 estimated female population aged 15–19 years in each area, percent.)

Area	2002	1991	Percent change, 1991–2002
United States [1]	43.0	61.8	-30
Alabama	54.5	73.6	-26
Alaska	39.5	66.0	-40
Arizona	61.2	79.7	-23
Arkansas	59.9	79.5	-25
California	41.1	73.8	-44
Colorado	47.0	58.3	-19
Connecticut	25.8	40.1	-36
Delaware	46.3	60.4	-23
District of Columbia	69.1	109.6	-37
Florida	44.5	67.9	-34
Georgia	55.7	76.0	-27
Hawaii	38.2	59.2	-35
Idaho	39.1	53.9	-27
Illinois	42.2	64.5	-35
Indiana	44.6	60.4	-26
Iowa	32.5	42.5	-24
Kansas	43.0	55.4	-22
Kentucky	51.0	68.8	-26
Louisiana	58.1	76.0	-24
Maine	25.4	43.5	-42
Maryland	35.4	54.1	-35
Massachusetts	23.3	37.5	-38
Michigan	34.8	58.9	-41
Minnesota	27.5	37.3	-26
Mississippi	64.7	85.3	-24
Missouri	44.1	64.4	-32
Montana	36.4	46.8	-22
Nebraska	37.0	42.4	-13
Nevada	53.9	74.5	-28
New Hampshire	20.0	33.1	-40
New Jersey	26.8	41.3	-35
New Mexico	62.4	79.5	-22
New York	29.5	45.5	-35
North Carolina	52.2	70.0	-25
North Dakota	27.2	35.5	-23
Ohio	39.5	60.5	-35
Oklahoma	58.0	72.1	-20
Oregon	36.8	54.8	-33
Pennsylvania	31.6	46.7	-32
Rhode Island	35.6	44.7	-20
South Carolina	53.0	72.5	-27
South Dakota	38.0	47.6	-20
Tennessee	54.3	74.8	-27
Texas	64.4	78.4	-18
Utah	36.8	48.0	-23
Vermont	24.2	39.2	-38
Virginia	37.6	53.4	-30
Washington	33.0	53.7	-39
West Virginia	45.5	58.0	-22
Wisconsin	32.3	43.7	-26
Wyoming	39.9	54.3	-27
Puerto Rico	62.2	72.4	-14
Virgin Islands	56.8	77.9	-27
Guam	64.7	95.7	-32
American Samoa	46.2	. . .	. . .
Northern Marianas	42.3	. . .	. . .

[1] Excludes data for the territories.
. . . = Not available.

TABLE I-11. LIVE BIRTHS BY RACE OF MOTHER FOR EACH STATE AND TERRITORY, 2002

The largest number of Black births took place in New York, followed by Florida, Georgia, and Texas. The largest number of American Indian births took place in Arizona, followed by Oklahoma and New Mexico. By far the largest number of Asian and Pacific Islanders births took place in California—about 30 percent of the total such births—followed by New York, Hawaii, and Texas.

Table I-11. Live Births by Race of Mother for each State and Territory, 2002

(Number.)

State	Number				
	All races	White	Black	American Indian [1]	Asian or Pacific Islander
United States [2]	4 021 726	3 174 760	593 691	42 368	210 907
Alabama	58 967	39 978	18 292	164	533
Alaska	9 938	6 377	432	2 438	691
Arizona	87 837	77 043	2 779	5 701	2 314
Arkansas	37 437	29 209	7 427	257	544
California	529 357	428 549	32 653	3 033	65 122
Colorado	68 418	62 425	2 934	662	2 397
Connecticut	42 001	34 654	5 195	224	1 928
Delaware	11 090	7 925	2 708	29	428
District of Columbia	7 498	2 677	4 620	7	194
Florida	205 579	152 855	46 238	1 075	5 411
Georgia	133 300	85 809	42 777	326	4 388
Hawaii	17 477	3 953	475	171	12 878
Idaho	20 970	20 151	103	377	339
Illinois	180 622	140 163	31 833	264	8 362
Indiana	85 081	74 309	9 332	157	1 283
Iowa	37 559	35 112	1 263	234	950
Kansas	39 412	34 904	2 890	448	1 170
Kentucky	54 233	48 399	4 943	103	788
Louisiana	64 872	36 757	26 659	412	1 044
Maine	13 559	13 049	174	116	220
Maryland	73 323	45 198	24 214	239	3 672
Massachusetts	80 645	66 689	8 344	191	5 421
Michigan	129 967	102 590	22 440	683	4 254
Minnesota	68 025	58 023	4 862	1 358	3 782
Mississippi	41 518	22 618	18 202	268	430
Missouri	75 251	62 374	11 028	353	1 496
Montana	11 049	9 512	37	1 395	105
Nebraska	25 383	22 980	1 442	403	558
Nevada	32 571	26 979	2 611	540	2 441
New Hampshire	14 442	13 691	225	48	478
New Jersey	114 751	84 493	19 952	172	10 134
New Mexico	27 753	23 281	511	3 548	413
New York	251 415	181 212	49 590	693	19 920
North Carolina	117 335	85 210	27 571	1 657	2 897
North Dakota	7 757	6 762	90	815	90
Ohio	148 720	122 887	22 547	288	2 998
Oklahoma	50 387	39 508	4 704	5 160	1 015
Oregon	45 192	41 047	941	804	2 400
Pennsylvania	142 850	117 817	20 265	353	4 415
Rhode Island	12 894	11 036	1 145	157	556
South Carolina	54 570	35 373	18 183	168	846
South Dakota	10 698	8 657	103	1 805	133
Tennessee	77 482	59 627	16 304	157	1 394
Texas	372 450	317 150	41 642	834	12 824
Utah	49 182	46 572	339	685	1 586
Vermont	6 387	6 239	44	10	94
Virginia	99 672	71 415	22 084	125	6 048
Washington	79 028	66 519	3 393	1 920	7 196
West Virginia	20 712	19 877	679	11	145
Wisconsin	68 560	58 979	6 418	1 051	2 112
Wyoming	6 550	6 147	54	279	70
Puerto Rico	52 747	47 811	4 925	. . .	. . .
Virgin Islands	1 634	326	1 222	86	-
Guam	3 212	260	41	2	2 909
American Samoa	1 627	8	-	-	1 619
Northern Marianas	1 290	17	1	-	1 272

[1]Includes births to Aleuts and Eskimos.
[2]Excludes data for the territories.
. . . = Not available.
- = Quantity zero.

TABLE I-12. LIVE BIRTHS BY SPECIFIED HISPANIC ORIGIN AND RACE OF MOTHER FOR EACH STATE AND TERRITORY, 2002

In California, Hispanic births accounted for virtually one-half of all births, in Texas they accounted for 48 percent, and in Arizona 43 percent. The mothers were predominantly of Mexican origin. In New Mexico, 53 percent of births were Hispanic, most tracing their origins to original Spanish settlers, rather than from Mexico. In Florida, about a fourth of all births came from Hispanic mothers, whose origin most frequently was Central or South America, followed by Mexico and then Cuba. Thus Cuban mothers accounted for 19 percent of Florida's Hispanic offspring. Outside of Florida, New Jersey had the highest number of births to mothers of Cuban origin. In New York, 22 percent of births were to Hispanic mothers, of which almost half were Central and South American, and almost a quarter were Puerto Rican.

Table I-12. Live Births by Specified Hispanic Origin and Race of Mother for each State and Territory, 2002

(Number.)

State	All origins	Origin of mother									Not stated
		Hispanic						Non-Hispanic			
		Total	Mexican	Puerto Rican	Cuban	Central and South American	Other and unknown Hispanic	Total [1]	White	Black	
United States [2]	4 021 726	876 642	627 505	57 465	14 232	125 981	51 459	3 119 944	2 298 156	578 335	25 140
Alabama	58 967	2 569	1 919	91	18	237	304	56 344	37 402	18 276	54
Alaska	9 938	799	276	61	6	72	384	6 902	4 427	267	2 237
Arizona	87 837	37 938	36 029	252	68	729	860	48 950	39 033	2 545	949
Arkansas	37 437	3 050	2 581	31	8	403	27	34 190	26 001	7 415	197
California	529 357	263 061	230 386	2 030	699	24 842	5 104	262 036	164 649	31 450	4 260
Colorado	68 418	21 029	16 581	283	63	767	3 335	47 382	41 858	2 761	7
Connecticut	42 001	6 982	803	4 054	74	1 723	328	34 680	27 685	4 932	339
Delaware	11 090	1 316	735	308	10	256	7	9 760	6 629	2 684	14
District of Columbia	7 498	954	94	9	3	782	66	6 494	1 733	4 563	50
Florida	205 579	51 619	12 689	8 866	10 093	18 677	1 294	153 581	102 294	45 257	379
Georgia	133 300	16 819	13 220	568	168	2 716	147	114 706	68 269	42 001	1 775
Hawaii	17 477	2 422	499	720	16	68	1 119	15 021	3 200	442	34
Idaho	20 970	2 788	2 219	15	5	76	473	17 849	17 074	100	333
Illinois	180 622	41 022	34 793	2 633	182	1 765	1 649	139 537	99 346	31 604	63
Indiana	85 081	6 169	5 347	296	13	413	100	78 545	67 894	9 271	367
Iowa	37 559	2 390	1 907	44	6	335	98	35 093	32 709	1 243	76
Kansas	39 412	5 023	4 192	92	21	305	413	33 994	29 563	2 846	395
Kentucky	54 233	1 630	1 187	107	75	218	43	52 573	46 811	4 923	30
Louisiana	64 872	1 383	528	70	62	139	584	63 452	35 428	26 611	37
Maine	13 559	167	37	27	2	36	65	13 347	12 852	167	45
Maryland	73 323	6 062	1 361	369	60	3 142	1 130	66 964	39 093	24 007	297
Massachusetts	80 645	9 592	430	4 505	66	4 395	196	70 530	58 313	6 635	523
Michigan	129 967	7 265	5 961	411	72	422	399	120 687	93 831	22 217	2 015
Minnesota	68 025	4 646	3 601	99	30	588	328	62 499	52 744	4 750	880
Mississippi	41 518	823	538	25	8	46	206	40 633	21 749	18 191	62
Missouri	75 251	3 267	2 462	123	44	415	223	71 879	59 079	10 983	105
Montana	11 049	382	160	8	1	23	190	10 471	8 967	32	196
Nebraska	25 383	3 313	2 630	33	13	467	170	21 487	19 121	1 416	583
Nevada	32 571	11 386	9 403	212	178	1 089	504	20 959	15 638	2 534	226
New Hampshire	14 442	503	113	96	9	203	82	13 383	12 690	182	556
New Jersey	114 751	24 664	4 695	6 659	810	12 304	196	89 944	61 741	18 010	143
New Mexico	27 753	14 623	7 556	61	44	173	6 789	13 125	8 759	475	5
New York	251 415	54 700	8 838	12 960	401	24 207	8 294	195 819	130 189	45 206	896
North Carolina	117 335	15 064	11 490	702	133	2 622	117	102 170	70 234	27 434	101
North Dakota	7 757	149	89	8	7	10	35	7 483	6 499	87	125
Ohio	148 720	4 817	2 706	1 255	49	522	285	143 560	117 990	22 354	343
Oklahoma	50 387	5 259	4 774	114	13	250	108	45 110	34 370	4 676	18
Oregon	45 192	8 040	7 417	81	45	326	171	36 870	32 949	897	282
Pennsylvania	142 850	8 696	1 771	5 255	101	527	1 042	133 040	108 620	19 727	1 114
Rhode Island	12 894	2 328	184	655	14	1 369	106	9 051	7 356	1 027	1 515
South Carolina	54 570	3 175	2 274	170	27	506	198	51 311	32 203	18 143	84
South Dakota	10 698	318	198	14	4	77	25	10 368	8 376	103	12
Tennessee	77 482	4 348	3 244	213	39	669	183	73 109	55 316	16 267	25
Texas	372 450	178 968	156 592	1 144	305	9 641	11 286	192 038	137 618	41 007	1 444
Utah	49 182	6 952	5 287	91	10	617	947	42 051	39 533	312	179
Vermont	6 387	32	7	8	3	8	6	6 276	6 131	43	79
Virginia	99 672	9 790	2 509	676	69	5 861	675	89 641	61 694	21 920	241
Washington	79 028	12 349	10 381	296	54	681	937	65 293	53 387	3 263	1 386
West Virginia	20 712	84	51	7	1	6	19	20 578	19 749	677	50
Wisconsin	68 560	5 295	4 204	653	28	249	161	63 252	53 820	6 349	13
Wyoming	6 550	622	557	5	2	7	51	5 927	5 540	53	1
Puerto Rico	52 747	. . .	. . .	. . .	. . .	. . .	. . .	. . .	. . .	. . .	52 747
Virgin Islands	1 634	323	11	262	2	-	48	1 240	83	1 086	71
Guam	3 212	54	22	16	1	7	8	3 069	220	39	89
American Samoa	1 627	. . .	. . .	. . .	. . .	. . .	. . .	. . .	. . .	. . .	1 627
Northern Marianas	1 290	. . .	. . .	. . .	. . .	. . .	. . .	. . .	. . .	. . .	1 289

Note: Race and Hispanic origin are reported separately on birth certificates. Persons of Hispanic origin may be of any race. In this table Hispanic women are classified only by place of origin; non-Hispanic women are classified by race.

[1]Includes races other than White and Black.
[2]Excludes data for the territories.
. . . = Not available.
- = Quantity zero.

TABLE I-13. TOTAL NUMBER OF BIRTHS, RATES (BIRTH, FERTILITY, AND TOTAL FERTILITY), AND PERCENT OF BIRTHS WITH SELECTED DEMOGRAPHIC CHARACTERISTICS, BY DETAILED RACE AND PLACE OF BIRTH OF MOTHER, 2002

This table summarizes by race the major rates shown in previous tables and newly presents the sex ratio and several demographics of interest about the mother, including marital status, teenage status, educational status, and whether the mother was born in the United States.

For Asians and Pacific Islanders more than one-half of births were recorded as "other" than Chinese, Japanese, Hawaiian, and Filipino. This suggests that information on origin was missing from birth certificates. Among the identified births, Filipino and Chinese were the largest.

The table shows the percentages of fourth- and higher-order births; this indicates that Chinese and Japanese mothers tend to have fewer children, although the immigrants in this group may be younger. American Indians and Blacks had higher percentages of high order births (19 percent and 15 percent, respectively), as well as higher percentages of births to unmarried mothers (60 percent and 68 percent, respectively).

Asian mothers had the highest rate of school completion of 12 years or more, a rate of 89.7 percent, and Japanese and Filipino mothers exceeded this average. These data on school completions were boosted by the very low Asian birth rate for mothers under the age of 20.

The data distinguish between mothers who were born in the United States and those who were born outside the 50 states and DC. Asian or Pacific Islander mothers were by far the least likely to have been born in the United States (16.9 percent), led by Chinese (10.0 percent) and those coded as "other" (11.6 percent). American Indian and Black mothers were the mostly likely to have been born in the United States (94 percent and 87 percent, respectively). For mothers born outside the states, a smaller percentage gave birth as teenagers in 2002 than those born in the United States. In part this could reflect entry into the United States of women who had already reached age 20.

The sex ratio of newborns is expressed as male live births per 1,000 female live births. The male count was 1,048 for all races, and the figure for Whites was close to this average. The ratio was somewhat lower for Blacks and American Indians, but it was higher for Asians. It is intriguing that persistent differences show up in the sex ratio between groups.

TABLE I-14. TOTAL NUMBER OF BIRTHS, RATES (BIRTH, FERTILITY, AND TOTAL FERTILITY), AND PERCENT OF BIRTHS WITH SELECTED DEMOGRAPHIC CHARACTERISTICS, BY SPECIFIED HISPANIC ORIGIN, RACE, AND PLACE OF BIRTH OF MOTHER, 2002

This table summarizes by ethnicity and race the major rates shown in previous tables and it presents the sex ratio and several demographics of interest about the mother, including marital status, teenage status,

Table I-13. Total Number of Births, Rates (Birth, Fertility, and Total Fertility), and Percent of Births with Selected Demographic Characteristics, by Detailed Race and Place of Birth of Mother, 2002

(Birth rates are live births per 1,000 population.)

Characteristic	All races	White	Black	American Indian [1]	Asian or Pacific Islander					
					Total	Chinese	Japanese	Hawaiian	Filipino	Other
Births (Number)	4 021 726	3 174 760	593 691	42 368	210 907	33 673	9 264	6 772	33 016	128 182
Birth rate	13.9	13.5	15.7	13.8	16.5	. . .	. . .	. . .	. . .	. . .
Fertility rate [2]	64.8	64.8	65.8	58.0	64.1	. . .	. . .	. . .	. . .	. . .
Total fertility rate [3]	2 013.0	2 027.5	1 991.0	1 735.0	1 819.5	. . .	. . .	. . .	. . .	. . .
Sex ratio [4]	1 048	1 050	1 032	1 023	1 065	1 070	1 089	1 075	1 070	1 059
All Births (Percent)										
Births to mothers under 20 years	10.8	9.8	18.0	18.5	3.8	0.9	1.7	14.6	4.5	4.0
4th- and higher-order births	10.8	10.2	15.3	19.1	6.5	2.1	3.9	16.3	7.3	7.1
Births to unmarried mothers	34.0	28.5	68.2	59.7	14.9	9.0	10.3	50.4	20.0	13.5
Mothers completing 12 years or more of school	78.5	78.4	75.6	69.2	89.7	88.7	97.8	85.7	94.7	88.4
Mothers born in the 50 states and DC	76.7	78.5	87.0	94.4	16.9	10.0	40.4	97.4	21.5	11.6
Mothers Born in the 50 States and DC (Percent)										
Births to mothers under 20 years	11.7	9.8	19.9	19.1	12.8	3.7	3.6	14.8	11.0	17.2
4th- and higher-order births	10.4	9.2	15.5	19.6	7.9	4.0	4.6	16.2	7.1	6.2
Births to unmarried mothers	34.9	26.8	72.1	61.4	33.0	12.2	17.8	50.7	34.5	33.0
Mothers completing 12 years or more of school	83.6	85.6	74.5	69.4	89.1	96.6	96.8	85.7	91.5	85.7
Mothers Born Outside the 50 States and DC (Percent)										
Births to mothers under 20 years	7.8	9.5	5.3	8.3	2.0	0.6	0.4	*	2.7	2.3
4th- and higher-order births	12.3	13.7	13.5	10.4	6.2	1.9	3.3	19.0	7.3	7.2
Births to unmarried mothers	30.7	34.5	41.4	30.3	11.1	8.5	5.2	40.2	16.0	10.9
Mothers completing 12 years or more of school	61.3	51.4	83.1	67.0	89.9	87.9	98.6	83.6	95.6	88.7

Note: Race and Hispanic origin are reported separately on birth certificates. In this table all women (including Hispanic women) are classified only according to their race.

[1]Includes births to Aleuts and Eskimos.
[2]Fertility rates are computed by relating total births, regardless of age of mother, to women aged 15–44 years.
[3]Total fertility rates are sums of birth rates for 5-year age groups multiplied by 5.
[4]Male live births per 1,000 female live births.
. . . = Not available.
* = Figure does not meet standards of reliability or precision; based on fewer than 20 births in numerator.

educational status, and whether the mother was born in the United States. Once again, this table highlights the big size of the Hispanic birth rate relative to the rate for non-Hispanics and particularly non-Hispanic Whites. It also shows the moderate proportion of Hispanic mothers in 2002, who were born in the United States (36.7 percent) in contrast to non-Hispanic White mothers, 94.2 percent of whom were native born. Among mothers born outside the United States, Hispanics were less likely to have completed 12 years or more of school compared with those born in the United States, while Blacks and Whites born outside the United States were more likely.

TABLE I-15. LIVE BIRTHS BY RACE OF MOTHER AND OBSERVED AND SEASONALLY ADJUSTED BIRTH AND FERTILITY RATES, BY MONTH, 2002

The monthly average number of births in 2002 was 335,144. The actual number of births per month ranged from about 304,000 (February) to 359,400 (August). Historically, the number of births tends to peak during the summer months and is at its lowest during the winter The observed birth rates, shown in the table, have been adjusted to the different number of days in the month. They peaked in August and September (14.7) and were lowest in November (13.4).

Table I-14. Total Number of Births, Rates (Birth, Fertility, and Total Fertility), and Percent of Births with Selected Demographic Characteristics, by Specified Hispanic Origin, Race, and Place of Birth of Mother, 2002

(Birth rates are live births per 1,000 population.)

Characteristic	All origins [1]	Hispanic						Non-Hispanic		
		Total	Mexican	Puerto Rican	Cuban	Central and South American	Other and unknown Hispanic	Total [2]	White	Black
Births (Number)	4 021 726	876 642	627 505	57 465	14 232	125 981	51 459	3 119 944	2 298 156	578 335
Birth rate [3]	13.9	22.6	24.2	16.5	10.0	22.4	([3])	12.6	11.7	16.1
Fertility rate [3,4]	64.8	94.4	102.8	65.4	59.0	86.1	([3])	59.6	57.4	67.4
Total fertility rate [3,5]	2 013.0	2 718.0	2 879.5	1 947.5	1 940.5	2 610.5	([3])	1 877.0	1 828.5	2 047.0
Sex ratio [6]	1 048	1 041	1 038	1 061	1 056	1 046	1 031	1 050	1 054	1 032
All Births (Percent)										
Births to mothers under 20 years	10.8	14.9	15.9	17.8	8.1	8.5	16.7	9.6	7.9	18.1
4th- and higher-order births	10.8	13.6	14.7	12.2	4.9	10.8	11.8	10.0	8.9	15.4
Births to unmarried mothers	34.0	43.5	42.1	59.1	29.8	44.8	44.4	31.3	23.0	68.4
Mothers completing 12 years or more of school	78.5	51.9	45.8	68.5	88.2	64.2	68.3	85.8	88.3	75.7
Mothers born in the 50 states and DC	76.7	36.7	36.1	65.8	45.0	11.5	71.7	87.9	94.2	88.0
Mothers Born in the 50 States and DC (Percent)										
Births to mothers under 20 years	11.7	21.4	22.3	19.6	13.1	19.6	19.6	10.5	8.2	19.9
4th- and higher-order births	10.4	11.8	12.5	11.2	5.3	5.7	11.5	10.2	8.8	15.6
Births to unmarried mothers	34.9	48.6	47.5	61.5	30.1	46.2	46.9	33.3	23.7	72.2
Mothers completing 12 years or more of school	83.6	68.5	66.6	68.6	86.3	80.7	72.1	85.4	88.1	74.5
Mothers Born Outside the 50 States and DC (Percent)										
Births to mothers under 20 years	7.8	11.1	12.2	14.1	4.0	7.1	9.3	2.9	3.0	5.0
4th- and higher-order births	12.3	14.7	15.9	14.0	4.6	11.5	12.5	8.7	9.4	13.8
Births to unmarried mothers	30.7	40.5	39.0	54.3	29.6	44.6	37.2	16.3	10.9	39.8
Mothers completing 12 years or more of school	61.3	42.2	33.8	68.2	89.8	62.0	58.7	89.3	90.9	84.5

Note: Race and Hispanic origin are reported separately on birth certificates. Persons of Hispanic origin may be of any race. In this table Hispanic women are classified only by place of origin; non-Hispanic women are classified by race.

[1]Includes origin not stated.
[2]Includes races other than White and Black.
[3]Rates for Central and South American include other and unknown Hispanic.
[4]Fertility rates are computed by relating total births, regardless of age of mother, to women aged 15–44 years.
[5]Total fertility rates are sums of birth rates for 5-year age groups multiplied by 5.
[6]Male live births per 1,000 female live births.

Table I-15. Live Births by Race of Mother and Observed and Seasonally Adjusted Birth and Fertility Rates, by Month, 2002

(Rates on an annual basis per 1,000 population for specified month. Monthly populations estimated as of the first of each month.)

Month	Number			Observed		Seasonally adjusted	
	All races [1]	White	Black	Birth rate	Fertility rate [2]	Birth rate	Fertility rate [2]
Total	4 021 726	3 174 760	593 691	13.9	64.8	X	X
January	330 674	258 390	51 703	13.6	62.8	14.0	65.1
February	303 977	238 937	46 328	13.8	63.9	14.0	65.0
March	331 505	262 328	48 469	13.6	63.0	13.8	63.9
April	324 432	257 867	46 088	13.7	63.7	14.0	65.1
May	339 007	270 311	47 702	13.9	64.4	13.9	64.8
June	327 588	259 483	47 598	13.8	64.3	13.7	63.8
July	357 669	283 120	52 831	14.6	67.9	14.1	65.6
August	359 417	284 227	52 733	14.7	68.2	14.0	64.9
September	348 814	276 618	50 361	14.7	68.4	14.0	65.2
October	345 814	273 288	50 396	14.1	65.6	14.0	65.2
November	318 573	249 042	48 284	13.4	62.4	13.8	64.4
December	334 256	261 149	51 198	13.6	63.4	13.9	64.8

[1]Includes races other than White and Black.
[2]Fertility rates are live births per 1,000 women aged 15–44 years.
X = Not applicable.

TABLE I-16. LIVE BIRTHS AND INDEX OF OCCURRENCE BY METHOD OF DELIVERY, DAY OF WEEK, AND RACE OF MOTHER, 2002

In 2002, there were on average 11,018 births per day. The number of births by day of week varied considerably; in 2002, the average ranged from a low of 7,526 on Sundays to a high of 12,581 on Tuesdays. In terms of index numbers, where 100 indicates the average, births varied from 68.3 on Sundays, to 77.8 on Saturdays, and 114.2 on Tuesdays. A weekend birth deficit exists for vaginal deliveries, but it is considerably greater for cesarean deliveries. These lows and highs were even more pronounced for repeat cesarean deliveries. For the latter, births occurred 33.5 percent on Sundays (relative to the average) and 134.1 percent (relative to the average) on Fridays. The weekend birth deficit for all cesarean births has grown noticeably since such data became available in 1989; at that time the Sunday index was 60.7 percent (data not shown), compared with 49.9 in 2001. This trend is not unexpected, as cesarean births, especially repeat cesareans, are increasingly "elective" surgeries decided on prior to labor.

Table I-16. Live Births and Index of Occurrence by Method of Delivery, Day of Week, and Race of Mother, 2002

Day of week and race of mother	Average number of births	Index of occurrence [1]				
		Total [2]	Method of delivery			
			Vaginal	Cesarean		
				Total	Primary	Repeat
All Races [3]	11 018	100.0	100.0	100.0	100.0	100.0
Sunday	7 526	68.3	75.0	49.9	60.5	33.5
Monday	11 453	103.9	101.6	110.3	100.1	126.0
Tuesday	12 823	116.4	113.7	123.8	121.0	128.3
Wednesday	12 083	109.7	107.8	114.9	113.1	117.5
Thursday	12 365	112.2	110.3	117.6	115.2	121.3
Friday	12 285	111.5	107.3	123.2	116.1	134.1
Saturday	8 573	77.8	84.2	60.1	73.7	39.0
White	8 698	100.0	100.0	100.0	100.0	100.0
Sunday	5 797	66.7	73.3	48.0	58.8	31.7
Monday	9 107	104.7	102.4	111.1	100.7	126.9
Tuesday	10 172	116.9	114.3	124.4	121.8	128.2
Wednesday	9 587	110.2	108.3	115.4	113.6	118.2
Thursday	9 826	113.0	111.1	118.3	116.0	121.8
Friday	9 751	112.1	107.7	124.4	116.8	135.8
Saturday	6 629	76.2	82.7	58.1	72.1	36.9
Black	1 627	100.0	100.0	100.0	100.0	100.0
Sunday	1 195	73.4	80.2	56.2	66.3	40.1
Monday	1 646	101.2	98.8	107.0	97.2	122.9
Tuesday	1 877	115.4	112.2	123.4	119.4	129.8
Wednesday	1 764	108.5	106.4	113.6	112.5	115.5
Thursday	1 784	109.7	107.6	115.1	112.4	119.4
Friday	1 771	108.9	105.5	117.9	113.1	125.7
Saturday	1 346	82.8	89.2	66.5	79.0	46.4

[1] Index is the ratio of the average number of births by a specified method of delivery on a given day of the week to the average daily number of births by a specified method of delivery for the year, multiplied by 100.
[2] Includes method of delivery not stated.
[3] Includes races other than White and Black.

TABLE I-17. NUMBER, BIRTH RATE, AND PERCENT OF BIRTHS TO UNMARRIED WOMEN BY AGE, RACE, AND HISPANIC ORIGIN OF MOTHER, 2002

Unmarried mothers accounted for 34.0 percent of all births in the United States in 2002. This percentage was highest for teenagers (80 percent for births at 15–19 years of age) and lowest for mothers in their 30s (about 14.5 percent). Unmarried mothers, aged 40 years and above, accounted for a somewhat higher percentage of total births (17.3 percent) than those in their 30s, reflecting perhaps a higher level of income or a push to beat the biological clock.

Births to unmarried mothers (as a percent of total births) were lowest for Asians and Pacific Islanders, and they were relatively low for non-Hispanic Whites (23.0 percent). For non-Hispanic Blacks the unmarried percentage was 68.4 (with the proportion remaining quite high through the mid-20s), and for Hispanics it was 43.5 percent.

Table I-17. Number, Birth Rate, and Percent of Births to Unmarried Women by Age, Race, and Hispanic Origin of Mother, 2002

Measure and age of mother	All races [1]	White		Black		American Indian [2,3]	Asian or Pacific Islander [2]	Hispanic [4]
		Total [2]	Non-Hispanic	Total [2]	Non-Hispanic			
Number								
All ages	1 365 966	904 461	528 535	404 864	395 538	25 297	31 344	381 466
Under 15 years	7 093	3 683	1 446	3 174	3 119	129	107	2 266
15–19 years	340 186	228 407	135 313	99 375	97 282	6 678	5 726	94 483
15 years	17 629	10 672	4 811	6 293	6 150	399	265	5 970
16 years	38 888	25 531	12 816	12 015	11 728	774	568	12 954
17 years	66 274	44 829	25 344	19 053	18 637	1 329	1 063	19 774
18 years	95 259	64 580	39 359	27 131	26 589	1 880	1 668	25 547
19 years	122 136	82 795	52 983	34 883	34 178	2 296	2 162	30 238
20–24 years	527 657	349 161	214 529	158 276	155 080	9 548	10 672	136 369
25–29 years	268 312	176 055	94 304	79 946	77 952	4 993	7 318	83 035
30–34 years	139 208	91 688	50 150	40 375	39 193	2 475	4 670	42 254
35–39 years	66 036	43 684	25 472	18 958	18 299	1 176	2 218	18 566
40 years and over	17 474	11 783	7 321	4 760	4 613	298	633	4 493
Rate per 1,000 Unmarried Women in Specified Group								
15–44 years [5]	43.7	38.9	27.8	66.2	...	...	21.3	87.9
15–19 years	35.4	30.4	22.1	64.8	...	...	13.4	66.1
15–17 years	20.8	17.5	11.5	39.9	...	...	7.5	43.0
18–19 years	58.6	51.0	38.8	104.1	...	...	22.2	105.3
20–24 years	70.5	61.6	46.1	119.2	...	...	26.5	131.4
25–29 years	61.5	56.8	38.5	85.9	...	...	27.5	123.1
30–34 years	40.8	38.3	26.0	49.9	...	...	28.6	88.1
35–39 years	20.8	19.4	13.5	24.9	...	...	18.7	51.3
40–44 years [6]	5.4	5.0	3.7	6.3	...	...	6.8	12.7
Percent of Births to Unmarried Women								
All ages	34.0	28.5	23.0	68.2	68.4	59.7	14.9	43.5
Under 15 years	97.0	94.8	96.9	99.6	99.6	97.0	97.3	93.6
15–19 years	80.0	74.6	75.4	95.7	95.9	86.6	71.5	73.9
15 years	94.3	91.5	93.8	99.3	99.3	98.3	89.5	89.8
16 years	90.1	86.5	88.9	99.0	99.0	96.1	81.3	84.6
17 years	86.2	82.0	83.9	98.2	98.2	91.5	80.5	79.8
18 years	80.3	75.1	76.9	96.0	96.1	86.9	73.6	72.7
19 years	72.7	66.7	67.4	92.7	92.8	79.7	63.2	65.7
20–24 years	51.6	44.6	41.3	81.3	81.5	66.6	35.5	51.4
25–29 years	25.3	20.7	15.3	58.5	58.7	49.2	11.7	35.2
30–34 years	14.6	11.8	8.1	42.5	42.5	39.1	6.6	26.8
35–39 years	14.5	11.8	8.6	39.2	39.1	39.5	6.8	26.0
40 years and over	17.3	14.5	11.2	39.6	39.6	40.7	8.9	28.8

Note: For 48 states and the District of Columbia, marital status is reported on the birth certificate; for Michigan and New York, mother's marital status is inferred. Rates cannot be computed for unmarried non-Hispanic Black women or for American Indian women because the necessary populations are not available.

[1]Includes races other than White and Black and origin not stated.
[2]Race and Hispanic origin are reported separately on the birth certificate. Race categories are consistent with the 1977 Office of Management and Budget guidelines. Data for persons of Hispanic origin are included in the data for each race group according to the mother's reported race.
[3]Includes births to Aleuts and Eskimos.
[4]Includes all persons of Hispanic origin of any race.
[5]Birth rates computed by relating total births to unmarried mothers, regardless of age of mother, to unmarried women aged 15–44 years.
[6]Birth rates computed by relating births to unmarried mothers aged 40 years and over to unmarried women aged 40–44 years.
. . . = Not available.

TABLE I-18. BIRTH RATES FOR UNMARRIED WOMEN BY AGE, RACE, AND HISPANIC ORIGIN OF MOTHER, SELECTED YEARS, 1970–2002

This table shows the number of births for unmarried women per an average 1,000 of unmarried women. For all childbearing ages this rate was 43.7 in 2002. This rate has been fairly stable since the beginning of the 1990s, but it increased sharply in the 1980s. For married and unmarried mothers, combined, the corresponding rate is 64.8.

Table I-18. Birth Rates for Unmarried Women by Age, Race, and Hispanic Origin of Mother, Selected Years, 1970–2002

(Rates are live births to unmarried women per 1,000 unmarried women.)

Year and race and Hispanic origin	Age of mother								
	15–44 years[1]	15–19 years			20–24 years	25–29 years	30–34 years	35–39 years	40–44 years[2]
		Total	15–17 years	18–19 years					
All Races[3]									
2002[4]	43.7	35.4	20.8	58.6	70.5	61.5	40.8	20.8	5.4
2001[4]	43.8	37.0	22.0	60.6	71.2	59.5	40.4	20.4	5.3
2000[4]	44.0	39.0	23.9	62.2	72.1	58.5	39.3	19.7	5.0
1999[4]	43.3	39.7	25.0	62.4	70.7	56.8	38.1	19.0	4.6
1998[4]	43.3	40.9	26.5	63.7	70.4	55.4	38.1	18.7	4.6
1997[4]	42.9	41.4	27.7	63.9	68.9	53.4	37.9	18.7	4.6
1996[4]	43.8	42.2	28.5	64.9	68.9	54.5	40.1	19.9	4.8
1995[4]	44.3	43.8	30.1	66.5	68.7	54.3	38.9	19.3	4.7
1994[4]	46.2	45.8	31.7	69.1	70.9	57.4	39.6	19.7	4.7
1993[4]	44.8	44.0	30.3	66.1	68.3	55.9	38.0	18.9	4.4
1992[4]	44.9	44.2	30.2	66.7	67.9	55.6	37.6	18.8	4.1
1991[4]	45.0	44.6	30.8	65.4	67.8	56.0	37.9	17.9	3.8
1990[4]	43.8	42.5	29.6	60.7	65.1	56.0	37.6	17.3	3.6
1989[4]	41.6	40.1	28.7	56.0	61.2	52.8	34.9	16.0	3.4
1988[4]	38.5	36.4	26.4	51.5	56.0	48.5	32.0	15.0	3.2
1987[4]	36.0	33.8	24.5	48.9	52.6	44.5	29.6	13.5	2.9
1986[4]	34.2	32.3	22.8	48.0	49.3	42.2	27.2	12.2	2.7
1985[4]	32.8	31.4	22.4	45.9	46.5	39.9	25.2	11.6	2.5
1984[4,5]	31.0	30.0	21.9	42.5	43.0	37.1	23.3	10.9	2.5
1983[4,5]	30.3	29.5	22.0	40.7	41.8	35.5	22.4	10.2	2.6
1982[4,5]	30.0	28.7	21.5	39.6	41.5	35.1	21.9	10.0	2.7
1981[4,5]	29.5	27.9	20.9	39.0	41.1	34.5	20.8	9.8	2.6
1980[4,5]	29.4	27.6	20.6	39.0	40.9	34.0	21.1	9.7	2.6
1980[5,6]	28.4	27.5	20.7	38.7	39.7	31.4	18.5	8.4	2.3
1975[5,6]	24.5	23.9	19.3	32.5	31.2	27.5	17.9	9.1	2.6
1970[6,7]	26.4	22.4	17.1	32.9	38.4	37.0	27.1	13.6	3.5
White Total									
2002[4]	38.9	30.4	17.5	51.0	61.6	56.8	38.3	19.4	5.0
2001[4]	38.5	31.3	18.1	52.1	61.7	54.6	37.2	18.6	4.9
2000[4]	38.2	32.7	19.7	53.1	61.6	52.9	35.9	17.9	4.5
1999[4]	37.4	33.2	20.7	52.8	60.2	50.8	34.9	17.3	4.1
1998[4]	36.9	33.6	21.5	53.0	59.5	48.6	34.1	16.9	4.1
1997[4]	36.3	33.6	22.0	52.8	57.9	47.0	33.6	16.6	3.9
1996[4]	37.0	34.0	22.3	53.4	57.9	48.1	35.3	17.7	4.3
1995[4]	37.0	35.0	23.3	54.6	57.2	47.4	33.7	16.8	4.2
1994[4]	37.8	35.8	23.9	55.7	57.5	48.6	33.8	17.2	4.3
1993[4]	35.6	33.3	21.9	51.9	53.8	46.0	31.9	16.3	3.9
1992[4]	35.0	32.7	21.5	51.1	52.4	44.8	31.3	16.1	3.6
1991[4]	34.5	32.7	21.7	49.4	51.4	44.3	30.9	15.2	3.2
1990[4]	32.9	30.6	20.4	44.9	48.2	43.0	29.9	14.5	3.2
1989[4]	30.2	28.0	19.3	40.2	43.8	39.1	26.8	13.1	2.9
1988[4]	27.4	25.3	17.6	36.8	39.2	35.4	24.2	12.1	2.7
1987[4]	25.3	23.2	16.2	34.5	36.6	32.0	22.3	10.7	2.4
1986[4]	23.9	21.8	14.9	33.5	34.2	30.5	20.1	9.7	2.2
1985[4]	22.5	20.8	14.5	31.2	31.7	28.5	18.4	9.0	2.0
1984[4,5]	20.6	19.3	13.7	27.9	28.5	25.5	16.8	8.4	2.0
1983[4,5]	19.8	18.7	13.6	26.4	27.1	23.8	15.9	7.8	2.0
1982[4,5]	19.3	18.0	13.1	25.3	26.5	23.1	15.3	7.4	2.1
1981[4,5]	18.6	17.2	12.6	24.6	25.8	22.3	14.2	7.2	1.9
1980[4,5]	18.1	16.5	12.0	24.1	25.1	21.5	14.1	7.1	1.8
Non-Hispanic White									
2002[4]	27.8	22.1	11.5	38.8	46.1	38.5	26.0	13.5	3.7
2001[4]	27.8	23.1	12.1	40.3	46.4	37.5	25.4	13.2	3.6
2000[4]	28.0	24.7	13.6	42.1	47.0	36.9	24.8	12.9	3.3
1999[4]	27.9	25.6	14.6	42.5	46.3	36.2	24.8	13.0	3.1
1998[4]	27.9	26.2	15.6	42.9	46.3	35.3	25.0	13.1	3.1
1997[4]	27.5	26.4	16.2	43.1	44.8	34.5	24.9	12.7	2.9
1996[4]	28.2	27.0	16.9	43.8	44.5	35.0	26.3	13.8	3.3
1995[4]	28.1	27.7	17.6	44.4	43.9	34.4	25.1	12.9	3.2
1994[4]	28.4	28.1	18.0	44.9	43.8	34.7	24.6	12.8	3.1
1993[4]	. . .	. . .	. . .	. . .	. . .	. . .	. . .	. . .	. . .
1992[4]	. . .	. . .	. . .	. . .	. . .	. . .	. . .	. . .	. . .
1991[4]	. . .	. . .	. . .	. . .	. . .	. . .	. . .	. . .	. . .
1990[4,8]	24.4	25.0	16.2	37.0	36.4	30.3	20.5	6.1	. . .

Note: Race and Hispanic origin are reported separately on the birth certificate. Data for persons of Hispanic origin are included in the data for each race group according to the mother's reported race. Rates cannot be computed for unmarried non-Hispanic Black women or for American Indian women because the necessary populations are not available.

[1]Fertility rates computed by relating total births to unmarried mothers, regardless of age of mother, to unmarried women aged 15–44 years.
[2]Rates computed by relating births to unmarried mothers aged 40 years and over to unmarried women aged 40–44 years.
[3]Includes races other than White, Black, and Asian or Pacific Islander.
[4]Data for states in which marital status was not reported have been inferred and included with data from the remaining states.
[5]Based on 100 percent of births in selected states and on a 50-percent sample of births in all other states.
[6]Births to unmarried women are estimated for the United States from data for registration areas in which marital status of mother was reported.
[7]Based on a 50-percent sample of births.
[8]Rates for 1990 based on data for 48 states and the District of Columbia that reported Hispanic origin on the birth certificate. Rates shown for 1990 for ages 35–39 years are based on births to unmarried women aged 35–44 years.
. . . = Not available.

Typically the birth rate is highest for unmarried women in their early 20s, while for all mothers combined it tends to peak in the 25–29 year age bracket. In 2002, the unmarried birth rate was 70.5 between 20 and 24 years of age, but it dropped down to 20.8 in the 35–39 year age bracket. For 18- and 19-year-old women, the rate was 58.6.

Since 1980 developments of the birth rate of unmarried mothers have been different for the major population groups. For Whites this birth rate has increased, for Hispanics it has been more or less steady, and for Blacks there has been a considerable decrease. This decrease is entirely accounted for by a sharp drop since 1991 in the unmarried birth rate of 15–19 year old Black women, especially those under 18 years. The older Black age groups also showed declines in the 1990s, but generally they were not enough to wipe out increases that had occurred in the 1980s.

Table I-18. Birth Rates for Unmarried Women by Age, Race, and Hispanic Origin of Mother, Selected Years, 1970–2002—*Continued*

(Rates are live births to unmarried women per 1,000 unmarried women.)

Year and race and Hispanic origin	Age of mother								
	15–44 years[1]	15–19 years			20–24 years	25–29 years	30–34 years	35–39 years	40–44 years[2]
		Total	15–17 years	18–19 years					
Black Total									
2002[4]	66.2	64.8	39.9	104.1	119.2	85.9	49.9	24.9	6.3
2001[4]	68.2	69.9	43.8	110.2	122.8	84.1	51.2	25.4	6.3
2000[4]	70.5	75.0	48.3	115.0	129.0	85.9	50.2	25.4	6.3
1999[4]	69.7	76.5	50.1	115.3	126.8	85.6	49.0	24.2	5.8
1998[4]	71.6	81.5	55.2	121.0	127.8	86.5	50.5	24.3	6.0
1997[4]	71.6	84.5	59.2	124.5	124.2	81.4	51.0	24.3	6.5
1996[4]	72.8	87.5	62.8	126.8	122.5	81.2	53.4	25.2	6.1
1995[4]	74.5	91.2	67.5	128.7	124.6	82.3	53.3	25.3	6.0
1994[4]	80.8	99.3	74.0	139.2	135.2	91.3	56.5	26.0	5.9
1993[4]	83.0	101.2	76.0	139.7	139.9	92.8	56.7	25.7	5.8
1992[4]	85.7	104.8	77.3	146.2	142.6	96.8	57.3	25.6	5.4
1991[4]	89.0	107.8	79.9	147.6	146.4	100.0	59.8	25.5	5.4
1990[4]	90.5	106.0	78.8	143.7	144.8	105.3	61.5	25.5	5.1
1989[4]	90.7	104.5	78.9	140.9	142.4	102.9	60.5	24.9	5.0
1988[4]	86.5	96.1	73.5	130.5	133.6	97.2	57.4	24.1	5.0
1987[4]	82.6	90.9	69.9	123.0	126.1	91.6	53.1	22.4	4.7
1986[4]	79.0	88.5	67.0	121.1	118.0	84.6	50.0	20.6	4.4
1985[4]	77.0	87.6	66.8	117.9	113.1	79.3	47.5	20.4	4.3
1984[4,5]	75.2	86.1	66.5	113.6	107.9	77.8	43.8	19.4	4.3
1983[4,5]	76.2	85.5	66.8	111.9	107.2	79.7	43.8	19.4	4.8
1982[4,5]	77.9	85.1	66.3	112.7	109.3	82.7	44.1	19.5	5.2
1981[4,5]	79.4	85.0	65.9	114.2	110.7	83.1	45.5	19.6	5.6
1980[4,5]	81.1	87.9	68.8	118.2	112.3	81.4	46.7	19.0	5.5
Asian or Pacific Islander Total									
2002[4]	21.3	13.4	7.5	22.2	26.5	27.5	28.6	18.7	6.8
2001[4]	21.2	14.6	8.7	23.0	25.2	26.7	29.4	19.7	6.3
2000[4]	20.9	15.2	9.6	23.2	24.2	25.4	29.7	18.4	6.9
Hispanic[9]									
2002[4]	87.9	66.1	43.0	105.3	131.4	123.1	88.1	51.3	12.7
2001[4]	87.8	67.1	44.2	104.3	132.3	120.7	91.4	49.5	12.2
2000[4]	87.2	68.5	47.0	102.2	130.3	121.6	89.4	45.9	12.2
1999[4]	84.9	68.6	48.6	100.1	126.1	119.6	84.4	42.3	11.2
1998[4]	82.8	69.3	49.7	101.4	120.5	115.7	78.1	38.8	12.0
1997[4]	83.2	69.2	50.5	100.9	122.7	114.8	78.8	40.6	12.1
1996[4]	86.1	69.3	49.6	102.7	131.6	121.7	84.6	41.1	12.3
1995[4]	88.7	73.2	52.5	109.6	135.7	122.2	84.2	42.2	12.1
1994[4]	95.8	77.6	55.5	116.0	144.5	131.7	91.2	47.4	13.9
1993[4]	91.4	71.1	49.5	109.1	134.3	130.4	87.8	47.1	14.1
1992[4]	92.8	70.3	49.2	106.6	138.2	133.4	89.9	47.8	14.6
1991[4]	92.5	71.0	49.5	107.4	134.2	135.1	88.2	47.5	14.1
1990[4,8]	89.6	65.9	45.9	98.9	129.8	131.7	88.1	50.8	13.7

Note: Race and Hispanic origin are reported separately on the birth certificate. Data for persons of Hispanic origin are included in the data for each race group according to the mother's reported race. Rates cannot be computed for unmarried non-Hispanic Black women or for American Indian women because the necessary populations are not available.

[1]Fertility rates computed by relating total births to unmarried mothers, regardless of age of mother, to unmarried women aged 15–44 years.
[2]Rates computed by relating births to unmarried mothers aged 40 years and over to unmarried women aged 40–44 years.
[4]Data for states in which marital status was not reported have been inferred and included with data from the remaining states.
[5]Based on 100 percent of births in selected states and on a 50-percent sample of births in all other states.
[8]Rates for 1990 based on data for 48 states and the District of Columbia that reported Hispanic origin on the birth certificate. Rates shown for 1990 for ages 35–39 years are based on births to unmarried women aged 35–44 years.
[9]Includes all persons of Hispanic origin of any race.

TABLE I-18A. NUMBER, RATE, AND PERCENT OF BIRTHS TO UNMARRIED WOMEN, AND BIRTH RATE FOR MARRIED WOMEN, SELECTED YEARS, 1980–2002

Since 1980, the number of births to unmarried women has increased from 665,747 to 1,365,966 and the share of these births to total births has increased from 18.4 percent to 34.0 percent. Most of this increase had taken place by the mid-1990s, when the rate of births by unmarried women reached a peak. The increase in nonmarital births since 1995 is due almost entirely to the 10 percent rise in the number of unmarried women of childbearing age. In 2002 this rate, that is, births to unmarried women per 1,000 unmarried women aged 15 to 44 years was 43.7, while the comparable rate for married women stood at 86.3.

Table I-18A. Number, Rate, and Percent of Births to Unmarried Women, and Birth Rate for Married Women, Selected Years, 1980–2002

(Number, rate, and percent.)

Year	Births to unmarried women			Birth rate for married women [3]
	Number	Rate [1]	Percent [2]	
2002	1 365 966	43.7	34.0	86.3
2001	1 349 249	43.8	33.5	86.7
2000	1 347 043	44.0	33.2	87.4
1999	1 308 560	43.3	33.0	84.8
1998	1 293 567	43.3	32.8	84.2
1997	1 257 444	42.9	32.4	82.7
1996	1 260 306	43.8	32.4	82.3
1995	1 253 976	44.3	32.2	82.6
1994	1 289 592	46.2	32.6	82.9
1993	1 240 172	44.8	31.0	86.1
1992	1 224 876	44.9	30.1	88.5
1991	1 213 769	45.0	29.5	89.6
1990	1 165 384	43.8	28.0	93.2
1989	1 094 169	41.6	27.1	91.9
1988	1 005 299	38.5	25.7	90.8
1987	933 013	36.0	24.5	90.0
1986	878 477	34.2	23.4	90.7
1985	828 174	32.8	22.0	93.3
1980	665 747	29.4	18.4	97.0

[1] Births to unmarried women per 1,000 unmarried women aged 15–44 years.
[2] Percent of all births to unmarried women.
[3] Births to married women per 1,000 married women aged 15–44 years.

TABLE I-19. NUMBER AND PERCENT OF BIRTHS TO UNMARRIED WOMEN BY RACE AND HISPANIC ORIGIN OF MOTHER FOR EACH STATE AND TERRITORY, 2002

The three states with the highest percentages of births by non-Hispanic White unmarried mothers were Maine (32.5 percent), Vermont (31.9 percent), and West Virginia (31.6 percent). For all states the average unmarried share of births for non-Hispanic Whites was 23.0 percent. For non-Hispanic Black mothers the highest unmarried percentages were 81.9 percent in Wisconsin, 76.8 percent in Illinois, and 76.4 percent in the District of Columbia. The average for non-Hispanic Blacks was 68.4 percent. For Hispanics, the highest relative count was in northeastern states where immigration from Puerto Rico has been important: Massachusetts (61.5 percent), Pennsylvania (60.9 percent), and Connecticut (60.8 percent). The average of births by unmarried mothers for all Hispanics was 43.5 percent.

Table I-19. Number and Percent of Births to Unmarried Women by Race and Hispanic Origin of Mother for each State and Territory, 2002

(Number, percent.)

State	Births to unmarried mothers						Percent unmarried					
	All races [1]	White		Black		Hispanic [3]	All races [1]	White		Black		Hispanic [3]
		Total [2]	Non-Hispanic	Total [2]	Non-Hispanic			Total [2]	Non-Hispanic	Total [2]	Non-Hispanic	
United States [4]	1 365 966	904 461	528 535	404 864	395 538	381 466	34.0	28.5	23.0	68.2	68.4	43.5
Alabama	20 515	7 879	7 238	12 511	12 506	629	34.8	19.7	19.4	68.4	68.4	24.5
Alaska	3 379	1 530	1 010	187	96	331	34.0	24.0	22.8	43.3	36.0	41.4
Arizona	35 494	29 419	9 752	1 712	1 564	19 705	40.4	38.2	25.0	61.6	61.5	51.9
Arkansas	13 891	8 054	6 880	5 633	5 626	1 138	37.1	27.6	26.5	75.8	75.9	37.3
California	174 681	143 842	33 110	20 394	19 690	111 099	33.0	33.6	20.1	62.5	62.6	42.2
Colorado	18 320	16 092	7 603	1 592	1 477	8 714	26.8	25.8	18.2	54.3	53.5	41.4
Connecticut	12 213	8 532	4 405	3 450	3 283	4 247	29.1	24.6	15.9	66.4	66.6	60.8
Delaware	4 505	2 549	1 823	1 907	1 889	740	40.6	32.2	27.5	70.4	70.4	56.2
District of Columbia	4 236	685	142	3 533	3 488	553	56.5	25.6	8.2	76.5	76.4	58.0
Florida	80 843	48 647	28 890	30 918	30 353	20 446	39.3	31.8	28.2	66.9	67.1	39.6
Georgia	50 401	21 583	14 420	28 149	27 786	7 243	37.8	25.2	21.1	65.8	66.2	43.1
Hawaii	5 870	686	537	88	78	1 061	33.6	17.4	16.8	18.5	17.6	43.8
Idaho	4 595	4 312	3 225	34	33	1 004	21.9	21.4	18.9	33.0	33.0	36.0
Illinois	62 890	37 769	20 367	24 448	24 287	17 550	34.8	26.9	20.5	76.8	76.8	42.8
Indiana	30 929	23 578	20 435	7 127	7 085	3 064	36.4	31.7	30.1	76.4	76.4	49.7
Iowa	11 022	9 758	8 774	935	918	986	29.3	27.8	26.8	74.0	73.9	41.3
Kansas	12 249	9 875	7 694	1 961	1 945	2 133	31.1	28.3	26.0	67.9	68.3	42.5
Kentucky	18 011	14 228	13 524	3 619	3 611	717	33.2	29.4	28.9	73.2	73.3	44.0
Louisiana	30 498	10 041	9 615	20 013	19 990	449	47.0	27.3	27.1	75.1	75.1	32.5
Maine	4 419	4 248	4 181	59	56	60	32.6	32.6	32.5	33.9	33.5	35.9
Maryland	25 530	10 806	8 027	14 359	14 248	2 753	34.8	23.9	20.5	59.3	59.3	45.4
Massachusetts	21 620	15 883	10 947	4 883	3 804	5 903	26.8	23.8	18.8	58.5	57.3	61.5
Michigan	44 366	27 056	23 649	16 546	16 437	3 053	34.1	26.4	25.2	73.7	74.0	42.0
Minnesota	18 668	13 763	11 299	2 840	2 769	2 357	27.4	23.7	21.4	58.4	58.3	50.7
Mississippi	19 553	5 458	5 108	13 841	13 836	345	47.1	24.1	23.5	76.0	76.1	41.9
Missouri	26 511	17 752	16 298	8 346	8 317	1 466	35.2	28.5	27.6	75.7	75.7	44.9
Montana	3 627	2 623	2 417	16	12	158	32.8	27.6	27.0	*	*	41.4
Nebraska	7 259	5 946	4 407	956	938	1 379	28.6	25.9	23.0	66.3	66.2	41.6
Nevada	12 183	9 451	4 438	1 826	1 775	5 054	37.4	35.0	28.4	69.9	70.0	44.4
New Hampshire	3 546	3 388	3 033	96	72	182	24.6	24.7	23.9	42.7	39.6	36.2
New Jersey	33 625	20 170	8 312	12 824	11 719	12 960	29.3	23.9	13.5	64.3	65.1	52.5
New Mexico	13 020	10 129	2 373	289	265	7 821	46.9	43.5	27.1	56.6	55.8	53.5
New York	89 840	53 571	23 558	32 589	29 714	32 571	35.7	29.6	18.1	65.7	65.7	59.5
North Carolina	40 669	21 232	14 102	18 140	18 066	7 189	34.7	24.9	20.1	65.8	65.9	47.7
North Dakota	2 250	1 624	1 518	32	30	59	29.0	24.0	23.4	35.6	34.5	39.6
Ohio	52 677	35 336	32 965	16 879	16 750	2 383	35.4	28.8	27.9	74.9	74.9	49.5
Oklahoma	18 357	12 238	10 085	3 308	3 287	2 229	36.4	31.0	29.3	70.3	70.3	42.4
Oregon	13 976	12 545	9 186	577	553	3 350	30.9	30.6	27.9	61.3	61.6	41.7
Pennsylvania	47 748	31 830	26 310	15 240	14 850	5 294	33.4	27.0	24.2	75.2	75.3	60.9
Rhode Island	4 604	3 576	1 881	721	652	1 379	35.7	32.4	25.6	63.0	63.5	59.2
South Carolina	22 026	8 784	7 425	13 032	13 005	1 374	40.4	24.8	23.1	71.7	71.7	43.3
South Dakota	3 744	2 265	2 143	39	39	155	35.0	26.2	25.6	37.9	37.9	48.7
Tennessee	28 039	15 859	13 880	11 890	11 872	1 999	36.2	26.6	25.1	72.9	73.0	46.0
Texas	120 704	93 441	29 676	25 828	25 483	63 723	32.4	29.5	21.6	62.0	62.1	35.6
Utah	8 465	7 660	4 980	159	151	2 670	17.2	16.4	12.6	46.9	48.4	38.4
Vermont	2 036	1 991	1 958	26	26	11	31.9	31.9	31.9	59.1	60.5	*
Virginia	30 233	15 976	12 079	13 733	13 659	3 928	30.3	22.4	19.6	62.2	62.3	40.1
Washington	22 780	18 428	13 130	1 806	1 731	5 122	28.8	27.7	24.6	53.2	53.0	41.5
West Virginia	6 811	6 299	6 248	489	488	29	32.9	31.7	31.6	72.0	72.1	34.5
Wisconsin	20 551	14 306	11 974	5 256	5 202	2 431	30.0	24.3	22.2	81.9	81.9	45.9
Wyoming	1 987	1 768	1 504	28	27	270	30.3	28.8	27.1	51.9	50.9	43.4
Puerto Rico	27 403	24 175	. . .	3 222	. . .	. . .	52.0	50.6	. . .	65.4	. . .	. . .
Virgin Islands	1 099	184	21	895	803	216	67.3	56.4	25.3	73.2	73.9	66.9
Guam	1 778	51	40	13	13	18	55.4	19.6	18.2	*	*	*
American Samoa	521	-	. . .	-	. . .	. . .	32.0	*	. . .	*	. . .	. . .
Northern Marianas	770	5	. . .	-	. . .	. . .	59.7	*	. . .	*	. . .	. . .

[1]Includes races other than White and Black and origin not stated.
[2]Race and Hispanic origin are reported separately on the birth certificate. Data for persons of Hispanic origin are included in the data for each race group according to the mother's reported race.
[3]Includes all persons of Hispanic origin of any race.
[4]Excludes data for the territories.
. . . = Not available.
* = Figure does not meet standards of reliability or precision; based on fewer than 20 births in numerator.
- = Quantity zero.

TABLE I-20. BIRTH RATES BY AGE AND RACE OF FATHER, 1980–2002

This table measures the frequency at which men in various age groups became fathers of a newborn in various years. The frequency is stated in terms of a 1,000 men. A fathers' "birth rate" of 48.4 per 1,000, for men between 15 and 54 years of age, is shown for 2002. This rate is less than the fertility rate of women, because the population of women is included only up to age 44.

The male rate has been declining since 1990 and it was at the lowest level ever reported in 2002. By age group it was highest in the 25–29 year bracket. For Whites in 2002, the fathers' birth rate between 20 and 24 years was 70.8, and between 25 and 29 years it was 104.8. Blacks had higher fathers' birth rates, except in their early 30s. The overall Black fathers' birth rate was 61.2, also sharply lower since 1990.

Information on the age of the father is often missing on

birth certificates of children born to mothers under 25 years of age and unmarried women. Such missing data are estimated according to the patterns prevailing for reporting mothers in each age group.

Table I-20. Birth Rates by Age and Race of Father, 1980–2002

(Rates are live births per 1,000 men in specified group. Figures for age of father not stated are distributed.)

Year and race of father	15–54 years [1]	Age of father								
		15–19 years [2]	20–24 years	25–29 years	30–34 years	35–39 years	40–44 years	45–49 years	50–54 years	55 years and over
All Races [3]										
2002	48.4	17.4	75.6	105.0	99.1	57.7	22.6	7.4	2.4	0.3
2001	49.0	18.5	78.5	105.8	99.6	57.0	22.3	7.3	2.4	0.3
2000	50.0	19.8	82.1	106.5	99.5	56.3	22.2	7.3	2.5	0.3
1999	49.2	20.6	81.1	105.3	95.9	53.9	21.1	7.0	2.4	0.3
1998	49.6	21.3	82.3	104.4	94.4	53.1	21.0	7.1	2.5	0.3
1997	49.4	21.9	82.1	102.6	92.0	51.5	20.7	7.0	2.5	0.3
1996	50.2	22.7	83.4	102.8	91.3	51.1	20.5	6.9	2.5	0.3
1995	51.0	23.9	83.9	103.2	90.7	50.4	20.3	7.0	2.5	0.3
1994	52.4	24.6	85.6	105.3	91.1	50.5	20.3	7.2	2.6	0.3
1993	53.7	24.4	86.0	108.1	91.7	50.7	20.2	7.3	2.7	0.4
1992	55.3	24.4	87.1	111.1	93.0	51.1	20.4	7.3	2.7	0.4
1991	56.8	24.7	87.9	113.5	94.3	51.6	20.2	7.4	2.7	0.4
1990	58.4	23.5	88.0	116.4	97.8	53.0	21.0	7.5	2.8	0.4
1989	57.2	21.9	85.4	114.3	94.8	51.3	20.4	7.4	2.7	0.6
1988	55.8	19.6	82.4	111.6	93.2	49.9	19.9	7.1	2.7	0.4
1987	55.0	18.3	80.5	109.9	91.2	48.6	19.0	6.9	2.6	0.4
1986	54.8	17.9	80.3	109.6	90.3	46.8	18.3	6.7	2.6	0.4
1985	55.6	18.0	81.2	112.3	91.1	47.3	18.1	6.6	2.5	0.4
1984 [4]	55.0	17.8	80.7	111.4	89.9	46.0	17.8	6.3	2.4	0.4
1983 [4]	55.1	18.2	82.6	113.0	89.1	45.2	17.4	6.4	2.3	0.4
1982 [4]	56.4	18.6	86.5	117.3	90.3	44.5	17.5	6.4	2.3	0.4
1981 [4]	56.3	18.4	88.4	119.1	88.7	43.3	17.0	6.2	2.3	0.4
1980 [4]	57.0	18.8	92.0	123.1	91.0	42.8	17.1	6.1	2.2	0.3
White										
2002	46.4	14.8	70.8	104.8	99.4	56.4	21.0	6.6	2.0	0.3
2001	46.9	15.5	73.1	105.4	99.9	55.7	20.8	6.5	2.0	0.3
2000	47.6	16.6	75.8	105.4	99.5	54.7	20.7	6.5	2.1	0.3
1999	46.9	17.3	74.7	104.1	96.2	52.7	19.8	6.3	2.1	0.3
1998	47.1	17.7	75.6	102.7	94.3	51.9	19.6	6.3	2.1	0.3
1997	46.8	18.0	75.3	100.9	91.7	50.2	19.3	6.2	2.1	0.3
1996	47.7	18.7	76.7	101.4	91.1	49.9	19.2	6.1	2.1	0.2
1995	48.4	19.4	77.0	101.7	90.4	49.1	19.1	6.2	2.1	0.2
1994	49.3	19.5	77.4	103.1	90.4	48.9	18.9	6.3	2.2	0.3
1993	50.3	18.9	77.2	105.5	90.7	48.9	18.7	6.4	2.2	0.2
1992	51.8	18.8	77.8	108.2	91.9	49.1	18.8	6.4	2.2	0.3
1991	53.1	19.0	78.4	110.2	92.8	49.6	18.5	6.5	2.2	0.3
1990	54.6	18.1	78.3	113.2	96.1	50.9	19.2	6.5	2.2	0.3
1989	53.3	16.7	75.9	110.8	93.0	49.1	18.7	6.3	2.1	0.4
1988	52.2	14.8	73.7	108.3	91.2	47.6	18.1	6.1	2.1	0.3
1987	51.6	13.9	72.8	107.0	89.5	46.2	17.3	5.9	2.0	0.3
1986	51.7	13.8	73.3	107.0	88.7	44.4	16.6	5.7	2.0	0.3
1985	52.6	14.0	74.7	109.9	89.5	44.8	16.3	5.6	1.9	0.3
1984 [4]	51.8	14.0	74.3	108.8	87.9	43.5	16.0	5.3	1.9	0.3
1983 [4]	52.0	14.4	76.3	110.2	86.8	42.6	15.5	5.3	1.8	0.3
1982 [4]	53.1	14.9	80.1	114.2	87.5	41.7	15.6	5.3	1.9	0.3
1981 [4]	52.9	15.0	81.7	115.8	85.8	40.3	15.0	5.2	1.8	0.3
1980 [4]	53.4	15.4	84.9	119.4	87.8	39.7	15.0	5.1	1.8	0.3
Black										
2002	61.2	33.3	116.2	123.6	94.0	57.8	28.5	12.0	4.7	0.9
2001	63.3	36.5	124.5	125.9	95.6	57.1	28.2	11.8	4.7	1.0
2000	66.2	39.6	135.5	131.0	95.2	56.9	28.4	11.7	5.0	1.0
1999	65.4	41.0	133.8	129.6	91.6	54.3	26.5	11.2	4.9	1.0
1998	66.8	42.8	137.0	130.3	90.9	54.0	26.7	11.6	5.0	1.0
1997	66.7	45.1	136.3	126.3	88.8	52.6	26.1	11.4	5.2	1.0
1996	67.2	46.7	137.6	123.9	87.0	51.8	25.7	11.3	5.3	1.1
1995	69.1	49.9	139.2	123.9	87.7	52.0	25.7	11.9	5.4	1.1
1994	74.0	54.1	149.1	129.6	91.4	53.8	26.4	12.8	5.8	1.1
1993	77.6	56.2	152.7	134.2	94.0	56.3	27.7	13.4	6.3	1.3
1992	80.4	57.0	157.1	138.6	95.8	56.7	28.4	13.7	6.1	1.4
1991	83.0	57.8	158.5	142.0	99.2	58.5	29.4	14.1	6.7	1.4
1990	84.9	55.2	158.2	144.9	103.2	60.4	31.1	15.0	7.1	1.4
1989	84.1	52.9	153.4	143.5	101.4	59.9	31.1	14.9	6.9	2.7
1988	80.7	48.1	144.1	137.9	100.0	58.0	30.6	14.3	6.9	1.4
1987	78.3	44.6	136.1	133.9	97.4	58.0	30.0	13.8	6.6	1.3
1986	77.2	42.6	131.4	131.6	97.4	58.0	29.1	13.5	6.7	1.3
1985	77.2	41.8	129.5	132.7	97.3	59.4	29.5	13.3	6.5	1.2
1984 [4]	76.7	40.9	128.0	132.2	98.3	58.4	29.3	13.3	6.1	1.2
1983 [4]	77.2	40.7	129.1	134.4	99.0	59.6	29.6	13.5	6.0	1.2
1982 [4]	79.5	40.3	133.4	141.2	103.6	61.1	29.6	13.9	6.0	1.2
1981 [4]	80.4	38.9	138.4	145.6	104.3	61.3	29.7	13.3	5.7	1.2
1980 [4]	83.0	40.1	145.3	152.8	109.6	62.0	31.2	13.6	5.9	1.1

Note: Race and Hispanic origin are reported separately on birth certificates. In this table all men (including Hispanic men) are classified only according to their race. Age of father was not stated for 13 percent of births in 2002.

[1] Rates computed by relating total births, regardless of age of father, to men aged 15–54 years.
[2] Rates computed by relating births of fathers under 20 years of age to men aged 15–19 years.
[3] Includes races other than White and Black.
[4] Based on 100 percent of births in selected states and on a 50-percent sample of births in all other states.

TABLE I-21. LIVE BIRTHS BY EDUCATIONAL ATTAINMENT, AGE, RACE, AND HISPANIC ORIGIN OF MOTHER, 2002

As many as 78.5 percent of mothers in 2002 had at least 12 years of education, and 25.9 percent had 16 years or more of schooling. These percentages were notably higher for Japanese mothers as well as for non-Hispanic Whites. Hispanics had 51.9 percent in the educational group of 12 years or more and 8.3 percent in the group with at least 16 years of education. The educational attainment of non-Hispanic Black mothers was considerably higher than that of Hispanics, but below that of non-Hispanic Whites: 75.7 percent had attained 12 years of school or more, and 12.7 percent were at the college graduate level.

Educational achievement has been positively correlated with timely prenatal care, behavior conducive to a healthy pregnancy, as well as a smaller number of children.

TABLE I-22. NUMBER OF LIVE BIRTHS AND PERCENT DISTRIBUTION BY WEIGHT GAIN OF MOTHER DURING PREGNANCY AND MEDIAN WEIGHT GAIN, ACCORDING TO PERIOD OF GESTATION, RACE, AND HISPANIC ORIGIN OF MOTHER: UNITED STATES EXCLUDING CALIFORNIA, 2002

The median weight gain of mothers during pregnancy was 30.5 pounds. Non-Hispanic Blacks had a slightly smaller median weight gain of 29.9 pounds, and Hispanics had a median weight gain of 28.7 pounds. To some extent, weight gain is related to the duration of gestation. Among mothers with a gestation period of 40 weeks and over, 13.8 percent had a weight gain of 46 pounds or more, while for those with gestation under 37 weeks only 11.3 percent had this large a weight gain. Conversely, among mothers with a gestation period under 37 weeks, 18.2 percent had weight gains of less than 16 pounds, while for mothers with a gestation period of 40 weeks and over, this category of weight gain occurred only for 10.8 percent of cases. In 2002 a large proportion of mothers had weight gains outside the guidelines recommended by the Institute of Medicine. In general, weight gain of less than 16 pounds is considered inadequate for most women, whereas weight gain of more than 40 pounds is considered excessive in most cases (underweight women are recommended to gain more weight, while overweight women are recommended to gain less weight). Inadequate maternal weight gain has been associated with an increased risk of intrauterine growth retardation, shortened period of gestation, low birthweight, and perinatal mortality. High weight gain has been linked with an elevated risk of a large-for-gestational-age infant, cesarean delivery, and long-term maternal weight retention.

Overall, the weight gain most often reported, with a frequency of 17.7 percent, was 26 to 30 pounds. But for mothers with gestation periods under 37 weeks, the most frequent weight gain was less than 16 pounds—with a frequency of 18.2 percent for all races and 25.1 percent for non-Hispanic Blacks.

Table I-21. Live Births by Educational Attainment, Age, Race, and Hispanic Origin of Mother, 2002

Age and race of mother	Total	Years of school completed by mother						Percent 12 years or more	Percent 16 years or more
		0–8 years	9–11 years	12 years	13–15 years	16 years or more	Not stated		
All Races [1]									
All ages	4 021 726	239 616	614 966	1 234 726	851 732	1 026 819	53 867	78.5	25.9
Under 15 years	7 315	5 525	1 543	-	-	-	247	*	*
15–19 years	425 493	36 451	213 684	146 680	21 817	-	6 861	40.2	*
15 years	18 703	5 577	12 568	-	-	-	558	*	*
16 years	43 147	5 577	35 418	1 294	-	-	858	3.1	*
17 years	76 881	6 594	56 429	12 333	220	-	1 305	16.6	*
18 years	118 651	8 260	54 402	50 621	3 606	-	1 762	46.4	*
19 years	168 111	10 443	54 867	82 432	17 991	-	2 378	60.6	*
20–24 years	1 022 106	65 557	212 778	445 622	227 368	57 339	13 442	72.4	5.7
25–29 years	1 060 391	62 102	105 296	315 029	271 475	293 155	13 334	84.0	28.0
30–34 years	951 219	41 916	54 355	206 783	211 563	424 862	11 740	89.8	45.2
35–39 years	453 927	21 569	22 380	98 280	98 235	207 087	6 376	90.2	46.3
40 years and over	101 275	6 496	4 930	22 332	21 274	44 376	1 867	88.5	44.6
White Total [2]									
All ages	3 174 760	216 927	461 280	937 991	664 942	854 863	38 757	78.4	27.3
Under 15 years	3 884	2 940	807	-	-	-	137	*	*
15–19 years	305 988	31 365	151 179	103 954	14 735	-	4 755	39.4	*
15 years	11 665	3 708	7 599	-	-	-	358	*	*
16 years	29 501	4 552	23 459	918	-	-	572	3.2	*
17 years	54 698	5 886	39 099	8 632	160	-	921	16.3	*
18 years	85 957	7 543	39 668	35 130	2 405	-	1 211	44.3	*
19 years	124 167	9 676	41 354	59 274	12 170	-	1 693	58.3	*
20–24 years	783 000	61 621	162 732	334 290	169 977	44 661	9 719	71.0	5.8
25–29 years	851 142	58 079	83 107	244 745	214 634	240 916	9 661	83.2	28.6
30–34 years	779 535	38 043	43 073	161 868	170 310	357 797	8 444	89.5	46.4
35–39 years	369 833	19 211	16 866	76 294	78 548	174 219	4 695	90.1	47.7
40 years and over	81 378	5 668	3 516	16 840	16 738	37 270	1 346	88.5	46.6
Non-Hispanic White									
All ages	2 298 156	37 286	230 461	680 848	550 543	781 617	17 401	88.3	34.3
Under 15 years	1 493	1 178	278	-	-	-	37	*	*
15–19 years	179 511	9 104	86 070	71 759	10 729	-	1 849	46.4	*
15 years	5 130	1 530	3 491	-	-	-	109	*	*
16 years	14 421	1 618	12 096	502	-	-	205	3.5	*
17 years	30 205	1 785	22 650	5 311	103	-	356	18.1	*
18 years	51 196	1 960	23 344	23 785	1 637	-	470	50.1	*
19 years	78 559	2 211	24 489	42 161	8 989	-	709	65.7	*
20–24 years	519 153	11 529	88 373	241 792	135 262	38 319	3 878	80.6	7.4
25–29 years	614 909	7 314	32 930	175 417	177 591	217 517	4 140	93.4	35.6
30–34 years	620 173	4 823	14 974	120 325	145 071	330 607	4 373	96.8	53.7
35–39 years	297 436	2 496	6 293	58 252	67 332	160 646	2 417	97.0	54.5
40 years and over	65 481	842	1 543	13 303	14 558	34 528	707	96.3	53.3
Black Total [2]									
All ages	593 691	13 912	128 422	231 835	135 545	73 837	10 140	75.6	12.7
Under 15 years	3 188	2 428	666	-	-	-	94	*	*
15–19 years	103 795	4 245	54 865	36 893	6 022	-	1 770	42.1	*
15 years	6 336	1 687	4 484	-	-	-	165	*	*
16 years	12 142	885	10 678	321	-	-	258	2.7	*
17 years	19 411	573	15 284	3 179	51	-	324	16.9	*
18 years	28 266	553	12 770	13 468	1 018	-	457	52.1	*
19 years	37 640	547	11 649	19 925	4 953	-	566	67.1	*
20–24 years	194 704	2 373	42 608	92 604	45 997	8 228	2 894	76.5	4.3
25–29 years	136 591	1 942	17 281	52 336	40 051	22 720	2 261	85.7	16.9
30–34 years	95 006	1 531	7 972	30 896	27 119	25 612	1 876	89.8	27.5
35–39 years	48 388	1 001	3 999	15 292	13 246	13 890	960	89.5	29.3
40 years and over	12 019	392	1 031	3 814	3 110	3 387	285	87.9	28.9
Non-Hispanic Black									
All ages	578 335	12 998	125 345	226 221	132 331	72 045	9 395	75.7	12.7
Under 15 years	3 132	2 390	653	-	-	-	89	*	*
15–19 years	101 494	4 124	53 654	36 180	5 880	-	1 656	42.1	*
15 years	6 191	1 654	4 380	-	-	-	157	*	*
16 years	11 846	856	10 432	309	-	-	249	2.7	*
17 years	18 980	557	14 978	3 094	47	-	304	16.8	*
18 years	27 656	535	12 462	13 248	990	-	421	52.3	*
19 years	36 821	522	11 402	19 529	4 843	-	525	67.1	*
20–24 years	190 241	2 165	41 675	90 716	44 919	8 035	2 731	76.6	4.3
25–29 years	132 824	1 730	16 801	50 951	39 086	22 165	2 091	85.8	17.0
30–34 years	92 155	1 352	7 696	29 917	26 484	25 009	1 697	90.0	27.6
35–39 years	46 831	881	3 867	14 770	12 919	13 524	870	89.7	29.4
40 years and over	11 658	356	999	3 687	3 043	3 312	261	88.1	29.1
Hispanic [3]									
All ages	876 642	180 513	233 253	260 237	115 398	71 042	16 199	51.9	8.3
Under 15 years	2 421	1 791	541	-	-	-	89	*	*
15–19 years	127 900	22 377	66 126	32 734	4 112	-	2 551	29.4	*
15 years	6 647	2 208	4 205	-	-	-	234	*	*
16 years	15 321	2 966	11 593	426	-	-	336	2.8	*
17 years	24 772	4 117	16 710	3 387	61	-	497	14.2	*
18 years	35 123	5 590	16 590	11 523	786	-	634	35.7	*
19 years	46 037	7 496	17 028	17 398	3 265	-	850	45.7	*
20–24 years	265 235	50 319	74 982	93 509	35 309	6 354	4 762	51.9	2.4
25–29 years	236 143	50 959	50 571	70 034	37 330	22 943	4 306	56.2	9.9
30–34 years	157 887	33 377	28 333	42 059	25 232	26 100	2 786	60.2	16.8
35–39 years	71 480	16 837	10 698	18 300	11 232	13 020	1 393	60.7	18.6
40 years and over	15 576	4 853	2 002	3 601	2 183	2 625	312	55.1	17.2

[1]Includes races other than White and Black.
[2]Race and Hispanic origin are reported separately on the birth certificate. Data for persons of Hispanic origin are included in the data for each race group according to the mother's reported race.
[3]Includes all persons of Hispanic origin of any race.
* = Figure does not meet standards of reliability or precision; based on fewer than 20 births in numerator.
- = Quantity zero.

Table I-22. Number of Live Births and Percent Distribution by Weight Gain of Mother During Pregnancy and Median Weight Gain, According to Period of Gestation, Race, and Hispanic Origin of Mother: United States Excluding California, 2002

Period of gestation [1] and race and Hispanic origin of mother	All births	Weight gain during pregnancy									
		Less than 16 pounds	16–20 pounds	21–25 pounds	26–30 pounds	31–35 pounds	36–40 pounds	41–45 pounds	46 pounds or more	Not stated	Median weight gain in pounds
NUMBER											
All Gestation Periods [2]											
All races [3]	3 492 369	400 890	359 358	449 914	578 289	445 368	411 316	220 001	407 787	219 446	X
White total [4]	2 746 211	291 163	272 174	355 666	463 693	364 439	334 119	180 258	325 615	159 084	X
Non-Hispanic White	2 133 507	207 156	195 340	272 669	364 708	296 096	272 852	150 096	272 828	101 762	X
Black total [4]	561 038	91 140	66 294	67 871	81 818	56 637	56 717	29 655	66 338	44 568	X
Non-Hispanic Black	546 885	89 531	64 790	66 090	79 649	54 872	55 077	28 757	64 559	43 560	X
Hispanic [5]	613 581	84 455	77 375	83 484	99 267	68 508	61 456	30 301	53 170	55 565	X
Under 37 Weeks											
All races [3]	430 200	71 942	53 676	55 638	63 227	43 789	40 357	21 448	44 530	35 593	X
White total [4]	310 427	46 530	37 271	40 994	47 052	33 790	30 745	16 803	33 985	23 257	X
Non-Hispanic White	236 654	32 664	26 957	31 307	36 495	27 035	24 890	13 984	28 656	14 666	X
Black total [4]	99 038	22 249	13 549	11 731	13 023	7 912	7 778	3 785	8 990	10 021	X
Non-Hispanic Black	97 147	21 919	13 329	11 476	12 761	7 714	7 617	3 696	8 812	9 823	X
Hispanic [5]	74 153	14 026	10 394	9 818	10 612	6 817	5 902	2 821	5 365	8 398	X
37–39 Weeks											
All races [3]	1 781 832	199 501	184 962	237 406	305 256	232 234	210 631	109 907	198 341	103 594	X
White total [4]	1 406 674	147 321	141 101	188 131	245 344	190 069	171 286	89 877	158 178	75 367	X
Non-Hispanic White	1 101 849	106 238	102 259	145 420	194 622	155 703	140 867	75 210	132 869	48 661	X
Black total [4]	277 750	42 724	32 653	34 818	42 061	29 228	28 678	14 918	32 367	20 303	X
Non-Hispanic Black	270 789	41 968	31 903	33 932	40 961	28 319	27 851	14 483	31 532	19 840	X
Hispanic [5]	305 281	41 231	39 107	42 886	50 858	34 430	30 501	14 744	25 516	26 008	X
40 Weeks and Over											
All races [3]	1 273 889	128 728	120 333	156 493	209 318	169 005	159 965	88 514	164 593	76 940	X
White total [4]	1 024 510	96 850	93 525	126 253	170 922	140 320	131 827	73 482	133 228	58 103	X
Non-Hispanic White	792 221	67 976	65 970	95 768	133 346	113 186	106 927	60 844	111 156	37 048	X
Black total [4]	182 821	25 936	19 992	21 250	26 649	19 441	20 177	10 922	24 896	13 558	X
Non-Hispanic Black	177 596	25 418	19 463	20 612	25 843	18 785	19 528	10 548	24 134	13 265	X
Hispanic [5]	232 658	29 023	27 752	30 669	37 670	27 176	24 960	12 702	22 216	20 490	X
PERCENT DISTRIBUTION											
All Gestation Periods [2]											
All races [3]	100.0	12.2	11.0	13.7	17.7	13.6	12.6	6.7	12.5	X	30.5
White total [4]	100.0	11.3	10.5	13.7	17.9	14.1	12.9	7.0	12.6	X	30.6
Non-Hispanic White	100.0	10.2	9.6	13.4	18.0	14.6	13.4	7.4	13.4	X	30.9
Black total [4]	100.0	17.6	12.8	13.1	15.8	11.0	11.0	5.7	12.8	X	30.0
Non-Hispanic Black	100.0	17.8	12.9	13.1	15.8	10.9	10.9	5.7	12.8	X	29.9
Hispanic [5]	100.0	15.1	13.9	15.0	17.8	12.3	11.0	5.4	9.5	X	28.7
Under 37 Weeks											
All races [3]	100.0	18.2	13.6	14.1	16.0	11.1	10.2	5.4	11.3	X	28.1
White total [4]	100.0	16.2	13.0	14.3	16.4	11.8	10.7	5.9	11.8	X	29.1
Non-Hispanic White	100.0	14.7	12.1	14.1	16.4	12.2	11.2	6.3	12.9	X	30.1
Black total [4]	100.0	25.0	15.2	13.2	14.6	8.9	8.7	4.3	10.1	X	25.5
Non-Hispanic Black	100.0	25.1	15.3	13.1	14.6	8.8	8.7	4.2	10.1	X	25.4
Hispanic [5]	100.0	21.3	15.8	14.9	16.1	10.4	9.0	4.3	8.2	X	25.7
37–39 Weeks											
All races [3]	100.0	11.9	11.0	14.1	18.2	13.8	12.6	6.5	11.8	X	30.5
White total [4]	100.0	11.1	10.6	14.1	18.4	14.3	12.9	6.8	11.9	X	30.6
Non-Hispanic White	100.0	10.1	9.7	13.8	18.5	14.8	13.4	7.1	12.6	X	30.8
Black total [4]	100.0	16.6	12.7	13.5	16.3	11.4	11.1	5.8	12.6	X	30.0
Non-Hispanic Black	100.0	16.7	12.7	13.5	16.3	11.3	11.1	5.8	12.6	X	30.0
Hispanic [5]	100.0	14.8	14.0	15.4	18.2	12.3	10.9	5.3	9.1	X	28.5
40 Weeks and Over											
All races [3]	100.0	10.8	10.1	13.1	17.5	14.1	13.4	7.4	13.8	X	30.9
White total [4]	100.0	10.0	9.7	13.1	17.7	14.5	13.6	7.6	13.8	X	31.0
Non-Hispanic White	100.0	9.0	8.7	12.7	17.7	15.0	14.2	8.1	14.7	X	32.0
Black total [4]	100.0	15.3	11.8	12.6	15.7	11.5	11.9	6.5	14.7	X	30.4
Non-Hispanic Black	100.0	15.5	11.8	12.5	15.7	11.4	11.9	6.4	14.7	X	30.4
Hispanic [5]	100.0	13.7	13.1	14.5	17.8	12.8	11.8	6.0	10.5	X	30.1

Note: Excludes data for California, which did not require reporting of weight gain during pregnancy.

[1]Expressed in completed weeks.
[2]Includes births with period of gestation not stated.
[3]Includes races other than White and Black and origin not stated.
[4]Race and Hispanic origin are reported separately on the birth certificate. Data for persons of Hispanic origin are included in the data for each race group according to the mother's reported race.
[5]Includes all persons of Hispanic origin of any race.
X = Not applicable.

TABLE I-23. PERCENT LOW BIRTHWEIGHT BY WEIGHT GAIN OF MOTHER DURING PREGNANCY, PERIOD OF GESTATION, AND RACE AND HISPANIC ORIGIN OF MOTHER: UNITED STATES EXCLUDING CALIFORNIA, 2002

This table focuses on the percent of newborns with low birthweight, defined as less than 5-1/2 pounds. The dominant factor is the length of the gestation period. For a gestation period under 37 weeks, 43.7 percent of births have low birthweight. For gestation periods of 37–39 weeks and of 40 weeks and over, this percentage falls to 4.0 percent and 1.6 percent, respectively. In addition to the length of gestation, maternal weight gain has a positive relationship to infant birthweight (although the trend begins to reverse for weight gains above 40 pounds). In general, Blacks have a markedly higher risk of low birthweight compared with non-Hispanic Whites, as well as with Hispanics.

Table I-23. Percent Low Birthweight by Weight Gain of Mother During Pregnancy, Period of Gestation, and Race and Hispanic Origin of Mother: United States Excluding California, 2002

(Low birthweight is defined as weight of less than 2,500 grams [5 lb 8 oz].)

Period of gestation [1] and race and Hispanic origin of mother	Total	Weight gain during pregnancy								
		Less than 16 pounds	16–20 pounds	21–25 pounds	26–30 pounds	31–35 pounds	36–40 pounds	41–45 pounds	46 pounds or more	Not stated
All Gestation Periods [2]										
All races [3]	8.0	13.9	10.5	8.1	6.7	5.7	5.6	5.5	6.0	12.1
White total [4]	7.0	11.7	9.1	7.2	5.9	5.2	5.0	5.1	5.6	10.3
Non-Hispanic White	7.0	12.0	9.6	7.3	6.0	5.2	5.1	5.2	5.7	10.7
Black total [4]	13.4	21.1	16.2	13.2	11.4	9.6	9.1	8.3	8.1	18.9
Non-Hispanic Black	13.5	21.2	16.4	13.3	11.5	9.6	9.2	8.4	8.2	19.0
Hispanic [5]	6.9	11.1	8.0	6.7	5.6	5.1	4.6	4.4	4.7	9.7
Mexican	6.5	10.2	7.2	6.1	5.1	4.7	4.5	4.0	4.4	8.7
Puerto Rican	9.8	17.1	13.3	10.5	7.9	7.7	6.5	5.7	6.0	15.4
Cuban	6.5	12.4	8.7	6.8	6.2	4.7	4.9	5.6	5.0	9.5
Central and South American	6.5	11.6	7.8	6.8	5.6	4.7	3.9	4.1	4.3	10.3
Other and unknown Hispanic	8.1	13.1	9.3	8.3	6.4	6.2	4.8	5.3	5.3	13.9
Under 37 Weeks										
All races [3]	43.7	55.7	47.8	42.5	38.5	36.9	36.2	36.6	36.9	54.2
White total [4]	41.3	52.3	45.3	40.4	36.7	35.7	35.1	36.1	36.6	51.5
Non-Hispanic White	42.7	54.8	48.1	42.0	38.2	36.9	36.5	37.6	37.8	55.0
Black total [4]	51.3	63.0	54.9	49.8	45.3	42.6	40.9	39.8	38.3	61.3
Non-Hispanic Black	51.4	63.1	55.0	49.9	45.3	42.7	41.0	40.1	38.3	61.4
Hispanic [5]	36.8	46.8	37.9	35.3	31.4	31.3	29.2	28.2	29.6	44.8
37–39 Weeks										
All races [3]	4.0	6.1	5.0	4.3	3.7	3.2	3.1	3.0	3.2	5.0
White total [4]	3.5	5.1	4.3	3.7	3.2	2.8	2.7	2.7	2.9	4.1
Non-Hispanic White	3.4	5.1	4.4	3.7	3.2	2.7	2.7	2.7	2.9	4.1
Black total [4]	6.7	9.3	7.8	7.0	6.4	5.4	5.2	5.0	4.6	8.1
Non-Hispanic Black	6.7	9.4	7.8	7.0	6.4	5.4	5.2	5.0	4.6	8.1
Hispanic [5]	3.7	5.2	4.2	3.9	3.3	3.1	2.9	2.8	2.8	4.3
40 Weeks and Over										
All races [3]	1.6	2.8	2.3	1.8	1.4	1.2	1.1	1.0	1.0	2.1
White total [4]	1.3	2.3	1.9	1.6	1.2	1.0	0.9	0.9	0.9	1.7
Non-Hispanic White	1.3	2.3	1.9	1.6	1.2	1.0	0.9	0.9	0.9	1.6
Black total [4]	3.0	4.7	3.8	3.3	2.8	2.3	2.3	1.9	1.8	3.7
Non-Hispanic Black	3.1	4.7	3.9	3.4	2.9	2.3	2.4	1.9	1.8	3.7
Hispanic [5]	1.5	2.3	1.9	1.6	1.4	1.1	1.0	0.9	0.9	2.0

Note: Excludes data for California, which did not require reporting of weight gain during pregnancy.

[1]Expressed in completed weeks.
[2]Includes births with period of gestation not stated.
[3]Includes races other than White and Black and origin not stated.
[4]Race and Hispanic origin are reported separately on the birth certificate. Data for persons of Hispanic origin are included in the data for each race group according to the mother's reported race.
[5]Includes all persons of Hispanic origin of any race.

TABLE I-24. PERCENT OF BIRTHS WITH SELECTED MEDICAL OR HEALTH CHARACTERISTICS BY DETAILED RACE AND PLACE OF BIRTH OF MOTHER, 2002

Health characteristics indicate that 11.4 percent of mothers in 2002 were smokers and 0.8 percent were drinkers. The National Center for Health Statistics, on the basis of national surveys of pregnant women, has concluded that alcohol use is substantially underreported (perhaps more than tenfold) on the birth certificates. The lowest prevalence of smoking and drinking was reported for Chinese mothers and the highest for American Indians, with 19.7 percent for smokers and 2.5 percent for drinkers. (The allegedly high incidence of fetal alcohol syndrome of American Indian children suggests that this estimate of drinking is also an underestimate.) While prenatal smoking is also believed to be somewhat underreported on birth certificates, the patterns of maternal smoking based on birth certificates have been largely corroborated by nationally representative surveys. See Tables 29 through 32 for more detail on maternal smoking.

The cesarean delivery rate averaged 26.1 percent, with a somewhat higher rate for Blacks and Filipinos, and a lower rate for American Indians, Native Hawaiians, and Chinese and Japanese.

The table also shows an early beginning of prenatal care for most prospective mothers, and it gives several gradations of birthweight, percentages of preterm births, and low 5-minute Apgar scores, an index of newborn physical signs of birth asphyxia. American Indians and Blacks have a lower proportion of first trimester prenatal care and higher proportion of inadequate maternal weight gain. Black infants are at higher risk of preterm delivery, low birthweight, and 5-minute Apgar score less than seven.

Table I-24. Percent of Births with Selected Medical or Health Characteristics by Detailed Race and Place of Birth of Mother, 2002

Characteristic	All races	White	Black	American Indian [1]	Asian or Pacific Islander					
					Total	Chinese	Japanese	Hawaiian	Filipino	Other
ALL BIRTHS										
Mother										
Prenatal care beginning in the first trimester	83.7	85.4	75.2	69.8	84.8	87.2	90.5	78.1	85.4	83.9
Late or no prenatal care	3.6	3.1	6.2	8.0	3.1	2.1	2.1	4.7	2.8	3.5
Smoker [2]	11.4	12.3	8.7	19.7	2.5	0.5	4.0	13.7	2.9	2.1
Drinker [3]	0.8	0.8	0.9	2.5	0.3	0.1	0.8	1.1	0.4	0.3
Weight gain of less than 16 lbs [4]	12.2	11.3	17.6	17.0	9.3	6.9	11.8	10.2	8.0	9.9
Median weight gain [4]	30.5	30.6	30.0	30.0	30.2	30.3	26.8	32.8	30.6	30.1
Cesarean delivery rate	26.1	25.9	27.6	23.1	25.0	23.9	20.8	22.4	28.5	24.8
Infant										
Preterm births [5]	12.1	11.1	17.5	13.1	10.4	7.7	9.2	13.5	12.7	10.5
Birthweight										
Very low birthweight [6]	1.5	1.2	3.1	1.3	1.1	0.7	1.0	1.6	1.3	1.2
Low birthweight [7]	7.8	6.8	13.3	7.2	7.8	5.5	7.6	8.1	8.6	8.2
4,000 grams or more [8]	9.2	10.2	5.0	11.2	5.5	6.2	4.7	8.3	5.5	5.2
5-minute Apgar score of less than 7 [9]	1.4	1.2	2.3	1.4	1.0	0.7	0.8	1.6	1.1	1.0
MOTHERS BORN IN THE 50 STATES AND DC										
Mother										
Prenatal care beginning in the first trimester	85.3	87.7	75.1	69.5	83.1	90.9	91.6	78.4	85.0	80.3
Late or no prenatal care	3.1	2.4	6.1	8.0	3.3	1.7	1.6	4.7	2.6	3.9
Smoker [2]	13.8	14.6	9.8	20.7	8.5	2.9	6.1	13.7	6.6	7.7
Drinker [3]	0.9	0.9	0.9	2.6	0.8	*	*	1.1	0.7	0.8
Weight gain of less than 16 lbs [4]	11.9	10.6	18.0	17.1	9.1	6.3	10.8	10.2	7.9	9.0
Median weight gain [4]	30.7	30.8	30.0	30.0	30.9	30.5	29.0	32.8	31.2	30.8
Cesarean delivery rate	26.3	26.1	27.4	22.9	22.1	22.3	25.1	22.4	22.8	20.8
Infant										
Preterm births [5]	12.4	11.3	18.1	13.1	12.3	10.7	11.7	13.6	13.4	11.8
Birthweight										
Very low birthweight [6]	1.5	1.2	3.2	1.3	1.4	1.1	1.2	1.6	1.4	1.5
Low birthweight [7]	8.2	7.0	13.8	7.1	8.6	7.9	8.3	8.2	9.5	8.5
4,000 grams or more [8]	9.4	10.4	4.6	11.5	6.5	5.6	5.4	8.3	5.5	6.7
5-minute Apgar score of less than 7 [9]	1.4	1.2	2.3	1.4	1.4	*	1.2	1.6	1.2	1.4
MOTHERS BORN OUTSIDE THE 50 STATES AND DC										
Mother										
Prenatal care beginning in the first trimester	78.6	77.2	76.3	75.5	85.1	86.8	89.7	69.7	85.5	84.4
Late or no prenatal care	5.2	5.5	6.7	6.8	3.1	2.2	2.4	*	2.9	3.4
Smoker [2]	1.8	2.0	1.3	3.8	1.3	0.2	2.6	14.6	1.8	1.4
Drinker [3]	0.4	0.4	0.3	*	0.3	*	1.1	*	0.3	0.2
Weight gain of less than 16 lbs [4]	13.7	14.5	15.0	14.9	9.4	6.9	12.5	*	8.1	10.0
Median weight gain [4]	28.8	28.5	29.2	27.7	30.0	30.3	25.8	32.0	30.5	29.8
Cesarean delivery rate	25.5	25.1	29.6	25.4	25.6	24.1	17.8	25.6	30.0	25.3
Infant										
Preterm births [5]	10.8	10.7	13.7	12.3	10.0	7.4	7.5	*	12.5	10.3
Birthweight										
Very low birthweight [6]	1.1	1.0	2.4	1.4	1.1	0.7	0.8	*	1.3	1.1
Low birthweight [7]	6.6	6.0	9.7	8.7	7.6	5.3	7.1	*	8.4	8.1
4,000 grams or more [8]	8.4	9.3	7.7	7.0	5.3	6.3	4.2	*	5.5	5.0
5-minute Apgar score of less than 7 [9]	1.0	1.0	1.7	1.3	0.9	0.7	*	*	1.0	1.0

Note: Race and Hispanic origin are reported separately on birth certificates. In this table all women (including Hispanic women) are classified only according to their race.

1Includes births to Aleuts and Eskimos.
2Excludes data for California, which did not report tobacco use on the birth certificate.
3Excludes data for California, which did not report alcohol use on the birth certificate.
4Excludes data for California, which did not report weight gain on the birth certificate. Median weight shown in pounds.
5Born prior to 37 completed weeks of gestation.
6Birthweight of less than 1,500 grams (3 lb 4 oz).
7Birthweight of less than 2,500 grams (5 lb 8 oz).
8Equivalent to 8 lb 14 oz.
9Excludes data for California and Texas, which did not report 5-minute Apgar score on the birth certificate.
* = Figure does not meet standards of reliability or precision; based on fewer than 20 births in numerator.

TABLE I-25. PERCENT OF BIRTHS WITH SELECTED MEDICAL OR HEALTH CHARACTERISTICS BY SPECIFIED HISPANIC ORIGIN, RACE, AND PLACE OF BIRTH OF MOTHER, 2002

Smoking and drinking were particularly low among Hispanics who originated in Central and South America. Non-Hispanic Whites, however, had a relatively high smoking rate of 15.0 percent. Preterm births—born prior to 37 weeks of gestation—averaged 12.1 percent for all mothers, but were especially high at 17.7 percent for non-Hispanic Blacks. This group also had a relatively high frequency of births with very low birthweight (less than 3-1/4 pounds), amounting to 3.1 percent. By and large, Hispanic infants had comparable or lower risk of preterm birth, low birthweight, or low Apgar score, compared with non-Hispanic Whites.

The cesarean delivery rate was particularly high among Cuban mothers; it amounted to 33.2 percent for Cubans born in the United States and 39.9 percent for foreign-born Cubans.

Table I-25. Percent of Births with Selected Medical or Health Characteristics by Specified Hispanic Origin, Race, and Place of Birth of Mother, 2002

Characteristic	All origins [1]	Origin of mother								
		Hispanic						Non-Hispanic		
		Total	Mexican	Puerto Rican	Cuban	Central and South American	Other and unknown Hispanic	Total [2]	White	Black
ALL BIRTHS										
Mother										
Prenatal care beginning in the first trimester	83.7	76.7	75.7	79.9	92.0	78.7	76.7	85.7	88.6	75.2
Late or no prenatal care	3.6	5.5	5.8	4.1	1.3	4.9	5.3	3.1	2.2	6.2
Smoker [3]	11.4	3.0	2.2	9.0	2.8	1.3	6.5	13.2	15.0	8.8
Drinker [4]	0.8	0.5	0.5	0.7	0.3	0.3	0.9	0.9	0.9	0.9
Weight gain of less than 16 lbs [5]	12.2	15.1	16.7	12.6	7.9	12.5	13.5	11.7	10.2	17.8
Median weight gain [5]	30.5	28.7	27.4	30.5	31.6	30.0	30.3	30.7	30.9	29.9
Cesarean delivery rate	26.1	25.2	24.5	26.2	36.9	27.0	24.6	26.4	26.2	27.7
Infant										
Preterm births [6]	12.1	11.6	11.4	14.0	10.5	11.2	12.8	12.2	11.0	17.7
Birthweight										
Very low birthweight [7]	1.5	1.2	1.1	2.0	1.2	1.2	1.4	1.5	1.2	3.1
Low birthweight [8]	7.8	6.5	6.2	9.7	6.5	6.5	7.9	8.2	6.9	13.4
4,000 grams or more [9]	9.2	8.5	8.8	6.8	8.5	8.5	6.8	9.3	10.8	4.9
5-minute Apgar score of less than 7 [10]	1.4	1.1	1.1	1.4	0.7	0.9	1.3	1.4	1.2	2.3
MOTHERS BORN IN THE 50 STATES AND DC										
Mother										
Prenatal care beginning in the first trimester	85.3	79.8	79.7	79.4	91.0	83.4	77.9	86.0	88.8	75.1
Late or no prenatal care	3.1	4.3	4.3	4.3	1.7	3.5	5.1	3.0	2.2	6.1
Smoker [3]	13.8	6.3	5.1	10.1	4.3	4.6	8.7	14.4	15.5	9.8
Drinker [4]	0.9	0.8	0.8	0.7	0.5	0.6	1.1	1.0	0.9	0.9
Weight gain of less than 16 lbs [5]	11.9	13.5	14.5	12.0	7.8	9.6	13.5	11.8	10.2	18.1
Median weight gain [5]	30.7	30.3	30.1	30.7	32.2	31.0	30.4	30.7	30.9	30.0
Cesarean delivery rate	26.3	25.1	25.0	25.6	33.2	25.1	24.3	26.4	26.3	27.4
Infant										
Preterm births [6]	12.4	12.6	12.4	13.9	10.1	11.3	13.3	12.4	11.1	18.2
Birthweight										
Very low birthweight [7]	1.5	1.4	1.2	1.9	1.2	1.5	1.5	1.6	1.2	3.2
Low birthweight [8]	8.2	7.5	7.1	9.6	6.8	7.5	8.3	8.2	6.9	13.8
4,000 grams or more [9]	9.4	7.7	8.0	6.8	7.7	8.4	6.5	9.6	10.8	4.5
5-minute Apgar score of less than 7 [10]	1.4	1.2	1.2	1.3	0.6	0.9	1.4	1.4	1.2	2.3
MOTHERS BORN OUTSIDE THE 50 STATES AND DC										
Mother										
Prenatal care beginning in the first trimester	78.6	75.0	73.5	80.9	92.8	78.1	74.1	83.9	86.2	76.2
Late or no prenatal care	5.2	6.1	6.6	3.8	1.0	5.1	5.7	3.8	3.2	6.8
Smoker [3]	1.8	1.1	0.7	6.8	1.6	0.8	1.6	2.7	5.1	1.2
Drinker [4]	0.4	0.3	0.2	0.8	*	0.2	0.3	0.5	0.9	0.3
Weight gain of less than 16 lbs [5]	13.7	16.1	17.9	13.8	8.0	12.9	13.6	10.6	9.2	15.4
Median weight gain [5]	28.8	27.3	25.9	30.2	31.1	29.6	29.6	30.2	30.7	28.9
Cesarean delivery rate	25.5	25.2	24.2	27.5	39.9	27.3	25.6	26.0	24.7	29.6
Infant										
Preterm births [6]	10.8	11.0	10.8	14.1	10.8	11.2	11.3	10.5	9.3	13.8
Birthweight										
Very low birthweight [7]	1.1	1.0	1.0	2.0	1.1	1.2	1.2	1.3	1.0	2.5
Low birthweight [8]	6.6	6.0	5.6	9.8	6.3	6.4	6.3	7.5	6.1	9.9
4,000 grams or more [9]	8.4	9.0	9.3	6.6	9.1	8.5	7.5	7.6	10.6	7.6
5-minute Apgar score of less than 7 [10]	1.0	1.0	1.0	1.4	0.8	0.9	1.1	1.1	0.9	1.7

Note: Race and Hispanic origin are reported separately on birth certificates. Persons of Hispanic origin may be of any race. In this table Hispanic women are classified only by place of origin; non-Hispanic women are classified by race.

1Includes origin not stated.
2Includes races other than White and Black.
3Excludes data for California, which did not report tobacco use on the birth certificate.
4Excludes data for California, which did not report alcohol use on the birth certificate.
5Excludes data for California, which did not report weight gain on the birth certificate. Median weight gain shown in pounds.
6Born prior to 37 completed weeks of gestation.
7Birthweight of less than 1,500 grams (3 lb 4 oz).
8Birthweight of less than 2,500 grams (5 lb 8 oz).
9Equivalent to 8 lb 14 oz.
10Excludes data for California and Texas, which did not report 5-minute Apgar score on the birth certificate.
* = Figure does not meet standards of reliability or precision; based on fewer than 20 births in numerator.

TABLE I-26. LIVE BIRTHS TO MOTHERS WITH SELECTED MEDICAL RISK FACTORS AND RATES BY AGE AND RACE OF MOTHER, 2002

Medical risk factors during pregnancy can contribute to serious complications and maternal and infant morbidity and mortality if not treated properly. Sixteen medical risk factors that can affect pregnancy outcome have been separately identified on birth certificates since 1989. The most frequently reported risk factors in 2002 were pregnancy-associated hypertension (37.8 per 1,000 births), diabetes (32.8 per 1,000), and anemia (25.7 per 1,000). These rates have risen considerably (about 40 percent each) since the onset of the 1990s.

Chronic hypertension, which is high blood pressure prior to pregnancy or before the 20th week of gestation, has increased about 30 percent since 1990. Eclampsia is coma and/or convulsions associated with hypertension, edema and proteinuria of unclear origin. It has declined 20 percent since 1990, most likely due to increased first trimester prenatal care (see Table 33A) and improved treatment of pre-eclampsia signs.

The rate of hydramnios/oligohydramnios (the excess or shortage of amniotic fluid) had doubled between 1990 and 2002 (from 5.9 to 13.9); this condition is associated with maternal diabetes. Acute or chronic lung disease, such as asthma or tuberculosis, has also risen very rapidly during this period (more than tripled), with most notable increases in younger women.

The risk of having a medical condition during pregnancy often differs by maternal age. Hypertension and lung disease is most prevalent for pregnant women both under 20 and over 40. Anemia is more frequent at a young age, especially teenage. Diabetes (as most other chronic conditions) occurs more often in the older age groups, with an incidence of 76.5 per 1,000 among pregnancies over 40 years. When Whites and Blacks are compared, Blacks report higher rates of pregnancy-associated hypertension both for the teenage and the older age groups; lower overall rates of diabetes (although higher rates of diabetes for older age groups); and higher rates of anemia for all age groups.

Table I-26. Live Births to Mothers with Selected Medical Risk Factors and Rates by Age and Race of Mother, 2002

(Rates are number of live births with specified medical risk factor per 1,000 live births in specified group.)

Medical risk factor and race of mother	All births [1]	Medical risk factor reported	Rate by age of mother							Not stated [2]
			All ages	Under 20 years	20–24 years	25–29 years	30–34 years	35–39 years	40–54 years	
All Races [3]										
Anemia	4 021 726	102 788	25.7	36.3	30.9	23.8	20.5	20.7	21.2	28 424
Cardiac disease	4 021 726	20 308	5.1	2.7	3.4	4.8	6.6	7.7	9.5	28 424
Acute or chronic lung disease	4 021 726	49 263	12.3	14.6	13.2	11.8	11.3	11.8	12.5	28 424
Diabetes	4 021 726	131 027	32.8	9.9	18.6	31.8	43.2	57.7	76.5	28 424
Genital herpes [4]	3 649 276	33 644	9.3	6.3	8.1	8.5	10.7	12.7	13.3	28 748
Hydramnios/oligohydramnios	4 021 726	55 590	13.9	14.7	13.9	13.5	13.5	14.4	17.8	28 424
Hemoglobinopathy	4 021 726	3 152	0.8	1.1	0.9	0.7	0.7	0.7	0.9	28 424
Hypertension, chronic	4 021 726	33 442	8.4	2.7	4.9	7.5	10.3	15.8	25.6	28 424
Hypertension, pregnancy-associated	4 021 726	150 854	37.8	41.3	37.5	37.3	35.6	38.6	47.8	28 424
Eclampsia	4 021 726	12 920	3.2	4.6	3.4	2.9	2.8	3.2	4.0	28 424
Incompetent cervix	4 021 726	11 703	2.9	1.4	2.1	2.8	3.7	4.5	4.9	28 424
Previous infant 4,000+ grams	4 021 726	42 102	10.5	1.3	5.8	10.8	14.9	18.1	20.0	28 424
Previous preterm or small-for-gestational-age infant	4 021 726	49 934	12.5	4.7	12.1	13.4	13.6	15.3	17.3	28 424
Renal disease	4 021 726	12 185	3.1	3.0	3.4	3.1	2.8	2.6	2.6	28 424
Rh sensitization [5]	3 982 314	26 648	6.7	5.3	6.2	6.8	7.5	7.4	7.5	30 643
Uterine bleeding [4]	3 649 276	20 666	5.7	4.1	5.0	5.7	6.4	6.8	8.0	28 748
White										
Anemia	3 174 760	72 058	22.9	32.2	27.2	21.2	18.8	19.2	19.1	22 500
Cardiac disease	3 174 760	17 148	5.4	2.8	3.5	5.1	7.0	8.3	10.1	22 500
Acute or chronic lung disease	3 174 760	37 660	11.9	13.1	12.5	11.6	11.3	11.8	12.5	22 500
Diabetes	3 174 760	99 472	31.6	10.2	18.2	30.2	40.3	53.7	71.2	22 500
Genital herpes [4]	2 857 610	26 134	9.2	5.7	7.3	8.2	10.9	13.5	14.6	22 774
Hydramnios/oligohydramnios	3 174 760	42 414	13.5	13.9	13.5	13.0	13.1	13.9	16.8	22 500
Hemoglobinopathy	3 174 760	1 109	0.4	0.3	0.3	0.3	0.4	0.4	0.5	22 500
Hypertension, chronic	3 174 760	22 796	7.2	2.3	4.3	6.5	8.8	13.0	20.4	22 500
Hypertension, pregnancy-associated	3 174 760	120 089	38.1	40.3	38.0	38.4	36.1	38.0	47.0	22 500
Eclampsia	3 174 760	9 418	3.0	4.0	3.1	2.7	2.6	3.1	3.8	22 500
Incompetent cervix	3 174 760	7 936	2.5	1.2	1.7	2.3	3.2	4.1	4.7	22 500
Previous infant 4,000+ grams	3 174 760	37 572	11.9	1.4	6.5	12.0	16.5	20.1	22.3	22 500
Previous preterm or small-for-gestational-age infant	3 174 760	38 518	12.2	4.3	11.5	12.9	13.5	15.0	17.1	22 500
Renal disease	3 174 760	10 438	3.3	3.4	3.7	3.4	3.1	2.8	2.8	22 500
Rh sensitization [5]	3 139 856	24 227	7.8	6.2	7.1	7.8	8.6	8.6	8.7	24 429
Uterine bleeding [4]	2 857 610	17 154	6.1	4.5	5.3	6.0	6.6	7.0	8.6	22 774
Black										
Anemia	593 691	24 029	40.7	45.1	44.1	39.7	34.9	32.8	35.9	3 045
Cardiac disease	593 691	2 356	4.0	2.4	3.3	4.3	5.3	6.1	8.0	3 045
Acute or chronic lung disease	593 691	9 466	16.0	18.7	16.5	15.5	13.7	14.2	15.3	3 045
Diabetes	593 691	17 952	30.4	8.3	18.0	33.3	50.5	67.8	86.4	3 045
Genital herpes [4]	552 049	6 286	11.5	8.0	11.6	12.6	13.5	11.9	8.5	3 126
Hydramnios/oligohydramnios	593 691	9 655	16.3	17.1	15.3	15.5	16.7	18.8	23.4	3 045
Hemoglobinopathy	593 691	1 885	3.2	3.5	3.1	3.1	3.0	3.2	3.7	3 045
Hypertension, chronic	593 691	9 245	15.7	4.0	7.9	15.6	25.9	41.1	62.0	3 045
Hypertension, pregnancy-associated	593 691	24 484	41.5	44.7	37.2	39.0	42.8	51.5	59.3	3 045
Eclampsia	593 691	2 922	4.9	6.2	4.5	4.6	4.8	4.9	6.1	3 045
Incompetent cervix	593 691	3 243	5.5	1.9	4.0	6.6	8.9	9.1	7.8	3 045
Previous infant 4,000+ grams	593 691	2 630	4.5	0.8	2.9	5.7	7.2	8.5	9.5	3 045
Previous preterm or small-for-gestational-age infant	593 691	8 882	15.0	5.9	14.9	18.7	17.6	18.8	20.6	3 045
Renal disease	593 691	1 246	2.1	2.0	2.2	2.0	2.2	2.0	*	3 045
Rh sensitization [5]	590 801	1 881	3.2	3.0	3.0	3.4	3.4	3.1	3.4	3 287
Uterine bleeding [4]	552 049	2 310	4.2	3.0	3.8	4.6	5.0	5.7	4.8	3 126

[1]Total number of births to residents of areas reporting specified medical risk factor.
[2]No response reported for the medical risk factor item.
[3]Includes races other than White and Black.
[4]Texas does not report this risk factor.
[5]Kansas does not report this risk factor.
* = Figure does not meet standards of reliability or precision; based on fewer than 20 births in numerator.

TABLE I-27. NUMBER AND RATE OF LIVE BIRTHS TO MOTHERS WITH SELECTED MEDICAL RISK FACTORS, COMPLICATIONS OF LABOR, AND OBSTETRIC PROCEDURES, BY DETAILED RACE OF MOTHER, 2002

American Indian women had the highest rates of pregnancy-associated hypertension and anemia, and particularly high rates of diabetes. High rates of diabetes were also reported for Asians, in particular Filipinos. Blacks have higher rates of pregnancy-associated hypertension and anemia, compared with Whites. Native Hawaiians had the highest rate of pregnancy-associated hypertension, as well as very high rates of anemia and diabetes. Chinese had the lowest rate of pregnancy-associated hypertension.

Blacks had higher rates of meconium staining and fetal distress, but lower rates of cephalopelvic disproportion and breech/malpresentation (leading risk factors for cesarean delivery), compared with Whites. Hawaiians also had elevated rates of meconium staining and fetal distress.

In regard to obstetric procedures, most mothers received electronic fetal monitoring (852 out of 1,000); despite some controversy over routine use of EFM, this rate has climbed continually since 1989. The use of ultrasound also has increased steadily, and it was received by 676 out of 1,000 mothers in 2002. Advances in this technology allow early screening of fetal abnormalities. Amniocentesis generally was used rarely, and its use has declined somewhat since 1989; it was obtained most frequently by Asians, and by Japanese women, in particular (79.4 out of 1,000).

Table I-27. Number and Rate of Live Births to Mothers with Selected Medical Risk Factors, Complications of Labor, and Obstetric Procedures, by Detailed Race of Mother, 2002

(Rates are number of live births with specified risk factors, complications, or procedures per 1,000 live births in specified group.)

Medical risk factor, complication, and obstetric procedure	All races	White	Black	American Indian [1]	Asian or Pacific Islander					
					Total	Chinese	Japanese	Hawaiian	Filipino	Other
NUMBER										
Medical Risk Factors										
Anemia	102 788	72 058	24 029	2 366	4 335	417	145	371	587	2 815
Diabetes	131 027	99 472	17 952	2 345	11 258	1 601	391	331	1 957	6 978
Hypertension, pregnancy-associated	150 854	120 089	24 484	1 941	4 340	389	155	324	1 008	2 464
Uterine bleeding [2]	20 666	17 154	2 310	237	965	126	62	42	184	551
Complications of Labor and/or Delivery										
Meconium, moderate/heavy	200 408	147 500	39 941	2 194	10 773	1 572	408	448	1 925	6 420
Premature rupture of membrane	92 550	70 750	15 458	1 403	4 939	696	274	187	863	2 919
Dysfunctional labor	114 479	90 497	15 555	1 538	6 889	1 427	338	269	1 128	3 727
Breech/malpresentation	152 425	124 541	18 370	1 584	7 930	1 305	374	344	1 205	4 702
Cephalopelvic disproportion	63 240	51 124	7 081	625	4 410	779	182	116	808	2 525
Fetal distress [3]	140 074	104 231	26 571	1 545	7 727	1 085	347	313	1 208	4 774
Obstetric Procedures										
Amniocentesis	79 035	64 891	7 133	566	6 445	1 606	733	206	1 032	2 868
Electronic fetal monitoring	3 413 557	2 697 901	508 334	35 649	171 673	27 541	7 308	5 254	26 187	105 383
Induction of labor	826 161	685 153	103 057	8 729	29 222	4 120	1 363	1 071	3 971	18 697
Ultrasound	2 707 007	2 168 878	367 325	26 382	144 422	24 224	6 798	4 566	22 339	86 495
Stimulation of labor	693 548	553 122	95 512	6 920	37 994	6 412	1 791	1 128	5 414	23 249
RATE										
Medical Risk Factors										
Anemia	25.7	22.9	40.7	56.7	20.8	12.5	15.8	55.2	17.9	22.2
Diabetes	32.8	31.6	30.4	56.2	54.0	47.9	42.5	49.2	59.8	55.1
Hypertension, pregnancy-associated	37.8	38.1	41.5	46.5	20.8	11.6	16.9	48.2	30.8	19.5
Uterine bleeding [2]	5.7	6.1	4.2	5.8	4.9	4.0	6.9	6.3	5.8	4.7
Complications of Labor and/or Delivery										
Meconium, Moderate/heavy	50.1	46.7	67.5	52.5	51.5	46.8	44.3	66.6	58.7	50.5
Premature rupture of membrane	23.1	22.4	26.1	33.6	23.6	20.7	29.7	27.8	26.3	23.0
Dysfunctional labor	28.6	28.6	26.3	36.8	32.9	42.5	36.7	40.0	34.4	29.3
Breech/malpresentation	38.1	39.4	31.0	37.9	37.9	38.8	40.6	51.2	36.8	37.0
Cephalopelvic disproportion	15.8	16.2	12.0	15.0	21.1	23.2	19.8	17.2	24.6	19.9
Fetal distress [3]	38.6	36.7	48.3	37.7	39.3	33.8	38.7	47.1	38.3	40.7
Obstetric Procedures										
Amniocentesis	19.7	20.5	12.0	13.5	30.7	47.8	79.4	30.6	31.4	22.5
Electronic fetal monitoring	851.9	853.0	858.2	851.8	818.7	819.1	791.9	779.9	797.2	828.2
Induction of labor	206.2	216.6	174.0	208.6	139.4	122.5	147.7	159.0	120.9	146.9
Ultrasound	675.6	685.7	620.1	630.3	688.8	720.4	736.7	677.7	680.1	679.7
Stimulation of labor	173.1	174.9	161.2	165.3	181.2	190.7	194.1	167.4	164.8	182.7

[1]Includes births to Aleuts and Eskimos.
[2]Texas does not report this risk factor.
[3]Texas does not report this complication.

TABLE I-28. NUMBER AND RATE OF LIVE BIRTHS TO MOTHERS WITH SELECTED MEDICAL RISK FACTORS, COMPLICATIONS OF LABOR, AND OBSTETRIC PROCEDURES, BY SPECIFIED HISPANIC ORIGIN AND RACE OF MOTHER, 2002

Compared with non-Hispanic White women, Hispanics had substantially less risk of pregnancy-induced hypertension, somewhat greater risk of anemia, and about the same frequency of diabetes. Hispanics—especially Mexicans and Puerto Ricans—seldom obtained tests for amniocentesis (1 percent of births), and they had a lower frequency of induced labor than non-Hispanic Whites.

Puerto Ricans had greater rates of medical risk factors and of complications of labor and/or delivery than Mexicans and Cubans, with the exception of high rates in Cubans of dysfunctional labor and breech/malpresentation (Cubans also have high rates of cesarean delivery, as shown in Table 25).

Compared with non-Hispanic Whites, non-Hispanic Blacks had higher rates of pregnancy-associated hypertension and anemia; in regard to complications of labor and/or delivery, they had a greater frequency of meconium (moderate/heavy) and of fetal distress.

Table I-28. Number and Rate of Live Births to Mothers with Selected Medical Risk Factors, Complications of Labor, and Obstetric Procedures, by Specified Hispanic Origin and Race of Mother, 2002

(Rates are number of live births with specified risk factors, complications, or procedures per 1,000 live births in specified group.)

Medical risk factor, complication, and obstetric procedure	All origins [1]	Origin of mother								
		Hispanic						Non-Hispanic		
		Total	Mexican	Puerto Rican	Cuban	Central and South American	Other and unknown Hispanic	Total [2]	White	Black
NUMBER										
Medical Risk Factors										
Anemia	102 788	23 234	15 511	2 328	344	2 700	2 351	78 895	49 136	23 403
Diabetes	131 027	28 417	19 857	2 406	388	3 908	1 858	101 884	71 263	17 417
Hypertension, pregnancy-associated	150 854	22 948	15 563	1 827	441	3 142	1 975	127 083	97 070	24 039
Uterine bleeding [3]	20 666	2 648	1 564	381	46	369	288	17 873	14 504	2 223
Complications of Labor and/or Delivery										
Meconium, moderate/heavy	200 408	48 221	33 975	3 177	505	7 789	2 775	151 008	99 674	38 934
Premature rupture of membrane	92 550	15 825	10 620	1 436	182	2 393	1 194	76 087	54 901	15 119
Dysfunctional labor	114 479	22 199	13 932	1 990	538	3 673	2 066	91 736	68 538	15 053
Breech/malpresentation	152 425	25 689	17 253	2 047	543	3 984	1 862	125 843	98 814	17 895
Cephalopelvic disproportion	63 240	10 169	7 152	772	172	1 475	598	52 647	40 830	6 926
Fetal distress [4]	140 074	19 858	12 054	2 211	366	3 838	1 389	119 502	84 546	26 023
Obstetric Procedures										
Amniocentesis	79 035	8 532	4 254	1 037	274	2 080	887	69 923	56 260	6 849
Electronic fetal monitoring	3 413 557	717 570	506 409	49 663	12 566	104 197	44 735	2 678 865	1 983 122	495 846
Induction of labor	826 161	120 472	80 600	9 490	2 909	16 780	10 693	701 317	563 923	100 839
Ultrasound	2 707 007	525 175	360 827	39 552	8 261	80 830	35 705	2 167 727	1 645 878	357 152
Stimulation of labor	693 548	135 841	93 036	11 685	1 833	20 971	8 316	553 698	417 475	92 724
RATE										
Medical Risk Factors										
Anemia	25.7	26.6	24.8	40.8	24.2	21.6	46.0	25.5	21.5	40.7
Diabetes	32.8	32.6	31.8	42.1	27.3	31.3	36.4	32.9	31.2	30.3
Hypertension, pregnancy-associated	37.8	26.3	24.9	32.0	31.1	25.1	38.6	41.0	42.6	41.8
Uterine bleeding [3]	5.7	3.8	3.3	6.8	3.3	3.2	7.2	6.2	6.8	4.2
Complications of Labor and/or Delivery										
Meconium, moderate/heavy	50.1	55.2	54.2	55.5	35.5	62.1	54.2	48.7	43.6	67.5
Premature rupture of membrane	23.1	18.1	17.0	25.1	12.8	19.1	23.3	24.5	24.0	26.2
Dysfunctional labor	28.6	25.4	22.2	34.8	37.9	29.3	40.4	29.6	30.0	26.1
Breech/malpresentation	38.1	29.4	27.5	35.8	38.2	31.7	36.4	40.5	43.2	31.0
Cephalopelvic disproportion	15.8	11.6	11.4	13.5	12.1	11.8	11.7	17.0	17.9	12.0
Fetal distress [4]	38.6	28.6	25.7	39.5	26.3	33.1	34.8	41.1	39.4	48.6
Obstetric Procedures										
Amniocentesis	19.7	9.8	6.8	18.1	19.3	16.6	17.3	22.5	24.6	11.9
Electronic fetal monitoring	851.9	820.2	808.2	867.5	883.8	829.8	872.4	862.0	866.5	859.2
Induction of labor	206.2	137.7	128.6	165.8	204.6	133.6	208.5	225.7	246.4	174.7
Ultrasound	675.6	600.3	575.9	690.9	581.0	643.7	696.3	697.5	719.1	618.9
Stimulation of labor	173.1	155.3	148.5	204.1	128.9	167.0	162.2	178.2	182.4	160.7

Note: Race and Hispanic origin are reported separately on birth certificates. Persons of Hispanic origin may be of any race. In this table Hispanic women are classified only by place of origin; non-Hispanic women are classified by race.

[1] Includes origin not stated.
[2] Includes races other than White and Black.
[3] Texas does not report this risk factor.
[4] Texas does not report this complication.

TABLE I-29. NUMBER OF LIVE BIRTHS BY SMOKING STATUS DURING PREGNANCY, PERCENT SMOKERS, AND CIGARETTE CONSUMPTION, ACCORDING TO AGE AND RACE OF MOTHER: UNITED STATES EXCLUDING CALIFORNIA, 2002

The negative consequences of prenatal smoking on birth outcomes are well documented, and include low birth-weight intrauterine growth retardation, miscarriage, and infant mortality, as well as negative impacts on child health and development. The impact on low birthweight is more severe with greater cigarette consumption, but studies also show that there is no safe level of smoking.

Smoking during pregnancy declined in 2002 to 11.4 percent of women giving birth, down from 19.5 percent in 1989, when this information first became available from the birth certificate. The highest incidence of 18.2 percent was among those aged 18–19. The frequency of smoking declined for higher ages through the mid-30s, with a low of 7.1 percent in the 30–34 year age group. However, as age increases, women who smoke are more likely to be heavier smokers. Smoking by Blacks is less frequent than by Whites in age groups below 30, but more frequent in age groups above 30.

Table I-29. Number of Live Births by Smoking Status During Pregnancy, Percent Smokers, and Cigarette Consumption, According to Age and Race of Mother: United States Excluding California, 2002

Smoking status, smoking measure, and race of mother	Age of mother									
	All ages	Under 15 years	15–19 years			20–24 years	25–29 years	30–34 years	35–39 years	40–54 years
			Total	15–17 years	18–19 years					
NUMBER										
All Races [1]										
Total	3 492 369	6 572	375 271	122 064	253 207	899 012	923 112	820 815	384 033	83 554
Smoker	397 197	382	62 193	16 272	45 921	149 563	90 493	57 699	29 868	6 999
Nonsmoker	3 077 173	6 150	311 090	105 074	206 016	744 930	828 111	758 982	351 935	75 975
Not stated	17 999	40	1 988	718	1 270	4 519	4 508	4 134	2 230	580
White										
Total	2 746 211	3 275	262 896	81 454	181 442	678 382	739 861	678 778	315 538	67 481
Smoker	337 312	301	53 447	13 877	39 570	127 352	76 740	49 145	24 737	5 590
Nonsmoker	2 394 739	2 954	208 045	67 103	140 942	547 640	659 555	626 188	288 933	61 424
Not stated	14 160	20	1 404	474	930	3 390	3 566	3 445	1 868	467
Black										
Total	561 038	3 084	99 384	36 447	62 937	185 104	128 823	88 671	44 989	10 983
Smoker	48 578	59	6 701	1 778	4 923	18 103	11 170	6 926	4 389	1 230
Nonsmoker	509 875	3 011	92 246	34 485	57 761	166 191	117 045	81 308	40 386	9 688
Not stated	2 585	14	437	184	253	810	608	437	214	65
PERCENT SMOKERS										
Total [1]	11.4	5.8	16.7	13.4	18.2	16.7	9.9	7.1	7.8	8.4
White	12.3	9.2	20.4	17.1	21.9	18.9	10.4	7.3	7.9	8.3
Black	8.7	1.9	6.8	4.9	7.9	9.8	8.7	7.8	9.8	11.3
PERCENT DISTRIBUTION [2]										
All Races [1]										
Smoker	100.0	100.0	100.0	100.0	100.0	100.0	100.0	100.0	100.0	100.0
1–5 cigarettes	32.2	52.3	40.2	43.8	38.9	33.2	29.6	28.0	27.5	26.9
6–10 cigarettes	42.0	35.6	42.3	40.7	42.8	42.9	42.1	40.8	39.5	37.5
11–15 cigarettes	5.5	*	4.0	3.5	4.2	5.0	6.0	6.8	6.9	7.5
16–20 cigarettes	17.8	9.7	12.2	10.8	12.7	16.9	19.5	20.8	21.9	23.1
21–30 cigarettes	1.9	*	1.0	0.9	1.0	1.5	2.0	2.6	2.9	3.3
31–40 cigarettes	0.6	*	0.3	0.3	0.3	0.5	0.7	0.8	1.1	1.6
41 cigarettes or more	0.1	*	0.1	*	0.1	0.1	0.1	0.1	0.1	*
White										
Smoker	100.0	100.0	100.0	100.0	100.0	100.0	100.0	100.0	100.0	100.0
1–5 cigarettes	29.0	45.1	36.6	39.8	35.5	29.6	26.5	25.5	24.6	23.5
6–10 cigarettes	43.1	41.0	44.3	43.3	44.7	44.4	42.9	41.1	39.5	37.1
11–15 cigarettes	6.0	*	4.4	3.8	4.6	5.4	6.5	7.3	7.6	8.5
16–20 cigarettes	19.2	11.3	13.3	11.8	13.8	18.3	21.0	22.3	23.7	25.2
21–30 cigarettes	2.0	*	1.1	1.0	1.1	1.7	2.2	2.9	3.2	3.8
31–40 cigarettes	0.7	*	0.3	0.3	0.3	0.5	0.8	0.9	1.2	1.8
41 cigarettes or more	0.1	*	0.1	*	0.1	0.1	0.1	0.1	0.2	*
Black										
Smoker	100.0	100.0	100.0	100.0	100.0	100.0	100.0	100.0	100.0	100.0
1–5 cigarettes	51.2	87.0	64.0	69.0	62.3	54.7	47.5	42.7	42.4	41.3
6–10 cigarettes	35.2	*	28.3	23.8	29.9	33.6	37.6	39.5	39.0	39.0
11–15 cigarettes	2.7	*	1.5	1.7	1.5	2.3	2.9	3.8	3.3	3.5
16–20 cigarettes	9.5	*	5.5	5.0	5.7	8.3	10.6	12.0	13.3	13.9
21–30 cigarettes	0.9	*	0.4	*	0.5	0.8	0.9	1.2	1.3	*
31–40 cigarettes	0.4	*	*	*	*	0.3	0.5	0.7	0.5	*
41 cigarettes or more	0.1	*	*	*	*	*	*	*	*	*

Note: Excludes data for California, which did not require reporting of tobacco use during pregnancy.

[1]Includes races other than White and Black.
[2]Excludes data for Indiana and New York state (but includes New York City), and South Dakota, which did not report average number of cigarettes smoked per day in standard categories.
* = Figure does not meet standards of reliability or precision; based on fewer than 20 births in numerator.

The trends and variations of smoking as reported here from birth certificates are corroborated by data from comparable surveys, but the extent of maternal smoking is believed to be somewhat underreported on the birth certificates. The results are in part due to the stigma associated with tobacco use, particularly in cases of poor birth outcome. The daily consumption of cigarettes by smokers, most frequently, was 6 to 10 cigarettes for Whites and 1 to 5 cigarettes by Blacks. Daily dosages of 16 to 20 cigarettes were reported by 18 percent of smokers. The proportion of smokers who smoked 11 cigarettes or more per day has fallen from 41 percent in 1989 to 26 percent in 2002.

TABLE I-30. NUMBER OF LIVE BIRTHS BY SMOKING STATUS DURING PREGNANCY AND PERCENT SMOKERS, BY AGE, SPECIFIED HISPANIC ORIGIN, AND RACE OF MOTHER: UNITED STATES EXCLUDING CALIFORNIA, 2002

This table presents numbers of smokers and percentages of smokers by originating country for Hispanics. It also shows the smoking pattern by age groups. On average only 3.0 percent of Hispanic mothers were smokers, as compared with 8.8 percent of non-Hispanic Blacks and 15.0 percent of non-Hispanic Whites. Hispanics show less variation in smoking frequency by age than non-Hispanic Whites and Blacks. Among the various Hispanic groups, the highest frequency of smoking was reported for Puerto Ricans (9.0 percent).

Table I-30. Number of Live Births by Smoking Status During Pregnancy and Percent Smokers, by Age, Specified Hispanic Origin, and Race of Mother: United States Excluding California, 2002

(Number, percent.)

Origin of mother	Smoking status (number)				Age of mother (percent smokers)									
	Total births	Smoker	Nonsmoker	Not stated	All ages	Under 15 years	15–19 years			20–24 years	25–29 years	30–34 years	35–39 years	40–54 years
							Total	15–17 years	18–19 years					
All origins [1]	3 492 369	397 197	3 077 173	17 999	11.4	5.8	16.7	13.4	18.2	16.7	9.9	7.1	7.8	8.4
Hispanic	613 581	18 488	592 559	2 534	3.0	1.8	3.7	3.1	4.0	3.7	2.5	2.3	2.8	3.1
Mexican	397 119	8 879	386 434	1 806	2.2	1.5	2.6	2.2	2.9	2.6	1.8	1.7	2.3	2.8
Puerto Rican	55 435	4 964	50 314	157	9.0	*	8.9	6.8	10.1	10.7	8.6	7.1	7.7	7.0
Cuban	13 533	378	13 142	13	2.8	*	5.3	5.8	5.1	4.4	1.9	2.2	2.1	*
Central and South American	101 139	1 265	99 626	248	1.3	*	1.9	1.6	2.1	1.5	1.0	0.9	1.3	1.5
Other and unknown Hispanic	46 355	3 002	43 043	310	6.5	*	7.6	6.4	8.4	7.8	5.5	5.1	5.4	5.9
Non-Hispanic [2]	2 857 908	375 979	2 467 692	14 237	13.2	7.4	21.0	17.5	22.5	20.2	11.4	7.8	8.5	9.1
White	2 133 507	317 665	1 805 177	10 665	15.0	18.6	29.4	27.2	30.2	24.6	12.6	8.2	8.8	9.2
Black	546 885	47 851	496 583	2 451	8.8	1.9	6.8	4.9	7.9	9.9	8.8	8.0	10.0	11.5

Note: Excludes data for California, which did not require reporting of tobacco use during pregnancy. Race and Hispanic origin are reported separately on birth certificates. Persons of Hispanic origin may be of any race. In this table Hispanic women are classified only by place of origin; non-Hispanic women are classified by race.

[1]Includes origin not stated.
[2]Includes races other than White and Black.
* = Figure does not meet standards of reliability or precision; based on fewer than 20 births in numerator.

TABLE I-31. NUMBER OF LIVE BIRTHS, PERCENT SMOKERS DURING PREGNANCY, AND CIGARETTE CONSUMPTION, ACCORDING TO EDUCATIONAL ATTAINMENT, RACE, AND HISPANIC ORIGIN OF MOTHER: UNITED STATES EXCLUDING CALIFORNIA, 2002

Mothers who completed only 9 to 11 years of education are most likely to be smokers, a pattern which holds for non-Hispanic Whites, Blacks, and Hispanics. For non-Hispanic Whites 42.7 percent of members of this group are smokers, but for non-Hispanic Blacks this percentage was 17.0 percent, and for Hispanics it was 4.7 percent. As educational attainment increases, the prevalence of smoking falls. Non-Hispanic White mothers with 16 years of education, or more, smoked in only 1.9 percent of these cases. For smokers, the number of cigarettes smoked per day did not change considerably according to educational attainment. Black and Hispanic smokers were more likely than non-Hispanic Whites to report fewer cigarettes per day, that is, a half pack per day or less.

Table I-31. Number of Live Births, Percent Smokers During Pregnancy, and Cigarette Consumption, According to Educational Attainment, Race, and Hispanic Origin of Mother: United States Excluding California, 2002

Smoking measure, and race and Hispanic origin of mother	Total	Years of school completed by mother					
		0–8 years	9–11 years	12 years	13–15 years	16 years or more	Not stated
ALL BIRTHS							
All races [1]	3 492 369	180 087	525 209	1 086 613	751 010	907 051	42 399
White total [2]	2 746 211	159 696	380 679	817 053	588 731	770 386	29 666
Non-Hispanic White	2 133 507	36 104	219 810	638 201	509 153	714 575	15 664
Black total [2]	561 038	13 563	123 552	219 071	126 718	69 096	9 038
Non-Hispanic Black	546 885	12 674	120 692	213 843	123 771	67 443	8 462
Hispanic [3]	613 581	124 460	162 837	181 202	80 062	53 534	11 486
PERCENT SMOKERS							
Total	11.4	8.3	24.1	15.8	8.7	1.7	11.2
White total [2]	12.3	8.2	26.7	18.2	9.8	1.9	11.9
Non-Hispanic White	15.0	31.2	42.7	22.3	10.9	1.9	18.0
Black total [2]	8.7	10.0	16.8	8.5	5.0	1.2	11.3
Non-Hispanic Black	8.8	10.4	17.0	8.6	5.0	1.2	11.3
Hispanic [3]	3.0	1.5	4.7	3.3	2.7	0.9	2.9
PERCENT DISTRIBUTION [4]							
All Races [1]							
Smoker	100.0	100.0	100.0	100.0	100.0	100.0	100.0
10 cigarettes or less	74.2	69.3	73.8	73.8	75.6	80.1	74.1
11–20 cigarettes	23.3	26.6	23.3	23.8	22.3	18.2	23.1
21 cigarettes or more	2.6	4.1	2.9	2.5	2.1	1.7	2.9
White Total [2]							
Smoker	100.0	100.0	100.0	100.0	100.0	100.0	100.0
10 cigarettes or less	72.1	67.3	70.9	71.8	74.1	79.4	71.2
11–20 cigarettes	25.2	28.2	25.9	25.5	23.7	18.8	25.5
21 cigarettes or more	2.8	4.5	3.2	2.6	2.2	1.8	3.3
Non-Hispanic White							
Smoker	100.0	100.0	100.0	100.0	100.0	100.0	100.0
10 cigarettes or less	71.3	64.0	69.7	71.2	73.7	79.2	70.0
11–20 cigarettes	25.9	30.9	27.0	26.1	24.0	19.0	26.6
21 cigarettes or more	2.9	5.1	3.3	2.7	2.2	1.8	3.4
Black Total [2]							
Smoker	100.0	100.0	100.0	100.0	100.0	100.0	100.0
10 cigarettes or less	86.4	83.0	86.0	86.8	87.5	89.8	81.4
11–20 cigarettes	12.2	15.6	12.4	12.0	11.4	9.1	17.0
21 cigarettes or more	1.4	*	1.6	1.3	1.0	*	*
Non-Hispanic Black							
Smoker	100.0	100.0	100.0	100.0	100.0	100.0	100.0
10 cigarettes or less	86.4	83.0	86.0	86.8	87.5	89.8	81.5
11–20 cigarettes	12.2	15.6	12.4	11.9	11.5	9.1	17.0
21 cigarettes or more	1.4	*	1.6	1.3	1.0	*	*
Hispanic [3]							
Smoker	100.0	100.0	100.0	100.0	100.0	100.0	100.0
10 cigarettes or less	86.3	87.9	87.2	85.7	84.3	85.5	83.4
11–20 cigarettes	12.5	11.3	11.4	13.1	14.7	13.1	15.4
21 cigarettes or more	1.2	*	1.4	1.2	*	*	*

Note: Excludes data for California, which did not require reporting of tobacco use during pregnancy.

[1]Includes races other than White and Black and origin not stated.
[2]Race and Hispanic origin are reported separately on the birth certificate. Data for persons of Hispanic origin are included in the data for each race group according to the mother's reported race.
[3]Includes all persons of Hispanic origin of any race.
[4]Excludes data for Indiana, New York state (but includes New York City), and South Dakota, which did not report average number of cigarettes smoked per day in standard categories.
* = Figure does not meet standards of reliability or precision; based on fewer than 20 births in numerator.

TABLE I-32. PERCENT LOW BIRTHWEIGHT BY SMOKING STATUS DURING PREGNANCY, AGE, RACE, AND HISPANIC ORIGIN OF MOTHER: UNITED STATES EXCLUDING CALIFORNIA, 2002

In 2002, the rate of low birthweight (less than 5-1/2 pounds) among babies born to smokers was 12.2 percent, as compared with 7.5 percent for nonsmokers. This gap tends to widen with advancing maternal age; for mothers 35–39 years old, smokers had a 16.7 percent incidence of low birthweight, while nonsmokers had a frequency of 8.1 percent. Apart from the influence of aging, this could reflect the fact that maternal smokers in the older age brackets tend to be somewhat heavier smokers (see Table 29). Black nonsmokers have roughly double the frequency of low birthweight as non-Hispanic White or Hispanic nonsmokers, reflecting other risk factors. However, the effect of smoking in Blacks is comparable with the other race and ethnicity groups.

Table I-32. Percent Low Birthweight by Smoking Status During Pregnancy, Age, Race, and Hispanic Origin of Mother: United States Excluding California, 2002

(Low birthweight is defined as weight of less than 2,500 grams [5 lb 8 oz].)

Smoking status and race of mother	All ages	Age of mother								
		Under 15 years	15–19 years			20–24 years	25–29 years	30–34 years	35–39 years	40–54 years
			Total	15–17 years	18–19 years					
All Races [1]										
Total	8.0	13.8	9.9	10.7	9.5	8.2	7.1	7.4	8.8	11.1
Smoker	12.2	12.9	11.5	12.2	11.2	11.0	11.4	13.7	16.7	19.3
Nonsmoker	7.5	13.7	9.5	10.5	9.1	7.6	6.6	6.9	8.1	10.3
Not stated	10.9	*	13.2	14.9	12.3	10.3	10.0	10.4	10.9	16.5
White Total [2]										
Total	7.0	12.5	8.4	9.1	8.0	6.9	6.2	6.6	7.8	10.1
Smoker	11.0	13.3	10.9	11.6	10.6	10.2	10.4	12.0	14.6	16.9
Nonsmoker	6.4	12.3	7.7	8.6	7.3	6.1	5.7	6.1	7.2	9.4
Not stated	9.5	*	11.1	13.6	9.8	8.6	8.9	9.5	9.4	15.6
Non-Hispanic White										
Total	7.0	12.7	8.5	9.4	8.2	7.0	6.2	6.5	7.7	10.1
Smoker	11.0	13.6	10.8	11.5	10.5	10.1	10.3	12.0	14.5	17.1
Nonsmoker	6.3	12.4	7.5	8.5	7.2	6.0	5.6	6.0	7.1	9.3
Not stated	9.2	*	11.3	14.9	9.5	8.2	8.9	9.2	8.5	15.2
Black Total [2]										
Total	13.4	15.2	14.0	14.5	13.7	13.0	12.5	13.4	15.3	17.6
Smoker	20.1	*	16.9	17.8	16.5	17.1	18.8	25.7	28.7	31.0
Nonsmoker	12.7	15.2	13.8	14.3	13.4	12.5	11.8	12.4	13.8	15.9
Not stated	18.6	*	21.1	20.8	21.3	17.3	17.8	18.1	19.2	*
Non-Hispanic Black										
Total	13.5	15.1	14.1	14.6	13.7	13.1	12.6	13.5	15.4	17.7
Smoker	20.1	*	17.0	18.1	16.6	17.2	18.8	25.6	28.8	31.1
Nonsmoker	12.8	15.1	13.8	14.4	13.5	12.6	11.9	12.5	13.9	16.0
Not stated	18.5	*	20.5	21.1	20.1	17.4	18.2	17.8	19.0	*
Hispanic [3]										
Total	6.9	12.5	8.2	8.8	7.8	6.5	6.0	6.7	8.4	10.1
Smoker	12.2	*	12.3	13.0	12.0	11.2	11.8	13.1	17.1	14.6
Nonsmoker	6.7	12.5	8.0	8.7	7.6	6.3	5.9	6.6	8.1	9.9
Not stated	9.6	*	12.1	*	12.4	8.8	8.4	9.5	11.7	*

Note: Excludes data for California, which did not require reporting of tobacco use during pregnancy.

[1]Includes races other than White and Black and origin not stated.
[2]Race and Hispanic origin are reported separately on the birth certificate. Data for persons of Hispanic origin are included in the data for each race group according to the mother's reported race.
[3]Includes all persons of Hispanic origin of any race.
* = Figure does not meet standards of reliability or precision; based on fewer than 20 births in numerator.

TABLE I-33. LIVE BIRTHS BY MONTH OF PREGNANCY PRENATAL CARE BEGAN, PERCENT OF MOTHERS BEGINNING CARE IN THE FIRST TRIMESTER, AND PERCENT WITH LATE OR NO CARE, BY AGE, RACE, AND HISPANIC ORIGIN OF MOTHER, 2002

Prenatal care began in the first trimester for 88.6 percent of non-Hispanic White mothers, 75.2 percent of non-Hispanic Black mothers, and 76.7 percent of Hispanic mothers. Young mothers, particularly teenagers, generally had a lower rate of medical care during the first trimester than those aged 25 and over. For all three race/ethnic groups, the frequency of first trimester increases with age until 30–34 years, decreasing slightly in older brackets. The frequency of late care (beginning in the third trimester) or no care was 2.2 percent for non-Hispanic Whites, 6.2 percent for non-Hispanic Blacks, and 5.5 percent for Hispanics.

First trimester care has increased 10 percent on average since 1990 and the gains for Blacks, American Indians, and Hispanics were closer to 20 percent. (These data are not shown in Table 33.) Recent studies suggest that the expansion of Medicaid for pregnant women in the late 1980s has contributed to these increases.

Table I-33. Live Births by Month of Pregnancy Prenatal Care Began, Percent of Mothers Beginning Care in the First Trimester, and Percent with Late or No Care, by Age, Race, and Hispanic Origin of Mother, 2002

(Number, percent.)

Age, race, and Hispanic origin of mother	All births	Month of pregnancy prenatal care began								Percent	
		1st trimester			2nd trimester	Late or no care			Not stated	1st trimester	Late or no care
		Total	1st and 2nd months	3rd month	4th–6th months	Total	7th–9th months	No care			
All Races [1]											
All ages	4 021 726	3 301 186	2 557 791	743 395	499 014	142 430	103 323	39 107	79 096	83.7	3.6
Under 15 years	7 315	3 396	2 110	1 286	2 614	1 042	764	278	263	48.2	14.8
15–19 years	425 493	290 790	199 445	91 345	96 864	27 573	20 198	7 375	10 266	70.0	6.6
15 years	18 703	10 349	6 612	3 737	5 774	1 985	1 471	514	595	57.2	11.0
16 years	43 147	26 550	17 303	9 247	11 884	3 512	2 600	912	1 201	63.3	8.4
17 years	76 881	50 977	34 280	16 697	18 669	5 354	3 905	1 449	1 881	68.0	7.1
18 years	118 651	82 284	56 482	25 802	26 348	7 325	5 329	1 996	2 694	71.0	6.3
19 years	168 111	120 630	84 768	35 862	34 189	9 397	6 893	2 504	3 895	73.5	5.7
20–24 years	1 022 106	786 656	581 528	205 128	167 285	46 539	34 005	12 534	21 626	78.6	4.7
25–29 years	1 060 391	898 647	708 929	189 718	111 438	31 219	22 878	8 341	19 087	86.3	3.0
30–34 years	951 219	839 344	679 553	159 791	73 659	21 613	15 631	5 982	16 603	89.8	2.3
35–39 years	453 927	397 004	319 130	77 874	36 847	11 178	7 635	3 543	8 898	89.2	2.5
40 years and over	101 275	85 349	67 096	18 253	10 307	3 266	2 212	1 054	2 353	86.3	3.3
White Total [2]											
All ages	3 174 760	2 664 117	2 078 838	585 279	357 575	96 924	71 672	25 252	56 144	85.4	3.1
Under 15 years	3 884	1 998	1 256	742	1 276	476	347	129	134	53.3	12.7
15–19 years	305 988	216 395	149 340	67 055	64 780	17 983	13 344	4 639	6 830	72.3	6.0
15 years	11 665	6 896	4 460	2 436	3 315	1 094	823	271	360	61.0	9.7
16 years	29 501	19 070	12 499	6 571	7 443	2 236	1 675	561	752	66.3	7.8
17 years	54 698	37 583	25 610	11 973	12 349	3 503	2 568	935	1 263	70.3	6.6
18 years	85 957	61 434	42 356	19 078	17 886	4 864	3 581	1 283	1 773	73.0	5.8
19 years	124 167	91 412	64 415	26 997	23 787	6 286	4 697	1 589	2 682	75.2	5.2
20–24 years	783 000	615 512	457 079	158 433	120 170	32 059	23 908	8 151	15 259	80.2	4.2
25–29 years	851 142	733 505	581 600	151 905	82 079	21 839	16 223	5 616	13 719	87.6	2.6
30–34 years	779 535	698 230	568 592	129 638	54 371	14 917	11 066	3 851	12 017	91.0	1.9
35–39 years	369 833	328 654	265 598	63 056	27 259	7 422	5 222	2 200	6 498	90.5	2.0
40 years and over	81 378	69 823	55 373	14 450	7 640	2 228	1 562	666	1 687	87.6	2.8
Non-Hispanic White											
All ages	2 298 156	2 006 365	1 593 447	412 918	206 537	50 563	37 992	12 571	34 691	88.6	2.2
Under 15 years	1 493	809	498	311	455	179	144	35	50	56.1	12.4
15–19 years	179 511	133 567	92 707	40 860	34 421	8 110	6 286	1 824	3 413	75.8	4.6
15 years	5 130	3 144	2 041	1 103	1 409	442	343	99	135	62.9	8.8
16 years	14 421	9 805	6 447	3 358	3 393	924	727	197	299	69.4	6.5
17 years	30 205	21 941	14 975	6 966	6 202	1 470	1 134	336	592	74.1	5.0
18 years	51 196	38 392	26 524	11 868	9 698	2 194	1 682	512	912	76.4	4.4
19 years	78 559	60 285	42 720	17 565	13 719	3 080	2 400	680	1 475	78.2	4.0
20–24 years	519 153	424 747	319 520	105 227	69 386	16 296	12 406	3 890	8 724	83.2	3.2
25–29 years	614 909	549 718	443 530	106 188	45 652	11 210	8 450	2 760	8 329	90.6	1.8
30–34 years	620 173	569 651	470 050	99 601	33 554	8 650	6 424	2 226	8 318	93.1	1.4
35–39 years	297 436	270 359	221 042	49 317	17 820	4 617	3 229	1 388	4 640	92.3	1.6
40 years and over	65 481	57 514	46 100	11 414	5 249	1 501	1 053	448	1 217	89.5	2.3
Black Total [2]											
All ages	593 691	434 083	323 778	110 305	107 393	35 768	23 756	12 012	16 447	75.2	6.2
Under 15 years	3 188	1 305	811	494	1 249	525	388	137	109	42.4	17.1
15–19 years	103 795	65 174	44 207	20 967	27 493	8 200	5 748	2 452	2 928	64.6	8.1
15 years	6 336	3 132	1 970	1 162	2 212	790	576	214	202	51.1	12.9
16 years	12 142	6 709	4 321	2 388	3 905	1 121	797	324	407	57.2	9.6
17 years	19 411	11 844	7 714	4 130	5 474	1 592	1 126	466	501	62.6	8.4
18 years	28 266	18 240	12 471	5 769	7 137	2 100	1 463	637	789	66.4	7.6
19 years	37 640	25 249	17 731	7 518	8 765	2 597	1 786	811	1 029	69.0	7.1
20–24 years	194 704	139 552	102 164	37 388	38 190	11 841	8 000	3 841	5 121	73.6	6.2
25–29 years	136 591	105 648	81 455	24 193	20 495	6 868	4 586	2 282	3 580	79.4	5.2
30–34 years	95 006	75 621	59 077	16 544	11 927	4 699	2 925	1 774	2 759	82.0	5.1
35–39 years	48 388	37 801	29 314	8 487	6 249	2 824	1 638	1 186	1 514	80.6	6.0
40 years and over	12 019	8 982	6 750	2 232	1 790	811	471	340	436	77.5	7.0
Non-Hispanic Black											
All ages	578 335	423 012	315 462	107 550	104 923	34 940	23 085	11 855	15 460	75.2	6.2
Under 15 years	3 132	1 285	796	489	1 226	518	383	135	103	42.4	17.1
15–19 years	101 494	63 745	43 230	20 515	26 934	8 012	5 601	2 411	2 803	64.6	8.1
15 years	6 191	3 070	1 938	1 132	2 154	773	561	212	194	51.2	12.9
16 years	11 846	6 545	4 205	2 340	3 818	1 092	773	319	391	57.1	9.5
17 years	18 980	11 575	7 535	4 040	5 375	1 548	1 091	457	482	62.6	8.4
18 years	27 656	17 846	12 195	5 651	7 004	2 050	1 425	625	756	66.3	7.6
19 years	36 821	24 709	17 357	7 352	8 583	2 549	1 751	798	980	68.9	7.1
20–24 years	190 241	136 456	99 882	36 574	37 354	11 578	7 791	3 787	4 853	73.6	6.2
25–29 years	132 824	102 801	79 281	23 520	19 985	6 700	4 446	2 254	3 338	79.4	5.2
30–34 years	92 155	73 422	57 354	16 068	11 610	4 586	2 833	1 753	2 537	81.9	5.1
35–39 years	46 831	36 598	28 374	8 224	6 070	2 756	1 581	1 175	1 407	80.6	6.1
40 years and over	11 658	8 705	6 545	2 160	1 744	790	450	340	419	77.5	7.0
Hispanic [3]											
All ages	876 642	657 240	484 277	172 963	152 458	46 691	34 096	12 595	20 253	76.7	5.5
Under 15 years	2 421	1 199	765	434	838	304	210	94	80	51.2	13.0
15–19 years	127 900	83 778	57 259	26 519	30 804	10 007	7 184	2 823	3 311	67.2	8.0
15 years	6 647	3 820	2 458	1 362	1 956	657	484	173	214	59.4	10.2
16 years	15 321	9 410	6 147	3 263	4 133	1 331	968	363	447	63.3	8.9
17 years	24 772	15 821	10 759	5 062	6 224	2 074	1 468	606	653	65.6	8.6
18 years	35 123	23 310	16 014	7 296	8 291	2 690	1 926	764	832	68.0	7.8
19 years	46 037	31 417	21 881	9 536	10 200	3 255	2 338	917	1 165	70.0	7.3
20–24 years	265 235	191 770	138 183	53 587	51 287	15 895	11 642	4 253	6 283	74.1	6.1
25–29 years	236 143	183 604	137 774	45 830	36 701	10 679	7 841	2 838	5 159	79.5	4.6
30–34 years	157 887	127 245	97 358	29 887	20 957	6 285	4 687	1 598	3 400	82.4	4.1
35–39 years	71 480	57 583	43 898	13 685	9 464	2 802	2 015	787	1 051	[illegible]	[illegible]
40 years and over	15 576	12 061	9 040	3 021	2 407	719	517	202	389	79.4	4.7

[1] Includes races other than White and Black and origin not stated.
[2] Race and Hispanic origin are reported separately on the birth certificate. Data for persons of Hispanic origin are included in the data for each race group according to the mother's reported race.
[3] Includes all persons of Hispanic origin of any race.

TABLE I-33A. FIRST TRIMESTER PRENATAL CARE, BY RACE AND HISPANIC ORIGIN OF MOTHER, SELECTED YEARS, 1980–2002

Timely initiation of prenatal care has improved slowly but steadily in recent years, rising 10 percent since 1990 (from 75.8 percent). For non-Hispanic Blacks, Hispanics, and American Indians, the disparity compared with non-Hispanic Whites was pronounced in 1990, but the gains over the decade were 21 percent to 27 percent.

Table I-33A. First Trimester Prenatal Care, by Race and Hispanic Origin of Mother, Selected Years, 1980–2002

(Percent.)

Year	All races [1]	Non-Hispanic		American Indian [2]	Asian or Pacific Islander [2]	Hispanic [3]
		White	Black			
2002	83.7	88.6	75.2	69.8	84.8	76.7
2001	83.4	88.5	74.5	69.3	84.0	75.7
2000	83.2	88.5	74.3	69.3	84.0	74.4
1995	81.3	87.1	70.4	66.7	79.9	70.8
1990	75.8	83.3	60.7	57.9	75.1	60.2
1985	76.2	. . .	. . .	57.5	74.1	. . .
1980	76.3	. . .	. . .	55.8	73.7	. . .

[1]Includes races other than White and Black and origin not stated.
[2]Includes persons of Hispanic and non-Hispanic origin.
[3]Includes all persons of Hispanic origin of any race.
. . . = Not available.

TABLE I-34. PERCENT OF MOTHERS BEGINNING PRENATAL CARE IN THE FIRST TRIMESTER VERSUS LATE OR NO PRENATAL CARE, BY RACE AND HISPANIC ORIGIN OF MOTHER, FOR EACH STATE AND TERRITORY, 2002

The New England states and Iowa reported the highest proportions of mothers with first trimester prenatal care for 2002. The percentages for this care ranged from 87.9 in Maine to 91.5 in New Hampshire. Nationwide non-Hispanic White mothers also received first trimester care within this range, specifically 88.6 percent. In contrast only 69.0 percent of mothers in New Mexico started care in the first trimester, and women with late or no care amounted to 7.9 percent in this state. Other states with low percentages of early care were Nevada (75.9 percent), Arizona (76.5 percent), Oklahoma (76.8 percent), and South Dakota (77.7 percent). Other states with high percentages of late or no care were the District of Columbia (7.4 percent), and Nevada (7.1 percent).

TABLE I-35. LIVE BIRTHS BY MONTH OF PREGNANCY PRENATAL CARE BEGAN, NUMBER OF PRENATAL VISITS, AND MEDIAN NUMBER OF VISITS, BY RACE AND HISPANIC ORIGIN OF MOTHER, 2002

This table summarizes the number of prenatal care visits by prospective mothers and classifies them by the time when care began. It also shows details by race and ethnicity. The median number of prenatal visits for all births was 12.2, and it was 12.8 for mothers who started care in the first two months of pregnancy. The frequency of visits drops for those who start later, and the median is 5.4 visits for women who start care in the seventh to ninth month. The median number of visits by non-Hispanic Blacks and by Hispanics is 11.6 and 11.7, respectively, somewhat less than the 12.5 median for non-Hispanic Whites.

Table I-34. Percent of Mothers Beginning Prenatal Care in the First Trimester Versus Late or No Prenatal Care, by Race and Hispanic Origin of Mother, for each State and Territory, 2002

(Percent.)

State	Percent beginning care in first trimester						Percent late [1] or no care					
	All races [2]	White		Black		Hispanic [4]	All races [2]	White		Black		Hispanic [4]
		Total [3]	Non-Hispanic	Total [3]	Non-Hispanic			Total [3]	Non-Hispanic	Total [3]	Non-Hispanic	
United States [5]	83.7	85.4	88.6	75.2	75.2	76.7	3.6	3.1	2.2	6.2	6.2	5.5
Alabama	83.3	87.7	90.0	73.6	73.6	53.9	3.8	2.9	1.8	5.7	5.7	19.0
Alaska	80.3	84.2	83.5	83.9	84.6	76.8	4.6	3.8	3.8	*	*	5.9
Arizona	76.5	77.1	87.2	76.4	77.4	66.7	6.6	6.5	2.7	5.5	5.1	10.4
Arkansas	79.6	82.1	83.7	70.5	70.4	68.4	4.9	4.2	3.6	7.5	7.5	9.4
California	86.4	86.6	90.4	83.0	83.0	84.1	2.6	2.6	1.8	3.4	3.4	3.1
Colorado	79.1	79.5	86.2	70.2	70.7	65.9	4.5	4.4	2.7	7.3	7.0	7.9
Connecticut	88.3	89.3	92.2	81.7	82.1	77.0	2.0	1.8	1.3	3.5	3.5	3.8
Delaware	87.1	88.5	91.1	82.3	82.3	75.2	3.5	2.9	2.3	5.7	5.6	5.9
District of Columbia	76.4	85.1	90.8	70.5	70.6	73.5	7.4	3.7	1.9	10.0	9.8	7.9
Florida	85.4	87.8	89.8	77.1	77.0	83.7	2.9	2.3	1.9	4.9	5.0	3.3
Georgia	84.7	87.3	90.5	79.1	78.9	73.6	3.4	2.9	1.9	4.5	4.6	7.1
Hawaii	83.9	88.4	89.1	94.2	94.7	82.5	3.5	2.5	2.2	*	*	3.2
Idaho	82.1	82.3	84.3	81.8	81.3	70.2	3.4	3.3	2.8	*	*	6.4
Illinois	84.9	87.2	90.8	74.3	74.2	78.4	2.9	2.1	1.6	6.7	6.7	3.5
Indiana	81.4	82.9	84.5	69.6	69.7	65.1	3.5	3.0	2.6	7.1	7.1	7.9
Iowa	88.8	89.3	90.4	78.1	78.3	74.7	2.2	2.0	1.8	5.5	5.4	5.6
Kansas	86.8	87.5	90.0	79.6	79.6	72.3	2.8	2.5	1.9	5.0	5.0	6.6
Kentucky	86.8	87.5	88.0	80.8	80.7	72.6	2.5	2.3	2.1	4.4	4.4	6.6
Louisiana	83.8	90.4	90.7	74.7	74.7	83.3	3.6	1.7	1.7	6.1	6.1	3.4
Maine	87.9	88.2	88.3	73.0	72.5	83.2	1.6	1.6	1.5	*	*	*
Maryland	84.1	88.1	90.8	76.5	76.5	70.6	3.6	2.3	1.8	6.2	6.2	5.9
Massachusetts	89.9	91.4	92.7	80.7	79.2	82.9	2.0	1.5	1.3	5.1	5.6	3.0
Michigan	85.6	88.6	89.4	71.3	71.2	77.4	3.5	2.5	2.4	7.9	7.9	4.7
Minnesota	85.5	88.0	89.8	70.1	70.1	68.0	2.3	1.7	1.4	6.2	6.3	5.8
Mississippi	83.8	90.1	90.7	76.1	76.1	74.7	3.1	1.8	1.6	4.6	4.6	6.6
Missouri	87.8	89.3	89.9	79.7	79.7	78.1	2.7	2.1	2.0	5.6	5.6	4.8
Montana	83.7	86.1	86.4	70.3	68.8	79.4	2.8	1.9	1.8	*	*	*
Nebraska	83.4	84.5	87.0	70.6	70.4	69.2	3.1	2.8	2.1	6.0	6.1	6.8
Nevada	75.9	76.2	85.3	70.6	70.5	63.4	7.1	7.0	4.0	8.6	8.6	11.1
New Hampshire	91.5	91.8	92.2	80.4	79.7	84.8	1.4	1.4	1.3	*	*	*
New Jersey	80.3	83.6	89.0	64.0	63.8	68.7	4.9	3.7	2.4	11.0	11.3	7.1
New Mexico	69.0	70.4	76.8	68.5	68.6	66.4	7.9	7.4	5.0	7.9	8.1	8.8
New York	81.6	84.5	88.2	72.1	71.9	75.0	4.7	3.7	2.8	8.4	8.5	5.9
North Carolina	84.3	87.3	90.9	75.5	75.5	70.2	2.9	2.2	1.4	5.1	5.1	6.1
North Dakota	86.1	88.6	88.9	82.2	81.6	83.6	2.7	1.9	1.8	*	*	*
Ohio	87.8	89.3	89.8	78.9	78.9	78.0	2.9	2.3	2.2	6.2	6.2	5.1
Oklahoma	76.8	78.6	80.8	69.4	69.3	63.6	5.4	4.9	4.4	7.5	7.5	8.3
Oregon	81.6	81.9	84.6	76.0	75.8	71.4	3.8	3.7	3.2	4.5	4.5	5.5
Pennsylvania	84.6	86.9	88.0	71.0	71.1	72.4	3.5	2.8	2.5	7.8	7.8	5.7
Rhode Island	89.6	91.0	92.4	79.9	79.2	86.1	1.5	1.2	1.0	3.4	3.2	2.1
South Carolina	78.4	82.5	84.7	70.3	70.3	60.4	4.7	3.4	2.7	7.1	7.1	10.4
South Dakota	77.7	81.8	82.4	61.2	61.2	64.8	4.4	2.7	2.4	*	*	10.8
Tennessee	82.8	85.7	87.8	72.3	72.3	58.7	3.9	2.9	2.2	7.5	7.5	12.2
Texas	80.5	80.6	87.8	76.7	76.7	75.1	5.0	5.0	2.8	5.8	5.8	6.6
Utah	79.5	80.5	83.6	59.7	58.4	62.4	4.9	4.5	3.6	14.2	15.1	9.5
Vermont	88.9	89.0	89.2	71.8	71.1	81.3	1.7	1.6	1.5	*	*	*
Virginia	85.2	87.9	90.7	76.6	76.6	70.1	3.5	2.7	2.0	5.9	5.9	7.5
Washington	83.4	83.9	86.0	78.1	77.8	74.2	3.1	2.9	2.4	4.6	4.5	5.2
West Virginia	85.9	86.3	86.4	75.6	75.5	75.0	2.2	2.1	2.0	4.4	4.4	*
Wisconsin	84.3	86.6	88.2	71.4	71.4	69.4	3.2	2.6	2.2	7.3	7.4	6.3
Wyoming	84.9	85.4	86.0	79.2	78.8	79.4	2.9	2.6	2.6	*	*	3.5
Puerto Rico	81.2	81.9	75.0	. . .	. . .	. . .	2.8	2.7	. . .	4.5	. . .	. . .
Virgin Islands	64.0	63.5	77.1	63.7	63.5	59.8	10.5	11.7	*	10.1	9.6	13.3
Guam	61.2	87.6	89.1	90.0	89.5	73.6	12.7	*	*	*	*	*
American Samoa	. . .	. . .	. . .	. . .	. . .	. . .	. . .	. . .	. . .	. . .	. . .	. . .
Northern Marianas	30.4	*	. . .	*	. . .	. . .	24.4	*	. . .	*	. . .	. . .

[1]Care beginning in 3rd trimester.
[2]Includes races other than White and Black and origin not stated.
[3]Race and Hispanic origin are reported separately on the birth certificate. Data for persons of Hispanic origin are included in the data for each race group according to the mother's reported race.
[4]Includes all persons of Hispanic origin of any race.
[5]Excludes data for the territories.
. . . = Not available.
* = Figure does not meet standards of reliability or precision; based on fewer than 20 births in numerator.

Table I-35. Live Births by Month of Pregnancy Prenatal Care Began, Number of Prenatal Visits, and Median Number of Visits, by Race and Hispanic Origin of Mother, 2002

Number of prenatal visits and race and Hispanic origin of mother	All births	Month of pregnancy prenatal care began								
		1st trimester			2nd trimester	Late or no care			Not stated
		Total	1st and 2nd months	3rd month	4th–6th months	Total	7th–9th months	No care	
All Races [1]	4 021 726	3 301 186	2 557 791	743 395	499 014	142 430	103 323	39 107	79 096
No visits	39 107	X	X	X	X	X	X	39 107	X
1–2 visits	41 246	11 385	7 585	3 800	9 852	18 404	18 404	X	1 605
3–4 visits	83 624	25 629	15 240	10 389	29 711	26 313	26 313	X	1 971
5–6 visits	182 825	82 083	48 702	33 381	72 569	25 345	25 345	X	2 828
7–8 visits	342 539	215 983	135 372	80 611	108 152	14 778	14 778	X	3 626
9–10 visits	789 703	636 162	437 214	198 948	137 630	8 810	8 810	X	7 101
11–12 visits	1 058 720	972 170	756 643	215 527	78 129	3 837	3 837	X	4 584
13–14 visits	659 936	627 664	522 947	104 717	28 222	1 608	1 608	X	2 442
15–16 visits	468 001	447 167	389 775	57 392	17 972	1 216	1 216	X	1 646
17–18 visits	102 974	98 766	84 906	13 860	3 526	247	247	X	435
19 visits or more	146 413	139 616	124 017	15 599	5 635	429	429	X	733
Not stated	106 638	44 561	35 390	9 171	7 616	2 336	2 336	X	52 125
Median number of visits	12.2	12.5	12.8	11.5	9.5	5.4	5.4	X	10.3
White Total [2]	3 174 760	2 664 117	2 078 838	585 279	357 575	96 924	71 672	25 252	56 144
No visits	25 252	X	X	X	X	X	X	25 252	X
1–2 visits	27 306	7 952	5 361	2 591	6 179	12 135	12 135	X	1 040
3–4 visits	55 943	17 259	10 344	6 915	19 309	17 980	17 980	X	1 395
5–6 visits	128 308	59 188	35 143	24 045	49 532	17 628	17 628	X	1 960
7–8 visits	257 423	166 502	104 914	61 588	77 628	10 523	10 523	X	2 770
9–10 visits	614 820	503 945	348 279	155 666	100 111	6 326	6 326	X	4 438
11–12 visits	861 082	796 219	622 870	173 349	58 366	2 908	2 908	X	3 589
13–14 visits	544 972	520 399	434 839	85 560	21 374	1 244	1 244	X	1 955
15–16 visits	380 123	364 605	318 581	46 024	13 306	926	926	X	1 286
17–18 visits	83 803	80 675	69 913	10 762	2 603	179	179	X	346
19 visits or more	118 402	113 565	101 549	12 016	3 986	303	303	X	548
Not stated	77 326	33 808	27 045	6 763	5 181	1 520	1 520	X	36 817
Median number of visits	12.3	12.6	12.8	11.6	9.6	5.5	5.5	X	10.3
Non-Hispanic White	2 298 156	2 006 365	1 593 447	412 918	206 537	50 563	37 992	12 571	34 691
No visits	12 571	X	X	X	X	X	X	12 571	X
1–2 visits	13 709	4 041	2 817	1 224	2 955	6 129	6 129	X	584
3–4 visits	30 281	10 498	6 436	4 062	10 029	8 963	8 963	X	791
5–6 visits	76 062	39 205	24 128	15 077	26 609	9 006	9 006	X	1 242
7–8 visits	168 867	117 287	76 134	41 153	43 994	5 729	5 729	X	1 857
9–10 visits	426 212	361 528	256 220	105 308	57 920	3 686	3 686	X	3 078
11–12 visits	659 377	618 155	490 646	127 509	36 755	1 879	1 879	X	2 588
13–14 visits	425 524	409 085	343 798	65 287	14 085	859	859	X	1 495
15–16 visits	282 596	273 618	240 904	32 714	7 393	614	614	X	971
17–18 visits	65 607	63 432	55 287	8 145	1 761	126	126	X	288
19 visits or more	91 752	88 594	79 824	8 770	2 540	201	201	X	417
Not stated	45 598	20 922	17 253	3 669	2 496	800	800	X	21 380
Median number of visits	12.5	12.6	12.8	11.8	9.8	5.7	5.7	X	10.6
Black Total [2]	593 691	434 083	323 778	110 305	107 393	35 768	23 756	12 012	16 447
No visits	12 012	X	X	X	X	X	X	12 012	X
1–2 visits	11 048	2 811	1 804	1 007	2 967	4 825	4 825	X	445
3–4 visits	21 552	6 608	3 888	2 720	8 163	6 342	6 342	X	439
5–6 visits	40 908	16 925	10 165	6 760	17 495	5 816	5 816	X	672
7–8 visits	60 063	33 584	20 515	13 069	22 736	3 119	3 119	X	624
9–10 visits	122 496	90 099	60 195	29 904	28 319	1 810	1 810	X	2 268
11–12 visits	129 196	113 224	84 942	28 282	14 647	647	647	X	678
13–14 visits	77 796	72 042	58 755	13 287	5 160	250	250	X	344
15–16 visits	61 955	57 732	49 592	8 140	3 756	218	218	X	249
17–18 visits	13 809	12 958	10 572	2 386	747	39	39	X	65
19 visits or more	21 185	19 598	16 782	2 816	1 366	94	94	X	127
Not stated	21 671	8 502	6 568	1 934	2 037	596	596	X	10 536
Median number of visits	11.6	12.4	12.6	11.1	9.1	5.1	5.1	X	10.2
Non-Hispanic Black	578 335	423 012	315 462	107 550	104 923	34 940	23 085	11 855	15 460
No visits	11 855	X	X	X	X	X	X	11 855	X
1–2 visits	10 819	2 748	1 765	983	2 905	4 734	4 734	X	432
3–4 visits	21 050	6 460	3 815	2 645	8 005	6 161	6 161	X	424
5–6 visits	39 958	16 516	9 935	6 581	17 132	5 658	5 658	X	652
7–8 visits	58 482	32 717	19 967	12 750	22 153	3 001	3 001	X	611
9–10 visits	118 764	87 504	58 364	29 140	27 683	1 739	1 739	X	1 838
11–12 visits	125 537	110 022	82 466	27 556	14 260	623	623	X	632
13–14 visits	75 879	70 286	57 364	12 922	5 018	241	241	X	334
15–16 visits	60 749	56 624	48 633	7 991	3 675	213	213	X	237
17–18 visits	13 513	12 674	10 333	2 341	738	37	37	X	64
19 visits or more	20 789	19 225	16 460	2 765	1 349	91	91	X	124
Not stated	20 940	8 236	6 360	1 876	2 005	587	587	X	10 112
Median number of visits	11.6	12.4	12.6	11.1	9.1	5.1	5.1	X	10.1
Hispanic [3]	876 642	657 240	484 277	172 963	152 458	46 691	34 096	12 595	20 253
No visits	12 595	X	X	X	X	X	X	12 595	X
1–2 visits	13 733	3 939	2 561	1 378	3 280	6 061	6 061	X	453
3–4 visits	25 994	6 841	3 934	2 907	9 388	9 160	9 160	X	605
5–6 visits	52 763	20 147	11 082	9 065	23 171	8 742	8 742	X	703
7–8 visits	89 332	49 503	28 936	20 567	34 042	4 884	4 884	X	903
9–10 visits	190 202	143 217	92 552	50 665	42 587	2 681	2 681	X	1 717
11–12 visits	201 780	177 962	132 054	45 908	21 775	1 028	1 028	X	1 015
13–14 visits	119 226	111 064	90 785	20 279	7 339	379	379	X	444
15–16 visits	96 764	90 199	76 944	13 255	5 951	306	306	X	308
17–18 visits	18 187	17 234	14 612	2 622	844	54	54	X	55
19 visits or more	26 285	24 626	21 374	3 252	1 440	99	99	X	120
Not stated	29 781	12 508	9 443	3 065	2 641	702	702	X	13 930
Median number of visits	11.7	12.4	12.7	11.0	9.3	5.3	5.3	X	10.0

[1]Includes races other than White and Black and origin not stated.
[2]Race and Hispanic origin are reported separately on the birth certificate. Data for persons of Hispanic origin are included in the data for each race group according to the mother's reported race.
[3]Includes all persons of Hispanic origin of any race.
X = Not applicable.

TABLE I-36. LIVE BIRTHS TO MOTHERS WITH SELECTED OBSTETRIC PROCEDURES AND RATES BY AGE AND RACE OF MOTHER, 2002

The frequency of some obstetric procedures declines moderately as the age of the mother increases. Examples are electronic fetal monitoring and induction of labor. The application of ultrasound, on the other hand, tends to increase in frequency with increasing age up to age 30–34. Amniocentesis, which can detect birth defects that are more prevalent in older mothers, rises very sharply in the 35–39 year age bracket and again for mothers between 40 and 54 years of age, an increase that is more pronounced in Whites than in Blacks. (See also the discussion of Tables 27 and 28.)

Table I-36. Live Births to Mothers with Selected Obstetric Procedures and Rates by Age and Race of Mother, 2002

(Rates are number of live births with specified procedure per 1,000 live births in specified group.)

Obstetric procedure and race of mother	All births [1]	Obstetric procedure reported	Rate by age of mother							Not stated [2]
			All ages	Under 20 years	20–24 years	25–29 years	30–34 years	35–39 years	40–54 years	
All Races [3]										
Amniocentesis	4 021 726	79 035	19.7	5.2	6.5	8.7	15.4	77.4	112.3	14 816
Electronic fetal monitoring	4 021 726	3 413 557	851.9	865.6	859.5	853.3	846.4	835.8	826.1	14 816
Induction of labor	4 021 726	826 161	206.2	201.2	209.3	213.7	205.2	192.7	187.3	14 816
Stimulation of labor	4 021 726	693 548	173.1	192.9	181.6	175.5	165.8	151.8	140.3	14 816
Tocolysis	4 021 726	83 695	20.9	23.4	22.7	20.8	19.3	18.3	18.6	14 816
Ultrasound	4 021 726	2 707 007	675.6	642.3	661.0	678.8	693.7	692.8	685.3	14 816
White										
Amniocentesis	3 174 760	64 891	20.5	5.2	6.6	8.6	15.6	79.5	117.6	11 743
Electronic fetal monitoring	3 174 760	2 697 901	853.0	865.0	859.8	855.0	848.8	837.8	827.3	11 743
Induction of labor	3 174 760	685 153	216.6	210.8	221.1	224.4	215.3	201.9	193.4	11 743
Stimulation of labor	3 174 760	553 122	174.9	196.8	184.8	177.2	167.1	153.6	142.8	11 743
Tocolysis	3 174 760	66 386	21.0	24.1	22.9	20.9	19.5	18.1	18.3	11 743
Ultrasound	3 174 760	2 168 878	685.7	656.4	673.3	686.9	700.9	700.3	693.0	11 743
Black										
Amniocentesis	593 691	7 133	12.0	5.3	6.2	8.7	13.3	44.3	64.7	1 337
Electronic fetal monitoring	593 691	508 334	858.2	872.1	864.6	854.5	846.4	840.9	833.6	1 337
Induction of labor	593 691	103 057	174.0	178.9	173.2	177.0	170.4	165.0	172.9	1 337
Stimulation of labor	593 691	95 512	161.2	183.7	169.4	156.9	145.1	133.3	119.7	1 337
Tocolysis	593 691	12 056	20.4	20.2	20.8	20.6	19.6	19.8	19.5	1 337
Ultrasound	593 691	367 325	620.1	601.2	612.2	628.2	636.8	635.6	631.1	1 337

[1]Total number of births to residents of areas reporting specified obstetric procedures.
[2]No response reported for the obstetric procedures item.
[3]Includes races other than White and Black.

TABLE I-36A. RATE OF INDUCTION OF LABOR, BY RACE AND HISPANIC ORIGIN OF MOTHER, SELECTED YEARS, 1989–2002

The rate of induction of labor has more than doubled since 1989. Even though each race/ethnic group increased, the absolute rate varied widely in 2002. For example the rate for non-Hispanic White women (24.6 percent) was considerably higher than that for non-Hispanic Black women (17.5 percent).

Table I-36A. Rate of Induction of Labor, by Race and Hispanic Origin of Mother, Selected Years, 1989–2002

(Percent.)

Race and Hispanic origin of mother	2002	2001	2000	1995	1989 [1]	Percent change, 1989–2002
All races [2]	20.6	20.5	19.9	16.0	9.0	129
Non-Hispanic White	24.6	24.4	23.6	18.9	10.5	134
Non-Hispanic Black	20.9	20.3	20.1	15.6	8.4	149
American Indian total [3]	17.5	17.1	16.5	11.7	6.7	161
Asian or Pacific Islander total [3]	13.9	14.0	13.3	10.8	5.5	153
Hispanic	13.8	13.5	13.2	10.2	5.3	160

[1]Excludes data for Louisiana, Nebraska, and Oklahoma, which did not report induction of labor in 1989.
[2]Includes races other than White and Black.
[3]Race and Hispanic origin are reported separately on the birth certificate. Data for persons of Hispanic origin are included in the data for each race group according to the mother's reported race.
[4]Includes all persons of Hispanic origin of any race.

TABLE I-37. LIVE BIRTHS TO MOTHERS WITH SELECTED COMPLICATIONS OF LABOR AND/OR DELIVERY AND RATES BY AGE AND RACE OF MOTHER, 2002

Depending on the severity of the condition, certain complications of labor and delivery may affect the health outcome of the infant. Many of these conditions are more common when the infant has low birthweight. The five most frequently reported complications of labor in 2002 were meconium (5 percent), fetal distress (4 percent), breech/malpresentation (4 percent), dysfunctional labor (3 percent), and premature rupture of membrane (2 percent). Complication rates frequently are high for mothers under 20 years of age (febrile, meconium, and fetal distress) and some rates also increase from age 35 and above (premature rupture of membrane, breech/malpresentation, and fetal distress). Complication rates vary between White and Black mothers, notably meconium and fetal distress occurring more commonly in Blacks and breech/malpresentation and cephalopelvic disproportion occurring more commonly in Whites.

Cord and placental complications are quite infrequent but they are among the top 10 causes of infant death.

Table I-37. Live Births to Mothers with Selected Complications of Labor and/or Delivery and Rates by Age and Race of Mother, 2002

(Rates are number of live births with specified complication per 1,000 live births in specified group.)

Complication and race of mother	All births[1]	Complication reported	Rate by age of mother							Not stated[2]
			All ages	Under 20 years	20–24 years	25–29 years	30–34 years	35–39 years	40–54 years	
All Races[3]										
Febrile	4 021 726	60 960	15.2	19.9	15.9	15.2	14.5	11.8	11.1	19 572
Meconium, moderate/heavy	4 021 726	200 408	50.1	57.2	51.4	48.9	47.6	48.3	50.6	19 572
Premature rupture of membrane	4 021 726	92 550	23.1	24.1	21.9	22.2	23.6	24.9	28.5	19 572
Abruptio placenta	4 021 726	21 409	5.3	5.3	5.1	5.0	5.4	6.4	7.7	19 572
Placenta previa	4 021 726	13 356	3.3	1.1	1.8	2.8	4.5	6.4	9.1	19 572
Other excessive bleeding	4 021 726	24 137	6.0	5.9	5.6	5.7	6.3	6.9	8.3	19 572
Seizures during labor	4 021 726	1 297	0.3	0.7	0.4	0.2	0.2	0.3	0.3	19 572
Precipitous labor	4 021 726	73 359	18.3	12.8	17.3	18.3	20.1	21.8	20.3	19 572
Prolonged labor	4 021 726	28 113	7.0	8.1	7.1	6.9	6.8	6.7	6.6	19 572
Dysfunctional labor	4 021 726	114 479	28.6	29.9	27.4	28.3	28.9	29.3	32.1	19 572
Breech/malpresentation	4 021 726	152 425	38.1	28.1	30.5	37.0	44.2	49.8	58.7	19 572
Cephalopelvic disproportion	4 021 726	63 240	15.8	16.8	14.8	16.1	16.2	15.6	15.8	19 572
Cord prolapse	4 021 726	6 941	1.7	1.3	1.5	1.6	1.9	2.3	2.6	19 572
Anesthetic complication[4]	3 649 276	2 354	0.6	0.5	0.6	0.6	0.7	0.8	0.9	22 204
Fetal distress[4]	3 649 276	140 074	38.6	43.4	38.3	36.6	37.2	40.4	47.3	22 204
White										
Febrile	3 174 760	46 285	14.7	19.7	15.7	14.6	13.7	11.4	10.3	15 519
Meconium, moderate/heavy	3 174 760	147 500	46.7	52.3	47.8	45.7	44.8	45.5	48.1	15 519
Premature rupture of membrane	3 174 760	70 750	22.4	23.0	21.0	21.6	23.0	24.2	28.1	15 519
Abruptio placenta	3 174 760	16 245	5.1	5.1	4.8	4.8	5.2	6.0	7.3	15 519
Placenta previa	3 174 760	10 294	3.3	1.1	1.8	2.8	4.3	5.9	8.6	15 519
Other excessive bleeding	3 174 760	19 239	6.1	6.2	5.9	5.7	6.3	6.7	8.0	15 519
Seizures during labor	3 174 760	914	0.3	0.6	0.4	0.2	0.2	0.3	*	15 519
Precipitous labor	3 174 760	57 191	18.1	11.7	16.7	18.0	20.1	21.9	20.3	15 519
Prolonged labor	3 174 760	22 862	7.2	8.5	7.5	7.1	6.9	6.8	6.9	15 519
Dysfunctional labor	3 174 760	90 497	28.6	30.0	27.7	28.3	28.9	29.0	31.7	15 519
Breech/malpresentation	3 174 760	124 541	39.4	29.6	31.6	38.1	45.3	50.4	59.7	15 519
Cephalopelvic disproportion	3 174 760	51 124	16.2	17.6	15.5	16.5	16.3	15.6	15.8	15 519
Cord prolapse	3 174 760	5 445	1.7	1.3	1.4	1.6	1.9	2.3	2.5	15 519
Anesthetic complication[4]	2 857 610	1 884	0.7	0.5	0.6	0.6	0.7	0.8	1.0	17 743
Fetal distress[4]	2 857 610	104 231	36.7	41.0	36.5	34.6	35.8	38.7	44.5	17 743
Black										
Febrile	593 691	9 013	15.2	20.1	15.9	13.5	13.6	10.7	11.0	1 906
Meconium, moderate/heavy	593 691	39 941	67.5	71.3	65.5	66.3	67.6	69.8	69.3	1 906
Premature rupture of membrane	593 691	15 458	26.1	26.4	24.5	24.9	28.2	29.8	32.4	1 906
Abruptio placenta	593 691	3 948	6.7	6.0	6.1	6.4	7.4	9.1	9.5	1 906
Placenta previa	593 691	1 842	3.1	1.0	2.0	3.1	4.9	7.3	9.3	1 906
Other excessive bleeding	593 691	2 646	4.5	3.9	3.6	4.4	5.4	6.3	8.6	1 906
Seizures during labor	593 691	300	0.5	1.0	0.4	0.4	0.4	0.5	*	1 906
Precipitous labor	593 691	11 373	19.2	14.8	18.9	20.6	21.2	21.9	21.2	1 906
Prolonged labor	593 691	3 189	5.4	6.6	5.1	5.1	5.1	5.5	5.3	1 906
Dysfunctional labor	593 691	15 555	26.3	28.5	25.1	25.2	26.7	27.6	29.3	1 906
Breech/malpresentation	593 691	18 370	31.0	23.5	25.9	31.3	38.9	47.2	50.6	1 906
Cephalopelvic disproportion	593 691	7 081	12.0	14.8	11.7	11.1	11.2	11.2	9.7	1 906
Cord prolapse	593 691	1 087	1.8	1.5	1.6	1.9	2.2	2.4	3.3	1 906
Anesthetic complication[4]	552 049	326	0.6	0.5	0.5	0.6	0.6	0.9	*	2 270
Fetal distress[4]	552 049	26 571	48.3	50.1	45.5	46.9	49.0	54.3	64.2	2 270

[1]Total number of births to residents of areas reporting specified complication.
[2]No response reported for the complications item.
[3]Includes races other than White and Black.
[4]Texas does not report this complication.
* = Figure does not meet standards of reliability or precision; based on fewer than 20 births in numerator.

TABLE I-38. LIVE BIRTHS BY ATTENDANT, PLACE OF DELIVERY, AND RACE AND HISPANIC ORIGIN OF MOTHER, 2002

The share of births delivered by physicians in hospitals has declined from 98.7 percent in 1975 to 91.3 percent in 2002. This reflected an increase in births attended by midwives from 1.0 percent to 8.1 percent over the same period (in fact midwife-attended births are believed to be underreported, so these are considered low estimates). The use of midwives was somewhat higher for Hispanic mothers, amounting to 9.6 percent. The great majority of midwives were certified nurse midwives. About 4.8 percent of the physicians were doctors of osteopathy.

Ninety-nine percent of births in 2002 were delivered in hospitals, essentially unchanged for the last several decades. Of the remainder, there were 23,000 deliveries in residences and nearly 10,000 in freestanding birthing centers.

Table I-38. Live Births by Attendant, Place of Delivery, and Race and Hispanic Origin of Mother, 2002

Place of delivery and race and Hispanic origin of mother	All births	Physician			Midwife			Other	Unspecified
		Total	Doctor of medicine	Doctor of osteopathy	Total	Certified nurse midwife	Other midwife		
All Races [1]									
Total	4 021 726	3 675 430	3 500 255	175 175	325 235	307 527	17 708	20 033	1 028
In hospital [2]	3 986 190	3 671 505	3 496 970	174 535	303 071	298 073	4 998	11 084	530
Not in hospital	35 416	3 890	3 252	638	22 142	9 437	12 705	8 906	478
Freestanding birthing center	9 683	1 040	658	382	8 415	5 689	2 726	221	7
Clinic or doctor's office	385	196	184	12	114	50	64	75	-
Residence	22 980	1 978	1 771	207	13 154	3 475	9 679	7 503	345
Other	2 368	676	639	37	459	223	236	1 107	126
Not specified	120	35	33	2	22	17	5	43	20
White Total [3]									
Total	3 174 760	2 897 429	2 747 527	149 902	260 665	244 223	16 442	15 964	702
In hospital [2]	3 143 215	2 894 518	2 745 214	149 304	239 717	235 508	4 209	8 627	353
Not in hospital	31 462	2 888	2 291	597	20 929	8 701	12 228	7 313	332
Freestanding birthing center	8 991	1 010	628	382	7 771	5 216	2 555	206	4
Clinic or doctor's office	330	161	149	12	111	47	64	58	-
Residence	20 458	1 328	1 150	178	12 627	3 251	9 376	6 251	252
Other	1 683	389	364	25	420	187	233	798	76
Not specified	83	23	22	1	19	14	5	24	17
Non-Hispanic White									
Total	2 298 156	2 110 332	1 990 524	119 808	175 818	161 824	13 994	11 533	473
In hospital [2]	2 270 789	2 107 809	1 988 575	119 234	157 381	154 075	3 306	5 354	245
Not in hospital	27 316	2 515	1 942	573	18 418	7 735	10 683	6 164	219
Freestanding birthing center	7 526	982	603	379	6 522	4 522	1 857	162	3
Clinic or doctor's office	300	149	138	11	101	42	59	50	-
Residence	18 210	1 110	944	166	11 568	3 024	8 544	5 353	179
Other	1 280	274	257	17	370	147	223	599	37
Not specified	51	8	7	1	19	14	5	15	9
Black Total [3]									
Total	593 691	547 701	529 610	18 091	42 939	42 213	726	2 810	241
In hospital [2]	590 829	546 857	528 799	18 058	42 218	41 722	496	1 619	135
Not in hospital	2 833	835	802	33	719	489	230	1 176	103
Freestanding birthing center	431	16	16	-	406	307	99	8	1
Clinic or doctor's office	16	11	11	-	2	2	-	3	-
Residence	1 868	573	551	22	282	152	130	949	64
Other	518	235	224	11	29	28	1	216	38
Not specified	29	9	9	-	2	2	-	15	3
Non-Hispanic Black									
Total	578 335	534 006	516 411	17 595	41 368	40 670	698	2 733	228
In hospital [2]	575 597	533 199	515 637	17 562	40 692	40 215	477	1 576	130
Not in hospital	2 722	802	769	33	675	454	221	1 147	98
Freestanding birthing center	406	16	16	-	382	286	96	7	1
Clinic or doctor's office	16	11	11	-	2	2	-	3	-
Residence	1 796	546	524	22	262	138	124	927	61
Other	504	229	218	11	29	28	1	210	36
Not specified	16	5	5	-	1	1	-	10	-
Hispanic [4]									
Total	876 642	787 769	757 795	29 974	84 492	82 337	2 155	4 203	178
In hospital [2]	872 929	787 406	757 456	29 950	82 210	81 439	771	3 218	95
Not in hospital	3 699	354	330	24	2 281	897	1 384	981	83
Freestanding birthing center	1 441	21	18	3	1 375	681	694	44	1
Clinic or doctor's office	29	11	10	1	9	4	5	9	-
Residence	1 854	211	199	12	853	175	678	737	53
Other	375	111	103	8	44	37	7	191	29
Not specified	14	9	9	-	1	1	-	4	-

[1]Includes races other than White and Black and origin not stated.
[2]Includes births occurring en route to or on arrival at hospital.
[3]Race and Hispanic origin are reported separately on the birth certificate. Data for persons of Hispanic origin are included in the data for each race group according to the mother's reported race.
[4]Includes all persons of Hispanic origin of any race.
- = Quantity zero.

TABLE I-39. LIVE BIRTHS BY METHOD OF DELIVERY AND RATES OF CESAREAN DELIVERY AND VAGINAL BIRTH AFTER PREVIOUS CESAREAN DELIVERY, BY RACE AND HISPANIC ORIGIN OF MOTHER, 1989–2002

In 2002 the rate of cesarean delivery rose to 26.1 percent of all births, the highest rate ever reported in the United States. The rise reflected an increase in the primary cesarean rate as well as a steep decrease in the rate of vaginal birth after a previous cesarean delivery. The primary cesarean rate, that is, the rate for mothers who did not have a previous cesarean, was 18.0 percent in 2002, 23 percent higher than the temporary low of 14.6 percent in 1996 and 1997. The rate of vaginal birth after previous cesarean delivery fell to 12.6 percent in 2002; it had attained a temporary high of 28.3 percent in 1996. The recent trends were shared by all age, racial, and ethnic groups; the dip in primary cesarean rates in the early 1990s was most pronounced in non-Hispanic Whites, moderate in Hispanics, and almost nonexistent in Blacks.

Figure I-9. Primary Cesarean Rates by Age, Race, and Hispanic Origin of Mother, 1989, 1996, and 2002

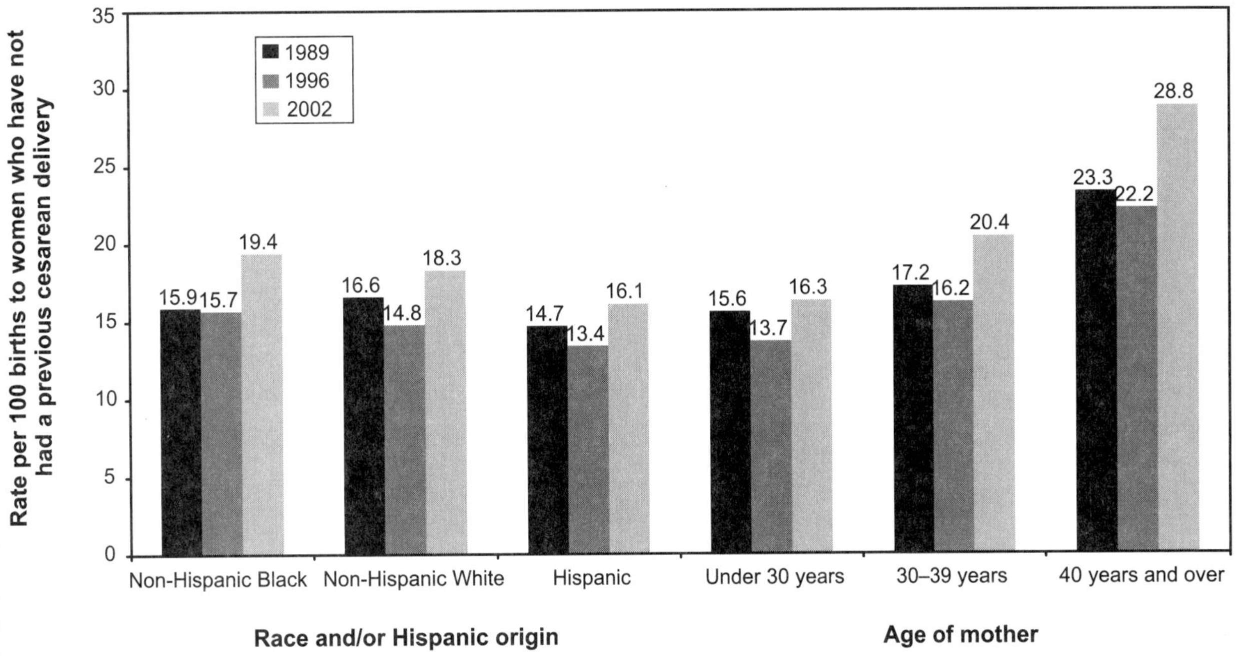

Table I-39. Live Births by Method of Delivery and Rates of Cesarean Delivery and Vaginal Birth after Previous Cesarean Delivery, by Race and Hispanic Origin of Mother, 1989–2002

Year and race and Hispanic origin of mother	Births by method of delivery							Cesarean delivery rate		Rate of vaginal birth after previous cesarean [3]
	All births	Vaginal		Cesarean			Not stated	Total [1]	Primary [2]	
		Total	After previous cesarean	Total	Primary	Repeat				
All Races [4]										
2002	4 021 726	2 958 423	59 248	1 043 846	634 426	409 420	19 457	26.1	18.0	12.6
2001	4 025 933	3 027 993	74 048	978 411	601 383	377 028	19 529	24.4	16.9	16.4
2000	4 058 814	3 108 188	89 978	923 991	577 638	346 353	26 635	22.9	16.1	20.6
1999	3 959 417	3 063 870	97 680	862 086	542 080	320 006	33 461	22.0	15.5	23.4
1998	3 941 553	3 078 537	108 903	825 870	519 975	305 895	37 146	21.2	14.9	26.3
1997	3 880 894	3 046 621	112 145	799 033	502 526	296 507	35 240	20.8	14.6	27.4
1996	3 891 494	3 061 092	116 045	797 119	503 724	293 395	33 283	20.7	14.6	28.3
1995	3 899 589	3 063 724	112 439	806 722	510 104	296 618	29 143	20.8	14.7	27.5
1994	3 952 767	3 087 576	110 341	830 517	520 647	309 870	34 674	21.2	14.9	26.3
1993	4 000 240	3 098 796	103 581	861 987	539 251	322 736	39 457	21.8	15.3	24.3
1992	4 065 014	3 100 710	97 549	888 622	554 662	333 960	75 682	22.3	15.6	22.6
1991	4 110 907	3 100 891	90 690	905 077	569 195	335 882	104 939	22.6	15.9	21.3
1990 [5]	4 110 563	3 111 421	84 299	914 096	575 066	339 030	85 046	22.7	16.0	19.9
1989 [6]	3 798 734	2 793 463	71 019	826 955	521 873	305 082	178 316	22.8	16.1	18.9
White Total [7]										
2002	3 174 760	2 340 512	46 004	818 347	493 002	325 345	15 901	25.9	17.7	12.4
2001	3 177 626	2 394 930	58 053	766 771	467 285	299 486	15 925	24.3	16.7	16.2
2000	3 194 005	2 449 264	70 414	723 209	449 161	274 048	21 532	22.8	15.9	20.4
1999	3 132 501	2 426 092	77 158	678 952	424 148	254 804	27 457	21.9	15.3	23.2
1998	3 118 727	2 440 113	86 495	649 987	406 439	243 548	28 627	21.0	14.7	26.2
1997	3 072 640	2 415 236	89 522	630 613	393 603	237 010	26 791	20.7	14.5	27.4
1996	3 093 057	2 434 079	93 783	631 409	395 851	235 558	27 569	20.6	14.5	28.5
1995	3 098 885	2 435 191	90 940	639 818	401 098	238 720	23 876	20.8	14.6	27.6
1994	3 121 004	2 435 965	88 471	656 400	407 946	248 454	28 639	21.2	14.8	26.3
1993	3 149 833	2 435 229	82 995	682 355	423 540	258 815	32 249	21.9	15.3	24.3
1992	3 201 678	2 434 959	77 977	705 841	437 398	268 443	60 878	22.5	15.7	22.5
1991	3 241 273	2 434 900	72 564	723 088	452 534	270 554	83 285	22.9	16.1	21.1
1990 [5]	3 252 473	2 453 857	67 191	732 713	458 656	274 057	65 903	23.0	16.1	19.7
1989 [6]	3 022 537	2 212 843	56 851	667 114	418 177	248 937	142 580	23.2	16.2	18.6
Non-Hispanic White										
2002	2 298 156	1 687 144	33 440	598 682	370 339	228 343	12 330	26.2	18.3	12.8
2001	2 326 578	1 746 551	43 215	567 488	353 977	213 511	12 539	24.5	17.2	16.8
2000	2 362 968	1 804 550	52 912	540 794	342 732	198 062	17 624	23.1	16.4	21.1
1999	2 346 450	1 810 682	59 480	514 051	327 106	186 945	21 717	22.1	15.7	24.1
1998	2 361 462	1 842 420	67 787	495 550	315 138	180 412	23 492	21.2	15.1	27.3
1997	2 333 363	1 829 213	70 284	481 982	305 605	176 377	22 168	20.9	14.8	28.5
1996	2 358 989	1 851 058	73 973	485 530	308 482	177 048	22 401	20.8	14.8	29.5
1995	2 382 638	1 867 024	72 124	496 103	313 933	182 170	19 511	21.0	14.9	28.4
1994	2 438 855	1 896 609	71 597	518 021	324 236	193 785	24 225	21.5	15.1	27.0
1993	2 472 031	1 902 433	67 536	542 013	338 236	203 777	27 585	22.2	15.6	24.9
1992 [9]	2 527 207	1 916 414	63 828	566 788	352 470	214 318	44 005	22.8	16.0	22.9
1991 [9]	2 589 878	1 941 726	60 174	587 802	368 721	219 081	60 350	23.2	16.4	21.5
1990 [5,10]	2 626 500	1 972 754	55 952	603 467	378 508	224 959	50 279	23.4	16.5	19.9
1989 [6,11]	2 526 367	1 806 753	47 559	556 585	349 858	206 727	163 029	23.6	16.6	18.7
Black Total [7]										
2002	593 691	427 801	9 567	163 295	100 621	62 674	2 595	27.6	19.4	13.2
2001	606 156	447 458	11 747	156 071	97 429	58 642	2 627	25.9	18.3	16.7
2000	622 598	468 497	14 382	150 401	94 767	55 634	3 700	24.3	17.3	20.5
1999	605 970	462 401	15 438	139 471	88 269	51 202	4 098	23.2	16.5	23.2
1998	609 902	470 088	17 062	135 727	86 438	49 289	4 087	22.4	16.0	25.7
1997	599 913	466 001	16 986	130 142	83 025	47 117	3 770	21.8	15.6	26.5
1996	594 781	462 378	16 866	128 357	82 646	45 711	4 046	21.7	15.6	27.0
1995	603 139	468 984	16 224	130 482	84 441	46 041	3 673	21.8	15.7	26.1
1994	636 391	493 879	16 970	138 067	88 636	49 431	4 445	21.8	15.7	25.6
1993	658 875	509 816	16 179	143 452	91 677	51 775	5 607	22.0	15.7	23.8
1992	673 633	514 929	15 382	146 480	93 165	53 315	12 224	22.1	15.7	22.4
1991	682 602	519 047	14 213	145 583	92 645	52 938	17 972	21.9	15.5	21.2
1990 [5]	679 236	516 581	13 496	146 472	93 476	52 996	16 183	22.1	15.7	20.3
1989 [6]	611 147	452 921	11 104	127 907	82 695	45 212	30 319	22.0	15.8	19.7

[1]Percent of all live births by cesarean delivery.
[2]Number of primary cesareans per 100 live births to women who have not had a previous cesarean.
[3]Number of vaginal births after previous cesarean delivery per 100 live births to women with a previous cesarean delivery.
[4]Includes races other than White and Black and origin not stated.
[5]Excludes data for Oklahoma, which did not report method of delivery on the birth certificate.
[6]Excludes data for Louisiana, Maryland, Nebraska, Nevada, and Oklahoma, which did not report method of delivery on the birth certificate.
[7]Race and Hispanic origin are reported separately on the birth certificate. Data for persons of Hispanic origin are included in the data for each race group according to the mother's reported race.
[9]Excludes data for New Hampshire, which did not report Hispanic origin.
[10]Excludes data for New Hampshire and Oklahoma, which did not report Hispanic origin.
[11]Excludes data for Louisiana, New Hampshire, and Oklahoma, which did not report Hispanic origin.

Table I-39. Live Births by Method of Delivery and Rates of Cesarean Delivery and Vaginal Birth after Previous Cesarean Delivery, by Race and Hispanic Origin of Mother, 1989–2002—*Continued*

Year and race and Hispanic origin of mother	Births by method of delivery							Cesarean delivery rate		Rate of vaginal birth after previous cesarean [3]
	All births	Vaginal		Cesarean			Not stated	Total [1]	Primary [2]	
		Total	After previous cesarean	Total	Primary	Repeat				
Non-Hispanic Black										
2002	578 335	416 516	9 317	159 297	98 245	61 052	2 522	27.7	19.4	13.2
2001	589 917	435 455	11 417	151 908	94 912	56 996	2 554	25.9	18.3	16.7
2000	604 346	454 736	13 910	146 042	92 044	53 998	3 568	24.3	17.3	20.5
1999	588 981	449 580	14 999	135 508	85 898	49 610	3 893	23.2	16.5	23.2
1998	593 127	457 186	16 510	131 999	84 169	47 830	3 942	22.4	16.0	25.7
1997	581 431	451 744	16 353	126 138	80 599	45 539	3 549	21.8	15.6	26.4
1996	578 099	449 544	16 322	124 836	80 457	44 379	3 719	21.7	15.7	26.9
1995	587 781	457 104	15 721	127 171	82 395	44 776	3 506	21.8	15.7	26.0
1994	619 198	480 551	16 478	134 526	86 411	48 115	4 121	21.9	15.7	25.5
1993	641 273	496 333	15 675	139 702	89 315	50 387	5 238	22.0	15.7	23.7
1992 [9]	657 450	502 669	14 950	143 153	91 086	52 067	11 628	22.2	15.7	22.3
1991 [9]	666 758	507 522	13 847	142 417	90 664	51 753	16 819	21.9	15.5	21.1
1990 [5,10]	661 701	503 720	13 157	142 838	91 175	51 663	15 143	22.1	15.7	20.3
1989 [6,11]	611 269	440 310	10 726	125 290	81 177	44 113	45 669	22.2	15.9	19.6
Hispanic [8]										
2002	876 642	653 516	12 610	219 777	122 603	97 174	3 349	25.2	16.1	11.5
2001	851 851	648 821	14 846	199 874	113 529	86 345	3 156	23.6	15.2	14.7
2000	815 868	633 220	17 062	179 583	104 597	74 986	3 065	22.1	14.5	18.5
1999	764 339	599 118	16 915	161 035	94 433	66 602	4 186	21.2	14.0	20.3
1998	734 661	580 143	17 803	150 317	88 763	61 554	4 201	20.6	13.6	22.4
1997	709 767	563 114	17 942	142 907	84 410	58 497	3 746	20.2	13.4	23.5
1996	701 339	558 105	18 491	139 554	83 392	56 162	3 680	20.0	13.4	24.8
1995	679 768	539 731	17 396	136 640	82 662	53 978	3 397	20.2	13.7	24.4
1994	665 026	525 928	16 206	135 569	81 961	53 608	3 529	20.5	13.9	23.2
1993	654 418	514 493	14 586	136 279	82 576	53 703	3 646	20.9	14.2	21.4
1992 [9]	643 271	494 338	13 111	133 369	81 211	52 158	15 564	21.2	14.4	20.1
1991 [9]	623 085	472 126	11 615	129 752	80 228	49 524	21 207	21.6	14.8	19.0
1990 [5,10]	595 073	458 242	10 395	122 969	76 027	46 942	13 862	21.2	14.5	18.1
1989 [6,11]	532 249	385 462	8 549	105 268	64 905	40 363	41 519	21.5	14.7	17.5

[1]Percent of all live births by cesarean delivery.
[2]Number of primary cesareans per 100 live births to women who have not had a previous cesarean.
[3]Number of vaginal births after previous cesarean delivery per 100 live births to women with a previous cesarean delivery.
[5]Excludes data for Oklahoma, which did not report method of delivery on the birth certificate.
[6]Excludes data for Louisiana, Maryland, Nebraska, Nevada, and Oklahoma, which did not report method of delivery on the birth certificate.
[8]Includes all persons of Hispanic origin of any race.
[9]Excludes data for New Hampshire, which did not report Hispanic origin.
[10]Excludes data for New Hampshire and Oklahoma, which did not report Hispanic origin.
[11]Excludes data for Louisiana, New Hampshire, and Oklahoma, which did not report Hispanic origin.

Figure I-10. Vaginal Birth after Cesarean (VBAC) by Age, Race, and Hispanic Origin of Mother, 1989, 1996, and 2002

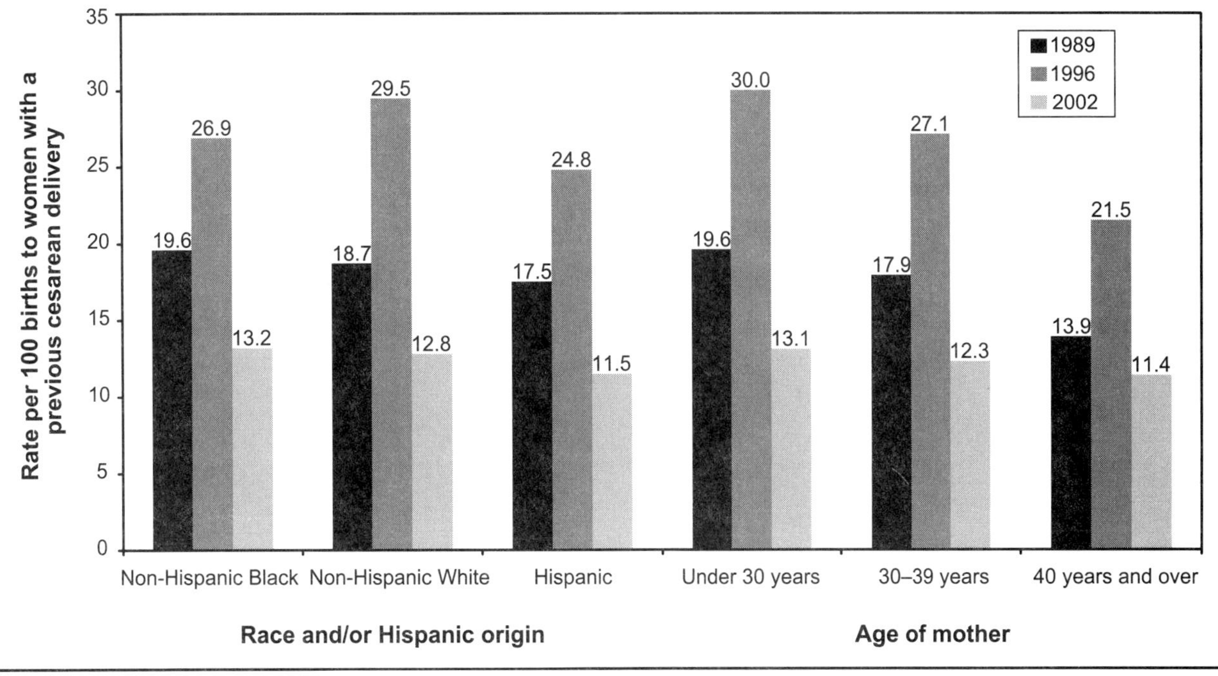

TABLE I-40. LIVE BIRTHS BY METHOD OF DELIVERY AND RATES OF CESAREAN DELIVERY AND VAGINAL BIRTH AFTER PREVIOUS CESAREAN DELIVERY, BY AGE, RACE, AND HISPANIC ORIGIN OF MOTHER, 2002

As in past years, overall cesarean rates rise as maternal age increases. The 2002 rate for mothers 40–54 years of age was 40.7 percent, while for mothers under age 20 years it was less than half (18.0 percent). Among other factors, the increased likelihood of cesareans for older mothers may be related to their high rate of multiple births, as well as increased patient or practitioner concerns and biological factors. In the older age groups cesareans for non-Hispanic Blacks as well as for Hispanics were more frequent than those for non-Hispanic Whites. At 40–54 years, non-Hispanic Blacks had a cesarean rate of 43.9 percent, Hispanics of 42.2 percent, and non-Hispanic Whites of 39.7 percent.

Table I-40. Live Births by Method of Delivery and Rates of Cesarean Delivery and Vaginal Birth after Previous Cesarean Delivery, by Age, Race, and Hispanic Origin of Mother, 2002

Age and race and Hispanic origin of mother	Births by method of delivery							Cesarean delivery rate		Rate of vaginal birth after previous cesarean [3]
	All births	Vaginal		Cesarean			Not stated	Total [1]	Primary [2]	
		Total	After previous cesarean	Total	Primary	Repeat				
All Races [4]	4 021 726	2 958 423	59 248	1 043 846	634 426	409 420	19 457	26.1	18.0	12.6
Under 20 years	432 808	353 653	1 506	77 563	67 741	9 822	1 592	18.0	16.1	13.3
20–24 years	1 022 106	799 537	10 910	218 239	144 947	73 292	4 330	21.4	15.5	13.0
25–29 years	1 060 391	788 776	16 294	266 452	159 869	106 583	5 163	25.3	17.1	13.3
30–34 years	951 219	663 566	17 976	282 576	156 159	126 417	5 077	29.9	19.5	12.4
35–39 years	453 927	293 251	10 278	158 005	82 497	75 508	2 671	35.0	22.6	12.0
40–54 years	101 275	59 640	2 284	41 011	23 213	17 798	624	40.7	28.8	11.4
White Total [5]	3 174 760	2 340 512	46 004	818 347	493 002	325 345	15 901	25.9	17.7	12.4
Under 20 years	309 872	254 659	936	54 046	47 427	6 619	1 167	17.5	15.7	12.4
20–24 years	783 000	615 786	7 740	163 799	109 359	54 440	3 415	21.0	15.2	12.4
25–29 years	851 142	635 069	12 493	211 779	126 723	85 056	4 294	25.0	16.9	12.8
30–34 years	779 535	545 608	14 479	229 600	125 761	103 839	4 327	29.6	19.1	12.2
35–39 years	369 833	240 983	8 489	126 667	65 511	61 156	2 183	34.5	22.0	12.2
40–54 years	81 378	48 407	1 867	32 456	18 221	14 235	515	40.1	28.1	11.6
Non-Hispanic White	2 298 156	1 687 144	33 440	598 682	370 339	228 343	12 330	26.2	18.3	12.8
Under 20 years	181 004	147 689	436	32 537	29 162	3 375	778	18.1	16.5	11.4
20–24 years	519 153	407 287	4 789	109 349	75 400	33 949	2 517	21.2	15.8	12.4
25–29 years	614 909	460 484	8 305	151 127	95 538	55 589	3 298	24.7	17.4	13.0
30–34 years	620 173	436 899	11 334	179 763	102 221	77 542	3 511	29.2	19.4	12.8
35–39 years	297 436	195 548	7 020	100 082	53 195	46 887	1 806	33.9	22.0	13.0
40–54 years	65 481	39 237	1 556	25 824	14 823	11 001	420	39.7	28.2	12.4
Black Total [5]	593 691	427 801	9 567	163 295	100 621	62 674	2 595	27.6	19.4	13.2
Under 20 years	106 983	85 354	517	21 268	18 344	2 924	361	19.9	17.8	15.0
20–24 years	194 704	147 381	2 712	46 562	29 756	16 806	761	24.0	17.1	13.9
25–29 years	136 591	97 412	2 825	38 552	21 633	16 919	627	28.4	18.6	14.3
30–34 years	95 006	62 182	2 165	32 364	17 624	14 740	460	34.2	22.7	12.8
35–39 years	48 388	28 766	1 062	19 297	10 242	9 055	325	40.1	27.0	10.5
40–54 years	12 019	6 706	286	5 252	3 022	2 230	61	43.9	32.0	11.4
Non-Hispanic Black	578 335	416 516	9 317	159 297	98 245	61 052	2 522	27.7	19.4	13.2
Under 20 years	104 626	83 420	507	20 849	17 963	2 886	357	20.0	17.8	14.9
20–24 years	190 241	143 871	2 664	45 627	29 144	16 483	743	24.1	17.1	13.9
25–29 years	132 824	94 614	2 746	37 605	21 108	16 497	605	28.4	18.7	14.3
30–34 years	92 155	60 277	2 092	31 434	17 140	14 294	444	34.3	22.8	12.8
35–39 years	46 831	27 829	1 026	18 687	9 949	8 738	315	40.2	27.1	10.5
40–54 years	11 658	6 505	282	5 095	2 941	2 154	58	43.9	32.1	11.6
Hispanic [6]	876 642	653 516	12 610	219 777	122 603	97 174	3 349	25.2	16.1	11.5
Under 20 years	130 321	108 166	515	21 782	18 513	3 269	373	16.8	14.7	13.6
20–24 years	265 235	209 624	2 976	54 748	34 148	20 600	863	20.7	14.2	12.6
25–29 years	236 143	174 488	4 215	60 708	31 151	29 557	947	25.8	15.5	12.5
30–34 years	157 887	107 505	3 139	49 628	23 323	26 305	754	31.6	18.3	10.7
35–39 years	71 480	44 773	1 460	26 376	12 150	14 226	331	37.1	21.9	9.3
40–54 years	15 576	8 960	305	6 535	3 318	3 217	81	42.2	27.7	8.7

[1]Percent of all live births by cesarean delivery.
[2]Number of primary cesareans per 100 live births to women who have not had a previous cesarean.
[3]Number of vaginal births after previous cesarean delivery per 100 live births to women with a previous cesarean delivery.

TABLE I-41. RATES OF CESAREAN DELIVERY AND VAGINAL BIRTH AFTER PREVIOUS CESAREAN DELIVERY, BY RACE AND HISPANIC ORIGIN OF MOTHER, FOR EACH STATE AND TERRITORY, 2002

Cesarean rates increased in all states from 2001 to 2002, as they also had in the previous year. The rates varied considerably by state in 2002, ranging from under 20 percent in New Mexico, Utah, Alaska, and Idaho to over 30 percent in Mississippi, New Jersey, and Louisiana. Puerto Rico had a cesarean delivery rate of 44.8 percent.

In contrast to a U.S. average rate of vaginal births after previous cesarean delivery of 12.6 percent, some states had frequencies above 20 percent in this category; these were Vermont, Utah, New Mexico, and Alaska. On the other hand, Louisiana and Mississippi only had 6.4 percent and 6.9 percent of such vaginal births, respectively, and Puerto Rico only had 3.4 percent.

Table I-41. Rates of Cesarean Delivery and Vaginal Birth after Previous Cesarean Delivery, by Race and Hispanic Origin of Mother, for each State and Territory, 2002

(Percent.)

State	Cesarean delivery rate [1]						Rate of vaginal births after previous cesarean [2]					
	All races [3]	White		Black		Hispanic [5]	All races [3]	White		Black		Hispanic [5]
		Total [4]	Non-Hispanic	Total [4]	Non-Hispanic			Total [4]	Non-Hispanic	Total [4]	Non-Hispanic	
United States [6]	26.1	25.9	26.2	27.6	27.7	25.2	12.6	12.4	12.8	13.2	13.2	11.5
Alabama	28.7	28.6	29.0	29.1	29.1	23.4	9.4	9.5	9.0	9.2	9.2	16.4
Alaska	19.5	22.0	21.2	22.5	20.2	17.9	20.8	14.5	12.7	*	*	*
Arizona	21.3	21.3	23.2	23.8	23.8	19.3	11.4	10.4	11.0	8.9	9.8	9.7
Arkansas	29.1	28.9	29.4	30.2	30.2	24.8	11.9	11.5	10.7	13.3	13.3	16.7
California	26.8	26.6	27.4	30.2	30.4	26.1	8.0	7.8	8.0	7.9	7.9	7.6
Colorado	21.1	21.1	21.6	21.2	21.3	19.9	18.6	18.5	17.7	19.0	18.3	20.2
Connecticut	26.1	26.2	27.1	26.4	26.4	22.4	13.3	13.4	12.9	13.6	13.8	15.9
Delaware	27.3	26.9	27.4	28.6	28.7	24.2	12.8	12.6	10.9	13.4	13.2	20.3
District of Columbia	26.5	24.9	26.7	27.6	27.7	21.6	8.4	10.4	*	7.4	7.5	*
Florida	28.5	28.9	27.9	27.8	27.8	30.8	8.3	7.8	8.9	9.7	9.8	5.9
Georgia	25.9	25.5	26.8	26.9	26.9	19.9	11.1	11.5	10.7	10.2	10.1	14.7
Hawaii	21.4	19.5	19.2	16.9	15.9	22.1	16.7	15.5	15.4	*	*	15.7
Idaho	19.7	19.5	19.4	22.3	22.0	20.5	17.4	17.6	16.6	*	*	22.0
Illinois	23.9	23.8	24.8	24.1	24.1	21.4	17.1	17.2	16.2	16.0	16.1	19.3
Indiana	24.8	24.7	24.7	25.5	25.5	24.5	12.2	11.7	11.4	15.3	15.3	14.1
Iowa	24.8	24.9	24.9	24.8	25.0	24.4	12.0	11.9	11.9	14.4	14.4	10.8
Kansas	24.8	24.8	25.1	27.3	27.3	23.4	11.7	11.6	11.3	12.0	11.9	13.5
Kentucky	28.0	28.0	28.1	28.6	28.6	27.3	9.4	9.1	9.0	11.6	11.5	13.0
Louisiana	30.4	31.3	31.3	29.3	29.2	30.0	6.4	4.6	4.6	8.8	8.9	*
Maine	25.7	25.6	25.6	24.1	24.6	28.9	10.4	10.6	10.5	*	*	*
Maryland	27.5	26.5	27.0	29.5	29.6	23.6	17.0	17.4	17.1	16.0	16.1	19.6
Massachusetts	28.0	28.1	28.7	29.2	29.3	24.5	14.9	14.7	14.3	15.2	16.1	16.6
Michigan	24.9	25.0	25.1	24.6	24.6	23.7	13.0	12.8	12.6	14.3	14.2	14.5
Minnesota	22.2	22.7	22.8	22.9	23.1	22.4	16.4	15.1	14.4	26.4	26.0	20.9
Mississippi	31.1	31.5	31.7	30.9	30.9	24.4	6.9	6.1	5.8	7.9	7.9	*
Missouri	25.7	25.9	26.0	25.0	25.0	24.1	14.1	13.3	13.2	17.5	17.6	14.4
Montana	22.9	22.5	22.5	*	*	23.4	16.1	16.4	16.2	*	*	*
Nebraska	26.7	26.9	27.4	24.4	24.2	23.8	9.8	9.3	9.5	15.3	15.8	9.3
Nevada	25.7	25.0	27.1	31.1	31.2	22.3	11.3	11.5	8.0	9.4	9.6	16.0
New Hampshire	24.2	24.2	24.4	25.6	28.9	21.1	15.7	15.6	15.5	*	*	*
New Jersey	30.9	30.9	31.5	31.7	31.4	29.7	16.8	16.1	16.0	19.7	20.6	16.2
New Mexico	19.1	19.6	21.2	20.7	21.1	18.7	21.5	20.4	21.5	*	*	19.9
New York	27.1	27.3	27.7	27.6	28.0	25.9	18.2	18.0	18.1	19.0	19.4	17.3
North Carolina	26.4	26.0	27.1	28.1	28.1	21.0	13.2	13.0	11.5	13.3	13.2	20.3
North Dakota	23.1	22.9	22.8	*	*	27.5	16.7	16.7	16.9	*	*	*
Ohio	23.6	23.5	23.5	24.4	24.4	23.2	18.9	18.3	18.3	22.0	21.9	18.5
Oklahoma	28.1	27.7	28.0	30.0	30.1	25.4	8.4	8.6	8.2	8.3	8.3	10.7
Oregon	23.4	23.3	23.6	25.9	26.2	21.7	16.3	16.5	15.2	*	*	20.8
Pennsylvania	24.8	24.8	24.9	25.2	25.2	23.6	18.4	17.7	17.6	22.6	22.8	18.1
Rhode Island	26.2	26.8	28.3	25.0	24.8	23.4	13.0	12.4	12.4	*	*	11.9
South Carolina	28.6	28.4	28.5	29.2	29.2	26.3	9.5	9.3	9.2	9.7	9.8	9.8
South Dakota	24.6	24.7	24.7	20.4	20.4	21.3	16.7	17.1	16.8	*	*	*
Tennessee	27.5	27.7	28.1	26.7	26.7	23.1	11.6	10.1	9.9	16.6	16.6	13.3
Texas	27.9	27.7	28.5	29.9	30.0	27.1	8.0	8.1	7.0	6.3	6.2	8.8
Utah	19.1	18.9	18.4	27.4	27.2	22.2	24.1	24.5	24.5	*	*	24.2
Vermont	20.9	20.9	21.0	*	*	*	25.6	25.6	25.4	*	*	*
Virginia	26.8	26.3	26.6	27.9	27.9	24.3	12.6	12.4	12.3	13.4	13.3	13.0
Washington	24.0	23.9	24.4	27.0	27.0	21.6	14.6	14.8	13.3	12.5	11.6	20.1
West Virginia	29.3	29.2	29.2	32.1	32.2	26.2	9.3	9.2	9.3	*	*	*
Wisconsin	20.6	21.1	21.3	18.1	18.1	19.6	18.3	17.5	17.0	24.6	24.2	21.9
Wyoming	21.1	21.1	20.8	*	*	23.3	18.0	18.0	18.4	*	*	*
Puerto Rico	44.8	-45.1	. . .	-42.7	. . .	. . .	3.7	3.7	. . .	3.4	. . .	. . .
Virgin Islands	23.4	26.2	26.8	23.2	22.7	25.9	18.3	*	*	18.2	18.5	*
Guam	20.6	20.5	20.5	*	*	*	21.1	*	*	*	*	*
American Samoa	. . .	. . .	. . .	. . .	. . .	. . .	. . .	. . .	. . .	. . .	. . .	. . .
Northern Marianas	24.8	*	. . .	*	. . .	. . .	*	*	. . .	*	. . .	. . .

[1]Percent of all live births by cesarean delivery.
[2]Number of vaginal births after previous cesarean delivery per 100 live births to women with a previous cesarean delivery.
[3]Includes races other than White and Black and origin not stated.
[4]Race and Hispanic origin are reported separately on the birth certificate. Data for persons of Hispanic origin are included in the data for each race group according to the mother's reported race.
[5]Includes all persons of Hispanic origin of any race.
[6]Excludes data for the territories.
. . . = Not available.
* = Figure does not meet standards of reliability or precision; based on fewer than 20 births in numerator.

Figure I-11. Rates of Cesarean Delivery, by State, 2002

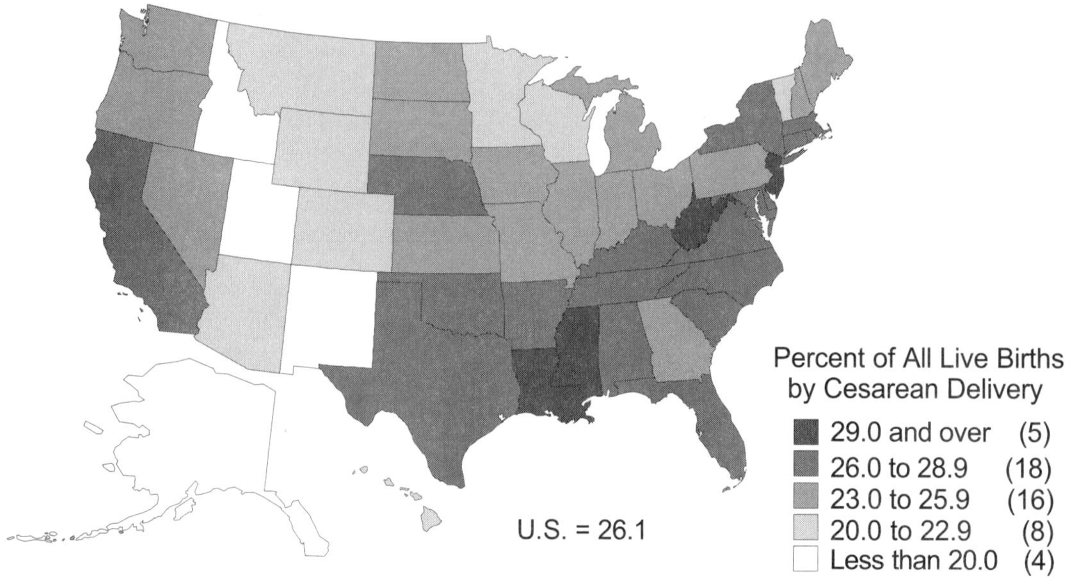

Figure I-12. Rates of Vaginal Birth after Previous Cesarean Delivery, by State, 2002

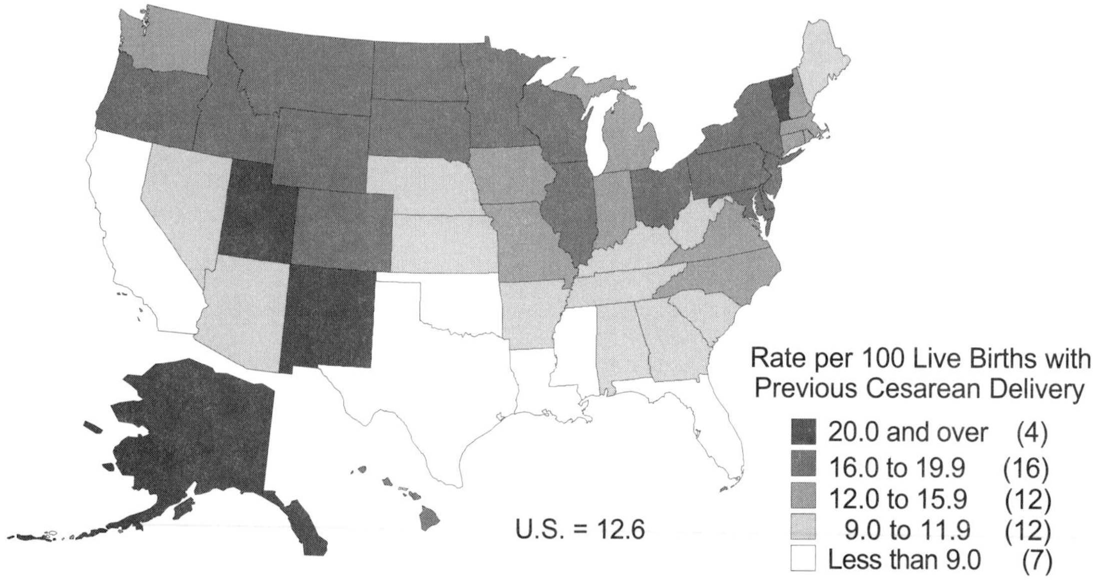

TABLE I-42. RATES OF CESAREAN DELIVERY AND VAGINAL BIRTH AFTER PREVIOUS CESAREAN DELIVERY, BY SELECTED MATERNAL MEDICAL RISK FACTORS AND COMPLICATIONS OF LABOR AND/OR DELIVERY, 2002

In the case of some medical risk factors, mothers are especially likely to have a cesarean delivery. High rates of cesareans in the neighborhood of 40 percent to 50 percent occurred for women with diabetes, chronic as well as pregnancy-associated hypertension, eclampsia, and other conditions. In the case of complications of labor and/or delivery, mothers with dysfunctional labor, breech/malpresentation, and cephalopelvic disproportion received a cesarean delivery in 71 percent, 87 percent, and 96 percent of the cases, respectively. However, the fairly frequent complication of meconium, moderate/heavy, was associated with a relatively low rate of cesarean delivery of 22.6 percent.

In cases of precipitous labor (less than three hours) the total cesarean rate was as low as 2.8 percent and the rate of vaginal birth after previous cesarean delivery surged to 72.8 percent.

Table I-42. Rates of Cesarean Delivery and Vaginal Birth after Previous Cesarean Delivery, by Selected Maternal Medical Risk Factors and Complications of Labor and/or Delivery, 2002

Medical risk factor and complication	All births to mothers with specified condition and/or procedure	Cesarean delivery rate		Rate of vaginal birth after previous cesarean [3]
		Total [1]	Primary [2]	
Medical Risk Factors				
Anemia	102 788	26.5	18.2	16.1
Cardiac disease	20 308	31.1	22.1	13.5
Acute or chronic lung disease	49 263	29.3	21.0	15.0
Diabetes	131 027	41.7	30.0	9.2
Genital herpes [4]	33 644	36.2	29.0	16.1
Hydramnios/oligohydramnios	55 590	40.2	34.1	11.3
Hemoglobinopathy	3 152	28.9	20.3	14.5
Hypertension, chronic	33 442	46.4	35.2	7.7
Hypertension, pregnancy-associated	150 854	41.6	35.7	8.6
Eclampsia	12 920	53.4	48.4	6.9
Incompetent cervix	11 703	39.2	31.3	15.3
Renal disease	12 185	28.4	19.9	14.9
Rh sensitization [5]	26 648	26.9	18.8	14.5
Uterine bleeding [4]	20 666	36.9	28.9	13.9
Complications of Labor and/or Delivery				
Febrile	60 960	31.9	30.6	41.1
Meconium, moderate/heavy	200 408	22.6	19.6	34.4
Premature rupture of membrane	92 550	29.2	25.4	25.0
Abruptio placenta	21 409	63.3	58.7	10.3
Placenta previa	13 356	81.3	77.0	2.3
Other excessive bleeding	24 137	32.7	25.6	18.0
Seizures during labor	1 297	57.1	55.0	*
Precipitous labor (less than 3 hours)	73 359	2.8	1.9	72.8
Prolonged labor (more than 20 hours)	28 113	37.4	36.6	43.5
Dysfunctional labor	114 479	70.7	69.4	12.1
Breech/malpresentation	152 425	86.9	85.4	3.4
Cephalopelvic disproportion	63 240	96.4	96.3	1.8
Cord prolapse	6 941	68.4	66.6	10.5
Anesthetic complication [4]	2 354	44.0	36.2	12.4
Fetal distress [4]	140 074	61.3	59.4	15.0

[1]Percent of all live births by cesarean delivery.
[2]Number of primary cesareans per 100 live births to women who have not had a previous cesarean.
[3]Number of vaginal births after previous cesarean delivery per 100 live births to women with a previous cesarean delivery.
[4]Texas does not report this risk factor or complication.
[5]Kansas does not report this risk factor.
* = Figure does not meet standards of reliability or precision; based on fewer than 20 births in numerator.

TABLE I-43. LIVE BIRTHS BY BIRTHWEIGHT AND PERCENT VERY LOW AND LOW BIRTHWEIGHT, BY PERIOD OF GESTATION, RACE, AND HISPANIC ORIGIN OF MOTHER, 2002

For all births, 1.5 percent resulted in very low birthweight (under 3-1/4 pounds) and 7.8 percent had low birthweight (under 5-1/2 pounds), the highest level reported in more than three decades. When the period of gestation becomes very short, these percentages increase dramatically. For all preterm deliveries, that is, deliveries with a period of gestation less than 37 weeks, the incidence of very low birthweight is 11.6 percent and that for low birthweight is 43.2 percent. For periods of gestation of 37–41 weeks, the frequency of very low birthweight drops to a mere 0.1 percent and that for low birthweight declines to 3.0 percent. For deliveries after 41 weeks of gestation, the low birthweight percentage drops further to 2.3 percent.

When comparing the same gestation periods between Black mothers and non-Hispanic White mothers, Blacks show a greater incidence of infants with very low birthweight and low birthweight, especially among near full-term and full-term births. The results for Hispanic mothers are similar to those for non-Hispanic Whites, except in the case of preterm babies; for these Hispanics report somewhat smaller percentages of underweight newborns.

Table I-43. Live Births by Birthweight and Percent Very Low and Low Birthweight, by Period of Gestation, Race, and Hispanic Origin of Mother, 2002

Birthweight, race, and Hispanic origin of mother	All births	Period of gestation [1]										Not stated
		Preterm					Term				Postterm	
		Total under 37 weeks	Under 28 weeks	28–31 weeks	32–35 weeks	36 weeks	Total 37–41 weeks	37–39 weeks	40 weeks	41 weeks	42 weeks and over	
ALL RACES [2]												
Total Number	4 021 726	480 812	29 425	48 420	224 364	178 603	3 231 563	2 029 753	807 896	393 914	268 095	41 256
Less than 500 grams	6 268	6 057	5 766	263	27	1	11	6	5	-	2	198
500–999 grams	22 845	22 190	16 361	5 210	594	25	191	116	44	31	30	434
1,000–1,499 grams	29 431	27 329	4 029	15 626	7 097	577	1 492	1 077	284	131	231	379
1,500–1,999 grams	61 652	51 203	908	11 917	33 412	4 966	8 833	7 348	1 021	464	820	796
2,000–2,499 grams	193 881	100 582	660	4 143	65 179	30 600	86 016	71 481	9 909	4 626	5 175	2 108
2,500–2,999 grams	688 630	127 482	944	4 082	56 940	65 516	520 030	396 514	86 416	37 100	34 306	6 812
3,000–3,499 grams	1 521 884	97 178	-	4 668	39 616	52 894	1 308 043	855 173	312 554	140 316	101 901	14 762
3,500–3,999 grams	1 125 959	38 612	-	2 440	17 011	19 161	985 884	544 969	292 587	148 328	90 665	10 798
4,000–4,499 grams	314 182	7 763	-	-	3 739	4 024	274 022	131 515	89 882	52 625	29 248	3 149
4,500–4,999 grams	48 606	1 211	-	-	524	687	41 799	18 864	13 711	9 224	5 071	525
5000 grams or more	5 396	211	-	-	103	108	4 524	2 239	1 320	965	556	105
Not stated	2 992	994	757	71	122	44	718	451	163	104	90	1 190
Percent												
Very low birthweight [3]	1.5	11.6	91.2	43.6	3.4	0.3	0.1	0.1	0.0	0.0	0.1	2.5
Low birthweight [4]	7.8	43.2	96.7	76.9	47.4	20.3	3.0	3.9	1.4	1.3	2.3	9.8
WHITE TOTAL [5]												
Total Number	3 174 760	350 269	17 582	32 733	164 401	135 553	2 577 102	1 605 486	651 241	320 375	214 605	32 784
Less than 500 grams	3 588	3 449	3 273	151	24	1	4	2	2	-	1	134
500–999 grams	13 909	13 439	9 775	3 252	395	17	144	85	33	26	24	302
1,000–1,499 grams	19 595	18 144	2 555	10 422	4 773	394	1 004	714	192	98	165	282
1,500–1,999 grams	43 074	35 822	584	8 477	23 364	3 397	6 102	5 070	721	311	562	588
2,000–2,499 grams	135 633	72 157	383	2 783	47 381	21 610	58 415	48 639	6 646	3 130	3 596	1 465
2,500–2,999 grams	495 046	94 371	562	2 561	42 248	49 000	371 206	283 904	60 827	26 475	24 492	4 977
3,000–3,499 grams	1 191 374	74 082	-	3 190	29 291	41 601	1 026 216	672 563	243 704	109 949	79 324	11 752
3,500–3,999 grams	947 951	30 624	-	1 850	13 279	15 495	832 272	460 609	246 546	125 117	75 912	9 143
4,000–4,499 grams	275 046	6 386	-	-	3 049	3 337	240 419	115 120	79 172	46 127	25 519	2 722
4,500–4,999 grams	42 751	1 005	-	-	430	575	36 851	16 507	12 132	8 212	4 435	460
5000 grams or more	4 630	169	-	-	80	89	3 872	1 906	1 126	840	504	85
Not stated	2 163	621	450	47	87	37	597	367	140	90	71	874
Percent												
Very low birthweight [3]	1.2	10.0	91.1	42.3	3.2	0.3	0.0	0.0	0.0	0.0	0.1	2.3
Low birthweight [4]	6.8	40.9	96.7	76.7	46.2	18.8	2.5	3.4	1.2	1.1	2.0	8.7
NON-HISPANIC WHITE												
Total Number	2 298 156	251 132	12 355	23 300	116 567	98 910	1 885 189	1 178 913	474 093	232 183	149 898	11 937
Less than 500 grams	2 545	2 496	2 369	104	23	-	2	1	1	-	1	46
500–999 grams	9 952	9 733	6 958	2 475	288	12	93	55	20	18	17	109
1,000–1,499 grams	14 403	13 498	1 754	7 889	3 578	277	704	502	130	72	102	99
1,500–1,999 grams	32 143	27 178	364	6 389	17 874	2 551	4 340	3 664	470	206	392	233
2,000–2,499 grams	99 549	54 306	244	1 847	35 972	16 243	42 126	35 383	4 599	2 144	2 552	565
2,500–2,999 grams	346 513	68 910	353	1 542	30 428	36 587	259 593	200 890	41 068	17 635	16 329	1 681
3,000–3,499 grams	842 347	50 072	-	1 883	18 241	29 948	734 586	486 658	171 619	76 309	53 615	4 074
3,500–3,999 grams	701 894	19 577	-	1 135	7 891	10 551	624 898	347 782	184 281	92 835	54 137	3 282
4,000–4,499 grams	210 885	4 196	-	-	1 917	2 279	186 654	89 454	61 448	35 752	18 959	1 076
4,500–4,999 grams	32 988	623	-	-	249	374	28 849	12 849	9 500	6 500	3 334	182
5000 grams or more	3 393	99	-	-	44	55	2 858	1 378	845	635	404	32
Not stated	1 544	444	313	36	62	33	486	297	112	77	56	558
Percent												
Very low birthweight [3]	1.2	10.3	92.0	45.0	3.3	0.3	0.0	0.0	0.0	0.0	0.1	2.2
Low birthweight [4]	6.9	42.8	97.1	80.4	49.6	19.3	2.5	3.4	1.1	1.1	2.0	9.2

[1]Expressed in completed weeks.
[2]Includes races other than White and Black and origin not stated.
[3]Birthweight of less than 1,500 grams (3 lb 4 oz).
[4]Birthweight of less than 2,500 grams (5 lb 8 oz).
[5]Race and Hispanic origin are reported separately on the birth certificate. Data for persons of Hispanic origin are included in the data for each race group according to the mother's reported race.
0.0 = Quantity more than zero but less than 0.05.
- = Quantity zero.

Table I-43. Live Births by Birthweight and Percent Very Low and Low Birthweight, by Period of Gestation, Race, and Hispanic Origin of Mother, 2002—*Continued*

Birthweight, race, and Hispanic origin of mother	All births	Period of gestation [1]										Not stated
		Preterm					Term				Postterm	
		Total under 37 weeks	Under 28 weeks	28–31 weeks	32–35 weeks	36 weeks	Total 37–41 weeks	37–39 weeks	40 weeks	41 weeks	42 weeks and over	
BLACK TOTAL [5]												
Total Number	593 691	103 437	10 495	13 143	47 212	32 587	448 002	292 061	105 733	50 208	37 956	4 296
Less than 500 grams	2 414	2 355	2 247	105	3	-	7	4	3	-	1	51
500–999 grams	7 842	7 704	5 838	1 694	165	7	35	22	8	5	4	99
1,000–1,499 grams	8 286	7 766	1 286	4 401	1 926	153	402	295	77	30	58	60
1,500–1,999 grams	15 146	12 594	286	2 894	8 162	1 252	2 192	1 812	255	125	221	139
2,000–2,499 grams	45 125	22 412	247	1 136	14 021	7 008	21 033	17 314	2 545	1 174	1 287	393
2,500–2,999 grams	140 506	25 614	326	1 258	11 517	12 513	106 500	80 315	18 351	7 834	7 447	945
3,000–3,499 grams	226 457	17 635	-	1 161	8 105	8 369	191 312	124 092	46 318	20 902	16 106	1 404
3,500–3,999 grams	117 795	5 904	-	478	2 728	2 698	101 110	55 658	30 185	15 267	10 020	761
4,000–4,499 grams	25 293	978	-	-	485	493	21 752	10 692	6 879	4 181	2 377	186
4,500–4,999 grams	3 739	139	-	-	60	79	3 177	1 604	971	602	391	32
5000 grams or more	475	22	-	-	12	10	410	202	128	80	33	10
Not stated	613	314	265	16	28	5	72	51	13	8	11	216
Percent												
Very low birthweight [3]	3.1	17.3	91.6	47.2	4.4	0.5	0.1	0.1	0.1	0.1	0.2	5.1
Low birthweight [4]	13.3	51.2	96.8	77.9	51.5	25.8	5.3	6.7	2.7	2.7	4.1	18.2
NON-HISPANIC BLACK												
Total Number	578 335	101 423	10 330	12 895	46 304	31 894	435 923	284 581	102 654	48 688	36 896	4 093
Less than 500 grams	2 365	2 311	2 205	103	3	-	7	4	3	-	1	46
500–999 grams	7 711	7 574	5 748	1 658	161	7	34	21	8	5	4	99
1,000–1,499 grams	8 118	7 610	1 268	4 302	1 888	152	393	289	76	28	58	57
1,500–1,999 grams	14 882	12 373	283	2 850	8 023	1 217	2 155	1 777	254	124	217	137
2,000–2,499 grams	44 300	22 020	240	1 120	13 776	6 884	20 635	16 989	2 500	1 146	1 270	375
2,500–2,999 grams	137 584	25 129	323	1 242	11 274	12 290	104 256	78 644	17 947	7 665	7 286	913
3,000–3,499 grams	220 468	17 241	-	1 136	7 951	8 154	186 194	120 872	45 016	20 306	15 688	1 345
3,500–3,999 grams	113 972	5 747	-	468	2 661	2 618	97 818	53 900	29 175	14 743	9 688	719
4,000–4,499 grams	24 310	951	-	-	470	481	20 920	10 294	6 609	4 017	2 262	177
4,500–4,999 grams	3 587	133	-	-	57	76	3 045	1 546	929	570	378	31
5000 grams or more	462	22	-	-	12	10	398	196	124	78	33	9
Not stated	576	312	263	16	28	5	68	49	13	6	11	185
Percent												
Very low birthweight [3]	3.1	17.3	91.6	47.1	4.4	0.5	0.1	0.1	0.1	0.1	0.2	5.2
Low birthweight [4]	13.4	51.3	96.8	77.9	51.5	25.9	5.3	6.7	2.8	2.7	4.2	18.3
HISPANIC [6]												
Total Number	876 642	99 510	5 202	9 529	47 998	36 781	692 314	426 983	177 190	88 141	64 997	19 821
Less than 500 grams	992	909	861	46	1	1	2	1	1	-	-	81
500–999 grams	3 984	3 738	2 826	799	110	3	52	31	13	8	7	187
1,000–1,499 grams	5 240	4 697	809	2 570	1 203	115	308	216	64	28	62	173
1,500–1,999 grams	10 943	8 649	217	2 082	5 489	861	1 780	1 426	248	106	172	342
2,000–2,499 grams	36 224	17 899	146	944	11 417	5 392	16 420	13 361	2 059	1 000	1 050	855
2,500–2,999 grams	149 219	25 562	211	1 032	11 892	12 427	112 260	83 468	19 921	8 871	8 255	3 142
3,000–3,499 grams	349 830	24 154	-	1 334	11 101	11 719	292 437	186 267	72 424	33 746	25 856	7 383
3,500–3,999 grams	245 222	11 075	-	711	5 408	4 956	206 679	112 477	62 003	32 199	21 830	5 638
4,000–4,499 grams	63 667	2 204	-	-	1 138	1 066	53 325	25 504	17 553	10 268	6 561	1 577
4,500–4,999 grams	9 691	384	-	-	180	204	7 947	3 640	2 601	1 706	1 091	269
5000 grams or more	1 217	70	-	-	36	34	995	522	276	197	101	51
Not stated	413	169	132	11	23	3	109	70	27	12	12	123
Percent												
Very low birthweight [3]	1.2	9.4	88.7	35.9	2.7	0.3	0.1	0.1	0.0	0.0	0.1	2.2
Low birthweight [4]	6.5	36.1	95.8	67.7	38.0	17.3	2.7	3.5	1.3	1.3	2.0	8.3

[1]Expressed in completed weeks.
[3]Birthweight of less than 1,500 grams (3 lb 4 oz).
[4]Birthweight of less than 2,500 grams (5 lb 8 oz).
[5]Race and Hispanic origin are reported separately on the birth certificate. Data for persons of Hispanic origin are included in the data for each race group according to the mother's reported race.
[6]Includes all persons of Hispanic origin of any race.
0.0 = Quantity more than zero but less than 0.05.
- = Quantity zero.

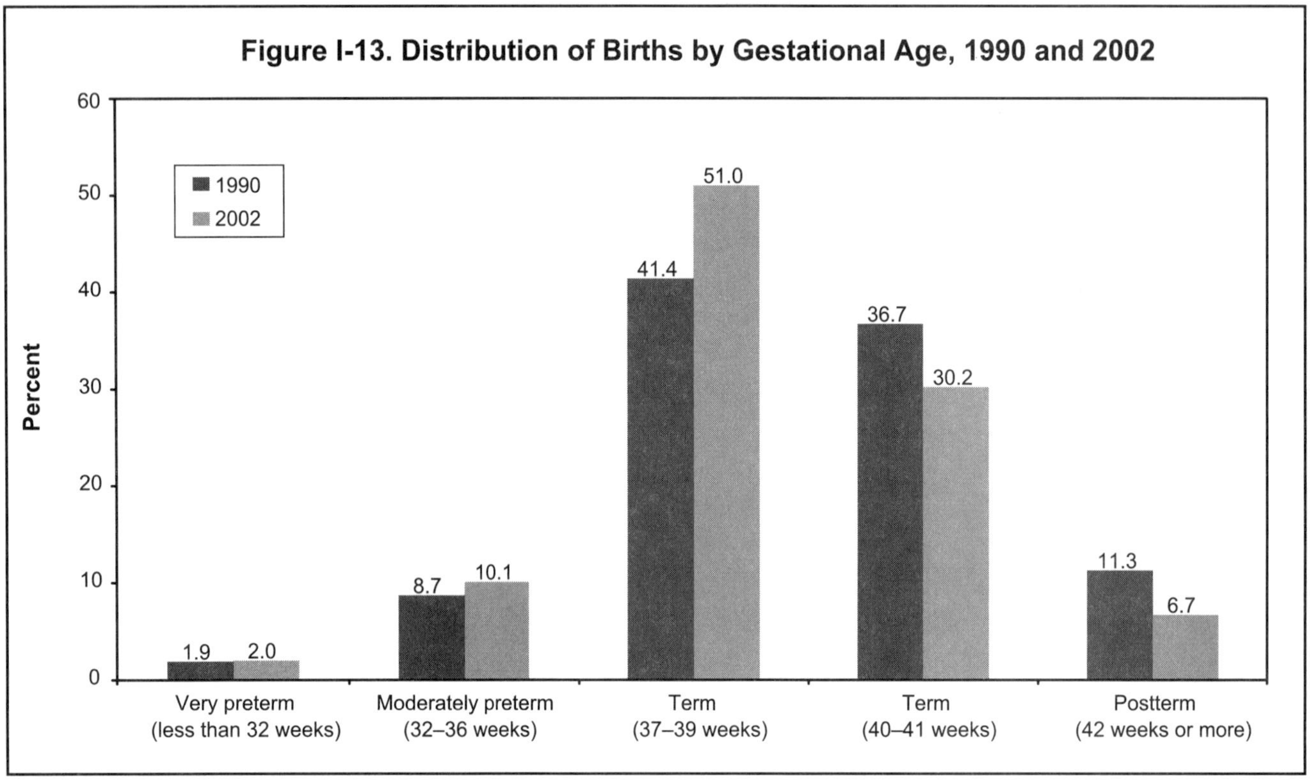
Figure I-13. Distribution of Births by Gestational Age, 1990 and 2002
Percent
1990
2002
1.9
2.0
8.7
10.1
41.4
51.0
36.7
30.2
11.3
6.7
Very preterm (less than 32 weeks)
Moderately preterm (32–36 weeks)
Term (37–39 weeks)
Term (40–41 weeks)
Postterm (42 weeks or more)

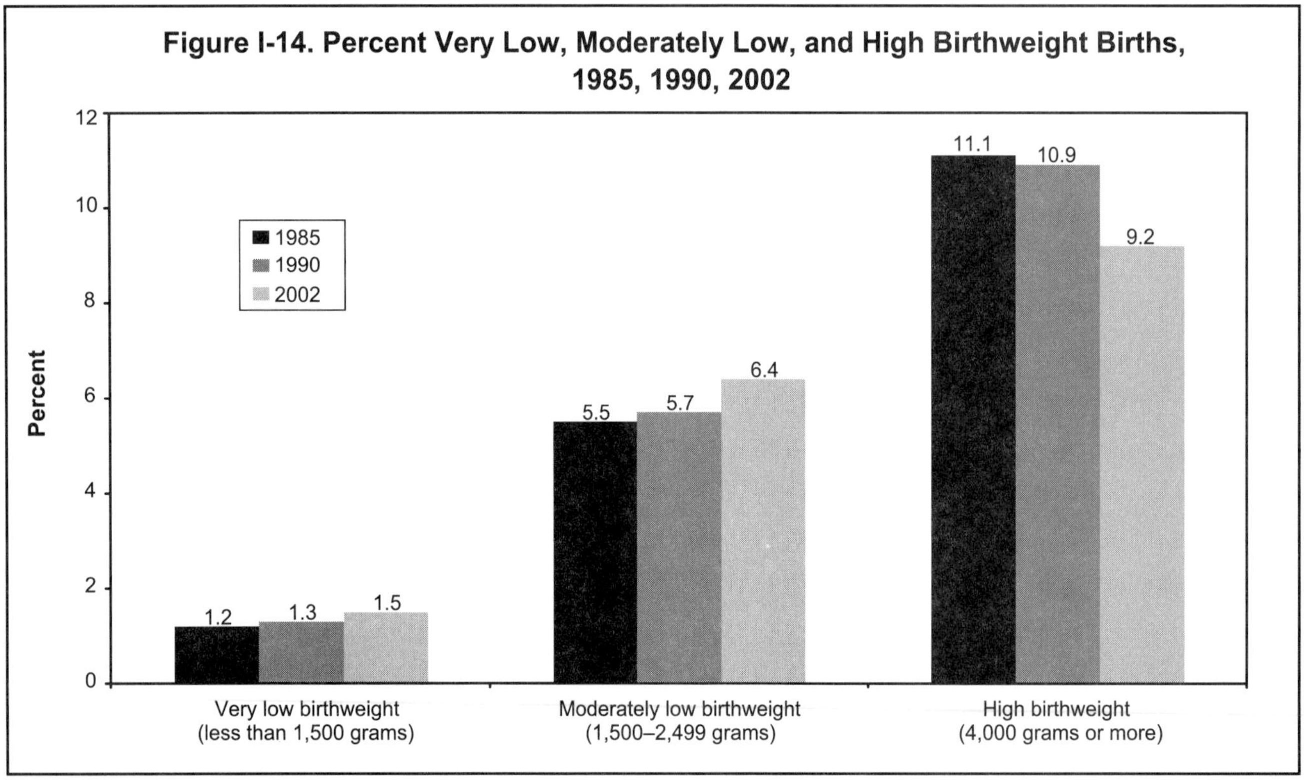
Figure I-14. Percent Very Low, Moderately Low, and High Birthweight Births, 1985, 1990, 2002
Percent
1985
1990
2002
1.2
1.3
1.5
5.5
5.7
6.4
11.1
10.9
9.2
Very low birthweight (less than 1,500 grams)
Moderately low birthweight (1,500–2,499 grams)
High birthweight (4,000 grams or more)

TABLE I-44. PERCENT OF LIVE BIRTHS THAT ARE VERY PRETERM, PRETERM, VERY LOW BIRTHWEIGHT, AND LOW BIRTHWEIGHT, BY RACE AND HISPANIC ORIGIN OF MOTHER, 1981–2002

The preterm birth rate (including births at less than 37 completed weeks of gestation) rose to 12.1 percent of all births for 2002, the highest level recorded in two decades. The "very preterm birth rate" (defined as less than 32 completed weeks of gestation) was 1.96 percent in 2002, about 8 percent higher than the 1981 level of 1.81 percent. Preterm birth is among the leading causes of infant death and is associated with nearly half of all congenital neurological defects, such as cerebral palsy. In 2001, 41 percent of all infants born earlier than 28 weeks of gestation did not survive their first year of life, as compared with 0.3 percent of those that had gestation periods of 37–41 weeks. The rise of multiple births over the past two decades has had an important influence on the preterm birth rate, as these births are much more likely to occur earlier.

The rising trend of preterm births has been concentrated among non-Hispanic Whites (29 percent increase since 1990), while rates for non-Hispanic Blacks have been declining. The preterm rate for non-Hispanic Blacks was

Table I-44. Percent of Live Births that are Very Preterm, Preterm, Very Low Birthweight, and Low Birthweight, by Race and Hispanic Origin of Mother, 1981–2002

Year	Very preterm [1]						Preterm [2]					
	All races [3]	White		Black		Hispanic [5]	All races [3]	White		Black		Hispanic [5]
		Total [4]	Non-Hispanic	Total [4]	Non-Hispanic			Total [4]	Non-Hispanic	Total [4]	Non-Hispanic	
2002	1.96	1.60	1.56	4.01	4.04	1.72	12.1	11.1	11.0	17.5	17.7	11.6
2001	1.95	1.59	1.55	4.02	4.05	1.69	11.9	11.0	10.8	17.5	17.6	11.4
2000	1.93	1.55	1.51	4.04	4.09	1.69	11.6	10.6	10.4	17.3	17.4	11.2
1999	1.96	1.57	1.54	4.13	4.18	1.68	11.8	10.7	10.5	17.5	17.6	11.4
1998	1.96	1.57	1.52	4.11	4.15	1.72	11.6	10.5	10.2	17.5	17.6	11.4
1997	1.94	1.53	1.49	4.17	4.19	1.68	11.4	10.2	9.9	17.5	17.6	11.2
1996	1.89	1.48	1.43	4.13	4.17	1.66	11.0	9.8	9.5	17.4	17.5	10.9
1995	1.89	1.46	1.41	4.25	4.29	1.66	11.0	9.7	9.4	17.7	17.8	10.9
1994	1.91	1.45	1.39	4.32	4.36	1.67	11.0	9.6	9.3	18.1	18.2	10.9
1993	1.93	1.45	1.39	4.41	4.45	1.67	11.0	9.5	9.1	18.5	18.6	11.0
1992 [8]	1.91	1.40	1.33	4.47	4.50	1.64	10.7	9.1	8.7	18.4	18.5	10.7
1991 [8]	1.94	1.41	1.35	4.62	4.65	1.65	10.8	9.1	8.7	18.9	19.0	11.0
1990 [9]	1.92	1.39	1.33	4.61	4.63	1.69	10.6	8.9	8.5	18.8	18.9	11.0
1989 [10]	1.95	1.41	1.34	4.64	4.68	1.76	10.6	8.8	8.4	18.9	19.0	11.1
1988	1.96	1.42	. . .	4.72	. . .	. . .	10.2	8.5	. . .	18.7	. . .	. . .
1987	1.96	1.44	. . .	4.61	. . .	. . .	10.2	8.5	. . .	18.4	. . .	. . .
1986	1.90	1.41	. . .	4.47	. . .	. . .	10.0	8.4	. . .	18.0	. . .	. . .
1985	1.88	1.42	. . .	4.37	. . .	. . .	9.8	8.2	. . .	17.8	. . .	. . .
1984	1.83	1.38	. . .	4.22	. . .	. . .	9.4	7.9	. . .	17.1	. . .	. . .
1983	1.86	1.40	. . .	4.34	. . .	. . .	9.6	8.0	. . .	17.7	. . .	. . .
1982	1.84	1.40	. . .	4.22	. . .	. . .	9.5	8.0	. . .	17.4	. . .	. . .
1981	1.81	1.37	. . .	4.13	. . .	. . .	9.4	7.9	. . .	17.3	. . .	. . .

Year	Very low birthweight [6]						Low birthweight [7]					
	All races [3]	White		Black		Hispanic [5]	All races [3]	White		Black		Hispanic [5]
		Total [4]	Non-Hispanic	Total [4]	Non-Hispanic			Total [4]	Non-Hispanic	Total [4]	Non-Hispanic	
2002	1.46	1.17	1.17	3.13	3.15	1.17	7.8	6.8	6.9	13.3	13.4	6.5
2001	1.44	1.16	1.17	3.04	3.08	1.14	7.7	6.7	6.8	13.0	13.1	6.5
2000	1.43	1.14	1.14	3.07	3.10	1.14	7.6	6.5	6.6	13.0	13.1	6.4
1999	1.45	1.15	1.15	3.14	3.18	1.14	7.6	6.6	6.6	13.1	13.2	6.4
1998	1.45	1.15	1.15	3.08	3.11	1.15	7.6	6.5	6.6	13.0	13.2	6.4
1997	1.42	1.13	1.12	3.04	3.05	1.13	7.5	6.5	6.5	13.0	13.1	6.4
1996	1.37	1.09	1.08	2.99	3.02	1.12	7.4	6.3	6.4	13.0	13.1	6.3
1995	1.35	1.06	1.04	2.97	2.98	1.11	7.3	6.2	6.2	13.1	13.2	6.3
1994	1.33	1.02	1.01	2.96	2.99	1.08	7.3	6.1	6.1	13.2	13.3	6.2
1993	1.33	1.01	1.00	2.96	2.99	1.06	7.2	6.0	5.9	13.3	13.4	6.2
1992 [8]	1.29	0.96	0.94	2.96	2.97	1.04	7.1	5.8	5.7	13.3	13.4	6.1
1991 [8]	1.29	0.96	0.94	2.96	2.97	1.02	7.1	5.8	5.7	13.6	13.6	6.1
1990 [9]	1.27	0.95	0.93	2.92	2.93	1.03	7.0	5.7	5.6	13.3	13.3	6.1
1989 [10]	1.28	0.95	0.93	2.95	2.97	1.05	7.0	5.7	5.6	13.5	13.6	6.2
1988	1.24	0.93	. . .	2.86	. . .	. . .	6.9	5.7	. . .	13.3	. . .	. . .
1987	1.24	0.94	. . .	2.79	. . .	. . .	6.9	5.7	. . .	13.0	. . .	. . .
1986	1.21	0.93	. . .	2.73	. . .	. . .	6.8	5.7	. . .	12.8	. . .	. . .
1985	1.21	0.93	. . .	2.71	. . .	. . .	6.8	5.7	. . .	12.6	. . .	. . .
1984	1.19	0.93	. . .	2.60	. . .	. . .	6.7	5.6	. . .	12.6	. . .	. . .
1983	1.19	0.92	. . .	2.60	. . .	. . .	6.8	5.7	. . .	12.8	. . .	. . .
1982	1.18	0.91	. . .	2.56	. . .	. . .	6.8	5.6	. . .	12.6	. . .	. . .
1981	1.16	0.91	. . .	2.52	. . .	. . .	6.8	5.7	. . .	12.7	. . .	. . .

[1]Births of less than 32 completed weeks of gestation.
[2]Births of less than 37 completed weeks of gestation.
[3]Includes races other than White and Black and origin not stated.
[4]Race and Hispanic origin are reported separately on the birth certificate. Data for persons of Hispanic origin are included in the data for each race group according to the mother's reported race.
[5]Includes all persons of Hispanic origin of any race.
[6]Less than 1,500 grams (3 lb 4 oz).
[7]Less than 2,500 grams (5 lb 8 oz).
[8]Data by Hispanic origin exclude New Hampshire, which did not report Hispanic origin.
[9]Data by Hispanic origin exclude New Hampshire and Oklahoma, which did not report Hispanic origin.
[10]Data by Hispanic origin exclude New Hampshire, Oklahoma, and Louisiana, which did not report Hispanic origin.
. . . = Not available.

19.0 percent in 1991, but declined to 17.7 percent in 2002. These risk levels, however, are considerably greater than the national average of 12.1 percent for preterm births. Preterm rates have risen 5 percent for Hispanics since 1990.

Trends in the rates of low birthweight correspond to developments in preterm births. Thus for all births since 1981, the frequency of low-weight infants has increased, reaching 7.8 percent. On the basis of detailed data available since 1989, the frequency of low-weight infants has increased for non-Hispanic Whites, and to a small extent for Hispanics. For non-Hispanic Blacks, the incidence of low weight has declined slightly since 1989. Nevertheless, the risk of low birthweight is high for non-Hispanic Blacks: singleton infants born to Black mothers continue to be more than twice as likely as non-Hispanic White or Hispanic infants to weigh less than 5-1/2 pounds at birth.

Among non-Hispanic Whites, there was a steep rise in multiple births, and infants born in multiple deliveries are approximately 10 times as likely to have low birthweight as are singletons. Also, according to a recent study, singletons conceived with assisted-reproductive technology are at greater risk of low birthweight than those conceived without such intervention.

TABLE I-44A. PRETERM BIRTH RATES, BY STATE, 1990 AND 2002

The proportion of births delivered at gestational ages of less than 37 weeks increased for all states except Georgia and the District of Columbia. Preterm rates in 2002 varied from greater than 15 percent for Mississippi, Alabama, and Louisiana, to less than 10 percent for Alaska, Minnesota, New Hampshire, Oregon, Vermont, and Washington.

Table I-44A. Preterm Birth Rates, by State, 1990 and 2002

(Percent.)

Area	2002	1990	Percent change, 1990–2002
United States	12.1	10.6	14
Alabama	15.8	13.1	20
Alaska	9.8	9.0	10
Arizona	12.6	10.2	23
Arkansas	12.7	12.5	2
California	10.2	9.8	4
Colorado	12.0	9.6	26
Connecticut	10.1	8.9	13
Delaware	13.6	11.1	23
District of Columbia	14.6	20.7	-29
Florida	13.0	11.6	12
Georgia	12.6	12.8	-1
Hawaii	13.7	10.0	37
Idaho	10.4	8.7	19
Illinois	12.6	11.4	10
Indiana	12.5	9.9	26
Iowa	11.6	8.7	33
Kansas	11.0	9.5	16
Kentucky	13.6	10.5	30
Louisiana	15.1	13.9	8
Maine	10.1	7.6	33
Maryland	12.9	11.2	15
Massachusetts	10.6	7.7	37
Michigan	11.9	10.7	11
Minnesota	9.8	8.1	21
Mississippi	17.2	15.1	14
Missouri	13.0	10.7	21
Montana	11.3	8.5	34
Nebraska	11.8	8.5	38
Nevada	13.0	11.1	17
New Hampshire	9.5	6.8	40
New Jersey	12.0	10.7	12
New Mexico	12.6	11.0	14
New York	11.4	10.7	7
North Carolina	13.3	12.5	7
North Dakota	11.3	8.3	37
Ohio	12.2	10.6	15
Oklahoma	12.6	10.1	25
Oregon	9.7	8.0	21
Pennsylvania	11.4	10.3	11
Rhode Island	11.3	9.6	18
South Carolina	14.2	12.3	16
South Dakota	11.3	8.9	26
Tennessee	13.8	12.6	10
Texas	13.3	11.2	18
Utah	10.5	8.7	21
Vermont	9.0	7.1	27
Virginia	11.8	11.0	7
Washington	9.6	8.4	15
West Virginia	13.4	9.9	35
Wisconsin	10.9	9.2	19
Wyoming	11.7	10.3	14

Note: Preterm is defined as less than 37 completed weeks of gestation.

Figure I-15. Preterm Birth Rates, by State, 2002

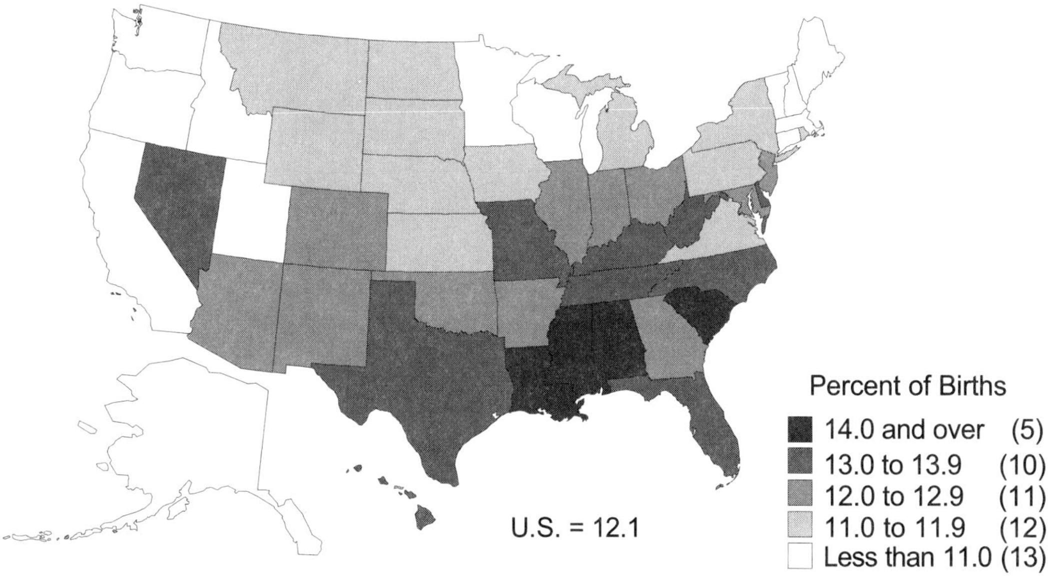

TABLE I-45. NUMBER AND PERCENT LOW BIRTHWEIGHT AND NUMBER OF LIVE BIRTHS BY BIRTHWEIGHT, BY AGE, RACE, AND HISPANIC ORIGIN OF MOTHER, 2002

For mothers between 25 and 29 years of age, the risk of low birthweight is smallest (6.9 percent). For younger mothers the risk increases (to 9.6 percent for 15–19 years of age), and for older mothers the risk increases rapidly to 10.4 percent for those 40–44 years old, and 20.2 percent for mothers 45–54 years of age. These patterns also apply to Hispanics and non-Hispanic Blacks, although the level of low birthweight risk is substantially greater for Blacks.

Much of the rise in risk among older mothers can be attributed to their higher multiple births rates. In 2002, multiples accounted for almost two-thirds of all low birthweight infants delivered to mothers aged 45 years and over. The percentage of low birthweight infants for singletons by mothers 45 years and over was just under 10 percent (data not shown).

Table I-45. Number and Percent Low Birthweight and Number of Live Births by Birthweight, by Age, Race, and Hispanic Origin of Mother, 2002

Age and race and Hispanic origin of mother	Low birthweight [1]		Birthweight												
	Number	Percent	Total	Less than 500 grams	500–999 grams	1,000–1,499 grams	1,500–1,999 grams	2,000–2,499 grams	2,500–2,999 grams	3,000–3,499 grams	3,500–3,999 grams	4,000–4,499 grams	4,500–4,999 grams	5000 grams or more	Not stated
All Races [2]															
All ages	314 077	7.8	4 021 726	6 268	22 845	29 431	61 652	193 881	688 630	1 521 884	1 125 959	314 182	48 606	5 396	2 992
Under 15 years	984	13.5	7 315	27	113	87	189	568	1 875	2 892	1 329	205	22	-	8
15–19 years	40 706	9.6	425 493	869	3 170	3 699	7 620	25 348	93 878	173 012	95 898	19 276	2 176	216	331
15 years	2 112	11.3	18 703	62	215	201	375	1 259	4 577	7 531	3 755	651	51	7	19
16 years	4 681	10.9	43 147	99	377	471	896	2 838	10 021	17 731	8 926	1 593	140	16	39
17 years	7 617	9.9	76 881	168	622	693	1 369	4 765	17 500	31 422	16 648	3 240	342	38	74
18 years	11 376	9.6	118 651	254	870	1 017	2 172	7 063	25 990	48 154	26 888	5 488	611	57	87
19 years	14 920	8.9	168 111	286	1 086	1 317	2 808	9 423	35 790	68 174	39 681	8 304	1 032	98	112
20–24 years	80 605	7.9	1 022 106	1 577	5 825	6 957	14 776	51 470	194 498	405 669	266 147	64 711	8 828	894	754
25–29 years	73 283	6.9	1 060 391	1 505	5 275	6 888	14 151	45 464	170 977	402 284	310 423	87 635	13 603	1 479	707
30–34 years	68 526	7.2	951 219	1 373	4 985	6 602	14 061	41 505	140 908	345 226	289 906	89 752	14 556	1 672	673
35–39 years	38 872	8.6	453 927	747	2 701	3 969	8 365	23 090	69 486	158 492	134 462	43 590	7 733	885	407
40–44 years	9 997	10.4	95 788	157	728	1 096	2 191	5 825	15 872	32 645	26 659	8 658	1 617	237	103
45–54 years	1 104	20.2	5 487	13	48	133	299	611	1 136	1 664	1 135	355	71	13	9
White Total [3]															
All ages	215 799	6.8	3 174 760	3 588	13 909	19 595	43 074	135 633	495 046	1 191 374	947 951	275 046	42 751	4 630	2 163
Under 15 years	471	12.1	3 884	10	53	47	99	262	864	1 605	800	126	13	-	5
15–19 years	24 909	8.1	305 988	460	1 816	2 232	4 702	15 699	61 962	125 588	75 420	15 927	1 786	178	218
15 years	1 115	9.6	11 665	37	121	97	204	656	2 556	4 757	2 697	491	35	4	10
16 years	2 745	9.3	29 501	52	199	299	524	1 671	6 353	12 278	6 730	1 256	101	12	26
17 years	4 636	8.5	54 698	95	347	423	854	2 917	11 386	22 665	12 995	2 658	275	32	51
18 years	7 050	8.2	85 957	127	509	610	1 349	4 455	17 364	35 197	21 212	4 528	505	47	54
19 years	9 363	7.5	124 167	149	640	803	1 771	6 000	24 303	50 691	31 786	6 994	870	83	77
20–24 years	52 231	6.7	783 000	839	3 327	4 258	9 551	34 256	136 322	311 153	219 008	55 414	7 587	756	529
25–29 years	51 339	6.0	851 142	880	3 228	4 673	10 110	32 448	125 163	319 516	264 198	77 159	11 990	1 250	527
30–34 years	50 296	6.5	779 535	821	3 223	4 746	10 540	30 966	105 800	278 636	249 861	79 947	13 035	1 463	497
35–39 years	28 422	7.7	369 833	471	1 743	2 779	6 195	17 234	52 276	127 529	115 069	38 621	6 847	768	301
40–44 years	7 236	9.4	76 928	97	480	754	1 636	4 269	11 770	26 027	22 635	7 543	1 434	205	78
45–54 years	895	20.1	4 450	10	39	106	241	499	889	1 320	960	309	59	10	8
Non-Hispanic White															
All ages	158 592	6.9	2 298 156	2 545	9 952	14 403	32 143	99 549	346 513	842 347	701 894	210 885	32 988	3 393	1 544
Under 15 years	188	12.6	1 493	5	24	21	37	101	295	608	343	50	5	-	4
15–19 years	15 139	8.4	179 511	286	1 131	1 442	2 878	9 402	35 029	71 535	45 933	10 412	1 212	114	137
15 years	512	10.0	5 130	21	63	53	96	279	1 025	1 985	1 317	261	25	1	4
16 years	1 421	9.9	14 421	30	109	168	262	852	2 945	5 737	3 504	726	65	7	16
17 years	2 683	8.9	30 205	60	211	265	511	1 636	6 007	12 194	7 488	1 616	167	16	34
18 years	4 437	8.7	51 196	77	337	410	846	2 767	10 076	20 396	12 930	2 968	331	29	29
19 years	6 086	7.8	78 559	98	411	546	1 163	3 868	14 976	31 223	20 694	4 841	624	61	54
20–24 years	36 028	6.9	519 153	554	2 343	2 928	6 648	23 555	88 644	201 070	147 901	39 178	5 472	522	338
25–29 years	37 851	6.2	614 909	622	2 294	3 434	7 573	23 928	88 603	226 821	193 568	57 869	8 930	883	384
30–34 years	40 119	6.5	620 173	640	2 454	3 761	8 485	24 779	82 414	219 065	201 351	65 159	10 550	1 126	389
35–39 years	22 688	7.6	297 436	358	1 293	2 135	4 995	13 907	41 411	101 569	93 641	31 723	5 587	592	225
40–44 years	5 824	9.4	61 853	74	385	594	1 321	3 450	9 386	20 626	18 380	6 242	1 186	148	61
45–54 years	755	20.8	3 628	6	28	88	206	427	731	1 053	777	252	46	8	6
Black Total [3]															
All ages	78 813	13.3	593 691	2 414	7 842	8 286	15 146	45 125	140 506	226 457	117 795	25 293	3 739	475	613
Under 15 years	483	15.2	3 188	15	56	35	87	290	958	1 197	472	70	5	-	3
15–19 years	14 406	13.9	103 795	396	1 261	1 344	2 687	8 718	28 425	41 118	16 871	2 576	272	28	99
15 years	928	14.7	6 336	25	88	97	162	556	1 848	2 495	908	131	14	3	9
16 years	1 774	14.6	12 142	46	165	159	337	1 067	3 331	4 855	1 858	276	32	3	13
17 years	2 747	14.2	19 411	70	264	250	476	1 687	5 479	7 679	2 996	439	45	5	21
18 years	3 948	14.0	28 266	124	334	374	754	2 362	7 655	11 191	4 622	745	70	6	29
19 years	5 009	13.3	37 640	131	410	464	958	3 046	10 112	14 898	6 487	985	111	11	27
20–24 years	24 963	12.8	194 704	691	2 301	2 446	4 643	14 882	48 572	76 346	36 691	6 964	881	93	194
25–29 years	16 803	12.3	136 591	557	1 780	1 867	3 199	9 400	30 147	52 372	29 259	6 692	1 053	136	129
30–34 years	12 710	13.4	95 006	481	1 466	1 421	2 552	6 790	19 413	34 603	21 610	5 560	878	125	107
35–39 years	7 347	15.2	48 388	231	784	898	1 536	3 898	10 302	16 777	10 518	2 779	532	68	65
40–44 years	1 993	17.4	11 443	40	190	262	417	1 084	2 541	3 850	2 289	624	107	23	16
45–54 years	108	18.8	576	3	4	13	25	63	148	194	85	28	11	2	-
Non-Hispanic Black															
All ages	77 376	13.4	578 335	2 365	7 711	8 118	14 882	44 300	137 584	220 468	113 972	24 310	3 587	462	576
Under 15 years	472	15.1	3 132	13	54	35	86	284	944	1 174	466	68	5	-	3
15–19 years	14 143	13.9	101 494	388	1 233	1 314	2 642	8 566	27 877	40 168	16 431	2 492	262	27	94
15 years	909	14.7	6 191	24	87	93	158	547	1 814	2 432	886	127	13	2	8
16 years	1 745	14.7	11 846	43	160	155	335	1 052	3 260	4 724	1 801	269	31	3	13
17 years	2 701	14.2	18 980	70	261	244	464	1 662	5 366	7 504	2 922	419	43	5	20
18 years	3 880	14.0	27 656	122	329	368	742	2 319	7 510	10 931	4 512	721	68	6	28
19 years	4 908	13.3	36 821	129	396	454	943	2 986	9 927	14 577	6 310	956	107	11	25
20–24 years	24 603	12.9	190 241	681	2 276	2 405	4 582	14 659	47 669	74 531	35 636	6 688	842	88	184
25–29 years	16 502	12.4	132 824	546	1 751	1 836	3 140	9 229	29 450	50 921	28 248	6 431	1 014	135	123
30–34 years	12 439	13.5	92 155	471	1 439	1 383	2 501	6 645	18 976	33 529	20 821	5 335	840	120	95
35–39 years	7 163	15.3	46 831	224	766	881	1 495	3 797	10 044	16 230	10 087	2 670	508	67	62
40–44 years	1 946	17.6	11 097	39	188	251	411	1 057	2 479	3 726	2 203	599	106	23	15
45–54 years	108	19.3	561	3	4	13	25	63	145	189	80	27	10	2	-
Hispanic [4]															
All ages	57 383	6.5	876 642	992	3 984	5 240	10 943	36 224	149 219	349 830	245 222	63 667	9 691	1 217	413
Under 15 years	291	12.0	2 421	5	32	25	62	167	580	1 006	457	78	8	-	1
15–19 years	9 967	7.8	127 900	171	700	809	1 855	6 432	27 298	54 651	29 716	5 553	581	65	69
15 years	617	9.3	6 647	15	59	48	111	384	1 557	2 828	1 392	233	10	4	6
16 years	1 348	8.8	15 321	22	92	133	263	838	3 471	6 643	3 273	534	39	5	8
17 years	1 995	8.1	24 772	34	139	164	354	1 304	5 471	10 556	5 541	1 063	112	17	17
18 years	2 655	7.6	35 123	50	175	205	506	1 719	7 389	14 966	8 340	1 563	172	17	21
19 years	3 352	7.3	46 037	50	235	259	621	2 187	9 410	19 658	11 170	2 160	248	22	17
20–24 years	16 313	6.2	265 235	283	995	1 353	2 922	10 760	48 075	110 786	71 345	16 227	2 116	235	138
25–29 years	13 551	5.7	236 143	251	949	1 254	2 549	8 548	36 722	92 791	70 406	19 182	3 042	357	92
30–34 years	10 100	6.4	157 887	161	763	981	2 046	6 149	23 245	59 263	47 859	14 567	2 452	335	66
35–39 years	5 662	7.9	71 480	96	449	642	1 185	3 290	10 796	25 737	21 108	6 724	1 245	168	40
40–44 years	1 384	9.3	14 809	21	86	164	297	816	2 354	5 336	4 161	1 279	234	55	6
45–54 years	115	15.0	767	4	10	12	27	62	149	260	170	57	13	2	1

[1]Less than 2,500 grams (5 lb 8 oz).
[2]Includes races other than White and Black and origin not stated.
[3]Race and Hispanic origin are reported separately on the birth certificate. Data for persons of Hispanic origin are included in the data for each race group according to the mother's reported race.
[4]Includes all persons of Hispanic origin of any race.
- = Quantity zero.

TABLE I-46. NUMBER AND PERCENT OF BIRTHS OF LOW BIRTHWEIGHT, BY RACE AND HISPANIC ORIGIN OF MOTHER, FOR EACH STATE AND TERRITORY, 2002

Low birthweight (newborns less than 5-1/2 pounds) was encountered least frequently in Alaska, Oregon, and Washington at rates between 5.8 percent and 5.9 percent. Rates of 10 percent or more were reported for the District of Columbia (11.6 percent), Mississippi (11.2 percent), Louisiana (10.4 percent), and South Carolina (10.0 percent), and also Puerto Rico and the Virgin Islands.

In every state where Blacks could be separately identified, their frequency of infants with low birthweight was greater than that for non-Hispanic Whites as well as that for Hispanics. The non-Hispanic Black rate had a low of 9.8 percent in Oregon, and a high of 15.2 percent in Mississippi. The Hispanic low weight birth rate ranged from 5.3 percent in Oregon and Washington to 9.2 percent in Pennsylvania and the District of Columbia.

Table I-46. Number and Percent of Births of Low Birthweight, by Race and Hispanic Origin of Mother, for each State and Territory, 2002

(Low birthweight is birthweight of less than 2,500 grams [5 lb 8 oz].)

State	Number						Percent					
	All races [1]	White		Black		Hispanic [3]	All races [1]	White		Black		Hispanic [3]
		Total [2]	Non-Hispanic	Total [2]	Non-Hispanic			Total [2]	Non-Hispanic	Total [2]	Non-Hispanic	
United States [4]	314 077	215 799	158 592	78 813	77 376	57 383	7.8	6.8	6.9	13.3	13.4	6.5
Alabama	5 825	3 140	2 957	2 626	2 620	189	9.9	7.9	7.9	14.4	14.3	7.4
Alaska	579	331	205	48	27	51	5.8	5.2	4.6	11.1	10.1	6.4
Arizona	5 938	5 039	2 608	326	310	2 415	6.8	6.5	6.7	11.7	12.2	6.4
Arkansas	3 204	2 136	1 957	1 008	1 007	169	8.6	7.3	7.5	13.6	13.6	5.5
California	33 824	25 064	9 838	3 765	3 658	15 134	6.4	5.8	6.0	11.5	11.6	5.8
Colorado	6 067	5 338	3 609	422	403	1 777	8.9	8.6	8.6	14.4	14.6	8.5
Connecticut	3 258	2 406	1 866	652	624	560	7.8	6.9	6.7	12.6	12.7	8.0
Delaware	1 102	651	553	403	403	97	9.9	8.2	8.3	14.9	15.0	7.4
District of Columbia	866	186	105	673	659	88	11.6	7.0	6.1	14.6	14.5	9.2
Florida	17 320	10 902	7 476	5 889	5 791	3 503	8.4	7.1	7.3	12.7	12.8	6.8
Georgia	11 915	5 840	4 799	5 672	5 564	1 015	8.9	6.8	7.0	13.3	13.3	6.0
Hawaii	1 450	278	219	49	48	218	8.3	7.0	6.9	10.3	10.9	9.0
Idaho	1 284	1 228	1 028	12	10	184	6.1	6.1	6.0	*	*	6.6
Illinois	14 725	9 451	6 944	4 545	4 517	2 530	8.2	6.7	7.0	14.3	14.3	6.2
Indiana	6 463	5 140	4 741	1 205	1 203	388	7.6	6.9	7.0	13.0	13.0	6.3
Iowa	2 489	2 277	2 131	130	127	149	6.6	6.5	6.5	10.3	10.2	6.2
Kansas	2 757	2 312	1 992	354	351	299	7.0	6.6	6.7	12.3	12.4	6.0
Kentucky	4 657	3 875	3 749	711	709	132	8.6	8.0	8.0	14.4	14.4	8.1
Louisiana	6 774	2 749	2 678	3 905	3 895	79	10.4	7.5	7.6	14.7	14.6	5.7
Maine	853	822	810	16	16	11	6.3	6.3	6.3	*	*	*
Maryland	6 607	3 145	2 723	3 178	3 163	418	9.0	7.0	7.0	13.1	13.2	6.9
Massachusetts	6 046	4 663	3 928	933	798	813	7.5	7.0	6.7	11.2	12.1	8.5
Michigan	10 363	6 869	6 306	3 120	3 100	458	8.0	6.7	6.7	13.9	14.0	6.3
Minnesota	4 251	3 343	3 022	520	511	284	6.3	5.8	5.7	10.7	10.8	6.1
Mississippi	4 635	1 824	1 776	2 767	2 767	47	11.2	8.1	8.2	15.2	15.2	5.7
Missouri	6 034	4 365	4 151	1 527	1 526	210	8.0	7.0	7.0	13.9	13.9	6.4
Montana	755	634	600	6	5	25	6.8	6.7	6.7	*	*	6.5
Nebraska	1 817	1 538	1 301	196	194	202	7.2	6.7	6.8	13.6	13.7	6.1
Nevada	2 445	1 843	1 093	379	370	738	7.5	6.8	7.0	14.5	14.6	6.5
New Hampshire	914	866	828	18	17	20	6.3	6.3	6.5	*	*	4.0
New Jersey	9 185	5 846	4 232	2 546	2 388	1 753	8.0	6.9	6.9	12.8	13.3	7.1
New Mexico	2 225	1 871	680	71	68	1 200	8.0	8.0	7.8	13.9	14.3	8.2
New York	19 802	12 276	8 513	5 945	5 552	4 077	7.9	6.8	6.5	12.0	12.3	7.5
North Carolina	10 514	6 238	5 332	3 875	3 860	923	9.0	7.3	7.6	14.1	14.1	6.1
North Dakota	486	412	391	12	11	11	6.3	6.1	6.0	*	*	*
Ohio	12 334	8 927	8 562	3 118	3 105	351	8.3	7.3	7.3	13.8	13.9	7.3
Oklahoma	4 019	2 946	2 593	652	646	367	8.0	7.5	7.5	13.9	13.8	7.0
Oregon	2 608	2 290	1 862	90	88	428	5.8	5.6	5.7	9.6	9.8	5.3
Pennsylvania	11 685	8 463	7 609	2 839	2 773	798	8.2	7.2	7.0	14.0	14.1	9.2
Rhode Island	1 019	823	543	123	114	173	7.9	7.5	7.4	10.8	11.1	7.5
South Carolina	5 455	2 672	2 456	2 679	2 674	216	10.0	7.6	7.6	14.7	14.8	6.8
South Dakota	765	604	582	12	12	24	7.2	7.0	7.0	*	*	7.5
Tennessee	7 106	4 593	4 347	2 370	2 359	253	9.2	7.7	7.9	14.6	14.5	5.8
Texas	28 646	22 264	9 698	5 301	5 242	12 534	7.7	7.0	7.1	12.7	12.8	7.0
Utah	3 164	2 938	2 478	49	49	451	6.4	6.3	6.3	14.5	15.7	6.5
Vermont	409	406	403	-	-	2	6.4	6.5	6.6	*	*	*
Virginia	7 888	4 615	4 025	2 781	2 763	605	7.9	6.5	6.5	12.6	12.6	6.2
Washington	4 604	3 646	2 937	344	335	651	5.9	5.5	5.5	10.2	10.4	5.3
West Virginia	1 855	1 741	1 726	96	96	7	9.0	8.8	8.7	14.2	14.2	*
Wisconsin	4 538	3 462	3 173	848	841	302	6.6	5.9	5.9	13.2	13.3	5.7
Wyoming	553	511	457	7	7	54	8.4	8.3	8.3	*	*	8.7
Puerto Rico	6 039	5 480	. . .	558	. . .	. . .	11.5	11.5	. . .	11.3	. . .	. . .
Virgin Islands	193	42	9	141	125	44	11.8	13.0	*	11.5	11.5	13.7
Guam	255	13	10	4	4	3	8.0	*	*	*	*	*
American Samoa	64	3	. . .	-	. . .	. . .	3.9	*	. . .	*	. . .	. . .
Northern Marianas	89	1	. . .	-	. . .	. . .	6.9	*	. . .	*	. . .	. . .

[1]Includes races other than White and Black and origin not stated.
[2]Race and Hispanic origin are reported separately on the birth certificate. Data for persons of Hispanic origin are included in the data for each race group according to the mother's reported race.
[3]Includes all persons of Hispanic origin of any race.
[4]Excludes data for the territories.
. . . = Not available.
* = Figure does not meet standards of reliability or precision; based on fewer than 20 births in numerator.
- = Quantity zero.

Figure I-16. Percent of Births of Low Birthweight, by State, 2002

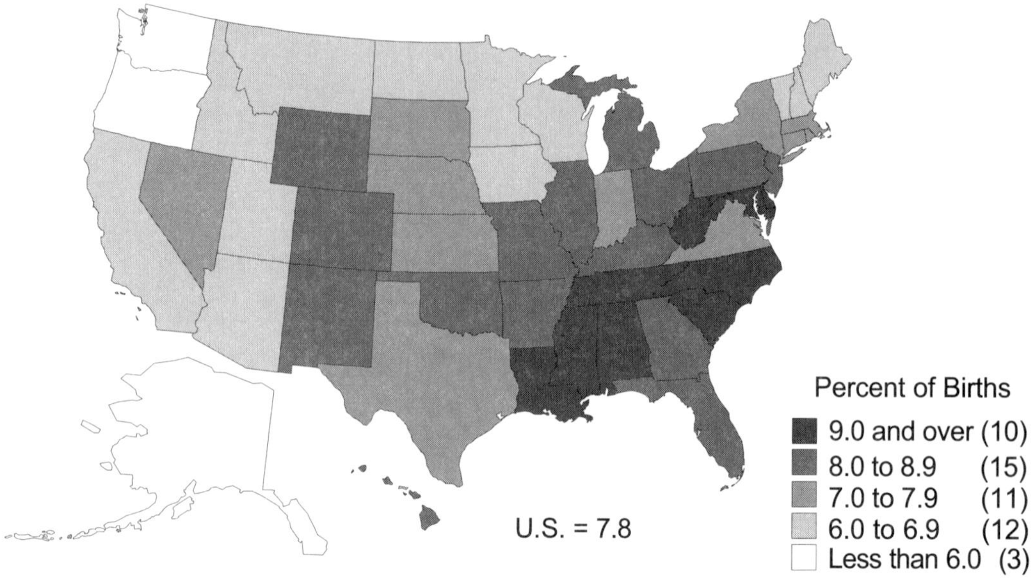

TABLE I-47. NUMBER AND PERCENT OF BIRTHS OF VERY LOW BIRTHWEIGHT, BY RACE AND HISPANIC ORIGIN OF MOTHER, FOR EACH STATE AND TERRITORY, 2002

For non-Hispanic White and also for Hispanic mothers, the rate of very low birthweight infants (less than 3-1/4 pounds) averaged 1.2 percent in 2002 in the United States; for non-Hispanic Blacks this rate was 3.1 percent. Overall, the highest rates, of 2.7 percent and 2.2 percent occurred in the District of Columbia and Mississippi, respectively. The lowest rate for all races was recorded in Alaska, Idaho, South Dakota, and Utah, all at 0.9 percent.

For non-Hispanic Whites, the smallest rate of very low birthweight was in Alaska (0.5 percent), and the highest rate was in Delaware (1.7 percent). For non-Hispanic Blacks the lowest rates were in Iowa and in the state of Washington (both 2.1 percent), and the highest rate was in North Carolina (3.6 percent). For Hispanics the lowest rate was recorded in Oklahoma (0.8 percent), and the highest rate was in Rhode Island (2.0 percent). Hispanics residing in New England generally had relatively high rates of very low birthweight, possibly associated with the sizable presence of residents of Puerto Rican origin.

Table I-47. Number and Percent of Births of Very Low Birthweight, by Race and Hispanic Origin of Mother, for each State and Territory, 2002

(Very low birthweight is birthweight of less than 1,500 grams [3 lb 4 oz].)

State	Number						Percent					
	All races [1]	White		Black		Hispanic [3]	All races [1]	White		Black		Hispanic [3]
		Total [2]	Non-Hispanic	Total [2]	Non-Hispanic			Total [2]	Non-Hispanic	Total [2]	Non-Hispanic	
United States [4]	58 544	37 092	26 900	18 542	18 194	10 216	1.5	1.2	1.2	3.1	3.1	1.2
Alabama	1 206	553	525	646	644	31	2.0	1.4	1.4	3.5	3.5	1.2
Alaska	92	45	23	6	4	8	0.9	0.7	0.5	*	*	*
Arizona	958	808	439	69	66	369	1.1	1.0	1.1	2.5	2.6	1.0
Arkansas	578	352	322	219	218	31	1.5	1.2	1.2	3.0	2.9	1.0
California	6 075	4 474	1 685	889	860	2 747	1.1	1.0	1.0	2.7	2.7	1.0
Colorado	885	754	493	80	74	272	1.3	1.2	1.2	2.7	2.7	1.3
Connecticut	664	469	337	167	162	132	1.6	1.4	1.2	3.2	3.3	1.9
Delaware	219	132	110	80	80	21	2.0	1.7	1.7	3.0	3.0	1.6
District of Columbia	200	33	17	166	161	17	2.7	1.2	*	3.6	3.5	*
Florida	3 295	1 798	1 184	1 426	1 401	624	1.6	1.2	1.2	3.1	3.1	1.2
Georgia	2 287	965	781	1 264	1 238	185	1.7	1.1	1.1	3.0	2.9	1.1
Hawaii	266	63	51	11	11	36	1.5	1.6	1.6	*	*	1.5
Idaho	195	180	141	3	3	35	0.9	0.9	0.8	*	*	1.3
Illinois	2 968	1 742	1 254	1 105	1 097	490	1.6	1.2	1.3	3.5	3.5	1.2
Indiana	1 172	906	833	246	246	73	1.4	1.2	1.2	2.6	2.7	1.2
Iowa	433	392	360	27	26	33	1.2	1.1	1.1	2.1	2.1	1.4
Kansas	505	403	341	85	84	59	1.3	1.2	1.2	2.9	3.0	1.2
Kentucky	878	697	676	169	168	21	1.6	1.4	1.4	3.4	3.4	1.3
Louisiana	1 312	417	402	870	868	16	2.0	1.1	1.1	3.3	3.3	*
Maine	159	153	150	6	6	3	1.2	1.2	1.2	*	*	*
Maryland	1 394	568	501	782	777	68	1.9	1.3	1.3	3.2	3.2	1.1
Massachusetts	1 089	788	635	224	185	175	1.4	1.2	1.1	2.7	2.8	1.8
Michigan	2 032	1 221	1 109	774	766	93	1.6	1.2	1.2	3.5	3.5	1.3
Minnesota	814	616	556	136	132	54	1.2	1.1	1.1	2.8	2.8	1.2
Mississippi	922	312	308	601	601	4	2.2	1.4	1.4	3.3	3.3	*
Missouri	1 162	772	722	369	368	51	1.5	1.2	1.2	3.3	3.4	1.6
Montana	122	98	94	2	2	4	1.1	1.0	1.0	*	*	*
Nebraska	327	269	229	46	46	32	1.3	1.2	1.2	3.2	3.2	1.0
Nevada	412	301	173	79	76	125	1.3	1.1	1.1	3.0	3.0	1.1
New Hampshire	157	152	144	1	1	2	1.1	1.1	1.1	*	*	*
New Jersey	1 760	987	681	657	620	337	1.5	1.2	1.1	3.3	3.4	1.4
New Mexico	343	276	105	12	11	172	1.2	1.2	1.2	*	*	1.2
New York	3 825	2 173	1 440	1 436	1 359	791	1.5	1.2	1.1	2.9	3.0	1.4
North Carolina	2 138	1 079	927	986	983	159	1.8	1.3	1.3	3.6	3.6	1.1
North Dakota	84	70	69	2	2	. . .	1.1	1.0	1.1	*	*	*
Ohio	2 391	1 618	1 535	725	724	78	1.6	1.3	1.3	3.2	3.2	1.6
Oklahoma	638	456	413	112	112	43	1.3	1.2	1.2	2.4	2.4	0.8
Oregon	463	406	334	16	16	72	1.0	1.0	1.0	*	*	0.9
Pennsylvania	2 299	1 560	1 391	691	677	152	1.6	1.3	1.3	3.4	3.4	1.7
Rhode Island	224	181	114	35	31	47	1.7	1.6	1.6	3.1	3.0	2.0
South Carolina	1 045	449	415	577	576	35	1.9	1.3	1.3	3.2	3.2	1.1
South Dakota	94	69	67	1	1	4	0.9	0.8	0.8	*	*	*
Tennessee	1 333	774	747	544	541	28	1.7	1.3	1.4	3.3	3.3	0.6
Texas	4 973	3 684	1 581	1 164	1 150	2 102	1.3	1.2	1.1	2.8	2.8	1.2
Utah	466	430	344	5	5	81	0.9	0.9	0.9	*	*	1.2
Vermont	68	68	67	-	-	1	1.1	1.1	1.1	*	*	*
Virginia	1 616	817	701	736	727	121	1.6	1.1	1.1	3.3	3.3	1.2
Washington	780	603	480	72	67	116	1.0	0.9	0.9	2.1	2.1	0.9
West Virginia	307	280	278	22	22	2	1.5	1.4	1.4	3.2	3.3	*
Wisconsin	852	616	558	201	199	59	1.2	1.0	1.0	3.1	3.1	1.1
Wyoming	67	63	58	-	-	5	1.0	1.0	1.0	*	*	*
Puerto Rico	741	670	. . .	71	. . .	. . .	1.4	1.4	. . .	1.4	. . .	. . .
Virgin Islands	32	7	1	25	21	7	2.0	*	*	2.0	1.9	*
Guam	28	2	1	-	-	1	0.9	*	*	*	*	*
American Samoa	6	-	. . .	-	. . .	. . .	*	*	. . .	*	. . .	. . .
Northern Marianas	7	-	. . .	-	. . .	. . .	*	*	. . .	*	. . .	. . .

[1]Includes races other than White and Black and origin not stated.
[2]Race and Hispanic origin are reported separately on the birth certificate. Data for persons of Hispanic origin are included in the data for each race group according to the mother's reported race.
[3]Includes all persons of Hispanic origin of any race.
[4]Excludes data for the territories.
. . . = Not available.
* = Figure does not meet standards of reliability or precision; based on fewer than 20 births in numerator.
- = Quantity zero.

TABLE I-48. LIVE BIRTHS WITH SELECTED ABNORMAL CONDITIONS OF THE NEWBORN AND RATES BY AGE AND RACE OF MOTHER, 2002

Eight abnormal conditions are reported on the birth certificate. Every year (since the beginning of collection in 1989) the three most frequently reported conditions have been assisted ventilation less than 30 minutes, assisted ventilation 30 minutes or longer, and hyaline membrane disease/respiratory distress syndrome. In 2002, the rate for assisted ventilation of less than 30 minutes was 21.2 per 1,000, nearly twice the 1989 rate. The rate of assisted ventilation 30 minutes or longer was 9.6 per 1,000, and it has been increasing gradually since 1989. The rate for hyaline membrane disease/respiratory distress syndrome (a frequent cause of morbidity in preterm infants) was 6.1 per 1,000 in 2002; this rate has been decreasing since the mid-1990s. Since 1989 the rate for meconium aspiration syndrome has slowly decreased, and the rate for anemia has decreased by 50 percent. Both these conditions are more common in Blacks than Whites, while birth injury is reported more frequently in Whites.

Table I-48. Live Births with Selected Abnormal Conditions of the Newborn and Rates by Age and Race of Mother, 2002

(Rates are number of live births with specified abnormal condition per 1,000 live births in specified group.)

Abnormal condition and race of mother	All births [1]	Abnormal condition reported	Rate by age of mother							Not stated [2]
			All ages	Under 20 years	20–24 years	25–29 years	30–34 years	35–39 years	40–54 years	
All Races [3]										
Anemia	4 021 726	4 080	1.0	1.0	1.0	1.0	1.0	1.1	1.5	27 558
Birth injury [4]	3 623 893	10 771	3.0	3.4	3.1	2.9	2.9	2.9	2.7	31 306
Fetal alcohol syndrome [5]	3 953 166	134	-	*	-	-	-	0.1	*	28 597
Hyaline membrane disease/RDS	4 021 726	24 258	6.1	6.8	6.2	5.9	5.6	6.1	7.3	27 558
Meconium aspiration syndrome	4 021 726	5 382	1.3	1.5	1.4	1.3	1.2	1.3	1.7	27 558
Assisted ventilation less than 30 minutes [6]	3 903 432	82 060	21.2	21.8	20.5	20.3	21.7	22.9	23.6	36 504
Assisted ventilation 30 minutes or longer [6]	3 903 432	37 284	9.6	11.1	9.5	8.9	9.3	10.4	13.2	36 504
Seizures	4 021 726	1 791	0.4	0.6	0.5	0.4	0.4	0.4	0.5	27 558
White										
Anemia	3 174 760	3 033	1.0	1.0	0.9	0.9	1.0	1.1	1.4	22 254
Birth injury [4]	2 834 630	8 737	3.1	3.5	3.3	3.0	3.0	2.9	2.8	25 539
Fetal alcohol syndrome [5]	3 115 781	100	-	*	-	*	-	0.1	*	23 236
Hyaline membrane disease/RDS	3 174 760	19 655	6.2	6.9	6.3	6.1	5.9	6.2	7.5	22 254
Meconium aspiration syndrome	3 174 760	4 046	1.3	1.4	1.3	1.3	1.2	1.2	1.7	22 254
Assisted ventilation less than 30 minutes [6]	3 106 697	66 278	21.5	21.9	20.6	20.6	22.1	23.6	24.0	29 251
Assisted ventilation 30 minutes or longer [6]	3 106 697	29 015	9.4	10.9	9.2	8.7	9.2	10.1	13.3	29 251
Seizures	3 174 760	1 446	0.5	0.6	0.5	0.4	0.4	0.4	0.5	22 254
Black										
Anemia	593 691	792	1.3	1.2	1.2	1.5	1.5	1.6	2.1	2 586
Birth injury [4]	550 607	1 015	1.9	2.1	1.9	1.8	1.8	1.7	*	2 956
Fetal alcohol syndrome [5]	587 273	26	-	*	*	*	*	*	*	2 617
Hyaline membrane disease/RDS	593 691	3 620	6.1	6.4	5.9	6.3	5.5	6.8	7.2	2 586
Meconium aspiration syndrome	593 691	1 007	1.7	1.7	1.5	1.7	2.0	1.9	2.1	2 586
Assisted ventilation less than 30 minutes [6]	558 610	11 532	20.8	21.2	20.0	20.4	21.7	21.3	24.5	3 632
Assisted ventilation 30 minutes or longer [6]	558 610	6 557	11.8	11.7	10.8	11.6	12.5	14.7	14.9	3 632
Seizures	593 691	250	0.4	0.6	0.4	0.4	0.4	*	*	2 586

[1]Total number of births to residents of areas reporting specified abnormal condition.
[2]No response reported for the abnormal condition item.
[3]Includes races other than White and Black.
[4]Nebraska and Texas do not report this condition.
[5]Wisconsin does not report this condition.
[6]New York City does not report this condition.
* = Figure does not meet standards of reliability or precision; based on fewer than 20 births in numerator.
- = Quantity zero.

TABLE I-49. LIVE BIRTHS WITH SELECTED CONGENITAL ANOMALIES AND RATES BY AGE AND RACE OF MOTHER: UNITED STATES EXCLUDING NEW MEXICO, 2002

Twenty-one types of malformations are reported on the birth certificate. Some of these are underreported due to difficulties of detection at birth. These anomalies are rare, but together they remain the leading cause of infant death. The most frequent among them are heart malformations (129.9 per 100,000), other circulatory/respiratory anomalies (131.7 per 100,000), and other musculoskeletal/integumental anomalies (228.9 per 100,000).

Studies have found a positive association between maternal smoking and certain birth defects, including cleft lip/palate and clubfoot. Since 1998 there has been mandatory fortification of cereal and grain products with folic acid in a successful effort to reduce spina bifida and anencephalus. Rates for Down's syndrome, heart malformations, and other circulatory/respiratory anomalies increase significantly for infants of mothers over 34 years of age.

Maternal age-specific rates for Down's syndrome are lower among Blacks than Whites. By contrast, rates for polydactyly/syndactyly/adactyly (digit abnormalities) and other musculoskeletal/integumental anomalies are markedly higher among Blacks.

Table I-49. Live Births with Selected Congenital Anomalies and Rates by Age and Race of Mother: United States Excluding New Mexico, 2002

(Rates are number of live births with specified congenital anomaly per 100,000 live births in specified group.)

Congenital anomaly and race of mother	All births [1]	Congenital anomaly reported	Rate by age of mother							Not stated [2]
			All ages	Under 20 years	20–24 years	25–29 years	30–34 years	35–39 years	40–54 years	
All Races [3]										
Anencephalus	3 993 973	391	9.9	13.4	10.6	7.9	9.7	8.9	*	28 613
Spina bifida/meningocele	3 993 973	793	20.0	21.6	22.6	19.1	19.0	17.4	*	28 613
Hydrocephalus	3 993 973	892	22.5	27.5	24.2	21.6	17.9	24.6	27.1	28 613
Microcephalus	3 993 973	219	5.5	6.8	5.3	5.0	5.0	5.8	*	28 613
Other central nervous system anomalies	3 993 973	881	22.2	27.5	21.6	22.2	19.9	21.5	31.1	28 613
Heart malformation	3 993 973	5 152	129.9	118.6	120.6	124.4	131.7	148.8	228.7	28 613
Other circulatory/respiratory anomalies	3 993 973	5 222	131.7	124.5	125.7	125.3	132.1	150.9	200.6	28 613
Rectal atresia/stenosis	3 993 973	329	8.3	8.7	8.3	6.4	8.0	12.1	*	28 613
Tracheo-esophageal fistula/esophageal atresia	3 993 973	430	10.8	8.5	9.8	9.5	12.3	14.8	*	28 613
Omphalocele/gastroschisis	3 993 973	1 203	30.3	81.3	42.5	19.6	13.4	16.8	23.1	28 613
Other gastrointestinal anomalies	3 993 973	1 432	36.1	38.3	38.6	33.5	32.2	37.8	59.2	28 613
Malformed genitalia	3 993 973	3 433	86.6	87.4	84.1	83.3	91.5	84.5	105.3	28 613
Renal agenesis	3 993 973	612	15.4	16.0	13.8	15.2	16.9	15.2	*	28 613
Other urogenital anomalies	3 993 973	4 036	101.8	94.2	96.7	97.8	108.1	110.6	128.4	28 613
Cleft lip/palate	3 993 973	3 114	78.5	86.0	86.9	77.2	68.5	72.0	99.3	28 613
Polydactyly/syndactyly/adactyly	3 993 973	3 259	82.2	102.6	89.9	77.4	73.8	74.0	82.3	28 613
Clubfoot	3 993 973	2 363	59.6	68.1	67.3	56.7	53.5	51.6	69.2	28 613
Diaphragmatic hernia	3 993 973	479	12.1	10.3	12.5	12.2	10.7	13.6	20.1	28 613
Other musculoskeletal/integumental anomalies	3 993 973	9 077	228.9	265.4	249.3	213.3	213.2	211.9	254.8	28 613
Down's syndrome	3 993 973	1 850	46.7	25.6	22.8	27.0	41.3	107.5	360.2	28 613
Other chromasomal anomalies	3 993 973	1 253	31.6	24.4	25.1	24.2	27.7	54.8	138.4	28 613
White										
Anencephalus	3 151 479	306	9.8	11.2	10.0	8.7	9.8	10.2	*	22 660
Spina bifida/meningocele	3 151 479	646	20.6	23.7	22.6	21.0	18.9	18.1	*	22 660
Hydrocephalus	3 151 479	705	22.5	23.0	25.0	22.5	18.2	25.0	27.5	22 660
Microcephalus	3 151 479	176	5.6	*	5.4	5.1	5.5	5.5	*	22 660
Other central nervous system anomalies	3 151 479	705	22.5	29.2	22.2	22.3	20.2	21.7	30.0	22 660
Heart malformation	3 151 479	4 069	130.0	121.3	120.7	122.2	132.6	148.5	227.3	22 660
Other circulatory/respiratory anomalies	3 151 479	4 139	132.3	131.1	130.0	125.5	129.9	144.3	197.3	22 660
Rectal atresia/stenosis	3 151 479	274	8.8	8.9	8.6	6.9	8.2	13.4	*	22 660
Tracheo-esophageal fistula/esophageal atresia	3 151 479	360	11.5	8.9	10.6	9.8	12.9	16.2	*	22 660
Omphalocele/gastroschisis	3 151 479	956	30.6	92.0	45.6	19.1	12.5	14.5	*	22 660
Other gastrointestinal anomalies	3 151 479	1 131	36.1	39.4	36.6	33.7	33.0	38.7	63.7	22 660
Malformed genitalia	3 151 479	2 906	92.9	96.9	91.9	89.9	96.5	87.8	106.2	22 660
Renal agenesis	3 151 479	514	16.4	15.8	15.9	16.2	17.3	15.6	*	22 660
Other urogenital anomalies	3 151 479	3 324	106.2	101.9	101.3	100.6	112.6	115.3	127.4	22 660
Cleft lip/palate	3 151 479	2 668	85.3	97.6	96.3	85.6	74.0	71.9	97.4	22 660
Polydactyly/syndactyly/adactyly	3 151 479	1 902	60.8	65.4	61.1	60.1	58.6	61.5	64.9	22 660
Clubfoot	3 151 479	2 048	65.5	78.9	73.6	63.2	58.1	55.2	76.2	22 660
Diaphragmatic hernia	3 151 479	394	12.6	9.5	13.9	13.5	10.1	14.3	*	22 660
Other musculoskeletal/integumental anomalies	3 151 479	5 891	188.3	211.6	204.2	174.1	183.6	175.9	196.1	22 660
Down's syndrome	3 151 479	1 602	51.2	29.6	24.2	29.7	45.4	115.3	383.4	22 660
Other chromasomal anomalies	3 151 479	1 029	32.9	24.6	26.7	25.4	28.6	55.2	142.4	22 660
Black										
Anencephalus	593 180	62	10.5	*	12.9	*	*	*	*	2 781
Spina bifida/meningocele	593 180	122	20.7	*	26.3	*	24.4	*	*	2 781
Hydrocephalus	593 180	145	24.6	39.4	21.2	19.1	*	*	*	2 781
Microcephalus	593 180	28	4.7	*	*	*	*	*	*	2 781
Other central nervous system anomalies	593 180	119	20.2	19.7	20.1	20.6	*	*	*	2 781
Heart malformation	593 180	788	133.5	108.9	113.6	145.8	147.2	178.9	242.8	2 781
Other circulatory/respiratory anomalies	593 180	673	114.0	90.2	94.5	114.9	136.6	183.1	175.8	2 781
Rectal atresia/stenosis	593 180	35	5.9	*	*	*	*	*	*	2 781
Tracheo-esophageal fistula/esophageal atresia	593 180	40	6.8	*	*	*	*	*	*	2 781
Omphalocele/gastroschisis	593 180	187	31.7	53.5	27.4	24.3	22.2	*	*	2 781
Other gastrointestinal anomalies	593 180	207	35.1	31.0	40.3	35.3	26.5	*	*	2 781
Malformed genitalia	593 180	317	53.7	57.3	52.1	47.1	57.2	58.3	*	2 781
Renal agenesis	593 180	71	12.0	*	*	*	*	*	*	2 781
Other urogenital anomalies	593 180	403	68.3	60.1	64.0	71.4	75.2	66.6	*	2 781
Cleft lip/palate	593 180	239	40.5	37.6	48.5	30.9	32.8	43.7	*	2 781
Polydactyly/syndactyly/adactyly	593 180	1 214	205.6	210.4	209.1	195.9	218.2	187.3	192.6	2 781
Clubfoot	593 180	224	37.9	42.3	47.0	28.7	30.7	*	*	2 781
Diaphragmatic hernia	593 180	69	11.7	*	*	*	*	*	*	2 781
Other musculoskeletal/integumental anomalies	593 180	2 529	428.4	401.0	416.7	417.5	458.6	459.8	619.6	2 781
Down's syndrome	593 180	163	27.6	*	16.0	15.5	23.3	79.1	293.0	2 781
Other chromasomal anomalies	593 180	145	24.6	23.5	20.7	16.9	22.2	47.9	*	2 781

Note: Excludes data for New Mexico, which did not report congenital anomalies.

[1]Total number of births to residents of areas reporting specified congenital anomaly.
[2]No response reported for the congenital anomalies item.
[3]Includes races other than White and Black.
* = Figure does not meet standards of reliability or precision; based on fewer than 20 births in numerator.

TABLE I-50. NUMBER AND RATE OF LIVE BIRTHS BY PLURALITY OF BIRTH, AGE, RACE, AND HISPANIC ORIGIN OF MOTHER, 2002

There has been an upsurge in multiple births over the last two decades associated with two related trends: greater access to fertility therapies, and an increase in childbearing among women in their 30s and 40s who are more likely to have multiple births even without the use of fertility therapies. In 2002, there were over 125,000 births in twin deliveries, 3.1 percent of total births. This rate was 65 percent higher than the rate in 1980. The most rapid increases have been for mothers over 40 years. Twinning rates increased among the three largest racial and ethnic groups. Since 1990, the likelihood of giving births to a twin has risen by more than 50 percent among non-Hispanic Whites, 30 percent among non-Hispanic Blacks, and 15 percent among Hispanics.

In contrast to the continued rise in twin birth rates, the remarkable growth of triplet and higher order multiples of the past two decades has ended at least temporarily. Such births numbered 7,401 in 2002, which represents a rate of 184 per 100,000 live births. The rate in 2002 was 5 percent less than the 1998 high. Refinements and less aggressive techniques in assisted reproductive technologies may be contributing to the current abatement in the frequency of higher order multiple births. The higher the plurality of a pregnancy, the greater the likelihood of a shorter gestation period, of very low birthweight, and of other adverse conditions.

Table I-50. Number and Rate of Live Births by Plurality of Birth, Age, Race, and Hispanic Origin of Mother, 2002

Plurality and race and Hispanic origin of mother	All ages	Age of mother									
		Under 15 years	15–19 years			20–24 years	25–29 years	30–34 years	35–39 years	40–44 years	45–54 years
			Total	15–17 years	18–19 years						
NUMBER											
All Live Births											
All races [1]	4 021 726	7 315	425 493	138 731	286 762	1 022 106	1 060 391	951 219	453 927	95 788	5 487
White total [2]	3 174 760	3 884	305 988	95 864	210 124	783 000	851 142	779 535	369 833	76 928	4 450
Non-Hispanic White	2 298 156	1 493	179 511	49 756	129 755	519 153	614 909	620 173	297 436	61 853	3 628
Black total [2]	593 691	3 188	103 795	37 889	65 906	194 704	136 591	95 006	48 388	11 443	576
Non-Hispanic Black	578 335	3 132	101 494	37 017	64 477	190 241	132 824	92 155	46 831	11 097	561
Hispanic [3]	876 642	2 421	127 900	46 740	81 160	265 235	236 143	157 887	71 480	14 809	767
Live Births in Single Deliveries											
All races [1]	3 889 191	7 226	418 654	136 793	281 861	998 739	1 028 038	911 431	430 447	90 378	4 278
White total [2]	3 069 915	3 833	301 654	94 666	206 988	766 890	825 777	746 091	350 039	72 238	3 393
Non-Hispanic White	2 212 453	1 473	176 872	49 119	127 753	507 680	594 703	591 239	280 164	57 649	2 673
Black total [2]	572 659	3 151	101 510	37 210	64 300	188 283	131 197	90 889	46 082	11 022	525
Non-Hispanic Black	557 680	3 095	99 254	36 352	62 902	183 895	127 535	88 122	44 584	10 685	510
Hispanic [3]	857 777	2 390	126 167	46 161	80 006	260 592	230 978	153 490	69 092	14 377	691
Live Births in Twin Deliveries											
All races [1]	125 134	89	6 746	1 921	4 825	22 856	30 725	36 956	21 637	5 033	1 092
White total [2]	98 304	51	4 277	1 187	3 090	15 717	23 942	30 849	18 172	4 350	946
Non-Hispanic White	79 949	20	2 605	631	1 974	11 183	18 962	26 581	15 846	3 889	863
Black total [2]	20 423	37	2 255	673	1 582	6 315	5 234	3 949	2 181	401	51
Non-Hispanic Black	20 064	37	2 210	659	1 551	6 240	5 135	3 871	2 128	392	51
Hispanic [3]	18 128	31	1 710	574	1 136	4 549	4 980	4 176	2 208	413	61
Live Births in Higher Order Multiple Deliveries [4]											
All races [1]	7 401	-	93	17	76	511	1 628	2 832	1 843	377	117
White total [2]	6 541	-	57	11	46	393	1 423	2 595	1 622	340	111
Non-Hispanic White	5 754	-	34	6	28	290	1 244	2 353	1 426	315	92
Black total [2]	609	-	30	6	24	106	160	168	125	20	-
Non-Hispanic Black	591	-	30	6	24	106	154	162	119	20	-
Hispanic [3]	737	-	23	5	18	94	185	221	180	19	15
RATE PER 1,000 LIVE BIRTHS											
All Multiple Births											
All races [1]	33.0	12.2	16.1	14.0	17.1	22.9	30.5	41.8	51.7	56.5	220.3
White total [2]	33.0	13.1	14.2	12.5	14.9	20.6	29.8	42.9	53.5	61.0	237.5
Non-Hispanic White	37.3	13.4	14.7	12.8	15.4	22.1	32.9	46.7	58.1	68.0	263.2
Black total [2]	35.4	11.6	22.0	17.9	24.4	33.0	39.5	43.3	47.7	36.8	88.5
Non-Hispanic Black	35.7	11.8	22.1	18.0	24.4	33.4	39.8	43.8	48.0	37.1	90.9
Hispanic [3]	21.5	12.8	13.5	12.4	14.2	17.5	21.9	27.8	33.4	29.2	99.1
Twin Births											
All races [1]	31.1	12.2	15.9	13.8	16.8	22.4	29.0	38.9	47.7	52.5	199.0
White total [2]	31.0	13.1	14.0	12.4	14.7	20.1	28.1	39.6	49.1	56.5	212.6
Non-Hispanic White	34.8	13.4	14.5	12.7	15.2	21.5	30.8	42.9	53.3	62.9	237.9
Black total [2]	34.4	11.6	21.7	17.8	24.0	32.4	38.3	41.6	45.1	35.0	88.5
Non-Hispanic Black	34.7	11.8	21.8	17.8	24.1	32.8	38.7	42.0	45.4	35.3	90.9
Hispanic [3]	20.7	12.8	13.4	12.3	14.0	17.2	21.1	26.4	30.9	27.9	79.5
RATE PER 100,000 LIVE BIRTHS											
Higher Order Multiple Births [4]											
All races [1]	184.0	*	21.9	*	26.5	50.0	153.5	297.7	406.0	393.6	2 132.3
White total [2]	206.0	*	18.6	*	21.9	50.2	167.2	332.9	438.6	442.0	2 494.4
Non-Hispanic White	250.4	*	18.9	*	21.6	55.9	202.3	379.4	479.4	509.3	2 535.8
Black total [2]	102.6	*	28.9	*	36.4	54.4	117.1	176.8	258.3	174.8	*
Non-Hispanic Black	103.2	*	29.6	*	37.2	55.7	115.9	175.8	254.1	180.2	*
Hispanic [3]	84.1	*	18.0	*	[illegible]	35.4	78.3	140.0	251.8	[illegible]	[illegible]

[1]Includes races other than White and Black and origin not stated.
[2]Race and Hispanic origin are reported separately on the birth certificate. Data for persons of Hispanic origin are included in the data for each race group according to the mother's reported race.
[3]Includes all persons of Hispanic origin of any race.
[4]Births in greater than twin deliveries.
* = Figure does not meet standards of reliability or precision; based on fewer than 20 births in numerator.
- = Quantity zero.

Figure I-17. Gestational Age and Birthweight Characteristics, by Plurality, 2002

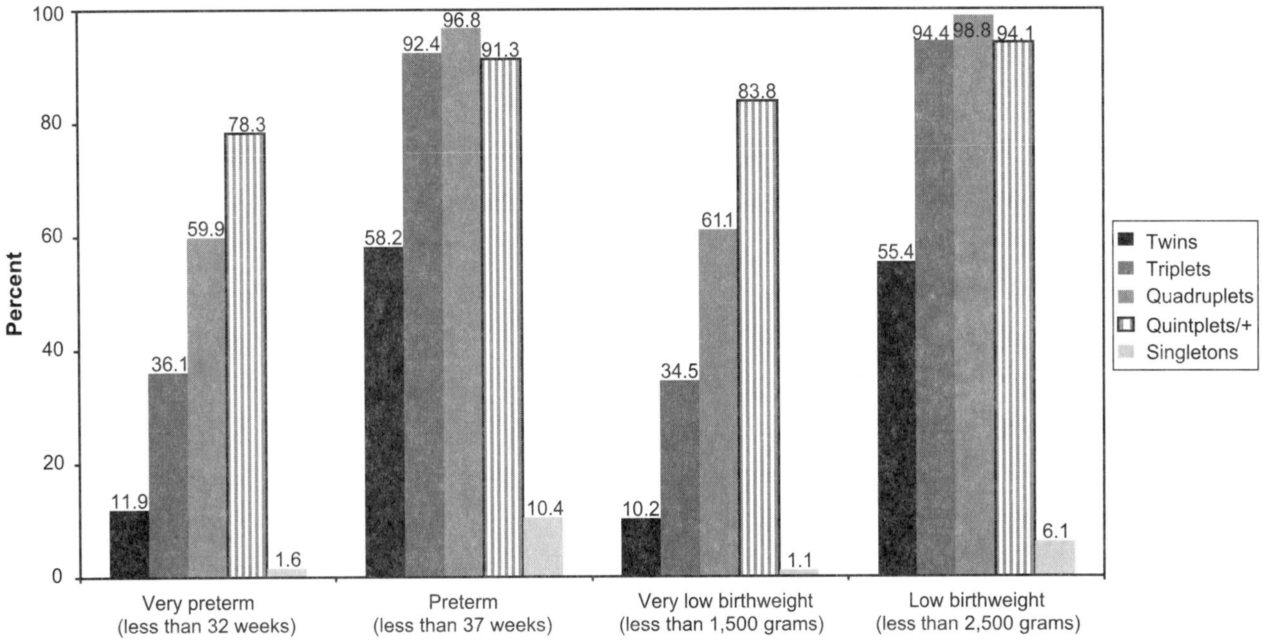

TABLE I-50A. GESTATIONAL AGE AND BIRTHWEIGHT CHARACTERISTICS, BY PLURALITY, 2002

The higher the plurality of a pregnancy, the greater the risk of poor perinatal outcome. The average twin in 2002 was delivered more than three weeks earlier than the average singleton (35.3 compared with 38.8 weeks) and weighed about 2 pounds less. The typical triplet weighed about half of its singleton counterpart (4 pounds) at birth; the average quintuplet less than one-third (only 3 pounds). As a result, twins are nearly five times as likely to die by their first birthday; for triplets or higher plurality the risk was 12 times higher.

Table I-50A. Gestational Age and Birthweight Characteristics, by Plurality, 2002

Characteristic	Twins	Triplets	Quadruplets	Quintuplets/+	Singletons
Number	125 134	6 898	434	69	3 889 191
Percent very preterm [1]	11.9	36.1	59.9	78.3	1.6
Percent preterm [2]	58.2	92.4	96.8	91.3	10.4
Mean gestational age (standard deviation) in weeks	35.3 (3.7)	32.2 (3.8)	29.9 (4.0)	28.5 (4.7)	38.8 (2.5)
Percent very low birthweight [3]	10.2	34.5	61.1	83.8	1.1
Percent low birthweight [4]	55.4	94.4	98.8	94.1	6.1
Mean birthweight (standard deviation) in grams	2 347 (645)	1 687 (561)	1 309 (522)	1 105 (777)	3 332 (573)

[1]Very preterm is less than 32 completed weeks of gestation.
[2]Preterm is less than 37 completed weeks of gestation.
[3]Very low birthweight is less than 1,500 grams.
[4]Low birthweight is less than 2,500 grams.

CHAPTER II: MORTALITY

This chapter republishes Final 2001 and Preliminary 2002 death data for the United States as originally derived from death certificates by the National Center for Health Statistics. The tables start by showing the number of deaths and death rates by sex, age, race, and Hispanic origin. Subsequent tables show life expectancy and its derivation. A major part of the chapter classifies deaths according to 113 selected causes with detail by sex, age, race and Hispanic origin. Death by certain mechanisms, such as injury by firearms and alcohol-induced causes, are shown separately. Statistics on infant mortality are presented toward the end of the chapter.

TABLE II-1. NUMBER OF DEATHS, DEATH RATES, AND AGE-ADJUSTED DEATH RATES, BY RACE AND SEX, SELECTED YEARS, 1940–2001

In 2001, a total of 2,416,425 resident deaths were registered in the United States. This compares with 4.026 million births. The number of deaths, of course, has increased considerably over the years as the population grew. The crude death rate, per 100,000 of population, was 848.5 in 2001. This death rate reflects the inherent health of the population as well as its age distribution. The younger the age, the lower the crude death rate will tend to be.

An age-adjusted death rate is also shown in the table; it attempts to eliminate the effects of changes in the age distribution. Age-adjusted rates are calculated to show what mortality would be if the age distribution of the population did not change from year to year and was the same for the various subgroups. Age-adjusted rates are constructed by measuring the age-specific death rates in a given population, but substituting a standard age distribution for the actual observed age distribution. The standard age distribution that is currently used is that of the total U.S. population in the year 2000. Age-adjusted rates are better suited than crude rates for examining trends over time or comparing relative risk across geographic areas or subgroups for gender and race.

For populations with a shorter than average life span, for example Blacks as well as White males, age adjustment tends to increase the death rate, while conversely it decreases the rate of White females who tend to live longer. Asians and Pacific Islanders have surprisingly low death rates, both crude and adjusted. Although there is some underreporting of their race on their death certificates (and also for American Indians), this deficiency is considered insufficient to explain their low levels of death rates. (The National Center for Health Statistics estimates a 21 percent underestimate for American Indians and 11 percent for Asians and Pacific Islanders.)

In 2001, the age-adjusted death rate for the Black population was 1.3 times that for Whites. That is, the average risk of death for Blacks was about 33 percent higher than that for Whites. For the total population, the age-adjusted death rate in the United States declined 17.8 percent between 1980 and 2001.

Figure II-1. Age-Adjusted Death Rates by Race, 1940–2001

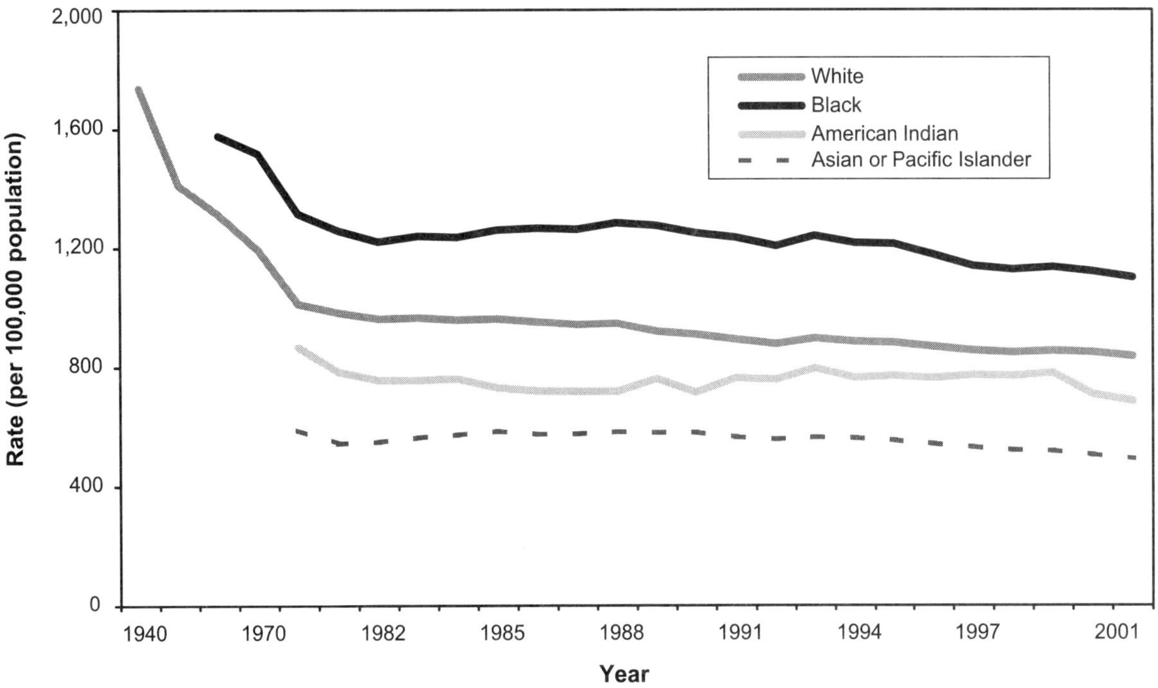

Table II-1. Number of Deaths, Death Rates, and Age-Adjusted Death Rates, by Race and Sex, Selected Years, 1940–2001

(Number, rate per 100,000 population.)

Year	All races [1]			White			Black			American Indian [2]			Asian or Pacific Islander [3]		
	Both sexes	Male	Female	Both sexes	Male	Female	Both sexes	Male	Female	Both sexes	Male	Female	Both sexes	Male	Female
Number of Deaths															
2001	2 416 425	1 183 421	1 233 004	2 079 691	1 011 218	1 068 473	287 709	145 908	141 801	11 977	6 466	5 511	37 048	19 829	17 219
2000	2 403 351	1 177 578	1 225 773	2 071 287	1 007 191	1 064 096	285 826	145 184	140 642	11 363	6 185	5 178	34 875	19 018	15 857
1999	2 391 399	1 175 460	1 215 939	2 061 348	1 005 335	1 056 013	285 064	145 703	139 361	11 312	6 092	5 220	33 675	18 330	15 345
1998	2 337 256	1 157 260	1 179 996	2 015 984	990 190	1 025 794	278 440	143 417	135 023	10 845	5 994	4 851	31 987	17 659	14 328
1997	2 314 245	1 154 039	1 160 206	1 996 393	986 884	1 009 509	276 520	144 110	132 410	10 576	5 985	4 591	30 756	17 060	13 696
1996	2 314 690	1 163 569	1 151 121	1 992 966	991 984	1 000 982	282 089	149 472	132 617	10 127	5 563	4 564	29 508	16 550	12 958
1995	2 312 132	1 172 959	1 139 173	1 987 437	997 277	990 160	286 401	154 175	132 226	9 997	5 574	4 423	28 297	15 933	12 364
1994	2 278 994	1 162 747	1 116 247	1 959 875	988 823	971 052	282 379	153 019	129 360	9 637	5 497	4 140	27 103	15 408	11 695
1993	2 268 553	1 161 797	1 106 756	1 951 437	988 329	963 108	282 151	153 502	128 649	9 579	5 434	4 145	25 386	14 532	10 854
1992	2 175 613	1 122 336	1 053 277	1 873 781	956 957	916 824	269 219	146 630	122 589	8 953	5 181	3 772	23 660	13 568	10 092
1991	2 169 518	1 121 665	1 047 853	1 868 904	956 497	912 407	269 525	147 331	122 194	8 621	4 948	3 673	22 173	12 727	9 446
1990	2 148 463	1 113 417	1 035 046	1 853 254	950 812	902 442	265 498	145 359	120 139	8 316	4 877	3 439	21 127	12 211	8 916
1989	2 150 466	1 114 190	1 036 276	1 853 841	950 852	902 989	267 642	146 393	121 249	8 614	5 066	3 548	20 042	11 688	8 354
1988	2 167 999	1 125 540	1 042 459	1 876 906	965 419	911 487	264 019	144 228	119 791	7 917	4 617	3 300	18 963	11 155	7 808
1987	2 123 323	1 107 958	1 015 365	1 843 067	953 382	889 685	254 814	139 551	115 263	7 602	4 432	3 170	17 689	10 496	7 193
1986	2 105 361	1 104 005	1 001 356	1 831 083	952 554	878 529	250 326	137 214	113 112	7 301	4 365	2 936	16 514	9 795	6 719
1985	2 086 440	1 097 758	988 682	1 819 054	950 455	868 599	244 207	133 610	110 597	7 154	4 181	2 973	15 887	9 441	6 446
1984	2 039 369	1 076 514	962 855	1 781 897	934 529	847 368	235 884	129 147	106 737	6 949	4 117	2 832	14 483	8 627	5 856
1983	2 019 201	1 071 923	947 278	1 765 582	931 779	833 803	233 124	127 911	105 213	6 839	4 064	2 775	13 554	8 126	5 428
1982	1 974 797	1 056 440	918 357	1 729 085	919 239	809 846	226 513	125 610	100 903	6 679	3 974	2 705	12 430	7 564	4 866
1981	1 977 981	1 063 772	914 209	1 731 233	925 490	805 743	228 560	127 296	101 264	6 608	4 016	2 592	11 475	6 908	4 567
1980	1 989 841	1 075 078	914 763	1 738 607	933 878	804 729	233 135	130 138	102 997	6 923	4 193	2 730	11 071	6 809	4 262
1970	1 921 031	1 078 478	842 553	1 682 096	942 437	739 659	225 647	127 540	98 107	5 675	3 391	. . .	. . .	. . .	. . .
1960	1 711 982	975 648	736 334	1 505 335	860 857	644 478	196 010	107 701	88 309	4 528	2 658	. . .	. . .	. . .	. . .
1950	1 452 454	827 749	624 705	1 276 085	731 366	544 719	169 606	92 004	77 602	4 440	2 497	. . .	. . .	. . .	. . .
1940	1 417 269	791 003	626 266	1 231 223	690 901	540 322	178 743	95 517	83 226	4 791	2 527	. . .	. . .	. . .	. . .
Death Rate															
2001	848.5	846.4	850.4	895.1	881.9	907.9	773.5	823.9	727.7	392.1	424.2	360.2	303.8	335.0	274.0
2000	854.0	853.0	855.0	900.2	887.8	912.3	781.1	834.1	733.0	380.8	415.6	346.1	296.6	332.9	262.3
1999	857.0	859.2	854.9	901.4	892.1	910.4	788.1	847.4	734.3	399.3	431.8	367.1	296.8	333.2	262.5
1998	847.3	856.4	838.5	889.5	887.3	891.6	782.3	848.2	722.6	397.8	441.9	354.2	293.8	335.4	254.9
1997	848.8	864.6	833.6	889.1	893.3	885.0	789.9	867.1	720.1	402.7	458.2	347.7	294.1	336.8	253.9
1996	859.2	882.8	836.7	896.0	907.1	885.3	819.7	915.3	733.3	399.5	441.5	358.0	294.4	340.2	251.1
1995	868.3	900.8	837.2	901.8	921.0	883.2	846.2	960.2	743.2	409.4	459.4	360.1	294.6	341.4	250.4
1994	866.1	904.2	829.7	897.8	922.6	873.8	849.0	970.2	739.7	408.2	468.8	348.3	294.6	344.0	247.7
1993	872.8	915.0	832.5	902.7	931.8	874.6	864.6	992.2	749.6	419.8	479.6	360.7	288.0	338.1	240.3
1992	848.1	896.1	802.4	875.8	912.2	840.8	841.8	967.6	728.6	406.6	474.1	340.0	282.1	331.1	235.3
1991	857.6	908.8	808.7	883.2	922.7	845.2	861.4	994.8	741.4	405.3	468.9	342.7	278.7	326.9	232.4
1990	863.8	918.4	812.0	888.0	930.9	846.9	871.0	1 008.0	747.9	402.8	476.4	330.4	283.3	334.3	234.3
1989	871.3	926.3	818.9	893.2	936.5	851.8	887.9	1 026.7	763.2	430.5	510.7	351.3	280.9	334.5	229.4
1988	886.7	945.1	831.2	910.5	957.9	865.3	888.3	1 026.1	764.6	411.7	485.0	339.9	282.0	339.0	227.4
1987	876.4	939.3	816.7	900.1	952.7	849.8	868.9	1 006.2	745.7	410.7	483.8	339.0	278.9	338.3	222.0
1986	876.7	944.7	812.3	900.1	958.6	844.3	864.9	1 002.6	741.5	409.5	494.9	325.9	276.2	335.1	219.9
1985	876.9	948.6	809.1	900.4	963.6	840.1	854.8	989.3	734.2	416.4	492.5	342.5	283.4	344.6	224.9
1984	864.8	938.8	794.7	887.8	954.1	824.6	836.1	968.5	717.4	419.6	502.7	338.4	275.9	336.5	218.1
1983	863.7	943.2	788.4	885.4	957.7	816.4	836.6	971.2	715.9	428.5	515.1	343.9	276.1	339.1	216.1
1982	852.4	938.4	771.2	873.1	951.8	798.2	823.4	966.2	695.5	434.5	522.9	348.1	271.3	338.3	207.4
1981	862.0	954.0	775.0	880.4	965.2	799.8	842.4	992.6	707.7	445.6	547.9	345.6	272.3	336.2	211.5
1980	878.3	976.9	785.3	892.5	983.3	806.1	875.4	1 034.1	733.3	487.4	597.1	380.1	296.9	375.3	222.5
1970	945.3	1 090.3	807.8	946.3	1 086.7	812.6	999.3	1 186.6	. . .	. . .	. . .	. . .	. . .	. . .	. . .
1960	954.7	1 104.5	809.2	947.8	1 098.5	800.9	1 038.6	1 181.7	. . .	. . .	. . .	. . .	. . .	. . .	. . .
1950	963.8	1 106.1	823.5	945.7	1 089.5	. . .	. . .	. . .	. . .	. . .	. . .	. . .	. . .	. . .	. . .
1940	1 076.4	1 197.4	954.6	1 041.5	1 162.2	. . .	. . .	. . .	. . .	. . .	. . .	. . .	. . .	. . .	. . .
Age-Adjusted Death Rate															
2001	854.5	1 029.1	721.8	836.5	1 006.1	706.7	1 101.2	1 375.0	912.5	686.7	798.9	594.0	492.1	597.4	412.0
2000	869.0	1 053.8	731.4	849.8	1 029.4	715.3	1 121.4	1 403.5	927.6	709.3	841.5	604.5	506.4	624.2	416.8
1999	875.6	1 067.0	734.0	854.6	1 040.0	716.6	1 135.7	1 432.6	933.6	780.9	925.9	668.2	519.7	641.2	427.5
1998	870.6	1 069.4	724.7	849.3	1 042.0	707.3	1 127.8	1 430.5	921.6	770.4	943.9	640.5	522.4	646.9	426.7
1997	878.1	1 088.1	725.6	855.7	1 059.1	707.8	1 139.8	1 458.8	922.1	774.0	974.8	625.3	531.8	660.2	432.6
1996	894.1	1 115.7	733.0	869.0	1 082.9	713.6	1 178.4	1 524.2	940.3	763.6	924.8	641.7	543.2	676.1	439.6
1995	909.8	1 143.9	739.4	882.3	1 107.5	718.7	1 213.9	1 585.7	955.9	771.2	932.0	643.9	554.8	693.4	446.7
1994	913.5	1 155.5	738.6	885.6	1 118.7	717.5	1 216.9	1 592.8	954.6	764.8	953.3	618.8	562.7	702.5	452.1
1993	926.1	1 177.3	745.9	897.0	1 138.9	724.1	1 241.2	1 632.2	969.5	796.4	1 006.3	641.6	565.8	709.9	450.4
1992	905.6	1 158.3	725.5	877.7	1 122.4	704.1	1 206.7	1 587.8	942.5	759.0	970.4	599.4	558.5	697.3	445.8
1991	922.3	1 180.5	738.2	893.2	1 143.1	716.1	1 235.4	1 626.1	963.3	763.9	970.6	608.3	566.2	703.4	453.2
1990	938.7	1 202.8	750.9	909.8	1 165.9	728.8	1 250.3	1 644.5	975.1	716.3	916.2	561.8	582.0	716.4	469.3
1989	950.5	1 215.0	761.8	920.2	1 176.6	738.8	1 275.5	1 670.1	998.1	761.6	999.8	586.3	581.3	729.6	458.4
1988	975.7	1 250.7	781.0	947.6	1 215.9	759.1	1 284.3	1 677.6	1 006.8	718.6	917.4	563.6	584.2	732.0	451.0
1987	970.0	1 246.1	774.2	943.4	1 213.4	753.3	1 263.1	1 650.3	989.7	719.8	899.3	583.7	577.3	732.4	448.1
1986	978.6	1 261.7	778.7	952.8	1 230.5	758.1	1 266.7	1 650.1	994.4	720.8	926.7	549.3	576.4	730.5	445.4
1985	988.1	1 278.1	784.5	963.6	1 249.8	764.3	1 261.2	1 634.5	994.4	731.7	926.1	577.2	586.5	755.4	456.7
1984	982.5	1 271.4	779.8	959.7	1 245.9	760.7	1 236.7	1 600.8	976.9	761.7	946.0	567.9	574.4	724.7	443.1
1983	990.0	1 284.5	783.3	967.3	1 259.4	763.9	1 240.5	1 600.7	980.7	757.3	945.0	605.5	565.1	718.8	428.8
1982	985.0	1 279.9	776.6	963.6	1 255.9	758.7	1 221.3	1 580.4	960.1	757.0	940.1	604.4	550.4	738.2	410.3
1981	1 007.1	1 308.2	792.7	984.0	1 282.2	773.6	1 258.4	1 626.6	986.6	784.6	1 030.2	588.0	544.7	710.3	405.3
1980	1 039.1	1 348.1	817.9	1 012.7	1 317.6	796.1	1 314.8	1 697.8	1 033.3	867.0	1 111.5	662.4	589.9	786.5	425.9
1970	1 222.6	1 542.1	971.4	1 193.3	1 513.7	944.0	1 518.1	1 873.9	1 228.7	. . .	. . .	. . .	. . .	. . .	. . .
1960	1 339.2	1 609.0	1 105.3	1 311.3	1 586.0	1 074.4	1 577.5	1 811.1	1 369.7	. . .	. . .	. . .	. . .	. . .	. . .
1950	1 446.0	1 674.2	1 236.0	1 410.8	1 642.5	1 198.0	. . .	. . .	. . .	. . .	. . .	. . .	. . .	. . .	. . .
1940	[illegible]	[illegible]	1 500.1	1 735.3	1 925.2	1 550.4		. . .	. . .	. . .	. . .	. . .	. . .	. . .	. . .

[1] For 1940–1991 includes deaths among races not shown separately; beginning in 1992 records coded as "other races" and records for which race was unknown, not stated, or not classifiable were assigned to the race of the previous record.
[2] Includes Aleuts and Eskimos.
[3] Includes Chinese, Filipino, Hawaiian, Japanese, and Other Asian or Pacific Islander.
. . . = Not available.

TABLE II-1A. DEATHS, AGE-ADJUSTED DEATH RATES, AND LIFE EXPECTANCY AT BIRTH, BY RACE AND SEX; AND INFANT AND MORTALITY RATES, BY RACE, 2001 AND PRELIMINARY 2002

The Division of Vital Statistics of the National Center for Health Statistics has issued preliminary mortality data for 2002. These data are based on comprehensive, but not complete information: 96 percent to 97 percent coverage of deaths in the United States was available, while reports on cause of death covered about 92 percent to 93 percent of cases.

Deaths totaled 2.448 million in 2002. The increase of 31,000 deaths since 2001 reflected a larger population, rather than a change in the crude death rate. The age-adjusted death rate, which calculates the risk of death on the basis of an unvarying age structure, reached a record low of 846.8 per 100,000, that is, 0.9 percent lower than in 2001. For White males and females, the age-adjusted rate declined 1.1 percent and 0.5 percent, respectively. The corresponding rates for Blacks declined 2.7 percent and 1.3 percent, respectively. But even with this larger decline, the risk of death among Blacks remained 30 percent higher than among Whites.

Life expectancy at birth reached a high of 77.4 years, an increase of 0.2 years relative to 2001. Male life expectancy was 74.7 years and female life expectancy reached 79.9 years. The gap between male and female life expectancy thus was 5.2 years, as compared with 5.4 years in 2001, and 7.8 years at its peak in 1979.

The infant mortality rate—deaths under one year per 1,000 live births—increased from 6.8 to 7.0 in 2002, the first increase of this rate in more than four decades. The neonatal period, the first 27 days of life, accounted for this increase, and it was particularly concentrated in the first six days after birth. Another puzzle was presented by a roughly offsetting decline in late fetal deaths (death of the fetus after 27 weeks of gestation). In general, neonatal deaths have been associated with short gestation periods and low birth weight (often occurring at multiple births). While the frequency of both of these factors increased somewhat further in 2002, they also had been increasing in previous years even when the infant mortality rate was moving lower (see Table I-44 and Figures I-13 and I-14). These trends may be associated with increased use of assisted reproductive therapies, such as in-vitro fertilization, as well as increased use of induction of labor and cesarean delivery, which occur more often at earlier gestational ages.

Table II-1A. Deaths, Age-Adjusted Death Rates, and Life Expectancy at Birth, by Race and Sex; and Infant and Mortality Rates, by Race, 2001 and Preliminary 2002

(Number, rate, years.)

Measure and sex	All races [1]		White		Black	
	2002	2001	2002	2001	2002	2001
All deaths	2 447 864	2 416 425	2 107 730	2 079 691	289 431	287 709
Age-adjusted death rate [2]	846.8	854.5	831.0	836.5	1 081.2	1 101.2
Male	1 015.3	1 029.1	995.1	1 006.1	1 338.4	1 375.0
Female	716.7	721.8	703.2	706.7	900.4	912.5
Life expectancy at birth [3]	77.4	77.2	77.8	77.7	72.5	72.2
Male	74.7	74.4	75.3	75.0	68.9	68.6
Female	79.9	79.8	80.3	80.2	75.7	75.5
All infant deaths	27 977	27 568	18 390	17 955	8 446	8 498
Infant mortality rate [4]	7.0	6.8	5.8	5.7	14.3	14.0

[1] Includes races other than White and Black.
[2] Age-adjusted death rates are per 100,000 U.S. standard population, based on the year 2000 standard.
[3] Life expectancy at birth stated in years.
[4] Infant mortality rates are deaths under 1 year per 1,000 live births in specified group.

TABLE II-1B. DEATHS, DEATH RATES, AND AGE-ADJUSTED DEATH RATES, BY AGE, SEX, RACE AND HISPANIC ORIGIN, 2001 AND PRELIMINARY 2002

This table compares the number and the rate of deaths in 2002 with those of 2001. The data are shown by age group, sex, race, and Hispanic origin. Overall, the age-adjusted death rate declined by 0.9 percent in 2002, and no age group had statistically significant increases in this year.

As was the case in previous years, surprisingly low death rates were reported for Asians and Pacific Islanders, Hispanics, and American Indians. Their preliminary age-adjusted rates were shown, respectively, as 44 percent, 25 percent, and 19 percent lower than those for the non-Hispanic White population. Mortality for races other than White and Black may be seriously understated due to underreporting for some race groups and Hispanic origin on death certificates.

Table II-1B. Deaths, Death Rates, and Age-Adjusted Death Rates, by Age, Sex, Race, and Hispanic Origin, 2001 and Preliminary 2002

(Number, rate per 100,000 population.)

Age, race, and sex	2002		2001	
	Number	Rate	Number	Rate
ALL RACES, BOTH SEXES				
All Ages	2 447 864	848.9	2 416 425	848.5
Under 1 year [1]	27 977	693.6	27 568	683.4
1–4 years	4 862	31.2	5 107	33.3
5–14 years	7 169	17.5	7 095	17.3
15–24 years	33 075	81.5	32 252	80.7
25–34 years	41 375	103.6	41 683	105.2
35–44 years	91 005	202.6	91 674	203.6
45–54 years	172 213	429.6	168 065	428.9
55–64 years	253 728	953.8	244 139	964.6
65–74 years	424 107	2 320.8	430 960	2 353.3
75–84 years	709 351	5 570.3	701 929	5 582.4
85 years and over	682 634	14 862.3	665 531	15 112.8
Not stated	367	X	422	X
Age-adjusted rate	X	846.8	X	854.5
ALL RACES, MALE				
All Ages	1 201 010	847.8	1 183 421	846.4
Under 1 year [1]	15 659	758.7	15 477	749.8
1–4 years	2 810	35.3	2 899	37.0
5–14 years	4 208	20.0	4 168	19.8
15–24 years	24 431	117.3	23 963	117.0
25–34 years	28 752	142.3	28 757	143.7
35–44 years	57 441	256.8	58 164	259.6
45–54 years	107 584	546.8	104 848	545.1
55–64 years	151 577	1 185.6	144 958	1 192.7
65–74 years	237 588	2 862.2	241 581	2 911.5
75–84 years	344 205	6 774.3	340 742	6 833.0
85 years and over	226 473	16 295.3	217 533	16 744.8
Not stated	283	X	331	X
Age-adjusted rate	X	1 015.3	X	1 029.1
ALL RACES, FEMALE				
All Ages	1 246 854	849.9	1 233 004	850.4
Under 1 year [1]	12 318	625.3	12 091	613.9
1–4 years	2 052	27.0	2 208	29.5
5–14 years	2 960	14.8	2 927	14.6
15–24 years	8 645	43.7	8 289	42.6
25–34 years	12 622	64.0	12 926	66.0
35–44 years	33 564	148.8	33 510	148.2
45–54 years	64 630	316.7	63 217	316.8
55–64 years	102 152	739.3	99 181	754.0
65–74 years	186 520	1 870.2	189 379	1 890.8
75–84 years	365 146	4 770.9	361 187	4 760.5
85 years and over	456 161	14 240.5	447 998	14 429.9
Not stated	84	X	91	X
Age-adjusted rate	X	716.7	X	721.8
WHITE, BOTH SEXES				
All Ages	2 107 730	897.9	2 079 691	895.1
Under 1 year [1]	18 390	587.4	17 955	570.9
1–4 years	3 412	28.1	3 670	30.7
5–14 years	5 150	16.2	5 170	16.2
15–24 years	24 677	77.2	23 748	75.5
25–34 years	29 618	93.6	29 994	95.3
35–44 years	68 036	186.5	68 137	185.6
45–54 years	131 756	395.1	128 699	393.2
55–64 years	205 843	904.4	197 853	913.5
65–74 years	361 277	2 275.3	367 128	2 298.9
75–84 years	634 144	5 559.9	627 752	5 560.9
85 years and over	625 129	15 056.2	609 265	15 287.5
Not stated	298	X	320	X
Age-adjusted rate	X	831.0	X	836.5
WHITE, MALE				
All Ages	1 027 391	885.9	1 011 218	881.9
Under 1 year [1]	10 444	651.6	10 099	627.6
1–4 years	1 961	31.6	2 094	34.2
5–14 years	3 010	18.4	3 014	18.4
15–24 years	18 106	109.9	17 529	108.0
25–34 years	20 842	128.5	20 978	130.3
35–44 years	43 868	238.8	44 243	239.7
45–54 years	83 597	505.0	81 403	501.3
55–64 years	123 771	1 120.6	118 353	1 127.5
65–74 years	204 390	2 804.4	207 793	2 842.3
75–84 years	309 444	6 756.0	306 278	6 799.7
85 years and over	207 725	16 519.4	199 179	16 935.4
Not stated	235	X	255	X
Age-adjusted rate	X	995.1	X	1 006.1

[1]Death rates for "Under 1 year" (based on population estimates) differ from infant mortality rates (based on live births).
X = Not applicable.

Table II-1B. Deaths, Death Rates, and Age-Adjusted Death Rates, by Age, Sex, Race, and Hispanic Origin, 2001 and Preliminary 2002—*Continued*

(Number, rate per 100,000 population.)

Age, race, and sex	2002		2001	
	Number	Rate	Number	Rate
WHITE, FEMALE				
All Ages	1 080 339	909.5	1 068 473	907.9
Under 1 year [1]	7 946	520.1	7 856	511.5
1–4 years	1 451	24.5	1 576	27.1
5–14 years	2 140	13.8	2 156	13.9
15–24 years	6 571	42.5	6 219	40.8
25–34 years	8 777	56.9	9 016	58.6
35–44 years	24 168	133.4	23 894	131.0
45–54 years	48 159	286.8	47 296	286.8
55–64 years	82 072	700.5	79 500	712.2
65–74 years	156 887	1 826.4	159 335	1 840.2
75–84 years	324 700	4 757.2	321 474	4 738.5
85 years and over	417 405	14 420.6	410 086	14 597.6
Not stated	63	X	65	X
Age-adjusted rate	X	703.2	X	706.7
NON-HISPANIC WHITE, BOTH SEXES				
All Ages	1 987 280	1 000.2	1 962 810	991.1
Under 1 year [1]	13 492	577.1	13 311	555.4
1–4 years	2 518	27.3	2 772	30.1
5–14 years	4 050	16.1	4 151	16.3
15–24 years	19 521	76.1	18 730	74.1
25–34 years	23 817	96.1	24 191	96.4
35–44 years	59 520	191.6	59 546	188.3
45–54 years	119 850	400.3	117 086	396.0
55–64 years	191 321	915.1	184 109	924.0
65–74 years	340 084	2 303.7	346 389	2 325.5
75–84 years	608 243	5 612.7	603 143	5 604.4
85 years and over	604 739	15 191.2	589 269	15 371.2
Not stated	126	X	113	X
Age-adjusted rate	X	839.8	X	842.9
NON-HISPANIC WHITE, MALE				
All Ages	959 958	986.3	945 967	975.6
Under 1 year [1]	7 722	644.8	7 508	611.6
1–4 years	1 437	30.4	1 592	33.7
5–14 years	2 375	18.4	2 428	18.6
15–24 years	14 009	107.1	13 542	105.1
25–34 years	16 397	131.4	16 613	131.5
35–44 years	38 018	244.7	38 213	241.6
45–54 years	75 751	510.1	73 731	502.6
55–64 years	114 758	1 128.6	109 993	1 136.3
65–74 years	192 421	2 832.6	196 158	2 869.4
75–84 years	296 544	6 819.7	294 013	6 851.5
85 years and over	200 424	16 691.0	192 084	17 055.3
Not stated	102	X	92	X
Age-adjusted rate	X	1 004.7	X	1 012.8
NON-HISPANIC WHITE, FEMALE				
All Ages	1 027 322	1 013.5	1 016 843	1 006.1
Under 1 year [1]	5 770	506.0	5 803	496.4
1–4 years	1 081	24.0	1 180	26.3
5–14 years	1 675	13.7	1 723	13.9
15–24 years	5 512	43.9	5 188	41.9
25–34 years	7 420	60.3	7 578	60.9
35–44 years	21 502	138.4	21 333	134.9
45–54 years	44 099	292.2	43 355	291.0
55–64 years	76 563	712.9	74 116	723.5
65–74 years	147 663	1 852.8	150 231	1 864.1
75–84 years	311 699	4 803.8	309 130	4 777.3
85 years and over	404 315	14 543.3	397 185	14 670.6
Not stated	24	X	21	X
Age-adjusted rate	X	712.0	X	713.5
BLACK [2], BOTH SEXES				
All Ages	289 431	766.8	287 709	773.5
Under 1 year [1]	8 446	1 252.0	8 498	1 304.5
1–4 years	1 192	46.9	1 181	47.5
5–14 years	1 678	24.7	1 587	23.3
15–24 years	7 040	113.6	7 147	117.7
25–34 years	10 144	186.3	10 028	185.0
35–44 years	20 190	347.8	20 687	356.5
45–54 years	35 516	763.5	34 518	771.9
55–64 years	41 264	1 562.5	39 976	1 576.8
65–74 years	53 213	3 153.3	54 379	3 257.6
75–84 years	62 785	6 510.9	62 233	6 585.6
85 years and over	47 911	14 227.8	47 387	14 550.8
Not stated	53	X	88	X
Age-adjusted rate	X	1 081.2	X	1 101.2

[1] Death rates for "Under 1 year" (based on population estimates) differ from infant mortality rates (based on live births).
[2] Race and Hispanic origin are reported separately on the death certificate. Data for persons of Hispanic origin are included in the data for each race group according to the decedent's reported race.
X = Not applicable.

Table II-1B. Deaths, Death Rates, and Age-Adjusted Death Rates, by Age, Sex, Race, and Hispanic Origin, 2001 and Preliminary 2002—*Continued*

(Number, rate per 100,000 population.)

Age, race, and sex	2002		2001	
	Number	Rate	Number	Rate
BLACK [2], MALE				
All Ages	146 447	814.6	145 908	823.9
Under 1 year [1]	4 595	1 334.9	4 764	1 426.4
1–4 years	701	54.3	672	53.2
5–14 years	1 006	29.1	960	27.7
15–24 years	5 361	172.5	5 479	180.7
25–34 years	6 818	263.3	6 673	259.2
35–44 years	11 804	433.1	12 135	445.0
45–54 years	21 055	979.9	20 594	996.1
55–64 years	23 969	2 036.6	22 905	2 025.3
65–74 years	28 106	4 011.4	28 775	4 166.6
75–84 years	28 426	8 154.7	28 395	8 355.9
85 years and over	14 572	15 622.6	14 489	16 439.9
Not stated	35	X	67	X
Age-adjusted rate	X	1 338.4	X	1 375.0
BLACK [2], FEMALE				
All Ages	142 984	723.3	141 801	727.7
Under 1 year [1]	3 851	1 165.7	3 734	1 176.3
1–4 years	491	39.3	509	41.7
5–14 years	672	20.1	627	18.7
15–24 years	1 679	54.3	1 668	54.9
25–34 years	3 326	116.5	3 355	117.9
35–44 years	8 386	272.3	8 552	278.0
45–54 years	14 461	577.8	13 924	579.1
55–64 years	17 295	1 181.4	17 071	1 215.5
65–74 years	25 107	2 544.1	25 604	2 616.1
75–84 years	34 359	5 580.3	33 838	5 591.6
85 years and over	33 339	13 693.5	32 898	13 849.9
Not stated	18	X	21	X
Age-adjusted rate	X	900.4	X	912.5
NON-HISPANIC BLACK, BOTH SEXES				
All Ages	285 941	791.1	284 343	798.1
Under 1 year [1]	8 213	1 269.3	8 277	1 354.4
1–4 years	1 155	48.1	1 156	49.5
5–14 years	1 648	25.5	1 557	24.0
15–24 years	6 899	116.8	7 024	121.4
25–34 years	9 982	193.4	9 859	191.5
35–44 years	19 905	356.9	20 436	365.8
45–54 years	35 096	779.4	34 098	786.9
55–64 years	40 808	1 590.2	39 523	1 602.5
65–74 years	52 646	3 198.6	53 803	3 300.7
75–84 years	62 099	6 575.9	61 592	6 645.3
85 years and over	47 452	14 352.5	46 968	14 660.8
Not stated	39	X	50	X
Age-adjusted rate	X	1 096.9	X	1 116.5
NON-HISPANIC BLACK, MALE				
All Ages	144 426	840.1	143 971	849.7
Under 1 year [1]	4 468	1 353.6	4 651	1 484.2
1–4 years	678	55.6	655	55.1
5–14 years	986	30.1	941	28.6
15–24 years	5 248	177.4	5 377	186.2
25–34 years	6 696	273.2	6 548	267.9
35–44 years	11 613	443.9	11 971	456.3
45–54 years	20 768	999.5	20 307	1 014.6
55–64 years	23 680	2 072.2	22 600	2 055.4
65–74 years	27 770	4 065.8	28 446	4 218.9
75–84 years	28 084	8 230.7	28 083	8 426.8
85 years and over	14 408	15 761.6	14 352	16 576.2
Not stated	27	X	40	X
Age-adjusted rate	X	1 357.5	X	1 393.7
NON-HISPANIC BLACK, FEMALE				
All Ages	141 515	746.6	140 372	751.2
Under 1 year [1]	3 745	1 181.6	3 626	1 217.8
1–4 years	477	40.4	501	43.6
5–14 years	662	20.8	616	19.3
15–24 years	1 651	56.0	1 647	56.8
25–34 years	3 286	121.3	3 311	122.4
35–44 years	8 292	280.1	8 465	285.7
45–54 years	14 328	590.9	13 791	591.5
55–64 years	17 128	1 203.2	16 923	1 238.2
65–74 years	24 876	2 583.4	25 357	2 653.0
75–84 years	34 014	5 639.6	33 509	5 645.1
85 years and over	33 045	13 814.5	32 616	13 951.4
Not stated	11	X	10	X
Age-adjusted rate	X	913.7	X	925.5

[1]Death rates for "Under 1 year" (based on population estimates) differ from infant mortality rates (based on live births).
[2]Race and Hispanic origin are reported separately on the death certificate. Data for persons of Hispanic origin are included in the data for each race group according to the decedent's reported race.
X = Not applicable.

Table II-1B. Deaths, Death Rates, and Age-Adjusted Death Rates, by Age, Sex, Race, and Hispanic Origin, 2001 and Preliminary 2002—*Continued*

(Number, rate per 100,000 population.)

Age, race, and sex	2002		2001	
	Number	Rate	Number	Rate
AMERICAN INDIAN [2, 3], BOTH SEXES				
All Ages	12 403	403.2	11 977	392.1
Under 1 year [1]	338	810.1	373	650.4
1–4 years	92	46.2	103	48.0
5–14 years	126	21.6	129	22.0
15–24 years	582	104.4	557	102.5
25–34 years	650	141.4	596	132.0
35–44 years	1 154	245.3	1 132	241.2
45–54 years	1 596	427.3	1 559	432.8
55–64 years	1 886	898.0	1 768	896.2
65–74 years	2 298	2 082.5	2 206	2 072.3
75–84 years	2 219	4 117.5	2 146	4 218.3
85 years and over	1 455	8 322.4	1 402	8 897.1
Not stated	7	X	6	X
Age-adjusted rate	X	677.6	X	686.7
AMERICAN INDIAN [2, 3], MALE				
All Ages	6 747	439.4	6 466	424.2
Under 1 year [1]	187	878.1	211	720.2
1–4 years	50	49.3	52	47.6
5–14 years	65	22.0	75	25.2
15–24 years	414	144.4	400	142.9
25–34 years	458	193.5	417	180.0
35–44 years	738	316.8	722	311.4
45–54 years	975	537.7	940	536.9
55–64 years	1 075	1 060.2	1 004	1 053.4
65–74 years	1 207	2 378.3	1 165	2 393.5
75–84 years	1 047	4 743.8	983	4 775.3
85 years and over	525	9 380.0	492	9 758.0
Not stated	6	X	5	X
Age-adjusted rate	X	796.1	X	798.9
AMERICAN INDIAN [2, 3], FEMALE				
All Ages	5 656	367.1	5 511	360.2
Under 1 year [1]	151	739.2	162	577.5
1–4 years	42	43.0	51	48.5
5–14 years	62	21.6	54	18.6
15–24 years	168	62.1	157	59.6
25–34 years	192	86.1	179	81.4
35–44 years	416	175.1	410	172.6
45–54 years	621	323.1	619	334.3
55–64 years	811	746.6	794	749.4
65–74 years	1 091	1 830.6	1 041	1 801.7
75–84 years	1 172	3 683.1	1 163	3 839.7
85 years and over	931	7 832.7	910	8 492.0
Not stated	1	X	1	X
Age-adjusted rate	X	580.7	X	594.0
ASIAN OR PACIFIC ISLANDER [2], BOTH SEXES				
All Ages	38 300	299.3	37 048	303.8
Under 1 year [1]	803	430.1	742	412.5
1–4 years	166	23.4	153	22.3
5–14 years	215	12.2	209	12.2
15–24 years	778	41.4	800	43.0
25–34 years	963	40.2	1 065	47.4
35–44 years	1 625	75.3	1 718	84.2
45–54 years	3 345	195.4	3 289	202.7
55–64 years	4 736	478.6	4 542	495.3
65–74 years	7 319	1 223.6	7 247	1 275.9
75–84 years	10 203	3 283.6	9 798	3 385.3
85 years and over	8 139	9 368.2	7 477	9 714.8
Not stated	9	X	8	X
Age-adjusted rate	X	473.9	X	492.1
ASIAN OR PACIFIC ISLANDER [2], MALE				
All Ages	20 424	330.5	19 829	335.0
Under 1 year [1]	433	453.5	403	438.8
1–4 years	98	27.4	81	23.5
5–14 years	127	14.1	119	13.6
15–24 years	550	58.2	555	59.2
25–34 years	635	54.6	689	62.4
35–44 years	1 031	99.1	1 064	108.2
45–54 years	1 956	246.6	1 911	253.6
55–64 years	2 762	599.7	2 696	625.5
65–74 years	3 886	1 486.7	3 848	1 556.0
75–84 years	5 288	4 063.1	5 086	4 168.9
85 years and over	3 652	10 908.7	3 373	11 308.9
Not stated	6	X	4	X
Age-adjusted rate	X	576.5	X	597.4

[1] Death rates for "Under 1 year" (based on population estimates) differ from infant mortality rates (based on live births).
[2] Race and Hispanic origin are reported separately on the death certificate. Data for persons of Hispanic origin are included in the data for each race group according to the decedent's reported race.
[3] Includes deaths among Aleuts and Eskimos.
X = Not applicable.

Table II-1B. Deaths, Death Rates, and Age-Adjusted Death Rates, by Age, Sex, Race, and Hispanic Origin, 2001 and Preliminary 2002—*Continued*

(Number, rate per 100,000 population.)

Age, race, and sex	2002		2001	
	Number	Rate	Number	Rate
ASIAN OR PACIFIC ISLANDER [2], FEMALE				
All Ages	17 875	270.1	17 219	274.4
Under 1 year [1]	370	405.6	339	385.0
1–4 years	68	19.3	72	21.1
5–14 years	88	10.1	90	10.8
15–24 years	227	24.3	245	26.6
25–34 years	328	26.6	376	32.9
35–44 years	594	53.1	654	61.9
45–54 years	1 389	151.2	1 378	158.5
55–64 years	1 973	372.9	1 846	379.9
65–74 years	3 434	1 019.7	3 399	1 059.9
75–84 years	4 915	2 721.9	4 712	2 814.4
85 years and over	4 487	8 402.5	4 104	8 706.2
Not stated	2	X	4	X
Age-adjusted rate	X	396.4	X	412.0
HISPANIC [4], BOTH SEXES				
All Ages	116 878	301.5	113 413	306.8
Under 1 year [1]	4 928	590.9	4 680	572.7
1–4 years	927	29.6	917	30.6
5–14 years	1	15.3	1 038	14.7
15–24 years	5 225	76.8	5 058	75.6
25–34 years	5 800	79.1	5 797	84.5
35–44 years	8 424	145.0	8 441	154.8
45–54 years	11 639	318.4	11 362	334.4
55–64 years	14 160	717.0	13 337	722.2
65–74 years	20 561	1 741.3	20 172	1 776.0
75–84 years	24 798	4 136.4	23 637	4 262.6
85 years and over	19 249	10 670.1	18 907	11 821.8
Not stated	55	X	67	X
Age-adjusted rate	X	628.4	X	658.7
HISPANIC [4], MALE				
All Ages	65 537	327.8	63 317	332.9
Under 1 year [1]	2 743	643.3	2 601	624.4
1–4 years	545	34.1	518	33.8
5–14 years	640	17.2	600	16.6
15–24 years	4 151	113.5	4 019	111.5
25–34 years	4 455	112.0	4 366	118.0
35–44 years	5 775	190.8	5 895	208.5
45–54 years	7 633	418.6	7 477	443.9
55–64 years	8 752	936.3	8 026	923.9
65–74 years	11 538	2 205.2	11 231	2 242.6
75–84 years	12 291	5 043.5	11 759	5 258.0
85 years and over	6 962	11 649.0	6 767	12 888.3
Not stated	52	X	58	X
Age-adjusted rate	X	765.8	X	802.5
HISPANIC [4], FEMALE				
All Ages	51 342	273.5	50 096	279.0
Under 1 year [1]	2 185	536.1	2 079	518.9
1–4 years	381	24.9	399	27.2
5–14 years	473	13.3	438	12.7
15–24 years	1 074	34.1	1 039	33.7
25–34 years	1 345	40.1	1 431	45.2
35–44 years	2 649	95.2	2 546	97.0
45–54 years	4 006	218.7	3 885	226.7
55–64 years	5 408	520.0	5 311	543.0
65–74 years	9 023	1 372.2	8 941	1 408.0
75–84 years	12 507	3 515.1	11 878	3 589.8
85 years and over	12 287	10 185.2	12 140	11 300.5
Not stated	3	X	9	X
Age-adjusted rate	X	517.4	X	544.2

[1] Death rates for "Under 1 year" (based on population estimates) differ from infant mortality rates (based on live births).
[2] Race and Hispanic origin are reported separately on the death certificate. Data for persons of Hispanic origin are included in the data for each race group according to the decedent's reported race.
[4] Includes all persons of Hispanic origin of any race.
X = Not applicable.

TABLE II-2. NUMBER OF DEATHS, DEATH RATES, AND AGE-ADJUSTED DEATH RATES, BY RACE, HISPANIC ORIGIN, AND SEX, 1997–2001

The age-adjusted death rate for Hispanics was 22 percent lower in 2001 than for non-Hispanic Whites. The Hispanic population has lower age-specific death rates at older ages (see Table 4). This could be caused in part by an underreporting of Hispanic origin on death certificates, and/or that some elderly immigrants return to their country of origin before their death (the so-called "salmon bias effect"). More generally, inconsistencies can arise between the reporting of Hispanic origin on death certificates and on censuses, which are used to estimate the size of the underlying populations.

Table II-2. Number of Deaths, Death Rates, and Age-Adjusted Death Rates, by Race, Hispanic Origin, and Sex, 1997–2001

(Number, rate per 100,000 population.)

Year	All origins[1]			Hispanic			Non-Hispanic[2]			Non-Hispanic White			Non-Hispanic Black		
	Both sexes	Male	Female	Both sexes	Male	Female	Both sexes	Male	Female	Both sexes	Male	Female	Both sexes	Male	Female
Number of Deaths															
2001	2 416 425	1 183 421	1 233 004	113 413	63 317	50 096	2 295 244	1 115 683	1 179 561	1 962 810	945 967	1 016 843	284 343	143 971	140 372
2000	2 403 351	1 177 578	1 225 773	107 254	60 172	47 082	2 287 846	1 112 704	1 175 142	1 959 919	944 781	1 015 138	282 676	143 297	139 379
1999	2 391 399	1 175 460	1 215 939	103 740	57 991	45 749	2 279 325	1 112 718	1 166 607	1 953 197	944 913	1 008 284	281 979	143 883	138 096
1998	2 337 256	1 157 260	1 179 996	98 406	55 821	42 585	2 230 127	1 096 677	1 133 450	1 912 802	931 844	980 958	275 264	141 627	133 637
1997	2 314 245	1 154 039	1 160 206	95 460	54 348	41 112	2 209 450	1 094 541	1 114 909	1 895 461	929 703	965 758	273 381	142 241	131 140
Death Rate															
2001	848.5	846.4	850.4	306.8	332.9	279.0	926.2	923.6	928.6	991.1	975.6	1 006.1	798.1	849.7	751.2
2000	854.0	853.0	855.0	303.8	331.3	274.6	929.6	928.1	931.0	993.2	978.5	1 007.3	805.5	859.5	756.7
1999	857.0	859.2	854.9	305.7	332.6	277.2	929.9	932.2	927.8	990.7	979.6	1 001.3	812.1	872.8	757.3
1998	847.3	856.4	838.5	303.9	336.0	270.0	916.0	925.3	907.1	972.9	969.2	976.5	805.6	873.7	744.1
1997	848.8	864.6	833.6	309.0	343.2	272.9	913.9	930.4	898.3	967.4	970.6	964.3	813.5	892.9	741.9
Age-Adjusted Death Rate															
2001	854.5	1 029.1	721.8	658.7	802.5	544.2	864.0	1 039.8	730.9	842.9	1 012.8	713.5	1 116.5	1 393.7	925.5
2000	869.0	1 053.8	731.4	665.7	818.1	546.0	877.9	1 063.8	740.0	855.5	1 035.4	721.5	1 137.0	1 422.0	941.2
1999	875.6	1 067.0	734.0	676.4	830.5	555.9	883.9	1 076.4	741.9	859.8	1 045.5	722.3	1 150.1	1 449.4	946.0
1998	870.6	1 069.4	724.7	665.4	833.6	536.9	878.4	1 078.2	732.4	854.1	1 046.7	712.8	1 141.8	1 448.2	932.9
1997	878.1	1 088.1	725.6	669.3	840.5	538.8	885.3	1 096.4	732.6	859.7	1 063.2	712.5	1 154.3	1 476.7	934.2

[1]Figures for origin not stated are included in "All origins" but are not distributed among specified origins.
[2]Includes races other than White and Black.

Figure II-2. Age-Adjusted Death Rates, by Sex, Race, and Hispanic Origin, 2001

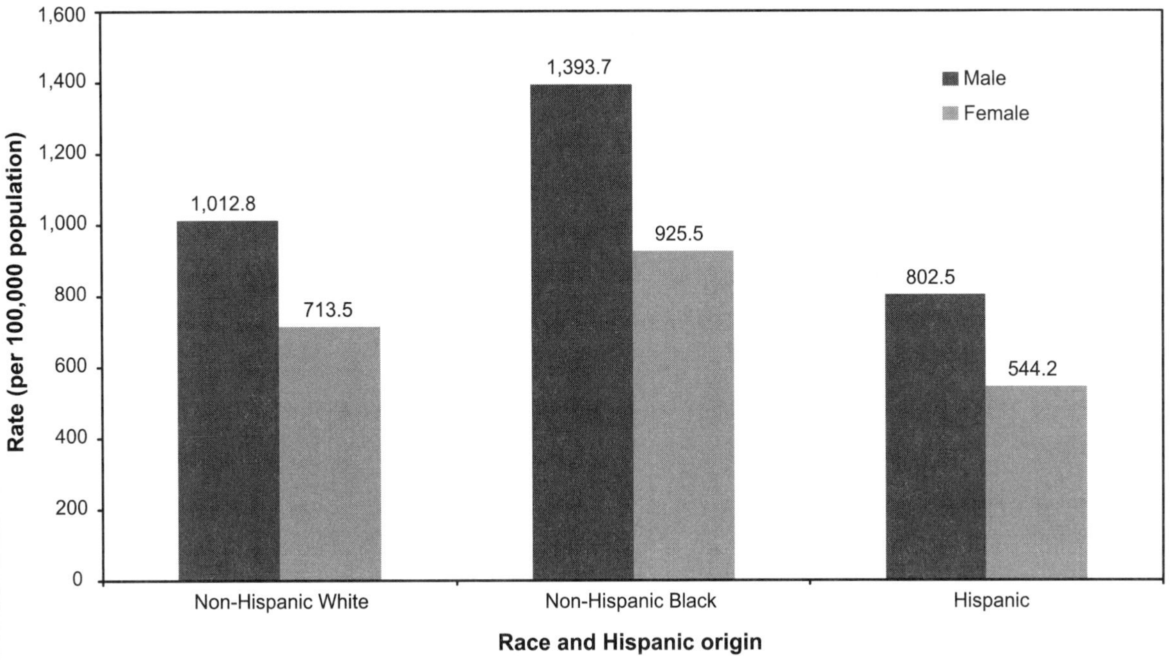

TABLE II-3. NUMBER OF DEATHS AND DEATH RATES, BY AGE, RACE, AND SEX, 2001

The death rate by age records a large figure of 683 per 100,000 for infant deaths (under 1 year). It drops as age increases and reaches a low of 15 per 100,000 in the 5–9 year group. It illustrates that a child proven "sea worthy" during its first year has excellent subsequent survival chances. Mortality rates, however, do increase for every successive age group after 5 to 9 years, first slowly, then more rapidly. The rate per 100,000 exceeds 1,200 by age 60–64, and exceeds 15,000 by age 85 and higher.

The low death rates for Asians and Pacific Islanders, already noted in the discussion of Table 1, are confirmed by their low specific-age death rates in this table. Table 4 will show that Hispanics also have relatively low age-specific death rates, but the rates for Asians and Pacific Islanders are still lower. Bias occurs if the coverage of Asians and Pacific Islanders in the census count differs from that reported on death certificates. The funeral director, who often fills out the death certificate, could mark the deceased as White instead of Asian. A recent study indicates net underreporting of deaths by Asians and Pacific Islanders of 11 percent. (Reference 16 on page 14, "Deaths Final Data 2001," National Vital Statistics Reports, Volume 52, Number 3.)

Table II-3. Number of Deaths and Death Rates, by Age, Race, and Sex, 2001

(Number, rate per 100,000 population.)

Age	All races			White			Black			American Indian[1]			Asian or Pacific Islander[2]		
	Both sexes	Male	Female	Both sexes	Male	Female	Both sexes	Male	Female	Both sexes	Male	Female	Both sexes	Male	Female
Number of Deaths															
All ages	2 416 425	1 183 421	1 233 004	2 079 691	1 011 218	1 068 473	287 709	145 908	141 801	11 977	6 466	5 511	37 048	19 829	17 219
Under 1 year	27 568	15 477	12 091	17 955	10 099	7 856	8 498	4 764	3 734	373	211	162	742	403	339
1–4 years	5 108	2 900	2 208	3 671	2 095	1 576	1 181	672	509	103	52	51	153	81	72
5–9 years	3 092	1 726	1 366	2 249	1 231	1 018	700	409	291	44	29	15	99	57	42
10–14 years	4 002	2 441	1 561	2 920	1 782	1 138	887	551	336	85	46	39	110	62	48
15–19 years	13 555	9 766	3 789	10 235	7 264	2 971	2 710	2 080	630	273	186	87	337	236	101
20–24 years	18 697	14 197	4 500	13 513	10 265	3 248	4 437	3 399	1 038	284	214	70	463	319	144
25–29 years	18 211	13 006	5 205	12 988	9 381	3 607	4 467	3 117	1 350	265	194	71	491	314	177
30–34 years	23 472	15 751	7 721	17 006	11 597	5 409	5 561	3 556	2 005	331	223	108	574	375	199
35–39 years	36 895	23 727	13 168	27 238	17 880	9 358	8 385	5 022	3 363	531	355	176	741	470	271
40–44 years	54 779	34 437	20 342	40 899	26 363	14 536	12 302	7 113	5 189	601	367	234	977	594	383
45–49 years	73 684	46 567	27 117	55 634	35 774	19 860	15 867	9 495	6 372	747	437	310	1 436	861	575
50–54 years	94 381	58 281	36 100	73 065	45 629	27 436	18 651	11 099	7 552	812	503	309	1 853	1 050	803
55–59 years	109 523	65 801	43 722	87 807	53 120	34 687	18 852	10 984	7 868	821	475	346	2 043	1 222	821
60–64 years	134 616	79 157	55 459	110 046	65 233	44 813	21 124	11 921	9 203	947	529	418	2 499	1 474	1 025
65–69 years	178 232	102 110	76 122	148 608	86 088	62 520	25 367	13 763	11 604	1 060	546	514	3 197	1 713	1 484
70–74 years	252 728	139 471	113 257	218 520	121 705	96 815	29 012	15 012	14 000	1 146	619	527	4 050	2 135	1 915
75–79 years	333 677	171 311	162 366	295 048	152 517	142 531	32 590	15 616	16 974	1 148	547	601	4 891	2 631	2 260
80–84 years	368 252	169 431	198 821	332 704	153 761	178 943	29 643	12 779	16 864	998	436	562	4 907	2 455	2 452
85 years and over	665 531	217 533	447 998	609 265	199 179	410 086	47 387	14 489	32 898	1 402	492	910	7 477	3 373	4 104
Not stated	422	331	91	320	255	65	88	67	21	6	5	1	8	4	4
Death Rate															
All ages[3]	848.5	846.4	850.4	895.1	881.9	907.9	773.5	823.9	727.7	392.1	424.2	360.2	303.8	335.0	274.4
Under 1 year[4]	683.4	749.8	613.9	570.9	627.6	511.5	1 304.5	1 426.4	1 176.3	650.4	720.2	577.5	412.5	438.8	385.0
1–4 years	33.3	37.0	29.5	30.7	34.2	27.1	47.5	53.2	41.7	48.0	47.6	48.5	22.3	23.5	21.1
5–9 years	15.3	16.7	13.9	14.3	15.3	13.3	20.7	23.8	17.5	15.5	20.2	. . .	11.6	13.1	10.1
10–14 years	19.2	22.8	15.3	17.9	21.3	14.4	25.8	31.6	19.8	28.0	29.8	26.0	12.8	14.0	11.5
15–19 years	66.9	93.7	38.5	64.2	88.3	38.5	86.3	130.4	40.8	94.5	125.7	61.7	38.0	52.1	23.3
20–24 years	95.0	141.1	46.8	87.1	128.2	43.2	151.3	236.3	69.4	111.7	162.2	57.2	47.6	65.8	29.5
25–29 years	96.2	135.6	55.8	87.0	122.4	49.6	168.8	247.0	97.5	117.1	165.9	65.0	43.9	57.3	31.0
30–34 years	113.5	151.2	75.2	102.7	137.4	66.6	200.5	271.0	137.3	146.8	194.4	97.5	50.8	67.4	34.7
35–39 years	165.9	213.7	118.2	151.2	196.7	104.9	286.0	364.1	216.6	222.9	298.4	147.6	69.9	91.3	49.7
40–44 years	240.5	304.8	177.2	218.8	281.4	155.9	428.4	527.8	340.6	260.0	325.0	197.8	99.7	126.9	74.8
45–49 years	354.8	455.4	257.2	322.8	417.1	229.4	644.1	830.2	482.8	377.0	454.4	304.1	164.2	211.1	123.3
50–54 years	512.4	646.8	383.7	471.4	595.5	350.1	928.5	1 201.4	696.2	500.9	637.3	371.5	247.6	303.7	199.4
55–59 years	771.8	958.4	596.9	723.2	896.0	558.4	1 328.9	1 718.2	1 009.5	718.6	859.2	586.7	395.5	503.9	299.6
60–64 years	1 210.7	1 496.8	951.3	1 156.1	1 427.9	905.3	1 891.7	2 424.6	1 472.5	1 140.8	1 321.5	972.5	624.1	782.0	483.7
65–69 years	1 869.7	2 315.6	1 485.9	1 805.8	2 237.6	1 426.7	2 738.8	3 497.3	2 178.4	1 728.7	1 924.2	1 560.3	1 012.3	1 219.5	846.3
70–74 years	2 878.3	3 587.4	2 314.8	2 823.2	3 513.9	2 263.9	3 904.2	5 053.2	3 138.8	2 539.2	3 049.6	2 122.0	1 606.0	1 998.2	1 317.6
75–79 years	4 494.0	5 603.2	3 717.6	4 446.8	5 543.5	3 669.9	5 660.2	7 255.7	4 707.8	3 607.9	4 104.4	3 250.1	2 683.3	3 391.3	2 158.7
80–84 years	7 151.9	8 781.9	6 175.1	7 149.4	8 771.1	6 169.3	8 028.9	10256.3	6 894.3	5 237.5	6 007.2	4 763.9	4 579.6	5 527.3	3 908.6
85 years and over	15 112.8	16 744.8	14 429.9	15 287.5	16 935.4	14 597.6	14550.8	16439.9	13849.9	8 897.1	9 758.0	8 492.0	9 714.8	11308.9	8 706.2

[1]Includes Aleuts and Eskimos.
[2]Includes Chinese, Filipino, Hawaiian, Japanese, and Other Asian or Pacific Islander.
[3]Figures for age not stated are included in "All ages" but not distributed among age groups.
[4]Death rates for "Under 1 year" (based on population estimates) differ from infant mortality rates (based on live births).

TABLE II-4. NUMBER OF DEATHS AND DEATH RATES, BY RACE, HISPANIC ORIGIN, AGE, AND SEX, 2001

Death rates for Hispanic men and women are considerably lower than those for non-Hispanic Whites (as well as for non-Hispanic Blacks) in all age groups above 24 years. The comments on Table 2 present some hypotheses for this finding.

TABLE II-5. NUMBER OF DEATHS AND DEATH RATES, BY AGE AND AGE-ADJUSTED DEATH RATES, ACCORDING TO SPECIFIED HISPANIC ORIGIN, RACE, AND SEX, 2001

Death rates for Cubans and Mexicans are particularly low relative to non-Hispanics. Data for Puerto Ricans are shown separately in this table, but Central and South American and other Hispanic deaths are combined for the computations of rates, due to small numbers.

Table II-4. Number of Deaths and Death Rates, by Race, Hispanic Origin, Age, and Sex, 2001

(Number, rate per 100,000 population.)

Age	All origins [1]			Hispanic			Non-Hispanic [2]			Non-Hispanic White			Non-Hispanic Black		
	Both sexes	Male	Female	Both sexes	Male	Female	Both sexes	Male	Female	Both sexes	Male	Female	Both sexes	Male	Female
Number of Deaths															
All ages	2 416 425	1 183 421	1 233 004	113 413	63 317	50 096	2 295 244	1 115 683	1 179 561	1 962 810	945 967	1 016 843	284 343	143 971	140 372
Under 1 year	27 568	15 477	12 091	4 680	2 601	2 079	22 629	12 739	9 890	13 311	7 508	5 803	8 277	4 651	3 626
1–4 years	5 108	2 900	2 208	917	518	399	4 172	2 372	1 800	2 773	1 593	1 180	1 156	655	501
5–9 years	3 092	1 726	1 366	480	281	199	2 598	1 437	1 161	1 776	956	820	685	400	285
10–14 years	4 002	2 441	1 561	558	319	239	3 434	2 116	1 318	2 374	1 471	903	872	541	331
15–19 years	13 555	9 766	3 789	1 998	1 530	468	11 508	8 197	3 311	8 264	5 756	2 508	2 662	2 041	621
20–24 years	18 697	14 197	4 500	3 060	2 489	571	15 551	11 639	3 912	10 466	7 786	2 680	4 362	3 336	1 026
25–29 years	18 211	13 006	5 205	2 863	2 223	640	15 252	10 715	4 537	10 130	7 168	2 962	4 388	3 056	1 332
30–34 years	23 472	15 751	7 721	2 934	2 143	791	20 404	13 510	6 894	14 061	9 445	4 616	5 471	3 492	1 979
35–39 years	36 895	23 727	13 168	3 859	2 739	1 120	32 857	20 859	11 998	23 334	15 102	8 232	8 287	4 959	3 328
40–44 years	54 779	34 437	20 342	4 582	3 156	1 426	49 906	31 059	18 847	36 212	23 111	13 101	12 149	7 012	5 137
45–49 years	73 684	46 567	27 117	5 397	3 655	1 742	67 927	42 656	25 271	50 151	32 045	18 106	15 657	9 354	6 303
50–54 years	94 381	58 281	36 100	5 965	3 822	2 143	87 989	54 159	33 830	66 935	41 686	25 249	18 441	10 953	7 488
55–59 years	109 523	65 801	43 722	6 114	3 760	2 354	102 952	61 716	41 236	81 505	49 212	32 293	18 646	10 845	7 801
60–64 years	134 616	79 157	55 459	7 223	4 266	2 957	126 860	74 498	52 362	102 604	60 781	41 823	20 877	11 755	9 122
65–69 years	178 232	102 110	76 122	8 910	5 129	3 781	168 796	96 644	72 152	139 503	80 805	58 698	25 103	13 617	11 486
70–74 years	252 728	139 471	113 257	11 262	6 102	5 160	240 704	132 895	107 809	206 886	115 353	91 533	28 700	14 829	13 871
75–79 years	333 677	171 311	162 366	12 560	6 535	6 025	320 213	164 283	155 930	282 018	145 725	136 293	32 243	15 429	16 814
80–84 years	368 252	169 431	198 821	11 077	5 224	5 853	356 299	163 796	192 503	321 125	148 288	172 837	29 349	12 654	16 695
85 years and over	665 531	217 533	447 998	18 907	6 767	12 140	645 020	210 256	434 764	589 269	192 084	397 185	46 968	14 352	32 616
Not stated	422	331	91	67	58	9	173	137	36	113	92	21	50	40	10
Death Rate															
All ages [3]	848.5	846.4	850.4	306.8	332.9	279.0	926.2	923.6	928.6	991.1	975.6	1 006.1	798.1	849.7	751.2
Under 1 year [4]	683.4	749.8	613.9	572.7	624.4	518.9	703.5	773.2	630.4	555.4	611.6	496.4	1 354.4	1 484.2	1 217.8
1–4 years	33.3	37.0	29.5	30.6	33.8	27.2	33.8	37.6	29.9	30.1	33.8	26.3	49.5	55.1	43.6
5–9 years	15.3	16.7	13.9	13.0	14.9	11.0	15.8	17.0	14.4	14.5	15.2	13.7	21.4	24.6	18.1
10–14 years	19.2	22.8	15.3	16.6	18.6	14.5	19.6	23.6	15.4	18.0	21.7	14.1	26.6	32.5	20.5
15–19 years	66.9	93.7	38.5	63.1	92.0	31.1	67.3	93.6	39.7	63.4	86.1	39.6	88.7	133.9	42.0
20–24 years	95.0	141.1	46.8	86.9	128.2	36.1	96.2	143.3	48.7	85.4	125.6	44.3	156.7	244.7	72.2
25–29 years	96.2	135.6	55.8	80.6	115.2	39.4	99.2	139.8	58.8	87.2	122.4	51.4	174.9	255.7	101.4
30–34 years	113.5	151.2	75.2	88.7	121.1	51.3	117.5	156.2	79.1	104.4	139.3	69.0	207.1	279.6	142.2
35–39 years	165.9	213.7	118.2	129.5	175.5	79.0	170.6	218.5	123.4	153.1	197.9	108.2	294.6	375.1	223.2
40–44 years	240.5	304.8	177.2	185.3	249.0	118.3	245.8	309.6	183.5	220.9	282.4	159.7	438.1	538.9	349.0
45–49 years	354.8	455.4	257.2	280.9	379.7	181.7	360.4	460.5	263.6	324.7	417.3	233.1	656.8	846.0	493.1
50–54 years	512.4	646.8	383.7	403.9	529.6	283.8	519.3	653.3	391.0	474.0	596.5	353.9	946.1	1 222.7	710.9
55–59 years	771.8	958.4	596.9	580.6	751.2	426.0	783.7	969.6	608.9	730.8	901.6	567.1	1 351.9	1 746.2	1 028.9
60–64 years	1 210.7	1 496.8	951.3	910.1	1 158.6	695.1	1 228.7	1 514.1	968.9	1 169.7	1 439.8	919.2	1 920.5	2 456.8	1 498.8
65–69 years	1 869.7	2 315.6	1 485.9	1 412.8	1 817.5	1 085.0	1 896.2	2 341.5	1 511.2	1 827.4	2 256.8	1 448.1	2 777.9	3 547.9	2 209.4
70–74 years	2 878.3	3 587.4	2 314.8	2 229.5	2 791.5	1 800.8	2 908.7	3 621.9	2 340.5	2 849.2	3 543.1	2 285.2	3 951.1	5 105.6	3 182.0
75–79 years	4 494.0	5 603.2	3 717.6	3 556.3	4 456.8	2 917.0	4 528.0	5 644.0	3 747.4	4 476.8	5 580.2	3 695.5	5 712.5	7 313.8	4 756.8
80–84 years	7 151.9	8 781.9	6 175.1	5 501.5	6 783.5	4 707.4	7 201.4	8 842.8	6 219.1	7 196.1	8 828.0	6 211.0	8 098.0	10 346.7	6 952.7
85 years and over	15 112.8	16 744.8	14 429.9	11 821.8	12 888.3	11 300.5	15 199.0	16 866.3	14 505.6	15 371.2	17 055.3	14 670.6	14 660.8	16 576.2	13 951.4

[1]Figures for origin not stated are included in "All origins" but not distributed among specified origins.
[2]Includes races other than White and Black.
[3]Figures for age not stated are included in "All ages" but not distributed among age groups.
[4]Death rates for "Under 1 year" (based on population estimates) differ from infant mortality rates (based on live births).

Table II-5. Number of Deaths and Death Rates, by Age, and Age-Adjusted Death Rates, According to Specified Hispanic Origin, Race, and Sex, 2001

(Number, rate per 100,000 population.)

Hispanic origin, race for non-Hispanic origin, and sex	All ages	Under 1 year[1]	1–4 years	5–14 years	15–24 years	25–34 years	35–44 years	45–54 years	55–64 years	65–74 years	75–84 years	85 years and over	Age not stated	Age-adjusted rate
NUMBER OF DEATHS														
All Origins	2 416 425	27 568	5 108	7 094	32 252	41 683	91 674	168 065	244 139	430 960	701 929	665 531	422	X
Male	1 183 421	15 477	2 900	4 167	23 963	28 757	58 164	104 848	144 958	241 581	340 742	217 533	331	X
Female	1 233 004	12 091	2 208	2 927	8 289	12 926	33 510	63 217	99 181	189 379	361 187	447 998	91	X
Hispanic	113 413	4 680	917	1 038	5 058	5 797	8 441	11 362	13 337	20 172	23 637	18 907	67	X
Male	63 317	2 601	518	600	4 019	4 366	5 895	7 477	8 026	11 231	11 759	6 767	58	X
Female	50 096	2 079	399	438	1 039	1 431	2 546	3 885	5 311	8 941	11 878	12 140	9	X
Mexican	63 496	3 312	680	743	3 592	3 709	4 751	6 419	7 397	11 133	12 674	9 061	25	X
Male	36 528	1 822	377	422	2 908	2 881	3 384	4 238	4 440	6 124	6 456	3 455	21	X
Female	26 968	1 490	303	321	684	828	1 367	2 181	2 957	5 009	6 218	5 606	4	X
Puerto Rican	14 920	467	60	103	442	664	1 359	1 914	2 206	2 767	2 833	2 102	3	X
Male	8 269	268	33	63	337	442	909	1 263	1 343	1 541	1 354	713	3	X
Female	6 651	199	27	40	105	222	450	651	863	1 226	1 479	1 389	-	X
Cuban	11 666	46	13	16	71	131	332	561	1 070	2 297	3 460	3 668	1	X
Male	5 892	25	11	11	48	96	235	378	711	1 383	1 752	1 241	1	X
Female	5 774	21	2	5	23	35	97	183	359	914	1 708	2 427	-	X
Central and South American	10 884	398	82	95	602	821	1 116	1 282	1 357	1 838	1 874	1 417	2	X
Male	5 727	234	46	56	464	611	734	773	748	909	760	390	2	X
Female	5 157	164	36	39	138	210	382	509	609	929	1 114	1 027	-	X
Other and unknown Hispanic	12 447	457	82	81	351	472	883	1 186	1 307	2 137	2 796	2 659	36	X
Male	6 901	252	51	48	262	336	633	825	784	1 274	1 437	968	31	X
Female	5 546	205	31	33	89	136	250	361	523	863	1 359	1 691	5	X
Non-Hispanic[2]	2 295 244	22 629	4 172	6 032	27 059	35 656	82 763	155 916	229 812	409 500	676 512	645 020	173	X
Male	1 115 683	12 739	2 372	3 553	19 836	24 225	51 918	96 815	136 214	229 539	328 079	210 256	137	X
Female	1 179 561	9 890	1 800	2 479	7 223	11 431	30 845	59 101	93 598	179 961	348 433	434 764	36	X
White	1 962 810	13 311	2 773	4 150	18 730	24 191	59 546	117 086	184 109	346 389	603 143	589 269	113	X
Male	945 967	7 508	1 593	2 427	13 542	16 613	38 213	73 731	109 993	196 158	294 013	192 084	92	X
Female	1 016 843	5 803	1 180	1 723	5 188	7 578	21 333	43 355	74 116	150 231	309 130	397 185	21	X
Black	284 343	8 277	1 156	1 557	7 024	9 859	20 436	34 098	39 523	53 803	61 592	46 968	50	X
Male	143 971	4 651	655	941	5 377	6 548	11 971	20 307	22 600	28 446	28 083	14 352	40	X
Female	140 372	3 626	501	616	1 647	3 311	8 465	13 791	16 923	25 357	33 509	32 616	10	X
Not stated[3]	7 768	259	19	24	135	230	470	787	990	1 288	1 780	1 604	182	X
Male	4 421	137	10	14	108	166	351	556	718	811	904	510	136	X
Female	3 347	122	9	10	27	64	119	231	272	477	876	1 094	46	X
DEATH RATE[4]														
All Origins[5]	848.5	683.4	33.3	17.3	80.7	105.2	203.6	428.9	964.6	2 353.3	5 582.4	15 112.8	X	854.5
Male	846.4	749.8	37.0	19.8	117.0	143.7	259.6	545.1	1 192.7	2 911.5	6 833.0	16 744.8	X	1 029.1
Female	850.4	613.9	29.5	14.6	42.6	66.0	148.2	316.8	754.0	1 890.8	4 760.5	14 429.9	X	721.8
Hispanic	306.8	572.7	30.6	14.7	75.6	84.5	154.8	334.4	722.2	1 776.0	4 262.6	11 821.8	X	658.7
Male	332.9	624.4	33.8	16.6	111.5	118.0	208.5	443.9	923.9	2 242.6	5 258.0	12 888.3	X	802.5
Female	279.0	518.9	27.2	12.7	33.7	45.2	97.0	226.7	543.0	1 408.0	3 589.8	11 300.5	X	544.2
Mexican	258.1	521.5	30.7	15.0	77.3	78.8	142.7	307.6	702.7	1 862.0	4 399.9	10 778.9	X	645.9
Male	282.8	550.0	33.7	16.5	115.2	111.4	189.0	396.6	882.1	2 228.9	5 262.6	*	X	779.0
Female	230.8	490.5	27.6	13.4	32.2	39.0	88.9	214.2	538.3	1 550.1	3 759.9	9 989.8	X	536.2
Puerto Rican	460.3	*	27.8	16.3	84.4	128.7	267.4	517.3	1 050.6	1 859.3	5 539.0	*	X	821.1
Male	537.6	*	*	20.3	127.0	183.4	373.2	767.3	1 424.5	2 671.7	*	*	X	1 029.1
Female	390.5	*	*	12.4	40.7	80.7	170.0	317.0	745.9	1 345.2	*	*	X	667.6
Cuban	860.5	*	*	*	44.8	74.8	154.9	318.7	693.1	1 482.8	3 727.8	*	X	590.9
Male	861.9	*	*	*	58.6	94.7	199.7	428.0	1 010.4	1 849.0	*	*	X	*
Female	859.1	*	*	*	*	*	100.3	208.7	427.3	1 140.9	3 047.2	*	X	450.6
Other Hispanic[6]	300.0	716.5	31.7	13.2	69.9	88.3	142.7	322.5	620.0	1 697.3	3 811.8	*	X	677.2
Male	325.4	865.5	35.3	15.8	99.0	122.9	202.2	440.4	763.4	2 333.0	*	*	X	747.6
Female	274.7	584.0	27.6	10.6	36.1	49.9	87.2	216.2	494.4	1 274.3	3 112.5	*	X	649.6
Non-Hispanic[2]	926.2	703.5	33.8	17.7	81.4	108.9	209.2	435.6	979.5	2 383.9	5 628.5	15 199.0	X	864.0
Male	923.6	773.2	37.6	20.4	117.5	148.5	265.2	551.6	1 207.0	2 944.1	6 888.0	16 866.3	X	1 039.8
Female	928.6	630.4	29.9	14.9	44.1	69.6	154.3	324.0	768.7	1 918.4	4 801.8	14 505.6	X	730.9
White	991.1	555.4	30.1	16.3	74.1	96.4	188.3	396.0	924.0	2 325.5	5 604.4	15 371.2	X	842.9
Male	975.6	611.6	33.8	18.6	105.1	131.5	241.6	502.6	1 136.3	2 869.4	6 851.5	17 055.3	X	1 012.8
Female	1 006.1	496.4	26.3	13.9	41.9	60.9	134.9	291.0	723.5	1 864.1	4 777.3	14 670.6	X	713.5
Black	798.1	1 354.4	49.5	24.0	121.4	191.5	365.8	786.9	1 602.5	3 300.7	6 645.3	14 660.8	X	1 116.5
Male	849.7	1 484.2	55.1	28.6	186.2	267.9	456.3	1 014.6	2 055.4	4 218.9	8 426.8	16 576.2	X	1 393.7
Female	751.2	1 217.8	43.6	19.3	56.8	122.4	285.7	591.5	1 238.2	2 653.0	5 645.1	13 951.4	X	925.5

[1]Death rates for "Under 1 year" (based on population estimates) differ from infant mortality rates (based on live births).
[2]Includes races other than White and Black.
[3]Includes deaths for which Hispanic origin was not reported on the death certificate.
[4]Figures for age not stated are included in "All ages" but not distributed among age groups.
[5]Figures for origin not stated are included in "All origins" but not distributed among specified origins.
[6]Includes Central and South American and Other and unknown Hispanic.
X = Not applicable.
- = Quantity zero.
* = Figure too small to meet standards of reliability or precision.

TABLE II-6. ABRIDGED LIFE TABLE FOR THE TOTAL POPULATION, 2001

Life expectancy at a given age represents the average number of years that a group of that age would live if they were to experience throughout the rest of life the age-specific death rates present in that year. Thus it does not reflect changes in mortality rates that may occur in the future. The U.S. life expectancy in 2001 was 77.2 years at birth, 58.1 years at age 20, and 14.6 years at age 70.

Column 2 – Probability of dying – shows the probability of dying between the ages shown in column 1. For example between the 20th and the 25th birthdays, the probability of dying is 0.00474. These probabilities are derived from vital statistics death rates up to age 85, as well as modeling from Medicare program data for ages over 85. All subsequent columns are derived from the calculations of the probability of dying.

Column 3 – Number surviving to age – the number of persons, from a theoretical cohort of 100,000 live births, who survive to the beginning of each age interval. For example, 98,682 would survive to age 20.

Column 4 – Number dying between ages – shows the number dying in each successive age interval out of the initial 100,000 live births. Each figure is the difference between two successive figures in column three.

Column 5 – Person-years lived between indicated ages – this column shows the total number of person-years lived between the beginning and the end of the given age bracket. Person-years represent the number of persons, multiplied by years lived within a bracket.

Column 6 – Total number of person-years lived – in this column all of the person-years shown in column five are cumulated from the bottom up. Starting with the cohort of 100,000, the column shows how many person-years of life are left at the beginning of each age bracket.

Column 7 – Expectancy of life – is derived by taking the total person-years that remain to be lived above a certain age and dividing this sum by the number of persons who survived to that age interval. Thus, it is the average number of years remaining to be lived by those who have survived to a given age.

Table II-6. Abridged Life Table for the Total Population, 2001

Age x to $x+n$	Probability of dying between ages x to $x+n$	Number surviving to age x	Number dying between ages x to $x+n$	Person-years lived between ages x to $x+n$	Total number of person-years lived above age x	Expectancy of life at age x
	$_nq_x$	l_x	$_nd_x$	$_nL_x$	T_x	e_x
0–1 year	0.00684	100 000	684	99 404	7 716 990	77.2
1–5 years	0.00133	99 316	132	396 950	7 617 586	76.7
5–10 years	0.00077	99 184	76	495 723	7 220 636	72.8
10–15 years	0.00096	99 108	95	495 329	6 724 913	67.9
15–20 years	0.00334	99 012	330	494 375	6 229 584	62.9
20–25 years	0.00474	98 682	468	492 242	5 735 209	58.1
25–30 years	0.00480	98 214	471	489 911	5 242 967	53.4
30–35 years	0.00567	97 743	554	487 395	4 753 057	48.6
35–40 years	0.00824	97 189	801	484 069	4 265 662	43.9
40–45 years	0.01198	96 388	1 154	479 237	3 781 593	39.2
45–50 years	0.01766	95 234	1 682	472 212	3 302 356	34.7
50–55 years	0.02537	93 552	2 373	462 194	2 830 144	30.3
55–60 years	0.03810	91 179	3 474	447 822	2 367 950	26.0
60–65 years	0.05913	87 705	5 186	426 352	1 920 129	21.9
65–70 years	0.08964	82 519	7 397	395 107	1 493 777	18.1
70–75 years	0.13455	75 122	10 107	351 564	1 098 670	14.6
75–80 years	0.20295	65 014	13 195	293 351	747 106	11.5
80–85 years	0.30639	51 820	15 877	220 319	453 756	8.8
85–90 years	0.44924	35 943	16 147	138 485	233 436	6.5
90–95 years	0.60144	19 796	11 906	66 838	94 952	4.8
95–100 years	0.74084	7 890	5 845	22 602	28 114	3.6
100 years and over	1.00000	2 045	2 045	5 512	5 512	2.7

TABLE II-7. EXPECTATION OF LIFE, BY AGE, RACE, AND SEX, 2001 AND PRELIMINARY 2002

Life expectancy at birth rose to a record level in 2002 of 77.4 years, up 0.2 years from 2001. New highs in life expectancy were reached for White and Black males (75.3 years and 68.9 years, respectively) and also for White and Black females (80.3 years and 75.7 years, respectively). On average, 18.2 years of life were remaining for a person who had reached age 65.

Table II-7. Expectation of Life, by Age, Race, and Sex, 2001 and Preliminary 2002

(Years of life remaining.)

Age (years) and race	Both sexes		Male		Female	
	2002	2001	2002	2001	2002	2001
All Races [1]						
0 years	77.4	77.2	74.7	74.4	79.9	79.8
1 year	76.9	76.7	74.3	74.0	79.4	79.3
5 years	73.0	72.8	70.4	70.1	75.5	75.4
10 years	68.1	67.9	65.4	65.2	70.6	70.4
15 years	63.1	52.9	60.5	60.2	65.6	65.5
20 years	58.3	58.1	55.8	55.5	60.7	60.6
25 years	53.6	53.4	51.1	50.9	55.9	55.7
30 years	48.8	48.6	46.5	46.2	51.0	50.9
35 years	44.1	43.9	41.8	41.5	46.2	46.0
40 years	39.4	39.2	37.2	37.0	41.5	41.3
45 years	34.9	34.7	32.7	32.5	36.8	36.6
50 years	30.4	30.3	28.4	28.2	32.2	32.1
55 years	26.2	26.0	24.2	24.0	27.8	27.7
60 years	22.0	21.9	20.3	20.1	23.5	23.4
65 years	18.2	18.1	16.6	16.4	19.5	19.4
70 years	14.7	14.6	13.3	13.1	15.8	15.7
75 years	11.6	11.5	10.4	10.2	12.5	12.4
80 years	8.9	8.8	8.0	7.7	9.5	9.4
85 years	6.7	6.5	5.9	5.7	7.0	6.9
90 years	4.9	4.8	4.4	4.2	5.1	5.0
95 years	3.7	3.6	3.3	3.2	3.8	3.7
100 years	2.8	2.7	2.6	2.5	2.8	2.8
White						
0 years	77.8	77.7	75.3	75.0	80.3	80.2
1 year	77.3	77.1	74.7	74.5	79.8	79.6
5 years	73.4	73.2	70.8	70.6	75.8	75.7
10 years	68.4	68.3	65.9	65.6	70.9	70.8
15 years	63.5	63.3	60.9	60.7	65.9	65.8
20 years	58.7	58.5	56.2	56.0	61.1	60.9
25 years	53.9	53.8	51.5	51.3	56.2	56.1
30 years	49.1	49.0	46.8	46.6	51.3	51.2
35 years	44.4	44.2	42.1	41.9	46.5	46.3
40 years	39.7	39.5	37.5	37.3	41.7	41.6
45 years	35.1	34.9	33.0	32.8	37.0	36.9
50 years	30.6	30.5	28.6	28.4	32.4	32.3
55 years	26.3	26.1	24.4	24.2	27.9	27.8
60 years	22.1	22.0	20.4	20.2	23.6	23.5
65 years	18.2	18.2	16.6	16.5	19.5	19.5
70 years	14.7	14.6	13.3	13.2	15.8	15.7
75 years	11.6	11.5	10.4	10.2	12.4	12.3
80 years	8.9	8.7	7.9	7.7	9.4	9.3
85 years	6.5	6.4	5.8	5.6	6.8	6.7
90 years	4.8	4.6	4.3	4.1	4.9	4.8
95 years	3.5	3.4	3.2	3.0	3.5	3.4
100 years	2.5	2.4	2.4	2.3	2.5	2.5
Black						
0 years	72.5	72.2	68.9	68.6	75.7	75.5
1 year	72.5	72.2	69.0	68.6	75.7	75.4
5 years	68.6	68.3	65.2	64.8	71.8	71.5
10 years	63.7	63.4	60.2	59.8	66.9	66.6
15 years	58.8	58.5	55.3	54.9	61.9	61.7
20 years	54.0	53.7	50.7	50.3	57.1	56.8
25 years	49.4	49.1	46.2	45.8	52.2	52.0
30 years	44.8	44.5	41.8	41.4	47.5	47.2
35 years	40.2	39.9	37.3	36.9	42.8	42.5
40 years	35.8	35.5	32.9	32.5	38.2	38.0
45 years	31.5	31.2	28.7	28.4	33.8	33.6
50 years	27.4	27.1	24.8	24.4	29.6	29.3
55 years	23.5	23.3	21.1	20.8	25.5	25.3
60 years	19.9	19.7	17.7	17.5	21.6	21.5
65 years	16.6	16.4	14.6	14.4	18.0	17.9
70 years	13.6	13.5	11.9	11.7	14.8	14.7
75 years	11.0	10.8	9.5	9.3	11.8	11.7
80 years	8.7	8.6	7.6	7.3	9.3	9.2
85 years	6.7	6.7	6.0	5.7	7.0	7.0
90 years	5.2	5.1	4.7	4.5	5.3	5.3
95 years	4.0	3.9	3.7	3.6	4.0	4.0
100 years	3.1	3.0	3.0	2.9	3.0	3.0

[1] Includes races other than White and Black.

Figure II-3. Life Expectancy at Selected Ages, by Sex, 2001

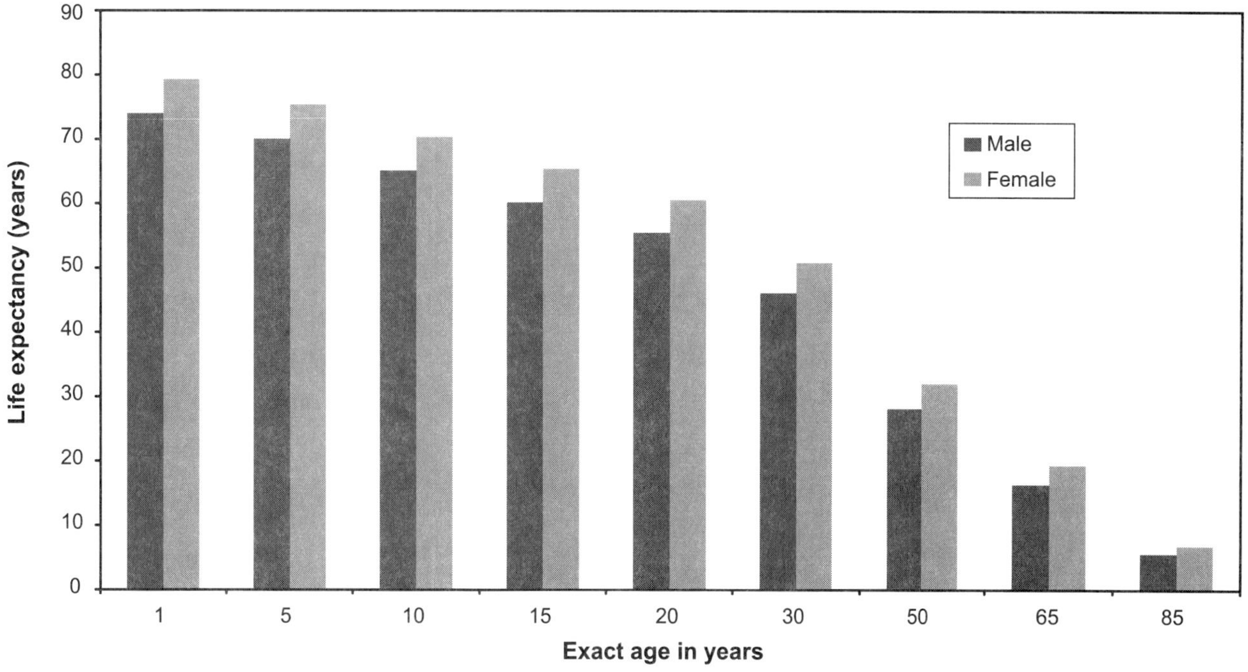

Figure II-4. Life Expectancy at Selected Ages, by Race, 2001

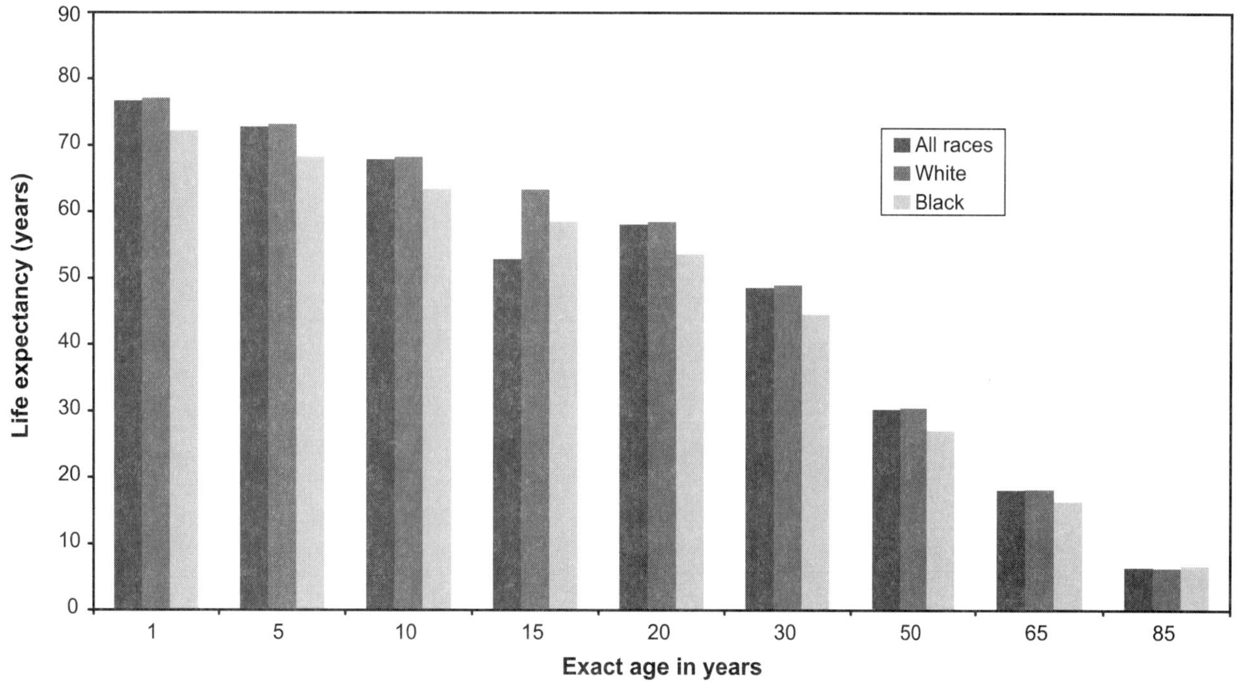

TABLE II-8. LIFE EXPECTANCY AT BIRTH, BY RACE AND SEX, SELECTED YEARS, 1940–2001

Life expectancy for the whole population has lengthened by 23 percent since 1940, a notable achievement. The 2001 life expectancy at birth of 77.2 years for the total population was a record high, as were life expectancies for all four race-sex groups. From 1900 (not shown in the table) to the late 1970s, the sex gap in life expectancy widened from 2.0 years to 7.8 years, but it has been narrowing since, down to 5.4 years in 2001.

Since 1970, when separate data for Blacks became available, Black expectancy has lengthened by 8.1 years, while the average life span of Whites increased 6.0 years. Nevertheless an average difference of 5.5 years of life span remains between the two races. The increase in life expectancy did not occur evenly: between 1984 and 1989 Black life expectancy, especially male, was decreasing.

Table II-8. Life Expectancy at Birth, by Race and Sex, Selected Years, 1940–2001

(Years.)

Year	All races [1]			White			Black		
	Both sexes	Male	Female	Both sexes	Male	Female	Both sexes	Male	Female
2001	77.2	74.4	79.8	77.7	75.0	80.2	72.2	68.6	75.5
2000	77.0	74.3	79.7	77.6	74.9	80.1	71.9	68.3	75.2
1999	76.7	73.9	79.4	77.3	74.6	79.9	71.4	67.8	74.7
1998	76.7	73.8	79.5	77.3	74.5	80.0	71.3	67.6	74.8
1997	76.5	73.6	79.4	77.1	74.3	79.9	71.1	67.2	74.7
1996	76.1	73.1	79.1	76.8	73.9	79.7	70.2	66.1	74.2
1995	75.8	72.5	78.9	76.5	73.4	79.6	69.6	65.2	73.9
1994	75.7	72.4	79.0	76.5	73.3	79.6	69.5	64.9	73.9
1993	75.5	72.2	78.8	76.3	73.1	79.5	69.2	64.6	73.7
1992	75.8	72.3	79.1	76.5	73.2	79.8	69.6	65.0	73.9
1991	75.5	72.0	78.9	76.3	72.9	79.6	69.3	64.6	73.8
1990	75.4	71.8	78.8	76.1	72.7	79.4	69.1	64.5	73.6
1989	75.1	71.7	78.5	75.9	72.5	79.2	68.8	64.3	73.3
1988	74.9	71.4	78.3	75.6	72.2	78.9	68.9	64.4	73.2
1987	74.9	71.4	78.3	75.6	72.1	78.9	69.1	64.7	73.4
1986	74.7	71.2	78.2	75.4	71.9	78.8	69.1	64.8	73.4
1985	74.7	71.1	78.2	75.3	71.8	78.7	69.3	65.0	73.4
1984	74.7	71.1	78.2	75.3	71.8	78.7	69.5	65.3	73.6
1983	74.6	71.0	78.1	75.2	71.6	78.7	69.4	65.2	73.5
1982	74.5	70.8	78.1	75.1	71.5	78.7	69.4	65.1	73.6
1981	74.1	70.4	77.8	74.8	71.1	78.4	68.9	64.5	73.2
1980	73.7	70.0	77.4	74.4	70.7	78.1	68.1	63.8	72.5
1979	73.9	70.0	77.8	74.6	70.8	78.4	68.5	64.0	72.9
1978	73.5	69.6	77.3	74.1	70.4	78.0	68.1	63.7	72.4
1977	73.3	69.5	77.2	74.0	70.2	77.9	67.7	63.4	72.0
1976	72.9	69.1	76.8	73.6	69.9	77.5	67.2	62.9	71.6
1975	72.6	68.8	76.6	73.4	69.5	77.3	66.8	62.4	71.3
1970	70.8	67.1	74.7	71.7	68.0	75.6	64.1	60.0	68.3
1960	69.7	66.6	73.1	70.6	67.4	74.1	. . .	. . .	. . .
1950	68.2	65.6	71.1	69.1	66.5	72.2	. . .	. . .	. . .
1940	62.9	60.8	65.2	64.2	62.1	66.6	. . .	. . .	. . .

[1]Includes races other than White and Black.
. . . = Not available.

Figure II-5. Life Expectancy at Birth, by Sex, 1940–2001

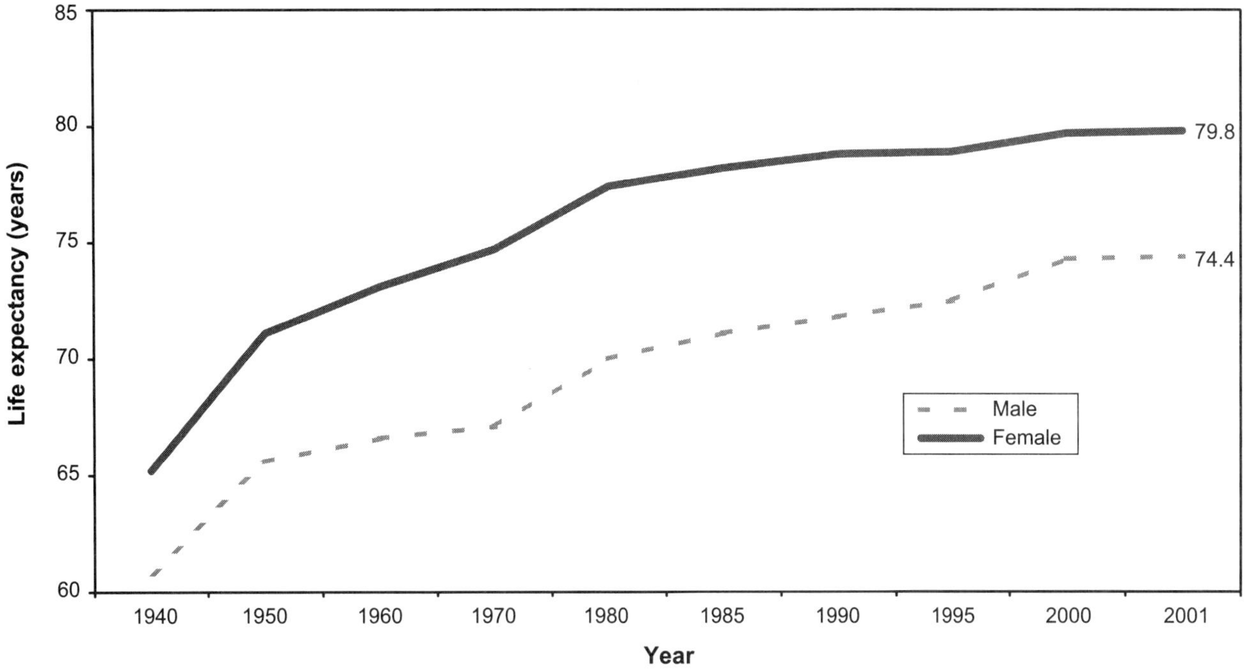

Figure II-6. Life Expectancy at Birth by Race, 1940–2001

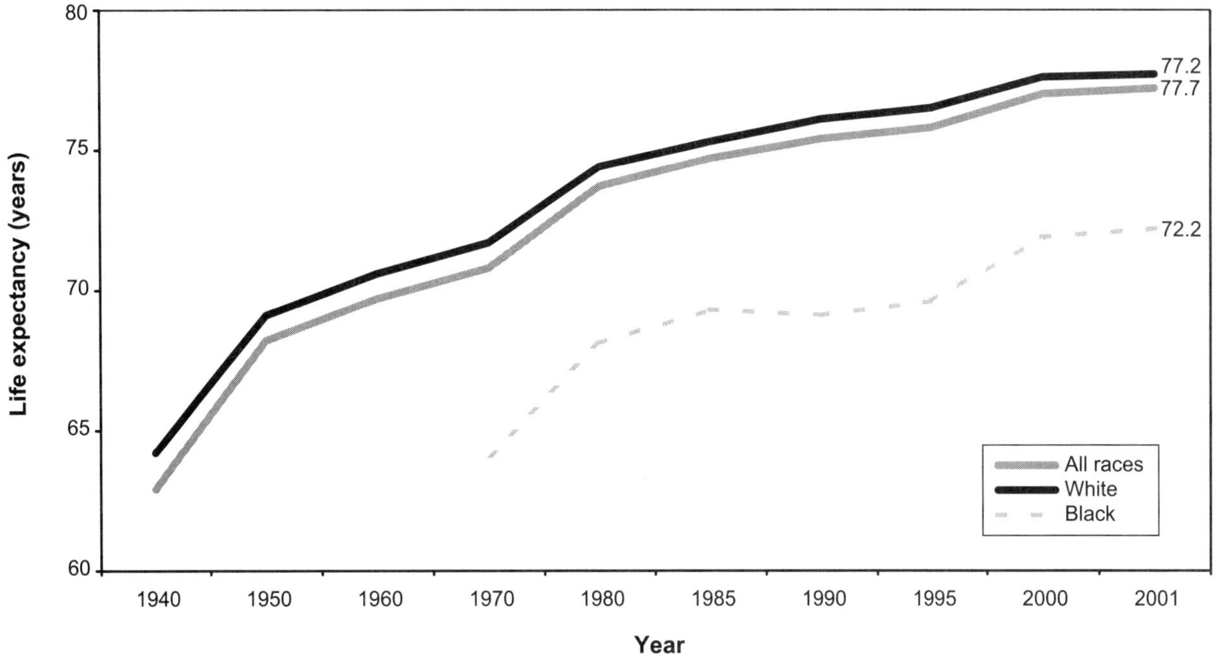

TABLE II-9. DEATH RATES BY AGE AND AGE-ADJUSTED DEATH RATES FOR THE 15 LEADING CAUSES OF DEATH, 1999–2001

Between 2000 and 2001 the age-adjusted death rate for all causes decreased by 1.7 percent, driven by decreases in such major causes as heart disease, cancer, stroke, influenza and pneumonia, and chronic lower respiratory disease. The last category has decreased recently despite a persistent upward trend for the last 50 years.

The 15 leading causes of death encompass 83.4 percent of all deaths in 2001. Diseases of the heart rank first as a cause of death, accounting for 29.0 percent of the total. Cancer is next with 22.9 percent, and stroke is third with 6.8 percent. (See Table 9A for further analysis of rates for the 15 leading causes of death.) The death rates from each of the three top causes declined between 1999 and 2001, especially in the case of heart diseases. The rankings of causes are not the same for each age group. Death rates from cancer exceed those from heart diseases through the 70–74-year age bracket, but thereafter the incidence of heart diseases grows sharply and jumps into first place. Influenza and pneumonia and Alzheimer's disease, which rank seventh and eighth in the general population jump, to the fourth and fifth rank beginning age 85.

Assault (homicide) ranks as number 13, somewhat lower than suicide, which has rank 11. Homicide jumped up sharply in 2001 due to the terrorist attacks on September 11. Death caused by HIV numbered 14,175, fortunately some 3,000 short of being included in the top 15 causes, but still representing a major public health problem. The age-adjusted death rate from HIV disease has declined for six consecutive years.

Figure II-7. Deaths by Leading Cause, 2001

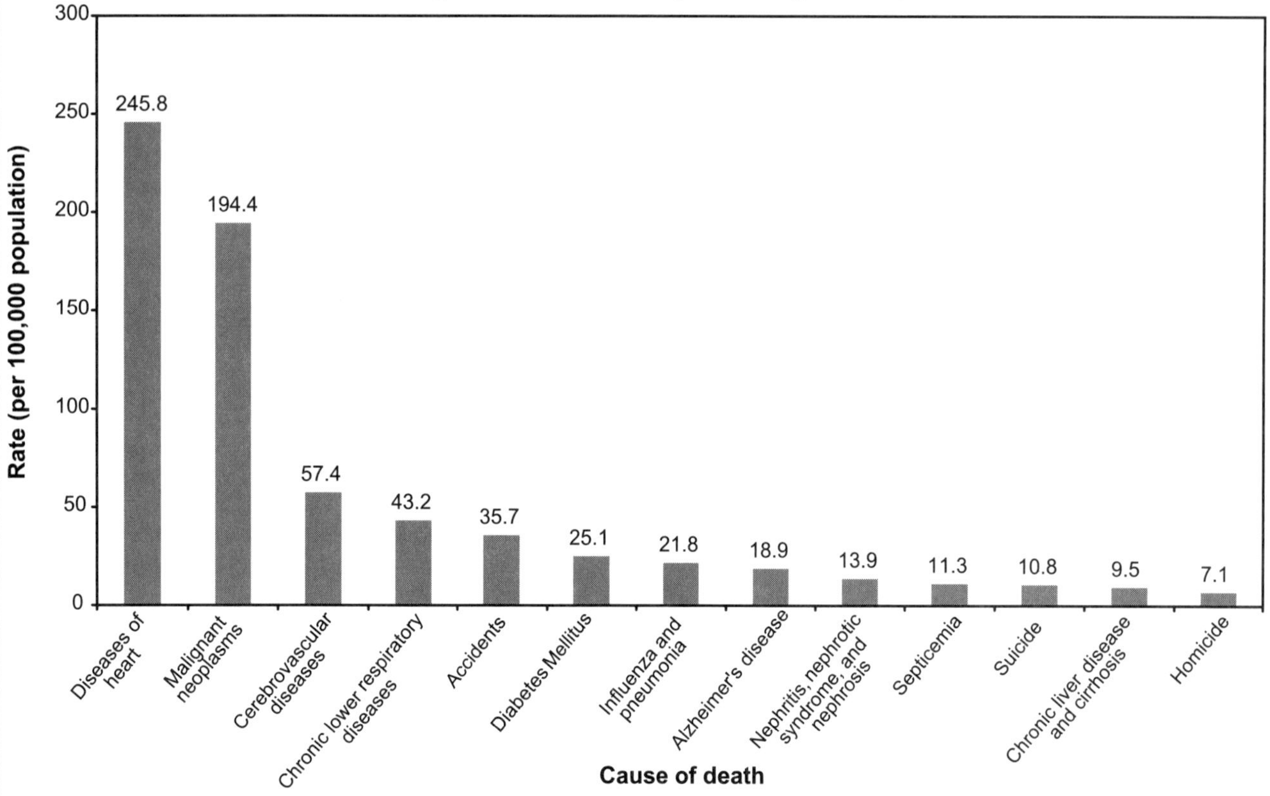

Table II-9. Death Rates by Age and Age-Adjusted Death Rates for the 15 Leading Causes of Death, 1999–2001

(Rate per 100,000 population.)

Cause of death and year [1]	All ages [2]	Under 1 year [3]	1–4 years	5–14 years	15–24 years	25–34 years	35–44 years	45–54 years	55–64 years	65–74 years	75–84 years	85 years and over	Age-adjusted rate
All Causes													
2001	848.5	683.4	33.3	17.3	80.7	105.2	203.6	428.9	964.6	2 353.3	5 582.4	15 112.8	854.5
2000	854.0	736.7	32.4	18.0	79.9	101.4	198.9	425.6	992.2	2 399.1	5 666.5	15 524.4	869.0
1999	857.0	736.0	34.2	18.6	79.3	102.2	198.0	418.2	1 005.0	2 457.3	5 714.5	15 554.6	875.6
Diseases of Heart (I00-I09, I11, I13, I20-I51)													
2001	245.8	11.9	1.5	0.7	2.5	8.0	29.6	92.9	246.9	635.1	1 725.7	5 664.2	247.8
2000	252.6	13.0	1.2	0.7	2.6	7.4	29.2	94.2	261.2	665.6	1 780.3	5 926.1	257.6
1999	259.9	13.8	1.2	0.7	2.8	7.6	30.2	95.7	269.9	701.7	1 849.9	6 063.0	266.5
Malignant Neoplasms (C00-C97)													
2001	194.4	1.6	2.7	2.5	4.3	10.1	36.8	126.5	356.5	802.8	1 315.8	1 765.6	196.0
2000	196.5	2.4	2.7	2.5	4.4	9.8	36.6	127.5	366.7	816.3	1 335.6	1 819.4	199.6
1999	197.0	1.8	2.7	2.5	4.5	10.0	37.1	127.6	374.6	827.1	1 331.5	1 805.8	200.8
Cerebrovascular Diseases (I60-I69)													
2001	57.4	2.7	0.4	0.2	0.5	1.5	5.5	15.1	38.0	123.4	443.9	1 500.2	57.9
2000	59.6	3.3	0.3	0.2	0.5	1.5	5.8	16.0	41.0	128.6	461.3	1 589.2	60.9
1999	60.0	2.7	0.3	0.2	0.5	1.4	5.7	15.2	40.6	130.8	469.8	1 614.8	61.6
Chronic Lower Respiratory Diseases (J40-J47)													
2001	43.2	1.0	0.3	0.3	0.4	0.7	2.2	8.5	44.1	167.9	379.8	644.7	43.7
2000	43.4	0.9	0.3	0.3	0.5	0.7	2.1	8.6	44.2	169.4	386.1	648.6	44.2
1999	44.5	0.9	0.4	0.3	0.5	0.8	2.0	8.5	47.5	177.2	397.8	646.0	45.4
Accidents (Unintentional Injuries) (V01-X59, Y85-Y86)													
2001	35.7	24.2	11.2	6.9	36.1	29.9	35.4	34.1	30.3	42.8	100.9	276.4	35.7
2000	34.8	23.1	11.9	7.3	36.0	29.5	34.1	32.6	30.9	41.9	95.1	273.5	34.9
1999	35.1	22.3	12.4	7.6	35.3	29.6	33.8	31.8	30.6	44.6	100.5	282.4	35.3
Diabetes Mellitus (E10-E14)													
2001	25.1	*	*	0.1	0.4	1.5	4.3	13.6	37.8	91.4	181.4	321.8	25.3
2000	24.6	*	*	0.1	0.4	1.6	4.3	13.1	37.8	90.7	179.5	319.7	25.0
1999	24.5	*	*	0.1	0.4	1.4	4.3	12.9	38.3	91.8	178.0	317.2	25.0
Influenza and Pneumonia (J10-J18)													
2001	21.8	7.4	0.7	0.2	0.5	0.9	2.2	4.6	10.7	36.3	148.5	685.6	22.0
2000	23.2	7.6	0.7	0.2	0.5	0.9	2.4	4.7	11.9	39.1	160.3	744.1	23.7
1999	22.8	8.4	0.8	0.2	0.5	0.8	2.4	4.6	11.0	37.2	157.0	751.8	23.5
Alzheimer's Disease (G30)													
2001	18.9	*	*	*	*	*	*	0.2	2.1	18.7	147.5	710.3	19.1
2000	17.6	*	*	*	*	*	*	0.2	2.0	18.7	139.6	667.7	18.1
1999	16.0	*	*	*	*	*	*	0.2	1.9	17.4	129.5	601.3	16.5
Nephritis, Nephrotic Syndrome and Nephrosis (N00-N07, N17-N19, N25-N27)													
2001	13.9	3.3	*	*	0.2	0.6	1.7	4.6	13.0	40.2	104.2	287.7	14.0
2000	13.2	4.3	*	0.1	0.2	0.6	1.6	4.4	12.8	38.0	100.8	277.8	13.5
1999	12.7	4.4	*	0.1	0.2	0.6	1.6	4.0	12.0	37.1	97.6	268.9	13.0
Septicemia (A40-A41)													
2001	11.3	7.7	0.7	0.2	0.3	0.7	1.8	5.0	12.3	32.8	82.3	205.9	11.4
2000	11.1	7.2	0.6	0.2	0.3	0.7	1.9	4.9	11.9	31.0	80.4	215.7	11.3
1999	11.0	7.5	0.6	0.2	0.3	0.7	1.8	4.6	11.4	31.2	79.4	220.7	11.3
Intentional Self-Harm (Suicide) (*U03, X60-X84, Y870)													
2001 [4]	10.8	X	X	0.7	9.9	12.8	14.7	15.2	13.1	13.3	17.4	17.5	10.7
2000	10.4	X	X	0.7	10.2	12.0	14.5	14.4	12.1	12.5	17.6	19.6	10.4
1999	10.5	X	X	0.6	10.1	12.7	14.3	13.9	12.2	13.4	18.1	19.3	10.5
Chronic Liver Disease and Cirrhosis (K70, K73-K74)													
2001	9.5	*	*	*	0.1	1.0	7.4	18.5	22.7	30.0	30.2	22.2	9.5
2000	9.4	*	*	*	0.1	1.0	7.5	17.7	23.8	29.8	31.0	23.1	9.5
1999	9.4	*	*	*	0.1	1.0	7.3	17.4	23.7	30.6	31.9	23.2	9.6
Assault (Homicide) (*U01-*U02, X85-Y09, Y871)													
2001 [4]	7.1	8.2	2.7	0.8	13.3	13.1	9.5	6.3	4.0	2.9	2.5	2.4	7.1
2000	6.0	9.2	2.3	0.9	12.6	10.4	7.1	4.7	3.0	2.4	2.4	2.4	5.9
1999	6.1	8.7	2.5	1.1	12.9	10.5	7.1	4.6	3.0	2.6	2.5	2.4	6.0
Essential (Primary) Hypertension and Hypertensive Renal Disease (I10, I12)													
2001	6.8	*	*	*	0.1	0.3	0.7	2.4	5.8	15.5	47.7	171.9	6.8
2000	6.4	*	*	*	*	0.2	0.8	2.3	5.9	15.1	45.5	162.9	6.5
1999	6.1	*	*	*	*	0.2	0.7	2.2	5.5	15.2	43.6	152.1	6.2
Pneumonitis Due to Solids and Liquids (J69)													
2001	6.1	*	*	*	0.1	0.2	0.4	1.0	2.6	10.0	45.8	189.4	6.1
2000	5.9	*	*	*	0.1	0.2	0.4	1.0	2.5	10.3	44.5	187.6	6.1
1999	5.5	*	*	*	0.1	0.2	0.4	0.8	2.5	9.5	41.1	175.6	5.6

[1]Based on codes from the "Tenth Revision, International Classification of Diseases, 1992." Codes *U01-*U03 have been added as of 2001 for deaths due to terrorism.
[2]Figures for age not stated are included in "All ages" but not distributed among age groups.
[3]Death rates for "Under 1 year" (based on population estimates) differ from infant mortality rates (based on live births).
[4]Figures include September 11, 2001 related deaths for which death certificates were filed as of 10/24/02.
* = Figure too small to meet standards of reliability or precision.
X = Not applicable.

Figure II-8. Leading Cause of Death and Death Rate by Age, 2001

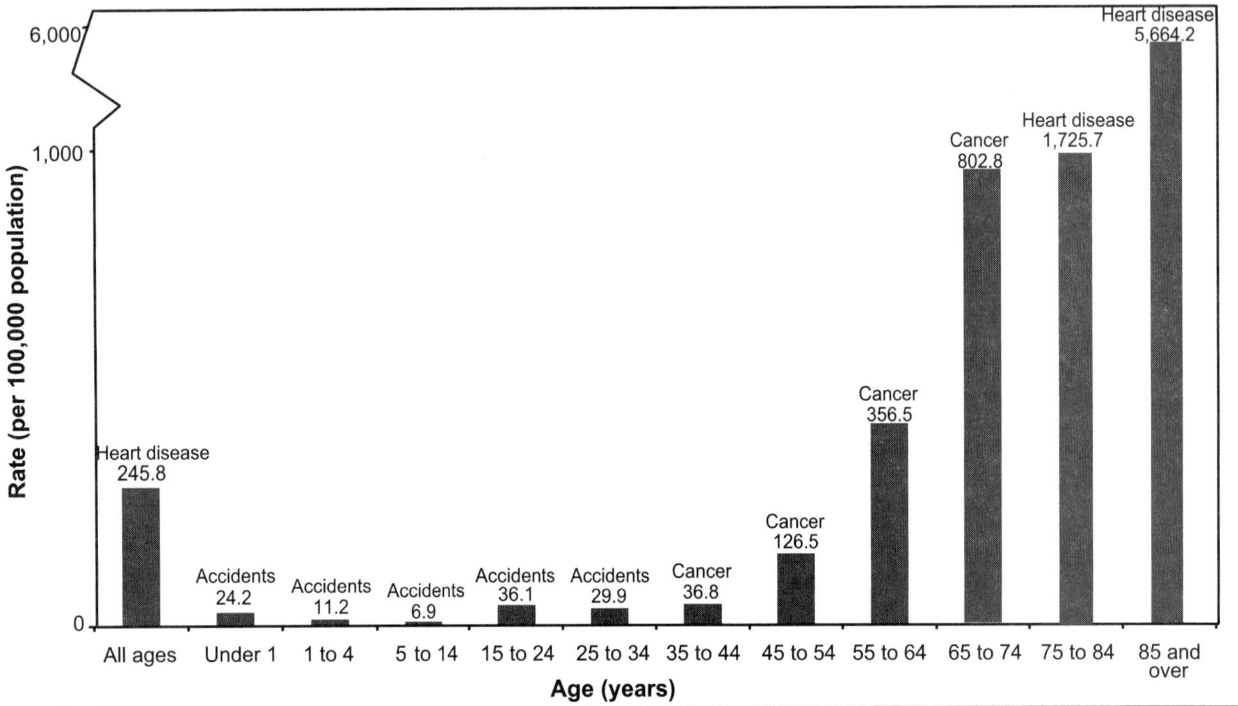

Figure II-9. Ratio of Age-Adjusted Death Rates, by Sex for the 15 Leading Causes, 2001

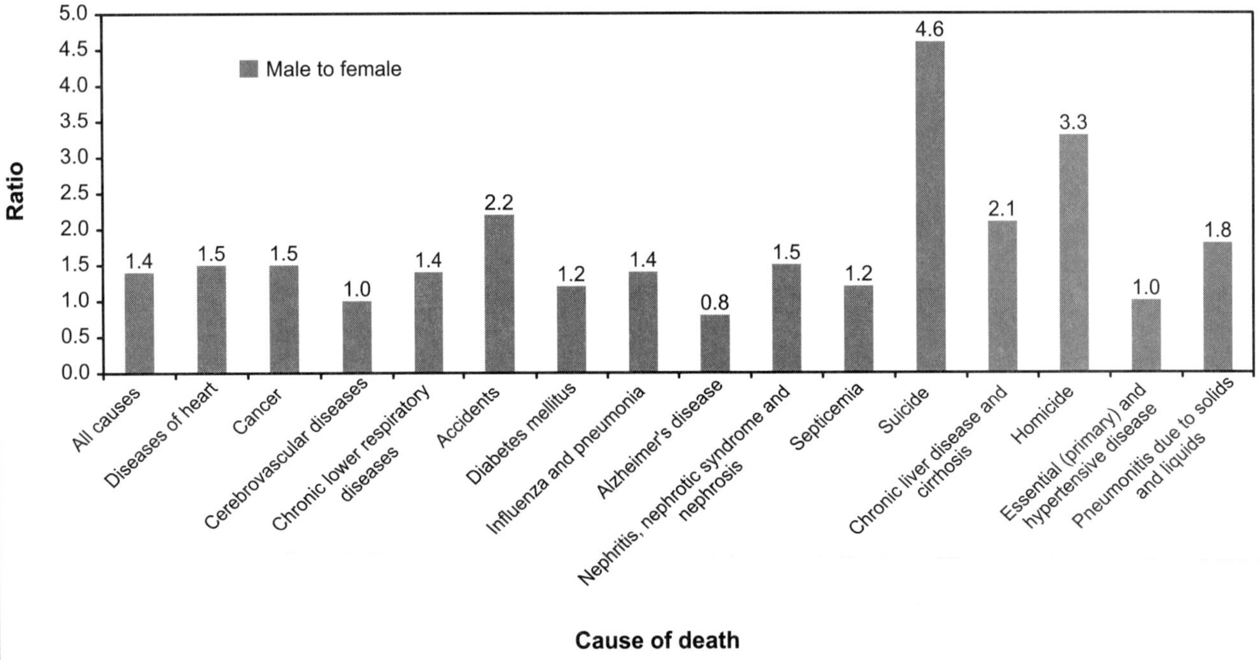

TABLE II-9A. COMPARISONS OF DEATHS, DEATH RATES, AND AGE-ADJUSTED DEATH RATES FOR 15 LEADING CAUSES OF DEATH, 2001

For each of the 15 leading causes of death in 2001, this table provides the proportion of total deaths, the percent change in age-adjusted rates between 2000 and 2001, and the ratio of age-adjusted rates between males and females, Blacks and Whites, and Hispanics and non-Hispanic Whites. Heart disease and cancer together accounted for nearly 52 percent of all deaths. The largest increases from 2000 to 2001 were homicide, Alzheimer's disease, and hypertension. The increase in homicide was mostly due to inclusion of deaths from the September 11 terrorist attacks, but the rate would have increased by 3.4 percent even without them. The rate for Alzheimer's disease has increased dramatically since 1979, due in large part to improvements in diagnosis and increased awareness of the condition in the medical community. The largest decreases were influenza and pneumonia (which fluctuates yearly), stroke, and heart disease. Rates for stroke have been generally declining since 1958 and for heart disease since 1993.

The relative risk of death in one population group compared with another can be expressed as a ratio. Ratios above the value of one indicate an increased risk in the first group, whereas ratios below one indicate an increased risk in the second group. Ratios of age-adjusted death rates show that males have higher risks than females for 12 of the 15 leading causes of death. The largest ratios were for suicide (4.6), homicide (3.3), accidents (2.2), and chronic liver disease and cirrhosis (2.1). For the largest killers, heart disease and cancer, men had a 50 percent greater risk of death than women. The only leading cause of death with a lower risk for men was Alzheimer's disease (0.8).

Age-adjusted rates for the Black population were at least 1.5 times greater than those for Whites for five of the leading causes of death. The largest ratios were for homicide (4.3), hypertension (2.9), kidney disease (2.4), septicemia (2.3), and diabetes (2.1). Rates were lower in Blacks for three leading causes of death. The lowest ratio was for suicide (0.5), indicating that Whites have double the risk of Blacks.

Age-adjusted rates for Hispanic were lower than in the non-Hispanic White population for 10 of the 15 leading causes of death. The lowest ratios were for chronic lower respiratory diseases (0.4), Alzheimer's disease (0.5), and suicide (0.5). The highest ratios were for homicide (2.1), chronic liver disease and cirrhosis (1.8), and diabetes (1.7).

Table II-9A. Comparisons of Deaths, Death Rates, and Age-Adjusted Death Rates for 15 Leading Causes of Death, 2001

(Number, percent, rate per 100,000 population.)

Rank [1]	Cause of death [2]	Number	Percent of total deaths	2001 crude death rate	Age-adjusted death rate: 2001	Age-adjusted death rate: Percent change, 2000–2001	Ratio: Male to female	Ratio: Black to White	Ratio: Hispanic to non-Hispanic White
	All causes	2 416 425	100.0	848.5	854.5	-1.7	1.4	1.3	0.8
1	Diseases of heart I00–I09, I11, I13, I20–I51	700 142	29.0	245.8	247.8	-3.8	1.5	1.3	0.8
2	Malignant neoplasms C00–C97	553 768	22.9	194.4	196.0	-1.8	1.5	1.3	0.7
3	Cerebrovascular diseases I60–I69	163 538	6.8	57.4	57.9	-4.9	1.0	1.4	0.8
4	Chronic lower respiratory diseases J40–J47	123 013	5.1	43.2	43.7	-1.1	1.4	0.7	0.4
5	Accidents (unintentional injuries) V01–X59, Y85–Y86	101 537	4.2	35.7	35.7	2.3	2.2	1.0	0.8
6	Diabetes mellitus E10–E14	71 372	3.0	25.1	25.3	1.2	1.2	2.1	1.7
7	Influenza and pneumonia J10–J18	62 034	2.6	21.8	22.0	-7.2	1.4	1.1	0.9
8	Alzheimer's disease G30	53 852	2.2	18.9	19.1	5.5	0.8	0.7	0.5
9	Nephritis,nephrotic syndrome and nephrosis N00–N07, N17–N19, N25–N27	39 480	1.6	13.9	14.0	3.7	1.5	2.4	1.0
10	Septicemia A40–A41	32 238	1.3	11.3	11.4	0.9	1.2	2.3	0.8
11	Intentional self-harm (suicide) *U03, X60–X84, Y87.0	30 622	1.3	10.8	10.7	2.9	4.6	0.5	0.5
12	Chronic liver disease and cirrhosis K70, K73–K74	27 035	1.1	9.5	9.5	0.0	2.1	1.0	1.8
13	Assault (homicide) *U01–U02, X85–Y09, Y87.1	20 308	0.8	7.1	7.1	20.3	3.3	4.3	2.1
14	Essential (primary) hypertension and hypertensive disease I10, I12	19 250	0.8	6.8	6.8	4.6	1.0	2.9	1.1
15	Pneumonitis due to solids and liquids J69	17 301	0.7	6.1	6.1	0.0	1.8	1.1	0.7
	All other causes (Residual)	400 935	16.6	140.8	X	X	X	X	X

[1]Rank based on number of deaths.
[2]Based on codes from the "Tenth Revision, International Classification of Diseases, 1992." Codes *U01-*U03 have been added as of 2001 for deaths due to terrorism.
X = Not applicable.

Figure II-10. Ratio of Age-Adjusted Death Rates by Race and Sex for the 15 Leading Causes, 2001

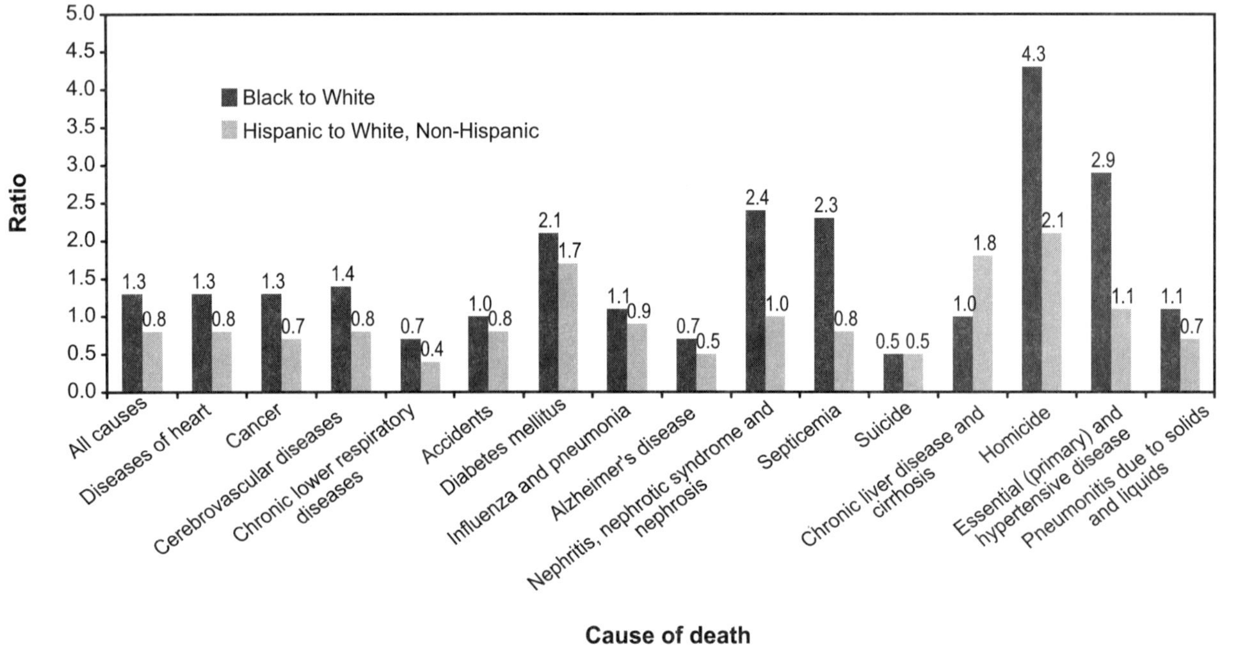

TABLE II-9B. DEATHS AND DEATH RATES FOR 2002 AND CHANGES IN AGE-ADJUSTED RATES FROM 2001 TO 2002 FOR THE 15 LEADING CAUSES OF DEATH, 2001 AND PRELIMINARY 2002

The three top killers—diseases of the heart, cancer, and stroke—all contributed to the overall decline of 0.9 percent in the 2002 age-adjusted death rates. Heart disease and cancer accounted for more than 1.2 million deaths, 51 percent of all deaths in 2002. The size of the decline ranged from 3.0 percent for heart disease to 1.0 percent for cancer. Chronic liver disease and cirrhosis also showed a reduction. A major decline occurred in homicides, and this reflected a return to baseline levels following the bulge of the terrorist attacks of 2001. (See also Table 29A.) Preliminary data have historically underestimated deaths for accidents, suicides, and homicides, because some of these causes were still under investigation at the state level.

Increases occurred for Alzheimer's disease (5.8 percent), influenza and pneumonia (3.2 percent), hypertension and hypertensive renal disease (2.9 percent), septicemia (2.6 percent), and nephritis (1.4 percent).

Table II-9B. Deaths and Death Rates for 2002 and Changes in Age-Adjusted Rates from 2001 to 2002 for the 15 Leading Causes of Death, 2001 and Preliminary 2002

(Number, rate, percent.)

Rank[1]	Cause of death[2]	Number	Death rate	Age-adjusted death rate		
				2002	2001	Percent change
	All causes	2 447 862	848.9	846.8	854.5	-0.9
1	Diseases of heart (I00-I09, I11, I13, I20-I51)	695 754	241.3	240.4	247.8	-3.0
2	Malignant neoplasms (C00-C97)	558 847	193.8	194.0	196.0	-1.0
3	Cerebrovascular diseases (I60-I69)	163 010	56.5	56.3	57.9	-2.8
4	Chronic lower respiratory diseases (J40-J47)	125 500	43.5	43.7	43.7	X
5	Accidents (unintentional injuries) (V01-X59, Y85-Y86)	102 303	35.5	35.3	35.7	-1.1
6	Diabetes mellitus (E10-E14)	73 119	25.4	25.4	25.3	0.4
7	Influenza and pneumonia (J10-J18)	65 984	22.9	22.7	22.0	3.2
8	Alzheimer's disease (G30)	58 785	20.4	20.2	19.1	5.8
9	Nephritis, nephrotic syndrome and nephrosis (N00-N07, N17-N19, N25-N27)	41 018	14.2	14.2	14.0	1.4
10	Septicemia (A40-A41)	33 881	11.7	11.7	11.4	2.6
11	Intentional self-harm (suicide) (*U03, X60-X84, Y87.0)	30 646	10.6	10.6	10.7	-0.9
12	Chronic liver disease and cirrhosis (K70, K73-K74)	27 045	9.4	9.3	9.5	-2.1
13	Essential (primary) hypertension and hypertensive renal disease	20 241	7.0	7.0	6.8	2.9
14	Pneumonitis due to solids and liquids	17 000	0.1	6.1	6.1	X
15	Assault (homicide)	17 045	5.9	5.9	7.1	-16.9
	All other causes (Residual)	416 991	144.6	X	X	X

[1]Rank based on number of deaths.
[2]Based on codes from the "Tenth Revision, International Classification of Diseases, 1992." Codes *U01-*U03 have been added as of 2001 for deaths due to terrorism.
X = Not applicable.

TABLE II-9C. DEATHS AND DEATH RATES FOR THE 10 LEADING CAUSES OF DEATH IN SPECIFIED AGE GROUPS, PRELIMINARY 2002

This table shows the 10 leading causes of death for six different age groups in 2002. Preliminary results tend to underestimate death and death rates for accidents, suicides, homicides, and sudden infant deaths, as well as misestimate seasonal causes such as influenza and pneumonia and chronic lower respiratory diseases. At young ages, 25–44 years, accidents ranked first as the cause of death, and motor vehicle accidents were a major

Table II-9C. Deaths and Death Rates for the 10 Leading Causes of Death in Specified Age Groups, Preliminary 2002

(Number, rate per 100,000 population.)

Rank[1] and age	Cause of death[2]	Number	Rate
All Ages[3]			
	All causes	2 447 862	848.9
1	Diseases of heart (I00-I09, I11, I13, I20-I51)	695 754	241.3
2	Malignant neoplasms (C00-C97)	558 847	193.8
3	Cerebrovascular diseases (I60-I69)	163 010	56.5
4	Chronic lower respiratory diseases (J40-J47)	125 500	43.5
5	Accidents (unintentional injuries) (V01-X59, Y85-Y86)	102 303	35.5
	Motor vehicle accidents (V02-V04, V09.0, V09.2, V12-V14, V19.0-V19.2, V19.4-V19.6, V20-V79, V80.3-V80.5, V81.0-V81.1, V82.0-V82.1, V83-V86, V87.0-V87.8, V88.0-V88.8, V89.0, V89.2)	44 572	15.5
	All other accidents (V01, V05-V06, V09.1, V09.3-V09.9, V10-V12, V15-V18, V19.3, V19.8-V19.9, V80.0-V80.2, V80.6-V80.9, V81.2-V81.9, V82.2-V82.9, V87.9, V88.9, V89.1, V89.3, V89.9, V90-V99, W00-X59, Y85-Y86)	57 731	20.0
6	Diabetes mellitus (E10-E14)	73 119	25.4
7	Influenza and pneumonia (J10-J18)	65 984	22.9
8	Alzheimer's disease (G30)	58 785	20.4
9	Nephritis, nephrotic syndrome and nephrosis (N00-N07, N17-N19, N25-N27)	41 018	14.2
10	Septicemia (A40-A41)	33 881	11.7
	All other causes. (Residual)	529 661	183.7
1–4 years			
	All causes	4 838	31.1
1	Accidents (unintentional injuries) (V01-X59, Y85-Y86)	1 609	10.3
	Motor vehicle accidents (V02-V04, V09.0, V09.2, V12-V14, V19.0-V19.2, V19.4-V19.6, V20-V79, V80.3-V80.5, V81.0-V81.1, V82.0-V82.1, V83-V86, V87.0-V87.8, V88.0-V88.8, V89.0, V89.2)	602	3.9
	All other accidents (V01, V05-V06, V09.1, V09.3-V09.9, V10-V12, V15-V18, V19.3, V19.8-V19.9, V80.0-V80.2, V80.6-V80.9, V81.2-V81.9, V82.2-V82.9, V87.9, V88.9, V89.1, V89.3, V89.9, V90-V99, W00-X59, Y85-Y86)	1 007	6.5
2	Congenital malformations, deformations and chromosomal abnormalities (Q00-Q99)	521	3.3
3	Malignant neoplasms (C00-C97)	401	2.6
4	Assault (homicide) (*U01-*U02, X85-Y09, Y87.1)	384	2.5
5	Diseases of heart (I00-I09, I11, I13, I20-I51)	164	1.1
6	Influenza and pneumonia (J10-J18)	107	0.7
7	Septicemia (A40-A41)	82	0.5
8	Certain conditions originating in the perinatal period (P00-P96)	63	0.4
9	Chronic lower respiratory diseases (J40-J47)	62	0.4
10	In situ neoplasms, benign neoplasms and neoplasms of uncertain or unknown behavior (D00-D48)	58	0.4
	All other causes. (Residual)	1 387	8.9
5–14 years			
	All causes	7 152	17.4
1	Accidents (unintentional injuries) (V01-X59, Y85-Y86)	2 692	6.6
	Motor vehicle accidents (V02-V04, V09.0, V09.2, V12-V14, V19.0-V19.2, V19.4-V19.6, V20-V79, V80.3-V80.5, V81.0-V81.1, V82.0-V82.1, V83-V86, V87.0-V87.8, V88.0-V88.8, V89.0, V89.2)	1 603	3.9
	All other accidents (V01, V05-V06, V09.1, V09.3-V09.9, V10-V12, V15-V18, V19.3, V19.8-V19.9, V80.0-V80.2, V80.6-V80.9, V81.2-V81.9, V82.2-V82.9, V87.9, V88.9, V89.1, V89.3, V89.9, V90-V99, W00-X59, Y85-Y86)	1 089	2.7
2	Malignant neoplasms (C00-C97)	1 061	2.6
3	Congenital malformations, deformations and chromosomal abnormalities (Q00-Q99)	395	1.0
4	Assault (homicide) (*U01-*U02, X85-Y09, Y87.1)	342	0.8
5	Intentional self-harm (suicide) (*U03, X60-X84, Y87.0)	259	0.6
6	Diseases of heart (I00-I09, I11, I13, I20-I51)	237	0.6
7	Chronic lower respiratory diseases (J40-J47)	118	0.3
8	Septicemia (A40-A41)	100	0.2
9	Influenza and pneumonia (J10-J18)	92	0.2
10	In situ neoplasms, benign neoplasms and neoplasms of uncertain or unknown behavior (D00-D48)	86	0.2
	All other causes. (Residual)	1 770	4.3
15–24 years			
	All causes	33 009	81.3
1	Accidents (unintentional injuries) (V01-X59, Y85-Y86)	15 026	37.0
	Motor vehicle accidents (V02-V04, V09.0, V09.2, V12-V14, V19.0-V19.2, V19.4-V19.6, V20-V79, V80.3-V80.5, V81.0-V81.1, V82.0-V82.1, V83-V86, V87.0-V87.8, V88.0-V88.8, V89.0, V89.2)	11 303	27.8
	All other accidents (V01, V05-V06, V09.1, V09.3-V09.9, V10-V12, V15-V18, V19.3, V19.8-V19.9, V80.0-V80.2, V80.6-V80.9, V81.2-V81.9, V82.2-V82.9, V87.9, V88.9, V89.1, V89.3, V89.9, V90-V99, W00-X59, Y85-Y86)	3 722	9.2
2	Assault (homicide) (*U01-*U02, X85-Y09, Y87.1)	5 070	12.5
3	Intentional self-harm (suicide) (*U03, X60-X84, Y87.0)	3 932	9.7
4	Malignant neoplasms (C00-C97)	1 728	4.3
5	Diseases of heart (I00-I09, I11, I13, I20-I51)	948	2.3
6	Congenital malformations, deformations and chromosomal abnormalities (Q00-Q99)	472	1.2
7	Human immunodeficiency virus (HIV) disease (B20-B24)	175	0.4
8	Chronic lower respiratory diseases (J40-J47)	172	0.4
9	Cerebrovascular diseases (I60-I69)	163	0.4
9	Diabetes mellitus (E10-E14)	163	0.4
	All other causes (Residual)	5 160	12.7

[1]Rank based on number of deaths.
[2]Based on codes from the "Tenth Revision, International Classification of Diseases, 1992." Codes *U01-*U03 have been added as of 2001 for deaths due to terrorism.
[3]Includes deaths under 1 year of age.

component. Between one and four years, congenital malformations ranked second.

For the age group of 15–24 years, homicide ranked second to accidents, and suicide ranked third. The death rate for HIV disease has declined slightly in the last few years. For the age group of 15–24 years, it showed up in seventh place, it moved to fifth place for the 25–44 year group, and thereafter its rank receded.

In the 45–64 year age group, cancer moved to rank one, followed by diseases of the heart. For persons 65 years and over, this ranking was reversed as deaths from diseases of the heart became more prevalent than cancer. Nevertheless, in the higher age group the cancer fatality rate was five times larger than for persons between 45 and 64 years old.

Table II-9C. Deaths and Death Rates for the 10 Leading Causes of Death in Specified Age Groups, Preliminary 2002—*Continued*

(Number, rate per 100,000 population.)

Rank[1] and age	Cause of death[2]	Number	Rate
25–44 years			
	All causes	132 052	155.6
1	Accidents (unintentional injuries) (V01-X59, Y85-Y86)	27 454	32.4
	Motor vehicle accidents (V02-V04, V09.0, V09.2, V12-V14, V19.0-V19.2, V19.4-V19.6, V20-V79, V80.3-V80.5, V81.0-V81.1, V82.0-V82.1, V83-V86, V87.0-V87.8, V88.0-V88.8, V89.0, V89.2)	13 902	16.4
	All other accidents (V01, V05-V06, V09.1, V09.3-V09.9, V10-V12, V15-V18, V19.3, V19.8-V19.9, V80.0-V80.2, V80.6-V80.9, V81.2-V81.9, V82.2-V82.9, V87.9, V88.9, V89.1, V89.3, V89.9, V90-V99, W00-X59, Y85-Y86)	13 552	16.0
2	Malignant neoplasms (C00-C97)	20 008	23.6
3	Diseases of heart (I00-I09, I11, I13, I20-I51)	16 155	19.0
4	Intentional self-harm (suicide) (*U03, X60-X84, Y87.0)	11 501	13.6
5	Human immunodeficiency virus (HIV) disease (B20-B24)	7 531	8.9
6	Assault (homicide) (*U01-*U02, X85-Y09, Y87.1)	7 505	8.8
7	Chronic liver disease and cirrhosis (K70, K73-K74)	3 476	4.1
8	Cerebrovascular diseases (I60-I69)	2 934	3.5
9	Diabetes mellitus (E10-E14)	2 747	3.2
10	Influenza and pneumonia (J10-J18)	1 294	1.5
	All other causes. (Residual)	31 447	37.1
45–64 years			
	All causes	425 412	637.9
1	Malignant neoplasms (C00-C97)	143 416	215.1
2	Diseases of heart (I00-I09, I11, I13, I20-I51)	100 378	150.5
3	Accidents (unintentional injuries) (V01-X59, Y85-Y86)	21 578	32.4
	Motor vehicle accidents (V02-V04, V09.0, V09.2, V12-V14, V19.0-V19.2, V19.4-V19.6, V20-V79, V80.3-V80.5, V81.0-V81.1, V82.0-V82.1, V83-V86, V87.0-V87.8, V88.0-V88.8, V89.0, V89.2)	9 522	14.3
	All other accidents (V01, V05-V06, V09.1, V09.3-V09.9, V10-V12, V15-V18, V19.3, V19.8-V19.9, V80.0-V80.2, V80.6-V80.9, V81.2-V81.9, V82.2-V82.9, V87.9, V88.9, V89.1, V89.3, V89.9, V90-V99, W00-X59, Y85-Y86)	12 056	18.1
4	Cerebrovascular diseases (I60-I69)	15 869	23.8
5	Diabetes mellitus (E10-E14)	15 452	23.2
6	Chronic lower respiratory diseases (J40-J47)	14 720	22.1
7	Chronic liver disease and cirrhosis (K70, K73-K74)	13 131	19.7
8	Intentional self-harm (suicide) (*U03, X60-X84, Y87.0)	9 517	14.3
9	Human immunodeficiency virus (HIV) disease (B20-B24)	5 729	8.6
10	Septicemia (A40-A41)	5 416	8.1
	All other causes. (Residual)	80 206	120.3
65 years and over			
	All causes	1 817 095	5 103.9
1	Diseases of heart (I00-I09, I11, I13, I20-I51)	577 353	1 621.7
2	Malignant neoplasms (C00-C97)	392 145	1 101.5
3	Cerebrovascular diseases (I60-I69)	143 780	403.9
4	Chronic lower respiratory diseases (J40-J47)	109 158	306.6
5	Influenza and pneumonia (J10-J18)	59 235	166.4
6	Alzheimer's disease (G30)	58 205	163.5
7	Diabetes mellitus (E10-E14)	54 717	153.7
8	Nephritis, nephrotic syndrome and nephrosis (N00-N07, N17-N19, N25-N27)	34 389	96.6
9	Accidents (unintentional injuries) (V01-X59, Y85-Y86)	32 973	92.6
	Motor vehicle accidents (V02-V04, V09.0, V09.2, V12-V14, V19.0-V19.2, V19.4-V19.6, V20-V79, V80.3-V80.5, V81.0-V81.1, V82.0-V82.1, V83-V86, V87.0-V87.8, V88.0-V88.8, V89.0, V89.2)	7 479	21.0
	All other accidents (V01, V05-V06, V09.1, V09.3-V09.9, V10-V12, V15-V18, V19.3, V19.8-V19.9, V80.0-V80.2, V80.6-V80.9, V81.2-V81.9, V82.2-V82.9, V87.9, V88.9, V89.1, V89.3, V89.9, V90-V99, W00-X59, Y85-Y86)	25 493	71.6
10	Septicemia (A40-A41)	26 688	75.0
	All other causes. (Residual)	328 452	922.6

[1]Rank based on number of deaths.
[2]Based on codes from the "Tenth Revision, International Classification of Diseases, 1992." Codes *U01-*U03 have been added as of 2001 for deaths due to terrorism.

TABLE II-10. NUMBER OF DEATHS FOR 113 SELECTED CAUSES, BY AGE, 2001

Of the 2.4 million deaths in 2001, some 550,000 were caused by cancer. Lung cancer, with 156,000 deaths, was the leading type of fatal cancer. Major cardiovascular diseases claimed 922,000 deaths in 2001; the principal component was diseases of the heart, with 700,000 fatalities. Stroke is the other major component of cardiovascular diseases. Accidents accounted for about 100,000 deaths.

This table has a very detailed listing of causes of death. For example, there are many detailed classifications for cancer and for heart diseases, as well as a number of categories for types of accidents. The classifications are based on the "Tenth Revision of International Classifications of Diseases (1992)."

Table II-10. Number of Deaths for 113 Selected Causes, by Age, 2001

(Number.)

Cause of death [1]	All ages	Under 1 year	1–4 years	5–14 years	15–24 years	25–34 years	35–44 years	45–54 years	55–64 years	65–74 years	75–84 years	85 years and over	Age not stated
All causes	2 416 425	27 568	5 108	7 094	32 252	41 683	91 674	168 065	244 139	430 960	701 929	665 531	422
Salmonella infections (A01-A02)	41	2	-	-	-	2	1	3	3	7	14	9	-
Shigellosis and amebiasis (A03, A06)	7	1	1	1	-	-	-	3	-	-	1	-	-
Certain other intestinal infections (A04, A07-A09)	1 537	10	18	4	-	4	12	42	81	235	566	565	-
Tuberculosis (A16-A19)	764	2	1	-	12	25	40	84	116	152	206	125	1
Respiratory tuberculosis (A16)	600	2	1	-	7	17	28	61	96	115	166	106	1
Other tuberculosis (A17-A19)	164	-	-	-	5	8	12	23	20	37	40	19	-
Whooping cough (A37)	17	16	1	-	-	-	-	-	-	-	-	-	-
Scarlet fever and erysipelas (A38, A46)	3	-	-	1	-	-	-	-	-	1	-	1	-
Meningococcal infection (A39)	199	16	19	26	55	10	13	13	15	15	8	9	-
Septicemia (A40-A41)	32 238	312	108	69	133	281	829	1 977	3 111	5 998	10 351	9 069	-
Syphilis (A50-A53)	36	1	-	1	-	1	3	3	4	5	13	5	-
Acute poliomyelitis (A80)	-	-	-	-	-	-	-	-	-	-	-	-	-
Arthropod-borne viral encephalitis (A83-A84, A85.2)	9	1	-	-	-	-	-	1	-	5	2	-	-
Measles (B05)	1	-	-	-	-	-	1	-	-	-	-	-	-
Viral hepatitis (B15-B19)	5 585	1	-	3	20	74	797	2 060	1 014	836	613	166	1
Human immunodeficiency virus (HIV) disease (B20-B24)	14 175	10	13	38	225	2 101	5 867	4 120	1 317	384	88	10	2
Malaria (B50-B54)	9	-	-	1	-	-	2	2	2	1	1	-	-
Other and unspecified infectious and parasitic diseases and their sequelae (A00, A05, A20-A36, A42-A44, A48-A49, A54-A79, A81-A82, A85.0-A85.1, A85.8, A86-B04, B06-B09, B25-B49, B55-B99)	5 715	191	64	64	86	147	369	663	771	1 143	1 330	887	-
Malignant neoplasms (C00-C97)	553 768	66	420	1 008	1 704	3 994	16 569	49 562	90 223	147 018	165 445	77 751	8
Malignant neoplasms of lip, oral cavity and pharynx (C00-C14)	7 702	-	-	4	29	56	283	1 076	1 740	1 912	1 787	814	1
Malignant neoplasm of esophagus (C15)	12 530	-	-	-	6	29	309	1 408	2 554	3 779	3 307	1 137	1
Malignant neoplasm of stomach (C16)	12 319	-	-	-	20	122	464	1 127	1 771	3 021	3 793	2 001	-
Malignant neoplasms of colon, rectum and anus (C18-C21)	56 887	-	-	-	46	299	1 433	4 548	7 984	13 791	17 527	11 259	-
Malignant neoplasms of liver and intrahepatic bile ducts (C22)	13 351	5	18	15	29	75	401	1 868	2 351	3 440	3 649	1 500	-
Malignant neoplasm of pancreas (C25)	29 803	-	1	2	2	55	574	2 410	4 868	8 158	9 322	4 410	1
Malignant neoplasm of larynx (C32)	3 797	-	-	-	-	3	86	419	913	1 170	910	296	-
Malignant neoplasms of trachea, bronchus and lung (C33-C34)	156 058	-	5	2	20	157	2 796	12 041	29 777	51 225	46 700	13 331	4
Malignant melanoma of skin (C43)	7 542	-	-	-	41	186	622	1 208	1 333	1 689	1 678	785	-
Malignant neoplasm of breast (C50)	41 809	-	-	-	9	463	2 823	6 593	7 632	8 715	9 663	5 911	-
Malignant neoplasm of cervix uteri (C53)	4 092	-	-	1	12	209	711	953	723	585	583	315	-
Malignant neoplasms of corpus uteri and uterus, part unspecified (C54-C55)	6 784	-	-	-	-	38	139	566	1 105	1 784	2 063	1 088	1
Malignant neoplasm of ovary (C56)	14 414	-	-	2	21	102	472	1 626	2 612	3 666	4 170	1 743	-
Malignant neoplasm of prostate (C61)	30 719	-	-	-	1	1	30	412	1 907	6 543	12 976	8 849	-
Malignant neoplasms of kidney and renal pelvis (C64-C65)	12 078	1	22	35	20	52	358	1 311	2 328	3 221	3 306	1 424	-
Malignant neoplasm of bladder (C67)	12 225	-	-	1	3	15	128	475	1 265	2 814	4 488	3 036	-
Malignant neoplasms of meninges, brain and other parts of central nervous system (C70-C72)	12 609	16	100	300	224	416	1 028	1 951	2 464	2 823	2 566	721	-
Malignant neoplasms of lymphoid, hematopoietic and related tissue (C81-C96)	56 341	27	157	367	705	988	1 857	4 064	7 455	13 659	18 257	8 805	-
Hodgkin's disease (C81)	1 323	-	-	6	78	167	161	182	143	211	278	97	-
Non-Hodgkin's lymphoma (C82-C85)	22 305	1	1	38	148	298	756	1 732	3 098	5 358	7 369	3 506	-
Leukemia (C91-C95)	21 451	25	155	323	479	513	779	1 475	2 535	4 886	6 696	3 585	-
Multiple myeloma and immunoproliferative neoplasms (C88, C90)	11 172	-	-	-	-	9	157	671	1 671	3 181	3 886	1 597	-
Other and unspecified malignant neoplasms of lymphoid, hematopoietic and related tissue (C96)	90	1	1	-	-	1	4	4	8	2 328	-	20	-
All other and unspecified malignant neoplasms (C17, C23-C24, C26-C31, C37-C41, C44-C49, C51-C52, C57-C60, C62-C63, C66, C68-C69, C73-C80, C97)	62 708	17	117	279	516	728	2 055	5 506	9 441	15 023	18 700	10 326	-
In situ neoplasms, benign neoplasms and neoplasms of uncertain or unknown behavior (D00-D48)	13 676	51	58	105	102	151	366	719	1 253	2 574	4 654	3 643	-
Anemias (D50-D64)	4 627	16	22	55	109	160	215	232	249	492	1 226	1 851	-
Diabetes mellitus (E10-E14)	71 372	2	6	39	151	595	1 958	5 343	9 570	16 731	22 805	14 171	1
Nutritional deficiencies (E40-E64)	3 704	4	4	2	10	14	44	110	153	374	1 100	1 889	-
Malnutrition (E40-E46)	3 454	2	-	2	9	13	39	99	146	347	1 025	1 772	-
Other nutritional deficiencies (E50-E64)	250	2	4	-	1	1	5	11	7	27	75	117	-
Meningitis (G00, G03)	727	78	26	24	40	41	69	105	113	94	96	41	-
Parkinson's disease (G20-G21)	16 544	-	-	-	-	2	15	45	301	2 147	8 125	5 909	-
Alzheimer's disease (G30)	53 852	-	-	-	-	2	3	73	529	3 422	18 542	31 281	-

[1]Based on codes from the "Tenth Revision, International Classification of Diseases, 1992." Codes *U01-*U03 have been added as of 2001 for deaths due to terrorism.
- = Quantity zero.

Table II-10. Number of Deaths for 113 Selected Causes, by Age, 2001—*Continued*

(Number.)

Cause of death [1]	All ages	Under 1 year	1–4 years	5–14 years	15–24 years	25–34 years	35–44 years	45–54 years	55–64 years	65–74 years	75–84 years	85 years and over	Age not stated
Major cardiovascular diseases (I00-I78)	922 334	600	285	365	1 280	4 038	16 686	44 450	76 224	148 409	291 995	337 930	72
Diseases of heart (I00-I09, I11, I13, I20-I51)	700 142	479	225	272	999	3 160	13 326	36 399	62 486	116 299	216 992	249 439	66
Acute rheumatic fever and chronic rheumatic heart diseases (I00-I09)	3 489	3	1	5	11	34	130	232	361	744	1 188	780	-
Hypertensive heart disease (I11)	24 689	-	-	1	24	246	1 091	2 464	2 769	3 619	6 119	8 345	11
Hypertensive heart and renal disease (I13)	2 826	1	-	-	7	28	99	146	247	417	855	1 026	-
Ischemic heart diseases (I20-I25)	502 189	26	9	20	163	1 116	7 534	25 073	46 574	87 463	159 854	174 309	48
Acute myocardial infarction (I21-I22)	184 757	15	2	7	64	475	3 176	10 817	19 830	35 380	59 254	55 727	10
Other acute ischemic heart diseases (I24)	3 351	3	-	2	7	14	94	307	463	645	871	945	-
Other forms of chronic ischemic heart disease (I20, I25)	314 081	8	7	11	92	627	4 264	13 949	26 281	51 438	99 729	117 637	38
Atherosclerotic cardiovascular disease, so described (I25.0)	68 103	-	-	-	20	179	1 457	5 047	8 438	12 005	19 196	21 736	25
All other forms of chronic ischemic heart disease (I20, I25.1-I25.9)	245 978	8	7	11	72	448	2 807	8 902	17 843	39 433	80 533	95 901	13
Other heart diseases (I26-I51)	166 949	449	215	246	794	1 736	4 472	8 484	12 535	24 056	48 976	64 979	7
Acute and subacute endocarditis (I33)	1 178	1	-	3	8	29	117	186	160	247	316	111	-
Diseases of pericardium and acute myocarditis (I30-I31, I40)	839	32	26	21	40	50	93	111	106	120	154	85	1
Heart failure (I50)	56 934	24	17	15	34	92	307	954	2 374	6 274	17 134	29 708	1
All other forms of heart disease (I26-I28, I34-I38, I42-I49, I51)	107 998	392	172	207	712	1 565	3 955	7 233	9 895	17 415	31 372	35 075	5
Essential (primary) hypertension and hypertensive renal disease (I10, I12)	19 250	-	1	1	23	104	331	931	1 462	2 832	5 993	7 572	-
Cerebrovascular diseases (I60-I69)	163 538	108	54	80	196	601	2 491	5 910	9 608	22 598	55 822	66 066	4
Atherosclerosis (I70)	14 086	1	1	1	1	3	40	186	473	1 335	4 156	7 888	1
Other diseases of circulatory system (I71-I78)	25 318	12	4	11	61	170	498	1 024	2 195	5 345	9 032	6 965	1
Aortic aneurysm and dissection (I71)	15 234	-	2	4	43	117	332	661	1 394	3 517	5 751	3 412	1
Other diseases of arteries, arterioles and capillaries (I72-I78)	10 084	12	2	7	18	53	166	363	801	1 828	3 281	3 553	-
Other disorders of circulatory system (I80-I99)	4 665	22	5	-	38	124	285	454	563	841	1 183	1 149	1
Influenza and pneumonia (J10-J18)	62 034	299	112	92	181	339	983	1 801	2 704	6 650	18 677	30 191	5
Influenza (J10-J11)	257	7	6	12	7	10	6	15	21	21	56	96	-
Pneumonia (J12-J18)	61 777	292	106	80	174	329	977	1 786	2 683	6 629	18 621	30 095	5
Other acute lower respiratory infections (J20-J22)	347	50	18	3	1	2	12	14	16	28	58	145	-
Acute bronchitis and bronchiolitis (J20-J21)	263	49	18	3	1	2	11	12	14	23	39	91	-
Unspecified acute lower respiratory infection (J22)	84	1	-	-	-	-	1	2	2	5	19	54	-
Chronic lower respiratory diseases (J40-J47)	123 013	41	43	104	171	290	969	3 324	11 166	30 751	47 762	28 391	1
Bronchitis, chronic and unspecified (J40-J42)	959	21	8	-	5	6	21	42	61	133	261	401	-
Emphysema (J43)	16 242	3	-	1	2	7	75	448	1 829	4 804	6 337	2 736	-
Asthma (J45-J46)	4 269	10	31	99	140	233	455	603	553	634	802	709	-
Other chronic lower respiratory diseases (J44, J47)	101 543	7	4	4	24	44	418	2 231	8 723	25 180	40 362	24 545	1
Pneumoconioses and chemical effects (J60-J66, J68)	1 136	-	1	-	-	1	5	11	64	266	516	272	-
Pneumonitis due to solids and liquids (J69)	17 301	12	14	15	34	79	179	378	659	1 830	5 759	8 342	-
Other diseases of respiratory system (J00-J06, J30-J39, J67, J70-J98)	26 178	294	93	78	141	216	570	1 356	2 432	5 510	9 202	6 283	3
Peptic ulcer (K25-K28)	4 491	-	1	5	6	32	118	284	418	718	1 438	1 470	1
Diseases of appendix (K35-K38)	428	1	8	6	15	15	26	40	45	89	90	93	-
Hernia (K40-K46)	1 522	43	2	3	3	15	34	82	131	213	443	553	-
Chronic liver disease and cirrhosis (K70, K73-K74)	27 035	3	-	2	26	387	3 336	7 259	5 750	5 486	3 801	979	6
Alcoholic liver disease (K70)	12 207	-	-	-	8	264	2 145	4 203	2 913	1 848	717	104	5
Other chronic liver disease and cirrhosis (K73-K74)	14 828	3	-	2	18	123	1 191	3 056	2 837	3 638	3 084	875	1
Cholelithiasis and other disorders of gallbladder (K80-K82)	2 965	1	-	1	7	19	44	137	242	526	945	1 043	-
Nephritis, nephrotic syndrome and nephrosis (N00-N07, N17-N19, N25-N27)	39 480	133	16	18	92	225	768	1 822	3 284	7 356	13 097	12 668	1
Acute and rapidly progressive nephritic and nephrotic syndrome (N00-N01, N04)	156	8	4	1	-	3	8	6	13	27	48	38	-
Chronic glomerulonephritis, nephritis and nephropathy, not specified as acute or chronic, and renal sclerosis, unspecified (N02-N03, N05-N07, N26)	511	-	2	1	8	12	27	41	40	111	147	122	-
Renal failure (N17-N19)	38 784	125	9	16	82	209	732	1 770	3 226	7 211	12 898	12 505	1
Other disorders of kidney (N25, N27)	29	-	1	-	2	1	1	5	5	7	4	3	-
Infections of kidney (N10-N12, N13.6, N15.1)	748	2	4	-	5	11	33	46	46	87	249	265	-
Hyperplasia of prostate (N40)	439	-	-	-	-	-	-	2	9	45	141	242	-
Inflammatory diseases of female pelvic organs (N70-N76)	127	1	-	-	2	3	10	12	12	18	31	38	-
Pregnancy, childbirth and the puerperium (O00-O99)	416	X	X	3	110	190	106	6	X	X	X	X	-
Pregnancy with abortive outcome (O00-O07)	38	X	X	2	13	14	9	-	X	X	X	X	-
Other complications of pregnancy, childbirth and the puerperium (O10-O99)	378	X	X	1	97	176	97	6	X	X	X	X	-
Certain conditions originating in the perinatal period (P00-P96)	13 887	13 734	72	36	17	6	9	5	2	3	1	1	1
Congenital malformations, deformations and chromosomal abnormalities (Q00-Q99)	10 442	5 513	558	375	505	458	586	636	581	448	465	317	-
Symptoms, signs and abnormal clinical and laboratory findings, not elsewhere classified (R00-R99)	32 326	3 491	272	181	949	1 530	3 126	3 504	2 292	2 774	5 003	9 089	115
All other diseases (Residual)	186 406	1 131	646	860	1 799	3 072	8 084	14 203	16 048	25 803	49 711	65 032	17

[1]Based on codes from the "Tenth Revision, International Classification of Diseases, 1992." Codes *U01-*U03 have been added as of 2001 for deaths due to terrorism.
- = Quantity zero.
X = Not applicable.

Table II-10. Number of Deaths for 113 Selected Causes, by Age, 2001—*Continued*

(Number.)

Cause of death [1]	All ages	Under 1 year	1–4 years	5–14 years	15–24 years	25–34 years	35–44 years	45–54 years	55–64 years	65–74 years	75–84 years	85 years and over	Age not stated
Accidents (unintentional injuries) (V01-X59, Y85-Y86)	101 537	976	1 714	2 836	14 411	11 839	15 945	13 344	7 658	7 835	12 688	12 171	120
Transport accidents (V01-V99, Y85)	47 288	151	678	1 823	11 139	7 454	7 827	6 251	3 861	3 282	3 507	1 272	43
Motor vehicle accidents (V02-V04, V09.0, V09.2, V12-V14, V19.0-V19.2, V19.4-V19.6, V20-V79, V80.3-80.5, V81.0-V81.1, V82.0-V82.1, V83-V86, V87.0-V87.8, V88.0-V88.8, V89.0, V89.2)	43 788	144	626	1 686	10 725	6 937	7 083	5 564	3 465	2 990	3 312	1 223	33
Other land transport accidents (V01, V05-V06, V09.1, V09.3-V09.9, V10-V11, V15-V18, V19.3, V19.8-19.9, V80.0-V80.2, V80.6-V80.9, V81.2-V81.9, V82.2-V82.9, V87.9, V88.9, V89.1, V89.3, V89.9)	1 405	3	28	80	198	177	285	247	133	117	98	29	10
Water, air and space, and other and unspecified transport accidents and their sequelae (V90-V99, Y85)	2 095	4	24	57	216	340	459	440	263	175	97	20	-
Nontransport accidents (W00-X59, Y86)	54 249	825	1 036	1 013	3 272	4 385	8 118	7 093	3 797	4 553	9 181	10 899	77
Falls (W00-W19)	15 019	23	32	66	256	340	647	1 024	1 004	1 833	4 440	5 350	4
Accidental discharge of firearms (W32-W34)	802	-	15	57	206	122	146	103	69	33	41	8	2
Accidental drowning and submersion (W65-W74)	3 281	68	458	333	596	374	462	359	206	159	157	75	34
Accidental exposure to smoke, fire and flames (X00-X09)	3 309	50	225	252	209	247	441	419	378	411	423	251	3
Accidental poisoning and exposure to noxious substances (X40-X49)	14 078	15	31	50	1 362	2 507	5 036	3 547	798	291	279	152	10
Other and unspecified nontransport accidents and their sequelae (W20-W31, W35-W64, W75-W99, X10-X39, X50-X59, Y86)	17 760	669	275	255	643	795	1 386	1 641	1 342	1 826	3 841	5 063	24
Intentional self-harm (suicide) (*U03, X60-X84, Y87.0) [2]	30 622	X	X	279	3 971	5 070	6 635	5 942	3 317	2 432	2 192	769	15
Intentional self-harm (suicide) by discharge of firearms (X72-X74)	16 869	X	X	90	2 130	2 564	3 030	3 023	2 083	1 758	1 648	537	6
Intentional self-harm (suicide) by other and unspecified means and their sequelae (*U03, X60-X71, X75-X84, Y87.0) [2]	13 753	X	X	189	1 841	2 506	3 605	2 919	1 234	674	544	232	9
Assault (homicide) (*U01-*U02, X85-Y09, Y87.1) [2]	20 308	332	415	326	5 297	5 204	4 268	2 467	1 018	532	312	105	32
Assault (homicide) by discharge of firearms (*U01.4, X93-X95)	11 348	11	55	180	4 200	3 308	1 978	934	364	192	99	16	11
Assault (homicide) by other and unspecified means and their sequelae (*U01.0-*U01.3, *U01.5-*U01.9, *U02, X85-X92, X96-Y09, Y87.1) [2]	8 960	321	360	146	1 097	1 896	2 290	1 533	654	340	213	89	21
Legal intervention (Y35, Y89.0)	396	-	-	-	88	107	115	51	24	6	4	1	-
Events of undetermined intent (Y10-Y34, Y87.2, Y89.9)	4 198	83	44	51	412	728	1 396	990	225	99	108	44	18
Discharge of firearms, undetermined intent (Y22-Y24)	231	-	-	6	64	48	50	29	12	10	11	1	-
Other and unspecified events of undetermined intent and their sequelae (Y10-Y21, Y25-Y34, Y87.2, Y89.9)	3 967	83	44	45	348	680	1 346	961	213	89	97	43	18
Operations of war and their sequelae (Y36, Y89.1)	17	-	-	-	1	-	1	2	-	4	8	1	-
Complications of medical and surgical care (Y40-Y84, Y88)	3 021	26	24	14	43	79	172	283	384	567	864	564	1

[1]Based on codes from the "Tenth Revision, International Classification of Diseases, 1992." Codes *U01-*U03 have been added as of 2001 for deaths due to terrorism.
[2]Figures include September 11, 2001 related deaths for which death certificates were filed as of 10/24/02.
- = Quantity zero.
X = Not applicable.

TABLE II-10A. DEATHS, DEATH RATES, AND AGE-ADJUSTED DEATH RATES FOR 113 SELECTED CAUSES, 2001 AND PRELIMINARY 2002

This table compares data for 113 selected causes of death for 2002 (preliminary) and 2001. Such comparisons for the 10 and 15 top causes have been discussed previously in connection with Table 9B and Table 9C. Preliminary results tend to underestimate deaths for accidents, suicides, homicides, and sudden infant death, as well as misestimate seasonal causes such as influenza, pneumonia, and chronic respiratory diseases.

In addition to the long list of causes of death, Table 10A has an addendum, which regroups certain causes of death. The addendum provides a different framework for classifying injury deaths. The focus is on mechanism, such as firearm or drug poisoning rather than intent, such as accident, suicide, or homicide. The categories are injury by firearms, drug-induced deaths, alcohol-induced deaths, and injury at work. The latter item is derived from a separate entry on the death certificate that asks the medical certifier whether the death resulted from an injury sustained at work. The entry thus provides for a possible additional cause of death on the death certificate.

Table II-10A. Deaths, Death Rates, and Age-Adjusted Death Rates for 113 Selected Causes, 2001 and Preliminary 2002

(Number, rate per 100,000 population.)

Cause of death[1]	2002			2001		
	Number	Rate	Age-adjusted rate	Number	Rate	Age-adjusted rate
All causes	2 447 862	848.9	846.8	2 416 425	848.5	854.5
Salmonella infections (A01-A02)	22	0.0	0.0	41	0.0	0.0
Shigellosis and amebiasis (A03, A06)	7	*	*	7	*	*
Certain other intestinal infections, (A04, A07-A09)	2 183	0.8	0.7	1 537	0.5	0.5
Tuberculosis (A16-A19)	802	0.3	0.3	764	0.3	0.3
Respiratory tuberculosis (A16)	642	0.2	0.2	600	0.2	0.2
Other tuberculosis (A17-A19)	160	0.1	0.0	164	0.1	0.1
Whooping cough (A37)	17	*	*	17	*	*
Scarlet fever and erysipelas (A38, A46)	2	*	*	3	*	*
Meningococcal infection (A39)	155	0.1	0.0	199	0.1	0.1
Septicemia, (A40-A41)	33 881	11.7	11.7	32 238	11.3	11.4
Syphilis (A50-A53)	42	0.0	0.0	36	0.0	0.0
Acute poliomyelitis (A80)	*	*	*	*	*	*
Arthropod-borne viral encephalitis (A83-A84, A85.2)	69	0.0	0.0	9	*	*
Measles (B05)	1	*	*	1	*	*
Viral hepatitis (B15-B19)	5 706	2.0	2.0	5 585	2.0	2.0
Human immunodeficiency virus (HIV) disease (B20-B24)	13 991	4.9	4.9	14 175	5.0	5.0
Malaria (B50-B54)	13	*	*	9	*	*
Other and unspecified infectious and parasitic diseases and their sequelae (A00, A05, A20-A36, A42-A44, A48-A49, A54-A79, A81-A82, A85.0-A85.1, A85.8, A86-B04, B06-B09, B25-B49, B55-B99)	6 641	2.3	2.3	5 715	2.0	2.0
Malignant neoplasms (C00-C97)	558 847	193.8	194.0	553 768	194.4	196.0
Malignant neoplasms of lip, oral cavity and pharynx (C00-C14)	7 732	2.7	2.7	7 702	2.7	2.7
Malignant neoplasm of esophagus (C15)	12 766	4.4	4.4	12 530	4.4	4.4
Malignant neoplasm of stomach (C16)	12 265	4.3	4.3	12 319	4.3	4.4
Malignant neoplasms of colon, rectum and anus (C18-C21)	56 866	19.7	19.7	56 887	20.0	20.1
Malignant neoplasms of liver and intrahepatic bile ducts (C22)	14 046	4.9	4.9	13 351	4.7	4.7
Malignant neoplasm of pancreas (C25)	30 323	10.5	10.5	29 803	10.5	10.5
Malignant neoplasm of larynx (C32)	3 731	1.3	1.3	3 797	1.3	1.3
Malignant neoplasms of trachea, bronchus and lung (C33-C34)	158 258	54.9	55.1	156 058	54.8	55.3
Malignant melanoma of skin (C43)	7 563	2.6	2.6	7 542	2.6	2.7
Malignant neoplasm of breast (C50)	41 989	14.6	14.5	41 809	14.7	14.8
Malignant neoplasm of cervix uteri (C53)	3 965	1.4	1.4	4 092	1.4	1.4
Malignant neoplasms of corpus uteri and uterus, part unspecified (C54-C55)	6 876	2.4	2.4	6 784	2.4	2.4
Malignant neoplasm of ovary (C56)	14 678	5.1	5.1	14 414	5.1	5.1
Malignant neoplasm of prostate (C61)	30 571	10.6	10.6	30 719	10.8	10.9
Malignant neoplasms of kidney and renal pelvis (C64-C65)	12 184	4.2	4.2	12 078	4.2	4.3
Malignant neoplasm of bladder (C67)	12 670	4.4	4.4	12 225	4.3	4.3
Malignant neoplasms of meninges, brain and other parts of central nervous system (C70-C72)	12 777	4.4	4.4	12 609	4.4	4.5
Malignant neoplasms of lymphoid, hematopoietic and related tissue (C81-C96)	56 377	19.6	19.6	56 341	19.8	20.0
Hodgkin's disease, (C81)	1 356	0.5	0.5	1 323	0.5	0.5
Non-Hodgkin's lymphoma (C82-C85)	21 998	7.6	7.6	22 305	7.8	7.9
Leukemia (C91-C95)	21 575	7.5	7.5	21 451	7.5	7.6
Multiple myeloma and immunoproliferative neoplasms, (C88, C90)	11 374	3.9	4.0	11 172	3.9	4.0
Other and unspecified malignant neoplasms of lymphoid, hematopoietic and related tissue (C96)	74	0.0	0.0	90	0.0	0.0
All other and unspecified malignant neoplasms (C17, C23-C24, C26-C31, C37-C41, C44-C49, C51-C52, C57-C60, C62-C63, C66, C68-C69, C73-C80, C97)	63 209	21.9	21.9	62 708	22.0	22.2
In situ neoplasms, benign neoplasms and neoplasms of uncertain or unknown behavior (D00-D48)	13 370	4.6	4.6	13 676	4.8	4.9
Anemias (D50-D64)	4 601	1.6	1.6	4 627	1.6	1.6
Diabetes mellitus (E10-E14)	73 119	25.4	25.4	71 372	25.1	25.3
Nutritional deficiencies (E40-E64)	3 785	1.3	1.3	3 704	1.3	1.3
Malnutrition (E40-E46)	3 515	1.2	1.2	3 454	1.2	1.2
Other nutritional deficiencies (E50-E64)	270	0.1	0.1	250	0.1	0.1
Meningitis (G00, G03)	694	0.2	0.2	727	0.3	0.3
Parkinson's disease (G20-G21)	16 961	5.9	5.9	16 544	5.8	5.9
Alzheimer's disease (G30)	58 785	20.4	20.2	53 852	18.9	19.1
Major cardiovascular diseases (I00-I78)	917 839	318.3	317.1	922 334	323.9	326.5
Diseases of heart (I00-I09, I11, I13, I20-I51)	695 754	241.3	240.4	700 142	245.8	247.8
Acute rheumatic fever and chronic rheumatic heart diseases (I00-I09)	3 743	1.3	1.3	3 489	1.2	1.2
Hypertensive heart disease (I11)	25 938	9.0	8.9	24 689	8.7	8.7
Hypertensive heart and renal disease (I13)	2 888	1.0	1.0	2 826	1.0	1.0
Ischemic heart diseases (I20-I25)	493 542	171.1	170.6	502 189	176.3	177.8
Acute myocardial infarction (I21-I22)	179 869	62.4	62.2	184 757	64.9	65.4
Other acute ischemic heart diseases (I24)	3 416	1.2	1.2	3 351	1.2	1.2
Other forms of chronic ischemic heart disease (I20, I25)	310 257	107.6	107.2	314 081	110.3	111.2
Atherosclerotic cardiovascular disease, so described (I25.0)	67 104	23.3	23.1	68 103	23.9	24.1
All other forms of chronic ischemic heart disease (I20, I25.1 I25.9)	243 153	84.3	84.0	245 978	86.4	87.1
Other heart diseases (I26-I51)	169 643	58.8	58.6	166 949	58.6	59.1
Acute and subacute endocarditis (I33)	1 149	0.4	0.4	1 178	0.4	0.4
Diseases of pericardium and acute myocarditis (I30-I31, I40)	802	0.3	0.3	839	0.3	0.3
Heart failure (I50)	56 790	19.7	19.6	56 934	20.0	20.2
All other forms of heart disease (I26-I28, I34-I38, I42-I49, I51)	110 902	38.5	38.3	107 998	37.9	38.2
Essential (primary) hypertension and hypertensive renal disease (I10, I12)	20 241	7.0	7.0	19 250	6.8	6.8
Cerebrovascular diseases (I60-I69)	163 010	56.5	56.3	163 538	57.4	57.9
Atherosclerosis (I70)	13 942	4.8	4.8	14 086	4.9	5.0
Other diseases of circulatory system (I71-I78)	24 891	8.6	8.6	25 318	8.9	9.0
Aortic aneurysm and dissection (I71)	14 746	5.1	5.1	15 234	5.3	5.4
Other diseases of arteries, arterioles and capillaries (I72-I78)	10 146	3.5	3.5	10 084	3.5	3.6
Other disorders of circulatory system (I80-I99)	4 079	1.0	1.0	4 000	1.0	1.0

[1]Based on codes from the "Tenth Revision, International Classification of Diseases, 1992." Codes *U01-*U03 have been added as of 2001 for deaths due to terrorism.
* = Figure too small to meet standards of reliability or precision.

Table II-10A. Deaths, Death Rates, and Age-Adjusted Death Rates for 113 Selected Causes, 2001 and Preliminary 2002—*Continued*

(Number, rate per 100,000 population.)

Cause of death[1]	2002			2001		
	Number	Rate	Age-adjusted rate	Number	Rate	Age-adjusted rate
Influenza and pneumonia (J10-J18)	65 984	22.9	22.7	62 034	21.8	22.0
Influenza (J10-J11)	753	0.3	0.2	257	0.1	0.1
Pneumonia (J12-J18)	65 231	22.6	22.5	61 777	21.7	21.9
Other acute lower respiratory infections (J20-J22)	361	0.1	0.1	347	0.1	0.1
Acute bronchitis and bronchiolitis (J20-J21)	257	0.1	0.1	263	0.1	0.1
Unspecified acute lower respiratory infection (J22)	103	0.0	0.0	84	0.0	0.0
Chronic lower respiratory diseases (J40-J47)	125 500	43.5	43.7	123 013	43.2	43.7
Bronchitis, chronic and unspecified (J40-J42)	961	0.3	0.3	959	0.3	0.3
Emphysema (J43)	15 568	5.4	5.4	16 242	5.7	5.8
Asthma (J45-J46)	4 157	1.4	1.5	4 269	1.5	1.5
Other chronic lower respiratory diseases (J44, J47)	104 814	36.3	36.5	101 543	35.7	36.1
Pneumoconioses and chemical effects (J60-J66, J68)	1 139	0.4	0.4	1 136	0.4	0.4
Pneumonitis due to solids and liquids (J69)	17 693	6.1	6.1	17 301	6.1	6.1
Other diseases of respiratory system (J00-J06, J30-J39, J67, J70-J98)	25 194	8.7	8.8	26 178	9.2	9.3
Peptic ulcer (K25-K28)	4 081	1.4	1.4	4 491	1.6	1.6
Diseases of appendix, (K35-K38)	474	0.2	0.2	428	0.2	0.1
Hernia (K40-K46)	1 595	0.6	0.5	1 522	0.5	0.5
Chronic liver disease and cirrhosis (K70, K73-K74)	27 045	9.4	9.3	27 035	9.5	9.5
Alcoholic liver disease (K70)	11 965	4.1	4.1	12 207	4.3	4.3
Other chronic liver disease and cirrhosis (K73-K74)	15 079	5.2	5.2	14 828	5.2	5.2
Cholelithiasis and other disorders of gallbladder (K80-K82)	2 965	1.0	1.0	2 965	1.0	1.0
Nephritis, nephrotic syndrome and nephrosis (N00-N07, N17-N19, N25-N27)	41 018	14.2	14.2	39 480	13.9	14.0
Acute and rapidly progressive nephritic and nephrotic syndrome (N00-N01, N04)	168	0.1	0.1	156	0.1	0.0
Chronic glomerulonephritis, nephrosis and nephropathy not specified as acute or chronic, and renal sclerosis unspecified (N02-N03, N05-N07, N26)	545	0.2	0.2	511	0.2	0.2
Renal failure (N17-N19)	40 272	14.0	13.9	38 784	13.6	13.7
Other disorders of kidney (N25, N27)	34	0.0	0.0	29	0.0	0.0
Infections of kidney (N10-N12, N13.6, N15.1)	792	0.3	0.3	748	0.3	0.3
Hyperplasia of prostate (N40)	431	0.1	0.1	439	0.2	0.1
Inflammatory diseases of female pelvic organs (N70-N76)	110	0.0	0.0	127	0.0	0.0
Pregnancy, childbirth and the puerperium (O00-O99)	358	0.1	0.1	416	0.1	0.1
Pregnancy with abortive outcome, (O00-O07)	22	0.0	0.0	38	0.0	0.0
Other complications of pregnancy, childbirth and the puerperium (O10-O99)	336	0.1	0.1	378	0.1	0.1
Certain conditions originating in the perinatal period (P00-P96)	14 148	4.9	4.8	13 887	4.9	4.7
Congenital malformations, deformations and chromosomal abnormalities (Q00-Q99)	10 635	3.7	3.7	10 442	3.7	3.6
Symptoms, signs and abnormal clinical and laboratory findings, not elsewhere classified (R00-R99)	40 532	14.1	13.9	32 326	11.4	11.3
All other diseases (Residual)	193 974	67.3	66.9	186 406	65.5	65.9
Accidents (unintentional injuries) (V01-X59, Y85-Y86)	102 303	35.5	35.3	101 537	35.7	35.7
Transport accidents (V01-V99, Y85)	47 478	16.5	16.4	47 288	16.6	16.6
Motor vehicle accidents (V02-V04, V09.0, V09.2, V12-V14, V19.0-V19.2, V19.4-V19.6, V20-V79, V80.3, V80.5, V81.0, V81.1, V82.0, V82.1, V83, V86, V87.0, V87.8, V88.0, V88.8, V89.0, V89.2)	44 572	15.5	15.4	43 788	15.4	15.3
Other land transport accidents (V01, V05-V06, V09.1, V09.3-V09.9, V10-V11, V15-V18, V19.3, V19.8-V19.9, V80.0-V80.2, V80.6-V80.9, V81.2-V81.9, V82.2-V82.9, V87.9, V88.9, V89.1, V89.3, V89.9)	1 064	0.4	0.4	1 405	0.5	0.5
Water, air and space, and other and unspecified transport accidents and their sequelae (V90-V99, Y85)	1 842	0.6	0.6	2 095	0.7	0.7
Nontransport accidents (W00-X59, Y86)	54 825	19.0	19.0	54 249	19.0	19.1
Falls (W00-W19)	15 848	5.5	5.5	15 019	5.3	5.3
Accidental discharge of firearms (W32-W34)	813	0.3	0.3	802	0.3	0.3
Accidental drowning and submersion (W65-W74)	3 399	1.2	1.2	3 281	1.2	1.1
Accidental exposure to smoke, fire and flames (X00-X09)	3 024	1.0	1.1	3 309	1.2	1.2
Accidental poisoning and exposure to noxious substances (X40-X49)	14 670	5.1	5.1	14 078	4.9	4.9
Other and unspecified nontransport accidents and their sequelae (W20-W31, W35-W64, W75-W99, X10-X39, X50-X59, Y86)	17 070	5.9	5.9	17 760	6.2	6.3
Intentional self-harm (suicide) (*U03, X60-X84, Y87.0)[2]	30 646	10.6	10.6	30 622	10.8	10.7
Intentional self-harm (suicide) by discharge of firearms (X72-X74)	16 882	5.9	5.8	16 869	5.9	5.9
Intentional self-harm (suicide) by other and unspecified means and their sequelae (*U03, X60-X71, X75-X84, Y87.0)[2]	13 763	4.8	4.8	13 753	4.8	4.8
Assault (homicide) (*U01-*U02, X85-Y09, Y87.1)[2]	17 045	5.9	5.9	20 308	7.1	7.1
Assault (homicide) by discharge of firearms (*U01.4, X93-X95)	11 546	4.0	4.0	11 348	4.0	3.9
Assault (homicide) by other and unspecified means and their sequelae (*U01.0-*U01.3, *U01.5-*U01.9, *U02, X85-X92, X96-Y09, Y87.1)[2]	5 500	1.9	1.9	8 960	3.1	3.1
Legal intervention (Y35, Y89.0)	365	0.1	0.1	396	0.1	0.1
Events of undetermined intent (Y10-Y34, Y87.2, Y89.9)	4 423	1.5	1.5	4 198	1.5	1.5
Discharge of firearms, undetermined intent (Y22-Y24)	214	0.1	0.1	231	0.1	0.1
Other and unspecified events of undetermined intent and their sequelae (Y10-Y21, Y25-Y34, Y87.2, Y89.9)	4 209	1.5	1.4	3 967	1.4	1.4
Operations of war and their sequelae (Y36, Y89.1)	21	0.0	0.0	17	*	*
Complications of medical and surgical care, (Y40-Y84, Y88)	2 820	1.0	1.0	3 021	1.1	1.1
ADDENDUM[3]						
Injury by firearms (W32-W34, X72-X74, X93-X95, Y22-Y24, Y35.0)[3]	29 737	10.3	10.3	29 573	10.4	10.3
Drug-induced deaths (F11.0, F11.5, F11.7, F11.9, F12.0, F12.5, F12.7, F12.9, F13.0, F13.5, F13.7 F13.9, F14.0-F14.5, F14.7-F14.9, F15.0-F15.5, F15.7-F15.9, F16.0-F16.5, F16.7-F16.9, F17.0, F17.3-F17.5, F17.7-F17.9, F18.0-F18.5, F18.7-F18.9, F19.0-F19.5, F19.7-F19.9, X40-X44, X60-X64, X85, Y10, Y14)[3]	22 296	7.7	7.7	21 683	7.6	7.6
Alcohol-induced deaths (F10, G31.2, G62.1, I42.6, K29.2, K70, R78.0, X45, X65, Y15)[3]	19 344	6.7	6.6	19 817	7.0	6.9
Injury at work[4]	5 252	2.3	2.3	8 303	3.7	3.7

[1]Based on codes from the "Tenth Revision, International Classification of Diseases, 1992." Codes *U01-*U03 have been added as of 2001 for deaths due to terrorism.
[2]Figures include September 11, 2001, related deaths for which death certificates were filed as of October 24, 2002.
[3]Classified under other headings above.
[4]Derived from a separate item on the death certificate.
* = Figure too small to meet standards of reliability or precision.

TABLE 11. DEATH RATES FOR 113 SELECTED CAUSES, BY AGE, 2001

The number of deaths, shown in Table 10, here are converted to rates per 100,000 of population. The information on deaths is obtained from death certificates, while population is based on updates of the 2000 Census. Rates are not calculated for items with relatively small numbers of deaths.

The typical pattern for most diseases is a spike during infancy, representing congenital disorders and susceptibility to infection, with a rapid drop in early childhood, followed by a steady increase as age progresses. Non-transportation accidents also follow this pattern. However, some causes are extremely rare in infancy. In older age, death rates from certain diseases, such as chronic liver disease, peak before the highest age brackets.

Table II-11. Death Rates for 113 Selected Causes, by Age, 2001

(Rate per 100,000.)

Cause of death [1]	All ages [2]	Under 1 year [3]	1–4 years	5–14 years	15–24 years	25–34 years	35–44 years	45–54 years	55–64 years	65–74 years	75–84 years	85 years and over
All causes	848.5	683.4	33.3	17.3	80.7	105.2	203.6	428.9	964.6	2 353.3	5 582.4	15 112.8
Salmonella infections (A01-A02)	0.0	*	*	*	*	*	*	*	*	*	*	*
Shigellosis and amebiasis (A03, A06)	9.9	*	*	*	*	*	*	*	*	*	*	*
Certain other intestinal infections (A04, A07-A09)	0.5	*	*	*	*	*	*	0.1	0.3	1.3	4.5	12.8
Tuberculosis (A16-A19)	0.3	*	*	*	*	0.1	0.1	0.2	0.5	0.8	1.6	2.8
Respiratory tuberculosis (A16)	0.2	*	*	*	*	*	0.1	0.2	0.4	0.6	1.3	2.4
Other tuberculosis (A17-A19)	0.1	*	*	*	*	*	*	0.1	0.1	0.2	0.3	*
Whooping cough (A37)	*	*	*	*	*	*	*	*	*	*	*	*
Scarlet fever and erysipelas (A38, A46)	*	*	*	*	*	*	*	*	*	*	*	*
Meningococcal infection (A39)	0.1	*	*	0.1	0.1	*	*	*	*	*	*	*
Septicemia (A40-A41)	11.3	7.7	0.7	0.2	0.3	0.7	1.8	5.0	12.3	32.8	82.3	205.9
Syphilis (A50-A53)	0.0	*	*	*	*	*	*	*	*	*	*	*
Acute poliomyelitis (A80)	*	*	*	*	*	*	*	*	*	*	*	*
Arthropod-borne viral encephalitis (A83-A84, A85.2)	9.9	*	*	*	*	*	*	*	*	*	*	*
Measles (B05)	*	*	*	*	*	*	*	*	*	*	*	*
Viral hepatitis (B15-B19)	2.0	*	*	*	0.1	0.2	1.8	5.3	4.0	4.6	4.9	3.8
Human immunodeficiency virus (HIV) disease (B20-B24)	5.0	*	*	0.1	0.6	5.3	13.0	10.5	5.2	2.1	0.7	*
Malaria (B50-B54)	*	*	*	*	*	*	*	*	*	*	*	*
Other and unspecified infectious and parasitic diseases and their sequelae (A00, A05, A20-A36, A42-A44, A48-A49, A54-A79, A81-A82, A85.0-A85.1, A85.8, A86-B04, B06-B09, B25-B49, B55-B99)	2.0	4.7	0.4	0.2	0.2	0.4	0.8	1.7	3.0	6.2	10.6	20.1
Malignant neoplasms (C00-C97)	194.4	1.6	2.7	2.5	4.3	10.1	36.8	126.5	356.5	802.8	1 315.8	1 765.6
Malignant neoplasms of lip, oral cavity and pharynx (C00-C14)	2.7	*	*	*	0.1	0.1	0.6	2.7	6.9	10.4	14.2	18.5
Malignant neoplasm of esophagus (C15)	4.4	*	*	*	*	0.1	0.7	3.6	10.1	20.6	26.3	25.8
Malignant neoplasm of stomach (C16)	4.3	*	*	*	0.1	0.3	1.0	2.9	7.0	16.5	30.2	45.4
Malignant neoplasms of colon, rectum and anus (C18-C21)	20.0	*	*	*	0.1	0.8	3.2	11.6	31.5	75.3	139.4	255.7
Malignant neoplasms of liver and intrahepatic bile ducts (C22)	4.7	*	*	*	0.1	0.2	0.9	4.8	9.3	18.8	29.0	34.1
Malignant neoplasm of pancreas (C25)	10.5	*	*	*	*	0.1	1.3	6.1	19.2	44.5	74.1	100.1
Malignant neoplasm of larynx (C32)	1.3	*	*	*	*	*	0.2	1.1	3.6	6.4	7.2	6.7
Malignant neoplasms of trachea, bronchus and lung (C33-C34)	54.8	*	*	*	0.1	0.4	6.2	30.7	117.7	279.7	371.4	302.7
Malignant melanoma of skin (C43)	2.6	*	*	*	0.1	0.5	1.4	3.1	5.3	9.2	13.3	17.8
Malignant neoplasm of breast (C50)	14.7	*	*	*	*	1.2	6.3	16.8	30.2	47.6	76.8	134.2
Malignant neoplasm of cervix uteri (C53)	1.4	*	*	*	*	0.5	1.6	2.4	2.9	3.2	4.6	7.2
Malignant neoplasms of corpus uteri and uterus, part unspecified (C54-C55)	2.4	*	*	*	*	0.1	0.3	1.4	4.4	9.7	16.4	24.7
Malignant neoplasm of ovary (C56)	5.1	*	*	*	0.1	0.3	1.0	4.1	10.3	20.0	33.2	39.6
Malignant neoplasm of prostate (C61)	10.8	*	*	*	*	*	0.1	1.1	7.5	35.7	103.2	200.9
Malignant neoplasms of kidney and renal pelvis (C64-C65)	4.2	*	0.1	0.1	0.1	0.1	0.8	3.3	9.2	17.6	26.3	32.3
Malignant neoplasm of bladder (C67)	4.3	*	*	*	*	*	0.3	1.2	5.0	15.4	35.7	68.9
Malignant neoplasms of meninges, brain and other parts of central nervous system (C70-C72)	4.4	*	0.7	0.7	0.6	1.1	2.3	5.0	9.7	15.4	20.4	16.4
Malignant neoplasms of lymphoid, hematopoietic and related tissue (C81-C96)	19.8	0.7	1.0	0.9	1.8	2.5	4.1	10.4	29.5	74.6	145.2	199.9
Hodgkin's disease (C81)	0.5	*	*	*	0.2	0.4	0.4	0.5	0.6	1.2	2.2	2.2
Non-Hodgkin's lymphoma (C82-C85)	7.8	*	*	0.1	0.4	0.8	1.7	4.4	12.2	29.3	58.6	79.6
Leukemia (C91-C95)	7.5	0.6	1.0	0.8	1.2	1.3	1.7	3.8	10.0	26.7	53.3	81.4
Multiple myeloma and immunoproliferative neoplasms (C88, C90)	3.9	*	*	*	*	*	0.3	1.7	6.6	17.4	30.9	36.3
Other and unspecified malignant neoplasms of lymphoid, hematopoietic and related tissue (C96)	0.0	*	*	*	*	*	*	*	*	0.1	0.2	0.5
All other and unspecified malignant neoplasms (C17, C23-C24, C26-C31, C37-C41, C44-C49, C51-C52, C57-C60, C62-C63, C66, C68-C69, C73-C80, C97)	22.0	*	0.8	0.7	1.3	1.8	4.6	14.1	37.3	82.0	148.7	234.5
In situ neoplasms, benign neoplasms and neoplasms of uncertain or unknown behavior (D00-D48)	4.8	1.3	0.4	0.3	0.3	0.4	0.8	1.8	5.0	14.1	37.0	82.7
Anemias (D50-D64)	1.6	*	0.1	0.1	0.3	0.4	0.5	0.6	1.0	2.7	9.8	42.0
Diabetes mellitus (E10-E14)	25.1	*	*	0.1	0.4	1.5	4.3	13.6	37.8	91.4	181.4	321.8
Nutritional deficiencies (E40-E64)	1.3	*	*	*	*	*	0.1	0.3	0.6	2.0	8.7	42.9
Malnutrition (E40-E46)	1.2	*	*	*	*	*	0.1	0.3	0.6	1.9	8.2	40.2
Other nutritional deficiencies (E50-E64)	0.1	*	*	*	*	*	*	*	*	0.1	0.6	2.7
Meningitis (G00, G03)	0.2	1.9	0.2	0.1	0.1	0.1	0.2	0.3	0.4	0.5	0.8	0.9
Parkinson's disease (G20-G21)	5.8	*	*	*	*	*	*	0.1	1.2	11.7	64.6	134.2
Alzheimer's disease (G30)	18.9	*	*	*	*	*	*	0.2	2.1	18.7	147.5	710.3

1Based on codes from the "Tenth Revision, International Classification of Diseases, 1992." Codes *U01-*U03 have been added as of 2001 for deaths due to terrorism.
2Figures for age not stated are included in "All ages" but not distributed among age groups.
3Death rates for "Under 1 year" (based on population estimates) differ from infant mortality rates (based on live births).
0.0 = Quantity more than zero but less than 0.05.
* = Figure too small to meet standards of reliability or precision.

Table II-11. Death Rates for 113 Selected Causes, by Age, 2001 —*Continued*

(Rate per 100,000.)

Cause of death [1]	All ages [2]	Under 1 year [3]	1–4 years	5–14 years	15–24 years	25–34 years	35–44 years	45–54 years	55–64 years	65–74 years	75–84 years	85 years and over
Major cardiovascular diseases (I00-I78)	323.9	14.9	1.9	0.9	3.2	10.2	37.1	113.4	301.2	810.4	2 322.2	7 673.7
Diseases of heart (I00-I09, I11, I13, I20-I51)	245.8	11.9	1.5	0.7	2.5	8.0	29.6	92.9	246.9	635.1	1 725.7	5 664.2
Acute rheumatic fever and chronic rheumatic heart diseases (I00-I09)	1.2	*	*	*	*	0.1	0.3	0.6	1.4	4.1	9.4	17.7
Hypertensive heart disease (I11)	8.7	*	*	*	0.1	0.6	2.4	6.3	10.9	19.8	48.7	189.5
Hypertensive heart and renal disease (I13)	1.0	*	*	*	*	0.1	0.2	0.4	1.0	2.3	6.8	23.3
Ischemic heart diseases (I20-I25)	176.3	0.6	*	0.0	0.4	2.8	16.7	64.0	184.0	477.6	1 271.3	3 958.2
Acute myocardial infarction (I21-I22)	64.9	*	*	*	0.2	1.2	7.1	27.6	78.4	193.2	471.2	1 265.4
Other acute ischemic heart diseases (I24)	1.2	*	*	*	*	*	0.2	0.8	1.8	3.5	6.9	21.5
Other forms of chronic ischemic heart disease (I20, I25)	110.3	*	*	*	0.2	1.6	9.5	35.6	103.8	280.9	793.1	2 671.3
Atherosclerotic cardiovascular disease, so described (I25.0)	23.9	*	*	*	0.1	0.5	3.2	12.9	33.3	65.6	152.7	493.6
All other forms of chronic ischemic heart disease (I20, I25.1-I25.9)	86.4	*	*	*	0.2	1.1	6.2	22.7	70.5	215.3	640.5	2 177.7
Other heart diseases (I26-I51)	58.6	11.1	1.4	0.6	2.0	4.4	9.9	21.6	49.5	131.4	389.5	1 475.5
Acute and subacute endocarditis (I33)	0.4	*	*	*	*	0.1	0.3	0.5	0.6	1.3	2.5	2.5
Diseases of pericardium and acute myocarditis (I30-I31, I40)	0.3	0.8	0.2	0.1	0.1	0.1	0.2	0.3	0.4	0.7	1.2	1.9
Heart failure (I50)	20.0	0.6	*	*	0.1	0.2	0.7	2.4	9.4	34.3	136.3	674.6
All other forms of heart disease (I26-I28, I34-I38, I42-I49, I51)	37.9	9.7	1.1	0.5	1.8	4.0	8.8	18.5	39.1	95.1	249.5	796.5
Essential (primary) hypertension and hypertensive renal disease (I10, I12)	6.8	*	*	*	0.1	0.3	0.7	2.4	5.8	15.5	47.7	171.9
Cerebrovascular diseases (I60-I69)	57.4	2.7	0.4	0.2	0.5	1.5	5.5	15.1	38.0	123.4	443.9	1 500.2
Atherosclerosis (I70)	4.9	*	*	*	*	*	0.1	0.5	1.9	7.3	33.1	179.1
Other diseases of circulatory system (I71-I78)	8.9	*	*	*	0.2	0.4	1.1	2.6	8.7	29.2	71.8	158.2
Aortic aneurysm and dissection (I71)	5.3	*	*	*	0.1	0.3	0.7	1.7	5.5	19.2	45.7	77.5
Other diseases of arteries, arterioles and capillaries (I72-I78)	3.5	*	*	*	*	0.1	0.4	0.9	3.2	10.0	26.1	80.7
Other disorders of circulatory system (I80-I99)	1.6	0.5	*	*	0.1	0.3	0.6	1.2	2.2	4.6	9.4	26.1
Influenza and pneumonia (J10-J18)	21.8	7.4	0.7	0.2	0.5	0.9	2.2	4.6	10.7	36.3	148.5	685.6
Influenza (J10-J11)	0.1	*	*	*	*	*	*	*	0.1	0.1	0.4	2.2
Pneumonia (J12-J18)	21.7	7.2	0.7	0.2	0.4	0.8	2.2	4.6	10.6	36.2	148.1	683.4
Other acute lower respiratory infections (J20-J22)	0.1	1.2	*	*	*	*	*	*	*	0.2	0.5	3.3
Acute bronchitis and bronchiolitis (J20-J21)	0.1	1.2	*	*	*	*	*	*	*	0.1	0.3	2.1
Unspecified acute lower respiratory infection (J22)	0.0	*	*	*	*	*	*	*	*	*	*	1.2
Chronic lower respiratory diseases (J40-J47)	43.2	1.0	0.3	0.3	0.4	0.7	2.2	8.5	44.1	167.9	379.8	644.7
Bronchitis, chronic and unspecified (J40-J42)	0.3	0.5	*	*	*	*	0.0	0.1	0.2	0.7	2.1	9.1
Emphysema (J43)	5.7	*	*	*	*	*	0.2	1.1	7.2	26.2	50.4	62.1
Asthma (J45-J46)	1.5	*	0.2	0.2	0.4	0.6	1.0	1.5	2.2	3.5	6.4	16.1
Other chronic lower respiratory diseases (J44, J47)	35.7	*	*	*	0.1	0.1	0.9	5.7	34.5	137.5	321.0	557.4
Pneumoconioses and chemical effects (J60-J66, J68)	0.4	*	*	*	*	*	*	*	0.3	1.5	4.1	6.2
Pneumonitis due to solids and liquids (J69)	6.1	*	*	*	0.1	0.2	0.4	1.0	2.6	10.0	45.8	189.4
Other diseases of respiratory system (J00-J06, J30-J39, J67, J70-J98)	9.2	7.3	0.6	0.2	0.4	0.5	1.3	3.5	9.6	30.1	73.2	142.7
Peptic ulcer (K25-K28)	1.6	*	*	*	*	0.1	0.3	0.7	1.7	3.9	11.4	33.4
Diseases of appendix (K35-K38)	0.2	*	*	*	*	*	0.1	0.1	0.2	0.5	0.7	2.1
Hernia (K40-K46)	0.5	1.1	*	*	*	*	0.1	0.2	0.5	1.2	3.5	12.6
Chronic liver disease and cirrhosis (K70, K73-K74)	9.5	*	*	*	0.1	1.0	7.4	18.5	22.7	30.0	30.2	22.2
Alcoholic liver disease (K70)	4.3	*	*	*	*	0.7	4.8	10.7	11.5	10.1	5.7	2.4
Other chronic liver disease and cirrhosis (K73-K74)	5.2	*	*	*	*	0.3	2.6	7.8	11.2	19.9	24.5	19.9
Cholelithiasis and other disorders of gallbladder (K80-K82)	1.0	*	*	*	*	*	0.1	0.3	1.0	2.9	7.5	23.7
Nephritis, nephrotic syndrome and nephrosis (N00-N07, N17-N19, N25-N27)	13.9	3.3	*	*	0.2	0.6	1.7	4.6	13.0	40.2	104.2	287.7
Acute and rapidly progressive nephritic and nephrotic syndrome (N00-N01, N04)	0.1	*	*	*	*	*	*	*	*	0.1	0.4	0.9
Chronic glomerulonephritis, nephritis and nephropathy, not specified as acute or chronic, and renal sclerosis, unspecified (N02-N03, N05-N07, N26)	0.2	*	*	*	*	*	0.1	0.1	0.2	0.6	1.2	2.8
Renal failure (N17-N19)	13.6	3.1	*	*	0.2	0.5	1.6	4.5	12.7	39.4	102.6	284.0
Other disorders of kidney (N25, N27)	0.0	*	*	*	*	*	*	*	*	*	*	*
Infections of kidney (N10-N12, N13.6, N15.1)	0.3	*	*	*	*	*	0.1	0.1	0.2	0.5	2.0	6.0
Hyperplasia of prostate (N40)	0.2	*	*	*	*	*	*	*	*	0.2	1.1	5.5
Inflammatory diseases of female pelvic organs (N70-N76)	0.0	*	*	*	*	*	*	*	*	*	0.2	0.9
Pregnancy, childbirth and the puerperium (O00-O99)	0.1	X	X	*	0.3	0.5	0.2	*	X	X	X	X
Pregnancy with abortive outcome (O00-O07)	0.0	X	X	*	*	*	*	*	X	X	X	X
Other complications of pregnancy, childbirth and the puerperium (O10-O99)	0.1	X	X	*	0.2	0.4	0.2	*	X	X	X	X
Certain conditions originating in the perinatal period (P00-P96)	4.9	340.5	0.5	0.1	*	*	*	*	*	*	*	*
Congenital malformations, deformations and chromosomal abnormalities (Q00-Q99)	3.7	136.7	3.6	0.9	1.3	1.2	1.3	1.6	2.3	2.4	3.7	7.2
Symptoms, signs and abnormal clinical and laboratory findings, not elsewhere classified (R00-R99)	11.4	86.5	1.8	0.4	2.4	3.9	6.9	8.9	9.1	15.1	39.8	206.4
All other diseases (Residual)	65.5	28.0	4.2	2.1	4.5	7.8	18.0	36.2	63.4	140.9	395.3	1 476.7

[1]Based on codes from the "Tenth Revision, International Classification of Diseases, 1992." Codes *U01-*U03 have been added as of 2001 for deaths due to terrorism.
[2]Figures for age not stated are included in "All ages" but not distributed among age groups.
[3]Death rates for "Under 1 year" (based on population estimates) differ from infant mortality rates (based on live births).
0.0 = Quantity more than zero but less than 0.05.
* = Figure too small to meet standards of reliability or precision.
X = Not applicable.

Table II-11. Death Rates for 113 Selected Causes, by Age, 2001—*Continued*

(Rate per 100,000.)

Cause of death [1]	All ages [2]	Under 1 year [3]	1–4 years	5–14 years	15–24 years	25–34 years	35–44 years	45–54 years	55–64 years	65–74 years	75–84 years	85 years and over
Accidents (unintentional injuries) (V01-X59, Y85-Y86)	35.7	24.2	11.2	6.9	36.1	29.9	35.4	34.1	30.3	42.8	100.9	276.4
Transport accidents (V01-V99, Y85)	16.6	3.7	4.4	4.4	27.9	18.8	17.4	16.0	15.3	17.9	27.9	28.9
Motor vehicle accidents (V02-V04, V09.0, V09.2, V12-V14, V19.0-V19.2, V19.4-V19.6, V20-V79, V80.3-80.5, V81.0-V81.1, V82.0-V82.1, V83-V86, V87.0-V87.8, V88.0-V88.8, V89.0, V89.2)	15.4	3.6	4.1	4.1	26.8	17.5	15.7	14.2	13.7	16.3	26.3	27.8
Other land transport accidents (V01, V05-V06, V09.1, V09.3-V09.9, V10-V11, V15-V18, V19.3, V19.8-19.9, V80.0-V80.2, V80.6-V80.9, V81.2-V81.9, V82.2-V82.9, V87.9, V88.9, V89.1, V89.3, V89.9)	0.5	*	0.2	0.2	0.5	0.4	0.6	0.6	0.5	0.6	0.8	0.7
Water, air and space, and other and unspecified transport accidents and their sequelae (V90-V99, Y85)	0.7	*	0.2	0.1	0.5	0.9	1.0	1.1	1.0	1.0	0.8	0.5
Nontransport accidents (W00-X59, Y86)	19.0	20.5	6.8	2.5	8.2	11.1	18.0	18.1	15.0	24.9	73.0	247.5
Falls (W00-W19)	5.3	0.6	0.2	0.2	0.6	0.9	1.4	2.6	4.0	10.0	35.3	121.5
Accidental discharge of firearms (W32-W34)	0.3	*	*	0.1	0.5	0.3	0.3	0.3	0.3	0.2	0.3	*
Accidental drowning and submersion (W65-W74)	1.2	1.7	3.0	0.8	1.5	0.9	1.0	0.9	0.8	0.9	1.2	1.7
Accidental exposure to smoke, fire and flames (X00-X09)	1.2	1.2	1.5	0.6	0.5	0.6	1.0	1.1	1.5	2.2	3.4	5.7
Accidental poisoning and exposure to noxious substances (X40-X49)	4.9	*	0.2	0.1	3.4	6.3	11.2	9.1	3.2	1.6	2.2	3.5
Other and unspecified nontransport accidents and their sequelae (W20-W31, W35-W64, W75-W99, X10-X39, X50-X59, Y86)	6.2	16.6	1.8	0.6	1.6	2.0	3.1	4.2	5.3	10.0	30.5	115.0
Intentional self-harm (suicide) (*U03, X60-X84, Y87.0) [4]	10.8	0.0	0.0	0.7	9.9	12.8	14.7	15.2	13.1	13.3	17.4	17.5
Intentional self-harm (suicide) by discharge of firearms (X72-X74)	5.9	X	X	0.2	5.3	6.5	6.7	7.7	8.2	9.6	13.1	12.2
Intentional self-harm (suicide) by other and unspecified means and their sequelae (*U03, X60-X71, X75-X84, Y87.0) [4]	4.8	X	X	0.5	4.6	6.3	8.0	7.4	4.9	3.7	4.3	5.3
Assault (homicide) (*U01-*U02, X85-Y09, Y87.1) [4]	7.1	8.2	2.7	0.8	13.3	13.1	9.5	6.3	4.0	2.9	2.5	2.4
Assault (homicide) by discharge of firearms (*U01.4, X93-X95)	4.0	*	0.4	0.4	10.5	8.4	4.4	2.4	1.4	1.0	0.8	*
Assault (homicide) by other and unspecified means and their sequelae (*U01.0-*U01.3, *U01.5-*U01.9, *U02, X85-X92, X96-Y09, Y87.1) [4]	3.1	8.0	2.3	0.4	2.7	4.8	5.1	3.9	2.6	1.9	1.7	2.0
Legal intervention (Y35, Y89.0)	0.1	*	*	*	0.2	0.3	0.3	0.1	0.1	*	*	*
Events of undetermined intent (Y10-Y34, Y87.2, Y89.9)	1.5	2.1	0.3	0.1	1.0	1.8	3.1	2.5	0.9	0.5	0.9	1.0
Discharge of firearms, undetermined intent (Y22-Y24)	0.1	*	*	*	0.2	0.1	0.1	0.1	*	*	*	*
Other and unspecified events of undetermined intent and their sequelae (Y10-Y21, Y25-Y34, Y87.2, Y89.9)	1.4	2.1	0.3	0.1	0.9	1.7	3.0	2.5	0.8	0.5	0.8	1.0
Operations of war and their sequelae (Y36, Y89.1)	*	*	*	*	*	*	*	*	*	*	*	*
Complications of medical and surgical care (Y40-Y84, Y88)	1.1	0.6	0.2	*	0.1	0.2	0.4	0.7	1.5	3.1	6.9	12.8

[1]Based on codes from the "Tenth Revision, International Classification of Diseases, 1992." Codes *U01-*U03 have been added as of 2001 for deaths due to terrorism.
[2]Figures for age not stated are included in "All ages" but not distributed among age groups.
[3]Death rates for "Under 1 year" (based on population estimates) differ from infant mortality rates (based on live births).
[4]Figures include September 11, 2001 related deaths for which death certificates were filed as of 10/24/02.
0.0 = Quantity more than zero but less than 0.05.
* = Figure too small to meet standards of reliability or precision.
X = Not applicable.

TABLE II-12. NUMBER OF DEATHS FOR 113 SELECTED CAUSES, BY RACE AND SEX, 2001

This table presents cause-specific death counts by sex for 1) all races, 2) Whites – including non-Black Hispanics – 3) total "all other" – including Blacks – and 4) Blacks. The same format is used for crude rates in Table 14 and age-adjusted rates in Table 16. In addition to their use in calculating death rates, the actual counts are useful for exploring the national burden of specific causes in race and sex groups.

Death from cardiovascular diseases is highly concentrated at old age, and thus females with their greater life expectancies register notably more fatalities from these diseases than males; this applies both to Whites and to Blacks. The opposite holds for cancer, which is less concentrated at the highest age levels.

Males more frequently are victims of accidents, whether motor vehicle accidents or other accidents. For Black men fatalities from accidents in 2001 were twice the number as Black females, and for Whites the fatal accident rate was 80 percent higher for males. In the case of suicide, men had four times greater fatalities than women. For both sexes combined, the White rate of suicide was twice as high as that for Blacks.

Table II-12. Number of Deaths for 113 Selected Causes, by Race and Sex, 2001

(Number.)

Cause of death [1]	All races			White			All other					
							Total			Black		
	Both sexes	Male	Female	Both sexes	Male	Female	Both sexes	Male	Female	Both sexes	Male	Female
All causes	2 416 425	1 183 421	1 233 004	2 079 691	1 011 218	1 068 473	336 734	172 203	164 531	287 709	145 908	141 801
Salmonella infections (A01-A02)	41	21	20	33	18	15	8	3	5	5	1	4
Shigellosis and amebiasis (A03, A06)	7	6	1	7	6	1	-	-	-	-	-	-
Certain other intestinal infections (A04, A07-A09)	1 537	572	965	1 417	518	899	120	54	66	101	43	58
Tuberculosis (A16-A19)	764	453	311	469	271	198	295	182	113	203	127	76
Respiratory tuberculosis (A16)	600	371	229	368	221	147	232	150	82	160	106	54
Other tuberculosis (A17-A19)	164	82	82	101	50	51	63	32	31	43	21	22
Whooping cough (A37)	17	7	10	16	7	9	1	-	1	1	-	1
Scarlet fever and erysipelas (A38, A46)	3	2	1	3	2	1	-	-	-	-	-	-
Meningococcal infection (A39)	199	94	105	158	72	86	41	22	19	36	18	18
Septicemia (A40-A41)	32 238	14 307	17 931	25 806	11 371	14 435	6 432	2 936	3 496	5 880	2 635	3 245
Syphilis (A50-A53)	36	19	17	21	9	12	15	10	5	14	9	5
Acute poliomyelitis (A80)	-	-	-	-	-	-	-	-	-	-	-	-
Arthropod-borne viral encephalitis (A83-A84, A85.2)	9	6	3	6	5	1	3	1	2	2	-	2
Measles (B05)	1	1	-	1	1	-	-	-	-	-	-	-
Viral hepatitis (B15-B19)	5 585	3 539	2 046	4 451	2 900	1 551	1 134	639	495	831	484	347
Human immunodeficiency virus (HIV) disease (B20-B24)	14 175	10 590	3 585	6 171	5 135	1 036	8 004	5 455	2 549	7 844	5 328	2 516
Malaria (B50-B54)	9	5	4	5	2	3	4	3	1	3	2	1
Other and unspecified infectious and parasitic diseases and their sequelae (A00, A05, A20-A36, A42-A44, A48-A49, A54-A79, A81-A82, A85.0-A85.1, A85.8, A86-B04, B06-B09, B25-B49, B55-B99)	5 715	2 931	2 784	4 688	2 405	2 283	1 027	526	501	825	427	398
Malignant neoplasms (C00-C97)	553 768	287 075	266 693	479 651	248 146	231 505	74 117	38 929	35 188	62 170	32 679	29 491
Malignant neoplasms of lip, oral cavity and pharynx (C00-C14)	7 702	5 156	2 546	6 369	4 183	2 186	1 333	973	360	1 065	782	283
Malignant neoplasm of esophagus (C15)	12 530	9 607	2 923	10 624	8 205	2 419	1 906	1 402	504	1 688	1 231	457
Malignant neoplasm of stomach (C16)	12 319	7 191	5 128	9 501	5 568	3 933	2 818	1 623	1 195	2 067	1 186	881
Malignant neoplasms of colon, rectum and anus (C18-C21)	56 887	28 225	28 662	48 748	24 335	24 413	8 139	3 890	4 249	6 905	3 251	3 654
Malignant neoplasms of liver and intrahepatic bile ducts (C22)	13 351	8 483	4 868	10 697	6 774	3 923	2 654	1 709	945	1 670	1 048	622
Malignant neoplasm of pancreas (C25)	29 803	14 467	15 336	25 769	12 641	13 128	4 034	1 826	2 208	3 394	1 523	1 871
Malignant neoplasm of larynx (C32)	3 797	3 018	779	3 049	2 417	632	748	601	147	678	541	137
Malignant neoplasms of trachea, bronchus and lung (C33-C34)	156 058	90 426	65 632	137 274	78 665	58 609	18 784	11 761	7 023	16 019	10 131	5 888
Malignant melanoma of skin (C43)	7 542	4 763	2 779	7 403	4 702	2 701	139	61	78	104	37	67
Malignant neoplasm of breast (C50)	41 809	415	41 394	35 422	344	35 078	6 387	71	6 316	5 571	65	5 506
Malignant neoplasm of cervix uteri (C53)	4 092	X	4 092	3 136	X	3 136	956	X	956	787	X	787
Malignant neoplasms of corpus uteri and uterus, part unspecified (C54-C55)	6 784	X	6 784	5 606	X	5 606	1 178	X	1 178	1 047	X	1 047
Malignant neoplasm of ovary (C56)	14 414	X	14 414	12 950	X	12 950	1 464	X	1 464	1 153	X	1 153
Malignant neoplasm of prostate (C61)	30 719	30 719	X	25 066	25 066	X	5 653	5 653	X	5 265	5 265	X
Malignant neoplasms of kidney and renal pelvis (C64-C65)	12 078	7 524	4 554	10 720	6 695	4 025	1 358	829	529	1 113	678	435
Malignant neoplasm of bladder (C67)	12 225	8 343	3 882	11 180	7 780	3 400	1 045	563	482	877	447	430
Malignant neoplasms of meninges, brain and other parts of central nervous system (C70-C72)	12 609	7 015	5 594	11 638	6 498	5 140	971	517	454	741	382	359
Malignant neoplasms of lymphoid, hematopoietic and related tissue (C81-C96)	56 341	30 060	26 281	49 989	26 767	23 222	6 352	3 293	3 059	5 227	2 676	2 551
Hodgkin's disease (C81)	1 323	725	598	1 170	644	526	153	81	72	137	73	64
Non-Hodgkin's lymphoma (C82-C85)	22 305	11 710	10 595	20 335	10 642	9 693	1 970	1 068	902	1 496	804	692
Leukemia (C91-C95)	21 451	11 850	9 601	19 239	10 672	8 567	2 212	1 178	1 034	1 765	928	837
Multiple myeloma and immunoproliferative neoplasms (C88, C90)	11 172	5 724	5 448	9 168	4 765	4 403	2 004	959	1 045	1 817	864	953
Other and unspecified malignant neoplasms of lymphoid, hematopoietic and related tissue (C96)	90	51	39	77	44	33	13	7	6	12	7	5
All other and unspecified malignant neoplasms (C17, C23-C24, C26-C31, C37-C41, C44-C49, C51-C52, C57-C60, C62-C63, C66, C68-C69, C73-C80, C97)	62 708	31 663	31 045	54 510	27 506	27 004	8 198	4 157	4 041	6 799	3 436	3 363
In situ neoplasms, benign neoplasms and neoplasms of uncertain or unknown behavior (D00-D48)	13 676	6 735	6 941	12 261	6 089	6 172	1 415	646	769	1 157	525	632
Anemias (D50-D64)	4 627	1 774	2 853	3 543	1 303	2 240	1 084	471	613	1 001	446	555
Diabetes mellitus (E10-E14)	71 372	32 841	38 531	57 180	26 917	30 263	14 192	5 924	8 268	12 305	5 049	7 256
Nutritional deficiencies (E40-E64)	3 704	1 233	2 471	3 152	1 000	2 152	552	233	319	475	204	271
Malnutrition (E40-E46)	3 454	1 148	2 306	2 939	932	2 007	515	216	299	441	188	253
Other nutritional deficiencies (E50-E64)	250	85	165	213	68	145	37	17	20	34	16	18
Meningitis (G00, G03)	727	394	333	526	275	251	201	119	82	175	105	70
Parkinson's disease (G20-G21)	16 544	9 405	7 139	15 757	8 982	6 775	787	423	364	551	294	257
Alzheimer's disease (G30)	53 852	15 762	38 090	50 348	14 714	35 634	3 504	1 048	2 456	3 114	910	2 204

[1]Based on codes from the "Tenth Revision, International Classification of Diseases, 1992." Codes *U01-*U03 have been added as of 2001 for deaths due to terrorism.
- = Quantity zero.
X = Not applicable.

Table II-12. Number of Deaths for 113 Selected Causes, by Race and Sex, 2001—*Continued*

(Number.)

Cause of death [1]	All races			White			All other					
	Both sexes	Male	Female	Both sexes	Male	Female	Total: Both sexes	Total: Male	Total: Female	Black: Both sexes	Black: Male	Black: Female
Major cardiovascular diseases (I00-I78)	922 334	428 065	494 269	801 052	370 918	430 134	121 282	57 147	64 135	104 316	48 237	56 079
Diseases of heart (I00-I09, I11, I13, I20-I51)	700 142	339 095	361 047	610 638	295 556	315 082	89 504	43 539	45 965	77 674	37 016	40 658
Acute rheumatic fever and chronic rheumatic heart diseases (I00-I09)	3 489	1 031	2 458	3 086	910	2 176	403	121	282	302	86	216
Hypertensive heart disease (I11)	24 689	10 451	14 238	18 232	7 395	10 837	6 457	3 056	3 401	5 970	2 813	3 157
Hypertensive heart and renal disease (I13)	2 826	1 169	1 657	1 922	762	1 160	904	407	497	832	376	456
Ischemic heart diseases (I20-I25)	502 189	254 005	248 184	442 536	224 464	218 072	59 653	29 541	30 112	50 933	24 593	26 340
Acute myocardial infarction (I21-I22)	184 757	95 875	88 882	162 943	85 280	77 663	21 814	10 595	11 219	18 581	8 717	9 864
Other acute ischemic heart diseases (I24)	3 351	1 720	1 631	2 821	1 431	1 390	530	289	241	454	249	205
Other forms of chronic ischemic heart disease (I20, I25)	314 081	156 410	157 671	276 772	137 753	139 019	37 309	18 657	18 652	31 898	15 627	16 271
Atherosclerotic cardiovascular disease, so described (I25.0)	68 103	35 440	32 663	56 414	29 039	27 375	11 689	6 401	5 288	10 321	5 578	4 743
All other forms of chronic ischemic heart disease (I20, I25.1-I25.9)	245 978	120 970	125 008	220 358	108 714	111 644	25 620	12 256	13 364	21 577	10 049	11 528
Other heart diseases (I26-I51)	166 949	72 439	94 510	144 862	62 025	82 837	22 087	10 414	11 673	19 637	9 148	10 489
Acute and subacute endocarditis (I33)	1 178	668	510	917	517	400	261	151	110	235	135	100
Diseases of pericardium and acute myocarditis (I30-I31, I40)	839	401	438	658	310	348	181	91	90	158	79	79
Heart failure (I50)	56 934	21 632	35 302	51 252	19 332	31 920	5 682	2 300	3 382	5 122	2 052	3 070
All other forms of heart disease (I26-I28, I34-I38, I42-I49, I51)	107 998	49 738	58 260	92 035	41 866	50 169	15 963	7 872	8 091	14 122	6 882	7 240
Essential (primary) hypertension and hypertensive renal disease (I10, I12)	19 250	7 254	11 996	14 752	5 351	9 401	4 498	1 903	2 595	4 038	1 715	2 323
Cerebrovascular diseases (I60-I69)	163 538	63 177	100 361	140 465	53 428	87 037	23 073	9 749	13 324	19 002	7 907	11 095
Atherosclerosis (I70)	14 086	5 143	8 943	12 961	4 696	8 265	1 125	447	678	970	370	600
Other diseases of circulatory system (I71-I78)	25 318	13 396	11 922	22 236	11 887	10 349	3 082	1 509	1 573	2 632	1 229	1 403
Aortic aneurysm and dissection (I71)	15 234	9 114	6 120	13 652	8 224	5 428	1 582	890	692	1 251	671	580
Other diseases of arteries, arterioles and capillaries (I72-I78)	10 084	4 282	5 802	8 584	3 663	4 921	1 500	619	881	1 381	558	823
Other disorders of circulatory system (I80-I99)	4 665	1 981	2 684	3 843	1 603	2 240	822	378	444	742	339	403
Influenza and pneumonia (J10-J18)	62 034	27 342	34 692	54 774	23 744	31 030	7 260	3 598	3 662	5 771	2 813	2 958
Influenza (J10-J11)	257	101	156	229	88	141	28	13	15	20	10	10
Pneumonia (J12-J18)	61 777	27 241	34 536	54 545	23 656	30 889	7 232	3 585	3 647	5 751	2 803	2 948
Other acute lower respiratory infections (J20-J22)	347	117	230	299	94	205	48	23	25	34	16	18
Acute bronchitis and bronchiolitis (J20-J21)	263	89	174	218	67	151	45	22	23	32	15	17
Unspecified acute lower respiratory infection (J22)	84	28	56	81	27	54	3	1	2	2	1	1
Chronic lower respiratory diseases (J40-J47)	123 013	59 697	63 316	113 819	54 561	59 258	9 194	5 136	4 058	7 589	4 187	3 402
Bronchitis, chronic and unspecified (J40-J42)	959	396	563	878	354	524	81	42	39	63	32	31
Emphysema (J43)	16 242	8 415	7 827	15 243	7 798	7 445	999	617	382	811	487	324
Asthma (J45-J46)	4 269	1 479	2 790	2 990	937	2 053	1 279	542	737	1 108	459	649
Other chronic lower respiratory diseases (J44, J47)	101 543	49 407	52 136	94 708	45 472	49 236	6 835	3 935	2 900	5 607	3 209	2 398
Pneumoconioses and chemical effects (J60-J66, J68)	1 136	1 095	41	1 073	1 034	39	63	61	2	57	55	2
Pneumonitis due to solids and liquids (J69)	17 301	8 576	8 725	15 425	7 705	7 720	1 876	871	1 005	1 599	741	858
Other diseases of respiratory system (J00-J06, J30-J39, J67, J70-J98)	26 178	12 495	13 683	23 133	11 085	12 048	3 045	1 410	1 635	2 491	1 131	1 360
Peptic ulcer (K25-K28)	4 491	2 074	2 417	3 942	1 751	2 191	549	323	226	409	243	166
Diseases of appendix (K35-K38)	428	252	176	342	198	144	86	54	32	73	47	26
Hernia (K40-K46)	1 522	587	935	1 343	502	841	179	85	94	157	76	81
Chronic liver disease and cirrhosis (K70, K73-K74)	27 035	17 393	9 642	23 408	15 048	8 360	3 627	2 345	1 282	2 775	1 843	932
Alcoholic liver disease (K70)	12 207	8 932	3 275	10 387	7 689	2 698	1 820	1 243	577	1 348	935	413
Other chronic liver disease and cirrhosis (K73-K74)	14 828	8 461	6 367	13 021	7 359	5 662	1 807	1 102	705	1 427	908	519
Cholelithiasis and other disorders of gallbladder (K80-K82)	2 965	1 341	1 624	2 567	1 157	1 410	398	184	214	310	130	180
Nephritis, nephrotic syndrome and nephrosis (N00-N07, N17-N19, N25-N27)	39 480	18 852	20 628	31 345	15 241	16 104	8 135	3 611	4 524	7 274	3 186	4 088
Acute and rapidly progressive nephritic and nephrotic syndrome (N00-N01, N04)	156	73	83	130	60	70	26	13	13	21	9	12
Chronic glomerulonephritis, nephritis and nephropathy, not specified as acute or chronic, and renal sclerosis, unspecified (N02-N03, N05-N07, N26)	511	264	247	417	216	201	94	48	46	69	37	32
Renal failure (N17-N19)	38 784	18 509	20 275	30 777	14 961	15 816	8 007	3 548	4 459	7 178	3 138	4 040
Other disorders of kidney (N25, N27)	29	6	23	21	4	17	8	2	6	6	2	4
Infections of kidney (N10-N12, N13.6, N15.1)	748	213	535	647	176	471	101	37	64	80	28	52
Hyperplasia of prostate (N40)	439	439	X	403	403	X	36	36	X	32	32	X
Inflammatory diseases of female pelvic organs (N70-N76)	127	X	127	103	X	103	24	X	24	23	X	23
Pregnancy, childbirth and the puerperium (O00-O99)	416	X	416	235	X	235	181	X	181	158	X	158
Pregnancy with abortive outcome (O00-O07)	38	X	38	20	X	20	18	X	18	16	X	16
Other complications of pregnancy, childbirth and the puerperium (O10-O99)	378	X	378	215	X	215	163	X	163	142	X	142
Certain conditions originating in the perinatal period (P00-P96)	13 887	7 868	6 019	8 551	4 846	3 705	5 336	3 022	2 314	4 847	2 746	2 101
Congenital malformations, deformations and chromosomal abnormalities (Q00-Q99)	10 442	5 448	4 994	8 338	4 376	3 962	2 104	1 072	1 032	1 720	882	838
Symptoms, signs and abnormal clinical and laboratory findings, not elsewhere classified (R00-R99)	32 326	15 779	16 547	26 877	12 868	14 009	5 449	2 911	2 538	4 691	2 465	2 226
All other diseases (Residual)	186 406	75 162	111 244	161 293	64 020	97 273	25 113	11 142	13 971	21 970	9 671	12 299

[1] Based on codes from the "Tenth Revision, International Classification of Diseases, 1992." Codes *U01-*U03 have been added as of 2001 for deaths due to terrorism.
X = Not applicable.

Table II-12. Number of Deaths for 113 Selected Causes, by Race and Sex, 2001—*Continued*

(Number.)

Cause of death [1]	All races			White			All other					
							Total			Black		
	Both sexes	Male	Female	Both sexes	Male	Female	Both sexes	Male	Female	Both sexes	Male	Female
Accidents (unintentional injuries) (V01-X59, Y85-Y86)	101 537	66 060	35 477	85 964	55 493	30 471	15 573	10 567	5 006	12 462	8 537	3 925
Transport accidents (V01-V99, Y85)	47 288	32 794	14 494	39 544	27 407	12 137	7 744	5 387	2 357	5 906	4 214	1 692
Motor vehicle accidents (V02-V04, V09.0, V09.2, V12-V14, V19.0-V19.2, V19.4-V19.6, V20-V79, V80.3-80.5, V81.0-V81.1, V82.0-V82.1, V83-V86, V87.0-V87.8, V88.0-V88.8, V89.0, V89.2)	43 788	30 064	13 724	36 576	25 084	11 492	7 212	4 980	2 232	5 506	3 900	1 606
Other land transport accidents (V01, V05-V06, V09.1, V09.3-V09.9, V10-V11, V15-V18, V19.3, V19.8-19.9, V80.0-V80.2, V80.6-V80.9, V81.2-V81.9, V82.2-V82.9, V87.9, V88.9, V89.1, V89.3, V89.9)	1 405	1 067	338	1 105	850	255	300	217	83	226	168	58
Water, air and space, and other and unspecified transport accidents and their sequelae (V90-V99, Y85)	2 095	1 663	432	1 863	1 473	390	232	190	42	174	146	28
Nontransport accidents (W00-X59, Y86)	54 249	33 266	20 983	46 420	28 086	18 334	7 829	5 180	2 649	6 556	4 323	2 233
Falls (W00-W19)	15 019	8 089	6 930	13 827	7 331	6 496	1 192	758	434	854	547	307
Accidental discharge of firearms (W32-W34)	802	690	112	675	580	95	127	110	17	112	99	13
Accidental drowning and submersion (W65-W74) ...	3 281	2 560	721	2 515	1 936	579	766	624	142	579	478	101
Accidental exposure to smoke, fire and flames (X00-X09)	3 309	1 998	1 311	2 387	1 436	951	922	562	360	846	511	335
Accidental poisoning and exposure to noxious substances (X40-X49)	14 078	9 885	4 193	11 662	8 232	3 430	2 416	1 653	763	2 169	1 490	679
Other and unspecified nontransport accidents and their sequelae (W20-W31, W35-W64, W75-W99, X10-X39, X50-X59, Y86)	17 760	10 044	7 716	15 354	8 571	6 783	2 406	1 473	933	1 996	1 198	798
Intentional self-harm (suicide) (*U03, X60-X84, Y87.0) [2]	30 622	24 672	5 950	27 710	22 328	5 382	2 912	2 344	568	1 957	1 627	330
Intentional self-harm (suicide) by discharge of firearms (X72-X74)	16 869	14 758	2 111	15 467	13 521	1 946	1 402	1 237	165	1 091	971	120
Intentional self-harm (suicide) by other and unspecified means and their sequelae (*U03, X60-X71, X75-X84, Y87.0) [2]	13 753	9 914	3 839	12 243	8 807	3 436	1 510	1 107	403	866	656	210
Assault (homicide) (*U01-*U02, X85-Y09, Y87.1) [2]	20 308	15 555	4 753	11 328	8 254	3 074	8 980	7 301	1 679	8 226	6 780	1 446
Assault (homicide) by discharge of firearms (*U01.4, X93-X95)	11 348	9 532	1 816	5 188	4 042	1 146	6 160	5 490	670	5 885	5 279	606
Assault (homicide) by other and unspecified means and their sequelae (*U01.0-*U01.3, *U01.5-*U01.9, *U02, X85-X92, X96-Y09, Y87.1) [2]	8 960	6 023	2 937	6 140	4 212	1 928	2 820	1 811	1 009	2 341	1 501	840
Legal intervention (Y35, Y89.0)	396	380	16	292	283	9	104	97	7	84	79	5
Events of undetermined intent (Y10-Y34, Y87.2, Y89.9) .	4 198	2 833	1 365	3 453	2 283	1 170	745	550	195	646	480	166
Discharge of firearms, undetermined intent (Y22-Y24)	231	190	41	188	150	38	43	40	3	32	29	3
Other and unspecified events of undetermined intent and their sequelae (Y10-Y21, Y25-Y34, Y87.2, Y89.9)	3 967	2 643	1 324	3 265	2 133	1 132	702	510	192	614	451	163
Operations of war and their sequelae (Y36, Y89.1)	17	16	1	16	15	1	1	1	-	1	1	-
Complications of medical and surgical care (Y40-Y84, Y88)	3 021	1 357	1 664	2 446	1 084	1 362	575	273	302	522	250	272

[1]Based on codes from the "Tenth Revision, International Classification of Diseases, 1992." Codes *U01-*U03 have been added as of 2001 for deaths due to terrorism.
[2]Figures include September 11, 2001 related deaths for which death certificates were filed as of 10/24/02.
- = Quantity zero.

TABLE II-13. NUMBER OF DEATHS FOR 113 SELECTED CAUSES, BY RACE, HISPANIC ORIGIN, AND SEX, 2001

This table presents cause-specific death counts by sex for 1) all ethnic origins, 2) all Hispanics, 3) non-Hispanics (all races), 4) non-Hispanic Whites, 5) non-Hispanic Blacks, and 6) origin not stated. The same format (without "origin not stated") is used for crude rates in Table 15 and age-adjusted rates in Table 17.

As discussed in Table 2, the Hispanic population has special qualities of good health. Overall, Hispanics account for 4.7 percent of United States deaths in 2001, but in terms of population they represent as much as 14 percent. A large part of this difference is due to the young age of the Hispanic population, but the discrepancy is also due to underreporting of Hispanic ethnicity on death certificates. Hispanic cancer fatalities account for 4.0 percent of the national total, and lung cancer cases account for 2.4 percent. When it comes to diabetes, however, Hispanic fatalities increase to 7.9 percent of the total population. Hispanics also account for a substantial share of accidental deaths, 9.4 percent. When an adjustment is made for the differences in the age distribution, the Hispanic mortality rate increases to 77 percent of the rate for the total population (see Table 17).

Table II-13. Number of Deaths for 113 Selected Causes, by Race, Hispanic Origin, and Sex 2001

(Number.)

Cause of death [1]	All origins			Hispanic			Non-Hispanic [2]		
	Both sexes	Male	Female	Both sexes	Male	Female	Both sexes	Male	Female
All causes	2 416 425	1 183 421	1 233 004	113 413	63 317	50 096	2 295 244	1 115 683	1 179 561
Salmonella infections (A01-A02)	41	21	20	5	5	-	36	16	20
Shigellosis and amebiasis (A03, A06)	7	6	1	3	3	-	4	3	1
Certain other intestinal infections (A04, A07-A09)	1 537	572	965	53	19	34	1 474	550	924
Tuberculosis (A16-A19)	764	453	311	97	67	30	660	379	281
Respiratory tuberculosis (A16)	600	371	229	77	56	21	517	309	208
Other tuberculosis (A17-A19)	164	82	82	20	11	9	143	70	73
Whooping cough (A37)	17	7	10	10	4	6	7	3	4
Scarlet fever and erysipelas (A38, A46)	3	2	1	-	-	-	3	2	1
Meningococcal infection (A39)	199	94	105	21	11	10	178	83	95
Septicemia (A40-A41)	32 238	14 307	17 931	1 386	684	702	30 754	13 573	17 181
Syphilis (A50-A53)	36	19	17	7	5	2	28	14	14
Acute poliomyelitis (A80)	-	-	-	-	-	-	-	-	-
Arthropod-borne viral encephalitis (A83-A84, A85.2)	9	6	3	-	-	-	9	6	3
Measles (B05)	1	1	-	-	-	-	1	1	-
Viral hepatitis (B15-B19)	5 585	3 539	2 046	791	487	304	4 776	3 038	1 738
Human immunodeficiency virus (HIV) disease (B20-B24)	14 175	10 590	3 585	1 853	1 437	416	12 141	9 011	3 130
Malaria (B50-B54)	9	5	4	-	-	-	9	5	4
Other and unspecified infectious and parasitic diseases and their sequelae (A00, A05, A20-A36, A42-A44, A48-A49, A54-A79, A81-A82, A85.0-A85.1, A85.8, A86-B04, B06-B09, B25-B49, B55-B99)	5 715	2 931	2 784	423	250	173	5 271	2 666	2 605
Malignant neoplasms (C00-C97)	553 768	287 075	266 693	22 371	11 825	10 546	530 058	274 492	255 566
Malignant neoplasms of lip, oral cavity and pharynx (C00-C14)	7 702	5 156	2 546	311	219	92	7 370	4 920	2 450
Malignant neoplasm of esophagus (C15)	12 530	9 607	2 923	401	316	85	12 085	9 254	2 831
Malignant neoplasm of stomach (C16)	12 319	7 191	5 128	1 189	680	509	11 099	6 492	4 607
Malignant neoplasms of colon, rectum and anus (C18-C21)	56 887	28 225	28 662	2 250	1 185	1 065	54 513	26 978	27 535
Malignant neoplasms of liver and intrahepatic bile ducts (C22)	13 351	8 483	4 868	1 256	812	444	12 053	7 640	4 413
Malignant neoplasm of pancreas (C25)	29 803	14 467	15 336	1 336	643	693	28 407	13 792	14 615
Malignant neoplasm of larynx (C32)	3 797	3 018	779	163	141	22	3 614	2 858	756
Malignant neoplasms of trachea, bronchus and lung (C33-C34)	156 058	90 426	65 632	3 816	2 495	1 321	151 882	87 706	64 176
Malignant melanoma of skin (C43)	7 542	4 763	2 779	145	78	67	7 387	4 678	2 709
Malignant neoplasm of breast (C50)	41 809	415	41 394	1 760	14	1 746	39 957	399	39 558
Malignant neoplasm of cervix uteri (C53)	4 092	X	4 092	412	X	412	3 669	X	3 669
Malignant neoplasms of corpus uteri and uterus, part unspecified (C54-C55)	6 784	X	6 784	319	X	319	6 449	X	6 449
Malignant neoplasm of ovary (C56)	14 414	X	14 414	603	X	603	13 783	X	13 783
Malignant neoplasm of prostate (C61)	30 719	30 719	X	1 226	1 226	X	29 407	29 407	X
Malignant neoplasms of kidney and renal pelvis (C64-C65)	12 078	7 524	4 554	640	426	214	11 415	7 084	4 331
Malignant neoplasm of bladder (C67)	12 225	8 343	3 882	336	223	113	11 858	8 100	3 758
Malignant neoplasms of meninges, brain and other parts of central nervous system (C70-C72)	12 609	7 015	5 594	617	335	282	11 967	6 667	5 300
Malignant neoplasms of lymphoid, hematopoietic and related tissue (C81-C96)	56 341	30 060	26 281	2 747	1 577	1 170	53 458	28 410	25 048
Hodgkin's disease (C81)	1 323	725	598	99	66	33	1 221	657	564
Non-Hodgkin's lymphoma (C82-C85)	22 305	11 710	10 595	1 008	579	429	21 246	11 105	10 141
Leukemia (C91-C95)	21 451	11 850	9 601	1 114	640	474	20 285	11 180	9 105
Multiple myeloma and immunoproliferative neoplasms (C88, C90)	11 172	5 724	5 448	524	291	233	10 618	5 418	5 200
Other and unspecified malignant neoplasms of lymphoid, hematopoietic and related tissue (C96)	90	51	39	2	1	1	88	50	38
All other and unspecified malignant neoplasms (C17, C23-C24, C26-C31, C37-C41, C44-C49, C51-C52, C57-C60, C62-C63, C66, C68-C69, C73-C80, C97)	62 708	31 663	31 045	2 844	1 455	1 389	59 685	30 107	29 578
In situ neoplasms, benign neoplasms and neoplasms of uncertain or unknown behavior (D00-D48)	13 676	6 735	6 941	543	276	267	13 098	6 442	6 656
Anemias (D50-D64)	4 627	1 774	2 853	185	82	103	4 424	1 687	2 737
Diabetes mellitus (E10-E14)	71 372	32 841	38 531	5 663	2 590	3 073	65 498	30 144	35 354
Nutritional deficiencies (E40-E64)	3 704	1 233	2 471	136	53	83	3 563	1 178	2 385
Malnutrition (E40-E46)	3 454	1 148	2 306	132	52	80	3 317	1 094	2 223
Other nutritional deficiencies (E50-E64)	250	85	165	4	1	3	246	84	162
Meningitis (G00, G03)	727	394	333	58	33	25	666	361	305
Parkinson's disease (G20-G21)	16 544	9 405	7 139	483	274	209	16 019	9 104	6 915
Alzheimer's disease (G30)	53 852	15 762	38 090	1 239	401	838	52 516	15 334	37 182

[1]Based on codes from the "Tenth Revision, International Classification of Diseases, 1992." Codes *U01-*U03 have been added as of 2001 for deaths due to terrorism.
[2]Includes races other than White and Black.
X = Not applicable.
- = Quantity zero.

Table II-13. Number of Deaths for 113 Selected Causes, by Race, Hispanic Origin, and Sex 2001—*Continued*

(Number.)

Cause of death [1]	Non-Hispanic White			Non-Hispanic Black			Origin not stated [3]		
	Both sexes	Male	Female	Both sexes	Male	Female	Both sexes	Male	Female
All causes	1 962 810	945 967	1 016 843	284 343	143 971	140 372	7 768	4 421	3 347
Salmonella infections (A01-A02)	28	13	15	5	1	4	-	-	-
Shigellosis and amebiasis (A03, A06)	4	3	1	-	-	-	-	-	-
Certain other intestinal infections (A04, A07-A09)	1 357	497	860	98	42	56	10	3	7
Tuberculosis (A16-A19)	368	200	168	200	124	76	7	7	-
Respiratory tuberculosis (A16)	288	162	126	157	103	54	6	6	-
Other tuberculosis (A17-A19)	8	38	42	43	21	22	1	1	-
Whooping cough (A37)	6	3	3	1	1	-	-	-	-
Scarlet fever and erysipelas (A38, A46)	3	2	1	-	-	-	-	-	-
Meningococcal infection (A39)	138	62	76	36	18	18	-	-	-
Septicemia (A40-A41)	24 389	10 668	13 721	5 829	2 611	3 218	98	50	48
Syphilis (A50-A53)	14	5	9	14	9	5	1	1	-
Acute poliomyelitis (A80)	-	-	-	-	-	-	-	-	-
Arthropod-borne viral encephalitis (A83-A84, A85.2)	6	5	1	2	2	-	-	-	-
Measles (B05)	1	1	-	-	-	-	-	-	-
Viral hepatitis (B15-B19)	3 663	2 410	1 253	819	479	340	18	14	4
Human immunodeficiency virus (HIV) disease (B20-B24)	4 286	3 669	617	7 710	5 227	2 483	181	142	39
Malaria (B50-B54)	5	2	3	3	2	1	-	-	-
Other and unspecified infectious and parasitic diseases and their sequelae (A00, A05, A20-A36, A42-A44, A48-A49, A54-A79, A81-A82, A85.0-A85.1, A85.8, A86-B04, B06-B09, B25-B49, B55-B99)	4 257	2 148	2 109	815	419	396	21	15	6
Malignant neoplasms (C00-C97)	456 709	236 004	220 705	61 580	32 335	29 245	1 339	758	581
Malignant neoplasms of lip, oral cavity and pharynx (C00-C14)	6 052	3 961	2 091	1 056	773	283	21	17	4
Malignant neoplasm of esophagus (C15)	10 193	7 863	2 330	1 676	1 222	454	44	37	7
Malignant neoplasm of stomach (C16)	8 308	4 884	3 424	2 051	1 177	874	31	19	12
Malignant neoplasms of colon, rectum and anus (C18-C21)	46 450	23 128	23 322	6 847	3 221	3 626	124	62	62
Malignant neoplasms of liver and intrahepatic bile ducts (C22)	9 433	5 956	3 477	1 649	1 036	613	42	31	11
Malignant neoplasm of pancreas (C25)	24 424	11 994	12 430	3 352	1 501	1 851	60	32	28
Malignant neoplasm of larynx (C32)	2 876	2 267	609	668	531	137	20	19	1
Malignant neoplasms of trachea, bronchus and lung (C33-C34)	133 266	76 066	57 200	15 883	10 030	5 853	360	225	135
Malignant melanoma of skin (C43)	7 250	4 618	2 632	102	36	66	10	7	3
Malignant neoplasm of breast (C50)	33 629	329	33 300	5 528	64	5 464	92	2	90
Malignant neoplasm of cervix uteri (C53)	2 725	X	2 725	778	X	778	11	X	-
Malignant neoplasms of corpus uteri and uterus, part unspecified (C54-C55)	5 283	X	5 283	1 037	X	1 037	16	X	-
Malignant neoplasm of ovary (C56)	12 338	X	12 338	1 140	X	1 140	28	X	-
Malignant neoplasm of prostate (C61)	23 811	23 811	X	5 215	5 215	X	86	86	X
Malignant neoplasms of kidney and renal pelvis (C64-C65)	10 073	6 263	3 810	1 105	673	432	23	14	9
Malignant neoplasm of bladder (C67)	10 823	7 541	3 282	870	444	426	31	20	11
Malignant neoplasms of meninges, brain and other parts of central nervous system (C70-C72)	11 004	6 154	4 850	736	378	358	25	13	12
Malignant neoplasms of lymphoid, hematopoietic and related tissue (C81-C96)	47 190	25 162	22 028	5 162	2 642	2 520	136	73	63
Hodgkin's disease (C81)	1 069	577	492	136	72	64	3	2	1
Non-Hodgkin's lymphoma (C82-C85)	19 300	10 052	9 248	1 478	795	683	51	26	25
Leukemia (C91-C95)	18 105	10 018	8 087	1 741	915	826	52	30	22
Multiple myeloma and immunoproliferative neoplasms (C88, C90)	8 641	4 472	4 169	1 795	853	942	30	15	15
Other and unspecified malignant neoplasms of lymphoid, hematopoietic and related tissue (C96)	75	43	32	12	7	5	-	-	-
All other and unspecified malignant neoplasms (C17, C23-C24, C26-C31, C37-C41, C44-C49, C51-C52, C57-C60, C62-C63, C66, C68-C69, C73-C80, C97)	51 581	26 007	25 574	6 725	3 392	3 333	179	101	78
In situ neoplasms, benign neoplasms and neoplasms of uncertain or unknown behavior (D00-D48)	11 701	5 801	5 900	1 143	522	621	35	17	18
Anemias (D50-D64)	3 350	1 220	2 130	993	443	550	18	5	13
Diabetes mellitus (E10-E14)	51 482	24 302	27 180	12 167	4 983	7 184	211	107	104
Nutritional deficiencies (E40-E64)	3 015	948	2 067	472	202	270	5	2	3
Malnutrition (E40-E46)	2 806	881	1 925	438	186	252	5	2	3
Other nutritional deficiencies (E50-E64)	209	67	142	34	16	18	-	-	-
Meningitis (G00, G03)	469	243	226	172	104	68	3	-	3
Parkinson's disease (G20-G21)	15 239	8 685	6 554	547	291	256	42	27	15
Alzheimer's disease (G30)	49 030	14 295	34 735	3 104	906	2 198	97	27	70

[1]Based on codes from the "Tenth Revision, International Classification of Diseases, 1992." Codes *U01-*U03 have been added as of 2001 for deaths due to terrorism.
[3]Includes deaths for which Hispanic origin was not reported on the death certificate.
X = Not applicable.
- = Quantity zero.

Table II-13. Number of Deaths for 113 Selected Causes, by Race, Hispanic Origin, and Sex 2001—*Continued*

(Number.)

Cause of death [1]	All origins			Hispanic			Non-Hispanic [2]		
	Both sexes	Male	Female	Both sexes	Male	Female	Both sexes	Male	Female
Major cardiovascular diseases (I00-I78)	922 334	428 065	494 269	35 565	18 138	17 427	883 949	408 437	475 512
Diseases of heart (I00-I09, I11, I13, I20-I51)	700 142	339 095	361 047	27 090	14 195	12 895	670 787	323 673	347 114
Acute rheumatic fever and chronic rheumatic heart diseases (I00-I09)	3 489	1 031	2 458	156	51	105	3 327	978	2 349
Hypertensive heart disease (I11)	24 689	10 451	14 238	1 211	648	563	23 343	9 725	13 618
Hypertensive heart and renal disease (I13)	2 826	1 169	1 657	160	76	84	2 650	1 086	1 564
Ischemic heart diseases (I20-I25)	502 189	254 005	248 184	20 664	11 019	9 645	479 842	242 054	237 788
Acute myocardial infarction (I21-I22)	184 757	95 875	88 882	7 403	3 972	3 431	176 895	91 655	85 240
Other acute ischemic heart diseases (I24)	3 351	1 720	1 631	63	42	21	3 284	1 675	1 609
Other forms of chronic ischemic heart disease (I20, I25)	314 081	156 410	157 671	13 198	7 005	6 193	299 663	148 724	150 939
Atherosclerotic cardiovascular disease, so described (I25.0)	68 103	35 440	32 663	3 150	1 928	1 222	64 534	33 247	31 287
All other forms of chronic ischemic heart disease (I20, I25.1-I25.9)	245 978	120 970	125 008	10 048	5 077	4 971	235 129	115 477	119 652
Other heart diseases (I26-I51)	166 949	72 439	94 510	4 899	2 401	2 498	161 625	69 830	91 795
Acute and subacute endocarditis (I33)	1 178	668	510	56	34	22	1 117	631	486
Diseases of pericardium and acute myocarditis (I30-I31, I40)	839	401	438	58	25	33	779	375	404
Heart failure (I50)	56 934	21 632	35 302	1 364	565	799	55 437	21 012	34 425
All other forms of heart disease (I26-I28, I34-I38, I42-I49, I51)	107 998	49 738	58 260	3 421	1 777	1 644	104 292	47 812	56 480
Essential (primary) hypertension and hypertensive renal disease (I10, I12)	19 250	7 254	11 996	889	365	524	18 306	6 856	11 450
Cerebrovascular diseases (I60-I69)	163 538	63 177	100 361	6 416	2 982	3 434	156 741	60 030	96 711
Atherosclerosis (I70)	14 086	5 143	8 943	416	172	244	13 626	4 952	8 674
Other diseases of circulatory system (I71-I78)	25 318	13 396	11 922	754	424	330	24 489	12 926	11 563
Aortic aneurysm and dissection (I71)	15 234	9 114	6 120	403	279	124	14 782	8 798	5 984
Other diseases of arteries, arterioles and capillaries (I72-I78)	10 084	4 282	5 802	351	145	206	9 707	4 128	5 579
Other disorders of circulatory system (I80-I99)	4 665	1 981	2 684	219	104	115	4 425	1 869	2 556
Influenza and pneumonia (J10-J18)	62 034	27 342	34 692	2 722	1 309	1 413	59 103	25 926	33 177
Influenza (J10-J11)	257	101	156	13	7	6	244	94	150
Pneumonia (J12-J18)	61 777	27 241	34 536	2 709	1 302	1 407	58 859	25 832	33 027
Other acute lower respiratory infections (J20-J22)	347	117	230	22	9	13	325	108	217
Acute bronchitis and bronchiolitis (J20-J21)	263	89	174	17	7	10	246	82	164
Unspecified acute lower respiratory infection (J22)	84	28	56	5	2	3	79	26	53
Chronic lower respiratory diseases (J40-J47)	123 013	59 697	63 316	2 832	1 482	1 350	119 839	58 028	61 811
Bronchitis, chronic and unspecified (J40-J42)	959	396	563	37	15	22	919	380	539
Emphysema (J43)	16 242	8 415	7 827	286	175	111	15 914	8 213	7 701
Asthma (J45-J46)	4 269	1 479	2 790	274	103	171	3 976	1 365	2 611
Other chronic lower respiratory diseases (J44, J47)	101 543	49 407	52 136	2 235	1 189	1 046	99 030	48 070	50 960
Pneumoconioses and chemical effects (J60-J66, J68)	1 136	1 095	41	26	25	1	1 108	1 068	40
Pneumonitis due to solids and liquids (J69)	17 301	8 576	8 725	555	283	272	16 700	8 270	8 430
Other diseases of respiratory system (J00-J06, J30-J39, J67, J70-J98)	26 178	12 495	13 683	1 275	643	632	24 835	11 818	13 017
Peptic ulcer (K25-K28)	4 491	2 074	2 417	166	99	67	4 308	1 966	2 342
Diseases of appendix (K35-K38)	428	252	176	30	16	14	398	236	162
Hernia (K40-K46)	1 522	587	935	92	32	60	1 425	552	873
Chronic liver disease and cirrhosis (K70, K73-K74)	27 035	17 393	9 642	3 301	2 410	891	23 629	14 914	8 715
Alcoholic liver disease (K70)	12 207	8 932	3 275	1 729	1 457	272	10 423	7 431	2 992
Other chronic liver disease and cirrhosis (K73-K74)	14 828	8 461	6 367	1 572	953	619	13 206	7 483	5 723
Cholelithiasis and other disorders of gallbladder (K80-K82)	2 965	1 341	1 624	197	92	105	2 764	1 248	1 516
Nephritis, nephrotic syndrome and nephrosis (N00-N07, N17-N19, N25-N27)	39 480	18 852	20 628	1 857	892	965	37 511	17 908	19 603
Acute and rapidly progressive nephritic and nephrotic syndrome (N00-N01, N04)	156	73	83	13	7	6	142	65	77
Chronic glomerulonephritis, nephritis and nephropathy, not specified as acute or chronic, and renal sclerosis, unspecified (N02-N03, N05-N07, N26)	511	264	247	34	17	17	477	247	230
Renal failure (N17-N19)	38 784	18 509	20 275	1 806	867	939	36 867	17 591	19 276
Other disorders of kidney (N25, N27)	29	6	23	4	1	3	25	5	20
Infections of kidney (N10-N12, N13.6, N15.1)	748	213	535	33	10	23	714	203	511
Hyperplasia of prostate (N40)	439	439	X	11	11	X	423	423	X
Inflammatory diseases of female pelvic organs (N70-N76)	127	X	127	5	X	5	122	X	122
Pregnancy, childbirth and the puerperium (O00-O99)	416	X	416	83	X	83	332	X	332
Pregnancy with abortive outcome (O00-O07)	38	X	38	1	X	1	37	X	37
Other complications of pregnancy, childbirth and the puerperium (O10-O99)	378	X	378	82	X	82	295	X	295
Certain conditions originating in the perinatal period (P00-P96)	13 887	7 868	6 019	2 227	1 268	959	11 500	6 511	4 989
Congenital malformations, deformations and chromosomal abnormalities (Q00-Q99)	10 442	5 448	4 994	1 632	892	740	8 748	4 527	4 221
Symptoms, signs and abnormal clinical and laboratory findings, not elsewhere classified (R00-R99)	32 326	15 779	16 547	2 190	1 410	780	29 931	14 246	15 685
All other diseases (Residual)	186 406	75 162	111 244	7 847	3 849	3 998	177 985	71 014	106 971

[1]Based on codes from the "Tenth Revision, International Classification of Diseases, 1992." Codes *U01-*U03 have been added as of 2001 for deaths due to terrorism.
[2]Includes races other than White and Black.
X = Not applicable.

Table II-13. Number of Deaths for 113 Selected Causes, by Race, Hispanic Origin, and Sex 2001—*Continued*

(Number.)

Cause of death [1]	Non-Hispanic White			Non-Hispanic Black			Origin not stated [3]		
	Both sexes	Male	Female	Both sexes	Male	Female	Both sexes	Male	Female
Major cardiovascular diseases (I00-I78)	764 021	352 013	412 008	103 213	47 646	55 567	2 820	1 490	1 330
Diseases of heart (I00-I09, I11, I13, I20-I51)	582 349	280 713	301 636	76 794	36 541	40 253	2 265	1 227	1 038
Acute rheumatic fever and chronic rheumatic heart diseases (I00-I09)	2 924	857	2 067	302	86	216	6	2	4
Hypertensive heart disease (I11)	16 991	6 728	10 263	5 880	2 760	3 120	135	78	57
Hypertensive heart and renal disease (I13)	1 759	687	1 072	821	370	451	16	7	9
Ischemic heart diseases (I20-I25)	420 959	212 928	208 031	50 295	24 255	26 040	1 683	932	751
Acute myocardial infarction (I21-I22)	155 312	81 175	74 137	18 399	8 628	9 771	459	248	211
Other acute ischemic heart diseases (I24)	2 756	1 386	1 370	453	249	204	4	3	1
Other forms of chronic ischemic heart disease (I20, I25)	262 891	130 367	132 524	31 443	15 378	16 065	1 220	681	539
Atherosclerotic cardiovascular disease, so described (I25.0)	53 031	26 960	26 071	10 156	5 476	4 680	419	265	154
All other forms of chronic ischemic heart disease (I20, I25.1-I25.9)	209 860	103 407	106 453	21 287	9 902	11 385	801	416	385
Other heart diseases (I26-I51)	139 716	59 513	80 203	19 496	9 070	10 426	425	208	217
Acute and subacute endocarditis (I33)	859	482	377	233	134	99	5	3	2
Diseases of pericardium and acute myocarditis (I30-I31, I40)	600	286	314	157	78	79	2	1	1
Heart failure (I50)	49 788	18 730	31 058	5 094	2 036	3 058	133	55	78
All other forms of heart disease (I26-I28, I34-I38, I42-I49, I51)	88 469	40 015	48 454	14 012	6 822	7 190	285	149	136
Essential (primary) hypertension and hypertensive renal disease (I10, I12)	13 837	4 970	8 867	4 012	1 698	2 314	55	33	22
Cerebrovascular diseases (I60-I69)	133 879	50 382	83 497	18 844	7 828	11 016	381	165	216
Atherosclerosis (I70)	12 517	4 511	8 006	957	365	592	44	19	25
Other diseases of circulatory system (I71-I78)	21 439	11 437	10 002	2 606	1 214	1 392	75	46	29
Aortic aneurysm and dissection (I71)	13 217	7 924	5 293	1 238	659	579	49	37	12
Other diseases of arteries, arterioles and capillaries (I72-I78)	8 222	3 513	4 709	1 368	555	813	26	9	17
Other disorders of circulatory system (I80-I99)	3 617	1 496	2 121	729	335	394	21	8	13
Influenza and pneumonia (J10-J18)	51 952	22 389	29 563	5 686	2 769	2 917	209	107	102
Influenza (J10-J11)	216	81	135	20	10	10	-	-	-
Pneumonia (J12-J18)	51 736	22 308	29 428	5 666	2 759	2 907	209	107	102
Other acute lower respiratory infections (J20-J22)	278	86	192	33	15	18	-	-	-
Acute bronchitis and bronchiolitis (J20-J21)	202	61	141	31	14	17	-	-	-
Unspecified acute lower respiratory infection (J22)	76	25	51	2	1	1	-	-	-
Chronic lower respiratory diseases (J40-J47)	110 753	52 952	57 801	7 504	4 143	3 361	342	187	155
Bronchitis, chronic and unspecified (J40-J42)	839	338	501	62	32	30	3	1	2
Emphysema (J43)	14 930	7 607	7 323	800	480	320	42	27	15
Asthma (J45-J46)	2 717	834	1 883	1 092	450	642	19	11	8
Other chronic lower respiratory diseases (J44, J47)	92 267	44 173	48 094	5 550	3 181	2 369	278	148	130
Pneumoconioses and chemical effects (J60-J66, J68)	1 046	1 008	38	56	54	2	2	2	-
Pneumonitis due to solids and liquids (J69)	14 843	7 411	7 432	1 584	731	853	46	23	23
Other diseases of respiratory system (J00-J06, J30-J39, J67, J70-J98)	21 835	10 433	11 402	2 462	1 116	1 346	68	34	34
Peptic ulcer (K25-K28)	3 766	1 649	2 117	403	238	165	17	9	8
Diseases of appendix (K35-K38)	313	182	131	72	47	25	-	-	-
Hernia (K40-K46)	1 248	469	779	155	74	81	5	3	2
Chronic liver disease and cirrhosis (K70, K73-K74)	20 065	12 609	7 456	2 733	1 819	914	105	69	36
Alcoholic liver disease (K70)	8 635	6 209	2 426	1 331	926	405	55	44	11
Other chronic liver disease and cirrhosis (K73-K74)	11 430	6 400	5 030	1 402	893	509	50	25	25
Cholelithiasis and other disorders of gallbladder (K80-K82)	2 368	1 065	1 303	309	130	179	4	1	3
Nephritis, nephrotic syndrome and nephrosis (N00-N07, N17-N19, N25-N27)	29 449	14 331	15 118	7 214	3 158	4 056	112	52	60
Acute and rapidly progressive nephritic and nephrotic syndrome (N00-N01, N04)	116	52	64	21	9	12	1	1	-
Chronic glomerulonephritis, nephritis and nephropathy, not specified as acute or chronic, and renal sclerosis, unspecified (N02-N03, N05-N07, N26)	384	200	184	68	36	32	-	-	-
Renal failure (N17-N19)	28 932	14 076	14 856	7 119	3 111	4 008	111	51	60
Other disorders of kidney (N25, N27)	17	3	14	6	2	4	-	-	-
Infections of kidney (N10-N12, N13.6, N15.1)	614	166	448	80	28	52	1	-	1
Hyperplasia of prostate (N40)	388	388	X	31	31	X	5	5	X
Inflammatory diseases of female pelvic organs (N70-N76)	98	X	98	23	X	23	-	X	-
Pregnancy, childbirth and the puerperium (O00-O99)	156	X	156	154	X	154	1	X	1
Pregnancy with abortive outcome (O00-O07)	19	X	19	16	X	16	-	X	-
Other complications of pregnancy, childbirth and the puerperium (O10-O99)	137	X	137	138	X	138	1	X	1
Certain conditions originating in the perinatal period (P00-P96)	6 323	3 574	2 749	4 713	2 674	2 039	160	89	71
Congenital malformations, deformations and chromosomal abnormalities (Q00-Q99)	6 709	3 487	3 222	1 674	861	813	62	29	33
Symptoms, signs and abnormal clinical and laboratory findings, not elsewhere classified (R00-R99)	24 596	11 404	13 192	4 606	2 414	2 192	205	123	82
All other diseases (Residual)	153 156	60 022	93 134	21 752	9 561	12 191	574	299	275

[1]Based on codes from the "Tenth Revision, International Classification of Diseases, 1992." Codes *U01-*U03 have been added as of 2001 for deaths due to terrorism.
[3]Includes deaths for which Hispanic origin was not reported on the death certificate.
X = Not applicable.
- = Quantity zero.

Table II-13. Number of Deaths for 113 Selected Causes, by Race, Hispanic Origin, and Sex 2001—*Continued*

(Number.)

Cause of death [1]	All origins			Hispanic			Non-Hispanic [2]		
	Both sexes	Male	Female	Both sexes	Male	Female	Both sexes	Male	Female
Accidents (unintentional injuries) (V01-X59, Y85-Y86)	101 537	66 060	35 477	9 523	7 157	2 366	91 522	58 528	32 994
Transport accidents (V01-V99, Y85)	47 288	32 794	14 494	5 774	4 334	1 440	41 339	28 327	13 012
Motor vehicle accidents (V02-V04, V09.0, V09.2, V12-V14, V19.0-V19.2, V19.4-V19.6, V20-V79, V80.3-80.5, V81.0-V81.1, V82.0-V82.1, V83-V86, V87.0-V87.8, V88.0-V88.8, V89.0, V89.2)	43 788	30 064	13 724	5 272	3 987	1 285	38 362	25 962	12 400
Other land transport accidents (V01, V05-V06, V09.1, V09.3-V09.9, V10-V11, V15-V18, V19.3, V19.8-19.9, V80.0-V80.2, V80.6-V80.9, V81.2-V81.9, V82.2-V82.9, V87.9, V88.9, V89.1, V89.3, V89.9)	1 405	1 067	338	200	173	27	1 187	878	309
Water, air and space, and other and unspecified transport accidents and their sequelae (V90-V99, Y85)	2 095	1 663	432	302	174	128	1 790	1 487	303
Nontransport accidents (W00-X59, Y86)	54 249	33 266	20 983	3 749	2 823	926	50 183	30 201	19 982
Falls (W00-W19)	15 019	8 089	6 930	757	525	232	14 204	7 527	6 677
Accidental discharge of firearms (W32-W34)	802	690	112	86	76	10	715	613	102
Accidental drowning and submersion (W65-W74)	3 281	2 560	721	421	348	73	2 832	2 188	644
Accidental exposure to smoke, fire and flames (X00-X09)	3 309	1 998	1 311	176	109	67	3 106	1 873	1 233
Accidental poisoning and exposure to noxious substances (X40-X49)	14 078	9 885	4 193	1 324	1 061	263	12 631	8 720	3 911
Other and unspecified nontransport accidents and their sequelae (W20-W31, W35-W64, W75-W99, X10-X39, X50-X59, Y86)	17 760	10 044	7 716	985	704	281	16 695	9 280	7 415
Intentional self-harm (suicide) (*U03, X60-X84, Y87.0) [4]	30 622	24 672	5 950	1 850	1 576	274	28 647	22 983	5 664
Intentional self-harm (suicide) by discharge of firearms (X72-X74)	16 869	14 758	2 111	798	731	67	16 020	13 978	2 042
Intentional self-harm (suicide) by other and unspecified means and their sequelae (*U03, X60-X71, X75-X84, Y87.0) [4]	13 753	9 914	3 839	1 052	845	207	12 627	9 005	3 622
Assault (homicide) (*U01-*U02, X85-Y09, Y87.1) [4]	20 308	15 555	4 753	3 331	2 756	575	16 726	12 608	4 118
Assault (homicide) by discharge of firearms (*U01.4, X93-X95)	11 348	9 532	1 816	2 123	1 892	231	9 134	7 559	1 575
Assault (homicide) by other and unspecified means and their sequelae (*U01.0-*U01.3, *U01.5-*U01.9, *U02, X85-X92, X96-Y09, Y87.1) [4]	8 960	6 023	2 937	1 208	864	344	7 592	5 049	2 543
Legal intervention (Y35, Y89.0)	396	380	16	65	65	-	330	314	16
Events of undetermined intent (Y10-Y34, Y87.2, Y89.9)	4 198	2 833	1 365	283	216	67	3 876	2 586	1 290
Discharge of firearms, undetermined intent (Y22-Y24)	231	190	41	21	16	5	208	172	36
Other and unspecified events of undetermined intent and their sequelae (Y10-Y21, Y25-Y34, Y87.2, Y89.9)	3 967	2 643	1 324	262	200	62	3 668	2 414	1 254
Operations of war and their sequelae (Y36, Y89.1)	17	16	1	2	2	-	15	14	1
Complications of medical and surgical care (Y40-Y84, Y88)	3 021	1 357	1 664	145	65	80	2 861	1 286	1 575

[1]Based on codes from the "Tenth Revision, International Classification of Diseases, 1992." Codes *U01-*U03 have been added as of 2001 for deaths due to terrorism.
[2]Includes races other than White and Black.
[4]Figures include September 11, 2001 related deaths for which death certificates were filed as of 10/24/02.
- = Quantity zero.

Table II-13. Number of Deaths for 113 Selected Causes, by Race, Hispanic Origin, and Sex 2001—*Continued*

(Number.)

Cause of death [1]	Non-Hispanic White			Non-Hispanic Black			Origin not stated [3]		
	Both sexes	Male	Female	Both sexes	Male	Female	Both sexes	Male	Female
Accidents (unintentional injuries) (V01-X59, Y85-Y86)	76 262	48 180	28 082	12 239	8 381	3 858	492	375	117
Transport accidents (V01-V99, Y85)	33 753	23 047	10 706	5 803	4 143	1 660	175	133	42
Motor vehicle accidents (V02-V04, V09.0, V09.2, V12-V14, V19.0-V19.2, V19.4-V19.6, V20-V79, V80.3-80.5, V81.0-V81.1, V82.0-V82.1, V83-V86, V87.0-V87.8, V88.0-V88.8, V89.0, V89.2)	31 283	21 073	10 210	5 423	3 842	1 581	154	115	39
Other land transport accidents (V01, V05-V06, V09.1, V09.3-V09.9, V10-V11, V15-V18, V19.3, V19.8-19.9, V80.0-V80.2, V80.6-V80.9, V81.2-V81.9, V82.2-V82.9, V87.9, V88.9, V89.1, V89.3, V89.9)	901	673	228	216	159	57	18	16	2
Water, air and space, and other and unspecified transport accidents and their sequelae (V90-V99, Y85)	1 569	1 301	268	164	142	22	3	2	1
Nontransport accidents (W00-X59, Y86)	42 509	25 133	17 376	6 436	4 238	2 198	317	242	75
Falls (W00-W19)	13 037	6 785	6 252	837	537	300	58	37	21
Accidental discharge of firearms (W32-W34)	591	506	85	110	97	13	1	1	-
Accidental drowning and submersion (W65-W74)	2 082	1 577	505	570	471	99	28	24	4
Accidental exposure to smoke, fire and flames (X00-X09)	2 204	1 322	882	829	501	328	27	16	11
Accidental poisoning and exposure to noxious substances (X40-X49)	10 269	7 111	3 158	2 122	1 452	670	123	104	19
Other and unspecified nontransport accidents and their sequelae (W20-W31, W35-W64, W75-W99, X10-X39, X50-X59, Y86)	14 326	7 832	6 494	1 968	1 180	788	80	60	20
Intentional self-harm (suicide) (*U03, X60-X84, Y87.0) [4]	25 813	20 710	5 103	1 904	1 578	326	125	113	12
Intentional self-harm (suicide) by discharge of firearms (X72-X74)	14 648	12 771	1 877	1 069	949	120	51	49	2
Intentional self-harm (suicide) by other and unspecified means and their sequelae (*U03, X60-X71, X75-X84, Y87.0) [4]	11 165	7 939	3 226	835	629	206	74	64	10
Assault (homicide) (*U01-*U02, X85-Y09, Y87.1) [4]	7 928	5 453	2 475	8 077	6 658	1 419	251	191	60
Assault (homicide) by discharge of firearms (*U01.4, X93-X95)	3 085	2 172	913	5 790	5 190	600	91	81	10
Assault (homicide) by other and unspecified means and their sequelae (*U01.0-*U01.3, *U01.5-*U01.9, *U02, X85-X92, X96-Y09, Y87.1) [4]	4 843	3 281	1 562	2 287	1 468	819	160	110	50
Legal intervention (Y35, Y89.0)	230	221	9	82	77	5	1	1	-
Events of undetermined intent (Y10-Y34, Y87.2, Y89.9)	3 155	2 054	1 101	629	467	162	39	31	8
Discharge of firearms, undetermined intent (Y22-Y24)	166	133	33	31	28	3	2	2	-
Other and unspecified events of undetermined intent and their sequelae (Y10-Y21, Y25-Y34, Y87.2, Y89.9)	2 989	1 921	1 068	598	439	159	37	29	8
Operations of war and their sequelae (Y36, Y89.1)	14	13	1	1	1	-	-	-	-
Complications of medical and surgical care (Y40-Y84, Y88)	2 294	1 016	1 278	515	247	268	15	6	9

[1]Based on codes from the "Tenth Revision, International Classification of Diseases, 1992." Codes *U01-*U03 have been added as of 2001 for deaths due to terrorism.
[3]Includes deaths for which Hispanic origin was not reported on the death certificate.
[4]Figures include September 11, 2001 related deaths for which death certificates were filed as of 10/24/02.
- = Quantity zero.

TABLE II-14. DEATH RATES FOR 113 SELECTED CAUSES, BY RACE AND SEX, 2001

Crude death rates can be utilized to characterize actual measures of mortality risk. However, due to differences in age distribution between sexes and races, age-adjusted rates (see Table 16) should be used to compare relative risk among groups. An example of the influence of age on crude rates is that death rates from major cardiovascular diseases are higher for women, even though age-adjusted rates for the same diseases are considerably higher for men.

Table II-14. Death Rates for 113 Selected Causes, by Race and Sex, 2001

(Rate per 100,000 population.)

Cause of death[1]	All races			White			All other					
							Total			Black		
	Both sexes	Male	Female	Both sexes	Male	Female	Both sexes	Male	Female	Both sexes	Male	Female
All causes	848.5	846.4	850.4	895.1	881.9	907.9	642.1	684.6	602.9	773.5	823.9	727.7
Salmonella infections (A01-A02)	0.0	0.0	0.0	0.0	*	*	*	*	*	*	*	*
Shigellosis and amebiasis (A03, A06)	*	*	*	*	*	*	*	*	*	*	*	*
Certain other intestinal infections (A04, A07-A09)	0.5	0.4	0.7	0.6	0.5	0.8	0.2	0.2	0.2	0.3	0.2	0.3
Tuberculosis (A16-A19)	0.3	0.3	0.2	0.2	0.2	0.2	0.6	0.7	0.4	0.5	0.7	0.4
Respiratory tuberculosis (A16)	0.2	0.3	0.2	0.2	0.2	0.1	0.4	0.6	0.3	0.4	0.6	0.3
Other tuberculosis (A17-A19)	0.1	0.1	0.1	0.0	0.0	0.0	0.1	0.1	0.1	0.1	0.1	0.1
Whooping cough (A37)	*	*	*	*	*	*	*	*	*	*	*	*
Scarlet fever and erysipelas (A38, A46)	*	*	*	*	*	*	*	*	*	*	*	*
Meningococcal infection (A39)	0.1	0.1	0.1	0.1	0.1	0.1	0.1	0.1	*	0.1	*	*
Septicemia (A40-A41)	11.3	10.2	12.4	11.1	9.9	12.3	12.3	11.7	12.8	15.8	14.9	16.7
Syphilis (A50-A53)	0.0	*	*	0.0	*	*	*	*	*	*	*	*
Acute poliomyelitis (A80)	*	*	*	*	*	*	*	*	*	*	*	*
Arthropod-borne viral encephalitis (A83-A84, A85.2)	*	*	*	*	*	*	*	*	*	*	*	*
Measles (B05)	*	*	*	*	*	*	*	*	*	*	*	*
Viral hepatitis (B15-B19)	2.0	2.5	1.4	1.9	2.5	1.3	2.2	2.5	1.8	2.2	2.7	1.8
Human immunodeficiency virus (HIV) disease (B20-B24)	5.0	7.6	2.5	2.7	4.5	0.9	15.3	21.7	9.3	21.1	30.1	12.9
Malaria (B50-B54)	*	*	*	*	*	*	*	*	*	*	*	*
Other and unspecified infectious and parasitic diseases and their sequelae (A00, A05, A20-A36, A42-A44, A48-A49, A54-A79, A81-A82, A85.0-A85.1, A85.8, A86-B04, B06-B09, B25-B49, B55-B99)	2.0	2.1	1.9	2.0	2.1	1.9	2.0	2.1	1.8	2.2	2.4	2.0
Malignant neoplasms (C00-C97)	194.4	205.3	183.9	206.4	216.4	196.7	141.3	154.8	128.9	167.1	184.5	151.3
Malignant neoplasms of lip, oral cavity and pharynx (C00-C14)	2.7	3.7	1.8	2.7	3.6	1.9	2.5	3.9	1.3	2.9	4.4	1.5
Malignant neoplasm of esophagus (C15)	4.4	6.9	2.0	4.6	7.2	2.1	3.6	5.6	1.8	4.5	7.0	2.3
Malignant neoplasm of stomach (C16)	4.3	5.1	3.5	4.1	4.9	3.3	5.4	6.5	4.4	5.6	6.7	4.5
Malignant neoplasms of colon, rectum and anus (C18-C21)	20.0	20.2	19.8	21.0	21.2	20.7	15.5	15.5	15.6	18.6	18.4	18.8
Malignant neoplasms of liver and intrahepatic bile ducts (C22)	4.7	6.1	3.4	4.6	5.9	3.3	5.1	6.8	3.5	4.5	5.9	3.2
Malignant neoplasm of pancreas (C25)	10.5	10.3	10.6	11.1	11.0	11.2	7.7	7.3	8.1	9.1	8.6	9.6
Malignant neoplasm of larynx (C32)	1.3	2.2	0.5	1.3	2.1	0.5	1.4	2.4	0.5	1.8	3.1	0.7
Malignant neoplasms of trachea, bronchus and lung (C33-C34)	54.8	64.7	45.3	59.1	68.6	49.8	35.8	46.8	25.7	43.1	57.2	30.2
Malignant melanoma of skin (C43)	2.6	3.4	1.9	3.2	4.1	2.3	0.3	0.2	0.3	0.3	0.2	0.3
Malignant neoplasm of breast (C50)	14.7	0.3	28.6	15.2	0.3	29.8	12.2	0.3	23.1	15.0	0.4	28.3
Malignant neoplasm of cervix uteri (C53)	1.4	X	2.8	1.3	X	2.7	1.8	X	3.5	2.1	X	4.0
Malignant neoplasms of corpus uteri and uterus, part unspecified (C54-C55)	2.4	X	4.7	2.4	X	4.8	2.2	X	4.3	2.8	X	5.4
Malignant neoplasm of ovary (C56)	5.1	X	9.9	5.6	X	11.0	2.8	X	5.4	3.1	X	5.9
Malignant neoplasm of prostate (C61)	10.8	22.0	X	10.8	21.9	X	10.8	22.5	X	14.2	29.7	X
Malignant neoplasms of kidney and renal pelvis (C64-C65)	4.2	5.4	3.1	4.6	5.8	3.4	2.6	3.3	1.9	3.0	3.8	2.2
Malignant neoplasm of bladder (C67)	4.3	6.0	2.7	4.8	6.8	2.9	2.0	2.2	1.8	2.4	2.5	2.2
Malignant neoplasms of meninges, brain and other parts of central nervous system (C70-C72)	4.4	5.0	3.9	5.0	5.7	4.4	1.9	2.1	1.7	2.0	2.2	1.8
Malignant neoplasms of lymphoid, hematopoietic and related tissue (C81-C96)	19.8	21.5	18.1	21.5	23.3	19.7	12.1	13.1	11.2	14.1	15.1	13.1
Hodgkin's disease (C81)	0.5	0.5	0.4	0.5	0.6	0.4	0.3	0.3	0.3	0.4	0.4	0.3
Non-Hodgkin's lymphoma (C82-C85)	7.8	8.4	7.3	8.8	9.3	8.2	3.8	4.2	3.3	4.0	4.5	3.6
Leukemia (C91-C95)	7.5	8.5	6.6	8.3	9.3	7.3	4.2	4.7	3.8	4.7	5.2	4.3
Multiple myeloma and immunoproliferative neoplasms (C88, C90)	3.9	4.1	3.8	3.9	4.2	3.7	3.8	3.8	3.8	4.9	4.9	4.9
Other and unspecified malignant neoplasms of lymphoid, hematopoietic and related tissue (C96)	0.0	0.0	0.0	0.0	0.0	0.0	*	*	*	*	*	*
All other and unspecified malignant neoplasms (C17, C23-C24, C26-C31, C37-C41, C44-C49, C51-C52, C57-C60, C62-C63, C66, C68-C69, C73-C80, C97)	22.0	22.6	21.4	23.5	24.0	22.9	15.6	16.5	14.8	18.3	19.4	17.3
In situ neoplasms, benign neoplasms and neoplasms of uncertain or unknown behavior (D00-D48)	4.8	4.8	4.8	5.3	5.3	5.2	2.7	2.6	2.8	3.1	3.0	3.2
Anemias (D50-D64)	1.6	1.3	2.0	1.5	1.1	1.9	2.1	1.9	2.2	2.7	2.5	2.8
Diabetes mellitus (E10-E14)	25.1	23.5	26.6	24.6	23.5	25.7	27.1	23.6	30.3	33.1	28.5	37.2
Nutritional deficiencies (E40-E64)	1.3	0.9	1.7	1.4	0.9	1.8	1.1	0.9	1.2	1.3	1.2	1.4
Malnutrition (E40-E46)	1.2	0.8	1.6	1.3	0.8	1.7	1.0	0.9	1.1	1.2	1.1	1.3
Other nutritional deficiencies (E50-E64)	0.1	0.1	0.1	0.1	0.1	0.1	0.1	*	0.1	0.1	*	*
Meningitis (G00, G03)	0.3	0.3	0.2	0.2	0.2	0.2	0.4	0.5	0.3	0.5	0.6	0.4
Parkinson's disease (G20-G21)	5.8	6.7	4.9	6.8	7.8	5.8	1.5	1.7	1.3	1.5	1.7	1.3
Alzheimer's disease (G30)	18.9	11.3	26.3	21.7	12.8	30.3	6.7	4.2	9.0	8.4	5.1	11.3

[1]Based on codes from the "Tenth Revision, International Classification of Diseases, 1992." Codes *U01-*U03 have been added as of 2001 for deaths due to terrorism.
0.0 = Quantity more than zero but less than 0.05.
* = Figure too small to meet standards of reliability or precision.

Table II-14. Death Rates for 113 Selected Causes, by Race and Sex, 2001—*Continued*

(Rate per 100,000 population.)

Cause of death [1]	All races			White			All other					
							Total			Black		
	Both sexes	Male	Female	Both sexes	Male	Female	Both sexes	Male	Female	Both sexes	Male	Female
Major cardiovascular diseases (I00-I78)	323.9	306.2	340.9	344.8	323.5	365.5	231.3	227.2	235.0	280.4	272.4	287.8
Diseases of heart (I00-I09, I11, I13, I20-I51)	245.8	242.5	249.0	262.8	257.8	267.7	170.7	173.1	168.4	208.8	209.0	208.6
Acute rheumatic fever and chronic rheumatic heart diseases (I00-I09)	1.2	0.7	1.7	1.3	0.8	1.8	0.8	0.5	1.0	0.8	0.5	1.1
Hypertensive heart disease (I11)	8.7	7.5	9.8	7.8	6.4	9.2	12.3	12.1	12.5	16.0	15.9	16.2
Hypertensive heart and renal disease (I13)	1.0	0.8	1.1	0.8	0.7	1.0	1.7	1.6	1.8	2.2	2.1	2.3
Ischemic heart diseases (I20-I25)	176.3	181.7	171.2	190.5	195.8	185.3	113.7	117.4	110.3	136.9	138.9	135.2
Acute myocardial infarction (I21-I22)	64.9	68.6	61.3	70.1	74.4	66.0	41.6	42.1	41.1	50.0	49.2	50.6
Other acute ischemic heart diseases (I24)	1.2	1.2	1.1	1.2	1.2	1.2	1.0	1.1	0.9	1.2	1.4	1.1
Other forms of chronic ischemic heart disease (I20, I25)	110.3	111.9	108.8	119.1	120.1	118.1	71.1	74.2	68.3	85.8	88.2	83.5
Atherosclerotic cardiovascular disease, so described (I25.0)	23.9	25.3	22.5	24.3	25.3	23.3	22.3	25.4	19.4	27.7	31.5	24.3
All other forms of chronic ischemic heart disease (I20, I25.1-I25.9)	86.4	86.5	86.2	94.8	94.8	94.9	48.9	48.7	49.0	58.0	56.7	59.2
Other heart diseases (I26-I51)	58.6	51.8	65.2	62.3	54.1	70.4	42.1	41.4	42.8	52.8	51.7	53.8
Acute and subacute endocarditis (I33)	0.4	0.5	0.4	0.4	0.5	0.3	0.5	0.6	0.4	0.6	0.8	0.5
Diseases of pericardium and acute myocarditis (I30-I31, I40)	0.3	0.3	0.3	0.3	0.3	0.3	0.3	0.4	0.3	0.4	0.4	0.4
Heart failure (I50)	20.0	15.5	24.3	22.1	16.9	27.1	10.8	9.1	12.4	13.8	11.6	15.8
All other forms of heart disease (I26-I28, I34-I38, I42-I49, I51)	37.9	35.6	40.2	39.6	36.5	42.6	30.4	31.3	29.6	38.0	38.9	37.2
Essential (primary) hypertension and hypertensive renal disease (I10, I12)	6.8	5.2	8.3	6.3	4.7	8.0	8.6	7.6	9.5	10.9	9.7	11.9
Cerebrovascular diseases (I60-I69)	57.4	45.2	69.2	60.5	46.6	74.0	44.0	38.8	48.8	51.1	44.6	56.9
Atherosclerosis (I70)	4.9	3.7	6.2	5.6	4.1	7.0	2.1	1.8	2.5	2.6	2.1	3.1
Other diseases of circulatory system (I71-I78)	8.9	9.6	8.2	9.6	10.4	8.8	5.9	6.0	5.8	7.1	6.9	7.2
Aortic aneurysm and dissection (I71)	5.3	6.5	4.2	5.9	7.2	4.6	3.0	3.5	2.5	3.4	3.8	3.0
Other diseases of arteries, arterioles and capillaries (I72-I78)	3.5	3.1	4.0	3.7	3.2	4.2	2.9	2.5	3.2	3.7	3.2	4.2
Other disorders of circulatory system (I80-I99)	1.6	1.4	1.9	1.7	1.4	1.9	1.6	1.5	1.6	2.0	1.9	2.1
Influenza and pneumonia (J10-J18)	21.8	19.6	23.9	23.6	20.7	26.4	13.8	14.3	13.4	15.5	15.9	15.2
Influenza (J10-J11)	0.1	0.1	0.1	0.1	0.1	0.1	0.1	*	*	0.1	*	*
Pneumonia (J12-J18)	21.7	19.5	23.8	23.5	20.6	26.2	13.8	14.3	13.4	15.5	15.8	15.1
Other acute lower respiratory infections (J20-J22)	0.1	0.1	0.2	0.1	0.1	0.2	0.1	0.1	0.1	0.1	*	*
Acute bronchitis and bronchiolitis (J20-J21)	0.1	0.1	0.1	0.1	0.1	0.1	0.1	0.1	0.1	0.1	*	*
Unspecified acute lower respiratory infection (J22)	0.0	0.0	0.0	0.0	0.0	0.0	*	*	*	*	*	*
Chronic lower respiratory diseases (J40-J47)	43.2	42.7	43.7	49.0	47.6	50.3	17.5	20.4	14.9	20.4	23.6	17.5
Bronchitis, chronic and unspecified (J40-J42)	0.3	0.3	0.4	0.4	0.3	0.4	0.2	0.2	0.1	0.2	0.2	0.2
Emphysema (J43)	5.7	6.0	5.4	6.6	6.8	6.3	1.9	2.5	1.4	2.2	2.7	1.7
Asthma (J45-J46)	1.5	1.1	1.9	1.3	0.8	1.7	2.4	2.2	2.7	3.0	2.6	3.3
Other chronic lower respiratory diseases (J44, J47)	35.7	35.3	36.0	40.8	39.7	41.8	13.0	15.6	10.6	15.1	18.1	12.3
Pneumoconioses and chemical effects (J60-J66, J68)	0.4	0.8	0.0	0.5	0.9	0.0	0.1	0.2	*	0.2	0.3	*
Pneumonitis due to solids and liquids (J69)	6.1	6.1	6.0	6.6	6.7	6.6	3.6	3.5	3.7	4.3	4.2	4.4
Other diseases of respiratory system (J00-J06, J30-J39, J67, J70-J98)	9.2	8.9	9.4	10.0	9.7	10.2	5.8	5.6	6.0	6.7	6.4	7.0
Peptic ulcer (K25-K28)	1.6	1.5	1.7	1.7	1.5	1.9	1.0	1.3	0.8	1.1	1.4	0.9
Diseases of appendix (K35-K38)	0.2	0.2	0.1	0.1	0.2	0.1	0.2	0.2	0.1	0.2	0.3	0.1
Hernia (K40-K46)	0.5	0.4	0.6	0.6	0.4	0.7	0.3	0.3	0.3	0.4	0.4	0.4
Chronic liver disease and cirrhosis (K70, K73-K74)	9.5	12.4	6.7	10.1	13.1	7.1	6.9	9.3	4.7	7.5	10.4	4.8
Alcoholic liver disease (K70)	4.3	6.4	2.3	4.5	6.7	2.3	3.5	4.9	2.1	3.6	5.3	2.1
Other chronic liver disease and cirrhosis (K73-K74)	5.2	6.1	4.4	5.6	6.4	4.8	3.4	4.4	2.6	3.8	5.1	2.7
Cholelithiasis and other disorders of gallbladder (K80-K82)	1.0	1.0	1.1	1.1	1.0	1.2	0.8	0.7	0.8	0.8	0.7	0.9
Nephritis, nephrotic syndrome and nephrosis (N00-N07, N17-N19, N25-N27)	13.9	13.5	14.2	13.5	13.3	13.7	15.5	14.4	16.6	19.6	18.0	21.0
Acute and rapidly progressive nephritic and nephrotic syndrome (N00-N01, N04)	0.1	0.1	0.1	0.1	0.1	0.1	0.0	*	*	0.1	*	*
Chronic glomerulonephritis, nephritis and nephropathy, not specified as acute or chronic, and renal sclerosis, unspecified (N02-N03, N05-N07, N26)	0.2	0.2	0.2	0.2	0.2	0.2	0.2	0.2	0.2	0.2	0.2	0.2
Renal failure (N17-N19)	13.6	13.2	14.0	13.2	13.0	13.4	15.3	14.1	16.3	19.3	17.7	20.7
Other disorders of kidney (N25, N27)	0.0	*	0.0	0.0	*	*	*	*	*	*	*	*
Infections of kidney (N10-N12, N13.6, N15.1)	0.3	0.2	0.4	0.3	0.2	0.4	0.2	0.1	0.2	0.2	0.2	0.3
Hyperplasia of prostate (N40)	0.2	0.3	X	0.2	0.4	X	0.1	0.1	X	0.1	0.2	X
Inflammatory diseases of female pelvic organs (N70-N76)	0.0	X	0.1	0.0	X	0.1	0.0	X	0.1	0.1	X	0.1
Pregnancy, childbirth and the puerperium (O00-O99)	0.1	X	0.3	0.1	X	0.2	0.3	X	0.7	0.4	X	0.8
Pregnancy with abortive outcome (O00-O07)	0.0	X	0.0	0.0	X	0.0	*	X	*	*	X	*
Other complications of pregnancy, childbirth and the puerperium (O10-O99)	0.1	X	0.3	0.1	X	0.2	0.3	X	0.6	0.4	X	0.7
Certain conditions originating in the perinatal period (P00-P96)	4.9	5.6	4.2	3.7	4.2	3.1	10.2	12.0	8.5	13.0	15.5	10.8
Congenital malformations, deformations and chromosomal abnormalities (Q00-Q99)	3.7	3.9	3.4	3.6	3.8	3.4	4.0	4.3	3.8	4.6	5.0	4.3
Symptoms, signs and abnormal clinical and laboratory findings, not elsewhere classified (R00-R99)	11.4	11.3	11.4	11.6	11.2	11.9	10.4	11.6	9.3	12.6	13.9	11.4
All other diseases (Residual)	65.5	53.8	76.7	69.4	55.8	82.7	47.9	44.3	51.2	59.1	54.6	63.1

[1]Based on codes from the "Tenth Revision, International Classification of Diseases, 1992." Codes *U01-*U03 have been added as of 2001 for deaths due to terrorism.
0.0 = Quantity more than zero but less than 0.05.
* = Figure too small to meet standards of reliability or precision.
X = Not applicable.

Table II-14. Death Rates for 113 Selected Causes, by Race and Sex, 2001—*Continued*

(Rate per 100,000 population.)

Cause of death [1]	All races			White			All other					
							Total			Black		
	Both sexes	Male	Female	Both sexes	Male	Female	Both sexes	Male	Female	Both sexes	Male	Female
Accidents (unintentional injuries) (V01-X59, Y85-Y86)	35.7	47.2	24.5	37.0	48.4	25.9	29.7	42.0	18.3	33.5	48.2	20.1
Transport accidents (V01-V99, Y85)	16.6	23.5	10.0	17.0	23.9	10.3	14.8	21.4	8.6	15.9	23.8	8.7
Motor vehicle accidents (V02-V04, V09.0, V09.2, V12-V14, V19.0-V19.2, V19.4-V19.6, V20-V79, V80.3-80.5, V81.0-V81.1, V82.0-V82.1, V83-V86, V87.0-V87.8, V88.0-V88.8, V89.0, V89.2)	15.4	21.5	9.5	15.7	21.9	9.8	13.8	19.8	8.2	14.8	22.0	8.2
Other land transport accidents (V01, V05-V06, V09.1, V09.3-V09.9, V10-V11, V15-V18, V19.3, V19.8-19.9, V80.0-V80.2, V80.6-V80.9, V81.2-V81.9, V82.2-V82.9, V87.9, V88.9, V89.1, V89.3, V89.9)	0.5	0.8	0.2	0.5	0.7	0.2	0.6	0.9	0.3	0.6	0.9	0.3
Water, air and space, and other and unspecified transport accidents and their sequelae (V90-V99, Y85)	0.7	1.2	0.3	0.8	1.3	0.3	0.4	0.8	0.2	0.5	0.8	0.1
Nontransport accidents (W00-X59, Y86)	19.0	23.8	14.5	20.0	24.5	15.6	14.9	20.6	9.7	17.6	24.4	11.5
Falls (W00-W19)	5.3	5.8	4.8	6.0	6.4	5.5	2.3	3.0	1.6	2.3	3.1	1.6
Accidental discharge of firearms (W32-W34)	0.3	0.5	0.1	0.3	0.5	0.1	0.2	0.4	*	0.3	0.6	*
Accidental drowning and submersion (W65-W74)	1.2	1.8	0.5	1.1	1.7	0.5	1.5	2.5	0.5	1.6	2.7	0.5
Accidental exposure to smoke, fire and flames (X00-X09)	1.2	1.4	0.9	1.0	1.3	0.8	1.8	2.2	1.3	2.3	2.9	1.7
Accidental poisoning and exposure to noxious substances (X40-X49)	4.9	7.1	2.9	5.0	7.2	2.9	4.6	6.6	2.8	5.8	8.4	3.5
Other and unspecified nontransport accidents and their sequelae (W20-W31, W35-W64, W75-W99, X10-X39, X50-X59, Y86)	6.2	7.2	5.3	6.6	7.5	5.8	4.6	5.9	3.4	5.4	6.8	4.1
Intentional self-harm (suicide) (*U03, X60-X84, Y87.0) [2]	10.8	17.6	4.1	11.9	19.5	4.6	5.6	9.3	2.1	5.3	9.2	1.7
Intentional self-harm (suicide) by discharge of firearms (X72-X74)	5.9	10.6	1.5	6.7	11.8	1.7	2.7	4.9	0.6	2.9	5.5	0.6
Intentional self-harm (suicide) by other and unspecified means and their sequelae (* U03, X60-X71, X75-X84, Y87.0) [2]	4.8	7.1	2.6	5.3	7.7	2.9	2.9	4.4	1.5	2.3	3.7	1.1
Assault (homicide) (*U01-*U02, X85-Y09, Y87.1) [2]	7.1	11.1	3.3	4.9	7.2	2.6	17.1	29.0	6.2	22.1	38.3	7.4
Assault (homicide) by discharge of firearms (*U01.4, X93-X95)	4.0	6.8	1.3	2.2	3.5	1.0	11.7	21.8	2.5	15.8	29.8	3.1
Assault (homicide) by other and unspecified means and their sequelae (*U01.0-*U01.3, *U01.5-*U01.9, *U02, X85-X92, X96-Y09, Y87.1) [2]	3.1	4.3	2.0	2.6	3.7	1.6	5.4	7.2	3.7	6.3	8.5	4.3
Legal intervention (Y35, Y89.0)	0.1	0.3	*	0.1	0.2	*	0.2	0.4	*	0.2	0.4	*
Events of undetermined intent (Y10-Y34, Y87.2, Y89.9)	1.5	2.0	0.9	1.5	2.0	1.0	1.4	2.2	0.7	1.7	2.7	0.9
Discharge of firearms, undetermined intent (Y22-Y24)	0.1	0.1	0.0	0.1	0.1	0.0	0.1	0.2	*	0.1	0.2	*
Other and unspecified events of undetermined intent and their sequelae (Y10-Y21, Y25-Y34, Y87.2, Y89.9)	1.4	1.9	0.9	1.4	1.9	1.0	1.3	2.0	0.7	1.7	2.5	0.8
Operations of war and their sequelae (Y36, Y89.1)	*	*	*	*	*	*	*	*	*	*	*	*
Complications of medical and surgical care (Y40-Y84, Y88)	1.1	1.0	1.1	1.1	0.9	1.2	1.1	1.1	1.1	1.4	1.4	1.4

[1]Based on codes from the "Tenth Revision, International Classification of Diseases, 1992." Codes *U01-*U03 have been added as of 2001 for deaths due to terrorism.
[2]Figures include September 11, 2001 related deaths for which death certificates were filed as of 10/24/02.
0.0 = Quantity more than zero but less than 0.05.
* = Figure too small to meet standards of reliability or precision.

TABLE II-15. DEATH RATES FOR 113 SELECTED CAUSES, BY RACE, HISPANIC ORIGIN, AND SEX, 2001

Crude death rates can be utilized to characterize actual measures of mortality risk. However, due to differences in age distribution between sexes and ethnic groups, age-adjusted rates (see Table 17) should be used to compare relative risk among groups.

Table II-15. Death Rates for 113 Selected Causes, by Race, Hispanic Origin, and Sex, 2001

(Rate per 100,000 population.)

Cause of death [1]	All origins [2]			Hispanic			Non-Hispanic [3]			Non-Hispanic White			Non-Hispanic Black		
	Both sexes	Male	Female	Both sexes	Male	Female	Both sexes	Male	Female	Both sexes	Male	Female	Both sexes	Male	Female
All causes	848.5	846.4	850.4	306.8	332.9	279.0	926.2	923.6	928.6	991.1	975.6	1006.1	798.1	849.7	751.2
Salmonella infections (A01-A02)	0.0	0.0	0.0	*	*	*	0.0	*	0.0	0.0	*	*	*	*	*
Shigellosis and amebiasis (A03, A06)	*	*	*	*	*	*	*	*	*	*	*	*	*	*	*
Certain other intestinal infections (A04, A07-A09)	0.5	0.4	0.7	0.1	*	0.2	0.6	0.5	0.7	0.7	0.5	0.9	0.3	0.2	0.3
Tuberculosis (A16-A19)	0.3	0.3	0.2	0.3	0.4	0.2	0.3	0.3	0.2	0.2	0.2	0.2	0.6	0.7	0.4
Respiratory tuberculosis (A16)	0.2	0.3	0.2	0.2	0.3	0.1	0.2	0.3	0.2	0.1	0.2	0.1	0.4	0.6	0.3
Other tuberculosis (A17-A19)	0.1	0.1	0.1	0.1	*	*	0.1	0.1	0.1	0.0	0.0	0.0	0.1	0.1	0.1
Whooping cough (A37)	*	*	*	*	*	*	*	*	*	*	*	*	*	*	*
Scarlet fever and erysipelas (A38, A46)	*	*	*	*	*	*	*	*	*	*	*	*	*	*	*
Meningococcal infection (A39)	0.1	0.1	0.1	0.1	*	*	0.1	0.1	0.1	0.1	0.1	0.1	0.1	*	*
Septicemia (A40-A41)	11.3	10.2	12.4	3.7	3.6	3.9	12.4	11.2	13.5	12.3	11.0	13.6	16.4	15.4	17.2
Syphilis (A50-A53)	0.0	*	*	*	*	*	0.0	*	*	*	*	*	*	*	*
Acute poliomyelitis (A80)	*	*	*	*	*	*	*	*	*	*	*	*	*	*	*
Arthropod-borne viral encephalitis (A83-A84, A85.2)	*	*	*	*	*	*	*	*	*	*	*	*	*	*	*
Measles (B05)	*	*	*	*	*	*	*	*	*	*	*	*	*	*	*
Viral hepatitis (B15-B19)	2.0	2.5	1.4	2.1	2.6	1.7	1.9	2.5	1.4	1.8	2.5	1.2	2.3	2.8	1.8
Human immunodeficiency virus (HIV) disease (B20-B24)	5.0	7.6	2.5	5.0	7.6	2.3	4.9	7.5	2.5	2.2	3.8	0.6	21.6	30.8	13.3
Malaria (B50-B54)	*	*	*	*	*	*	*	*	*	*	*	*	*	*	*
Other and unspecified infectious and parasitic diseases and their sequelae (A00, A05, A20-A36, A42-A44, A48-A49, A54-A79, A81-A82, A85.0-A85.1, A85.8, A86-B04, B06-B09, B25-B49, B55-B99)	2.0	2.1	1.9	1.1	1.3	1.0	2.1	2.2	2.1	2.1	2.2	2.1	2.3	2.5	2.1
Malignant neoplasms (C00-C97)	194.4	205.3	183.9	60.5	62.2	58.7	213.9	227.2	201.2	230.6	243.4	218.4	172.8	190.8	156.5
Malignant neoplasms of lip, oral cavity and pharynx (C00-C14)	2.7	3.7	1.8	0.8	1.2	0.5	3.0	4.1	1.9	3.1	4.1	2.1	3.0	4.6	1.5
Malignant neoplasm of esophagus (C15)	4.4	6.9	2.0	1.1	1.7	0.5	4.9	7.7	2.2	5.1	8.1	2.3	4.7	7.2	2.4
Malignant neoplasm of stomach (C16)	4.3	5.1	3.5	3.2	3.6	2.8	4.5	5.4	3.6	4.2	5.0	3.4	5.8	6.9	4.7
Malignant neoplasms of colon, rectum and anus (C18-C21)	20.0	20.2	19.8	6.1	6.2	5.9	22.0	22.3	21.7	23.5	23.9	23.1	19.2	19.0	19.4
Malignant neoplasms of liver and intrahepatic bile ducts (C22)	4.7	6.1	3.4	3.4	4.3	2.5	4.9	6.3	3.5	4.8	6.1	3.4	4.6	6.1	3.3
Malignant neoplasm of pancreas (C25)	10.5	10.3	10.6	3.6	3.4	3.9	11.5	11.4	11.5	12.3	12.4	12.3	9.4	8.9	9.9
Malignant neoplasm of larynx (C32)	1.3	2.2	0.5	0.4	0.7	0.1	1.5	2.4	0.6	1.5	2.3	0.6	1.9	3.1	0.7
Malignant neoplasms of trachea, bronchus and lung (C33-C34)	54.8	64.7	45.3	10.3	13.1	7.4	61.3	72.6	50.5	67.3	78.4	56.6	44.6	59.2	31.3
Malignant melanoma of skin (C43)	2.6	3.4	1.9	0.4	0.4	0.4	3.0	3.9	2.1	3.7	4.8	2.6	0.3	0.2	0.4
Malignant neoplasm of breast (C50)	14.7	0.3	28.6	4.8	*	9.7	16.1	0.3	31.1	17.0	0.3	32.9	15.5	0.4	29.2
Malignant neoplasm of cervix uteri (C53)	1.4	X	2.8	1.1	X	2.3	1.5	X	2.9	1.4	X	2.7	2.2	X	4.2
Malignant neoplasms of corpus uteri and uterus, part unspecified (C54-C55)	2.4	X	4.7	0.9	X	1.8	2.6	X	5.1	2.7	X	5.2	2.9	X	5.5
Malignant neoplasm of ovary (C56)	5.1	X	9.9	1.6	X	3.4	5.6	X	10.9	6.2	X	12.2	3.2	X	6.1
Malignant neoplasm of prostate (C61)	10.8	22.0	X	3.3	6.4	X	11.9	24.3	X	12.0	24.6	X	14.6	30.8	X
Malignant neoplasms of kidney and renal pelvis (C64-C65)	4.2	5.4	3.1	1.7	2.2	1.2	4.6	5.9	3.4	5.1	6.5	3.8	3.1	4.0	2.3
Malignant neoplasm of bladder (C67)	4.3	6.0	2.7	0.9	1.2	0.6	4.8	6.7	3.0	5.5	7.8	3.2	2.4	2.6	2.3
Malignant neoplasms of meninges, brain and other parts of central nervous system (C70-C72)	4.4	5.0	3.9	1.7	1.8	1.6	4.8	5.5	4.2	5.6	6.3	4.8	2.1	2.2	1.9
Malignant neoplasms of lymphoid, hematopoietic and related tissue (C81-C96)	19.8	21.5	18.1	7.4	8.3	6.5	21.6	23.5	19.7	23.8	25.9	21.8	14.5	15.6	13.5
Hodgkin's disease (C81)	0.5	0.5	0.4	0.3	0.3	0.2	0.5	0.5	0.4	0.5	0.6	0.5	0.4	0.4	0.3
Non-Hodgkin's lymphoma (C82-C85)	7.8	8.4	7.3	2.7	3.0	2.4	8.6	9.2	8.0	9.7	10.4	9.2	4.1	4.7	3.7
Leukemia (C91-C95)	7.5	8.5	6.6	3.0	3.4	2.6	8.2	9.3	7.2	9.1	10.3	8.0	4.9	5.4	4.4
Multiple myeloma and immunoproliferative neoplasms (C88, C90)	3.9	4.1	3.8	1.4	1.5	1.3	4.3	4.5	4.1	4.4	4.6	4.1	5.0	5.0	5.0
Other and unspecified malignant neoplasms of lymphoid, hematopoietic and related tissue (C96)	0.0	0.0	0.0	*	*	*	0.0	0.0	0.0	0.0	0.0	0.0	*	*	*
All other and unspecified malignant neoplasms (C17, C23-C24, C26-C31, C37-C41, C44-C49, C51-C52, C57-C60, C62-C63, C66, C68-C69, C73-C80, C97)	22.0	22.6	21.4	7.7	7.7	7.7	24.1	24.9	23.3	26.0	26.8	25.3	18.9	20.0	17.8
In situ neoplasms, benign neoplasms and neoplasms of uncertain or unknown behavior (D00-D48)	4.8	4.8	4.8	1.5	1.5	1.5	5.3	5.3	5.2	5.9	6.0	5.8	3.2	3.1	3.3
Anemias (D50-D64)	1.6	1.3	2.0	0.5	0.4	0.6	1.8	1.4	2.2	1.7	1.3	2.1	2.8	2.6	2.9
Diabetes mellitus (E10-E14)	25.1	23.5	26.6	15.3	13.6	17.1	26.4	25.0	27.8	26.0	25.1	26.9	34.1	29.4	38.4
Nutritional deficiencies (E40-E64)	1.3	0.9	1.7	0.4	0.3	0.5	1.4	1.0	1.9	1.5	1.0	2.0	1.3	1.2	1.4
Malnutrition (E40-E46)	1.2	0.8	1.6	0.4	0.3	0.4	1.3	0.9	1.7	1.4	0.9	1.9	1.2	1.1	1.3
Other nutritional deficiencies (E50-E64)	0.1	0.1	0.1	*	*	*	0.1	0.1	0.1	0.1	0.1	0.1	0.1	*	*
Meningitis (G00, G03)	0.3	0.3	0.2	0.2	0.2	0.1	0.3	0.3	0.2	0.2	0.3	0.2	0.5	0.6	0.4
Parkinson's disease (G20-G21)	5.8	6.7	4.9	1.3	1.4	1.2	6.5	7.5	5.4	7.7	9.0	6.5	1.5	1.7	1.4
Alzheimer's disease (G30)	18.9	11.3	26.3	3.4	2.1	4.7	21.2	12.7	29.3	24.8	14.7	34.4	8.7	5.3	11.8

[1] Based on codes from the "Tenth Revision, International Classification of Diseases, 1992." Codes *U01-*U03 have been added as of 2001 for deaths due to terrorism.
[2] Figures for origin not stated are included in "All origins" but not distributed among specified origins.
[3] Includes races other than White and Black.
0.0 = Quantity more than zero but less than 0.05.
* = Figure too small to meet standards of reliability or precision.
X = Not applicable.

Table II-15. Death Rates for 113 Selected Causes, by Race, Hispanic Origin, and Sex, 2001—*Continued*

(Rate per 100,000 population.)

Cause of death [1]	All origins [2]			Hispanic			Non-Hispanic [3]			Non-Hispanic White			Non-Hispanic Black		
	Both sexes	Male	Female	Both sexes	Male	Female	Both sexes	Male	Female	Both sexes	Male	Female	Both sexes	Male	Female
Major cardiovascular diseases (I00-I78)	323.9	306.2	340.9	96.2	95.4	97.1	356.7	338.1	374.3	385.8	363.0	407.6	289.7	281.2	297.4
Diseases of heart (I00-I09, I11, I13, I20-I51)	245.8	242.5	249.0	73.3	74.6	71.8	270.7	268.0	273.3	294.1	289.5	298.4	215.5	215.7	215.4
Acute rheumatic fever and chronic rheumatic heart diseases (I00-I09)	1.2	0.7	1.7	0.4	0.3	0.6	1.3	0.8	1.8	1.5	0.9	2.0	0.8	0.5	1.2
Hypertensive heart disease (I11)	8.7	7.5	9.8	3.3	3.4	3.1	9.4	8.1	10.7	8.6	6.9	10.2	16.5	16.3	16.7
Hypertensive heart and renal disease (I13)	1.0	0.8	1.1	0.4	0.4	0.5	1.1	0.9	1.2	0.9	0.7	1.1	2.3	2.2	2.4
Ischemic heart diseases (I20-I25)	176.3	181.7	171.2	55.9	57.9	53.7	193.6	200.4	187.2	212.6	219.6	205.8	141.2	143.2	139.4
Acute myocardial infarction (I21-I22)	64.9	68.6	61.3	20.0	20.9	19.1	71.4	75.9	67.1	78.4	83.7	73.4	51.6	50.9	52.3
Other acute ischemic heart diseases (I24)	1.2	1.2	1.1	0.2	0.2	0.1	1.3	1.4	1.3	1.4	1.4	1.4	1.3	1.5	1.1
Other forms of chronic ischemic heart disease (I20, I25)	110.3	111.9	108.8	35.7	36.8	34.5	120.9	123.1	118.8	132.7	134.4	131.1	88.2	90.8	86.0
Atherosclerotic cardiovascular disease, so described (I25.0)	23.9	25.3	22.5	8.5	10.1	6.8	26.0	27.5	24.6	26.8	27.8	25.8	28.5	32.3	25.0
All other forms of chronic ischemic heart disease (I20, I25.1-I25.9)	86.4	86.5	86.2	27.2	26.7	27.7	94.9	95.6	94.2	106.0	106.6	105.3	59.7	58.4	60.9
Other heart diseases (I26-I51)	58.6	51.8	65.2	13.3	12.6	13.9	65.2	57.8	72.3	70.6	61.4	79.4	54.7	53.5	55.8
Acute and subacute endocarditis (I33)	0.4	0.5	0.4	0.2	0.2	0.1	0.5	0.5	0.4	0.4	0.5	0.4	0.7	0.8	0.5
Diseases of pericardium and acute myocarditis (I30-I31, I40)	0.3	0.3	0.3	0.2	0.1	0.2	0.3	0.3	0.3	0.3	0.3	0.3	0.4	0.5	0.4
Heart failure (I50)	20.0	15.5	24.3	3.7	3.0	4.5	22.4	17.4	27.1	25.1	19.3	30.7	14.3	12.0	16.4
All other forms of heart disease (I26-I28, I34-I38, I42-I49, I51)	37.9	35.6	40.2	9.3	9.3	9.2	42.1	39.6	44.5	44.7	41.3	47.9	39.3	40.3	38.5
Essential (primary) hypertension and hypertensive renal disease (I10, I12)	6.8	5.2	8.3	2.4	1.9	2.9	7.4	5.7	9.0	7.0	5.1	8.8	11.3	10.0	12.4
Cerebrovascular diseases (I60-I69)	57.4	45.2	69.2	17.4	15.7	19.1	63.2	49.7	76.1	67.6	52.0	82.6	52.9	46.2	59.0
Atherosclerosis (I70)	4.9	3.7	6.2	1.1	0.9	1.4	5.5	4.1	6.8	6.3	4.7	7.9	2.7	2.2	3.2
Other diseases of circulatory system (I71-I78)	8.9	9.6	8.2	2.0	2.2	1.8	9.9	10.7	9.1	10.8	11.8	9.9	7.3	7.2	7.4
Aortic aneurysm and dissection (I71)	5.3	6.5	4.2	1.1	1.5	0.7	6.0	7.3	4.7	6.7	8.2	5.2	3.5	3.9	3.1
Other diseases of arteries, arterioles and capillaries (I72-I78)	3.5	3.1	4.0	0.9	0.8	1.1	3.9	3.4	4.4	4.2	3.6	4.7	3.8	3.3	4.4
Other disorders of circulatory system (I80-I99)	1.6	1.4	1.9	0.6	0.5	0.6	1.8	1.5	2.0	1.8	1.5	2.1	2.0	2.0	2.1
Influenza and pneumonia (J10-J18)	21.8	19.6	23.9	7.4	6.9	7.9	23.8	21.5	26.1	26.2	23.1	29.2	16.0	16.3	15.6
Influenza (J10-J11)	0.1	0.1	0.1	*	*	*	0.1	0.1	0.1	0.1	0.1	0.1	0.1	*	*
Pneumonia (J12-J18)	21.7	19.5	23.8	7.3	6.8	7.8	23.8	21.4	26.0	26.1	23.0	29.1	15.9	16.3	15.6
Other acute lower respiratory infections (J20-J22)	0.1	0.1	0.2	0.1	*	*	0.1	0.1	0.2	0.1	0.1	0.2	0.1	*	*
Acute bronchitis and bronchiolitis (J20-J21)	0.1	0.1	0.1	*	*	*	0.1	0.1	0.1	0.1	0.1	0.1	0.1	*	*
Unspecified acute lower respiratory infection (J22)	0.0	0.0	0.0	*	*	*	0.0	0.0	0.0	0.0	0.0	0.1	*	*	*
Chronic lower respiratory diseases (J40-J47)	43.2	42.7	43.7	7.7	7.8	7.5	48.4	48.0	48.7	55.9	54.6	57.2	21.1	24.5	18.0
Bronchitis, chronic and unspecified (J40-J42)	0.3	0.3	0.4	0.1	*	0.1	0.4	0.3	0.4	0.4	0.3	0.5	0.2	0.2	0.2
Emphysema (J43)	5.7	6.0	5.4	0.8	0.9	0.6	6.4	6.8	6.1	7.5	7.8	7.2	2.2	2.8	1.7
Asthma (J45-J46)	1.5	1.1	1.9	0.7	0.5	1.0	1.6	1.1	2.1	1.4	0.9	1.9	3.1	2.7	3.4
Other chronic lower respiratory diseases (J44, J47)	35.7	35.3	36.0	6.0	6.3	5.8	40.0	39.8	40.1	46.6	45.6	47.6	15.6	18.8	12.7
Pneumoconioses and chemical effects (J60-J66, J68)	0.4	0.8	0.0	0.1	0.1	*	0.4	0.9	0.0	0.5	1.0	0.0	0.2	0.3	*
Pneumonitis due to solids and liquids (J69)	6.1	6.1	6.0	1.5	1.5	1.5	6.7	6.8	6.6	7.5	7.6	7.4	4.4	4.3	4.6
Other diseases of respiratory system (J00-J06, J30-J39, J67, J70-J98)	9.2	8.9	9.4	3.4	3.4	3.5	10.0	9.8	10.2	11.0	10.8	11.3	6.9	6.6	7.2
Peptic ulcer (K25-K28)	1.6	1.5	1.7	0.4	0.5	0.4	1.7	1.6	1.8	1.9	1.7	2.1	1.1	1.4	0.9
Diseases of appendix (K35-K38)	0.2	0.2	0.1	0.1	*	*	0.2	0.2	0.1	0.2	0.2	0.1	0.2	0.3	0.1
Hernia (K40-K46)	0.5	0.4	0.6	0.2	0.2	0.3	0.6	0.5	0.7	0.6	0.5	0.8	0.4	0.4	0.4
Chronic liver disease and cirrhosis (K70, K73-K74)	9.5	12.4	6.7	8.9	12.7	5.0	9.5	12.3	6.9	10.1	13.0	7.4	7.7	10.7	4.9
Alcoholic liver disease (K70)	4.3	6.4	2.3	4.7	7.7	1.5	4.2	6.2	2.4	4.4	6.4	2.4	3.7	5.5	2.2
Other chronic liver disease and cirrhosis (K73-K74)	5.2	6.1	4.4	4.3	5.0	3.4	5.3	6.2	4.5	5.8	6.6	5.0	3.9	5.3	2.7
Cholelithiasis and other disorders of gallbladder (K80-K82)	1.0	1.0	1.1	0.5	0.5	0.6	1.1	1.0	1.2	1.2	1.1	1.3	0.9	0.8	1.0
Nephritis, nephrotic syndrome and nephrosis (N00-N07, N17-N19, N25-N27)	13.9	13.5	14.2	5.0	4.7	5.4	15.1	14.8	15.4	14.9	14.8	15.0	20.2	18.6	21.7
Acute and rapidly progressive nephritic and nephrotic syndrome (N00-N01, N04)	0.1	0.1	0.1	*	*	*	0.1	0.1	0.1	0.1	0.1	0.1	0.1	*	*
Chronic glomerulonephritis, nephritis and nephropathy, not specified as acute or chronic, and renal sclerosis, unspecified (N02-N03, N05-N07, N26)	0.2	0.2	0.2	0.1	*	*	0.2	0.2	0.2	0.2	0.2	0.2	0.2	0.2	0.2
Renal failure (N17-N19)	13.6	13.2	14.0	4.9	4.6	5.2	14.9	14.6	15.2	14.6	14.5	14.7	20.0	18.4	21.4
Other disorders of kidney (N25, N27)	0.0	*	0.0	*	*	*	0.0	*	0.0	*	*	*	*	*	*
Infections of kidney (N10-N12, N13.6, N15.1)	0.3	0.2	0.4	0.1	*	0.1	0.3	0.2	0.4	0.3	0.2	0.4	0.2	0.2	0.3
Hyperplasia of prostate (N40)	0.2	0.3	X	*	*	X	0.2	0.4	X	0.2	0.4	X	0.1	0.2	X
Inflammatory diseases of female pelvic organs (N70-N76)	0.0	X	0.1	*	X	*	0.0	X	0.1	0.0	X	0.1	0.1	X	0.1
Pregnancy, childbirth and the puerperium (O00-O99)	0.1	X	0.3	0.2	X	0.5	0.1	X	0.3	0.1	X	0.2	0.4	X	0.8
Pregnancy with abortive outcome (O00-O07)	0.0	X	0.0	*	X	*	0.0	X	0.0	*	X	*	*	X	*
Other complications of pregnancy, childbirth and the puerperium (O10-O99)	0.1	X	0.3	0.2	X	0.5	0.1	X	0.2	0.1	X	0.1	0.4	X	0.7
Certain conditions originating in the perinatal period (P00-P96)	4.9	5.6	4.2	6.0	6.7	5.3	4.6	5.4	3.9	3.2	3.7	2.7	13.2	15.8	10.9
Congenital malformations, deformations and chromosomal abnormalities (Q00-Q99)	3.7	3.9	3.4	4.4	4.7	4.1	3.5	3.7	3.3	3.4	3.6	3.2	4.7	5.1	4.4
Symptoms, signs and abnormal clinical and laboratory findings, not elsewhere classified (R00-R99)	11.4	11.3	11.4	5.9	7.4	4.3	12.1	11.8	12.3	12.4	11.8	13.1	12.9	14.2	11.7
All other diseases (Residual)	65.5	53.8	76.7	21.2	20.2	22.3	71.8	58.8	84.2	77.3	61.9	92.1	61.1	56.4	65.2

[1]Based on codes from the "Tenth Revision, International Classification of Diseases, 1992." Codes *U01-*U03 have been added as of 2001 for deaths due to terrorism.
[2]Figures for origin not stated are included in "All origins" but not distributed among specified origins.
[3]Includes races other than White and Black.
0.0 = Quantity more than zero but less than 0.05.
* = Figure too small to meet standards of reliability or precision.
X = Not applicable.

Table II-15. Death Rates for 113 Selected Causes, by Race, Hispanic Origin, and Sex, 2001—*Continued*

(Rate per 100,000 population.)

Cause of death[1]	All origins[2]			Hispanic			Non-Hispanic[3]			Non-Hispanic White			Non-Hispanic Black		
	Both sexes	Male	Female	Both sexes	Male	Female	Both sexes	Male	Female	Both sexes	Male	Female	Both sexes	Male	Female
Accidents (unintentional injuries) (V01-X59, Y85-Y86)	35.7	47.2	24.5	25.8	37.6	13.2	36.9	48.5	26.0	38.5	49.7	27.8	34.4	49.5	20.6
Transport accidents (V01-V99, Y85)	16.6	23.5	10.0	15.6	22.8	8.0	16.7	23.5	10.2	17.0	23.8	10.6	16.3	24.5	8.9
Motor vehicle accidents (V02-V04, V09.0, V09.2, V12-V14, V19.0-V19.2, V19.4-V19.6, V20-V79, V80.3-80.5, V81.0-V81.1, V82.0-V82.1, V83-V86, V87.0-V87.8, V88.0-V88.8, V89.0, V89.2)	15.4	21.5	9.5	14.3	21.0	7.2	15.5	21.5	9.8	15.8	21.7	10.1	15.2	22.7	8.5
Other land transport accidents (V01, V05-V06, V09.1, V09.3-V09.9, V10-V11, V15-V18, V19.3, V19.8-19.9, V80.0-V80.2, V80.6-V80.9, V81.2-V81.9, V82.2-V82.9, V87.9, V88.9, V89.1, V89.3, V89.9)	0.5	0.8	0.2	0.5	0.9	0.2	0.5	0.7	0.2	0.5	0.7	0.2	0.6	0.9	0.3
Water, air and space, and other and unspecified transport accidents and their sequelae (V90-V99, Y85)	0.7	1.2	0.3	0.8	0.9	0.7	0.7	1.2	0.2	0.8	1.3	0.3	0.5	0.8	0.1
Nontransport accidents (W00-X59, Y86)	19.0	23.8	14.5	10.1	14.8	5.2	20.2	25.0	15.7	21.5	25.9	17.2	18.1	25.0	11.8
Falls (W00-W19)	5.3	5.8	4.8	2.0	2.8	1.3	5.7	6.2	5.3	6.6	7.0	6.2	2.3	3.2	1.6
Accidental discharge of firearms (W32-W34)	0.3	0.5	0.1	0.2	0.4	*	0.3	0.5	0.1	0.3	0.5	0.1	0.3	0.6	*
Accidental drowning and submersion (W65-W74)	1.2	1.8	0.5	1.1	1.8	0.4	1.1	1.8	0.5	1.1	1.6	0.5	1.6	2.8	0.5
Accidental exposure to smoke, fire and flames (X00-X09)	1.2	1.4	0.9	0.5	0.6	0.4	1.3	1.6	1.0	1.1	1.4	0.9	2.3	3.0	1.8
Accidental poisoning and exposure to noxious substances (X40-X49)	4.9	7.1	2.9	3.6	5.6	1.5	5.1	7.2	3.1	5.2	7.3	3.1	6.0	8.6	3.6
Other and unspecified nontransport accidents and their sequelae (W20-W31, W35-W64, W75-W99, X10-X39, X50-X59, Y86)	6.2	7.2	5.3	2.7	3.7	1.6	6.7	7.7	5.8	7.2	8.1	6.4	5.5	7.0	4.2
Intentional self-harm (suicide) (*U03, X60-X84, Y87.0)[4]	10.8	17.6	4.1	5.0	8.3	1.5	11.6	19.0	4.5	13.0	21.4	5.0	5.3	9.3	1.7
Intentional self-harm (suicide) by discharge of firearms (X72-X74)	5.9	10.6	1.5	2.2	3.8	0.4	6.5	11.6	1.6	7.4	13.2	1.9	3.0	5.6	0.6
Intentional self-harm (suicide) by other and unspecified means and their sequelae (*U03, X60-X71, X75-X84, Y87.0)[4]	4.8	7.1	2.6	2.8	4.4	1.2	5.1	7.5	2.9	5.6	8.2	3.2	2.3	3.7	1.1
Assault (homicide) (*U01-*U02, X85-Y09, Y87.1)[4]	7.1	11.1	3.3	9.0	14.5	3.2	6.7	10.4	3.2	4.0	5.6	2.4	22.7	39.3	7.6
Assault (homicide) by discharge of firearms (*U01.4, X93-X95)	4.0	6.8	1.3	5.7	9.9	1.3	3.7	6.3	1.2	1.6	2.2	0.9	16.3	30.6	3.2
Assault (homicide) by other and unspecified means and their sequelae (*U01.0-*U01.3, *U01.5-*U01.9, *U02, X85-X92, X96-Y09, Y87.1)[4]	3.1	4.3	2.0	3.3	4.5	1.9	3.1	4.2	2.0	2.4	3.4	1.5	6.4	8.7	4.4
Legal intervention (Y35, Y89.0)	0.1	0.3	*	0.2	0.3	*	0.1	0.3	*	0.1	0.2	*	0.2	0.5	*
Events of undetermined intent (Y10-Y34, Y87.2, Y89.9)	1.5	2.0	0.9	0.8	1.1	0.4	1.6	2.1	1.0	1.6	2.1	1.1	1.8	2.8	0.9
Discharge of firearms, undetermined intent (Y22-Y24)	0.1	0.1	0.0	0.1	*	*	0.1	0.1	0.0	0.1	0.1	0.0	0.1	0.2	*
Other and unspecified events of undetermined intent and their sequelae (Y10-Y21, Y25-Y34, Y87.2, Y89.9)	1.4	1.9	0.9	0.7	1.1	0.3	1.5	2.0	1.0	1.5	2.0	1.1	1.7	2.6	0.9
Operations of war and their sequelae (Y36, Y89.1)	*	*	*	*	*	*	*	*	*	*	*	*	*	*	*
Complications of medical and surgical care (Y40-Y84, Y88)	1.1	1.0	1.1	0.4	0.3	0.4	1.2	1.1	1.2	1.2	1.0	1.3	1.4	1.5	1.4

[1]Based on codes from the "Tenth Revision, International Classification of Diseases, 1992." Codes *U01-*U03 have been added as of 2001 for deaths due to terrorism.
[2]Figures for origin not stated are included in "All origins" but not distributed among specified origins.
[3]Includes races other than White and Black.
[4]Figures include September 11, 2001 related deaths for which death certificates were filed as of 10/24/02.
0.0 = Quantity more than zero but less than 0.05.
* = Figure too small to meet standards of reliability or precision.

TABLE II-16. AGE-ADJUSTED DEATH RATES FOR 113 SELECTED CAUSES, BY RACE AND SEX, 2001

These age-adjusted data allow a comparison of the susceptibility of various groups to different health risks. Adjusted for age, the overall male death rate is 43 percent higher than the female rate. As a result males in general are expected to have higher age-adjusted death rates from most specific causes. Two cases of interest that do not fit this pattern are Alzheimer's disease, for which women have a higher age-adjusted rate, and cerebrovascular diseases (stroke), which are closely matched between the sexes.

Blacks have a significantly higher risk of cancer death than Whites. For cancer, White males have a 46 percent increased relative risk of death compared with females, but for Black males this excess of cancer mortality risk is 73 percent compared with Black females. Lung cancer among Black males is a major contributor to this excess. For fatalities from diabetes, which are much smaller, the incidence for Blacks is more than twice that of Whites; for Blacks the rate for men and women is about the same. Pneumonia hits Black men particularly heavily.

Fatal accidents occur more than twice as often for men compared with women, both in the case of Whites as for Blacks. Suicide is also primarily a male thing, and has twice the frequency among Whites than Blacks. Homicide victims are primarily men and much more frequently Black men.

Table II-16. Age-Adjusted Death Rates for 113 Selected Causes, by Race and Sex, 2001

(Rate per 100,000 U.S. standard population.)

Cause of death [1]	All races			White			All other					
							Total			Black		
	Both sexes	Male	Female	Both sexes	Male	Female	Both sexes	Male	Female	Both sexes	Male	Female
All causes	854.5	1 029.1	721.8	836.5	1 006.1	706.7	939.9	1 151.4	788.5	1 101.2	1 375.0	912.5
Salmonella infections (A01-A02)	0.0	0.0	0.0	0.0	*	*	*	*	*	*	*	*
Shigellosis and amebiasis (A03, A06)	*	*	*	*	*	*	*	*	*	*	*	*
Certain other intestinal infections (A04, A07-A09)	0.5	0.5	0.5	0.6	0.5	0.6	0.4	0.4	0.3	0.4	0.5	0.4
Tuberculosis (A16-A19)	0.3	0.4	0.2	0.2	0.2	0.1	0.8	1.3	0.6	0.8	1.2	0.5
Respiratory tuberculosis (A16)	0.2	0.3	0.1	0.1	0.2	0.1	0.7	1.0	0.4	0.6	1.0	0.3
Other tuberculosis (A17-A19)	0.1	0.1	0.0	0.0	0.0	0.0	0.2	0.2	0.1	0.2	0.2	0.1
Whooping cough (A37)	*	*	*	*	*	*	*	*	*	*	*	*
Scarlet fever and erysipelas (A38, A46)	*	*	*	*	*	*	*	*	*	*	*	*
Meningococcal infection (A39)	0.1	0.1	0.1	0.1	0.0	0.1	0.1	0.1	*	0.1	*	*
Septicemia (A40-A41)	11.4	12.7	10.5	10.3	11.5	9.5	18.9	21.7	17.0	23.6	27.7	21.2
Syphilis (A50-A53)	0.0	*	*	0.0	*	*	*	*	*	*	*	*
Acute poliomyelitis (A80)	*	*	*	*	*	*	*	*	*	*	*	*
Arthropod-borne viral encephalitis (A83-A84, A85.2)	*	*	*	*	*	*	*	*	*	*	*	*
Measles (B05)	*	*	*	*	*	*	*	*	*	*	*	*
Viral hepatitis (B15-B19)	2.0	2.6	1.3	1.8	2.5	1.2	2.7	3.2	2.2	2.7	3.4	2.1
Human immunodeficiency virus (HIV) disease (B20-B24)	5.0	7.5	2.5	2.6	4.4	0.9	16.1	23.7	9.5	22.8	33.8	13.4
Malaria (B50-B54)	*	*	*	*	*	*	*	*	*	*	*	*
Other and unspecified infectious and parasitic diseases and their sequelae (A00, A05, A20-A36, A42-A44, A48-A49, A54-A79, A81-A82, A85.0-A85.1, A85.8, A86-B04, B06-B09, B25-B49, B55-B99)	2.0	2.4	1.7	1.9	2.3	1.6	2.6	3.0	2.2	2.8	3.4	2.4
Malignant neoplasms (C00-C97)	196.0	243.7	164.7	193.9	239.2	163.9	208.7	276.1	167.0	243.1	330.9	191.3
Malignant neoplasms of lip, oral cavity and pharynx (C00-C14)	2.7	4.1	1.6	2.6	3.9	1.5	3.4	5.8	1.7	3.9	6.7	1.8
Malignant neoplasm of esophagus (C15)	4.4	7.8	1.8	4.3	7.6	1.7	5.2	9.0	2.4	6.4	11.3	3.0
Malignant neoplasm of stomach (C16)	4.4	6.1	3.1	3.8	5.4	2.7	8.2	11.5	5.9	8.3	11.9	5.8
Malignant neoplasms of colon, rectum and anus (C18-C21)	20.1	24.2	17.2	19.6	23.7	16.6	23.5	27.8	20.7	27.6	33.1	24.1
Malignant neoplasms of liver and intrahepatic bile ducts (C22)	4.7	6.9	3.0	4.3	6.3	2.7	7.1	10.6	4.6	6.2	9.1	4.1
Malignant neoplasm of pancreas (C25)	10.5	12.1	9.3	10.4	12.0	9.1	11.7	12.6	10.9	13.6	15.0	12.5
Malignant neoplasm of larynx (C32)	1.3	2.5	0.5	1.2	2.3	0.5	2.0	3.9	0.7	2.5	5.0	0.9
Malignant neoplasms of trachea, bronchus and lung (C33-C34)	55.3	75.2	41.0	55.6	74.2	42.1	52.8	81.1	33.8	62.5	99.1	38.7
Malignant melanoma of skin (C43)	2.7	3.9	1.8	3.0	4.3	2.0	0.4	0.4	0.4	0.4	0.3	0.4
Malignant neoplasm of breast (C50)	14.8	0.3	26.0	14.4	0.3	25.5	16.7	0.5	28.5	20.4	0.6	34.4
Malignant neoplasm of cervix uteri (C53)	1.4	X	2.7	1.3	X	2.5	2.4	X	4.2	2.8	X	4.8
Malignant neoplasms of corpus uteri and uterus, part unspecified (C54-C55)	2.4	X	4.2	2.3	X	4.0	3.4	X	5.7	4.2	X	7.0
Malignant neoplasm of ovary (C56)	5.1	X	9.0	5.2	X	9.3	4.1	X	6.9	4.5	X	7.5
Malignant neoplasm of prostate (C61)	10.9	29.1	X	10.0	26.6	X	18.3	50.3	X	23.0	66.1	X
Malignant neoplasms of kidney and renal pelvis (C64-C65)	4.3	6.2	2.8	4.3	6.3	2.8	3.7	5.5	2.5	4.2	6.4	2.8
Malignant neoplasm of bladder (C67)	4.3	7.5	2.3	4.5	7.9	2.2	3.2	4.5	2.4	3.7	5.1	2.9
Malignant neoplasms of meninges, brain and other parts of central nervous system (C70-C72)	4.5	5.5	3.6	4.8	5.9	3.9	2.4	3.0	2.0	2.6	3.1	2.2
Malignant neoplasms of lymphoid, hematopoietic and related tissue (C81-C96)	20.0	25.5	16.0	20.2	25.9	16.2	17.5	21.8	14.5	20.0	25.3	16.5
Hodgkin's disease (C81)	0.5	0.6	0.4	0.5	0.6	0.4	0.3	0.4	0.3	0.4	0.5	0.4
Non-Hodgkin's lymphoma (C82-C85)	7.9	9.9	6.4	8.2	10.2	6.7	5.4	6.9	4.3	5.6	7.2	4.5
Leukemia (C91-C95)	7.6	10.1	5.9	7.8	10.4	6.0	5.9	7.6	4.8	6.7	8.7	5.4
Multiple myeloma and immunoproliferative neoplasms (C88, C90)	4.0	4.9	3.3	3.7	4.6	3.0	5.8	6.8	5.1	7.3	8.8	6.4
Other and unspecified malignant neoplasms of lymphoid, hematopoietic and related tissue (C96)	0.0	0.0	0.0	0.0	0.0	0.0	*	*	*	*	*	*
All other and unspecified malignant neoplasms (C17, C23-C24, C26-C31, C37-C41, C44-C49, C51-C52, C57-C60, C62-C63, C66, C68-C69, C73-C80, C97)	22.2	26.7	18.9	22.0	26.5	18.8	22.8	27.8	19.3	26.2	32.8	21.9
In situ neoplasms, benign neoplasms and neoplasms of uncertain or unknown behavior (D00-D48)	4.9	6.0	4.1	4.9	6.2	4.1	4.1	4.9	3.7	4.6	5.6	4.1
Anemias (D50-D64)	1.6	1.6	1.6	1.4	1.4	1.4	2.8	2.7	2.8	3.5	3.6	3.4
Diabetes mellitus (E10-E14)	25.3	28.1	23.1	23.0	26.2	20.5	41.3	41.3	40.7	49.2	49.9	48.1
Nutritional deficiencies (E40-E64)	1.3	1.2	1.4	1.2	1.1	1.3	1.8	1.9	1.7	2.1	2.4	1.8
Malnutrition (E40-E46)	1.2	1.1	1.3	1.2	1.0	1.2	1.7	1.8	1.6	1.9	2.2	1.7
Other nutritional deficiencies (E50-E64)	0.1	0.1	0.1	0.1	0.1	0.1	0.1	*	0.1	0.1	*	*
Meningitis (G00, G03)	0.3	0.3	0.2	0.2	0.3	0.2	0.4	0.6	0.3	0.5	0.8	0.4
Parkinson's disease (G20-G21)	5.9	9.1	4.0	6.2	9.7	4.3	2.7	4.0	1.9	2.5	3.9	1.8
Alzheimer's disease (G30)	19.1	16.1	20.5	19.8	16.7	21.3	12.4	10.8	13.1	14.6	13.1	15.1

[1]Based on codes from the "Tenth Revision, International Classification of Diseases, 1992." Codes *U01-*U03 have been added as of 2001 for deaths due to terrorism.
0.0 = Quantity more than zero but less than 0.05.
* = Figure too small to meet standards of reliability or precision.
X = Not applicable

Table II-16. Age-Adjusted Death Rates for 113 Selected Causes, by Race and Sex, 2001—*Continued*

(Rate per 100,000 U.S. standard population.)

Cause of death [1]	All races: Both sexes	All races: Male	All races: Female	White: Both sexes	White: Male	White: Female	All other: Total: Both sexes	All other: Total: Male	All other: Total: Female	All other: Black: Both sexes	All other: Black: Male	All other: Black: Female
Major cardiovascular diseases (I00-I78)	326.5	388.1	278.8	319.2	381.1	270.8	365.9	426.3	321.2	427.5	505.3	372.4
Diseases of heart (I00-I09, I11, I13, I20-I51)	247.8	305.4	203.9	243.5	301.8	198.7	268.5	321.5	229.9	316.9	384.5	269.8
Acute rheumatic fever and chronic rheumatic heart diseases (I00-I09)	1.2	0.9	1.5	1.2	0.9	1.5	1.1	0.8	1.3	1.1	0.7	1.3
Hypertensive heart disease (I11)	8.7	8.9	8.1	7.3	7.3	6.8	18.3	20.2	16.5	23.2	26.3	20.5
Hypertensive heart and renal disease (I13)	1.0	1.1	1.0	0.8	0.8	0.7	2.6	2.8	2.5	3.3	3.6	3.0
Ischemic heart diseases (I20-I25)	177.8	228.5	139.9	176.5	228.4	137.4	182.3	223.4	152.7	211.6	262.0	176.7
Acute myocardial infarction (I21-I22)	65.4	84.9	50.9	65.1	85.4	49.9	66.2	80.2	56.6	76.8	93.1	66.0
Other acute ischemic heart diseases (I24)	1.2	1.5	0.9	1.1	1.4	0.9	1.5	1.9	1.2	1.8	2.3	1.3
Other forms of chronic ischemic heart disease (I20, I25)	111.2	142.1	88.1	110.2	141.6	86.7	114.5	141.3	95.0	133.0	166.7	109.3
Atherosclerotic cardiovascular disease, so described (I25.0)	24.1	30.8	18.5	22.5	28.8	17.3	34.6	45.3	26.6	41.8	56.0	31.7
All other forms of chronic ischemic heart disease (I20, I25.1-I25.9)	87.1	111.3	69.6	87.6	112.8	69.4	80.0	96.0	68.4	91.1	110.6	77.6
Other heart diseases (I26-I51)	59.1	66.1	53.4	57.7	64.4	52.3	64.3	74.5	56.9	77.8	91.8	68.2
Acute and subacute endocarditis (I33)	0.4	0.5	0.3	0.4	0.5	0.3	0.6	0.9	0.5	0.8	1.1	0.6
Diseases of pericardium and acute myocarditis (I30-I31, I40)	0.3	0.3	0.3	0.3	0.3	0.3	0.4	0.4	0.4	0.5	0.5	0.4
Heart failure (I50)	20.2	21.2	19.2	20.2	21.2	19.3	18.3	19.5	17.3	22.1	24.4	20.6
All other forms of heart disease (I26-I28, I34-I38, I42-I49, I51)	38.2	44.1	33.6	36.8	42.4	32.4	44.9	53.7	38.7	54.4	65.8	46.5
Essential (primary) hypertension and hypertensive renal disease (I10, I12)	6.8	6.7	6.8	5.8	5.6	5.9	13.6	14.2	13.0	16.6	17.9	15.4
Cerebrovascular diseases (I60-I69)	57.9	59.0	56.4	55.8	56.5	54.5	70.7	75.4	66.9	78.8	85.4	73.7
Atherosclerosis (I70)	5.0	5.1	4.8	5.1	5.2	5.0	3.7	4.0	3.5	4.3	4.6	4.1
Other diseases of circulatory system (I71-I78)	9.0	12.0	6.9	8.9	12.0	6.7	9.3	11.2	7.9	10.8	12.8	9.4
Aortic aneurysm and dissection (I71)	5.4	8.0	3.6	5.5	8.2	3.6	4.6	6.2	3.5	5.0	6.4	3.9
Other diseases of arteries, arterioles and capillaries (I72-I78)	3.6	3.9	3.3	3.4	3.8	3.2	4.7	5.0	4.5	5.9	6.4	5.5
Other disorders of circulatory system (I80-I99)	1.6	1.7	1.6	1.5	1.6	1.5	2.2	2.4	2.1	2.7	3.1	2.5
Influenza and pneumonia (J10-J18)	22.0	26.6	19.2	21.7	26.0	19.1	22.8	30.1	18.4	24.1	*	*
Influenza (J10-J11)	0.1	0.1	0.1	0.1	0.1	0.1	0.1	*	*	0.1	*	*
Pneumonia (J12-J18)	21.9	26.5	19.1	21.6	25.9	19.0	22.7	30.0	18.4	24.1	32.2	19.4
Other acute lower respiratory infections (J20-J22)	0.1	0.1	0.1	0.1	0.1	0.1	0.1	0.1	0.1	0.1	*	*
Acute bronchitis and bronchiolitis (J20-J21)	0.1	0.1	0.1	0.1	0.1	0.1	0.1	0.1	0.1	0.1	*	*
Unspecified acute lower respiratory infection (J22)	0.0	0.0	0.0	0.0	0.0	0.0	*	*	*	*	*	*
Chronic lower respiratory diseases (J40-J47)	43.7	54.0	37.6	45.6	55.5	39.8	27.6	40.9	19.9	30.9	46.3	22.4
Bronchitis, chronic and unspecified (J40-J42)	0.3	0.4	0.3	0.3	0.4	0.3	0.2	0.3	0.2	0.2	0.3	0.2
Emphysema (J43)	5.8	7.4	4.7	6.1	7.8	5.1	3.0	4.8	1.9	3.3	5.3	2.2
Asthma (J45-J46)	1.5	1.2	1.7	1.2	0.9	1.5	3.0	2.8	3.2	3.6	3.2	3.8
Other chronic lower respiratory diseases (J44, J47)	36.1	45.0	30.8	37.9	46.4	32.9	21.3	32.9	14.7	23.7	37.5	16.2
Pneumoconioses and chemical effects (J60-J66, J68)	0.4	1.0	0.0	0.4	1.1	0.0	0.2	0.5	*	0.2	0.7	*
Pneumonitis due to solids and liquids (J69)	6.1	8.5	4.8	6.1	8.5	4.8	6.1	7.9	5.2	7.0	9.4	5.7
Other diseases of respiratory system (J00-J06, J30-J39, J67, J70-J98)	9.3	11.0	8.1	9.3	11.1	8.1	8.5	9.6	7.7	9.4	10.7	8.6
Peptic ulcer (K25-K28)	1.6	1.8	1.4	1.6	1.8	1.4	1.6	2.2	1.1	1.6	2.3	1.1
Diseases of appendix (K35-K38)	0.1	0.2	0.1	0.1	0.2	0.1	0.2	0.3	0.1	0.2	0.4	0.2
Hernia (K40-K46)	0.5	0.5	0.5	0.5	0.5	0.5	0.5	0.6	0.4	0.6	0.8	0.5
Chronic liver disease and cirrhosis (K70, K73-K74)	9.5	13.2	6.2	9.6	13.3	6.3	8.5	12.2	5.5	9.3	13.8	5.6
Alcoholic liver disease (K70)	4.3	6.6	2.2	4.3	6.6	2.2	4.1	6.3	2.4	4.4	6.9	2.4
Other chronic liver disease and cirrhosis (K73-K74)	5.2	6.6	4.0	5.3	6.7	4.1	4.4	5.9	3.2	4.9	6.9	3.2
Cholelithiasis and other disorders of gallbladder (K80-K82)	1.0	1.2	0.9	1.0	1.2	0.9	1.2	1.4	1.1	1.3	1.4	1.2
Nephritis, nephrotic syndrome and nephrosis (N00-N07, N17-N19, N25-N27)	14.0	17.4	12.0	12.5	16.0	10.5	24.2	27.2	22.3	29.5	33.8	27.1
Acute and rapidly progressive nephritic and nephrotic syndrome (N00-N01, N04)	0.0	0.1	0.0	0.0	0.1	0.0	0.1	*	*	0.1	*	*
Chronic glomerulonephritis, nephritis and nephropathy, not specified as acute or chronic, and renal sclerosis, unspecified (N02-N03, N05-N07, N26)	0.2	0.2	0.2	0.2	0.2	0.1	0.2	0.3	0.2	0.2	0.3	0.2
Renal failure (N17-N19)	13.7	17.1	11.8	12.3	15.8	10.2	23.8	26.8	22.0	29.2	33.4	26.7
Other disorders of kidney (N25, N27)	0.0	*	0.0	0.0	*	*	*	*	*	*	*	*
Infections of kidney (N10-N12, N13.6, N15.1)	0.3	0.2	0.3	0.3	0.2	0.3	0.3	0.3	0.3	0.3	0.3	0.3
Hyperplasia of prostate (N40)	0.1	0.5	X	0.2	0.5	X	0.1	0.3	X	0.1	0.4	X
Inflammatory diseases of female pelvic organs (N70-N76)	0.0	X	0.1	0.0	X	0.1	0.1	X	0.1	0.1	X	0.1
Pregnancy, childbirth and the puerperium (O00-O99)	0.1	X	0.3	0.1	X	0.2	0.3	X	0.6	0.4	X	0.8
Pregnancy with abortive outcome (O00-O07)	0.0	X	0.0	0.0	X	0.0	*	X	*	*	X	*
Other complications of pregnancy, childbirth and the puerperium (O10-O99)	0.1	X	0.3	0.1	X	0.2	0.3	X	0.5	0.4	X	0.7
Certain conditions originating in the perinatal period (P00-P96)	4.7	5.3	4.2	3.7	4.1	3.3	8.3	9.2	7.4	10.3	11.3	9.1
Congenital malformations, deformations and chromosomal abnormalities (Q00-Q99)	3.6	3.8	3.4	3.6	3.8	3.4	3.6	3.8	3.5	4.1	4.3	3.9
Symptoms, signs and abnormal clinical and laboratory findings, not elsewhere classified (R00-R99)	11.3	12.7	9.8	11.0	12.2	9.5	12.7	14.6	10.8	15.1	17.7	12.8
All other diseases (Residual)	65.9	66.1	64.0	64.7	64.5	63.0	70.6	74.9	66.7	84.6	92.0	78.5

[1]Based on codes from the "Tenth Revision, International Classification of Diseases, 1992." Codes *U01-*U03 have been added as of 2001 for deaths due to terrorism.
0.0 = Quantity more than zero but less than 0.05.
* = Figure too small to meet standards of reliability or precision.
X = Not applicable.

Table II-16. Age-Adjusted Death Rates for 113 Selected Causes, by Race and Sex, 2001—*Continued*

(Rate per 100,000 U.S. standard population.)

Cause of death [1]	All races			White			All other					
							Total			Black		
	Both sexes	Male	Female	Both sexes	Male	Female	Both sexes	Male	Female	Both sexes	Male	Female
Accidents (unintentional injuries) (V01-X59, Y85-Y86)	35.7	50.2	22.5	36.0	50.3	22.8	33.4	49.3	20.3	37.6	56.9	22.0
Transport accidents (V01-V99, Y85)	16.6	23.7	9.8	16.8	24.0	10.0	15.3	22.7	8.9	16.5	25.5	8.9
Motor vehicle accidents (V02-V04, V09.0, V09.2, V12-V14, V19.0-V19.2, V19.4-V19.6, V20-V79, V80.3-80.5, V81.0-V81.1, V82.0-V82.1, V83-V86, V87.0-V87.8, V88.0-V88.8, V89.0, V89.2)	15.3	21.8	9.3	15.6	22.0	9.5	14.2	20.9	8.4	15.4	23.5	8.5
Other land transport accidents (V01, V05-V06, V09.1, V09.3-V09.9, V10-V11, V15-V18, V19.3, V19.8-19.9, V80.0-V80.2, V80.6-V80.9, V81.2-V81.9, V82.2-V82.9, V87.9, V88.9, V89.1, V89.3, V89.9)	0.5	0.8	0.2	0.5	0.7	0.2	0.6	0.9	0.3	0.6	1.0	0.3
Water, air and space, and other and unspecified transport accidents and their sequelae (V90-V99, Y85)	0.7	1.2	0.3	0.8	1.3	0.3	0.5	0.8	0.2	0.5	1.0	0.1
Nontransport accidents (W00-X59, Y86)	19.1	26.4	12.7	19.2	26.4	12.8	18.2	26.6	11.4	21.1	31.4	13.1
Falls (W00-W19)	5.3	7.2	3.9	5.6	7.5	4.1	3.4	5.2	2.1	3.3	5.1	2.0
Accidental discharge of firearms (W32-W34)	0.3	0.5	0.1	0.3	0.5	0.1	0.2	0.4	*	0.3	0.5	*
Accidental drowning and submersion (W65-W74)	1.1	1.8	0.5	1.1	1.7	0.5	1.4	2.4	0.5	1.5	2.6	0.5
Accidental exposure to smoke, fire and flames (X00-X09)	1.2	1.5	0.9	1.0	1.3	0.7	2.1	2.9	1.5	2.7	3.8	1.9
Accidental poisoning and exposure to noxious substances (X40-X49)	4.9	7.0	2.9	5.0	7.1	2.9	4.9	7.2	2.9	6.3	9.5	3.7
Other and unspecified nontransport accidents and their sequelae (W20-W31, W35-W64, W75-W99, X10-X39, X50-X59, Y86)	6.3	8.4	4.5	6.2	8.3	4.5	6.1	8.5	4.3	7.0	9.8	4.9
Intentional self-harm (suicide) (*U03, X60-X84, Y87.0) [1]	10.7	18.2	4.0	11.7	19.6	4.5	5.7	9.9	2.1	5.5	9.8	1.8
Intentional self-harm (suicide) by discharge of firearms (X72-X74)	5.9	11.0	1.4	6.5	12.0	1.6	2.7	5.2	0.6	3.1	6.0	0.6
Intentional self-harm (suicide) by other and unspecified means and their sequelae (*U03, X60-X71, X75-X84, Y87.0) [1]	4.8	7.1	2.6	5.2	7.6	2.9	3.0	4.6	1.5	2.4	3.8	1.1
Assault (homicide) (*U01-*U02, X85-Y09, Y87.1) [1]	7.1	10.8	3.3	4.9	7.1	2.6	16.2	27.0	6.1	21.2	36.2	7.4
Assault (homicide) by discharge of firearms (*U01.4, X93-X95)	3.9	6.6	1.3	2.2	3.5	1.0	10.8	19.7	2.3	14.8	27.4	3.0
Assault (homicide) by other and unspecified means and their sequelae (*U01.0-*U01.3, *U01.5-*U01.9, *U02, X85-X92, X96-Y09, Y87.1) [1]	3.1	4.3	2.0	2.6	3.6	1.6	5.4	7.3	3.7	6.4	8.8	4.3
Legal intervention (Y35, Y89.0)	0.1	0.3	*	0.1	0.2	*	0.2	0.4	*	0.2	0.5	*
Events of undetermined intent (Y10-Y34, Y87.2, Y89.9)	1.5	2.0	0.9	1.5	2.0	1.0	1.4	2.2	0.7	1.8	2.9	0.9
Discharge of firearms, undetermined intent (Y22-Y24)	0.1	0.1	0.0	0.1	0.1	0.0	0.1	0.1	*	0.1	0.2	*
Other and unspecified events of undetermined intent and their sequelae (Y10-Y21, Y25-Y34, Y87.2, Y89.9)	1.4	1.9	0.9	1.4	1.8	1.0	1.4	2.1	0.7	1.7	2.7	0.9
Operations of war and their sequelae (Y36, Y89.1)	*	*	*	*	*	*	*	*	*	*	*	*
Complications of medical and surgical care (Y40-Y84, Y88)	1.1	1.1	1.0	1.0	1.1	0.9	1.6	1.8	1.4	2.0	2.3	1.7

[1]Based on codes from the "Tenth Revision, International Classification of Diseases, 1992." Codes *U01-*U03 have been added as of 2001 for deaths due to terrorism.
[2]Figures include September 11, 2001 related deaths for which death certificates were filed as of 10/24/02.
0.0 = Quantity more than zero but less than 0.05.
* = Figure too small to meet standards of reliability or precision.

TABLE II-17. AGE-ADJUSTED DEATH RATES FOR 113 SELECTED CAUSES, BY RACE, HISPANIC ORIGIN, AND SEX, 2001

Overall, age-adjusted Hispanic death rates are about 0.8 relative to non-Hispanic rates. Rates for Hispanic males are 47 percent greater than for females, as compared with a 42 percent difference for non-Hispanic males. Cancer mortality is significantly lower among Hispanics compared with non-Hispanics. Hispanics are more affected by diabetes than are non-Hispanic Whites, but an even larger death rate for this disease occurs among non-Hispanic Blacks.

Suicide occurs only about one half as often among Hispanics as among non-Hispanics. The low Hispanic rate is about the same as that for non-Hispanic Blacks. Hispanics are victims of homicide at about twice the rate as non-Hispanic Whites, but still much less frequently than non-Hispanic Blacks.

Table II-17. Age-Adjusted Death Rates for 113 Selected Causes, by Race, Hispanic Origin, and Sex, 2001

(Rate per 100,000 U.S. standard population.)

Cause of death [1]	All origins [2]			Hispanic			Non-Hispanic [3]			Non-Hispanic White			Non-Hispanic Black		
	Both sexes	Male	Female	Both sexes	Male	Female	Both sexes	Male	Female	Both sexes	Male	Female	Both sexes	Male	Female
All causes	854.5	1029.1	721.8	658.7	802.5	544.2	864.0	1039.8	730.9	842.9	1012.8	713.5	1116.5	1393.7	925.5
Salmonella infections (A01-A02)	0.0	0.0	0.0	*	*	*	0.0	*	0.0	0.0	*	*	*	*	*
Shigellosis and amebiasis (A03, A06)	*	*	*	*	*	*	*	*	*	*	*	*	*	*	*
Certain other intestinal infections (A04, A07-A09)	0.5	0.5	0.5	0.4	*	0.5	0.5	0.5	*	0.6	0.6	0.6	0.4	0.5	0.4
Tuberculosis (A16-A19)	0.3	0.4	0.2	0.5	0.8	0.3	0.2	0.3	0.2	0.1	0.2	0.1	0.8	1.2	0.5
Respiratory tuberculosis (A16)	0.2	0.3	0.1	0.4	0.7	0.2	0.2	0.3	0.1	0.1	0.2	0.1	0.6	1.0	0.3
Other tuberculosis (A17-A19)	0.1	0.1	0.0	0.1	*	0.1	0.0	*	*	0.0	0.0	0.0	0.2	0.2	0.1
Whooping cough (A37)	*	*	*	*	*	*	*	*	*	*	*	*	*	*	*
Scarlet fever and erysipelas (A38, A46)	*	*	*	*	*	*	*	*	*	*	*	*	*	*	*
Meningococcal infection (A39)	0.1	0.1	0.1	0.0	*	0.1	0.1	*	*	0.1	0.1	0.1	0.1	*	*
Septicemia (A40-A41)	11.4	12.7	10.5	8.8	9.9	7.9	11.5	12.9	10.7	10.4	11.5	9.6	24.0	28.1	21.5
Syphilis (A50-A53)	0.0	*	*	*	*	*	0.0	*	*	*	*	*	*	*	*
Acute poliomyelitis (A80)	*	*	*	*	*	*	*	*	*	*	*	*	*	*	*
Arthropod-borne viral encephalitis (A83-A84, A85.2)	*	*	*	*	*	*	*	*	*	*	*	*	*	*	*
Measles (B05)	*	*	*	*	*	*	*	*	*	*	*	*	*	*	*
Viral hepatitis (B15-B19)	2.0	2.6	1.3	3.7	4.5	2.8	1.8	2.5	1.2	1.7	2.3	1.0	2.8	3.5	2.1
Human immunodeficiency virus (HIV) disease (B20-B24)	5.0	7.5	2.5	6.2	9.7	2.7	4.8	7.3	2.5	2.1	3.6	0.6	23.2	34.4	13.8
Malaria (B50-B54)	*	*	*	*	*	*	*	*	*	*	*	*	*	*	*
Other and unspecified infectious and parasitic diseases and their sequelae (A00, A05, A20-A36, A42-A44, A48-A49, A54-A79, A81-A82, A85.0-A85.1, A85.8, A86-B04, B06-B09, B25-B49, B55-B99)	2.0	2.4	1.7	2.0	2.4	1.5	2.0	2.4	1.7	1.9	2.2	1.6	2.9	3.4	2.5
Malignant neoplasms (C00-C97)	196.0	243.7	164.7	132.3	168.2	108.6	199.7	247.9	168.0	197.4	243.1	167.2	246.7	335.3	194.4
Malignant neoplasms of lip, oral cavity and pharynx (C00-C14)	2.7	4.1	1.6	1.8	2.9	1.0	2.8	4.2	1.6	2.6	3.9	1.6	3.9	6.8	1.8
Malignant neoplasm of esophagus (C15)	4.4	7.8	1.8	2.4	4.4	1.0	4.5	8.0	1.8	4.4	7.9	1.7	6.5	11.5	3.0
Malignant neoplasm of stomach (C16)	4.4	6.1	3.1	6.9	9.2	5.3	4.2	5.9	2.9	3.6	5.0	2.5	8.4	12.1	5.9
Malignant neoplasms of colon, rectum and anus (C18-C21)	20.1	24.2	17.2	14.1	17.5	11.8	20.5	24.6	17.5	19.9	24.0	16.9	28.0	33.6	24.5
Malignant neoplasms of liver and intrahepatic bile ducts (C22)	4.7	6.9	3.0	7.3	10.4	4.9	4.6	6.7	2.9	4.1	6.0	2.6	6.3	9.2	4.1
Malignant neoplasm of pancreas (C25)	10.5	12.1	9.3	8.4	9.0	7.8	10.7	12.3	9.4	10.5	12.2	9.2	13.8	15.1	12.6
Malignant neoplasm of larynx (C32)	1.3	2.5	0.5	1.0	1.9	0.2	1.4	2.5	0.5	1.2	2.3	0.5	2.6	5.0	0.9
Malignant neoplasms of trachea, bronchus and lung (C33-C34)	55.3	75.2	41.0	23.8	36.8	14.5	57.2	77.4	42.7	57.7	76.6	44.0	63.4	100.5	39.4
Malignant melanoma of skin (C43)	2.7	3.9	1.8	0.8	0.9	0.7	2.8	4.1	1.8	3.2	4.7	2.1	0.4	0.3	0.5
Malignant neoplasm of breast (C50)	14.8	0.3	26.0	9.1	*	15.1	0.4	26.6	*	14.7	0.3	26.0	20.8	0.6	35.1
Malignant neoplasm of cervix uteri (C53)	1.4	X	2.7	1.8	X	3.4	1.4	X	2.6	1.2	X	2.3	2.8	X	4.9
Malignant neoplasms of corpus uteri and uterus, part unspecified (C54-C55)	2.4	X	4.2	1.9	X	3.3	2.4	X	4.2	2.3	X	4.0	4.3	X	7.0
Malignant neoplasm of ovary (C56)	5.1	X	9.0	3.5	X	6.1	5.2	X	9.2	5.4	X	9.5	4.6	X	7.6
Malignant neoplasm of prostate (C61)	10.9	29.1	X	9.1	23.5	X	11.0	29.3	X	10.0	26.7	X	23.3	66.8	X
Malignant neoplasms of kidney and renal pelvis (C64-C65)	4.3	6.2	2.8	3.7	5.6	2.3	4.3	6.2	2.8	4.4	6.3	2.9	4.3	6.5	2.9
Malignant neoplasm of bladder (C67)	4.3	7.5	2.3	2.4	4.0	1.4	4.4	7.7	2.3	4.6	8.1	2.3	3.8	5.1	2.9
Malignant neoplasms of meninges, brain and other parts of central nervous system (C70-C72)	4.5	5.5	3.6	2.9	3.4	2.4	4.6	5.7	3.7	4.9	6.1	4.0	2.6	3.2	2.2
Malignant neoplasms of lymphoid, hematopoietic and related tissue (C81-C96)	20.0	25.5	16.0	14.9	19.3	11.6	20.2	25.8	16.2	20.4	26.1	16.3	20.3	25.7	16.8
Hodgkin's disease (C81)	0.5	0.6	0.4	0.5	0.7	0.3	0.5	0.6	0.4	0.5	0.6	0.4	0.4	0.5	0.4
Non-Hodgkin's lymphoma (C82-C85)	7.9	9.9	6.4	5.9	7.5	4.6	8.0	10.0	6.5	8.3	10.4	6.8	5.7	7.4	4.5
Leukemia (C91-C95)	7.6	10.1	5.9	5.2	6.7	4.2	7.7	10.2	5.9	7.9	10.5	6.1	6.7	8.8	5.4
Multiple myeloma and immunoproliferative neoplasms (C88, C90)	4.0	4.9	3.3	3.3	4.3	2.6	4.0	4.9	3.4	3.7	4.6	3.1	7.4	8.9	6.4
Other and unspecified malignant neoplasms of lymphoid, hematopoietic and related tissue (C96)	0.0	0.0	0.0	*	*	*	0.0	0.0	0.0	0.0	0.0	0.0	*	*	*
All other and unspecified malignant neoplasms (C17, C23-C24, C26-C31, C37-C41, C44-C49, C51-C52, C57-C60, C62-C63, C66, C68-C69, C73-C80, C97)	22.2	26.7	18.9	16.6	19.3	14.6	22.5	27.2	19.1	22.3	26.9	19.0	26.6	33.2	22.2
In situ neoplasms, benign neoplasms and neoplasms of uncertain or unknown behavior (D00-D48)	4.9	6.0	4.1	3.3	4.0	2.8	4.9	6.1	4.2	5.0	6.3	4.2	4.7	5.7	4.1
Anemias (D50-D64)	1.6	1.6	1.6	1.2	1.1	1.2	1.7	1.6	1.7	1.4	1.4	1.4	3.6	3.7	3.4
Diabetes mellitus (E10-E14)	25.3	28.1	23.1	36.7	37.8	35.5	24.6	27.5	22.4	22.1	25.4	19.5	49.8	50.5	48.7
Nutritional deficiencies (E40-E64)	1.3	1.2	1.4	1.1	1.2	1.1	1.3	1.2	1.4	1.3	1.1	1.3	2.1	2.4	1.9
Malnutrition (E40-E46)	1.2	1.1	1.3	1.1	1.1	1.1	1.2	1.1	1.3	1.2	1.0	1.2	1.9	2.2	1.7
Other nutritional deficiencies (E50-E64)	0.1	0.1	0.1	*	*	*	0.1	0.1	0.1	0.1	0.1	0.1	0.1	*	*
Meningitis (G00, G03)	0.3	0.3	0.2	0.2	0.3	0.2	0.3	0.3	0.2	0.2	0.2	0.2	0.6	0.8	0.4
Parkinson's disease (G20-G21)	5.9	9.1	4.0	3.8	5.6	2.7	5.9	9.3	4.1	6.3	9.9	4.3	2.5	4.0	1.8
Alzheimer's disease (G30)	19.1	16.1	20.5	10.7	9.2	11.5	19.4	16.4	20.8	20.2	17.0	21.7	14.8	13.3	15.3

[1]Based on codes from the "Tenth Revision, International Classification of Diseases, 1992." Codes *U01-*U03 have been added as of 2001 for deaths due to terrorism.
[2]Figures for origin not stated are included in "All origins" but not distributed among specified origins.
[3]Includes races other than White and Black.
0.0 = Quantity more than zero but less than 0.05.
* = Figure too small to meet standards of reliability or precision.
X = Not applicable.

Table II-17. Age-Adjusted Death Rates for 113 Selected Causes, by Race, Hispanic Origin, and Sex, 2001—*Continued*

(Rate per 100,000 U.S. standard population.)

Cause of death [1]	All origins [2]			Hispanic			Non-Hispanic [3]			Non-Hispanic White			Non-Hispanic Black		
	Both sexes	Male	Female	Both sexes	Male	Female	Both sexes	Male	Female	Both sexes	Male	Female	Both sexes	Male	Female
Major cardiovascular diseases (I00-I78)	326.5	388.1	278.8	252.0	298.0	216.3	329.8	392.2	281.3	321.6	384.4	272.5	432.5	511.0	377.0
Diseases of heart (I00-I09, I11, I13, I20-I51)	247.8	305.4	203.9	192.2	232.6	161.0	250.4	309.0	205.7	245.6	304.8	200.0	320.4	388.7	272.9
Acute rheumatic fever and chronic rheumatic heart diseases (I00-I09)	1.2	0.9	1.5	0.9	0.7	1.1	1.2	0.9	1.5	1.2	0.9	1.4	1.1	0.7	1.4
Hypertensive heart disease (I11)	8.7	8.9	8.1	7.7	8.8	6.7	8.7	8.9	8.2	7.2	7.1	6.8	23.4	26.5	20.7
Hypertensive heart and renal disease (I13)	1.0	1.1	1.0	1.1	1.3	1.0	1.0	1.0	0.9	0.7	0.8	0.7	3.4	3.6	3.1
Ischemic heart diseases (I20-I25)	177.8	228.5	139.9	149.9	185.8	122.6	179.0	230.5	140.5	177.5	230.2	137.7	213.5	264.4	178.4
Acute myocardial infarction (I21-I22)	65.4	84.9	50.9	52.8	66.2	43.0	66.1	86.0	51.2	65.8	86.5	50.1	77.8	94.3	66.8
Other acute ischemic heart diseases (I24)	1.2	1.5	0.9	0.4	0.7	0.2	1.2	1.5	1.0	1.2	1.5	0.9	1.8	2.4	1.4
Other forms of chronic ischemic heart disease (I20, I25)	111.2	142.1	88.1	96.6	118.9	79.4	111.7	143.0	88.3	110.5	142.3	86.7	134.0	167.8	110.2
Atherosclerotic cardiovascular disease, so described (I25.0)	24.1	30.8	18.5	21.0	28.5	15.1	24.1	30.8	18.6	22.5	28.6	17.3	42.1	56.4	32.0
All other forms of chronic ischemic heart disease (I20, I25.1-I25.9)	87.1	111.3	69.6	75.6	90.5	64.3	87.6	112.2	69.7	88.0	113.7	69.4	91.8	111.4	78.2
Other heart diseases (I26-I51)	59.1	66.1	53.4	32.6	36.2	29.6	60.4	67.6	54.6	58.9	65.8	53.3	79.1	93.3	69.3
Acute and subacute endocarditis (I33)	0.4	0.5	0.3	0.3	0.3	0.2	0.4	0.6	0.3	0.4	0.5	0.3	0.8	1.1	0.6
Diseases of pericardium and acute myocarditis (I30-I31, I40)	0.3	0.3	0.3	0.3	0.2	0.3	0.3	0.3	0.3	0.3	0.3	0.3	0.5	0.5	0.5
Heart failure (I50)	20.2	21.2	19.2	10.7	10.9	10.4	20.5	21.6	19.6	20.6	21.7	19.7	22.4	24.7	20.9
All other forms of heart disease (I26-I28, I34-I38, I42-I49, I51)	38.2	44.1	33.6	21.4	24.7	18.7	39.1	45.2	34.3	37.6	43.4	33.1	55.3	66.9	47.3
Essential (primary) hypertension and hypertensive renal disease (I10, I12)	6.8	6.7	6.8	6.5	6.2	6.5	6.8	6.7	6.8	5.8	5.6	5.8	16.8	18.2	15.7
Cerebrovascular diseases (I60-I69)	57.9	59.0	56.4	44.9	48.9	41.6	58.4	59.2	56.9	56.0	56.5	54.8	79.9	86.5	74.8
Atherosclerosis (I70)	5.0	5.1	4.8	3.5	3.7	3.3	5.0	5.1	4.9	5.2	5.3	5.0	4.3	4.7	4.1
Other diseases of circulatory system (I71-I78)	9.0	12.0	6.9	5.0	6.5	3.9	9.2	12.2	7.0	9.1	12.3	6.9	10.9	12.9	9.5
Aortic aneurysm and dissection (I71)	5.4	8.0	3.6	2.5	4.1	1.5	5.5	8.2	3.7	5.6	8.4	3.7	5.0	6.5	4.0
Other diseases of arteries, arterioles and capillaries (I72-I78)	3.6	3.9	3.3	2.4	2.4	2.4	3.6	4.0	3.3	3.4	3.8	3.2	5.9	6.5	5.6
Other disorders of circulatory system (I80-I99)	1.6	1.7	1.6	1.4	1.5	1.3	1.7	1.7	1.6	1.6	1.6	1.5	2.8	3.1	2.5
Influenza and pneumonia (J10-J18)	22.0	26.6	19.2	20.5	24.6	17.8	22.0	26.6	19.2	21.7	26.0	19.1	24.3	32.5	19.6
Influenza (J10-J11)	0.1	0.1	0.1	*	*	*	0.1	0.1	0.1	0.1	0.1	0.1	0.1	*	*
Pneumonia (J12-J18)	21.9	26.5	19.1	20.4	24.5	17.8	21.9	26.5	19.1	21.6	25.9	19.0	24.3	32.4	19.5
Other acute lower respiratory infections (J20-J22)	0.1	0.1	0.1	0.1	*	*	0.1	0.1	0.1	0.1	0.1	0.1	0.1	*	*
Acute bronchitis and bronchiolitis (J20-J21)	0.1	0.1	0.1	*	*	*	0.1	0.1	0.1	0.1	0.1	0.1	0.1	*	*
Unspecified acute lower respiratory infection (J22)	0.0	0.0	0.0	*	*	*	0.0	0.0	0.0	0.0	0.0	0.0	*	*	*
Chronic lower respiratory diseases (J40-J47)	43.7	54.0	37.6	20.7	27.6	16.5	44.9	55.4	38.7	47.0	56.9	41.1	31.3	47.0	22.6
Bronchitis, chronic and unspecified (J40-J42)	0.3	0.4	0.3	0.2	*	0.2	0.3	0.4	0.3	0.4	0.4	0.3	0.2	0.3	0.2
Emphysema (J43)	5.8	7.4	4.7	2.1	3.3	1.4	6.0	7.7	4.9	6.4	8.0	5.3	3.4	5.4	2.2
Asthma (J45-J46)	1.5	1.2	1.7	1.4	1.0	1.7	1.5	1.2	1.8	1.2	0.9	1.4	3.6	3.3	3.9
Other chronic lower respiratory diseases (J44, J47)	36.1	45.0	30.8	16.9	23.1	13.2	37.1	46.1	31.7	39.0	47.7	34.0	24.0	38.0	16.3
Pneumoconioses and chemical effects (J60-J66, J68)	0.4	1.0	0.0	0.2	0.4	*	0.4	1.0	0.0	0.4	1.1	0.0	0.2	0.7	*
Pneumonitis due to solids and liquids (J69)	6.1	8.5	4.8	4.4	6.0	3.5	6.2	8.6	4.9	6.2	8.7	4.8	7.1	9.5	5.8
Other diseases of respiratory system (J00-J06, J30-J39, J67, J70-J98)	9.3	11.0	8.1	8.2	9.8	7.1	9.3	11.1	8.1	9.3	11.2	8.1	9.6	10.9	8.8
Peptic ulcer (K25-K28)	1.6	1.8	1.4	1.2	1.6	0.8	1.6	1.8	1.4	1.6	1.8	1.4	1.6	2.3	1.1
Diseases of appendix (K35-K38)	0.1	0.2	0.1	0.1	*	*	0.1	0.2	0.1	0.1	0.2	0.1	0.2	0.4	0.2
Hernia (K40-K46)	0.5	0.5	0.5	0.6	0.4	0.7	0.5	0.5	0.5	0.5	0.5	0.5	0.6	0.8	0.5
Chronic liver disease and cirrhosis (K70, K73-K74)	9.5	13.2	6.2	15.8	23.7	8.7	9.0	12.3	6.1	9.0	12.2	6.1	9.4	14.1	5.7
Alcoholic liver disease (K70)	4.3	6.6	2.2	7.6	13.6	2.3	4.0	6.0	2.2	3.9	5.9	2.2	4.5	7.1	2.4
Other chronic liver disease and cirrhosis (K73-K74)	5.2	6.6	4.0	8.2	10.1	6.4	5.0	6.3	3.9	5.0	6.3	4.0	5.0	7.0	3.2
Cholelithiasis and other disorders of gallbladder (K80-K82)	1.0	1.2	0.9	1.3	1.5	1.2	1.0	1.2	0.9	1.0	1.2	0.9	1.3	1.4	1.2
Nephritis, nephrotic syndrome and nephrosis (N00-N07, N17-N19, N25-N27)	14.0	17.4	12.0	12.4	14.5	11.1	14.0	17.5	12.0	12.4	16.0	10.3	30.0	34.3	27.4
Acute and rapidly progressive nephritic and nephrotic syndrome (N00-N01, N04)	0.0	0.1	0.0	*	*	*	0.0	0.1	0.0	0.0	0.1	0.0	0.1	*	*
Chronic glomerulonephritis, nephritis and nephropathy, not specified as acute or chronic, and renal sclerosis, unspecified (N02-N03, N05-N07, N26)	0.2	0.2	0.2	0.2	*	*	0.2	0.2	0.2	0.2	0.2	0.1	0.2	0.3	0.2
Renal failure (N17-N19)	13.7	17.1	11.8	12.1	14.2	10.8	13.8	17.2	11.8	12.2	15.7	10.1	29.6	33.9	27.1
Other disorders of kidney (N25, N27)	0.0	*	0.0	*	*	*	0.0	*	0.0	*	*	*	*	*	*
Infections of kidney (N10-N12, N13.6, N15.1)	0.3	0.2	0.3	0.2	*	0.2	0.3	0.2	0.3	0.3	0.2	0.3	0.3	0.3	0.3
Hyperplasia of prostate (N40)	0.1	0.5	X	*	*	X	0.2	0.5	X	0.2	0.5	X	0.1	0.4	X
Inflammatory diseases of female pelvic organs (N70-N76)	0.0	X	0.1	*	X	*	0.0	X	0.1	0.0	X	0.1	0.1	X	0.1
Pregnancy, childbirth and the puerperium (O00-O99)	0.1	X	0.3	0.2	X	0.4	0.1	X	0.3	0.1	X	0.2	0.4	X	0.8
Pregnancy with abortive outcome (O00-O07)	0.0	X	0.0	*	X	*	0.0	X	0.0	*	X	*	*	X	*
Other complications of pregnancy, childbirth and the puerperium (O10-O99)	0.1	X	0.3	0.2	X	0.4	0.1	X	0.2	0.1	X	0.2	0.4	X	0.7
Certain conditions originating in the perinatal period (P00-P96)	4.7	5.3	4.2	3.8	4.2	3.3	4.9	5.4	4.4	3.6	4.0	3.2	10.7	11.8	9.5
Congenital malformations, deformations and chromosomal abnormalities (Q00-Q99)	3.6	3.8	3.4	3.4	3.8	3.0	3.6	3.8	3.4	3.5	3.8	3.3	4.2	4.4	4.1
Symptoms, signs and abnormal clinical and laboratory findings, not elsewhere classified (R00-R99)	11.3	12.7	9.8	7.9	9.9	5.8	11.5	12.8	10.0	11.0	12.2	9.7	15.4	18.0	13.1
All other diseases (Residual)	65.9	66.1	64.0	46.8	49.2	44.0	66.8	67.0	65.0	65.4	65.1	63.9	85.9	93.5	79.7

[1]Based on codes from the "Tenth Revision, International Classification of Diseases, 1992." Codes *U01-*U03 have been added as of 2001 for deaths due to terrorism.
[2]Figures for origin not stated are included in "All origins" but not distributed among specified origins.
[3]Includes races other than White and Black.
0.0 = Quantity more than zero but less than 0.05.
* = Figure too small to meet standards of reliability or precision.
X = Not applicable.

Table II-17. Age-Adjusted Death Rates for 113 Selected Causes, by Race, Hispanic Origin, and Sex, 2001—*Continued*

(Rate per 100,000 U.S. standard population.)

Cause of death [1]	All origins [2]			Hispanic			Non-Hispanic [3]			Non-Hispanic White			Non-Hispanic Black		
	Both sexes	Male	Female	Both sexes	Male	Female	Both sexes	Male	Female	Both sexes	Male	Female	Both sexes	Male	Female
Accidents (unintentional injuries) (V01-X59, Y85-Y86)	35.7	50.2	22.5	30.7	44.7	16.8	35.9	50.4	23.0	36.2	50.4	23.3	38.3	57.9	22.4
Transport accidents (V01-V99, Y85)	16.6	23.7	9.8	16.7	24.6	8.9	16.5	23.6	10.0	16.7	23.7	10.1	16.9	26.1	9.1
Motor vehicle accidents (V02-V04, V09.0, V09.2, V12-V14, V19.0-V19.2, V19.4-V19.6, V20-V79, V80.3-80.5, V81.0-V81.1, V82.0-V82.1, V83-V86, V87.0-V87.8, V88.0-V88.8, V89.0, V89.2)	15.3	21.8	9.3	15.0	22.2	7.8	15.3	21.7	9.5	15.5	21.7	9.7	15.7	24.1	8.6
Other land transport accidents (V01, V05-V06, V09.1, V09.3-V09.9, V10-V11, V15-V18, V19.3, V19.8-19.9, V80.0-V80.2, V80.6-V80.9, V81.2-V81.9, V82.2-V82.9, V87.9, V88.9, V89.1, V89.3, V89.9)	0.5	0.8	0.2	0.7	1.1	0.2	0.5	0.7	0.2	0.4	0.7	0.2	0.7	1.0	0.3
Water, air and space, and other and unspecified transport accidents and their sequelae (V90-V99, Y85)	0.7	1.2	0.3	1.0	1.2	0.9	0.7	1.2	0.2	0.8	1.3	0.3	0.5	1.0	0.1
Nontransport accidents (W00-X59, Y86)	19.1	26.4	12.7	14.0	20.1	7.9	19.4	26.8	13.1	19.5	26.7	13.2	21.4	31.8	13.3
Falls (W00-W19)	5.3	7.2	3.9	4.1	5.4	2.8	5.3	7.3	3.9	5.6	7.5	4.2	3.3	5.2	2.0
Accidental discharge of firearms (W32-W34)	0.3	0.5	0.1	0.2	0.4	*	0.3	0.5	0.1	0.3	0.5	0.1	0.3	0.6	*
Accidental drowning and submersion (W65-W74)	1.1	1.8	0.5	1.0	1.6	0.3	1.2	1.8	0.5	1.1	1.7	0.5	1.5	2.7	0.5
Accidental exposure to smoke, fire and flames (X00-X09)	1.2	1.5	0.9	0.6	0.8	0.5	1.2	1.6	0.9	1.0	1.4	0.8	2.7	3.8	1.9
Accidental poisoning and exposure to noxious substances (X40-X49)	4.9	7.0	2.9	3.9	6.1	1.6	5.0	7.1	3.0	5.1	7.2	3.1	6.4	9.6	3.8
Other and unspecified nontransport accidents and their sequelae (W20-W31, W35-W64, W75-W99, X10-X39, X50-X59, Y86)	6.3	8.4	4.5	4.2	5.8	2.6	6.4	8.5	4.6	6.3	8.5	4.6	7.1	10.0	5.0
Intentional self-harm (suicide) (*U03, X60-X84, Y87.0) [4]	10.7	18.2	4.0	5.7	10.1	1.6	11.3	19.2	4.3	12.5	21.0	4.9	5.5	9.9	2.0
Intentional self-harm (suicide) by discharge of firearms (X72-X74)	5.9	11.0	1.4	2.5	4.8	0.4	6.3	11.7	1.6	7.0	12.9	1.8	3.1	6.1	1.0
Intentional self-harm (suicide) by other and unspecified means and their sequelae (*U03, X60-X71, X75-X84, Y87.0) [4]	4.8	7.1	2.6	3.2	5.3	1.2	5.0	7.4	2.8	5.5	8.1	3.1	2.4	3.8	1.0
Assault (homicide) (*U01-*U02, X85-Y09, Y87.1) [4]	7.1	10.8	3.3	8.3	12.9	3.1	6.8	10.3	3.3	4.0	5.6	2.5	21.7	37.2	8.0
Assault (homicide) by discharge of firearms (*U01.4, X93-X95)	3.9	6.6	1.3	4.9	8.2	1.2	3.7	6.2	1.3	1.6	2.2	0.9	15.2	28.2	3.0
Assault (homicide) by other and unspecified means and their sequelae (*U01.0-*U01.3, *U01.5-*U01.9, *U02, X85-X92, X96-Y09, Y87.1) [4]	3.1	4.3	2.0	3.3	4.7	1.9	3.1	4.1	2.0	2.4	3.3	1.5	6.5	9.0	4.0
Legal intervention (Y35, Y89.0)	0.1	0.3	*	0.1	0.3	*	0.1	0.3	*	0.1	0.2	*	0.2	0.4	*
Events of undetermined intent (Y10-Y34, Y87.2, Y89.9)	1.5	2.0	0.9	0.9	1.3	0.4	1.5	2.1	1.0	1.6	2.1	1.1	1.8	2.9	1.0
Discharge of firearms, undetermined intent (Y22-Y24)	0.1	0.1	0.0	0.1	*	*	0.1	0.1	0.0	0.1	0.1	0.0	0.1	0.2	*
Other and unspecified events of undetermined intent and their sequelae (Y10-Y21, Y25-Y34, Y87.2, Y89.9)	1.4	1.9	0.9	0.8	1.2	0.4	1.5	2.0	1.0	1.5	1.9	1.0	1.7	2.8	1.0
Operations of war and their sequelae (Y36, Y89.1)	*	*	*	*	*	*	*	*	*	*	*	*	*	*	*
Complications of medical and surgical care (Y40-Y84, Y88)	1.1	1.1	1.0	0.8	0.9	0.7	1.1	1.2	1.0	1.0	1.1	0.9	2.0	2.3	2.0

[1]Based on codes from the "Tenth Revision, International Classification of Diseases, 1992." Codes *U01-*U03 have been added as of 2001 for deaths due to terrorism.
[2]Figures for origin not stated are included in "All origins" but not distributed among specified origins.
[3]Includes races other than White and Black.
[4]Figures include September 11, 2001 related deaths for which death certificates were filed as of 10/24/02.
0.0 = Quantity more than zero but less than 0.05.
* = Figure too small to meet standards of reliability or precision.

Figure II-11. Age-Adjusted Death Rates for 8 Leading Causes of Death, by Race and Hispanic Origin, 2001

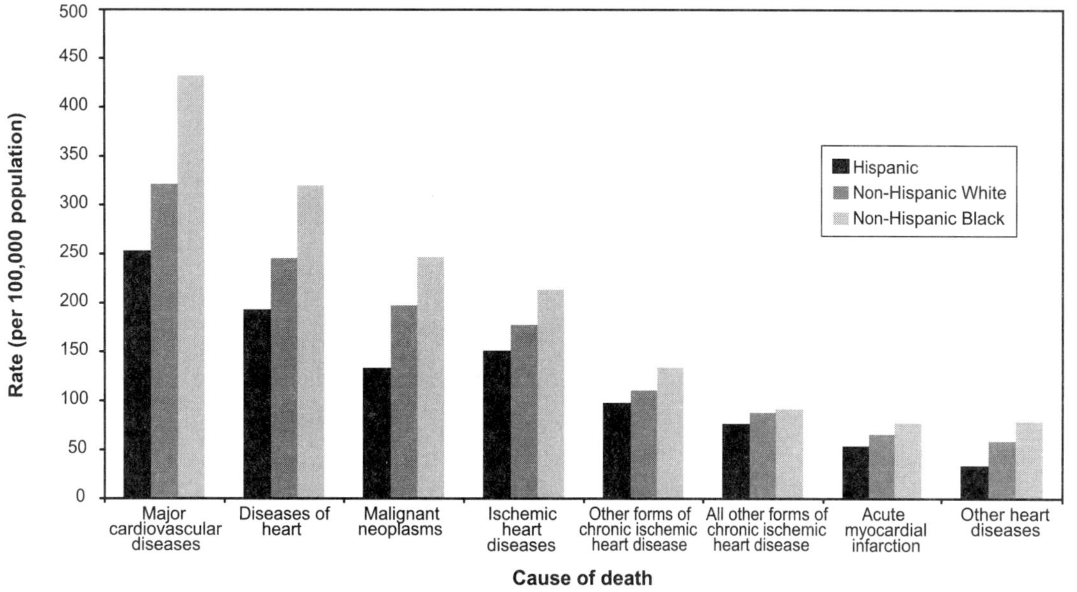

TABLE II-18. NUMBER OF DEATHS, DEATH RATES, AND AGE-ADJUSTED DEATH RATES FOR INJURY DEATHS ACCORDING TO MECHANISM AND INTENT OF DEATH, 2001

This table classifies the 157,078 deaths from injury in 2001 in two different ways: by type of hazard (drowning, firearms, etc.) and whether the injury was inflicted intentionally—homicide or suicide—or occurred accidentally (or could not be determined). Four major mechanisms of injury in 2001 accounted for 70 percent of all injury deaths: motor-vehicle traffic, firearm, poisoning, and falls.

For all injuries combined, 101,537 deaths (65 percent) were unintentional, 30,622 (19 percent) were classified as suicides, 20,308 (13 percent) were homicides, 4,198 (3 percent) were undetermined, and 413 were caused by legal intervention or war. Motor-vehicle traffic caused 42,443 deaths, including those of 2,976 motorcyclists, 585 bicyclists, and 4,822 pedestrians. Firearms caused almost 30,000 fatalities, and 16,869 of these were suicides.

The next largest category was poisoning, which claimed 22,242 deaths; while the majority of these fatalities were unintentional, almost one-fourth represented suicides. Falls, mostly unintentional, were also a major source of deaths, claiming 15,764 lives.

TABLE II-19. NUMBER OF DEATHS, DEATH RATES, AND AGE-ADJUSTED DEATH RATES FOR INJURY BY FIREARMS, BY RACE AND SEX, 1999–2001

Firearm injuries accounted for 18.8 percent of all injury deaths in 2001. Fatality rates from injury by firearms were 10.4 persons per 100,000 for the entire population in 2001, a slight uptick from the previous two years. However, these rates differed according to sex and race. For White females the death rate was 2.7 per 100,000, while White males had a rate of 16.2. For Blacks, the female fatality rate was 3.8, and the male rate was 36.4, all per 100,000.

TABLE II-20. NUMBER OF DEATHS, DEATH RATES, AND AGE-ADJUSTED DEATH RATES FOR INJURY BY FIREARMS, BY RACE, HISPANIC ORIGIN, AND SEX, 1999–2001

For Hispanics, the fatality rate from firearms was lower than that of the whole population, and also somewhat lower than the rate for non-Hispanic Whites. For Hispanic males the firearm death rate in 2001 was 14.6 per 100,000 and for females it was 1.7.

Table II-18. Number of Deaths, Death Rates, and Age-Adjusted Death Rates for Injury Deaths According to Mechanism and Intent of Death, 2001

(Number, rate per 100,000 population.)

Mechanism and intent of death [1]	Number	Rate	Age-adjusted rate
ALL INJURY (*U01-*U03, V01-Y36, Y85-Y87, Y89) [2]	157 078	55.2	55.1
Unintentional (V01-X59, Y85-Y86)	101 537	35.7	35.7
Suicide (*U03, X60-X84, Y87.0) [2]	30 622	10.8	10.7
Homicide (*U01-*U02, X85-Y09, Y87.1) [2]	20 308	7.1	7.1
Undetermined (Y10-Y34, Y87.2, Y89.9) 4,198	4 198	1.5	1.5
Legal intervention/ war (Y35-Y36, Y89 [.0, .1])	413	0.1	0.2
Cut/Pierce (W25-W29, W45, X78, X99, Y28, Y35.4)	2 532	0.9	0.9
Unintentional (W25-W29, W45)	85	0.0	0.0
Suicide (X78)	458	0.2	0.2
Homicide (X99)	1 971	0.7	0.7
Undetermined (Y28)	18	*	*
Legal intervention/ war (Y35.4)	-	*	*
Drowning (W65-W74, X71, X92, Y21)	3 923	1.4	1.4
Unintentional (W65-W74)	3 281	1.2	1.1
Suicide (X71)	339	0.1	0.1
Homicide (X92)	68	0.0	0.0
Undetermined (Y21)	235	0.1	0.1
Fall (W00-W19, X80, Y01, Y30)	15 764	5.5	5.6
Unintentional (W00-W19)	15 019	5.3	5.3
Suicide (X80)	651	0.2	0.2
Homicide (Y01)	17	*	*
Undetermined (Y30)	77	0.0	0.0
Fire/Hot Object or Substance (*U01.3, X00-X19, X76-X77, X97-X98, Y26-Y27, Y36.3) [3]	3 796	1.3	1.3
Unintentional (X00-X19)	3 423	1.2	1.2
Suicide (X76-X77)	148	0.1	0.1
Homicide (*U01.3, X97-X98)	148	0.1	0.0
Undetermined (Y26-Y27)	77	0.0	0.0
Legal intervention/war (Y36.3)	-	*	*
Fire/Flame (X00-X09, X76, X97, Y26)	3 673	1.3	1.3
Unintentional (X00-X09)	3 309	1.2	1.2
Suicide (X76)	147	0.1	0.1
Homicide (X97)	141	0.0	0.0
Undetermined (Y26)	76	0.0	0.0
Hot Object/Substance (X10-X19, X77, X98, Y27)	123	0.0	0.0
Unintentional (X10-X19)	114	0.0	0.0
Suicide (X77)	1	*	*
Homicide (X98)	7	*	*
Undetermined (Y27)	1	*	*
Firearm (*U01.4, W32-W34, X72-X74, X93-X95, Y22-Y24, Y35.0)	29 573	10.4	10.3
Unintentional (W32-W34)	802	0.3	0.3
Suicide (X72-X74)	16 869	5.9	5.9
Homicide (*U01.4, X93-X95)	11 348	4.0	3.9
Undetermined (Y22-Y24)	231	0.1	0.1
Legal intervention/ war (Y35.0)	323	0.1	0.1
Machinery (W24, W30-W31)	648	0.2	0.2
All Transport (*U01.1, V01-V99, X82, Y03, Y32, Y36.1) [2]	249 827	17.5	17.4
Unintentional (V01-V99)	46 706	16.4	16.3
Suicide (X82)	91	0.0	0.0
Homicide (*U01.1, Y03) [2]	3 008	1.1	1.0
Undetermined (Y32)	22	0.0	0.0
Legal intervention/ war (Y36.1)	-	*	*
Motor vehicle traffic (V02-V04 [.1, .9], V09.2, V12-V14 [.3-.9], V19 [.4-.6], V20-V28 [.3-.9], V29-V79 [.4-.9], V80 [.3-.5], V81.1, V82.1, V83-V86 [.0-.3], V87 [.0-.8], V89.2) [4]	42 443	14.9	14.9
Occupant (V30-V79 [.4-.9], V83-V86 [.0-.3]) [4]	19 270	6.8	6.8
Motorcyclist (V20-V28 [.3-.9], V29 [.4-.9]) [4]	2 976	1.0	1.0
Pedal cyclist (V12-V14 [.3-.9], V19 [.4-.6]) [4]	585	0.2	0.2
Pedestrian (V02-V04 [.1, .9], V09.2) [4]	4 822	1.7	1.7
Other (V80 [.3-.5], V81.1, V82.1) [4]	15	*	*
Unspecified (V87 [.0-.8], V89.2) [4]	14 775	5.2	5.2
Pedal cyclist, other (V10-V11, V12-V14 [.0-.2], V15-V18, V19 [.0-.3, .8, .9]) [4]	207	0.1	0.1
Pedestrian, other (V01, V02-V04 [.0], V05, V06, V09 [.0, .1, .3, .9]) [4]	1 249	0.4	0.4
Other land transport (V20-V28 [.0-.2], V29-V79 [.0-.3], V80 [.0-.2, .6-.9], V81-V82 [.0, .2-.9], V83-V86 [.4-.9], V87.9, V88 [.0-.9], V89 [.0, .1, .3, .9], X82, Y03, Y32) V83-V86 [.4-.9], V87.9, V88 [.0-.9], V89 [.0, .1, .3, .9])	1 493	0.5	0.5
Unintentional (V20-V28 [.0-.2], V29-V79 [.0-.3], V80 [.0-.2, .6-.9], V81-V82 [.0, .2-.9])	1 294	0.5	0.5
Suicide (X82)	91	0.0	0.0
Homicide (Y03)	86	0.0	0.0
Undetermined (Y32)	22	0.0	0.0
Other transport (*U01.1, V90-V99, Y36.1) [2]	4 435	1.6	1.5
Unintentional (V90-V99)	1 513	0.5	0.5
Homicide (*U01.1) [2]	2 922	1.0	1.0
Legal intervention/ war (Y36.1) .	-	*	*

[1]Based on codes from the "Tenth Revision, International Classification of Diseases, 1992." Codes *U01-*U03 have been added as of 2001 for deaths due to terrorism.
[2]Figures include September 11, 2001 related deaths for which certificates were filed as of 10/24/02.
[3]Subcategories may not add to the total.
[4]Intent of death is unintentional.
0.0 = Quantity more than zero but less than 0.05.
* = Figure too small to meet standards of reliability or precision.
- = Quantity zero.

Table II-18. Number of Deaths, Death Rates, and Age-Adjusted Death Rates for Injury Deaths According to Mechanism and Intent of Death, 2001—*Continued*

(Number, rate per 100,000 population.)

Mechanism and intent of death [1]	Number	Rate	Age-adjusted rate
Natural/Environmental (W42-W43, W53-W64, W92-W99, X20-X39, X51-X57) [4]	1 427	0.5	0.5
Overexertion (X50) [4]	8	*	*
Poisoning (*U01 [.6-.7], X40-X49, X60-X69, X85-X90, Y10-Y19, Y35.2)	22 242	7.8	7.8
Unintentional (X40-X49)	14 078	4.9	4.9
Suicide (X60-X69)	5 191	1.8	1.8
Homicide (*U01 [.6-.7], X85-X90)	64	0.0	0.0
Undetermined (Y10-Y19)	2 909	1.0	1.0
Legal intervention/ war (Y35.2)	-	*	*
Struck By or Against (W20-W22, W50-W52, X79, Y00, Y04, Y29, Y35.3)	1 244	0.4	0.4
Unintentional (W20-W22, W50-W52)	898	0.3	0.3
Suicide (X79)	2	*	*
Homicide (Y00, Y04)	341	0.1	0.1
Undetermined (Y29)	3	*	*
Legal intervention/ war (Y35.3)	-	*	*
Suffocation (W75-W84, X70, X91, Y20)	12 574	4.4	4.4
Unintentional (W75-W84)	5 555	2.0	2.0
Suicide (X70)	6 198	2.2	2.2
Homicide (X91)	690	0.2	0.2
Undetermined (Y20)	131	0.0	0.0
Other Specified, Classifiable (*U01 [.0, .2, .5], *U03.0, W23, W35-W41, W44, W49, W85-W91, X75, X81, X96, Y02, Y05-Y07, Y25, Y31, Y35 [.1, .5], Y36 [.0, .2, .4-.8], Y85) [2]	2 061	0.7	0.7
Unintentional (W23, W35-W41, W44, W49, W85-W91, Y85)	1 355	0.5	0.5
Suicide (*U03, X75, X81) [2]	283	0.1	0.1
Homicide (*U01 [.0, .2, .5], X96, Y02, Y05-Y07)	316	0.1	0.1
Undetermined (Y25, Y31)	42	0.0	0.0
Legal intervention/ war (Y35 [.1, .5], Y36 [.0, .2, .4-.8])	65	0.0	0.0
Other Specified, Not Elsewhere Classified (*U01.8, *U02, X58, X83, Y08, Y33, Y35.6, Y86-Y87, Y89 [.0-.1])	2 299	0.8	0.8
Unintentional (X58, Y86)	1 034	0.4	0.4
Suicide (X83, Y87.0)	246	0.1	0.1
Homicide (*U01.8, * U02, Y08, Y87.1)	831	0.3	0.3
Undetermined (Y33, Y87.2)	163	0.1	0.1
Legal intervention/ war (Y35.6, Y89 [.0, .1])	25	0.0	0.0
Unspecified (*U01.9, *U03.9, X59, X84, Y09, Y34, Y35.7, Y36.9, Y89.9)	9 160	3.2	3.2
Unintentional (X59)	7 218	2.5	2.5
Suicide (*U03.9, X84)	146	0.1	0.1
Homicide (*U01.9, Y09)	1 506	0.5	0.5
Undetermined (Y34, Y89.9)	290	0.1	0.1
Legal intervention/ war (Y35.7, Y36.9)	-	*	*

[1]Based on codes from the "Tenth Revision, International Classification of Diseases, 1992." Codes *U01-*U03 have been added as of 2001 for deaths due to terrorism.
[2]Figures include September 11, 2001 related deaths for which certificates were filed as of 10/24/02.
[4]Intent of death is unintentional.
0.0 = Quantity more than zero but less than 0.05.
* = Figure too small to meet standards of reliability or precision.
- = Quantity zero.

Table II-19. Number of Deaths, Death Rates, and Age-Adjusted Death Rates for Injury by Firearms, by Race and Sex, 1999–2001

(Number, rate per 100,000 population.)

Year	All races			White			All other					
							Total			Black		
	Both sexes	Male	Female	Both sexes	Male	Female	Both sexes	Male	Female	Both sexes	Male	Female
Number												
2001	29 573	25 480	4 093	21 760	18 527	3 233	7 813	6 953	860	7 184	6 438	746
2000	28 663	24 582	4 081	20 945	17 750	3 195	7 718	6 832	886	7 054	6 284	770
1999	28 874	24 700	4 174	21 143	17 942	3 201	7 731	6 758	973	7 017	6 184	833
Rate												
2001	10.4	18.2	2.8	9.4	16.2	2.7	14.9	27.6	3.2	19.3	36.4	3.8
2000	10.2	17.8	2.8	9.1	15.6	2.7	15.0	27.8	3.3	19.3	36.1	4.0
1999	10.3	18.1	2.9	9.2	15.8	[illegible]	15.1	[illegible]	3.7	19.4	36.0	4.4
Age-Adjusted Rate												
2001	10.3	18.5	2.8	9.2	16.3	2.7	14.0	25.9	3.1	18.4	34.5	3.8
2000	10.2	18.1	2.8	9.0	15.9	2.7	14.1	26.0	3.2	18.4	34.2	3.9
1999	10.3	18.4	2.9	9.1	16.2	2.7	14.4	26.3	3.6	18.4	34.1	4.3

Table II-20. Number of Deaths, Death Rates, and Age-Adjusted Death Rates for Injury by Firearms, by Race, Hispanic Origin, and Sex, 1999–2001

(Number, rate per 100,000 population.)

Year	All origins [1]			Hispanic			Non-Hispanic [2]			Non-Hispanic White			Non-Hispanic Black		
	Both sexes	Male	Female	Both sexes	Male	Female	Both sexes	Male	Female	Both sexes	Male	Female	Both sexes	Male	Female
Number															
2001	29 573	25 480	4 093	3 087	2 774	313	26 341	22 573	3 768	18 676	15 760	2 916	7 063	6 323	740
2000	28 663	24 582	4 081	2 891	2 582	309	25 637	21 881	3 756	18 042	15 160	2 882	6 958	6 193	765
1999	28 874	24 700	4 174	2 878	2 549	329	25 877	22 050	3 827	18 260	15 384	2 876	6 933	6 114	819
Rate															
2001	10.4	18.2	2.8	8.3	14.6	1.7	10.6	18.7	3.0	9.4	16.3	2.9	19.8	37.3	4.0
2000	10.2	17.8	2.8	8.2	14.2	1.8	10.4	18.3	3.0	9.1	15.7	2.9	19.8	37.1	4.2
1999	10.3	18.1	2.9	8.5	14.6	2.0	10.6	18.5	3.0	9.3	15.9	2.9	20.0	37.1	4.5
Age-Adjusted Rate															
2001	10.3	18.5	2.8	7.8	13.7	1.7	10.5	18.8	3.0	9.1	16.0	2.8	18.9	35.4	3.9
2000	10.2	18.1	2.8	7.8	13.6	1.8	10.3	18.4	3.0	8.8	15.5	2.8	18.9	35.2	4.1
1999	10.3	18.4	2.9	8.2	14.2	2.0	10.5	18.7	3.0	8.9	15.8	2.8	19.0	35.2	4.4

[1]Figures for origin not stated are included in "All origins" but are not distributed among specified origins.
[2]Includes races other than White and Black.

TABLE II-21. NUMBER OF DEATHS, DEATH RATES, AND AGE-ADJUSTED DEATH RATES FOR DRUG-INDUCED CAUSES, BY RACE AND SEX, 1999–2001

Drug-induced deaths totaled 21,683 in 2001, a rate of 7.6 per 100,000. This category consists of accidental poisoning by drugs (legal and illegal use), including medically prescribed drugs, and biological substances, as well as the direct use of these items for intentional self-poisoning and for homicidal assaults; the latter two categories are classified under intentional self-harm and assault in the tables on causes of death. Newborn deaths associated with the mother's drug use are not included.

Drug-induced deaths increased in both 2000 and 2001. Between 2000 and 2001, the age-adjusted death rate increased by 8.5 percent for White males and by 12.8 percent for White females. The rate did not change significantly for either Black males or females or for Hispanic males or females (see also Table 22). Fatalities for White males were about twice as frequent as for females in 2001. While the rates for Black and White females were quite similar, the fatality rate for Black males (at 13.6 per 100,000, age-adjusted) was considerably higher than the White male rate of 10.2 per 100,000.

Table II-21. Number of Deaths, Death Rates, and Age-Adjusted Death Rates for Drug-Induced Causes, by Race and Sex, 1999–2001

(Number, rate per 100,000 population.)

Year	All races			White			All other					
							Total			Black		
	Both sexes	Male	Female	Both sexes	Male	Female	Both sexes	Male	Female	Both sexes	Male	Female
Number												
2001	21 683	14 244	7 439	18 176	11 873	6 303	3 507	2 371	1 136	3 163	2 163	1 000
2000	19 698	13 125	6 573	16 371	10 849	5 522	3 327	2 276	1 051	3 032	2 093	939
1999	19 102	12 873	6 229	15 694	10 497	5 197	3 408	2 376	1 032	3 094	2 188	906
Rate												
2001	7.6	10.2	5.1	7.8	10.4	5.4	6.7	9.4	4.2	8.5	12.2	5.1
2000	7.0	9.5	4.6	7.1	9.6	4.7	6.5	9.2	3.9	8.3	12.0	4.9
1999	6.8	9.4	4.4	6.9	9.3	4.5	6.8	9.9	3.9	8.6	12.7	4.8
Age-Adjusted Rate												
2001	7.6	10.1	5.1	7.8	10.2	5.3	7.1	10.3	4.3	9.2	13.6	5.4
2000	7.0	9.5	4.6	7.1	9.4	4.7	6.9	10.2	4.1	9.0	13.5	5.2
1999	6.8	9.4	4.4	6.8	9.2	4.4	7.2	10.8	4.1	9.3	14.3	5.0

TABLE II-22. NUMBER OF DEATHS, DEATH RATES, AND AGE-ADJUSTED DEATH RATES FOR DRUG-INDUCED CAUSES, BY RACE, HISPANIC ORIGIN, AND SEX, 1999–2001

See Table 21 comment for definition of drug-induced causes. Drug-induced deaths among Hispanics are considerably less than among non-Hispanic Whites. The rate per 100,000 was 2.5 for Hispanic women in 2001, as compared with 5.7 for non-Hispanic White women (age-adjusted). For men, the Hispanic rate per 100,000 was 8.0, while the non-Hispanic White rate was 10.4. Between 2000 and 2001, drug-induced death rates increased significantly for non-Hispanic Whites, but not for Hispanics.

TABLE II-23. NUMBER OF DEATHS, DEATH RATES, AND AGE-ADJUSTED DEATH RATES FOR ALCOHOL-INDUCED CAUSES, BY RACE AND SEX, 1999–2001

In 2001 a total of 19,817 persons died of alcohol-induced causes in the United States. This category includes deaths from the direct use of alcohol, including disease due to alcoholism and accidental poisoning by alcohol. But it excludes causes, such as accidents, that are indirectly related to alcohol, as well as newborn deaths associated with maternal alcohol use.

The age-adjusted death rate for alcohol-induced causes for males in 2001 was 3.3 times the rate for females, which compares with a 2 to 1 male excess in the case of drug-induced fatalities. The rate for alcohol-induced causes for Black men was almost 40 percent higher than the rate for White men. Between 2000 and 2001 the age-adjusted rate for the total population remained unchanged.

Table II-22. Number of Deaths, Death Rates, and Age-Adjusted Death Rates for Drug-Induced Causes, by Race, Hispanic Origin, and Sex, 1999–2001

(Number, rate per 100,000 population.)

Year	All origins [1]			Hispanic			Non-Hispanic [2]			Non-Hispanic White			Non-Hispanic Black		
	Both sexes	Male	Female	Both sexes	Male	Female	Both sexes	Male	Female	Both sexes	Male	Female	Both sexes	Male	Female
Number															
2001	21 683	14 244	7 439	1 731	1 335	396	19 778	12 769	7 009	16 349	10 456	5 893	3 097	2 113	984
2000	19 698	13 125	6 573	1 700	1 348	352	17 813	11 644	6 169	14 568	9 431	5 137	2 975	2 049	926
1999	19 102	12 873	6 229	1 965	1 605	360	16 940	11 124	5 816	13 624	8 822	4 802	3 024	2 131	893
Rate															
2001	7.6	10.2	5.1	4.7	7.0	2.2	8.0	10.6	5.5	8.3	10.8	5.8	8.7	12.5	5.3
2000	7.0	9.5	4.6	4.8	7.4	2.1	7.2	9.7	4.9	7.4	9.8	5.1	8.5	12.3	5.0
1999	6.8	9.4	4.4	5.8	9.2	2.2	6.9	9.3	4.6	6.9	9.1	4.8	8.7	12.9	4.9
Age-Adjusted Rate															
2001	7.6	10.1	5.1	5.3	8.0	2.5	7.9	10.4	5.4	8.1	10.6	5.7	9.3	13.8	5.5
2000	7.0	9.5	4.6	5.4	8.3	2.4	7.1	9.5	4.8	7.2	9.6	4.9	9.1	13.6	5.3
1999	6.8	9.4	4.4	6.4	10.3	2.5	6.8	9.2	4.5	6.8	8.9	4.6	9.4	14.4	5.1

[1] Figures for origin not stated are included in "All origins" but are not distributed among specified origins.
[2] Includes races other than White and Black.

Table II-23. Number of Deaths, Death Rates, and Age-Adjusted Death Rates for Alcohol-Induced Causes, by Race and Sex, 1999–2001

(Number, rate per 100,000 population.)

Year	All races			White			All other					
							Total			Black		
	Both sexes	Male	Female	Both sexes	Male	Female	Both sexes	Male	Female	Both sexes	Male	Female
Number												
2001	19 817	14 923	4 894	16 428	12 426	4 002	3 389	2 497	892	2 643	1 988	655
2000	19 358	14 770	4 588	16 019	12 349	3 670	3 339	2 421	918	2 636	1 933	703
1999	19 171	14 665	4 506	15 686	12 110	3 576	3 485	2 555	930	2 757	2 042	715
Rate												
2001	7.0	10.7	3.4	7.1	10.8	3.4	6.5	9.9	3.3	7.1	11.2	3.4
2000	6.9	10.7	3.2	7.0	10.9	3.1	6.5	9.8	3.4	7.2	11.1	3.7
1999	6.9	10.7	3.2	6.9	10.7	3.1	6.9	10.6	3.5	7.6	11.9	3.8
Age-Adjusted Rate												
2001	6.9	11.0	3.3	6.8	10.8	3.2	7.7	12.6	3.6	8.6	14.7	3.8
2000	6.9	11.2	3.2	6.8	10.9	3.0	7.8	12.8	4.0	8.8	14.8	4.2
1999	7.0	11.3	3.1	6.7	10.9	2.9	8.5	14.1	4.1	9.6	16.2	4.4

TABLE II-24. NUMBER OF DEATHS, DEATH RATES, AND AGE-ADJUSTED DEATH RATES FOR ALCOHOL-INDUCED CAUSES, BY RACE, HISPANIC ORIGIN, AND SEX, 1999–2001

See Table 23 comment for definition of alcohol-induced fatalities. On an age-adjusted basis, the alcohol-induced fatality rate for Hispanics was 1.6 times the rate for the non-Hispanic White population. The Hispanic rate also exceeded the rate for Blacks, but by a smaller margin.

Table II-24. Number of Deaths, Death Rates, and Age-Adjusted Death Rates for Alcohol-Induced Causes, by Race, Hispanic Origin, and Sex, 1999–2001

(Number, rate per 100,000 population.)

Year	All origins [1]			Hispanic			Non-Hispanic [2]			Non-Hispanic White			Non-Hispanic Black		
	Both sexes	Male	Female	Both sexes	Male	Female	Both sexes	Male	Female	Both sexes	Male	Female	Both sexes	Male	Female
Number															
2001	19 817	14 923	4 894	2 363	2 009	354	17 315	12 801	4 514	13 991	10 351	3 640	2 598	1 957	641
2000	19 358	14 770	4 588	2 290	1 994	296	16 929	12 654	4 275	13 645	10 280	3 365	2 598	1 901	697
1999	19 171	14 665	4 506	2 166	1 849	317	16 863	12 691	4 172	13 434	10 185	3 249	2 719	2 008	711
Rate															
2001	7.0	10.7	3.4	6.4	10.6	2.0	7.0	10.6	3.6	7.1	10.7	3.6	7.3	11.6	3.4
2000	6.9	10.7	3.2	6.5	11.0	1.7	6.9	10.6	3.4	6.9	10.6	3.3	7.4	11.4	3.8
1999	6.9	10.7	3.2	6.4	10.6	1.9	6.9	10.6	3.3	6.8	10.6	3.2	7.8	12.2	3.9
Age-Adjusted Rate															
2001	6.9	11.0	3.3	10.0	18.0	2.9	6.6	10.3	3.3	6.4	9.9	3.2	8.7	14.9	3.8
2000	6.9	11.2	3.2	10.4	19.2	2.5	6.6	10.5	3.2	6.4	10.0	3.0	9.0	15.0	4.3
1999	7.0	11.3	3.1	10.3	18.5	2.9	6.7	10.6	3.1	6.3	10.0	2.9	9.7	16.4	4.5

[1]Figures for origin not stated are included in "All origins" but are not distributed among specified origins.
[2]Includes races other than White and Black.

TABLE II-25. NUMBER OF DEATHS, DEATH RATES, AND AGE-ADJUSTED DEATH RATES FOR AGES 15 YEARS AND OVER, BY RACE, SEX, AND MARITAL STATUS, 2001

For those aged 15 years and over, the number of deaths of persons who were currently married was 942,458 in 2001; for widowed persons it was 918,655; for divorced, 262,695; and for never married, 242,936. On an age-adjusted basis, the death rate was lowest for currently married persons. It was 75 percent higher for widowed persons, and 82 percent higher for divorced persons. Individuals who were never married had an age-adjusted rate of death 2.1 times higher than those currently married.

For women of all races, the age-adjusted death rate of those who were never married was 2.6 times higher than that for currently married women. In the case of men, the age-adjusted rate for those who were never married was 1.9 times larger than that for currently married men.

For all age groups 15 years and over, death rates for married persons were much lower than those for never-married persons. For ages 25–34 years, widowed persons had the highest death rates; but beginning at age 35, those who never married had the highest death rates.

Table II-25. Number of Deaths, Death Rates, and Age-Adjusted Death Rates for Ages 15 Years and Over, by Race, Sex, and Marital Status, 2001

(Number, rate per 100,000 population.)

Race, sex, and marital status	15 years and over [1]	15–24 years	25–34 years	35–44 years	45–54 years	55–64 years	65–74 years	75 years and over	Age-adjusted rate [3]
NUMBER									
All Races, Both Sexes [2]	2 376 233	32 252	41 683	91 674	168 065	244 139	430 960	1 367 460	X
Never married	242 936	29 380	23 262	31 775	32 278	24 379	28 261	73 601	X
Ever married	2 123 808	2 774	18 179	59 066	134 196	217 928	400 803	1 290 862	X
Married	942 458	2 387	13 437	37 828	83 125	137 318	236 133	432 230	X
Widowed	918 655	61	289	1 673	6 576	25 229	103 912	780 915	X
Divorced	262 695	326	4 453	19 565	44 495	55 381	60 758	77 717	X
Not stated	9 489	98	242	833	1 591	1 832	1 896	2 997	X
All Races, Male [2]	1 160 546	23 963	28 757	58 164	104 848	144 958	241 581	558 275	X
Never married	145 795	22 138	17 169	22 498	22 258	15 926	17 751	28 055	X
Ever married	1 007 910	1 743	11 395	35 014	81 281	127 526	222 371	528 580	X
Married	642 223	1 514	8 502	22 351	50 649	87 181	158 788	313 238	X
Widowed	223 913	33	134	606	2 233	7 152	29 790	183 965	X
Divorced	141 774	196	2 759	12 057	28 399	33 193	33 793	31 377	X
Not stated	6 841	82	193	652	1 309	1 506	1 459	1 640	X
All Races, Female [2]	1 215 687	8 289	12 926	33 510	63 217	99 181	189 379	809 185	X
Never married	97 141	7 242	6 093	9 277	10 020	8 453	10 510	45 546	X
Ever married	1 115 898	1 031	6 784	24 052	52 915	90 402	178 432	762 282	X
Married	300 235	873	4 935	15 477	32 476	50 137	77 345	118 992	X
Widowed	694 742	28	155	1 067	4 343	18 077	74 122	596 950	X
Divorced	120 921	130	1 694	7 508	16 096	22 188	26 965	46 340	X
Not stated	2 648	16	49	181	282	326	437	1 357	X
White, Both Sexes	2 052 576	23 748	29 994	68 137	128 699	197 853	367 128	1 237 017	X
Never married	182 138	21 331	15 133	20 107	20 998	17 117	21 912	65 540	X
Ever married	1 863 680	2 348	14 689	47 459	106 623	179 443	343 853	1 169 265	X
Married	832 059	2 002	10 627	30 014	66 752	115 749	208 397	398 518	X
Widowed	811 615	46	221	1 159	4 372	18 580	84 720	702 517	X
Divorced	220 006	300	3 841	16 286	35 499	45 114	50 736	68 230	X
Not stated	6 758	69	172	571	1 078	1 293	1 363	2 212	X
White, Male	995 756	17 529	20 978	44 243	81 403	118 353	207 793	505 457	X
Never married	107 706	15 998	11 574	15 224	15 070	11 431	13 956	24 453	X
Ever married	883 123	1 471	9 257	28 559	65 423	105 835	192 789	479 789	X
Married	567 633	1 270	6 760	17 874	40 704	73 205	139 907	287 913	X
Widowed	196 664	23	107	440	1 541	5 252	24 439	164 862	X
Divorced	118 826	178	2 390	10 245	23 178	27 378	28 443	27 014	X
Not stated	4 927	60	147	460	910	1 087	1 048	1 215	X
White, Female	1 056 820	6 219	9 016	23 894	47 296	79 500	159 335	731 560	X
Never married	74 432	5 333	3 559	4 883	5 928	5 686	7 956	41 087	X
Ever married	980 557	877	5 432	18 900	41 200	73 608	151 064	689 476	X
Married	264 426	732	3 867	12 140	26 048	42 544	68 490	110 605	X
Widowed	614 951	23	114	719	2 831	13 328	60 281	537 655	X
Divorced	101 180	122	1 451	6 041	12 321	17 736	22 293	41 216	X
Not stated	1 831	9	25	111	168	206	315	997	X
Black, Both Sexes	276 355	7 147	10 028	20 687	34 518	39 976	54 379	109 620	X
Never married	55 190	6 801	7 141	10 755	10 586	6 723	5 840	7 344	X
Ever married	218 620	321	2 822	9 695	23 451	32 755	48 041	101 535	X
Married	87 369	289	2 245	6 362	13 318	17 451	22 155	25 549	X
Widowed	92 618	11	60	461	1 995	5 993	16 800	67 298	X
Divorced	38 633	21	517	2 872	8 138	9 311	9 086	8 688	X
Not stated	2 545	25	65	237	481	498	498	741	X
Black, Male	139 445	5 479	6 673	12 135	20 594	22 905	28 775	42 884	X
Never married	34 289	5 241	4 874	6 606	6 726	4 177	3 485	3 180	X
Ever married	103 375	218	1 755	5 352	13 495	18 338	24 906	39 311	X
Married	59 072	194	1 425	3 634	8 139	11 361	15 268	19 051	X
Widowed	23 632	8	21	142	633	1 728	4 785	16 315	X
Divorced	20 671	16	309	1 576	4 723	5 249	4 853	3 945	X
Not stated	1 781	20	44	177	373	390	384	393	X
Black, Female	136 910	1 668	3 355	8 552	13 924	17 071	25 604	66 736	X
Never married	20 901	1 560	2 267	4 149	3 860	2 546	2 355	4 164	X
Ever married	115 245	103	1 067	4 343	9 956	14 417	23 135	62 224	X
Married	28 297	95	820	2 728	5 179	6 090	6 887	6 498	X
Widowed	68 986	3	39	319	1 362	4 265	12 015	50 983	X
Divorced	17 962	5	208	1 296	3 415	4 062	4 233	4 743	X
Not stated	764	5	21	60	108	108	114	348	X

[1]Excludes figures for age not stated.
[2]Includes races other than White and Black.

Table II-25. Number of Deaths, Death Rates, and Age-Adjusted Death Rates for Ages 15 Years and Over, by Race, Sex, and Marital Status, 2001—*Continued*

(Number, rate per 100,000 population.)

Race, sex, and marital status	15 years and over [1]	15–24 years	25–34 years	35–44 years	45–54 years	55–64 years	65–74 years	75 years and over	Age-adjusted rate [3]
RATE [4]									
All Races, Both Sexes [2]	1 059.1	80.7	105.2	203.6	428.9	964.6	2 353.3	8 054.4	1 284.8
Never married	382.3	83.1	159.3	442.3	898.8	1 662.0	4 126.4	10 949.6	1 999.6
Ever married	1 320.6	60.2	72.7	156.1	377.0	914.1	2 273.6	7 916.7	1 227.5
Married	756.3	56.1	60.4	119.0	291.6	752.0	1 934.0	5 873.6	950.5
Widowed	6 051.7	*	203.4	401.5	714.3	1 418.0	2 754.3	9 614.0	1 662.4
Divorced	1 249.6	98.7	169.7	347.9	720.9	1 456.9	3 690.3	9 431.3	1 727.0
All Races, Male [2]	1 065.9	117.0	143.7	259.6	545.1	1 192.7	2 911.5	8 881.5	1 496.2
Never married	418.2	117.6	202.0	525.8	1 177.4	2 046.9	5 227.0	11 107.6	2 266.7
Ever married	1 361.9	105.2	99.0	193.2	468.6	1 121.0	2 794.4	8 761.1	1 421.3
Married	1 030.1	98.6	81.5	144.3	350.0	915.9	2 409.4	7 256.6	1 173.2
Widowed	8 098.6	*	*	739.6	1 263.1	2 141.6	4 155.2	12 955.0	2 492.6
Divorced	1 593.1	177.6	258.8	472.0	1 052.5	2 179.5	5 194.0	10 577.1	2 204.0
All Races, Female [2]	1 052.7	42.6	66.0	148.2	316.8	754.0	1 890.8	7 568.2	1 118.3
Never married	338.7	43.9	99.8	319.3	589.1	1 227.2	3 043.9	10 854.6	1 713.5
Ever married	1 285.4	34.9	50.3	122.0	289.9	725.2	1 845.1	7 420.8	1 080.5
Married	482.1	32.1	41.8	94.9	231.4	573.5	1 376.5	3 911.3	663.9
Widowed	5 595.9	*	130.4	318.8	583.9	1 250.8	2 425.7	8 906.1	1 476.9
Divorced	997.5	59.1	108.8	244.6	463.4	973.9	2 707.9	8 786.8	1 408.9
White, Both Sexes	1 107.7	75.5	95.3	185.6	393.2	913.5	2 298.9	8 098.8	1 262.5
Never married	380.2	77.5	144.2	398.7	827.0	1 494.5	3 882.8	10 876.5	1 916.2
Ever married	1 356.4	59.3	70.0	149.9	353.1	874.7	2 232.1	7 969.7	1 215.7
Married	776.4	54.6	56.9	112.2	272.6	724.1	1 918.7	5 911.4	942.3
Widowed	6 330.7	*	207.2	380.8	651.4	1 362.3	2 693.7	9 743.4	1 643.2
Divorced	1 264.0	108.1	174.7	352.9	705.4	1 425.7	3 627.4	9 480.0	1 720.0
White, Male	1 099.9	108.0	130.3	239.7	501.3	1 127.5	2 842.2	8 898.2	1 464.9
Never married	399.7	108.0	182.0	475.4	1 061.3	1 788.0	4 907.7	10 656.8	2 116.0
Ever married	1 389.0	103.3	95.0	187.2	441.5	1 073.7	2 743.7	8 801.9	1 405.6
Married	1 055.9	95.6	76.6	137.2	327.2	878.7	2 387.1	7 299.9	1 162.4
Widowed	8 466.1	*	*	760.0	1 179.0	2 206.8	3 983.0	13 146.4	2 470.0
Divorced	1 584.3	201.5	266.1	471.6	1 031.0	2 124.9	5 153.5	10 684.1	2 199.4
White, Female	1 115.1	40.8	58.6	131.0	286.8	712.2	1 840.2	7 625.5	1 100.7
Never married	355.2	42.0	86.1	265.2	529.8	1 123.6	2 841.8	11 011.7	1 664.7
Ever married	1 328.3	34.6	48.3	115.2	268.0	690.7	1 803.0	7 477.6	1 070.2
Married	495.0	31.4	39.2	88.5	216.2	555.8	1 369.7	3 953.7	659.4
Widowed	5 858.2	*	134.8	291.8	523.8	1 183.8	2 381.2	9 026.9	1 456.2
Divorced	1 021.4	64.5	111.6	247.4	442.4	945.4	2 632.6	8 827.9	1 398.1
Black, Both Sexes	1 014.4	117.7	185.0	356.5	771.9	1 576.8	3 257.5	8 627.1	1 639.9
Never married	465.8	120.3	236.3	591.6	1 153.9	2 421.0	5 645.8	12 558.8	2 514.2
Ever married	1 420.2	76.3	117.7	243.3	659.7	1 450.9	3 068.0	8 376.3	1 514.2
Married	818.3	76.4	109.1	205.2	554.9	1 228.7	2 423.8	6 280.2	1 189.6
Widowed	5 114.0	*	*	489.5	1 053.9	1 949.1	3 642.3	9 281.5	1 885.8
Divorced	1 329.7	*	166.6	363.3	843.3	1 757.2	4 767.6	10 821.9	2 036.1
Black, Male	1 102.3	180.7	259.2	445.0	996.1	2 025.3	4 166.5	10 020.7	2 007.9
Never married	591.8	182.5	331.1	760.4	1 668.8	3 675.0	*	*	3 503.3
Ever married	1 507.5	135.1	159.2	288.0	810.8	1 802.7	3 866.6	9 593.1	1 808.2
Married	1 094.4	134.5	146.7	236.0	654.3	1 497.0	3 109.0	7 664.4	1 461.3
Widowed	6 819.6	*	*	*	*	2 353.5	5 563.4	12 614.2	*
Divorced	1 857.1	*	236.2	525.1	1 233.7	2 838.0	7 239.2	*	2 715.8
Black, Female	938.3	54.9	117.9	278.0	579.1	1 215.5	2 616.1	7 919.4	1 385.6
Never married	345.2	56.1	146.3	437.1	750.5	1 552.1	4 135.2	*	1 892.9
Ever married	1 350.0	39.7	82.3	204.2	526.7	1 162.3	2 509.9	7 754.9	1 315.3
Married	536.0	40.6	75.5	174.8	447.9	920.8	1 628.2	4 106.2	826.9
Widowed	4 710.4	*	*	420.6	898.5	1 822.2	3 202.0	8 558.0	1 691.8
Divorced	1 002.1	*	115.8	264.2	586.5	1 177.7	3 426.4	*	1 636.4

[1]Excludes figures for age not stated.
[2]Includes races other than White and Black.
[3]Calculated based on ages 25 years and over.
[4]Figures for marital status are included in totals but not distributed among specified marital status groups.
* = Figure too small to meet standards of reliability or precision.

TABLE II-26. NUMBER OF DEATHS, DEATH RATES, AND AGE-ADJUSTED DEATH RATES FOR AGES 15 YEARS AND OVER, BY RACE, HISPANIC ORIGIN, SEX, AND MARITAL STATUS, 2001

Hispanic persons showed the same patterns of mortality by marital status as the non-Hispanic population but, as was encountered in previous tables, they had significantly lower death rates. This difference prevailed for each marital status. Non-Hispanic Blacks also exhibited the same patterns of mortality by marital status as the general population, but in their case the age-adjusted death rates were higher for each category.

Table II-26. Number of Deaths, Death Rates, and Age-Adjusted Death Rates for Ages 15 Years and Over, by Race, Hispanic Origin, Sex, and Marital Status, 2001

(Number, rate per 100,000 population.)

Race, sex, and marital status	15 years and over [1]	15–24 years	25–34 years	35–44 years	45–54 years	55–64 years	65–74 years	75 years and over	Age-adjusted rate [4]
NUMBER									
All Origins, Both Sexes [2]	2 376 233	32 252	41 683	91 674	168 065	244 139	430 960	136 746	X
Never married	242 936	29 380	23 262	31 775	32 278	24 379	28 261	73 601	X
Ever married	2 123 808	2 774	18 179	59 066	134 196	217 928	400 803	129 086	X
Married	942 458	2 387	13 437	37 828	83 125	137 318	236 133	432 230	X
Widowed	918 655	61	289	1 673	6 576	25 229	103 912	780 915	X
Divorced	262 695	326	4 453	19 565	44 495	55 381	60 758	77 717	X
Not stated	9 489	98	242	833	1 591	1 832	1 896	2 997	X
All Origins, Male [2]	1 160 546	23 963	28 757	58 164	104 848	144 958	241 581	558 275	X
Never married	145 795	22 138	17 169	22 498	22 258	15 926	17 751	28 055	X
Ever married	1 007 910	1 743	11 395	35 014	81 281	127 526	222 371	528 580	X
Married	642 223	1 514	8 502	22 351	50 649	87 181	158 788	313 238	X
Widowed	223 913	33	134	606	2 233	7 152	29 790	183 965	X
Divorced	141 774	196	2 759	12 057	28 399	33 193	33 793	31 377	X
Not stated	6 841	82	193	652	1 309	1 506	1 459	1 640	X
All Origins, Female [2]	1 215 687	8 289	12 926	33 510	63 217	99 181	189 379	809 185	X
Never married	97 141	7 242	6 093	9 277	10 020	8 453	10 510	45 546	X
Ever married	1 115 898	1 031	6 784	24 052	52 915	90 402	178 432	762 282	X
Married	300 235	873	4 935	15 477	32 476	50 137	77 345	118 992	X
Widowed	694 742	28	155	1 067	4 343	18 077	74 122	596 950	X
Divorced	120 921	130	1 694	7 508	16 096	22 188	26 965	46 340	X
Not stated	2 648	16	49	181	282	326	437	1 357	X
Hispanic, Both Sexes	106 711	5 058	5 797	8 441	11 362	13 337	20 172	42 544	X
Never married	18 939	4 394	3 080	3 030	2 343	1 617	1 685	2 790	X
Ever married	86 989	641	2 649	5 302	8 862	11 569	18 354	39 612	X
Married	45 475	595	2 221	3 833	5 979	7 606	11 085	14 156	X
Widowed	28 794	10	46	170	448	1 343	4 551	22 226	X
Divorced	12 720	36	382	1 299	2 435	2 620	2 718	3 230	X
Not stated	783	23	68	109	157	151	133	142	X
Hispanic, Male	59 540	4 019	4 366	5 895	7 477	8 026	11 231	18 526	X
Never married	13 118	3 552	2 434	2 324	1 723	1 072	970	1 043	X
Ever married	45 771	445	1 870	3 474	5 612	6 828	10 150	17 392	X
Married	31 125	420	1 616	2 572	3 875	4 889	7 442	10 311	X
Widowed	7 603	4	21	59	165	395	1 258	5 701	X
Divorced	7 043	21	233	843	1 572	1 544	1 450	1 380	X
Not stated	651	22	62	97	142	126	111	91	X
Hispanic, Female	47 171	1 039	1 431	2 546	3 885	5 311	8 941	24 018	X
Never married	5 821	842	646	706	620	545	715	1 747	X
Ever married	41 218	196	779	1 828	3 250	4 741	8 204	22 220	X
Married	14 350	175	605	1 261	2 104	2 717	3 643	3 845	X
Widowed	21 191	6	25	111	283	948	3 293	16 525	X
Divorced	5 677	15	149	456	863	1 076	1 268	1 850	X
Not stated	132	1	6	12	15	25	22	51	X
Non-Hispanic, Both Sexes [3]	2 262 238	27 059	35 656	82 763	155 916	229 812	409 500	132 153	X
Never married	222 809	24 876	20 065	28 604	29 784	22 613	26 422	70 445	X
Ever married	2 032 304	2 131	15 470	53 594	124 995	205 826	381 628	124 866	X
Married	895 149	1 790	11 166	33 886	76 946	129 429	224 616	417 316	X
Widowed	887 864	51	243	1 503	6 104	23 800	99 126	757 037	X
Divorced	249 291	290	4 061	18 205	41 945	52 597	57 886	74 307	X
Not stated	7 125	52	121	565	1 137	1 373	1 450	2 427	X
Non-Hispanic, Male [3]	10 968 824	19 836	24 225	51 918	96 815	136 214	229 539	538 335	X
Never married	131 924	18 499	14 645	20 069	20 434	14 742	16 674	26 861	X
Ever married	959 955	1 296	9 489	31 424	75 464	120 359	211 763	510 160	X
Married	609 847	1 092	6 854	19 705	46 653	82 100	151 053	302 390	X
Widowed	215 768	29	113	547	2 060	6 722	28 446	177 851	X
Divorced	134 340	175	2 522	11 172	26 751	31 537	32 264	29 919	X
Not stated	5 003	41	91	425	917	1 113	1 102	1 314	X

[1] Excludes figures for age not stated.
[2] Figures for origin not stated are included in "All origins" but not distributed among specified groups.
[3] Includes races other than White and Black.

Table II-26. Number of Deaths, Death Rates, and Age-Adjusted Death Rates for Ages 15 Years and Over, by Race, Hispanic Origin, Sex, and Marital Status, 2001—*Continued*

(Number, rate per 100,000 population.)

Race, sex, and marital status	15 years and over [1]	15–24 years	25–34 years	35–44 years	45–54 years	55–64 years	65–74 years	75 years and over	Age-adjusted rate [4]
NUMBER—*Continued*									
Non-Hispanic, Female [3]	1 165 356	7 223	11 431	30 845	59 101	93 598	179 961	783 197	X
Never married	90 885	6 377	5 420	8 535	9 350	7 871	9 748	43 584	X
Ever married	1 072 349	835	5 981	22 170	49 531	85 467	169 865	738 500	X
Married	285 302	698	4 312	14 181	30 293	47 329	73 563	114 926	X
Widowed	672 096	22	130	956	4 044	17 078	70 680	579 186	X
Divorced	114 951	115	1 539	7 033	15 194	21 060	25 622	44 388	X
Not stated	2 122	11	30	140	220	260	348	1 113	X
Non-Hispanic White, Both Sexes	1 942 463	18 730	24 191	59 546	117 086	184 109	346 389	119 241	X
Never married	162 914	16 982	12 070	17 058	18 650	15 441	20 169	62 544	X
Ever married	1 774 669	1 714	12 046	42 131	97 711	167 728	325 221	112 811	X
Married	785 894	1 413	8 408	26 163	60 745	108 078	197 163	383 924	X
Widowed	781 694	36	176	996	3 926	17 220	80 081	679 259	X
Divorced	207 081	265	3 462	14 972	33 040	42 430	47 977	64 935	X
Not stated	4 880	34	75	357	725	940	999	1 750	X
Non-Hispanic White, Male	934 347	13 542	16 613	38 213	73 731	109 993	196 158	486 097	X
Never married	94 444	12 486	9 162	12 878	13 351	10 310	12 932	23 325	X
Ever married	836 452	1 028	7 392	25 058	59 783	98 898	182 472	461 821	X
Married	536 036	852	5 144	15 289	36 817	68 263	132 365	277 306	X
Widowed	188 768	19	86	384	1 378	4 847	23 142	158 912	X
Divorced	111 648	157	2 162	9 385	21 588	25 788	26 965	25 603	X
Not stated	3 451	28	59	277	597	785	754	951	X
Non-Hispanic White, Female	1 008 116	5 188	7 578	21 333	43 355	74 116	150 231	706 315	X
Never married	68 470	4 496	2 908	4 180	5 299	5 131	7 237	39 219	X
Ever married	938 217	686	4 654	17 073	37 928	68 830	142 749	666 297	X
Married	249 858	561	3 264	10 874	23 928	39 815	64 798	106 618	X
Widowed	592 926	17	90	612	2 548	12 373	56 939	520 347	X
Divorced	95 433	108	1 300	5 587	11 452	16 642	21 012	39 332	X
Not stated	1 429	6	16	80	128	155	245	799	X
Non-Hispanic Black, Both Sexes	273 303	7 024	9 859	20 436	34 098	39 523	53 803	108 560	X
Never married	54 447	6 692	7 039	10 658	10 460	6 645	5 752	7 201	X
Ever married	216 753	316	2 778	9 592	23 248	32 474	47 630	100 715	X
Married	86 536	284	2 207	6 293	13 192	17 296	21 951	25 313	X
Widowed	91 938	11	59	456	1 978	5 943	16 686	66 805	X
Divorced	38 279	21	512	2 843	8 078	9 235	8 993	8 597	X
Not stated	2 103	16	42	186	390	404	421	644	X
Non-Hispanic Black, Male	137 684	5 377	6 548	11 971	20 307	22 600	28 446	42 435	X
Never married	33 798	5 149	4 790	6 542	6 635	4 121	3 436	3 125	X
Ever married	102 434	216	1 728	5 294	13 371	18 169	24 686	38 970	X
Married	58 515	192	1 401	3 590	8 060	11 263	15 130	18 879	X
Widowed	23 425	8	21	140	626	1 706	4 744	16 180	X
Divorced	20 494	16	306	1 564	4 685	5 200	4 812	3 911	X
Not stated	1 452	12	30	135	301	310	324	340	X
Non-Hispanic Black, Female	135 619	1 647	3 311	8 465	13 791	16 923	25 357	66 125	X
Never married	20 649	1 543	2 249	4 116	3 825	2 524	2 316	4 076	X
Ever married	114 319	100	1 050	4 298	9 877	14 305	22 944	61 745	X
Married	28 021	92	806	2 703	5 132	6 033	6 821	6 434	X
Widowed	68 513	3	38	316	1 352	4 237	11 942	50 625	X
Divorced	17 785	5	206	1 279	3 393	4 035	4 181	4 686	X
Not stated	651	4	12	51	89	94	97	304	X

[1]Excludes figures for age not stated.
[2]Figures for origin not stated are included in "All origins" but not distributed among specified groups.

able II-26. Number of Deaths, Death Rates, and Age-Adjusted Death Rates for Ages 15 Years and Over, by Race, Hispanic Origin, Sex, and Marital Status, 2001—*Continued*

(Number, rate per 100,000 population.)

Race, sex, and marital status	15 years and over [1]	15–24 years	25–34 years	35–44 years	45–54 years	55–64 years	65–74 years	75 years and over	Age-adjusted rate [4]
RATE [5]									
All Origins, Both Sexes [2]	1 059.1	80.7	105.2	203.6	428.9	964.6	2 353.3	8 054.4	1 284.8
Never married	382.3	83.1	159.3	442.3	898.8	1 662.0	4 126.4	10 949.6	1 999.6
Ever married	1 320.6	60.2	72.7	156.1	377.0	914.1	2 273.6	7 916.7	1 227.5
Married	756.3	56.1	60.4	119.0	291.6	752.0	1 934.0	5 873.6	950.5
Widowed	6 051.7	*	203.4	401.5	714.3	1 418.0	2 754.3	9 614.0	1 662.4
Divorced	1 249.6	98.7	169.7	347.9	720.9	1 456.9	3 690.3	9 431.3	1 727.0
All Origins, Male [2]	1 065.9	117.0	143.7	259.6	545.1	1 192.7	2 911.5	8 881.5	1 496.2
Never married	418.2	117.6	202.0	525.8	1 177.4	2 046.9	5 227.0	11 107.6	2 266.7
Ever married	1 361.9	105.2	99.0	193.2	468.6	1 121.0	2 794.4	8 761.1	1 421.3
Married	1 030.1	98.6	81.5	144.3	350.0	915.9	2 409.4	7 256.6	1 173.2
Widowed	8 098.6	*	*	739.6	1 263.1	2 141.6	4 155.2	12 955.0	2 492.6
Divorced	1 593.1	177.6	258.8	472.0	1 052.5	2 179.5	5 194.0	10 577.1	2 204.0
All Origins, Female [2]	1 052.7	42.6	66.0	148.2	316.8	754.0	1 890.8	7 568.2	1 118.3
Never married	338.7	43.9	99.8	319.3	589.1	1 227.2	3 043.9	10 854.6	1 713.5
Ever married	1 285.4	34.9	50.3	122.0	289.9	725.2	1 845.1	7 420.8	1 080.5
Married	482.1	32.1	41.8	94.9	231.4	573.5	1 376.5	3 911.3	663.9
Widowed	5 595.9	*	130.4	318.8	583.9	1 250.8	2 425.7	8 906.1	1 476.9
Divorced	997.5	59.1	108.8	244.6	463.4	973.9	2 707.9	8 786.8	1 408.9
Hispanic, Both Sexes	408.9	75.6	84.5	154.8	334.4	722.2	1 776.0	5 954.8	960.9
Never married	203.1	79.6	132.6	349.0	661.3	1 042.4	2 257.7	*	1 449.0
Ever married	518.6	54.7	58.4	115.7	291.2	683.9	1 729.6	5 800.9	912.3
Married	321.5	52.5	53.2	95.5	239.7	582.7	1 597.9	4 350.9	733.0
Widowed	3 206.0	*	*	*	480.9	846.4	1 834.0	6 945.2	1 178.4
Divorced	734.4	*	112.1	250.5	534.5	1 150.9	2 278.4	*	1 390.2
Hispanic, Male	442.3	111.5	118.0	208.5	443.9	923.9	2 242.6	6 708.9	1 149.5
Never married	242.3	113.3	169.1	439.0	898.5	1 342.8	*	*	*
Ever married	568.7	94.8	82.8	151.2	376.0	865.5	2 145.4	6 524.1	1 078.5
Married	433.6	91.6	76.4	124.9	298.9	736.9	1 931.4	5 158.7	887.9
Widowed	4 649.0	*	*	*	*	*	*	10 610.9	1 948.8
Divorced	995.9	*	165.8	364.8	863.8	1 695.5	*	*	*
Hispanic, Female	373.3	33.7	45.2	97.0	226.7	543.0	1 408.0	5 479.7	809.6
Never married	148.8	35.3	73.1	208.4	381.4	723.8	*	*	1 140.6
Ever married	472.4	27.9	34.2	80.0	209.5	525.2	1 395.1	5 337.8	782.5
Married	206.0	25.9	29.3	64.6	175.6	423.3	1 181.2	3 064.2	522.7
Widowed	2 884.7	*	*	*	357.3	762.5	1 637.0	6 205.6	1 018.9
Divorced	553.9	*	74.4	158.7	315.4	787.8	1 615.6	*	1 097.1
Non-Hispanic, Both Sexes [3]	1 141.0	81.4	108.9	209.2	435.6	979.5	2 383.9	8 125.9	1 300.2
Never married	411.4	83.5	163.7	452.9	920.6	1 729.1	4 335.7	10 997.1	2 042.5
Ever married	1 410.3	61.4	75.5	161.2	383.9	929.1	2 303.4	7 992.6	1 242.9
Married	809.8	56.8	61.7	122.0	295.8	763.3	1 950.1	5 936.4	961.4
Widowed	6 218.9	*	199.1	413.4	738.3	1 470.6	2 816.9	9 698.7	1 690.9
Divorced	1 292.8	98.2	178.8	356.6	734.1	1 469.8	3 781.1	9 438.2	1 745.5
Non-Hispanic, Male [3]	1 149.6	117.5	148.5	265.2	551.6	1 207.0	2 944.1	8 957.8	1 512.3
Never married	448.5	118.0	207.9	535.4	1 202.4	2 119.1	5 349.4	11 016.3	2 289.3
Ever married	1 454.5	107.3	102.4	198.5	476.0	1 136.6	2 829.2	8 848.0	1 438.7
Married	1 104.6	99.5	82.3	146.7	354.1	927.1	2 435.1	7 352.0	1 187.9
Widowed	8 287.1	*	*	747.2	1 267.0	2 243.9	4 243.0	12 995.8	2 524.3
Divorced	1 641.9	174.8	274.8	481.4	1 064.5	2 198.0	5 277.2	10 524.7	2 218.3
Non-Hispanic, Female [3]	1 133.1	44.1	69.6	154.3	324.0	768.7	1 918.4	7 638.3	1 133.4
Never married	367.3	45.2	103.9	332.5	608.7	1 285.9	3 274.4	10 985.3	1 765.4
Ever married	1 372.9	36.9	53.3	127.2	296.5	739.0	1 870.1	7 492.3	1 094.9
Married	515.6	34.0	44.2	98.8	236.0	584.2	1 384.1	3 940.2	671.2
Widowed	5 757.6	*	126.4	329.3	608.9	1 294.9	2 481.3	8 997.7	1 504.1
Divorced	1 035.5	58.9	113.7	252.5	474.7	982.4	2 786.3	8 824.2	1 426.9
Non-Hispanic White, Both Sexes	1 206.7	74.1	96.4	188.3	396.0	924.0	2 325.5	8 169.7	1 274.7
Never married	415.2	75.8	144.8	402.2	843.8	1 549.2	4 069.9	10 949.2	1 953.9
Ever married	1 457.9	59.4	71.9	153.8	357.2	886.2	2 258.5	8 044.0	1 229.2
Married	836.3	53.8	56.8	113.8	274.1	732.2	1 931.8	5 970.9	951.0
Widowed	6 531.1	*	201.6	383.8	672.9	1 422.3	2 755.1	9 834.8	1 670.2
Divorced	1 311.3	106.9	185.3	362.1	716.7	1 435.7	3 727.7	9 440.9	1 734.8
Non-Hispanic White, Male	1 198.8	105.1	131.5	241.6	502.6	1 136.3	2 869.4	8 972.8	1 476.9
Never married	431.6	105.1	183.0	474.8	1 073.5	1 831.1	5 006.8	10 582.0	2 127.5
Ever married	1 492.2	102.1	96.9	191.2	445.3	1 084.8	2 774.0	8 886.2	1 420.2
Married	1 139.1	92.4	75.1	137.8	327.9	885.4	2 407.9	7 392.0	1 174.0
Widowed	8 721.5	*	*	778.5	1 173.2	2 393.7	4 079.8	13 190.8	2 510.1
Divorced	1 633.9	195.9	284.5	479.9	1 038.8	2 140.2	5 250.7	10 627.8	2 213.8

[1]Excludes figures for age not stated.
[2]Figures for origin not stated are included in "All origins" but not distributed among specified groups.
[3]Includes races other than White and Black.
[4]Calculated based on ages 25 years and over.
[5]Figures for marital status not stated are included in totals but not distributed among specified marital status groups.
* = Figure too small to meet standards of reliability or precision.

Table II-26. Number of Deaths, Death Rates, and Age-Adjusted Death Rates for Ages 15 Years and Over, by Race, Hispanic Origin, Sex, and Marital Status, 2001—*Continued*

(Number, rate per 100,000 population.)

Race, sex, and marital status	15 years and over [1]	15–24 years	25–34 years	35–44 years	45–54 years	55–64 years	65–74 years	75 years and over	Age-adjusted rate [4]
RATE [5]—*Continued*									
Non-Hispanic White, Female	1 214.1	41.9	60.9	134.9	291.0	723.5	1 864.1	7 695.6	1 113.6
Never married	394.5	42.8	87.4	273.3	548.2	1 183.2	3 050.0	11 180.0	1 715.9
Ever married	1 428.5	36.5	51.0	119.6	272.2	701.6	1 825.0	7 548.1	1 083.0
Married	532.6	32.9	41.1	91.4	218.8	564.6	1 376.1	3 980.5	665.4
Widowed	6 047.6	*	130.2	291.2	546.8	1 227.2	2 433.9	9 125.7	1 480.3
Divorced	1 065.3	64.4	117.3	256.3	452.3	950.7	2 716.5	8 801.2	1 410.4
Non-Hispanic Black, Both Sexes	1 043.2	121.4	191.5	365.8	786.9	1 602.5	3 300.7	8 704.2	1 661.8
Never married	477.6	123.8	242.2	602.6	1 168.9	2 443.5	5 953.2	12 380.3	2 539.1
Ever married	1 464.8	82.6	123.9	251.3	676.2	1 479.9	3 106.1	8 470.3	1 537.5
Married	849.0	82.7	115.5	212.1	571.6	1 261.4	2 457.2	6 355.7	1 211.0
Widowed	5 164.4	*	*	517.8	1 064.9	1 951.9	3 666.1	9 362.0	1 905.5
Divorced	1 355.1	*	169.5	373.2	855.3	1 780.1	4 862.5	11 136.9	2 083.8
Non-Hispanic Black, Male	1 133.1	186.2	267.9	456.3	1 014.6	2 055.4	4 218.9	10 107.5	2 033.9
Never married	607.4	187.6	339.3	776.4	1 690.9	3 725.4	*	*	3 517.6
Ever married	1 555.3	150.6	167.3	297.3	830.9	1 837.3	3 922.7	9 699.3	1 836.8
Married	1 133.8	150.4	154.7	243.7	672.4	1 536.1	3 162.2	7 736.1	1 486.1
Widowed	6 830.1	*	*	*	*	2 315.0	5 548.7	12 759.5	*
Divorced	1 894.0	*	241.1	540.6	1 254.4	2 857.4	7 363.4	*	2 765.3
Non-Hispanic Black, Female	965.4	56.8	122.4	285.7	591.5	1 238.2	2 653.0	7 992.2	1 404.7
Never married	353.8	58.0	150.4	444.4	761.3	1 564.6	4 481.5	*	1 919.2
Ever married	1 392.1	41.8	86.8	211.0	540.0	1 186.7	2 537.7	7 843.0	1 335.1
Married	556.9	42.6	80.2	180.9	462.6	945.8	1 644.1	4 171.5	843.6
Widowed	4 767.0	*	*	453.0	908.8	1 836.0	3 230.6	8 627.8	1 713.2
Divorced	1 020.5	*	117.6	270.7	594.2	1 198.1	3 495.9	*	1 680.7

[1]Excludes figures for age not stated.
[4]Calculated based on ages 25 years and over.
[5]Figures for marital status not stated are included in totals but not distributed among specified marital status groups.
* = Figure too small to meet standards of reliability or precision.

Figure II-12. Age-Adjusted Death Rates for Ages 15 Years and Over, by Marital Status, Sex, Race and Hispanic Origin, 2001

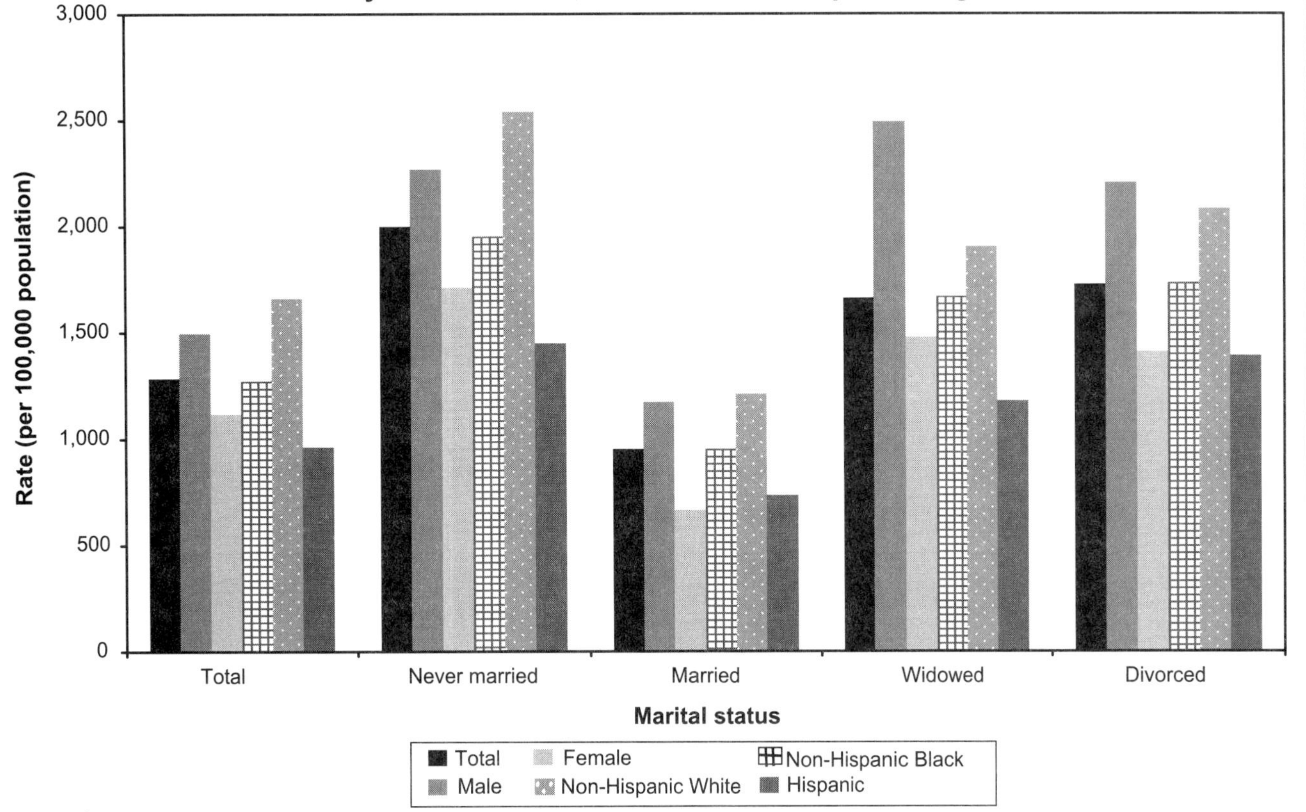

TABLE II-27. NUMBER OF DEATHS, DEATH RATES, AND AGE-ADJUSTED DEATH RATES FOR AGES 25–64 YEARS, BY EDUCATIONAL ATTAINMENT AND SEX, 2001

The higher the educational attainment, the lower are the death rates. In this table three levels of educational achievement are compared: under 12 years, 12 years, and 13 years or more. The drop in death rates is particularly sharp between 12 years and 13 years or more.

The table is limited to the age between 25 and 64 years, because educational attainment may not be completed under age 25, and because educational attainment reported on the death certificates of older persons is less reliable than the reports at younger ages.

Data were unavailable for Georgia, Rhode Island, and South Dakota.

Table II-27. Number of Deaths, Death Rates, and Age-Adjusted Death Rates for Ages 25–64 Years, by Educational Attainment and Sex, 2001

(Number, rate per 100,000 population.)

Years of school completed and sex	25–64 years [1]	25–34 years	35–44 years	45–54 years	55–64 years	Age-adjusted rate [2]
NUMBER						
Both Sexes	524 207	39 953	87 925	161 460	234 869	X
Under 12 years	112 654	9 063	18 304	29 269	56 018	X
12 years	228 103	17 698	40 605	68 928	100 872	X
13 years or more	165 251	12 023	26 035	57 719	69 474	X
Not stated	18 199	1 169	2 981	5 544	8 505	X
Male	323 545	27 627	55 868	100 683	139 367	X
Under 12 years	72 128	6 564	12 281	18 988	34 295	X
12 years	139 070	12 627	26 359	42 986	57 098	X
13 years or more	99 929	7 599	15 185	34 823	42 322	X
Not stated	12 418	837	2 043	3 886	5 652	X
Female	200 662	12 326	32 057	60 777	95 502	X
Under 12 years	40 526	2 499	6 023	10 281	21 723	X
12 years	89 033	5 071	14 246	25	43 774	X
13 years or more	65 322	4 424	10 850	22 896	27 152	X
Not stated	5 781	332	938	1 658	2 853	X
RATE [3]						
Both Sexes	364.6	105.0	202.6	426.6	959.6	362.2
Under 12 years	628.4	192.7	367.8	716.9	1 345.7	576.6
12 years	500.1	160.3	283.9	587.0	1 181.3	480.9
13 years or more	206.0	53.9	108.0	261.9	590.1	214.6
Male	454.7	143.5	259.1	542.1	1 184.7	457.5
Under 12 years	784.6	254.8	473.0	934.0	1 726.6	745.8
12 years	617.4	209.3	357.9	788.0	1 553.2	631.2
13 years or more	253.4	71.4	130.9	314.1	693.7	257.3
Female	276.4	65.6	146.8	315.3	751.3	270.7
Under 12 years	464.0	117.5	253.0	501.6	998.0	407.1
12 years	385.7	101.3	205.3	413.0	900.1	348.6
13 years or more	160.1	38.0	86.7	209.1	478.7	171.5

[1] Excludes figures for age not stated.
[2] Calculated based on ages 25–64 years.
[3] Figures for education not stated are included in totals for "Both sexes," "Male," and "Female" but not distributed among specified years of education.
X = Not applicable.

TABLE II-28. NUMBER OF DEATHS, DEATH RATES, AND AGE-ADJUSTED DEATH RATES FOR AGES 15 YEARS AND OVER, BY INJURY AT WORK, RACE, AND SEX, 2001

Deaths due to injury at work totaled 8,303 in 2001, an unusually high number that included 2,826 victims of the September 11 terrorist attacks (See also Table 29). The risk of work-related death was much higher for males than for females. White males had a 6.8 age-adjusted death rate, while the female rate was 0.9 per 100,000. For Black males the corresponding rate was 5.5 per 100,000, and for Black females is was 1.1. By age group the highest risk occurred in the 35 to 44 year group.

Table II-28. Number of Deaths, Death Rates, and Age-Adjusted Death Rates for Ages 15 Years and Over, by Injury at Work, Race, and Sex, 2001

(Number, rate per 100,000 population.)

Race and sex	15 years and over [1]	15–24 years	25–34 years	35–44 years	45–54 years	55–64 years	65 years and over [2]	Age-adjusted rate [2]
NUMBER								
All Races, Both Sexes [3]	8 303	650	1 813	2 355	1 827	972	686	X
Male	7 181	585	1 527	2 037	1 577	851	604	X
Female	1 122	65	286	318	250	121	82	X
White, Both Sexes	7 093	567	1 520	1 991	1 549	834	632	X
Male	6 211	517	1 301	1 752	1 350	735	556	X
Female	882	50	219	239	199	99	76	X
Black, Both Sexes	849	49	184	271	204	99	42	X
Male	680	40	146	213	160	83	38	X
Female	169	9	38	58	44	16	4	X
RATE								
All Races, Both Sexes [3]	3.7	1.6	4.6	5.2	4.7	3.8	1.9	3.7
Male	6.6	2.9	7.6	9.1	8.2	7.0	4.1	6.6
Female	1.0	0.3	1.5	1.4	1.3	0.9	0.0	1.0
White, Both Sexes	3.8	1.8	4.8	5.4	4.7	3.9	2.0	3.8
Male	6.9	3.2	8.1	9.5	8.3	7.0	4.3	6.8
Female	0.9	0.3	1.4	1.3	1.2	0.9	0.0	0.9
Black, Both Sexes	3.1	0.8	3.4	4.7	4.6	3.9	1.4	3.1
Male	5.4	1.3	5.7	7.8	7.7	7.0	3.4	5.5
Female	1.2	*	1.3	1.9	1.8	*	*	1.1

[1]Excludes figures for age not stated.
[2]Calculated based on ages 15 years and over.
[3]Includes races other than White and Black.
X = Not applicable.
* = Figure too small to meet standards of reliability or precision.

Figure II-13. Age-Adjusted Death Rates and (Number of Deaths) by Injury at Work, Race, and Sex, 1999–2001

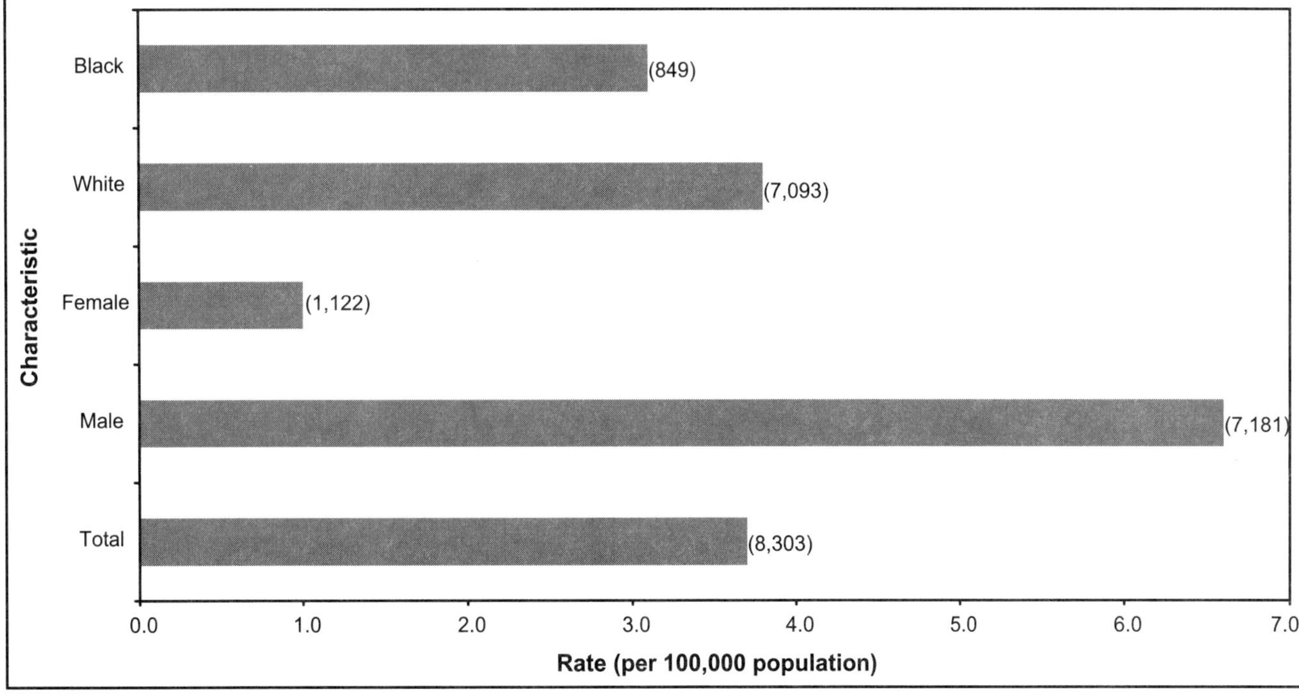

TABLE II-29. NUMBER OF DEATHS, DEATH RATES, AND AGE-ADJUSTED DEATH RATES FOR INJURY AT WORK, BY RACE AND SEX, 1993–2001

In the years prior to the terrorist attacks in September 2001, the rate of deaths from injury at work held fairly steady, although there was a downward drift over the years since 1993. For the total population, the age-adjusted rates were 2.9 per 100,000 in 1993 and 2.5 in 2000. The unadjusted rates were the same. In 2001 these rates jumped to 3.7, due to inclusion of September 11 deaths on the job.

In most years, rates for males were about 11 times higher than those for females. Until 1996, death rates for Blacks were somewhat greater than those for Whites.

TABLE II-29A. NUMBER OF DEATHS AND AGE-ADJUSTED DEATH RATES FOR HOMICIDE AND INJURY AT WORK, 2000, 2001, AND PRELIMINARY 2002

Changes in deaths from homicides and from injury at work between 2000, 2001, and 2002 were dominated by the effects of the terrorist attacks in 2001. Homicides increased by 3,543 in 2001, and declined by 3,263 in 2002. The age-adjusted rate was the same in 2002 as in 2000, thus the increased number of death over those two years reflected an increase in population. Injury at work is a separate category that can be reported along with another cause of death, such as homicide (see discussion of Table 10A). The number of deaths at work increased by 2,873 in 2001, and fell by 3,051 in 2002.

Table II-29. Number of Deaths, Death Rates, and Age-Adjusted Death Rates for Injury at Work, by Race and Sex, 1993–2001

(Number, rate per 100,000 population.)

Year	All races			White			All other					
							Total			Black		
	Both sexes	Male	Female	Both sexes	Male	Female	Both sexes	Male	Female	Both sexes	Male	Female
Number												
2001	8 303	7 181	1 122	7 093	6 211	882	1 210	970	240	849	680	169
2000	5 430	4 969	461	4 657	4 270	387	773	699	74	591	536	55
1999	5 651	5 152	499	4 805	4 385	420	846	767	79	659	598	61
1998	5 543	5 036	507	4 804	4 366	438	739	670	69	587	535	52
1997	5 666	5 144	522	4 785	4 352	433	881	792	89	684	626	58
1996	5 778	5 280	498	4 940	4 535	405	838	745	93	649	582	67
1995	5 872	5 334	538	5 007	4 550	457	865	784	81	692	627	65
1994	5 987	5 425	562	5 103	4 642	461	884	783	101	710	632	78
1993	5 847	5 352	495	4 979	4 581	398	868	771	97	677	608	69
Rate												
2001	3.7	6.6	1.0	3.8	6.9	0.9	3.1	5.3	1.2	3.1	5.4	1.2
2000	2.5	4.6	0.4	2.5	4.8	0.4	2.0	3.7	0.4	2.2	4.3	0.4
1999	2.6	4.9	0.4	2.6	4.9	0.5	2.3	4.4	0.4	2.5	4.9	0.4
1998	2.6	4.8	0.5	2.7	5.0	0.5	2.0	3.9	0.4	2.3	4.5	0.4
1997	2.7	5.0	0.5	2.7	5.0	0.5	2.5	4.8	0.5	2.7	5.3	0.4
1996	2.7	5.2	0.5	2.8	5.3	0.4	2.4	4.6	0.5	2.6	5.1	0.5
1995	2.8	5.3	0.5	2.9	5.4	0.5	2.6	5.0	0.5	2.8	5.5	0.5
1994	2.9	5.5	0.5	3.0	5.5	0.5	2.7	5.1	0.6	3.0	5.7	0.6
1993	2.9	5.5	0.5	2.9	5.5	0.5	2.7	5.2	0.6	2.9	5.6	0.5
Age-Adjusted Rate [1]												
2001	3.7	6.6	1.0	3.8	6.8	0.9	3.1	5.3	1.1	3.1	5.5	1.1
2000	2.5	4.6	0.4	2.5	4.8	0.4	2.1	3.9	0.4	2.3	4.6	0.4
1999	2.6	4.9	0.4	2.6	4.9	0.4	2.3	4.5	0.4	2.6	5.1	0.4
1998	2.6	4.8	0.5	2.7	5.0	0.5	2.1	4.1	0.4	2.3	4.7	0.4
1997	2.7	5.0	0.5	2.7	5.0	0.5	2.6	5.0	0.5	2.8	5.5	0.4
1996	2.8	5.2	0.5	2.8	5.3	0.4	2.5	4.8	0.5	2.6	5.3	0.5
1995	2.8	5.3	0.5	2.9	5.4	0.5	2.7	5.3	0.5	3.0	6.0	0.5
1994	2.9	5.5	0.5	3.0	5.6	0.5	2.8	5.4	0.6	3.1	6.0	0.6
1993	2.9	5.5	0.5	2.9	5.5	0.5	2.8	5.4	0.6	3.0	6.0	0.5

[1]Calculated based on ages 15 years and over.

Table II-29A. Number of Deaths and Age-Adjusted Death Rates for Homicide and Injury at Work, 2000, 2001 and Preliminary 2002

(Number, age-adjusted death rates per 100,000 U.S. standard population based on year 2000 population.)

Cause of death and measure	2002	2001 [1]	2000
Assault (Homicide)			
Number of deaths	17 045	20 908	16 765
Age-adjusted death rate	5.9	7.1	5.9
Injury at Work			
Number of deaths	5 252	8 303	5 430
Age-adjusted death rate	2.3	3.7	2.5

[1]Includes deaths from September 11, 2001, terrorist attacks.

TABLE II-30. NUMBER OF DEATHS, DEATH RATES, AND AGE-ADJUSTED DEATH RATES FOR MAJOR CAUSES OF DEATH FOR EACH STATE AND TERRITORY, 2001

This table includes statistics for all causes of death combined as well as details on the leading 15 causes of death in 2001, with the exception of septicemia, hypertension, and pneumonitis due to solids and liquids. In addition, death rates are given for HIV disease, motor vehicle accidents, and injury by firearms, causes with significant public health policy implications (HIV disease was among the 15 leading cause of death until 1997).

The 2.4 million deaths in the United States in 2001 represented an age-adjusted death rate of 854.5 per 100,000. Age-adjusted rates differed considerably among states. The four states with the lowest rates were Hawaii (652.6 per 100,000), Minnesota (744.9), Connecticut (767.8), and North Dakota (775.9). The four highest rates were reported for Mississippi (1,023.2), Louisiana (1,005.9), West Virginia (994.9), and Alabama (992.9). The District of Columbia, which as a city is not generally comparable to the states, showed an even higher rate of 1,038.2.

Table II-30. Number of Deaths, Death Rates, and Age-Adjusted Death Rates for Major Causes of Death for each State and Territory, 2001

(Number, rate per 100,000 population.)

Area	All causes			Human inmmunodeficiency virus (HIV) disease (B20-B24)			Malignant neoplasms (C00-C97)			Diabetes mellitus (E1-E14)		
	Number	Rate	Age-adjusted rate [1]	Number	Rate	Age-adjusted rate [1]	Number	Rate	Age-adjusted rate [1]	Number	Rate	Age-adjusted rate [1]
United States [2]	2 416 425	848.5	854.5	14 175	5.0	5.0	553 768	194.4	196.0	71 372	25.1	25.3
Alabama	45 316	1 014.0	992.9	214	4.8	4.9	9 801	219.3	212.2	1 344	30.1	29.4
Alaska	2 974	469.4	825.8	7	*	*	689	108.7	192.6	80	12.6	22.6
Arizona	41 058	773.7	787.4	156	2.9	3.1	9 143	172.3	172.9	1 058	19.9	20.1
Arkansas	27 759	1 030.1	948.1	64	2.4	2.5	6 092	226.1	208.0	756	28.1	25.8
California	234 044	676.4	775.1	1 505	4.3	4.4	53 924	155.8	179.2	6 395	18.5	21.4
Colorado	28 294	638.5	787.8	93	2.1	2.0	6 145	138.7	169.6	667	15.1	18.7
Connecticut	29 827	868.4	767.8	193	5.6	5.4	7 093	206.5	187.2	759	22.1	19.6
Delaware	7 112	892.8	891.9	67	8.4	8.3	1 786	224.2	219.3	218	27.4	27.2
District of Columbia	5 951	1 037.1	1 038.2	223	38.9	38.1	1 327	231.3	235.3	209	36.4	37.0
Florida	167 269	1 021.6	799.7	1 664	10.2	10.4	39 090	238.7	187.2	4 631	28.3	22.0
Georgia	64 485	767.2	956.5	722	8.6	8.5	13 818	164.4	202.1	1 475	17.5	21.9
Hawaii	8 394	684.1	652.6	26	2.1	2.1	2 024	165.0	155.9	173	14.1	13.3
Idaho	9 753	738.5	798.0	12	*	*	2 093	158.5	172.2	318	24.1	26.2
Illinois	105 430	842.1	857.3	518	4.1	4.2	24 778	197.9	204.5	3 092	24.7	25.4
Indiana	55 198	900.9	908.3	119	1.9	2.0	12 831	209.4	211.6	1 677	27.4	27.7
Iowa	27 791	947.9	777.4	18	*	*	6 404	218.4	188.7	709	24.2	20.1
Kansas	24 647	912.1	839.8	39	1.4	1.5	5 448	201.6	193.3	721	26.7	25.0
Kentucky	39 861	979.7	987.7	90	2.2	2.2	9 358	230.0	227.9	1 099	27.0	27.1
Louisiana	41 757	934.1	1 005.9	404	9.0	9.3	9 517	212.9	227.4	1 734	38.8	41.7
Maine	12 421	967.0	843.2	14	*	*	3 045	237.1	207.4	398	31.0	26.9
Maryland	43 839	813.9	881.4	572	10.6	10.1	10 322	191.6	204.6	1 458	27.1	29.2
Massachusetts	56 754	886.6	801.4	249	3.9	3.7	13 750	214.8	199.3	1 422	22.2	20.2
Michigan	86 424	863.7	877.6	255	2.5	2.5	19 689	196.8	199.0	2 655	26.5	26.9
Minnesota	37 735	757.0	744.9	53	1.1	1.0	8 967	179.9	183.1	1 213	24.3	24.3
Mississippi	28 259	988.2	1 023.2	202	7.1	7.4	5 958	208.3	216.3	661	23.1	24.1
Missouri	54 982	975.3	909.5	147	2.6	2.7	12 365	219.3	206.8	1 535	27.2	25.5
Montana	8 265	912.9	840.3	5	*	*	1 956	216.0	198.8	229	25.3	23.1
Nebraska	15 174	882.2	793.5	33	1.9	2.0	3 389	197.0	185.0	400	23.3	21.2
Nevada	16 285	776.3	922.6	76	3.6	3.6	3 815	181.9	205.9	322	15.3	17.6
New Hampshire	9 815	779.4	798.5	9	*	*	2 398	190.4	193.9	291	23.1	23.6
New Jersey	74 710	877.8	831.9	785	9.2	8.9	18 165	213.4	203.2	2 556	30.0	28.5
New Mexico	14 129	771.7	825.4	39	2.1	2.2	2 900	158.4	167.0	538	29.4	31.4
New York	159 240	834.4	800.9	2 085	10.9	10.7	36 975	193.7	187.7	3 844	20.1	19.4
North Carolina	70 934	864.4	911.0	464	5.7	5.6	16 065	195.8	202.5	2 181	26.6	27.8
North Dakota	6 048	950.1	775.9	4	*	*	1 391	218.5	188.4	196	30.8	25.7
Ohio	108 027	948.5	905.7	207	1.8	1.9	24 804	217.8	207.8	3 750	32.9	31.4
Oklahoma	34 682	999.6	959.7	99	2.9	2.9	7 411	213.6	205.2	1 065	30.7	29.5
Oregon	30 158	868.2	825.6	63	1.8	1.9	7 057	203.2	196.1	1 011	29.1	27.9
Pennsylvania	129 729	1 054.4	869.9	496	4.0	4.0	29 914	243.1	203.3	3 826	31.1	25.7
Rhode Island	10 021	945.7	806.4	30	2.8	2.8	2 398	226.3	200.6	265	25.0	21.7
South Carolina	36 612	901.3	939.1	310	7.6	7.6	8 272	203.6	206.8	1 088	26.8	27.5
South Dakota	6 923	912.9	784.8	4	*	*	1 590	209.7	187.9	212	28.0	24.2
Tennessee	55 151	959.2	972.9	277	*	*	12 229	212.7	212.5	1 746	30.4	30.6
Texas	152 779	714.9	877.8	1 055	4.9	5.1	33 488	156.7	191.0	5 456	25.5	31.6
Utah	12 662	555.7	776.8	12	*	*	2 313	101.5	143.4	509	22.3	31.9
Vermont	5 201	848.5	811.2	4	*	*	1 249	203.8	194.8	155	25.3	24.1
Virginia	56 280	782.0	862.2	275	3.8	3.7	13 345	185.4	199.9	1 613	22.4	24.5
Washington	44 642	744.9	792.9	113	1.9	1.8	10 802	180.2	193.3	1 403	23.4	25.1
West Virginia	20 967	1 164.2	994.9	29	1.6	1.7	4 685	260.1	218.7	802	44.5	37.6
Wisconsin	46 628	862.5	806.5	73	1.4	1.3	10 777	199.4	190.9	1 337	24.7	23.3
Wyoming	4 029	816.0	851.7	2	*	*	923	186.9	192.9	120	24.3	25.5
Puerto Rico [3]	29 082	757.7	841.7	715	18.6	20.3	4 713	122.8	135.5	2 475	64.5	71.4
Virgin Islands [3]	605	556.3	751.6	3	*	*	114	104.8	142.6	31	28.5	41.1
Guam [3]	663	418.7	789.9	3	*	*	97	61.3	112.9	19	*	*
American Samoa [3]	239	415.4	1 229.5	-	*	*	34	59.1	147.2	21	36.5	134.4
Northern Marianas [3]	148	205.9	1 015.4	-	*	*	27	37.6	195.7	9	*	*

[1]Death rates are affected by the population composition of the area. Age-adjusted death rates should be used for comparisons between areas.
[2]Excludes data for Puerto Rico, Virgin Islands, Guam, American Samoa and Northern Marianas.
[3]Age-adjusted death rates for Puerto Rico, Virgin Islands, Guam, American Samoa, and Northern Marianas are calculated using different age groups in the weighting procedure.
* = Figure too small to meet standards of reliability or precision.
- = Quantity zero.

Turning to specific causes of death, HIV age-adjusted death rates ranked highest in the District of Columbia and Puerto Rico and they amounted to more than 10 per 100,000 in New York, Florida, and Maryland. Death rates from cancer were lowest in Utah, Hawaii, New Mexico and Colorado. Apart from the District of Columbia, cancer death rates were highest in Kentucky, Louisiana, Delaware, West Virginia and Mississippi. Diabetes death rates were highest in Louisiana, West Virginia, and the District of Columbia.

The state of Washington, Maine, Oregon, Colorado, and North Dakota ranked highest for Alzheimer death rates. Minnesota, Hawaii, Colorado, Utah, and Alaska had the smallest death rates of heart diseases, while Mississippi, Oklahoma, West Virginia, and also the District of Columbia had the highest rates.

Table II-30. Number of Deaths, Death Rates, and Age-Adjusted Death Rates for Major Causes of Death for each State and Territory, 2001—*Continued*

(Number, rate per 100,000 population.)

Area	Alzheimer's disease (G30)			Diseases of heart (I00-I09, I11, I13, I20-I51)			Cerebrovascular diseases (I60-I69)			Influenza and pneumonia (J10-J18)		
	Number	Rate	Age-adjusted rate [1]	Number	Rate	Age-adjusted rate [1]	Number	Rate	Age-adjusted rate [1]	Number	Rate	Age-adjusted rate [1]
United States [2]	53 852	18.9	19.1	700 142	245.8	247.8	163 538	57.4	57.9	62 034	21.8	22.0
Alabama	1 103	24.7	24.6	13 207	295.5	289.3	2 998	67.1	66.0	1 105	24.7	24.4
Alaska	44	6.9	18.8	603	95.2	187.7	158	24.9	57.4	36	5.7	12.9
Arizona	1 108	20.9	22.1	10 588	199.5	204.5	2 480	46.7	48.2	1 133	21.3	22.1
Arkansas	452	16.8	15.1	8 263	306.6	279.1	2 256	83.7	75.9	735	27.3	24.8
California	4 935	14.3	16.9	68 234	197.2	230.4	18 088	52.3	61.2	8 129	23.5	27.6
Colorado	852	19.2	25.9	6 293	142.0	181.0	1 825	41.2	53.9	650	14.7	19.0
Connecticut	573	16.7	13.8	8 582	249.9	216.4	2 003	58.3	49.6	854	24.9	20.9
Delaware	135	16.9	17.6	2 033	255.2	257.2	383	48.1	49.2	181	22.7	23.4
District of Columbia	99	17.3	17.0	1 761	306.9	308.5	265	46.2	46.4	86	15.0	15.1
Florida	3 643	22.2	16.1	50 629	309.2	233.2	10 414	63.6	47.6	3 317	20.3	15.2
Georgia	1 419	16.9	23.5	17 478	207.9	268.0	4 312	51.3	67.3	1 515	18.0	24.0
Hawaii	122	9.9	9.6	2 310	188.3	179.5	766	62.4	59.8	207	16.9	16.1
Idaho	294	22.3	24.4	2 489	188.5	205.1	781	59.1	64.7	239	18.1	19.6
Illinois	2 163	17.3	17.3	30 990	247.5	251.5	7 230	57.7	58.4	2 659	21.2	21.3
Indiana	1 231	20.1	20.2	15 682	256.0	258.0	3 877	63.3	63.7	1 184	19.3	19.4
Iowa	863	29.4	21.7	8 250	281.4	224.0	2 218	75.6	58.8	877	29.9	22.7
Kansas	642	23.8	20.3	6 716	248.5	224.0	1 848	68.4	60.3	561	20.8	18.2
Kentucky	966	23.7	24.7	11 808	290.2	294.0	2 557	62.8	64.4	1 114	27.4	28.2
Louisiana	957	21.4	24.2	11 474	256.7	280.1	2 638	59.0	64.8	941	21.0	23.4
Maine	471	36.7	30.9	3 272	254.7	218.8	822	64.0	54.7	269	20.9	17.8
Maryland	948	17.6	20.2	12 310	228.6	251.6	2 882	53.5	59.6	928	17.2	19.3
Massachusetts	1 533	23.9	20.5	15 144	236.6	210.4	3 535	55.2	48.6	1 770	27.7	24.0
Michigan	1 803	18.0	18.5	26 896	268.8	273.8	5 701	57.0	58.2	2 076	20.7	21.2
Minnesota	1 173	23.5	22.0	8 760	175.7	171.0	2 727	54.7	52.5	828	16.6	15.5
Mississippi	449	15.7	16.4	9 050	316.5	329.0	1 935	67.7	70.5	754	26.4	27.4
Missouri	1 106	19.6	17.6	16 633	295.1	271.9	3 796	67.3	61.6	1 604	28.5	25.8
Montana	249	27.5	24.7	1 970	217.6	197.9	578	63.8	57.6	193	21.3	19.2
Nebraska	430	25.0	20.8	4 150	241.3	210.9	1 130	65.7	57.3	296	17.2	14.5
Nevada	211	10.1	14.7	4 393	209.4	257.1	913	43.5	55.1	352	16.8	21.9
New Hampshire	305	24.2	25.0	2 835	225.1	230.9	633	50.3	52.0	216	17.2	17.7
New Jersey	1 376	16.2	15.0	22 704	266.8	250.8	4 007	47.1	44.2	1 893	22.2	20.8
New Mexico	297	16.2	18.3	3 423	187.0	203.4	824	45.0	49.6	398	21.7	24.0
New York	1 705	8.9	8.4	56 643	296.8	282.2	7 706	40.4	38.4	5 085	26.6	25.2
North Carolina	1 786	21.8	24.2	18 792	229.0	244.0	5 401	65.8	71.3	1 780	21.7	23.7
North Dakota	224	35.2	25.3	1 700	267.1	210.9	507	79.6	60.4	162	25.4	18.8
Ohio	2 367	20.8	19.7	32 453	284.9	271.0	6 891	60.5	57.4	2 401	21.1	20.0
Oklahoma	682	19.7	18.6	10 840	312.4	298.1	2 384	68.7	65.6	913	26.3	25.1
Oregon	1 030	29.7	27.2	7 075	203.7	191.6	2 588	74.5	69.5	575	16.6	15.3
Pennsylvania	2 753	22.4	17.2	39 438	320.6	257.8	8 619	70.1	55.5	2 735	22.2	17.6
Rhode Island	251	23.7	18.4	3 076	290.3	240.5	614	57.9	47.4	295	27.8	22.3
South Carolina	881	21.7	24.4	9 471	233.2	245.5	2 832	69.7	75.0	754	18.6	20.1
South Dakota	156	20.6	15.7	1 985	261.8	218.1	490	64.6	52.5	188	24.8	19.5
Tennessee	1 148	20.0	20.9	15 688	272.9	278.2	4 037	70.2	72.3	1 617	28.1	29.1
Texas	3 439	16.1	21.4	43 199	202.1	255.1	10 612	49.7	63.8	3 608	16.9	21.7
Utah	314	13.8	20.7	2 896	127.1	185.2	870	38.2	56.4	411	18.0	26.7
Vermont	149	24.3	22.9	1 429	233.1	221.4	323	52.7	50.2	133	21.7	20.4
Virginia	1 184	16.5	19.4	14 913	207.2	231.9	4 129	57.4	65.4	1 481	20.6	23.7
Washington	2 050	34.2	36.5	11 281	188.2	201.2	3 765	62.8	67.2	965	16.1	17.1
West Virginia	398	22.1	18.6	6 325	351.2	296.0	1 272	70.6	59.4	483	26.8	22.7
Wisconsin	1 269	23.5	20.8	13 023	240.9	222.3	3 658	67.7	61.7	1 236	22.9	20.5
Wyoming	114	23.1	25.0	985	199.5	209.5	260	52.7	56.1	112	22.7	24.1
Puerto Rico [3]	842	21.9	26.0	6 409	167.0	188.2	1 700	44.3	50.6	1 091	28.4	32.8
Virgin Islands [3]	7	*	*	172	158.2	228.6	47	43.2	61.4	9	*	*
Guam [3]	2	*	*	204	128.8	276.1	62	39.2	87.2	17	*	*
American Samoa [3]	-	*	*	51	88.7	318.9	19	*	*	4	*	*
Northern Marianas [3]	-	*	*	23	32.0	204.5	15	*	*	1	*	*

[1]Death rates are affected by the population composition of the area. Age-adjusted death rates should be used for comparisons between areas.
[2]Excludes data for Puerto Rico, Virgin Islands, Guam, American Samoa and Northern Marianas.
[3]Age-adjusted death rates for Puerto Rico, Virgin Islands, Guam, American Samoa, and Northern Marianas are calculated using different age groups in the weighting procedure.
* = Figure too small to meet standards of reliability or precision.
\- = Quantity zero.

The effect of age on accidents, suicide, homicide, and firearm deaths (which can be examined in Table 11) is less pronounced than in disease causes, but still enough to warrant comparison of age-adjusted rates. For motor vehicle accidents, the states with the lowest fatality rates were New York, Massachusetts, Rhode Island, New Jersey, and the District of Columbia. Those with the highest rates were Mississippi, Wyoming, and Arkansas. When all types of accidents are considered Alaska stands out with the highest rate of fatalities, followed by New Mexico, Mississippi, and Wyoming. The lowest rates of completed suicides are found in Massachusetts, New York, the District of Columbia, and New Jersey. The highest rates of suicide occurred in New Mexico, Montana, Nevada, Colorado, Wyoming, and Idaho. Alaska changed from the second highest suicide rate in 2000 (and the highest crude rate) to the seventh highest rate in 2001.

Table II-30. Number of Deaths, Death Rates, and Age-Adjusted Death Rates for Major Causes of Death for each State and Territory, 2001—*Continued*

(Number, rate per 100,000 population.)

Area	Chronic lower respiratory diseases (J40-J47)			Chronic liver disease and cirrhosis (K70, K73-K74)			Nephritis, nephrotic syndrome, and nephrosis (N00-N07, N17-N19, N25-N27)			Accidents (V01-X59, Y85-Y86)		
	Number	Rate	Age-adjusted rate [1]	Number	Rate	Age-adjusted rate [1]	Number	Rate	Age-adjusted rate [1]	Number	Rate	Age-adjusted rate [1]
United States [2]	123 013	43.2	43.7	27 035	9.5	9.5	39 480	13.9	14.0	101 537	35.7	35.7
Alabama	2 204	49.3	48.0	445	10.0	9.6	974	21.8	21.4	2 211	49.5	49.2
Alaska	147	23.2	49.6	57	9.0	10.7	24	3.8	7.2	345	54.4	60.4
Arizona	2 499	47.1	47.3	661	12.5	12.8	644	12.1	12.3	2 475	46.6	47.3
Arkansas	1 368	50.8	46.2	233	8.6	8.3	521	19.3	17.6	1 277	47.4	46.4
California	12 965	37.5	43.9	3 692	10.7	11.7	1 994	5.8	6.7	8 132	23.5	24.4
Colorado	1 836	41.4	53.5	420	9.5	10.2	391	8.8	11.2	1 720	38.8	41.5
Connecticut	1 486	43.3	38.0	294	8.6	8.0	585	17.0	15.0	1 056	30.7	29.5
Delaware	306	38.4	37.9	75	9.4	9.1	108	13.6	13.7	290	36.4	36.3
District of Columbia	150	26.1	26.5	63	11.0	11.1	79	13.8	13.7	223	38.9	38.3
Florida	8 954	54.7	40.9	2 013	12.3	10.8	2 172	13.3	10.0	6 969	42.6	40.5
Georgia	3 097	36.8	47.9	654	7.8	8.7	1 362	16.2	20.9	3 396	40.4	43.6
Hawaii	280	22.8	21.6	84	6.8	6.4	123	10.0	9.5	372	30.3	29.7
Idaho	581	44.0	48.4	105	8.0	8.5	95	7.2	7.9	568	43.0	44.0
Illinois	4 776	38.1	39.3	1 103	8.8	9.1	2 178	17.4	17.7	4 077	32.6	32.7
Indiana	3 136	51.2	51.9	459	7.5	7.5	1 149	18.8	18.9	2 185	35.7	35.7
Iowa	1 552	52.9	43.5	214	7.3	6.8	220	7.5	6.0	1 051	35.8	32.3
Kansas	1 444	53.4	49.9	208	7.7	7.7	498	18.4	16.4	1 156	42.8	41.3
Kentucky	2 265	55.7	56.2	387	9.5	9.2	807	19.8	20.2	1 990	48.9	48.7
Louisiana	1 763	39.4	43.0	327	7.3	7.6	993	22.2	24.1	2 029	45.4	46.4
Maine	800	62.3	54.1	111	8.6	7.6	208	16.2	13.9	490	38.1	36.0
Maryland	1 907	35.4	39.2	481	8.9	9.0	593	11.0	12.0	1 339	24.9	25.7
Massachusetts	2 810	43.9	39.7	556	8.7	8.3	1 207	18.9	16.8	1 506	23.5	22.0
Michigan	4 148	41.5	42.3	1 058	10.6	10.5	1 533	15.3	15.6	3 291	32.9	33.1
Minnesota	1 909	38.3	38.7	359	7.2	7.3	729	14.6	14.2	1 797	36.1	35.2
Mississippi	1 333	46.6	48.7	265	9.3	9.6	602	21.1	21.9	1 571	54.9	55.6
Missouri	2 882	51.1	47.7	439	7.8	7.5	991	17.6	16.2	2 455	43.5	42.4
Montana	583	64.4	59.1	104	11.5	10.6	128	14.1	12.8	468	51.7	49.9
Nebraska	882	51.3	46.5	103	6.0	5.9	289	16.8	14.8	633	36.8	34.7
Nevada	1 136	54.2	65.6	297	14.2	14.5	376	17.9	21.6	739	35.2	37.2
New Hampshire	616	48.9	50.8	121	9.6	9.6	136	10.8	11.2	374	29.7	29.9
New Jersey	2 911	34.2	32.4	778	9.1	8.7	1 495	17.6	16.6	2 405	28.3	27.7
New Mexico	775	42.3	45.6	302	16.5	16.7	172	9.4	10.2	1 020	55.7	57.6
New York	6 906	36.2	34.8	1 488	7.8	7.6	2 421	12.7	12.2	4 982	26.1	25.6
North Carolina	3 514	42.8	45.4	725	8.8	8.9	1 332	16.2	17.3	3 438	41.9	42.6
North Dakota	308	48.4	39.5	62	9.7	9.3	50	7.9	6.2	238	37.4	33.8
Ohio	5 881	51.6	49.0	1 027	9.0	8.7	1 912	16.8	16.0	3 857	33.9	33.3
Oklahoma	1 922	55.4	53.0	363	10.5	10.2	567	16.3	15.6	1 700	49.0	48.3
Oregon	1 726	49.7	47.8	339	9.8	9.4	331	9.5	9.0	1 313	37.8	36.8
Pennsylvania	5 845	47.5	38.2	1 134	9.2	8.1	2 838	23.1	18.5	4 552	37.0	34.4
Rhode Island	511	48.2	40.8	96	9.1	8.4	118	11.1	9.3	293	27.7	25.1
South Carolina	1 731	42.6	44.4	461	11.3	11.2	632	15.6	16.3	1 960	48.3	48.7
South Dakota	362	47.7	41.5	77	10.2	10.1	103	13.6	11.1	382	50.4	47.0
Tennessee	2 941	51.2	52.0	613	10.7	10.4	624	10.9	11.1	2 709	47.1	47.1
Texas	7 739	36.2	46.1	2 213	10.4	11.8	2 269	10.6	13.4	7 920	37.1	39.5
Utah	525	23.0	33.6	138	6.1	8.1	178	7.8	11.4	643	28.2	32.3
Vermont	305	49.8	48.1	37	6.0	5.6	86	14.0	13.3	230	37.5	36.4
Virginia	2 744	38.1	42.8	531	7.4	7.5	1 128	15.7	17.6	2 432	33.8	35.1
Washington	2 634	43.9	47.8	571	9.5	9.8	279	4.7	4.9	2 072	34.6	35.1
West Virginia	1 273	70.7	58.9	212	11.8	10.3	391	21.7	18.2	834	46.3	44.6
Wisconsin	2 375	43.9	41.4	431	8.0	7.8	791	14.6	13.4	2 100	38.8	37.3
Wyoming	270	54.7	57.5	49	9.9	9.7	54	10.9	11.6	272	55.1	55.4
Puerto Rico [3]	1 226	31.9	36.9	283	7.4	7.9	804	20.9	23.2	1 290	33.6	34.8
Virgin Islands [3]	8	*	*	9	*	*	5	*	*	40	36.8	39.6
Guam [3]	22	13.9	34.3	10	*	*	15	*	*	45	28.4	36.5
American Samoa [3]	9	*	*	3	*	*	7	*	*	14	*	*
Northern Marianas [3]	5	*	*	2	*	*	7	*	*	6	*	*

[1]Death rates are affected by the population composition of the area. Age-adjusted death rates should be used for comparisons between areas.
[2]Excludes data for Puerto Rico, Virgin Islands, Guam, American Samoa and Northern Marianas.
[3]Age-adjusted death rates for Puerto Rico, Virgin Islands, Guam, American Samoa, and Northern Marianas are calculated using different age groups in the weighting procedure.
* = Figure too small to meet standards of reliability or precision.

The highest rate of homicides occurred in the District of Columbia, followed by New York, New Jersey, and Louisiana. The lowest rates were recorded in Maine, Iowa, and New Hampshire. However, there were four states with absolute numbers of homicides too low to calculate reliable rates, and of those, North and South Dakota and Vermont appear to be in the same range as the lowest rates.

Age-adjusted death rates were high in American Samoa and the Northern Marianas, they were close to the U.S. average in Puerto Rico, and below that in the Virgin Islands and Guam. In regard to specific causes of death, rates for diabetes were very high in American Samoa and Puerto Rico, as well as relatively high in the Virgin Islands. Also noticeable were a high death rate for nephritis in Puerto Rico and a high homicide rate in the Virgin Island.

Table II-30. Number of Deaths, Death Rates, and Age-Adjusted Death Rates for Major Causes of Death for each State and Territory, 2001—*Continued*

(Number, rate per 100,000 population.)

Area	Motor vehicle accidents [4]			Intentional self-harm (suicide) (*U03, X60-X84, Y87.0) [5]			Assault (homicide) (*U01-*U02, X85-Y09, Y87.1) [5]			Injury by firearms [6]		
	Number	Rate	Age-adjusted rate [1]	Number	Rate	Age-adjusted rate [1]	Number	Rate	Age-adjusted rate [1]	Number	Rate	Age-adjusted rate [1]
United States [2]	43 788	15.4	15.3	30 622	10.8	10.7	20 308	7.1	7.1	29 573	10.4	10.3
Alabama	1 037	23.2	23.1	512	11.5	11.3	425	9.5	9.5	737	16.5	16.4
Alaska	103	16.3	17.4	102	16.1	15.9	38	6.0	5.7	95	15.0	14.9
Arizona	1 020	19.2	19.4	767	14.5	14.8	494	9.3	9.3	842	15.9	16.0
Arkansas	661	24.5	24.4	382	14.2	14.2	183	6.8	6.9	412	15.3	15.3
California	3 932	11.4	11.5	2 831	8.2	8.5	2 223	6.4	6.2	3 200	9.2	9.3
Colorado	758	17.1	17.2	722	16.3	16.5	169	3.8	3.7	510	11.5	11.7
Connecticut	317	9.2	9.4	283	8.2	8.2	172	5.0	5.1	191	5.6	5.6
Delaware	122	15.3	15.2	108	13.6	13.3	33	4.1	4.1	76	9.5	9.4
District of Columbia	54	9.4	9.2	40	7.0	6.5	197	34.3	31.3	169	29.5	26.4
Florida	3 094	18.9	18.6	2 314	14.1	13.5	969	5.9	6.1	1 816	11.1	10.7
Georgia	1 628	19.4	19.7	935	11.1	11.4	681	8.1	7.8	1 124	13.4	13.5
Hawaii	128	10.4	10.3	136	11.1	10.9	34	2.8	2.8	47	3.8	3.8
Idaho	256	19.4	19.3	210	15.9	16.3	41	3.1	3.1	175	13.3	13.5
Illinois	1 585	12.7	12.7	1 139	9.1	9.1	1 087	8.7	8.6	1 289	10.3	10.2
Indiana	970	15.8	15.8	715	11.7	11.7	437	7.1	7.1	729	11.9	11.9
Iowa	447	15.2	14.9	304	10.4	10.1	61	2.1	2.1	191	6.5	6.3
Kansas	539	19.9	19.6	293	10.8	10.9	143	5.3	5.3	267	9.9	9.9
Kentucky	861	21.2	20.9	495	12.2	11.9	221	5.4	5.3	524	12.9	12.6
Louisiana	990	22.1	22.2	493	11.0	11.2	541	12.1	11.9	781	17.5	17.5
Maine	200	15.6	15.2	161	12.5	12.0	21	1.6	1.6	98	7.6	7.2
Maryland	703	13.1	13.3	454	8.4	8.4	555	10.3	10.4	607	11.3	11.4
Massachusetts	565	8.8	8.7	426	6.7	6.5	246	3.8	3.8	194	3.0	3.0
Michigan	1 431	14.3	14.3	1 051	10.5	10.5	689	6.9	6.9	1 077	10.8	10.8
Minnesota	628	12.6	12.6	480	9.6	9.6	127	2.5	2.5	324	6.5	6.5
Mississippi	825	28.8	28.9	328	11.5	11.7	326	11.4	11.4	502	17.6	17.6
Missouri	1 105	19.6	19.4	725	12.9	12.7	437	7.8	7.8	744	13.2	13.1
Montana	212	23.4	23.2	175	19.3	19.1	32	3.5	3.6	160	17.7	17.6
Nebraska	271	15.8	15.5	187	10.9	10.9	44	2.6	2.6	140	8.1	8.1
Nevada	331	15.8	16.2	387	18.4	18.8	175	8.3	8.4	345	16.4	16.7
New Hampshire	136	10.8	10.8	167	13.3	13.1	28	2.2	2.2	91	7.2	7.2
New Jersey	755	8.9	9.0	588	6.9	6.8	1 051	12.3	12.2	370	4.3	4.4
New Mexico	430	23.5	23.8	362	19.8	20.2	130	7.1	7.1	274	15.0	15.2
New York	1 678	8.8	8.7	1 253	6.6	6.5	2 818	14.8	14.5	1 068	5.6	5.6
North Carolina	1 647	20.1	20.1	997	12.1	12.1	599	7.3	7.1	1 093	13.3	13.2
North Dakota	115	18.1	17.5	79	12.4	12.7	9	*	*	47	7.4	7.5
Ohio	1 470	12.9	12.9	1 219	10.7	10.6	524	4.6	4.7	1 029	9.0	9.0
Oklahoma	728	21.0	20.8	515	14.8	14.9	210	6.1	6.0	489	14.1	14.0
Oregon	506	14.6	14.4	505	14.5	14.4	100	2.9	2.9	358	10.3	10.2
Pennsylvania	1 574	12.8	12.5	1 276	10.4	10.1	671	5.5	5.7	1 172	9.5	9.5
Rhode Island	97	9.2	8.8	88	8.3	8.0	37	3.5	3.5	46	4.3	4.3
South Carolina	999	24.6	24.3	467	11.5	11.3	350	8.6	8.5	564	13.9	13.7
South Dakota	184	24.3	24.0	105	13.8	13.9	18	*	*	54	7.1	7.1
Tennessee	1 305	22.7	22.5	711	12.4	12.2	454	7.9	7.8	845	14.7	14.5
Texas	3 964	18.5	18.9	2 225	10.4	10.9	1 415	6.6	6.4	2 350	11.0	11.2
Utah	299	13.1	13.6	321	14.1	15.4	77	3.4	3.4	229	10.0	10.9
Vermont	88	14.4	14.2	72	11.7	11.3	14	*	*	51	8.3	8.0
Virginia	961	13.4	13.5	797	11.1	11.0	509	7.1	6.9	771	10.7	10.6
Washington	745	12.4	12.5	712	11.9	11.8	195	3.3	3.2	510	8.5	8.5
West Virginia	373	20.7	20.6	286	15.9	15.0	68	3.8	3.7	251	13.9	13.1
Wisconsin	821	15.2	15.0	639	11.8	11.7	214	4.0	3.9	475	8.8	8.7
Wyoming	140	28.4	28.0	83	16.8	16.4	16	*	*	70	14.2	13.9
Puerto Rico [3]	499	13.0	13.0	289	7.5	7.8	760	19.8	19.0	738	19.2	18.4
Virgin Islands [3]	17	*	*	4	*	*	28	25.7	27.0	27	24.8	26.8
Guam [3]	20	12.6	14.2	23	14.5	13.8	8	*	*	8	*	*
American Samoa [3]	2	*	*	4	*	*	*	*	*	-	*	*
Northern Marianas [3]	2	*	*	12	*	*	4	*	*	1	*	*

[1]Death rates are affected by the population composition of the area. Age-adjusted death rates should be used for comparisons between areas.
[2]Excludes data for Puerto Rico, Virgin Islands, Guam, American Samoa and Northern Marianas.
[3]Age-adjusted death rates for Puerto Rico, Virgin Islands, Guam, American Samoa, and Northern Marianas are calculated using different age groups in the weighting procedure.
[4]ICD-10 codes for Motor vehicle accidents are V02-V04, V09.0, V09.2, V12-V14, V19.0-V19.2, V19.4-V19.6, V20-V79, V80.3-V80.5, V81.0-V81.1, V82.0-V82.1, V83-V86, V87.0-V87.8, V00.0 V00.0, V00.0, V00.0.
[5]Figures include September 11, 2001 related deaths for which death certificates were filed as of 10/24/02.
[6]ICD-10 codes for Injury by firearms are *U01.4, W32-W34, X72-X74, X93-X95, Y22-Y24, Y35.0.
* = Figure too small to meet standards of reliability or precision.
\- = Quantity zero.

Figure II-14. Age-Adjusted Death Rate, 2001

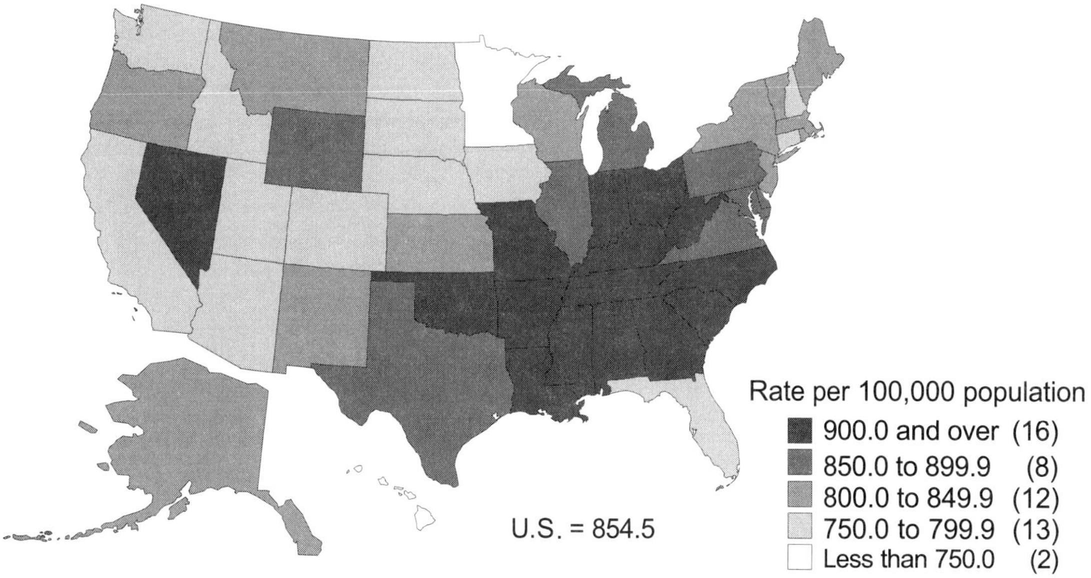

Figure II-15. Age-Adjusted Cancer Death Rate, 2001

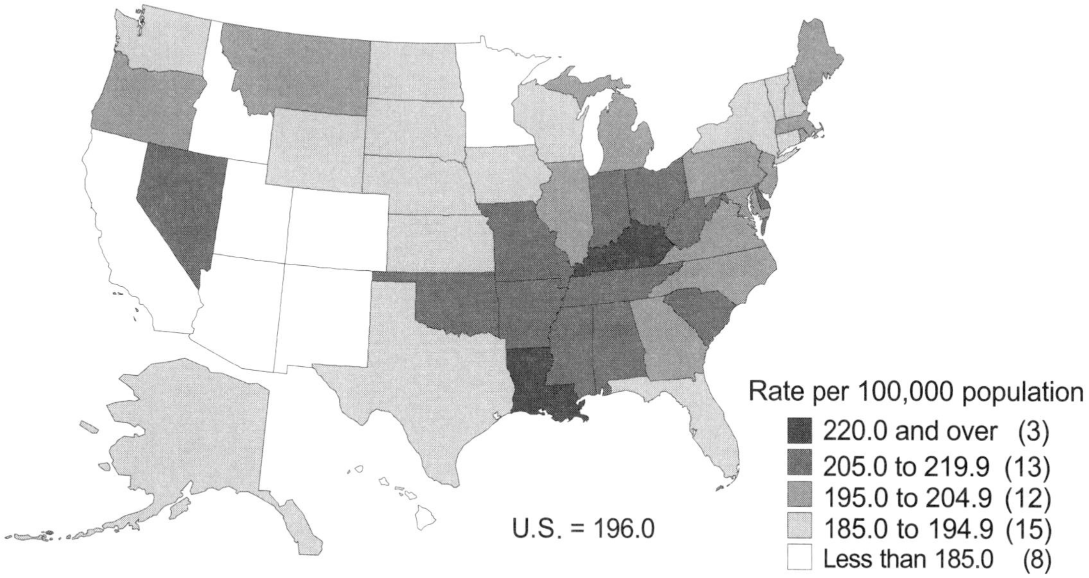

Table II-30A. Number of Deaths, Death Rates, and Age-Adjusted Death Rates for each State and Territory, 2001 and Preliminary 2002

(Number, rate per 100,000 population.)

Area	2002			2001		
	Number	Rate	Age-adjusted rate	Number	Rate	Age-adjusted rate
United States [1]	2 447 864	848.9	846.8	2 416 425	848.5	854.5
Alabama	46 055	1 026.5	997.8	45 316	1 014.0	992.9
Alaska	3 033	471.1	790.5	2 974	469.4	825.8
Arizona	42 807	784.5	795.5	41 058	773.7	787.4
Arkansas	28 517	1 052.3	964.6	27 759	1 030.1	948.1
California	234 660	668.2	758.4	234 044	676.4	775.1
Colorado	29 209	648.1	790.2	28 294	638.5	787.8
Connecticut	30 129	870.7	762.5	29 827	868.4	767.8
Delaware	6 858	849.4	837.7	7 112	892.8	891.9
District of Columbia	5 844	1 023.7	1 021.1	5 951	1 037.1	1 038.2
Florida	167 812	1 004.1	786.4	167 269	1 021.6	799.7
Georgia	65 463	764.7	949.3	64 485	767.2	956.5
Hawaii	8 802	707.0	660.5	8 394	684.1	652.6
Idaho	9 928	740.3	793.3	9 753	738.5	798.0
Illinois [2]	. . .	. . .	. . .	105 430	842.1	857.3
Indiana	55 352	898.7	898.8	55 198	900.9	908.3
Iowa	27 978	952.7	774.5	27 791	947.9	777.4
Kansas	25 015	921.1	843.3	24 647	912.1	839.8
Kentucky	41 218	1 007.1	1 006.6	39 861	979.7	987.7
Louisiana	42 055	938.2	1 002.2	41 757	934.1	1 005.9
Maine	12 695	980.7	846.6	12 421	967.0	843.2
Maryland	43 959	805.4	863.9	43 839	813.9	881.4
Massachusetts	56 949	886.0	792.2	56 754	886.6	801.4
Michigan	87 798	873.6	876.2	86 424	863.7	877.6
Minnesota	38 557	768.1	748.3	37 735	757.0	744.9
Mississippi	28 852	1 004.7	1 036.2	28 259	988.2	1 023.2
Missouri	55 952	986.4	916.9	54 982	975.3	909.5
Montana	8 514	936.2	850.5	8 265	912.9	840.3
Nebraska	15 737	910.1	814.7	15 174	882.2	793.5
Nevada	16 924	778.7	916.3	16 285	776.3	922.6
New Hampshire	9 847	772.3	781.2	9 815	779.4	798.5
New Jersey	73 971	861.1	808.3	74 710	877.8	831.9
New Mexico	14 365	774.4	816.2	14 129	771.7	825.4
New York	158 110	825.3	783.2	159 240	834.4	800.9
North Carolina	72 103	866.6	907.0	70 934	864.4	911.0
North Dakota	5 890	928.9	749.4	6 048	950.1	775.9
Ohio	109 547	959.1	906.4	108 027	948.5	905.7
Oklahoma	35 548	1 017.5	974.4	34 682	999.6	959.7
Oregon	31 128	883.9	834.4	30 158	868.2	825.6
Pennsylvania	130 235	1 055.8	862.1	129 729	1 054.4	869.9
Rhode Island	10 250	958.2	809.7	10 021	945.7	806.4
South Carolina	37 732	918.7	946.7	36 612	901.3	939.1
South Dakota	6 900	906.6	771.6	6 923	912.9	784.8
Tennessee	56 606	976.4	981.5	55 151	959.2	972.9
Texas	159 273	731.3	891.5	152 779	714.9	877.8
Utah	13 117	566.3	781.8	12 662	555.7	776.8
Vermont	5 075	823.1	774.9	5 201	848.5	811.2
Virginia	57 197	784.2	856.6	56 280	782.0	862.2
Washington	45 318	746.7	785.0	44 642	744.9	792.9
West Virginia	21 014	1 166.2	991.7	20 967	1 164.2	994.9
Wisconsin	46 990	863.6	800.0	46 628	862.5	806.5
Wyoming	4 171	836.4	863.6	4 029	816.0	851.7
Puerto Rico	27 927	723.7	791.7	29 082	757.7	841.7
Virgin Islands	623	572.6	743.0	605	556.3	751.6
Guam	639	396.8	738.7	663	418.7	789.9
American Samoa	294	509.4	1 471.4	239	415.4	1 229.5
Northern Marianas	163	220.3	1 058.4	148	205.9	1 015.4

[1] Excludes data for U.S. territories.
[2] Illinois data are not shown separately but are included in U.S. totals.
. . . = Not available.

Figure II-16. Age-Adjusted Death Rate for Heart Disease, 2001

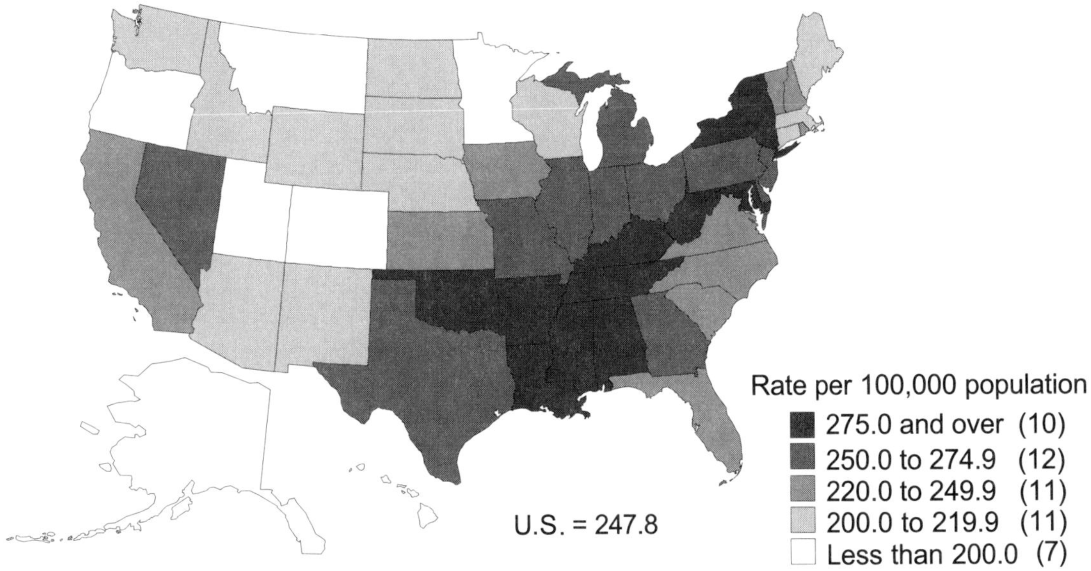

TABLE II-30A. NUMBER OF DEATHS, DEATH RATES, AND AGE-ADJUSTED DEATH RATES FOR EACH STATE AND TERRITORY, 2001 AND PRELIMINARY 2002

Most states had small changes in age-adjusted death rates between 2001 and 2002. For California, a 2.2 percent decline exceeded the average reduction of 0.9 percent for the national age-adjusted rate. For Texas there was an age-adjusted rate increase of 1.5 percent. New York and New Jersey residents had both been affected by the terrorist attacks of 2001; in 2002 the age-adjusted death rates of these states declined by 2.2 percent and 2.8 percent, respectively. Florida, the fourth most populous state, had a 1.7 percent reduction in its age-adjusted death rate.

TABLE II-31. INFANT, NEONATAL, AND POSTNEONATAL MORTALITY RATES, BY RACE AND SEX, SELECTED YEARS, 1940–2001

Infant death rates, by age, that were presented prior to this table use as denominators the estimated mid-year population under one year of age, based on the 2000 Census (for example, Tables 9 and 11). However, the most commonly used index for measuring the risk of dying during the first year of life is the "Infant mortality rate", which is calculated by dividing the number of infant deaths in a calendar year by the number of live births registered in that year. Because of the different denominators, infant mortality rates may differ from infant death rates.

Infant mortality rates declined from 47.0, per 1,000 live births in 1940, to 12.6 in 1980, and further to 6.8 in 2001. The rate in 2001 was the lowest ever recorded for the United States. For babies of White mothers the infant mortality rate was 5.7 per 1,000 in 2001; but for those of Black mothers it was 14.0, which represented a fall from 72.9 in 1940. The ratio of the Black to White infant mortality rates was 2.5 in 2001. The infant mortality rate tends to be 20 percent to 25 percent greater for male babies than for female infants. The infant mortality rate for Hispanic infants (not shown in tables) was 5.5 per 1,000 in 2001.

Neonatal deaths refer to infants who die before they are 28 days old. Post-neonatal comprises fatalities between 28 days of life and the end of the 12th month. Between 1980 and 2001, the White neonatal mortality rate fell by 49 percent, while the rate for Black mothers fell by 37 percent. The post-neonatal mortality rate remained constant between 1999 and 2001 at 2.3 per 1,000 for all races combined, 1.9 per 1,000 for Whites, and about 4.8 per 1,000 for Blacks.

In 2001, neonatal deaths accounted for two-thirds of all infant deaths within the first year, and post-neonatal deaths accounted for one third of this total.

Table II-31. Infant, Neonatal, and Postneonatal Mortality Rates, by Race and Sex, Selected Years, 1940–2001

(Death rates per 1,000 live births in specified group.)

Year	All races			White			All other					
							Total			Black		
	Both sexes	Male	Female	Both sexes	Male	Female	Both sexes	Male	Female	Both sexes	Male	Female
INFANT MORTALITY RATE												
Race of Mother [1]												
2001	6.8	7.5	6.1	5.7	6.2	5.1	11.3	12.4	10.2	14.0	15.5	12.5
2000	6.9	7.6	6.2	5.7	6.2	5.1	11.4	12.6	10.3	14.1	15.5	12.6
1999	7.1	7.7	6.4	5.8	6.4	5.2	11.9	12.9	10.9	14.6	15.9	13.2
1998	7.2	7.8	6.5	6.0	6.5	5.4	11.9	13.0	10.8	14.3	15.7	12.8
1997	7.2	8.0	6.5	6.0	6.7	5.4	11.8	12.8	10.7	14.2	15.5	12.8
1996	7.3	8.0	6.6	6.1	6.7	5.4	12.2	13.3	11.0	14.7	16.0	13.3
1995	7.6	8.3	6.8	6.3	7.0	5.6	12.6	13.5	11.6	15.1	16.3	13.9
1994	8.0	8.8	7.2	6.6	7.2	5.9	13.5	14.8	12.1	15.8	17.5	14.1
1993	8.4	9.3	7.4	6.8	7.6	6.0	14.1	15.6	12.5	16.5	18.3	14.7
1992	8.5	9.4	7.6	6.9	7.7	6.1	14.4	15.7	13.1	16.8	18.4	15.3
1991	8.9	10.0	7.8	7.3	8.3	6.3	15.1	16.5	13.6	17.6	19.4	15.7
1990	9.2	10.3	8.1	7.6	8.5	6.6	15.5	17.0	14.0	18.0	19.6	16.2
1989	9.8	10.8	8.8	8.1	9.0	7.1	16.3	17.6	15.0	18.6	20.0	17.2
1988	10.0	11.0	8.9	8.4	9.4	7.3	16.1	17.3	14.8	18.5	20.0	17.0
1987	10.1	11.2	8.9	8.5	9.5	7.5	16.5	18.1	14.8	18.8	20.6	16.8
1986	10.4	11.5	9.1	8.8	9.9	7.7	16.7	18.5	14.9	18.9	20.9	16.8
1985	10.6	11.9	9.3	9.2	10.4	7.9	16.8	18.3	15.3	19.0	20.8	17.2
1984	10.8	11.9	9.6	9.3	10.4	8.2	17.1	18.4	15.7	19.2	20.7	17.6
1983	11.2	12.3	10.0	9.6	10.7	8.5	17.8	19.4	16.1	20.0	22.0	18.0
1982	11.5	12.8	10.2	9.9	11.1	8.7	18.3	20.1	16.5	20.5	22.5	18.4
1981	11.9	13.1	10.7	10.3	11.5	9.1	18.8	20.4	17.2	20.8	22.5	19.0
1980	12.6	13.9	11.2	10.9	12.1	9.5	20.2	21.9	18.4	22.2	24.2	20.2
Race of Child [2]												
1980	12.6	13.9	11.2	11.0	12.3	9.6	19.1	20.7	17.5	21.4	23.3	19.4
1979	13.1	14.5	11.6	11.4	12.8	9.9	19.8	21.5	18.1	21.8	23.7	19.8
1978	13.8	15.3	12.2	12.0	13.4	10.6	21.1	23.1	18.9	23.1	25.4	20.8
1977	14.1	15.8	12.4	12.3	13.9	10.7	21.7	23.7	19.6	23.6	25.9	21.3
1976	15.2	16.8	13.6	13.3	14.8	11.7	23.5	25.5	21.4	25.5	27.8	23.2
1975	16.1	17.9	14.2	14.2	15.9	12.3	24.2	26.2	22.2	26.2	28.3	24.0
1970	20.0	22.4	17.5	17.8	20.0	15.4	30.9	34.2	27.5	32.6	36.2	29.0
1960	26.0	29.3	22.6	22.9	26.0	19.6	43.2	47.9	38.5	44.3	49.1	39.4
1950	29.2	32.8	25.5	26.8	30.2	23.1	44.5	48.9	39.9	43.9	48.3	39.4
1940	47.0	52.5	41.3	43.2	48.3	37.8	73.8	82.2	65.2	72.9	81.1	64.6
NEONATAL MORTALITY RATE												
Race of Mother [1]												
2001	4.5	5.0	4.1	3.8	4.2	3.4	7.4	8.1	6.6	9.2	10.2	8.2
2000	4.6	5.1	4.2	3.8	4.2	3.5	7.6	8.4	6.8	9.4	10.4	8.3
1999	4.7	5.1	4.3	3.9	4.2	3.6	7.9	8.6	7.3	9.8	10.7	8.8
1998	4.8	5.2	4.4	4.0	4.3	3.6	7.9	8.6	7.2	9.5	10.5	8.6
1997	4.8	5.2	4.3	4.0	4.4	3.6	7.7	8.4	7.1	9.4	10.1	8.6
1996	4.8	5.2	4.3	4.0	4.3	3.6	7.9	8.6	7.1	9.6	10.4	8.7
1995	4.9	5.4	4.4	4.1	4.5	3.6	8.1	8.7	7.5	9.8	10.6	9.0
1994	5.1	5.6	4.6	4.2	4.5	3.8	8.6	9.5	7.7	10.2	11.3	9.1
1993	5.3	5.7	4.8	4.3	4.6	3.9	9.0	9.9	8.1	10.7	11.8	9.6
1992	5.4	5.8	4.9	4.3	4.7	4.0	9.2	10.0	8.3	10.8	11.8	9.8
1991	5.6	6.2	5.0	4.5	5.0	4.0	9.5	10.5	8.5	11.2	12.6	9.9
1990	5.8	6.5	5.2	4.8	5.4	4.2	9.9	10.8	8.9	11.6	12.7	10.4
1989	6.2	6.8	5.6	5.1	5.7	4.6	10.3	11.1	9.5	11.9	12.8	11.0
1988	6.3	6.9	5.7	5.3	5.8	4.7	10.3	11.2	9.4	12.1	13.1	10.9
1987	6.5	7.1	5.8	5.4	6.0	4.8	10.7	11.7	9.6	12.3	13.5	11.1
1986	6.7	7.4	6.0	5.7	6.3	5.1	10.8	11.8	9.7	12.3	13.6	11.0
1985	7.0	7.8	6.1	6.0	6.8	5.2	11.0	12.0	10.0	12.6	13.8	11.0
1984	7.0	7.7	6.3	6.1	6.7	5.4	10.9	11.7	10.1	12.3	13.2	11.4
1983	7.3	8.0	6.5	6.3	7.0	5.6	11.4	12.5	10.3	12.9	14.2	11.6
1982	7.7	8.5	6.9	6.7	7.4	5.9	12.0	13.2	10.9	13.6	14.9	12.3
1981	8.0	8.8	7.2	7.0	7.7	6.2	12.5	13.5	11.5	14.0	15.2	12.8
1980	8.5	9.3	7.6	7.4	8.2	6.5	13.2	14.3	12.1	14.6	15.9	13.3
Race of Child [2]												
1980	8.5	9.3	7.6	7.5	8.3	6.6	12.5	13.5	11.5	14.1	15.3	12.8
1979	8.9	9.8	7.9	7.9	8.8	6.9	12.9	13.9	11.8	14.3	15.5	13.1
1978	9.5	10.5	8.4	8.4	9.3	7.4	14.0	15.5	12.4	15.5	17.2	13.7
1977	9.9	11.0	8.7	8.7	9.8	7.6	14.7	16.0	13.3	16.1	17.6	14.5
1976	10.9	12.0	9.7	9.7	10.7	8.5	16.3	17.7	14.9	17.9	19.5	16.3
1975	11.6	12.9	10.2	10.4	11.7	9.0	16.8	18.2	15.3	18.3	19.8	16.8
1970	15.1	17.0	13.1	13.8	15.5	11.9	21.4	23.9	18.9	22.8	25.4	20.1
1960	18.7	21.2	16.1	17.2	19.7	14.7	26.9	30.0	23.6	27.8	31.1	24.5
1950	20.5	23.3	17.5	19.4	22.2	16.4	27.5	30.8	24.2	27.8	31.1	24.4
1940	28.8	32.6	24.7	27.2	30.9	23.3	39.7	44.9	34.5	39.9	44.8	34.9

Note: Infant age: under 1 year; neonatal age: under 28 days; and postneonatal age: 28 days to 1 year

[1]Infant deaths based on race of child as stated on the death certificate; live births based on race of mother as stated on the birth certificate.
[2]Infant deaths based on race of child as stated on the death certificate; live births based on race of parents as stated on the birth certificate.

Table II-31. Infant, Neonatal, and Postneonatal Mortality Rates, by Race and Sex, Selected Years, 1940–2001—*Continued*

(Death rates per 1,000 live births in specified group.)

Year	All races			White			All other					
							Total			Black		
	Both sexes	Male	Female	Both sexes	Male	Female	Both sexes	Male	Female	Both sexes	Male	Female
POSTNEONATAL MORTALITY RATE												
Race of Mother [1]												
2001	2.3	2.5	2.1	1.9	2.1	1.7	4.0	4.4	3.5	4.8	5.3	4.3
2000	2.3	2.5	2.0	1.9	2.1	1.7	3.8	4.2	3.5	4.7	5.1	4.3
1999	2.3	2.6	2.0	1.9	2.2	1.6	4.0	4.3	3.6	4.8	5.2	4.4
1998	2.4	2.6	2.2	2.0	2.2	1.8	4.0	4.4	3.6	4.8	5.2	4.3
1997	2.5	2.8	2.1	2.0	2.3	1.8	4.0	4.5	3.6	4.8	5.3	4.2
1996	2.5	2.8	2.2	2.1	2.4	1.8	4.3	4.7	3.9	5.1	5.6	4.6
1995	2.7	3.0	2.4	2.2	2.5	1.9	4.5	4.8	4.1	5.3	5.7	4.8
1994	2.9	3.2	2.6	2.4	2.7	2.1	4.9	5.3	4.4	5.6	6.2	5.0
1993	3.1	3.5	2.6	2.5	2.9	2.1	5.1	5.7	4.4	5.8	6.6	5.1
1992	3.1	3.5	2.7	2.6	3.0	2.2	5.2	5.7	4.8	6.0	6.5	5.5
1991	3.4	3.8	2.9	2.8	3.2	2.3	5.6	6.0	5.1	6.3	6.8	5.8
1990	3.4	3.8	3.0	2.8	3.1	2.4	5.7	6.2	5.1	6.4	6.9	5.9
1989	3.6	4.0	3.1	2.9	3.4	2.5	6.0	6.5	5.5	6.7	7.2	6.2
1988	3.6	4.0	3.2	3.1	3.5	2.6	5.7	6.1	5.4	6.5	6.9	6.1
1987	3.6	4.1	3.2	3.1	3.5	2.6	5.8	6.3	5.2	6.4	7.1	5.8
1986	3.6	4.1	3.1	3.1	3.5	2.6	5.9	6.6	5.2	6.6	7.3	5.8
1985	3.7	4.2	3.2	3.2	3.6	2.7	5.8	6.3	5.3	6.4	7.0	5.8
1984	3.8	4.2	3.3	3.2	3.7	2.8	6.2	6.7	5.6	6.8	7.5	6.2
1983	3.9	4.3	3.4	3.3	3.7	2.9	6.4	7.0	5.8	7.0	7.8	6.3
1982	3.8	4.3	3.3	3.2	3.7	2.8	6.3	6.9	5.6	6.9	7.6	6.1
1981	3.9	4.3	3.5	3.4	3.8	2.9	6.3	6.8	5.8	6.8	7.4	6.3
1980	4.1	4.6	3.6	3.5	3.9	3.0	7.0	7.6	6.3	7.6	8.3	6.9
Race of Child [2]												
1980	4.1	4.6	3.6	3.5	4.0	3.0	6.6	7.2	6.0	7.3	7.9	6.6
1979	4.2	4.7	3.7	3.5	4.0	3.0	6.9	7.6	6.3	7.5	8.2	6.7
1978	4.3	4.7	3.9	3.6	4.0	3.2	7.0	7.6	6.5	7.6	8.2	7.0
1977	4.2	4.8	3.7	3.6	4.1	3.1	7.0	7.7	6.3	7.6	8.3	6.8
1976	4.3	4.8	3.8	3.6	4.1	3.2	7.2	7.8	6.5	7.6	8.4	6.9
1975	4.5	4.9	4.0	3.8	4.2	3.3	7.5	8.0	6.9	7.9	8.5	7.2
1970	4.9	5.4	4.4	4.0	4.4	3.5	9.5	10.3	8.6	9.9	10.8	8.9
1960	7.3	8.1	6.5	5.7	6.3	4.9	16.4	17.8	14.8	16.5	18.0	14.9
1950	8.7	9.4	8.0	7.4	8.0	6.7	16.9	18.1	15.7	16.1	17.2	15.0
1940	18.3	19.9	16.6	16.0	17.5	14.5	34.1	37.3	30.7	33.0	36.4	29.7

Note: Infant age: under 1 year; neonatal age: under 28 days; and postneonatal age: 28 days to 1 year.

[1]Infant deaths based on race of child as stated on the death certificate; live births based on race of mother as stated on the birth certificate.
[2]Infant deaths based on race of child as stated on the death certificate; live births based on race of parents as stated on the birth certificate.

Figure II-17. Infant Death Mortality Rates, by Race, 1940–2001

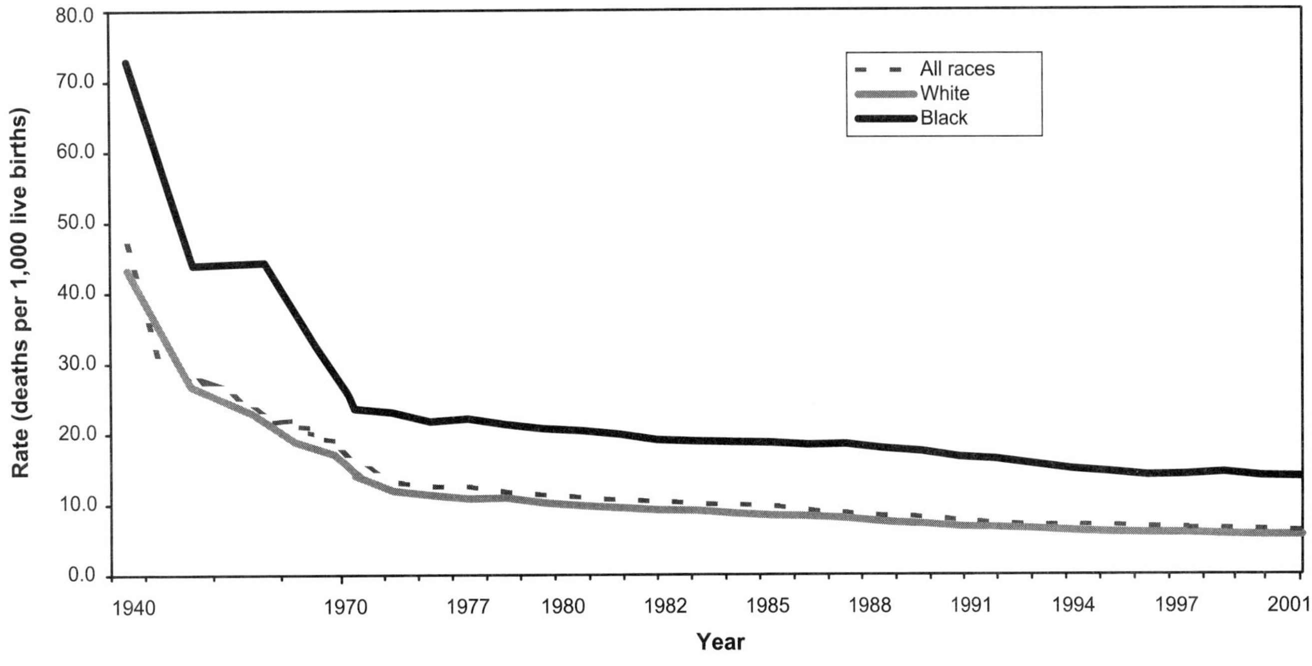

TABLE II-31A. NUMBER OF INFANT DEATHS AND INFANT MORTALITY RATES, BY AGE AND RACE AND HISPANIC ORIGIN, 2001 AND PRELIMINARY 2002

The increase in the infant mortality rate (per 1,000 live births) from 6.8 in 2001 to 7.0 in 2002 was previously discussed in the text for Table 1A. This was the first time in 44 years that the rate did not decline or stay unchanged. The increase in fatalities came entirely in neonatal births, that is, under 28 days of life, and it was concentrated in the first six days after birth. There also was a decline of roughly offsetting magnitude in late fetal deaths—after 27 weeks of gestation.

TABLE II-31B. NUMBER OF INFANT, NEONATAL, AND POSTNEONATAL DEATHS AND MORTALITY RATES, BY RACE AND SEX, 2001

This table summarizes infant deaths for 2001 by race and sex. Infants are defined as children under 1 year old, neonates as less than 28 days old, and post-neonates as 28 days until the end of 12 months old. The infant mortality rate is number of infant deaths in a calendar year divided by the number of live births registered in that year, with analogous definitions for neonatal and postneonatal mortality rates (see comments for Table 31 for further discussion).

In 2001 there were 27,568 infant deaths, 467 fewer than in 2000. Roughly two-thirds of infant deaths occur during the neonatal period. Rates for black infants are more than twice as high as for White infants. Male infants have higher rates than females, and the gap between males and females is more pronounced among Black infants.

Table II-31A. Number of Infant Deaths and Infant Mortality Rates, by Age and Race and Hispanic Origin, 2001 and Preliminary 2002

(Number, rate per 1,000 live births.)

Age and race and Hispanic origin	2002		2001	
	Number	Rate	Number	Rate
All Races [1]				
Under 1 year	27 977	7.0	27 568	6.8
Under 28 days	18 749	4.7	18 265	4.5
28 days–11 months	9 229	2.3	9 303	2.3
White [2]				
Under 1 year	18 390	5.8	17 955	5.7
Under 28 days	12 419	3.9	12 013	3.8
28 days–11 months	5 971	1.9	5 942	1.9
Non-Hispanic White				
Under 1 year	13 492	5.9	13 311	5.7
Under 28 days	9 064	3.9	8 888	3.8
28 days–11 months	4 428	1.9	4 423	1.9
Black [2]				
Under 1 year	8 446	14.3	8 498	14.0
Under 28 days	5 580	9.4	5 585	9.2
28 days–11 months	2 866	4.9	2 913	4.8
Hispanic [3]				
Under 1 year	4 928	5.6	4 680	5.5
Under 28 days	3 345	3.8	3 112	3.7
28 days–11 months	1 583	1.8	1 568	1.8

[1]Includes races other than White or Black.
[2]Race and Hispanic origin are reported separately on both the birth and death certificates. Data for persons of Hispanic origin are included in the data for each race group.
[3]Includes all persons of Hispanic origin.

Table II-31B. Number of Infant, Neonatal, and Postneonatal Deaths and Mortality Rates, by Race and Sex, 2001

(Number, rate per 1,000 live births.)

Race and sex	Infant		Neonatal		Postneonatal	
	Number	Rate	Number	Rate	Number	Rate
All Races [1]	27 568	6.8	18 265	4.5	9 303	2.3
Male	15 477	7.5	10 237	5.0	5 240	2.5
Female	12 091	6.1	8 028	4.1	4 063	2.1
White	17 955	5.7	12 013	3.8	5 942	1.9
Male	10 099	6.2	6 750	4.2	3 349	2.1
Female	7 856	5.1	5 263	3.4	2 593	1.7
Black	8 498	14.0	5 585	9.2	2 913	4.8
Male	4 764	15.5	3 125	10.2	1 639	5.3
Female	3 734	12.5	2 460	8.2	1 274	4.3

Note: Infant age: under 1 year; neonatal age: under 28 days; and postneonatal age: 28 days to 1 year.

[1]Includes races other than White and Black.

TABLE II-32. NUMBER OF INFANT DEATHS AND INFANT MORTALITY RATES FOR 130 SELECTED CAUSES, BY RACE, 2001

This table presents numbers of deaths and mortality rates by race for 130 selected causes. Specific causes are indented under more general summary groupings. Rates are not considered reliable or precise enough when less than 20 deaths occur and are therefore not presented. See Table 32A for a summarization of the 10 leading causes of death, which accounted for 67.2 percent of all infant deaths.

Table II-32. Number of Infant Deaths and Infant Mortality Rates for 130 Selected Causes, by Race, 2001

(Number, rates are infant (under 1 year) deaths per 100,000 live births in specified group.)

Cause of death [1]	Number			Rate		
	All races [2]	White	Black	All races [2]	White	Black
All causes	27 568	17 955	8 498	684.8	565.0	1 401.9
Certain infectious and parasitic diseases (A00-B99)	563	334	203	14.0	10.5	33.5
Certain intestinal infectious diseases (A00-A08)	12	9	2	*	*	*
Diarrhea and gastroenteritis of infectious origin (A09)	2	-	2	*	*	*
Tuberculosis (A16-A19)	2	1	-	*	*	*
Tetanus (A33, A35)	-	-	-	*	*	*
Diphtheria (A36)	-	-	-	*	*	*
Whooping cough (A37)	16	15	1	*	*	*
Meningococcal infection (A39)	16	14	2	*	*	*
Septicemia (A40-A41)	312	171	126	7.7	5.4	20.8
Congenital syphilis (A50)	1	1	-	*	*	*
Gonococcal infection (A54)	-	-	-	*	*	*
Viral diseases (A80-B34)	121	72	43	3.0	2.3	7.1
Acute poliomyelitis (A80)	-	-	-	*	*	*
Varicella (chickenpox) (B01)	1	-	-	*	*	*
Measles (B05)	-	-	-	*	*	*
Human immunodeficiency virus (HIV) disease (B20-B24)	10	3	7	*	*	*
Mumps (B26)	-	-	-	*	*	*
Other and unspecified viral diseases (A81-B00, B02-B04, B06-B19, B25, B27-B34)	110	69	36	2.7	2.2	5.9
Candidiasis (B37)	21	13	7	0.5	*	*
Malaria (B50-B54)	-	-	-	*	*	*
Pneumocystosis (B59)	2	2	-	*	*	*
All other and unspecified infectious and parasitic diseases (A20-A32, A38, A42-A49, A51-A53, A55-A79, B35-B36, B38-B49, B55-B58, B60-B99)	58	36	20	1.4	1.1	3.3
Neoplasms (C00-D48)	117	91	17	2.9	2.9	*
Malignant neoplasms (C00-C97)	66	54	6	1.6	1.7	*
Hodgkin's disease and non-Hodgkin s lymphomas (C81-C85)	1	1	-	*	*	*
Leukemia (C91-C95)	25	20	4	0.6	0.6	*
Other and unspecified malignant neoplasms (C00-C80, C88, C90, C96-C97)	40	33	2	1.0	1.0	*
In situ neoplasms, benign neoplasms and neoplasms of uncertain or unknown behavior (D00-D48)	51	37	11	1.3	1.2	*
Diseases of the blood and blood-forming organs and certain disorders involving the immune mechanism (D50-D89)	106	72	23	2.6	2.3	3.8
Anemias (D50-D64)	16	6	6	*	*	*
Hemorrhagic conditions and other diseases of blood and blood-forming organs (D65-D76)	59	42	11	1.5	1.3	*
Certain disorders involving the immune mechanism (D80-D89)	31	24	6	0.8	0.8	*
Endocrine, nutritional and metabolic diseases (E00-E88)	292	213	67	7.3	6.7	11.1
Short stature, not elsewhere classified (E34.3)	26	21	5	0.6	0.7	*
Nutritional deficiencies (E40-E64)	4	3	1	*	*	*
Cystic fibrosis (E84)	6	6	-	*	*	*
Volume depletion, disorders of fluid, electrolyte and acid-base balance (E86-E87)	90	54	33	2.2	1.7	5.4
All other endocrine, nutritional and metabolic diseases (E00-E32, E34.0-E34.2, E34.4-E34.9, E65-E83, E85, E88)	166	129	28	4.1	4.1	4.6
Diseases of the nervous system (G00-G98)	360	249	94	8.9	7.8	15.5
Meningitis (G00, G03)	78	47	29	1.9	1.5	4.8
Infantile spinal muscular atrophy, type I (Werdnig-Hoffman) (G12.0)	21	18	2	0.5	*	*
Infantile cerebral palsy (G80)	19	7	9	*	*	*
Anoxic brain damage, not elsewhere classified (G93.1)	26	15	9	0.6	*	*
Other diseases of nervous system (G04, G06-G11, G12.1-G12.9, G20-G72, G81-G92, G93.0, G93.2-G93.9, G95-G98)	216	162	45	5.4	5.1	7.4
Diseases of the ear and mastoid process (H60-H93)	3	2	1	*	*	*
Diseases of the circulatory system (I00-I99)	622	400	191	15.4	12.6	31.5
Pulmonary heart disease and diseases of pulmonary circulation (I26-I28)	170	114	44	4.2	3.6	7.3
Pericarditis, endocarditis and myocarditis (I30, I33, I40)	21	9	10	0.5	*	*
Cardiomyopathy (I42)	137	88	44	3.4	2.8	7.3
Cardiac arrest (I46)	24	18	4	0.6	*	*
Cerebrovascular diseases (I60-I69)	108	60	43	2.7	1.9	7.1
All other diseases of circulatory system (I00-I25, I31, I34-I38, I44-I45, I47-I51, I70-I99)	162	111	46	4.0	3.5	7.6
Diseases of the respiratory system (J00-J98)	696	386	258	17.3	12.1	42.6
Acute upper respiratory infections (J00-J06)	9	5	3	*	*	*
Influenza and pneumonia (J10-J18)	299	168	106	7.4	5.3	17.5
Influenza (J10-J11)	7	7	-	*	*	*
Pneumonia (J12-J18)	292	161	106	7.3	5.1	17.5
Acute bronchitis and acute bronchiolitis (J20-J21)	49	30	15	1.2	0.9	*
Bronchitis, chronic and unspecified (J40-J42)	21	11	10	0.5	*	*
Asthma (J45-J46)	10	5	4	*	*	*
Pneumonitis due to solids and liquids (J69)	12	9	3	*	*	*
Other and unspecified diseases of respiratory system (J22, J30-J39, J43-J44, J47-J68, J70-J98)	296	158	117	7.4	5.0	19.3
Diseases of the digestive system (K00-K92)	478	274	186	11.9	8.6	30.7
Gastritis, duodenitis, and noninfective enteritis and colitis (K29, K50-K55)	246	126	114	6.1	4.0	18.8
Hernia of abdominal cavity and intestinal obstruction without hernia (K40-K46, K56)	62	44	17	1.5	1.4	*
All other and unspecified diseases of digestive system (K00-K28, K30-K38, K57-K92)	170	104	55	4.2	3.3	9.1
Diseases of the genitourinary system (N00-N95)	157	94	60	3.9	3.0	9.9
Renal failure and other disorders of kidney (N17-N19, N25, N27)	125	70	53	3.1	2.2	8.7
Other and unspecified diseases of genitourinary system (N00-N15, N20-N23, N26, N28-N95)	32	24	7	0.8	0.8	*

[1]Based on codes from the "Tenth Revision, International Classification of Diseases, 1992." Codes *U01-*U03 have been added as of 2001 for deaths due to terrorism.
[2]Includes races other than White and Black.
- = Quantity zero.
* = Figure too small to meet standards of reliability or precision.

Table II-32. Number of Infant Deaths and Infant Mortality Rates for 130 Selected Causes, by Race, 2001—*Continued*

(Number, rates are infant (under 1 year) deaths per 100,000 live births in specified group.)

Cause of death [1]	Number			Rate		
	All races [2]	White	Black	All races [2]	White	Black
Certain conditions originating in the perinatal period (P00-P96)	13 734	8 444	4 808	341.1	265.7	793.2
Newborn affected by maternal factors and by complications of pregnancy, labor and delivery (P00-P04)	2 832	1 806	930	70.3	56.8	153.4
Newborn affected by maternal hypertensive disorders (P00.0)	60	31	28	1.5	1.0	4.6
Newborn affected by other maternal conditions which may be unrelated to present pregnancy (P00.1-P00.9)	70	41	24	1.7	1.3	4.0
Newborn affected by maternal complications of pregnancy (P01)	1 499	915	537	37.2	28.8	88.6
Newborn affected by incompetent cervix (P01.0)	403	223	166	10.0	7.0	27.4
Newborn affected by premature rupture of membranes (P01.1)	663	393	244	16.5	12.4	40.3
Newborn affected by multiple pregnancy (P01.5)	270	183	85	6.7	5.8	14.0
Newborn affected by other maternal complications of pregnancy (P01.2-P01.4, P01.6-P01.9)	163	116	42	4.0	3.7	6.9
Newborn affected by complications of placenta, cord and membranes (P02)	1 018	690	294	25.3	21.7	48.5
Newborn affected by complications involving placenta (P02.0-P02.3)	520	397	106	12.9	12.5	17.5
Newborn affected by complications involving cord (P02.4-P02.6)	51	36	12	1.3	1.1	*
Newborn affected by chorioamnionitis (P02.7)	447	257	176	11.1	8.1	29.0
Newborn affected by other and unspecified abnormalities of membranes (P02.8-P02.9)	-	-	-	*	*	*
Newborn affected by other complications of labor and delivery (P03)	148	110	29	3.7	3.5	4.8
Newborn affected by noxious influences transmitted via placenta or breast milk (P04)	37	19	18	0.9	*	*
Disorders related to length of gestation and fetal malnutrition (P05-P08)	4 473	2 492	1 854	111.1	78.4	305.9
Slow fetal growth and fetal malnutrition (P05)	63	41	19	1.6	1.3	*
Disorders related to short gestation and low birth weight, not elsewhere classified (P07)	4 410	2 451	1 835	109.5	77.1	302.7
Extremely low birth weight or extreme immaturity (P07.0, P07.2)	3 316	1 813	1 411	82.4	57.1	232.8
Other low birth weight or preterm (P07.1, P07.3)	1 094	638	424	27.2	20.1	69.9
Disorders related to long gestation and high birth weight (P08)	-	-	-	*	*	*
Birth trauma (P10-P15)	145	95	45	3.6	3.0	7.4
Intrauterine hypoxia and birth asphyxia (P20-P21)	534	385	129	13.3	12.1	21.3
Intrauterine hypoxia (P20)	120	82	33	3.0	2.6	5.4
Birth asphyxia (P21)	414	303	96	10.3	9.5	15.8
Respiratory distress of newborn (P22)	1 011	630	353	25.1	19.8	58.2
Other respiratory conditions originating in the perinatal period (P23-P28)	1 503	941	511	37.3	29.6	84.3
Congenital pneumonia (P23)	72	39	32	1.8	1.2	5.3
Neonatal aspiration syndromes (P24)	71	46	23	1.8	1.4	3.8
Interstitial emphysema and related conditions originating in the perinatal period (P25)	204	142	56	5.1	4.5	9.2
Pulmonary hemorrhage originating in the perinatal period (P26)	279	157	109	6.9	4.9	18.0
Chronic respiratory disease originating in the perinatal period (P27)	310	172	130	7.7	5.4	21.4
Atelectasis (P28.0-P28.1)	501	341	143	12.4	10.7	23.6
All other respiratory conditions originating in the perinatal period (P28.2-P28.9)	66	44	18	1.6	1.4	*
Infections specific to the perinatal period (P35-P39)	869	552	280	21.6	17.4	46.2
Bacterial sepsis of newborn (P36)	696	444	226	17.3	14.0	37.3
Omphalitis of newborn with or without mild hemorrhage (P38)	3	3	*	*	*	*
All other infections specific to the perinatal period (P35, P37, P39)	170	105	54	4.2	3.3	8.9
Hemorrhagic and hematological disorders of newborn (P50-P61)	600	435	143	14.9	13.7	23.6
Neonatal hemorrhage (P50-P52, P54)	453	332	103	11.3	10.4	17.0
Hemorrhagic disease of newborn (P53)	-	-	-	*	*	*
Hemolytic disease of newborn due to isoimmunization and other perinatal jaundice (P55-P59)	14	11	3	*	*	*
Hematological disorders (P60-P61)	133	92	37	3.3	2.9	6.1
Syndrome of infant of a diabetic mother and neonatal diabetes mellitus (P70.0-P70.2)	7	2	4	*	*	*
Necrotizing enterocolitis of newborn (P77)	380	217	140	9.4	6.8	23.1
Hydrops fetalis not due to hemolytic disease (P83.2)	166	141	16	4.1	4.4	*
Other perinatal conditions (P29, P70.3-P70.9, P71-P76, P78-P81, P83.0-P83.1, P83.3-P83.9, P90-P96)	1 214	748	403	30.2	23.5	66.5
Congenital malformations, deformations and chromosomal abnormalities (Q00-Q99)	5 513	4 247	1 032	136.9	133.7	170.3
Anencephaly and similar malformations (Q00)	263	218	34	6.5	6.9	5.6
Congenital hydrocephalus (Q03)	105	83	21	2.6	2.6	3.5
Spina bifida (Q05)	19	15	3	*	*	*
Other congenital malformations of nervous system (Q01-Q02, Q04, Q06-Q07)	286	220	52	7.1	6.9	8.6
Congenital malformations of heart (Q20-Q24)	1 556	1 175	311	38.6	37.0	51.3
Other congenital malformations of circulatory system (Q25-Q28)	192	144	39	4.8	4.5	6.4
Congenital malformations of respiratory system (Q30-Q34)	594	446	127	14.8	14.0	21.0
Congenital malformations of digestive system (Q35-Q45)	106	81	18	2.6	2.5	*
Congenital malformations of genitourinary system (Q50-Q64)	356	281	63	8.8	8.8	10.4
Congenital malformations and deformations of musculoskeletal system, limbs and integument (Q65-Q85)	477	362	94	11.8	11.4	15.5
Down's syndrome (Q90)	119	96	15	3.0	3.0	*
Edward's syndrome (Q91.0-Q91.3)	446	356	75	11.1	11.2	12.4
Patau's syndrome (Q91.4-Q91.7)	276	223	44	6.9	7.0	7.3
Other congenital malformations and deformations (Q10-Q18, Q86-Q89)	526	407	98	13.1	12.8	16.2
Other chromosomal abnormalities, not elsewhere classified (Q92-Q99)	192	140	38	4.8	4.4	6.3
Symptoms, signs and abnormal clinical and laboratory findings, not elsewhere classified (R00-R99)	3 491	2 231	1 104	86.7	70.2	182.1
Sudden infant death syndrome (R95)	2 234	1 406	745	55.5	44.2	122.9
Other symptoms, signs and abnormal clinical and laboratory findings, not elsewhere classified (R00-R53, R55-R94, R96-R99)	1 257	825	359	31.2	26.0	59.2
All other diseases (Residual)	19	13	5	*	*	*

[1]Based on codes from the "Tenth Revision, International Classification of Diseases, 1992." Codes *U01-*U03 have been added as of 2001 for deaths due to terrorism.
[2]Includes races other than White and Black.
- = Quantity zero.
* = Figure too small to meet standards of reliability or precision.

Table II-32. Number of Infant Deaths and Infant Mortality Rates for 130 Selected Causes, by Race, 2001—*Continued*

(Number, rates are infant (under 1 year) deaths per 100,000 live births in specified group.)

Cause of death[1]	Number			Rate		
	All races[2]	White	Black	All races[2]	White	Black
External causes of mortality (*U01, V01-Y84)[3]	1 417	905	449	35.2	28.5	74.1
Accidents (unintentional injuries) (V01-X59)	976	647	281	24.2	20.4	46.4
Transport accidents (V01-V99)	151	115	19	3.8	3.6	*
Motor vehicle accidents (V02-V04, V09.0, V09.2, V12-V14, V19.0-V19.2, V19.4-V19.6, V20-V79, V80.3-V80.5, V81.0-V81.1, V82.0-V82.1, V83-V86, V87.0-V87.8, V88.0-V88.8, V89.0, V89.2)	144	109	19	3.6	3.4	*
Other and unspecified transport accidents (V01, V05-V06, V09.1, V09.3-V09.9, V10-V11, V15-V18, V19.3, V19.8-V19.9, V80.0-V80.2, V80.6-V80.9, V81.2-V81.9, V82.2-V82.9, V87.9, V88.9, V89.1, V89.3, V89.9, V90-V99)	7	6	-	*	*	*
Falls (W00-W19)	23	15	6	0.6	*	*
Accidental discharge of firearms (W32-W34)	-	-	-	*	*	*
Accidental drowning and submersion (W65-W74)	68	52	12	1.7	1.6	*
Accidental suffocation and strangulation in bed (W75)	390	250	128	9.7	7.9	21.1
Other accidental suffocation and strangulation (W76-W77, W81-W84)	162	91	63	4.0	2.9	10.4
Accidental inhalation and ingestion of food or other objects causing obstruction of respiratory tract (W78-W80)	62	41	20	1.5	1.3	3.3
Accidents caused by exposure to smoke, fire and flames (X00-X09)	50	34	15	1.2	1.1	*
Accidental poisoning and exposure to noxious substances (X40-X49)	15	9	6	*	*	*
Other and unspecified accidents (W20-W31, W35-W64, W85-W99, X10-X39, X50-X59)	55	40	12	1.4	1.3	*
Assault (homicide) (*U01, X85-Y09)[3]	332	196	123	8.2	6.2	20.3
Assault (homicide) by hanging, strangulation and suffocation (X91)	40	23	16	1.0	0.7	*
Assault (homicide) by discharge of firearms (*U01.4, X93-X95)	11	8	2	*	*	*
Neglect, abandonment and other maltreatment syndromes (Y06-Y07)	116	69	43	2.9	2.2	7.1
Assault (homicide) by other and unspecified means (*U01.0-*U01.3, *U01.5-*U01.9, X85-X90, X92, X96-X99, Y00-Y05, Y08-Y09)[3]	165	96	62	4.1	3.0	10.2
Complications of medical and surgical care (Y40-Y84)	26	17	9	0.6	*	*
Other external causes (X60-X84, Y10-Y36)	83	45	36	2.1	1.4	5.9

[1]Based on codes from the "Tenth Revision, International Classification of Diseases, 1992." Codes *U01-*U03 have been added as of 2001 for deaths due to terrorism.
[2]Includes races other than White and Black.
[3]Figures include September 11, 2001 related deaths for which death certificates were filed as of 10/24/02.
- = Quantity zero.
* = Figure too small to meet standards of reliability or precision.

TABLE II-32A. NUMBER OF INFANT DEATHS AND INFANT MORTALITY RATES FOR 130 SELECTED CAUSES, 2001 AND PRELIMINARY 2002

This table parses the increase of 406 infant deaths in 2002, from 27,568 to 27,974, into a multitude of causes. To simplify, consider the 10 leading causes of infant death, which re shown in the following tables, Table 32B for 2001 and Table 32C for 2002. Ranking one and two as causes of death are "congenital malformations, deformations and chromosomal abnormalities;" and "disorders related to short gestation and low birth weight, not elsewhere classified." Rank four is "newborn affected by maternal complications of pregnancy." Historically, the majority of these causes of deaths have been among infants born with low birth weight. Turning back to Table 32A, the 2002 increase in these three causes is indicated to be 554. It thus appears that these three causes may account for the 2002 increase in infant deaths. (Further information is in "Supplemental Analyses of Recent Trends in Infant Mortality" by K. D. Kochanek and J. A. Martin of the National Center for Health Statistics of the CDC.)

Table II-32A. Number of Infant Deaths and Infant Mortality Rates for 130 Selected Causes, 2001 and Preliminary 2002

(Number, rate per 100,000 live births.)

Cause of death[1]	2002		2001	
	Number	Rate	Number	Rate
All causes	27 974	696.0	27 568	684.8
Certain infectious and parasitic diseases (A00-B99)	573	14.3	563	14.0
Certain intestinal infectious diseases (A00-A08)	10	*	12	*
Diarrhea and gastroenteritis of infectious origin (A09)	3	*	2	*
Tuberculosis (A16-A19)	1	*	2	*
Tetanus (A33, A35)	-	*	-	*
Diphtheria (A36)	-	*	-	*
Whooping cough (A37)	17	*	16	*
Meningococcal infection (A39)	16	*	16	*
Septicemia (A40-A41)	295	7.3	312	7.7
Congenital syphilis (A50)	1	*	1	*
Gonococcal infection (A54)	1	*	-	*
Viral diseases (A80-B34)	159	4.0	121	3.0
Acute poliomyelitis (A80)	-	*	-	*
Varicella (chickenpox) (B01)	1	*	1	*
Measles (B05)	-	*	-	*
Human immunodeficiency virus (HIV) disease (B20-B24)	8	*	10	*
Mumps (B26)	-	*	-	*
Other and unspecified viral diseases (A81-B00, B02-B04, B06-B19, B25, B27-B34)	149	3.7	110	2.7
Candidiasis (B37)	21	0.5	21	0.5
Malaria (B50 B54)	-	*	-	*
Pneumocystosis (B59)	2	*	2	*
All other and unspecified infectious and parasitic diseases (A20-A32, A38, A42-A49, A51-A53, A55-A79, B35-B36, B38-B49, B55-B58, B60-B99)	47	1.2	58	1.4
Neoplasms (C00-D48)	150	3.7	117	2.9
Malignant neoplasms (C00-C97)	81	2.0	66	1.6
Hodgkin's disease and non-Hodgkin's lymphomas (C81-C85)	2	*	1	*
Leukemia (C91-C95)	20	0.5	25	0.6
Other and unspecified malignant neoplasms (C00-C80, C88-C90, C96-C97)	58	1.4	40	1.0
In situ neoplasms, benign neoplasms and neoplasms of uncertain or unknown behavior (D00-D48)	69	1.7	51	1.3
Diseases of the blood and blood-forming organs and certain disorders involving the immune mechanism (D50-D89)	86	2.1	106	2.6
Anemias (D50-D64)	18	*	16	*
Hemorrhagic conditions and other diseases of blood and blood-forming organs (D65-D76)	45	1.1	59	1.5
Certain disorders involving the immune mechanism (D80-D89)	22	0.5	31	0.8
Endocrine, nutritional and metabolic diseases (E00-E88)	255	6.3	292	7.3
Short stature, not elsewhere classified (E34.3)	26	0.6	26	0.6
Nutritional deficiencies (E40-E64)	4	*	4	*
Cystic fibrosis (E84)	9	*	6	*
Volume depletion, disorders of fluid, electrolyte and acid-base balance (E86-E87)	74	1.8	90	2.2
All other endocrine, nutritional and metabolic diseases (E00-E32, E34.0 E34.2, E34.4 E34.9, E65-E83, E85, E88)	142	3.5	166	4.1
Diseases of the nervous system (G00-G98)	376	9.4	360	8.9
Meningitis (G00, G03)	71	1.8	78	1.9
Infantile spinal muscular atrophy, type I (Werdnig-Hoffman) (G12.0)	24	0.6	21	0.5
Infantile cerebral palsy (G80)	9	*	19	*
Anoxic brain damage, not elsewhere classified (G93.1)	41	1.0	26	0.6
Other diseases of nervous system (G04, G06-G11, G12.1-G12.9, G20-G72, G81-G92, G93.0, G93.2-G93.9, G95-G98)	231	5.7	216	5.4
Diseases of the ear and mastoid process (H60-H93)	4	*	3	*
Diseases of the circulatory system (I00-I99)	649	16.1	622	15.4
Pulmonary heart disease and diseases of pulmonary circulation (I26-I28)	177	4.4	170	4.2
Pericarditis, endocarditis and myocarditis (I30, I33, I40)	19	*	21	0.5
Cardiomyopathy (I42)	115	2.9	137	3.4
Cardiac arrest (I46)	36	0.9	24	0.6
Cerebrovascular diseases (I60-I69)	121	3.0	108	2.7
All other diseases of circulatory system (I00-I25, I31, I34-I38, I44-I45, I47-I51, I70-I99)	181	4.5	162	4.0
Diseases of the respiratory system (J00-J98)	586	14.6	696	17.3
Acute upper respiratory infections (J00-J06)	8	*	9	*
Influenza and pneumonia (J10-J18)	232	5.8	299	7.4
Influenza (J10-J11)	7	*	7	*
Pneumonia (J12-J18)	226	5.6	292	7.3
Acute bronchitis and acute bronchiolitis (J20-J21)	45	1.1	49	1.2
Bronchitis, chronic and unspecified (J40-J42)	20	0.5	21	0.5
Asthma (J45-J46)	2	*	10	*
Pneumonitis due to solids and liquids (J69)	2	*	12	*
Other and unspecified diseases of respiratory system (J22, J30-J39, J43-J44, J47-J68, J70-J98)	276	6.9	296	7.4
Diseases of the digestive system (K00-K92)	482	12.0	478	11.9
Gastritis, duodenitis, and noninfective enteritis and colitis (K29, K50-K55)	251	6.2	246	6.1
Hernia of abdominal cavity and intestinal obstruction without hernia (K40-K46, K56)	46	1.1	62	1.5
All other and unspecified diseases of digestive system (K00-K28, K30-K38, K57-K92)	185	4.6	170	4.2

[1]Based on codes from the "Tenth Revision, International Classification of Diseases, 1992." Codes *U01-*U03 have been added as of 2001 for deaths due to terrorism.
- = Quantity zero.
* = Figure too small to meet standards of reliability or precision.

Table II-32A. Number of Infant Deaths and Infant Mortality Rates for 130 Selected Causes, 2001 and Preliminary 2002—*Continued*

(Number, rate per 100,000 live births.)

Cause of death[1]	2002		2001	
	Number	Rate	Number	Rate
Diseases of the genitourinary system (N00-N98)	183	4.6	157	3.9
Renal failure and other disorders of kidney (N17-N19, N25, N27)	162	4.0	125	3.1
Other and unspecified diseases of genitourinary system (N00-N15, N20-N23, N26, N28-N98)	21	0.5	32	0.8
Certain conditions originating in the perinatal period (P00-P96)	13 987	348.0	13 734	341.1
Newborn affected by maternal factors and by complications of pregnancy, labor and delivery (P00-P04)	3 058	76.1	2 832	70.3
Newborn affected by maternal hypertensive disorders (P00.0)	68	1.7	60	1.5
Newborn affected by other maternal conditions which may be unrelated to present pregnancy (P00.1 P00.9)	68	1.7	70	1.7
Newborn affected by maternal complications of pregnancy (P01)	1 723	42.9	1 499	37.2
Newborn affected by incompetent cervix (P01.0)	486	12.1	403	10.0
Newborn affected by premature rupture of membranes (P01.1)	764	19.0	663	16.5
Newborn affected by multiple pregnancy (P01.5)	254	6.3	270	6.7
Newborn affected by other maternal complications of pregnancy (P01.2 P01.4, P01.6 P01.9)	219	5.4	163	4.0
Newborn affected by complications of placenta, cord and membranes (P02)	1 017	25.3	1 018	25.3
Newborn affected by complications involving placenta (P02.0 P02.3)	508	12.6	520	12.9
Newborn affected by complications involving cord (P02.4 P02.6)	54	1.3	51	1.3
Newborn affected by chorioamnionitis (P02.7)	453	11.3	447	11.1
Newborn affected by other and unspecified abnormalities of membranes (P02.8 P02.9)	1	*	-	*
Newborn affected by other complications of labor and delivery (P03)	151	3.8	-	3.7
Newborn affected by noxious influences transmitted via placenta or breast milk (P04)	31	0.8	37	0.9
Disorders related to length of gestation and fetal malnutrition (P05 P08)	4 673	116.3	4 473	111.1
Slow fetal growth and fetal malnutrition (P05)	75	1.9	63	1.6
Disorders related to short gestation and low birth weight, not elsewhere classified (P07)	4 598	114.4	4 410	109.5
Extremely low birth weight or extreme immaturity (P07.0, P07.2)	3 531	87.9	3 316	82.4
Other low birth weight or preterm (P07.1, P07.3)	1 067	26.5	1 094	27.2
Disorders related to long gestation and high birth weight (P08)	-	*	-	*
Birth trauma (P10-P15)	-	8.7	-	3.6
Birth trauma (P10-P15)	578	14.4	534	13.3
Intrauterine hypoxia (P20)	106	2.6	120	3.0
Birth asphyxia (P21)	472	11.7	414	10.3
Respiratory distress of newborn (P22)	958	23.8	1 011	25.1
Other respiratory conditions originating in the perinatal period (P23-P28)	1 205	30.0	1 503	37.3
Congenital pneumonia (P23)	71	1.8	72	1.8
Neonatal aspiration syndromes (P24)	50	1.2	71	1.8
Interstitial emphysema and related conditions originating in the perinatal period (P25)	157	3.9	204	5.1
Pulmonary hemorrhage originating in the perinatal period (P26)	189	4.7	279	6.9
Chronic respiratory disease originating in the perinatal period (P27)	311	7.7	310	7.7
Atelectasis (P28.0-P28.1)	387	9.6	501	12.4
All other respiratory conditions originating in the perinatal period (P28.2-P28.9)	40	1.0	66	1.6
Infections specific to the perinatal period (P35-P39)	927	23.1	869	21.6
Infections specific to the perinatal period (P35-P39)	734	18.3	696	17.3
Omphalitis of newborn with or without mild hemorrhage (P38)	1	*	3	*
All other infections specific to the perinatal period (P35, P37, P39)	192	4.8	170	4.2
Hemorrhagic and hematological disorders of newborn (P50-P61)	492	12.2	600	14.9
Hemorrhagic and hematological disorders of newborn (P50-P61)	374	9.3	453	11.3
Hemorrhagic disease of newborn (P53)	-	*	-	*
Hemolytic disease of newborn due to isoimmunization and other perinatal jaundice (P55-P59)	-	*	-	*
Hematological disorders (P60-P61)	100	2.5	133	3.3
Syndrome of infant of a diabetic mother and neonatal diabetes mellitus (P70.0-P70.2)	8	*	7	*
Necrotizing enterocolitis of newborn (P77)	348	8.7	380	9.4
Hydrops fetalis not due to hemolytic disease (P83.2)	187	4.7	166	4.1
Other perinatal conditions (P29, P70.3-P76, P78-P81, P83.0-P83.1, P83.3-P96)	1 205	30.0	1 214	30.2
Congenital malformations, deformations and chromosomal abnormalities (Q00-Q99)	5 655	140.7	5 513	136.9
Anencephaly and similar malformations (Q00)	304	7.6	263	6.5
Congenital hydrocephalus (Q03)	90	2.2	105	2.6
Spina bifida (Q05)	20	0.5	19	*
Other congenital malformations of nervous system (Q01-Q02, Q04, Q06-Q07)	293	7.3	286	7.1
Congenital malformations of heart (Q20-Q24)	1 508	37.5	1 556	38.6
Congenital malformations of heart (Q20-Q24)	217	5.4	192	4.8
Congenital malformations of respiratory system (Q30-Q34)	639	15.9	594	14.8
Congenital malformations of digestive system (Q35-Q45)	99	2.5	106	2.6
Congenital malformations of genitourinary system (Q50-Q64)	354	8.8	356	8.8
Congenital malformations and deformations of musculoskeletal system, limbs and integument (Q65-Q85)	472	11.7	477	11.8
Down's syndrome (Q90)	119	3.0	119	3.0
Edward's syndrome (Q91.0-Q91.3)	505	12.6	446	11.1
Patau's syndrome (Q91.4-Q91.7)	306	7.6	276	6.9
Other congenital malformations and deformations (Q10-Q18, Q86-Q89)	542	13.5	526	13.1
Other chromosomal abnormalities, not elsewhere classified (Q92-Q99)	189	4.7	192	4.8
Symptoms, signs and abnormal clinical and laboratory findings, not elsewhere classified (R00-R99)	3 702	92.1	3 491	86.7

[1]Based on codes from the "Tenth Revision, International Classification of Diseases, 1992." Codes *U01-*U03 have been added as of 2001 for deaths due to terrorism.
- = Quantity zero.
* = Figure too small to meet standards of reliability or precision.

Table II-32A. Number of Infant Deaths and Infant Mortality Rates for 130 Selected Causes, 2001 and Preliminary 2002*—Continued*

(Number, rate per 100,000 live births.)

Cause of death[1]	2002 Number	2002 Rate	2001 Number	2001 Rate
Sudden infant death syndrome (R95)	2 035	50.6	2 234	55.5
Other symptoms, signs and abnormal clinical and laboratory findings, not elsewhere classified (R00-R53, R55-R94, R96-R99)	1 667	41.5	1 257	31.2
All other diseases (Residual)	26	0.6	19	*
External causes of mortality (*U01, V01-Y84)	1 261	31.4	1 417	35.2
External causes of mortality (*U01, V01-Y84)	893	22.2	976	24.2
Transport accidents (V01-V99)	129	3.2	151	3.8
Motor vehicle accidents (V02-V04, V09.0, V09.2, V12-V14, V19.0-V19.2, V19.4-V19.6, V20-V79, V80.3-V80.5, V81.0-V81.1, V82.0-V82.1, V83-V86, V87.0-V87.8, V88.0-V88.8, V89.0, V89.2)	124	3.1	144	3.6
Other and unspecified transport accidents (V01, V05-V06, V09.1, V09.3-V09.9, V10-V11, V15-V18, V19.3, V19.8, V19.9, V80.0-V80.2, V80.6-V80.9, V81.2-V81.9, V82.2-V82.9, V87.9, V88.9, V89.1, V89.3, V89.9, V90-V99)	5	*	7	*
Falls (W00-W19)	24	0.6	23	0.6
Accidental discharge of firearms (W32-W34)	1	*	-	*
Accidental drowning and submersion (W65-W74)	61	1.5	68	1.7
Accidental suffocation and strangulation in bed (W75)	380	9.5	390	9.7
Other accidental suffocation and strangulation (W76-W77, W81-W84)	147	3.7	162	4.0
Accidental inhalation and ingestion of food or other objects causing obstruction of respiratory tract (W78-W80)	51	1.3	62	1.5
Accidents caused by exposure to smoke, fire and flames (X00-X09)	35	0.9	50	1.2
Accidental poisoning and exposure to noxious substances (X40-X49)	27	0.7	15	*
Other and unspecified accidents (W20-W31, W35-W64, W85-W99, X10-X39, X50-X59)	38	0.9	55	1.4
Assault (homicide) (*U01, X85-Y09)	274	6.8	332	8.2
Assault (homicide) by hanging, strangulation and suffocation (X91)	28	0.7	40	1.0
Assault (homicide) by discharge of firearms (*U01.4, X93-X95)	9	*	11	*
Neglect, abandonment and other maltreatment syndromes (Y06-Y07)	85	2.1	116	2.9
Assault (homicide) by other and unspecified means (*U01.0-*U01.3, *U01.5-*U01.9, X85-X90, X92, X96-X99, Y00-Y05, Y08-Y09)	152	3.8	165	4.1
Complications of medical and surgical care (Y40-Y84)	14	*	26	0.6
Other external causes and their sequelae (X60-X84, Y10-Y36)	80	2.0	83	2.1

[1]Based on codes from the "Tenth Revision, International Classification of Diseases, 1992." Codes *U01-*U03 have been added as of 2001 for deaths due to terrorism.
- = Quantity zero.
* = Figure too small to meet standards of reliability or precision.

TABLE II-32B. NUMBER AND PERCENT DISTRIBUTION OF INFANT DEATHS, AND INFANT MORTALITY RATES FOR 10 LEADING CAUSES OF INFANT DEATH, 2001

The infant mortality rates in the table are expressed as deaths per 100,00 live births. Of the 130 causes of infant death, the seven leading causes accounted for 60.4 percent. The top cause, representing 20.0 percent of infant deaths in 2001, was congenital malformations, deformations and chromosomal abnormalities. Disorders related to short gestation and low birth weight, not elsewhere classified, was next in importance, affecting 16.0 percent. Sudden infant death syndrome ranked third, with 8.1 percent. Newborn affected by maternal complications of pregnancy accounted for 5.4 percent. Newborn affected by complications of placenta, cord and membranes represented 3.7 percent. Respiratory distress of newborn caused 3.7 percent of infant deaths. Accidents ranked seventh, accounting for 2.5 percent.

Changes in infant mortality rates between 2000 and 2001 among the 10 leading causes were statistically significant for three causes: sudden infant death syndrome decreased by 10.8 percent, intrauterine hypoxia and birth asphyxia decreased by 14.2 percent, and accidents increased by 11.5 percent. The list of the 10 leading causes of infant death did not change from 2000 to 2001.

Table II-32B. Number and Percent Distribution of Infant Deaths, and Infant Mortality Rates for 10 Leading Causes of Infant Death, 2001

(Number, percent, rate per 100,000 live births.)

Rank[1]	Cause of death[2]	Number	Percent of total deaths	Rate	Percent change, 2000–2001
	All causes	27 568	100.0	684.8	-0.9
1	Congenital malformations,deformations and chromosomal abnormalities Q00–Q99	5 513	20.0	136.9	-3.3
2	Disorders related to short gestation and low birth weight,not elsewhere classified P07	4 410	16.0	109.5	1.1
3	Sudden infant death syndrome R95	2 234	8.1	55.5	-10.8
4	Newborn affected by maternal complications of pregnancy P01	1 499	5.4	37.2	7.5
5	Newborn affected by complications of placenta,cord and membranes P02	1 018	3.7	25.3	-3.4
6	Respiratory distress of newborn P22	1 011	3.7	25.1	2.0
7	Accidents (unintentional injuries) V01–X59	976	3.5	24.2	11.5
8	Bacterial sepsis of newborn	696	2.5	17.3	-8.5
9	Diseases of the circulatory system I00–I99	622	2.3	15.4	-5.5
10	Intrauterine hypoxia and birth asphyxia P20–P21	534	1.9	13.3	-14.2
	All other causes (Residual)	9 055	32.8	224.9	X

[1]Rank based on number of deaths.
[2]Based on codes from the "Tenth Revision, International Classification of Diseases, 1992." Codes *U01-*U03 have been added as of 2001 for deaths due to terrorism.
X = Not applicable.

TABLE II-32C. INFANT DEATHS AND INFANT MORTALITY RATES FOR 10 LEADING CAUSES OF INFANT DEATH, BY RACE AND HISPANIC ORIGIN, PRELIMINARY 2002

For the most part the rankings of causes of infant death among Whites, Blacks, and Hispanics were similar. Blacks, however, had a much higher rate of infant mortality than Whites or Hispanics. A related factor, shown in Table I-44, is that Black babies were more frequently born preterm and thus with lower birth weight than White babies. For Black infant mortality, "Disorders related to short gestation and low birth weight, not elsewhere classified" was the first ranking cause of death, while it ranked second for the other groups in the table.

Table II-32C. Infant Deaths and Infant Mortality Rates for 10 Leading Causes of Infant Death, by Race and Hispanic Origin, Preliminary 2002

(Number, rate per 100,000 live births.)

Rank [1] and race	Cause of death [2]	Number	Rate
All Races [3]			
	All causes	27 974	696.0
1	Congenital malformations, deformations and chromosomal abnormalities (Q00-Q99)	5 655	140.7
2	Disorders related to short gestation and low birth weight,not elsewhere classified (P07)	4 598	114.4
3	Sudden infant death syndrome (R95)	2 035	50.6
4	Newborn affected by maternal complications of pregnancy (P01)	1 723	42.9
5	Newborn affected by complications of placenta,cord and membranes (P02)	1 017	25.3
6	Respiratory distress of newborn (P22)	958	23.8
7	Accidents (unintentional injuries) (V01-X59)	893	22.2
8	Bacterial sepsis of newborn (P36)	734	18.3
9	Diseases of the circulatory system (I00-I99)	649	16.1
10	Intrauterine hypoxia and birth asphyxia (P20-P21)	578	14.4
	All other causes (Residual)	9 134	227.2
White [4]			
	All causes	18 447	580.8
1	Congenital malformations, deformations and chromosomal abnormalities (Q00-Q99)	4 384	138.0
2	Disorders related to short gestation and low birth weight,not elsewhere classified (P07)	2 573	81.0
3	Sudden infant death syndrome (R95)	1 343	42.3
4	Newborn affected by maternal complications of pregnancy (P01)	1 070	33.7
5	Newborn affected by complications of placenta,cord and membranes (P02)	672	21.2
6	Respiratory distress of newborn (P22)	598	18.8
7	Accidents (unintentional injuries) (V01-X59)	558	17.6
8	Bacterial sepsis of newborn (P36)	485	15.3
9	Diseases of the circulatory system (I00-I99)	426	13.4
10	Intrauterine hypoxia and birth asphyxia (P20-P21)	411	12.9
	All other causes (Residual)	5 927	186.6
Non-Hispanic White			
	All causes	13 542	587.9
1	Congenital malformations, deformations and chromosomal abnormalities (Q00-Q99)	3 110	135.0
2	Disorders related to short gestation and low birth weight,not elsewhere classified (P07)	1 811	78.6
3	Sudden infant death syndrome (R95)	1 112	48.3
4	Newborn affected by maternal complications of pregnancy (P01)	821	35.6
5	Newborn affected by complications of placenta,cord and membranes (P02)	493	21.4
6	Respiratory distress of newborn (P22)	446	19.4
7	Accidents (unintentional injuries) (V01-X59)	437	19.0
8	Bacterial sepsis of newborn (P36)	361	15.7
9	Diseases of the circulatory system (I00-I99)	320	13.9
10	Intrauterine hypoxia and birth asphyxia (P20-P21)	311	13.5
	All other causes (Residual)	4 321	187.6
Black [4]			
	All causes	8 380	1 419.0
1	Disorders related to short gestation and low birth weight,not elsewhere classified (P07)	1 836	310.9
2	Congenital malformations, deformations and chromosomal abnormalities (Q00-Q99)	1 023	173.2
3	Sudden infant death syndrome (R95)	603	102.1
4	Newborn affected by maternal complications of pregnancy (P01)	575	97.3
5	Respiratory distress of newborn (P22)	336	57.0
6	Newborn affected by complications of placenta,cord and membranes (P02)	315	53.3
7	Accidents (unintentional injuries) (V01-X59)	303	51.4
8	Bacterial sepsis of newborn (P36)	231	39.1
9	Diseases of the circulatory system (I00-I99)	189	32.1
10	Intrauterine hypoxia and birth asphyxia (P20-P21)	147	24.8
	All other causes (Residual)	2 822	477.8
Hispanic [5]			
	All causes	4 933	565.5
1	Congenital malformations, deformations and chromosomal abnormalities (Q00-Q99)	1 269	145.5
2	Disorders related to short gestation and low birth weight,not elsewhere classified (P07)	769	88.1
3	Newborn affected by maternal complications of pregnancy (P01)	246	28.2
4	Sudden infant death syndrome (R95)	241	27.7
5	Newborn affected by complications of placenta,cord and membranes (P02)	186	21.3
6	Respiratory distress of newborn (P22)	156	17.9
7	Accidents (unintentional injuries) (V01-X59)	128	14.6
8	Bacterial sepsis of newborn (P36)	126	14.5
9	Diseases of the circulatory system (I00-I99)	106	12.1
10	Intrauterine hypoxia and birth asphyxia (P20-P21)	98	11.2
	All other causes (Residual)	1 610	184.6

[1]Rank based on number of deaths.
[2]Based on codes from the "Tenth Revision, International Classification of Diseases, 1992." Codes *U01-*U03 have been added as of 2001 for deaths due to terrorism.
[3]Includes races other than White and Black.
[4]Race and Hispanic origin are reported separately on both the birth and death certificate.
[5]Includes all persons of Hispanic origin of any race.

TABLE II-33. NUMBER OF INFANT AND NEONATAL DEATHS AND MORTALITY RATES, BY RACE FOR EACH STATE AND TERRITORY, 2001

Delaware recorded the highest infant mortality rate of 10.7 per 1,000 in 2001, as compared with the United States average of 6.8. The Delaware rate represents a 16 percent increase from year 2000, and may be aberrant. The rate for the District of Columbia, usually the highest nationwide by far, was 10.6 per 1,000 in 2001, a 12 percent decrease from 2000. Rates in Mississippi, and Louisiana also were quite high. On the low side, New Hampshire had an astoundingly low infant mortality rate of 3.8 per 1,000, Utah had a rate of 4.8, and Massachusetts' rate was 5.0.

The Black infant mortality rate, which averaged 14.0 for the United States, was near 20, or above, in three states: Arizona, Kansas, and Delaware. The rates, however, appear to fluctuate considerably from year to year, as none of these states stood out in terms of Black infant mortality rate in year 2000. In Arizona and Kansas, White infant mortality rates were only moderately above the national average. States with the lowest Black infant mortality rates were Massachusetts, Minnesota, New York, and Kentucky.

Table II-33. Number of Infant and Neonatal Deaths and Mortality Rates, by Race for each State and Territory, 2001

(Number, rate per 1,000 live births.)

Area	Infant deaths						Neonatal deaths					
	All races [1]		White		Black		All races [1]		White		Black	
	Number	Rate	Number	Rate	Number	Rate	Number	Rate	Number	Rate	Number	Rate
United States [2]	27 568	6.8	17 955	5.7	8 498	14.0	18 265	4.5	12 013	3.8	5 585	9.2
Male	15 477	7.5	10 099	6.2	4 764	15.5	10 237	5.0	6 750	4.2	3 125	10.2
Female	12 091	6.1	7 856	5.1	3 734	12.5	8 028	4.1	5 263	3.4	2 460	8.2
Alabama	567	9.4	276	6.8	290	15.1	355	5.9	171	4.2	184	9.6
Alaska	81	8.1	37	5.8	4	*	38	3.8	24	3.8	2	*
Arizona	592	6.9	458	6.1	63	22.8	389	4.5	317	4.2	42	15.2
Arkansas	309	8.3	196	6.8	108	14.5	173	4.7	105	3.6	66	8.9
California	2 830	5.4	2 131	5.0	435	12.9	1 896	3.6	1 438	3.4	295	8.7
Colorado	388	5.8	336	5.5	38	12.8	288	4.3	251	4.1	29	9.8
Connecticut	260	6.1	167	4.7	87	16.9	195	4.6	127	3.6	63	12.3
Delaware	115	10.7	61	8.0	54	19.9	82	7.6	44	5.7	38	14.0
District of Columbia	81	10.6	7	*	73	15.0	59	7.7	5	*	53	10.9
Florida	1 495	7.3	844	5.5	635	13.5	981	4.8	544	3.6	426	9.0
Georgia	1 146	8.6	535	6.2	583	13.3	780	5.8	357	4.2	400	9.1
Hawaii	106	6.2	23	6.0	4	*	68	4.0	17	*	4	*
Idaho	129	6.2	125	6.3	1	*	84	4.1	82	4.1	1	*
Illinois	1 413	7.7	854	6.0	515	15.5	954	5.2	607	4.3	315	9.5
Indiana	650	7.5	515	6.8	132	13.7	407	4.7	325	4.3	80	8.3
Iowa	212	5.6	195	5.5	14	*	130	3.5	117	3.3	13	*
Kansas	287	7.4	225	6.5	57	20.5	181	4.7	144	4.2	34	12.2
Kentucky	325	5.9	275	5.6	48	9.7	196	3.6	169	3.5	26	5.3
Louisiana	643	9.8	239	6.5	390	14.4	417	6.4	151	4.1	256	9.5
Maine	84	6.1	76	5.7	6	*	65	4.7	59	4.4	6	*
Maryland	594	8.1	248	5.5	333	13.7	439	6.0	191	4.2	237	9.8
Massachusetts	405	5.0	312	4.6	79	9.6	308	3.8	240	3.5	59	7.2
Michigan	1 069	8.0	646	6.1	401	17.0	731	5.5	447	4.2	268	11.3
Minnesota	361	5.3	281	4.8	46	9.6	236	3.5	190	3.3	23	4.8
Mississippi	445	10.5	160	7.0	280	14.9	278	6.6	97	4.3	180	9.6
Missouri	558	7.4	371	5.9	184	16.5	370	4.9	241	3.9	126	11.3
Montana	74	6.7	63	6.7	-	*	48	4.4	43	4.6	-	*
Nebraska	168	6.8	144	6.4	14	*	119	4.8	106	4.7	8	*
Nevada	180	5.7	133	5.1	43	17.1	93	3.0	69	2.6	24	9.5
New Hampshire	56	3.8	53	3.8	-	*	38	2.6	35	2.5	-	*
New Jersey	747	6.5	430	5.1	293	14.2	501	4.3	303	3.6	187	9.1
New Mexico	174	6.4	143	6.3	7	*	107	3.9	92	4.0	4	*
New York	1 482	5.8	926	5.1	507	9.7	1 028	4.0	664	3.6	337	6.5
North Carolina	1 009	8.5	522	6.1	449	15.8	703	5.9	352	4.1	320	11.3
North Dakota	67	8.8	53	8.0	1	*	44	5.8	35	5.3	1	*
Ohio	1 161	7.7	773	6.2	377	16.4	760	5.0	516	4.1	237	10.3
Oklahoma	366	7.3	254	6.5	69	15.0	205	4.1	146	3.7	34	7.4
Oregon	246	5.4	228	5.5	8	*	157	3.5	147	3.6	2	*
Pennsylvania	1 033	7.2	725	6.1	291	14.4	728	5.1	531	4.5	186	9.2
Rhode Island	86	6.8	67	6.1	12	*	70	5.5	54	4.9	10	*
South Carolina	496	8.9	210	5.9	284	15.0	337	6.0	137	3.8	199	10.5
South Dakota	78	7.4	52	6.1	1	*	37	3.5	27	3.2	1	*
Tennessee	681	8.7	403	6.7	269	16.2	437	5.6	255	4.2	176	10.6
Texas	2 171	5.9	1 637	5.2	491	12.0	1 345	3.7	1 020	3.3	292	7.2
Utah	232	4.8	222	4.9	3	*	149	3.1	146	3.2	1	*
Vermont	35	5.5	35	5.6	-	*	23	3.6	23	3.7	-	*
Virginia	747	7.6	380	5.4	351	15.8	486	4.9	236	3.3	238	10.7
Washington	459	5.8	383	5.7	43	12.9	292	3.7	245	3.6	27	8.1
West Virginia	148	7.2	140	7.2	8	*	110	5.4	102	5.2	8	*
Wisconsin	491	7.1	352	5.9	117	17.8	324	4.7	245	4.1	67	10.2
Wyoming	36	5.9	34	5.9	-	*	24	3.9	24	4.2	*	*
Puerto Rico	517	9.3	506	9.9	11	*	394	7.1	385	7.5	9	*
Virgin Islands	13	*	0	*	10	*	11	*	2	*	9	*
Guam	35	9.8	1	*	-	*	25	7.0	1	*	-	*
American Samoa	14	*	-	*	-	*	8	*	-	*	-	*
Northern Marianas	11	*	-	*	-	*	9	*	-	*	-	*

Note: Infant deaths based on race of decedent; live births based on race of mother.
[1]Includes races other than white and black.
[2]Excludes data for Puerto Rico, Virgin Islands, Guam, American Samoa and Northern Marianas.
* = Figure too small to meet standards of reliability or precision.
\- = Quantity zero.

Figure II-18. Infant Death Rate, 2001

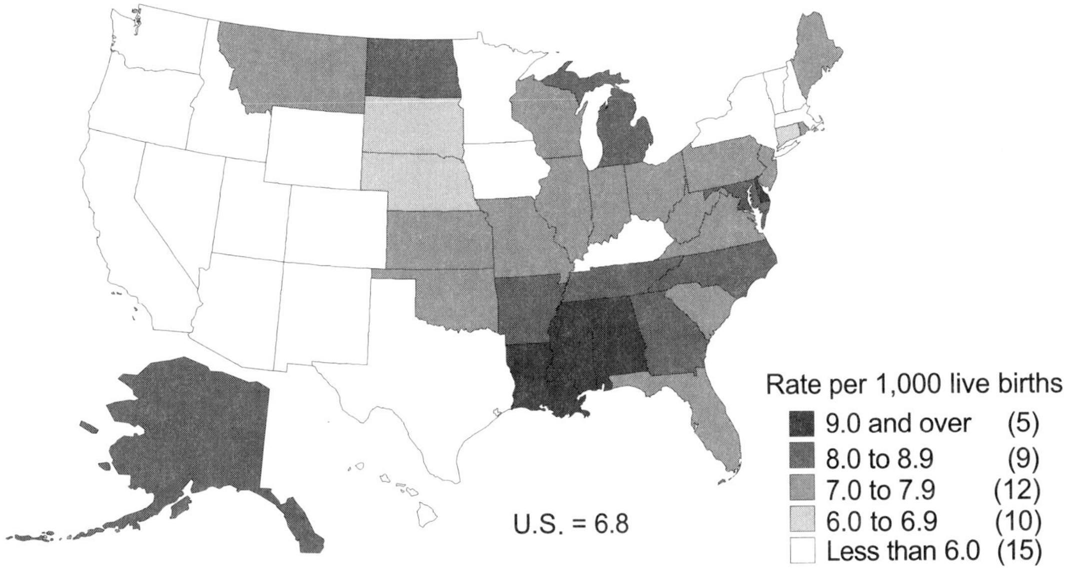

TABLE II-34. NUMBER OF MATERNAL DEATHS AND MATERNAL MORTALITY RATES FOR SELECTED CAUSES, BY RACE, 2001

The maternal mortality rate represents the risk of a pregnant woman dying of maternal causes, and is therefore also computed on the basis of the number of live births. In 2001, a total of 399 women were reported to have died of maternal causes. Included here are the deaths that were assigned, on the death certificate, to causes related to or aggravated by pregnancy (or pregnancy management). Deaths occurring more than 42 days after the termination of pregnancy are excluded, as well as accidental or intentional deaths.

The maternal mortality rate for 2001 was 9.9 deaths per 100,000 live births. There was a big difference between the rate for White and for Black pregnancies. The rate for Whites was 7.2 deaths per 100,000, while that for Blacks was 24.7.

TABLE II-35. NUMBER OF MATERNAL DEATHS AND MATERNAL MORTALITY RATES FOR SELECTED CAUSES, BY RACE AND HISPANIC ORIGIN, 2001

The maternal mortality rate for Hispanic women was 9.5 deaths per 100,000 live births. The corresponding rate for non-Hispanic Whites was 6.5. As with other statistics involving Hispanic origin, these data should be interpreted with caution because of inconsistencies between reporting Hispanic origin on death certificates and on data for live births. The latter data are needed to obtain the denominators in the calculation of rates.

Table II-34. Number of Maternal Deaths and Maternal Mortality Rates for Selected Causes, by Race, 2001

(Number, rate per 100,000 live births.)

Cause of death [1]	Number				Rate			
	All races	White	All other		All races	White	All other	
			Total	Black			Total	Black
Maternal causes (A34, O00-O95, O98-O99)	399	228	171	150	10.0	7.2	20.0	24.7
Pregnancy with abortive outcome (O00-O07)	38	20	18	16	1.0	0.6	*	*
Ectopic pregnancy (O00)	23	11	12	10	1.0	*	*	*
Spontaneous abortion (O03)	7	5	2	2	*	*	*	*
Medical abortion (O04)	1	1	-	-	*	*	*	*
Other abortion (O05)	-	-	-	-	*	*	*	*
Other and unspecified pregnancy with abortive outcome (O01-O02, O06-O07)	7	3	4	4	*	*	*	*
Other direct obstetric causes (A34, O10-O92)	309	179	130	114	8.0	5.6	15.0	18.8
Eclampsia and pre-eclampsia (O11, O13-O16)	72	43	29	26	2.0	1.4	3.0	4.3
Hemorrhage of pregnancy and childbirth and placenta previa (O20, O44-O46, O67, O72)	44	26	18	15	1.0	0.8	*	*
Complications predominantly related to the puerperium (A34, O85-O92)	101	55	46	39	3.0	1.7	5.0	6.4
Obstetrical tetanus (A34)	-	-	-	-	*	*	*	*
Obstetric embolism (O88)	69	39	30	24	2.0	1.2	4.0	4.0
Other complications predominantly related to the puerperium (O85-O87, O89-O92)	32	16	16	15	1.0	*	*	*
All other direct obstetric causes (O10, O12, O21-O43, O47-O66, O68-O71, O73-O75)	92	55	37	34	2.0	1.7	4.0	5.6
Obstetric death of unspecifed cause (O95)	1	1	-	-	*	*	*	*
Indirect obstetric causes (O98-O99)	51	28	23	20	1.0	0.9	3.0	3.3
Maternal causes more than 42 days after delivery or termination of pregnancy (O96-O97)	17	7	10	8	*	*	*	*
Death from any obstetric cause occurring more than 42 days but less than one year after delivery (O96)	9	4	5	4	*	*	*	*
Death from sequelae of direct obstetric causes (O97)	8	3	5	4	*	*	*	*

[1]Based on codes from the Tenth Revision, International Classification of Diseases, 1992. Codes *U01-*U03 have been added as of 2001 for deaths due to terrorism.
* = Figure too small to meet standards of reliability or precision.
- = Quantity zero.

Table II-35. Number of Maternal Deaths and Maternal Mortality Rates for Selected Causes, by Race and Hispanic Origin, 2001

(Number, rate per 100,000 live births.)

Cause of death [1]	Number					Rate				
	All origins [2]	Hispanic	Non-Hispanic [3]	Non-Hispanic White	Non-Hispanic Black	All origins [2]	Hispanic	Non-Hispanic [3]	Non-Hispanic White	Non-Hispanic Black
Maternal causes (A34, O00-O95, O98-O99)	399	81	317	151	146	9.9	10.0	10.1	7.0	24.7
Pregnancy with abortive outcome (O00-O07)	38	1	37	19	16	0.9	*	1.2	*	*
Ectopic pregnancy (O00)	23	1	22	10	10	0.6	*	0.7	*	*
Spontaneous abortion (O03)	7	-	7	5	2	*	*	*	*	*
Medical abortion (O04)	1	-	1	1	-	*	*	*	*	*
Other abortion (O05)	-	-	-	-	-	*	*	*	*	*
Other and unspecified pregnancy with abortive outcome (O01-O02, O06-O07)	7	*	7	3	4	*	*	*	*	*
Other direct obstetric causes (A34, O10-O92)	309	75	233	108	110	7.7	9.0	7.4	5.0	18.6
Eclampsia and pre-eclampsia (O11, O13-O16)	72	25	47	18	26	1.8	3.0	1.5	*	4.4
Hemorrhage of pregnancy and childbirth and placenta previa (O20, O44-O46, O67, O72)	44	10	33	17	13	1.1	*	1.0	*	*
Complications predominantly related to the puerperium (A34, O85-O92)	101	17	84	40	38	2.5	*	2.7	1.0	6.4
Obstetrical tetanus (A34)	-	-	-	-	-	*	*	*	*	*
Obstetric embolism (O88)	69	13	56	28	23	1.7	*	1.8	1.0	3.9
Other complications predominantly related to the puerperium (O85-O87, O89-O92)	32	4	28	12	15	0.8	*	0.9	*	*
All other direct obstetric causes (O10, O12, O21-O43, O47-O66, O68-O71, O73-O75)	92	23	69	33	33	2.3	3.0	2.2	1.0	5.6
Obstetric death of unspecifed cause (O95)	1	1	-	-	-	*	*	*	*	*
Indirect obstetric causes (O98-O99)	51	4	47	24	20	1.3	*	1.5	1.0	3.4
Maternal causes more than 42 days after delivery or termination of pregnancy (O96-O97)	17	2	15	5	8	*	*	*	*	*
Death from any obstetric cause occurring more than 42 days but less than one year after delivery (O96)	9	1	8	3	4	*	*	*	*	*
Death from sequelae of direct obstetric causes (O97)	8	1	7	2	4	*	*	*	*	*

[1]Based on codes from the Tenth Revision, International Classification of Diseases, 1992. Codes *U01-*U03 have been added as of 2001 for deaths due to terrorism.
[2]All origins includes origin not stated; specified origins exclude origin not stated.
[3]Includes races other than White and Black.
* = Figure too small to meet standards of reliability or precision.
- = Quantity zero.

CHAPTER III: HEALTH

The first section of the **Health** Chapter is **Measures of Health**. It provides information on the self-assessment of their health status by survey respondents, habits of exercise, health care, and overweight frequencies.

The section on **Selected Chronic Diseases** deals with heart diseases, cancer, AIDS, migraines, and a number of other diseases.

Disabilities examines individuals who need help for personal care needs and those who need assistance to live independently. Lost workdays from occupational injuries are also reported.

Use of Addictive Substances covers statistics on alcohol, cigarettes, and illicit drugs.

Ambulatatory Care deals with vaccinations of children, visits to doctors and emergency departments, dental visits, and medical tests. Statistics are shown on patients of special treatment organizations, such as substance abuse units, mental health organizations, home health care services, and hospices.

Inpatient Care covers statistics on hospitals, such as the volume of discharges and admissions by diagnosis and length of stay, and the frequency of medical procedures. There also is a section on nursing homes.

MEASURES OF HEALTH

TABLE III-1. HEALTHY PEOPLE 2010 LEADING HEALTH INDICATORS, SELECTED YEARS, 1990–2002, AND 2010 TARGET

This table presents measures of health for the U.S. population and for selected age groups. The data are given for various years, up to 2002 when available. Improvement targets are presented for 2010, in the way of a road map developed by the Department of Health and Human Services, and specifically by Healthy People 2010, operating from the Centers for Disease Control and Prevention's (CDC) Wonder Web site.

The indicators that are covered are physical activity, obesity, tobacco use, mental health, environmental quality, access to health care, and a number of other topics.

TABLE III-1A. HEALTH INSURANCE COVERAGE STATUS AND TYPE OF COVERAGE BY SELECTED CHARACTERISTICS, 2001

This table shows data for the civilian noninstitutionalized population in 2001. The health insurance status for seniors, those 65 years and over, is quite different from that of younger persons. Some 97 percent of seniors are covered by Medicare and more than 60 percent of those also carry private insurance, usually as secondary policies.

Among persons under 65 years, 41 million, or 16.5 percent, were not covered by health insurance at any time during 2001. Young adults between 18 and 24 years old were most likely to be uninsured (28 percent). Among children the uninsured percentage of was 11.7.

About two-thirds of all persons under 65 years of age were covered by employment-based health insurance. Medicaid insurance, which is available mainly for low income households, covered another 11.4 percent. The table permits more than one type of insurance coverage for a respondent.

Table III-1. Healthy People 2010 Leading Health Indicators, Selected Years, 1990–2002, and 2010 Target

(Percent.)

Leading health indicators	2010 Target	2002	2001	2000	1999	1998	1995	1990
Physical Activity								
Adolescents in grades 9–12: percent who engaged in 20 minutes or more of vigorous activity 3 or more days a week	85	...	65	...	65	...	64	...
Adults age 18 years and over: age-adjusted percent who engaged in moderate activity (at least 30 minutes, 5 days per week) or vigorous activity (at least 20 minutes, 3 days per week) [1]	50	33	32	32	30	30	...	...
Overweight and Obesity [2]								
Overweight or obese children and adolescents, age 6–19 years: percent who are at or above the sex- and age-specific 95th percentile of Body Mass Index (BMI) based on CDC Growth Charts	5	...	...	15	...	...	...	11
Obese adults age 20 years and over: age-adjusted percent with BMI of 30 Kg/m2/ or more	15	...	...	31	...	...	...	23
Tobacco Use								
Adolescents in grades 9–12: percent who smoked cigarettes one or more days in the past 30 days	16	...	28	...	35	...	35	...
Adults age 18 years and over: age-adjusted percent who smoked more than 100 cigarettes in their lifetime and now report smoking on some days or every day [1]	12	22	23	23	23	24	25	25
Substance Abuse [3]								
Adolescents age 12–17 years: percent who reported no use of alcohol or illicit drugs in the past 30 days	89	...	...	80	80	...	...	...
Adults age 18 years and over: percent who reported illicit drug use in the past 30 days	2	...	7	6	6	...	...	...
Adults age 18 years and over: percent who reported binge drinking in the past 30 days	6	...	22	22	21	...	...	...
Responsible Sexual Behavior								
Adolescents in grades 9–12: percent who are not sexually active or sexually active and used condoms	95	...	86	...	85	...	83	...
Never had intercourse	X	...	54	...	50	...	47	...
Used a condom at last intercourse	X	...	19	...	21	...	21	...
No sexual intercourse in past 3 months	X	...	12	...	14	...	15	...
Sexually active unmarried women age 18-44 years: percent who reported condom use by partners	50	...	...	...	...	...	23	...
New cases of gonorrhea per 100,000 population	19	...	129	129	132	132	149	277
Mental Health								
Adults age 18 years and over: percent with recognized depression who received treatment [4]	50	...	...	...	...	...	23	...
Age-adjusted suicide rate per 100,000 standard population	5.0	...	10.7	10.4	10.5	11.1	11.8	12.5
Injury and Violence								
Age-adjusted death rate for motor vehicle traffic-related injuries per 100,000 standard population [5]	9.2	...	14.9	14.9	14.7	15.3	15.9	18.0
Age-adjusted homicide rate per 100,000 standard population	3.0	...	7.1	5.9	6.0	6.4	8.3	9.4
Environmental Quality								
Percent of population exposed to ozone above EPA standard	0	...	41	43	43	43	...	...
Persons age 4 years and over: age-adjusted percent of nonsmokers exposed to environmental tobacco smoke [2]	45	...	...	54	...	...	...	88
Persons under age 18 years: hospital admissions for asthma per 10,000 population	17	...	21	...	26	...	...	...
Immunization								
Children age 19–35 months: percent who received all DTaP, polio, MMR, Hib, and HepB vaccines	80	75	74	73	73	73	...	...
Adults age 65 years and over: age-adjusted percent who received influenza vaccine in the past 12 months [1]	90	66	63	65	66	64	59	...
Adults age 65 years and over: age-adjusted percent who ever received pneumococcal vaccine [1]	90	55	54	53	50	46	35	...
Access to Health Care								
Persons under age 65 years: age-adjusted percent with health insurance [1]	100	84	84	83	84	83	84	...
Persons of all ages: age-adjusted percent with a specific source of ongoing primary care [1]	96	88	88	87	86	87	87	...
Pregnant women: percent who received prenatal care in the first trimester	90	...	83	83	83	83	81	76

[1]Data for 1997 and later years are not strictly comparable with data for earlier years due to the 1997 questionnaire redesign. Data for 2002 are provisional.
[2]National Health and Nutrition Examination Survey (NHANES) data for 1990 are for the period 1988–1994. NHANES data for 2000 are for the period 1999–2000, and are based on a smaller sample size and therefore are subject to greater sampling error than the estimate for 1988–1994.
[3]In a major redesign of the survey in 1999, the sample size, mode of administration, and survey content changed. Data are not shown for 1998 and earlier years because only limited comparisons can be made between data from the redesigned surveys (1999 onward) and data obtained from surveys prior to 1999.
[4]Baseline year 1997 and baseline value 23 percent are shown in 1995.
[5]Motor vehicle traffic-related injuries.
. . . = Not available.
X = Not applicable.

Figure III-1. Health Insurance Coverage Among Persons Under 65 Years of Age, 1994–2001

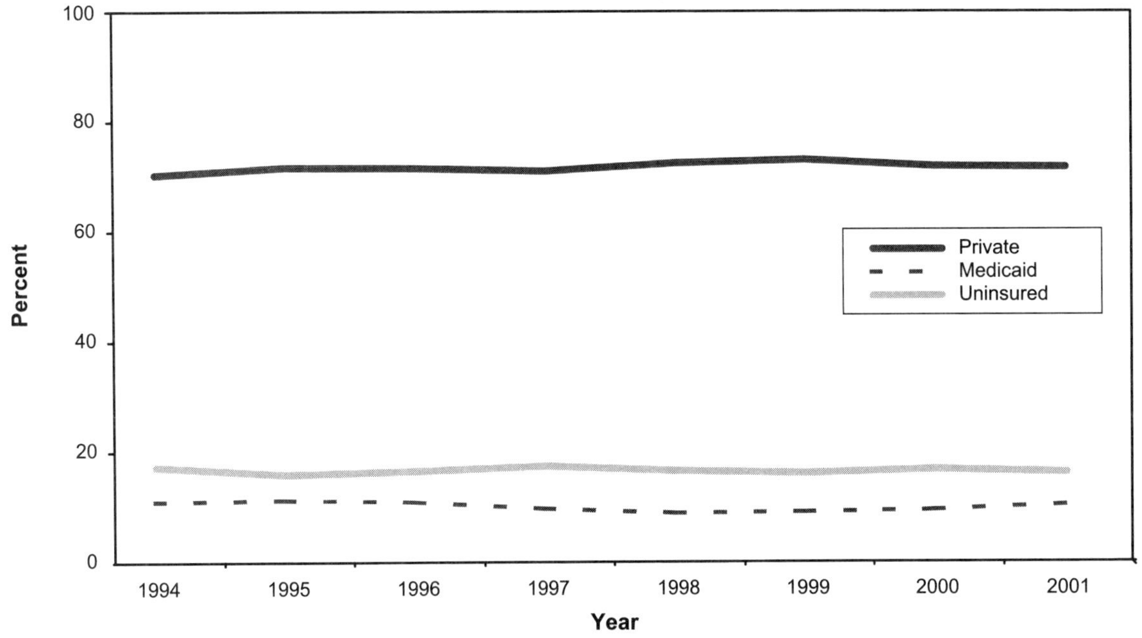

Table III-1A. Health Insurance Coverage Status and Type of Coverage, by Selected Characteristics, 2001

(Number in thousands, percent.)

Characteristic	Total	Not covered at any time during the year	Covered by some type of insurance during the year	Covered by private insurance	Covered by employment based insurance	Covered by own employment based insurance	Covered by government health plan	Covered by Medicaid	Covered by Medicaid or Indian Health Service	Covered by Medicare
NUMBER										
Total	282 082	41 207	240 875	199 860	176 551	93 470	71 295	31 601	31 601	38 043
Age										
Under 18 years	72 628	8 509	64 118	49 647	46 439	290	18 822	16 502	16 502	423
Under 6 years	23 373	2 503	20 869	15 062	14 264	0	7 322	6 546	6 546	156
6–11 years	24 626	2 769	21 857	16 885	16 057	0	6 490	5 676	5 676	126
12–17 years	24 629	3 237	21 392	17 699	16 118	290	5 010	4 280	4 280	140
18–24 years	27 312	7 673	19 640	17 012	13 766	5 776	3 642	2 831	2 831	180
25–34 years	38 670	9 051	29 619	26 905	25 306	19 537	3 653	2 587	2 587	489
35–44 years	44 284	7 131	37 153	34 315	32 386	23 427	4 003	2 532	2 532	860
45–54 years	39 545	5 179	34 365	31 649	29 487	22 020	3 990	2 071	2 071	1 331
55–64 years	25 874	3 392	22 482	19 581	17 521	13 349	4 567	1 807	1 807	2 301
Under 65 years	248 312	40 935	207 377	179 108	164 906	84 399	38 677	28 330	28 330	5 585
65 years and over	33 769	272	33 498	20 751	11 645	9 071	32 618	3 270	3 270	32 458
Sex										
Male	137 871	21 722	116 149	98 150	87 960	52 563	31 764	13 769	13 769	16 360
Female	144 211	19 485	124 726	101 709	88 591	40 908	39 531	17 832	17 832	21 683
Nativity										
Native	249 629	30 364	219 265	182 556	161 168	83 359	65 204	28 522	28 522	34 577
Foreign-born	32 453	10 843	21 610	17 303	15 383	10 111	6 091	3 079	3 079	3 466
Naturalized citizen	11 962	2 060	9 902	7 844	6 892	4 914	3 270	1 086	1 086	2 480
Not a citizen	20 491	8 782	11 708	9 459	8 491	5 197	2 821	1 994	1 994	986
PERCENT										
Total	100.0	14.6	85.4	70.8	62.6	33.1	25.3	11.2	11.2	13.5
Age										
Under 18 years	100.0	11.7	88.3	68.4	63.9	0.4	25.9	22.7	22.7	0.6
Under 6 years	100.0	10.7	89.3	64.4	61.0	0.0	31.3	28.0	28.0	0.7
6–11 years	100.0	11.2	88.8	68.6	65.2	0.0	26.4	23.0	23.0	0.5
12–17 years	100.0	13.1	86.9	71.9	65.4	1.2	20.3	17.4	17.4	0.6
18–24 years	100.0	28.1	71.9	62.3	50.4	21.2	13.3	10.4	10.4	0.7
25–34 years	100.0	23.4	76.6	69.6	65.4	50.5	9.4	6.7	6.7	1.3
35–44 years	100.0	16.1	83.9	77.5	73.1	52.9	9.0	5.7	5.7	1.9
45–54 years	100.0	13.1	86.9	80.0	74.6	55.7	10.1	5.2	5.2	3.4
55–64 years	100.0	13.1	86.9	75.7	67.7	51.6	17.6	7.0	7.0	8.9
Under 65 years	100.0	16.5	83.5	72.1	66.4	34.0	15.6	11.4	11.4	2.2
65 years and over	100.0	0.8	99.2	61.4	34.5	26.9	96.6	9.7	9.7	96.1
Sex										
Male	100.0	15.8	84.2	71.2	63.8	38.1	23.0	10.0	10.0	11.9
Female	100.0	13.5	86.5	70.5	61.4	28.4	27.4	12.4	12.4	15.0
Nativity										
Native	100.0	12.2	87.8	73.1	64.6	33.4	26.1	11.4	11.4	13.8
Foreign-born	100.0	33.4	66.6	53.3	47.4	31.2	18.8	9.5	9.5	10.7
Naturalized citizen	100.0	17.2	82.8	65.6	57.6	41.1	27.3	9.1	9.1	20.7
Not a citizen	100.0	42.9	57.1	46.2	41.4	25.4	13.8	9.7	9.7	4.8

TABLE III-2. RESPONDENT ASSESSED HEALTH STATUS BY AGE, SEX, RACE, AND HISPANIC ORIGIN, 1997–2003

These health status data were obtained by asking respondents to assess their own health, and that of family members living in the same household, as excellent, very good, good, fair, or poor. In early 2003, the percent of persons whose health was assessed as excellent or very good was 67.3 percent. This percentage had declined somewhat since it reached a high of 69.1 percent in 1998.

Compared with men, the health status of women was less likely to be reported as excellent (37.1 percent versus 34.0 percent) and more likely to be reported as good (22.4 percent versus 24.1 percent) or fair (6.4 versus 7.3 percent).

For both sexes, the percentages reporting excellent or very good health were highest under 18 years of age (82.6 percent), they declined to 66.6 percent for the 18–64 year age group, and they amounted to 38.6 percent for respondents 65 years and over. The health assessment of non-Hispanic Whites was more than 10 percentage points higher than that of either Hispanics or of non-Hispanic Blacks.

Table III-2. Respondent Assessed Health Status by Age, Sex, Race, and Hispanic Origin, 1997–2003

(Percent.)

Characteristic	Percent
Excellent or Very Good Health Status	
2003 (January–June)	67.3
2002	67.4
2001	68.4
2000	68.3
1999	68.7
1998	69.1
1997	68.5
JANUARY 2003–JUNE 2003	
Health Status and Sex	
Excellent	
Both sexes	35.9
Men	37.4
Women	34.5
Very good	
Both sexes	31.4
Men	31.5
Women	31.4
Good	
Both sexes	23.5
Men	22.7
Women	24.2
Fair	
Both sexes	6.8
Men	6.3
Women	7.4
Poor	
Both sexes	2.4
Men	2.2
Women	2.5
Health Assessed as Excellent or Very Good by Age, Sex, Race, and Hispanic Origin	
Under 18 years	
Both sexes	82.6
Male	82.6
Female	82.6
18–64 years	
Both sexes	66.6
Male	67.9
Female	65.3
65 years and over	
Both sexes	38.6
Male	38.7
Female	38.5
Race and Hispanic Origin	
Hispanic or Latino [1]	57.8
White, single race	70.8
Black, single race	56.5

Note: Health status data were obtained by asking respondents to assess their own health and that of family members living in the same household as excellent, very good, good, or poor. The analyses excluded persons with unknown health status.

TABLE III-3. LIFE EXPECTANCIES AND EXPECTED YEARS IN GOOD OR BETTER HEALTH AND EXCELLENT HEALTH FOR SELECTED AGES, 1995

This table shows life expectancies that prevailed in 1995 at various age levels. A newborn baby, at that time, would be expected to live 75.8 years and, taking an average for the population, 66.5 of these years would be lived in "good or better" health, while only 26.4 years would be expected to be lived in "excellent" health. Naturally, as age advances the expected additional years of healthy life tend to shrink. At age 75, for example, life expectancy was 11.1 years, expected years in "good or better" health had diminished to 7.4 years, and expected years in "excellent" health averaged only 1.4 years. Not only did the remaining years of health decline, the percentage of healthy life relative to total life expectancy also fell. Females showed significantly more expected years of life as well as expected years lived in good or better health, but there were no significant gender differences in expected years of life in excellent health.

Table III-3. Life Expectancies and Expected Years in Good or Better Health and Excellent Health for Selected Ages, 1995

(Percent.)

Age and sex	Expected years of life	Expected years in good or better health	Years in good or better health as a percentage of life expectancy	Expected years in excellent health	Years in excellent health as a percentage of life expectancy
Both Sexes					
Birth	75.8	66.5	87.7	26.4	34.8
20 years	56.9	48.0	84.3	16.5	29.0
30 years	47.5	39.0	82.1	12.7	26.6
45 years	33.8	26.2	77.5	7.3	21.7
65 years	17.5	12.4	71.0	2.6	14.9
70 years	14.1	9.8	69.7	2.0	13.9
75 years	11.1	7.4	67.4	1.4	13.1
80 years	8.3	5.4	65.3	1.1	13.0
Male					
Birth	72.8	64.6	88.7	27.1	37.3
20 years	54.0	46.2	85.5	17.2	31.9
30 years	44.9	37.3	83.1	13.0	29.0
45 years	31.5	24.6	78.1	7.4	23.3
65 years	15.9	11.2	70.4	2.4	14.9
70 years	12.8	8.8	68.9	1.7	13.6
75 years	10.1	6.7	66.5	1.2	11.8
80 years	7.7	4.9	64.2	0.8	10.8
Female					
Birth	78.8	68.4	86.8	25.7	32.6
20 years	59.8	49.8	83.3	15.8	26.4
30 years	50.1	40.6	81.2	12.3	24.6
45 years	35.9	27.6	76.9	7.3	20.2
65 years	18.8	13.4	71.3	2.8	14.8
70 years	15.1	10.6	70.1	2.1	14.0
75 years	11.7	8.0	67.9	1.6	13.7
80 years	8.7	5.7	65.9	1.2	14.2
Gender Difference [1]					
Birth	+6.1	+3.9	X	-1.4	X
20 years	+5.7	+3.6	X	-1.4	X
30 years	+5.2	+3.4	X	-0.7	X
45 years	+4.3	+3.0	X	-0.1	X
65 years	+2.9	+2.2	X	0.4	X
70 years	+2.3	+1.8	X	0.4	X
75 years	+1.7	+1.3	X	0.4	X
80 years	+1.0	+0.8	X	0.4	X

[1]Calculated by subtracting expected years for males from expected years for females.
X = Not applicable.

TABLE III-4. PERCENT DISTRIBUTIONS OF RESPONDENT-ASSESSED HEALTH STATUS AMONG PERSONS 18 YEARS OF AGE AND OVER, BY SELECTED CHARACTERISTICS, 2001

Distributions into three classes of health status are shown by sex, age, race, Hispanic origin, education, family income, poverty status, health insurance coverage, marital status, and residence location. American Indians and Blacks reported fair or poor health more frequently than Asians, Whites, or Hispanics. Older persons, as would be expected, reported fair or poor health more frequently than younger persons. Respondents in higher income brackets or with high educational attainment reported better health conditions than those in lower brackets. In regard to health insurance coverage, persons with private insurance reported the best health conditions. Health assessments by uninsured persons, however, were considerably more positive than those given by Medicaid patients. On a regional basis, the South reported slightly lower health assessments than the other regions, and living in a larger urban area was associated with reporting better health status.

Table III-4. Percent Distributions of Respondent-Assessed Health Status Among Persons 18 Years of Age and Over, by Selected Characteristics, 2001

(Percent.)

Characteristic	Current health status among persons 18 years of age and over [1]						
	Total	Excellent or very good	Standard error for excellent or very good	Good	Standard error for good	Fair or poor	Standard error for fair or poor
Total [2]	100.0	63.7	0.36	24.5	0.29	11.9	0.22
Sex							
Male	100.0	65.4	0.49	23.6	0.41	11.0	0.30
Female	100.0	62.1	0.44	25.3	0.38	12.6	0.28
Age							
18–44 years	100.0	74.4	0.44	20.0	0.39	5.6	0.21
45–64 years	100.0	57.5	0.61	27.1	0.53	15.4	0.43
65–74 years	100.0	43.8	1.04	34.2	0.91	22.0	0.88
75 years and over	100.0	35.2	1.06	34.6	0.99	30.2	0.98
Race							
1 race [3]	100.0	63.8	0.36	24.4	0.29	11.8	0.22
White	100.0	65.1	0.39	23.8	0.32	11.1	0.25
Black or African American	100.0	54.7	1.04	27.8	0.88	17.5	0.69
American Indian or Alaska Native	100.0	52.9	3.74	25.6	2.83	21.5	4.25
Asian	100.0	69.2	1.85	23.3	1.71	7.6	1.01
Native Hawaiian or other Pacific Islander	100.0	65.8	8.37	27.0	7.13	*7.1	3.61
2 or more races [4]	100.0	52.6	2.98	30.7	2.65	16.7	2.24
Black or African American, white	100.0	60.5	8.70	27.2	7.96	*12.4	5.42
American Indian or Alaska Native, white	100.0	46.2	4.49	35.1	4.18	18.7	3.42
Hispanic or Latino Origin and Race [5]							
Hispanic or Latino	100.0	60.3	0.83	26.8	0.70	12.9	0.54
Mexican or Mexican American	100.0	59.5	1.04	28.7	0.91	11.8	0.66
Not Hispanic or Latino	100.0	64.1	0.39	24.2	0.32	11.7	0.24
White, single race	100.0	65.5	0.42	23.6	0.34	10.9	0.26
Black or African American, single race	100.0	54.5	1.05	27.7	0.88	17.7	0.70
Education [6]							
Less than a high school diploma	100.0	38.7	0.76	31.9	0.69	29.4	0.67
High school diploma or GED [7]	100.0	56.3	0.66	29.7	0.58	14.1	0.44
Some college	100.0	65.8	0.64	24.4	0.58	9.8	0.37
Bachelor's degree or higher	100.0	78.4	0.59	17.3	0.53	4.3	0.26
Family Income [8]							
Less than $20,000	100.0	46.6	0.87	29.0	0.61	24.4	0.66
$20,000 or more	100.0	68.5	0.39	23.1	0.34	8.4	0.22
$20,000–$34,999	100.0	55.9	0.87	29.1	0.77	15.0	0.59
$35,000–$54,999	100.0	66.0	0.74	25.1	0.67	8.9	0.47
$55,000–$74,999	100.0	73.2	0.91	21.5	0.82	5.3	0.42
$75,000 or more	100.0	79.3	0.67	17.0	0.62	3.7	0.31
Poverty Status [9]							
Poor	100.0	48.8	1.36	27.0	0.93	24.3	1.02
Near poor	100.0	50.8	0.92	29.7	0.80	19.5	0.65
Not poor	100.0	70.5	0.42	22.1	0.38	7.3	0.23

[1]The data in this table are based on a question in the survey that asked respondents, "Would you say [subject name's] health in general was excellent, very good, good, fair, or poor?" This information was obtained during a part of the interview that allowed proxy responses, such that a knowledgeable adult family member could respond on behalf of adults not taking part in the interview (however, the sample in this table is based on the reported health status for the Sample Adult only). "Excellent" and "very good "are combined in this table, as are "fair" and "poor."

[2]Total includes other races not shown separately and persons with unknown poverty status.

[3]The category "1 race" refers to people who indicated only a single race group. Persons who indicated a single race other than the groups shown are included in the total for "1 race" but not shown separately due to small sample sizes. Therefore the frequencies for the category "1 race" will will be greater than the sum of the frequencies for the specific groups shown separately.

[4]The category "2 or more races" refers to all persons who indicated more than one race group. Only two combinations of race groups are shown due to small sample sizes for other combinations. Therefore, the frequencies for the category "2 or more races" will be greater than the sum of the frequencies for the specific combinations shown separately.

[5]Persons of Hispanic or Latino origin may be of any race or any combination of races. Similarly, the category "Not Hispanic or Latino" refers to all persons who are not of Hispanic or Latino origin, regardless of race.

[6]Education is shown only for persons aged 25 years and over.

[7]GED is General Educational Development high school equivalency diploma.

[8]The categories "less than $20,000"and "$20,000 or more" include both persons reporting dollar amounts and persons reporting only that their incomes were within one of these two categories. The indented categories include only those persons who reported dollard amounts.

[9]Poverty status is based on family income and family size using the U.S. Census Bureau's poverty thresholds for the previous calendar year. "Poor" persons are defined as below the poverty threshold. "Near poor" persons have incomes of 100 percent to less than 200 percent of the poverty threshold. "Not poor" persons have incomes that are 200 percent of the poverty threshold or greater.

* = Figure does not meet standards of reliability or precision.

Table III-4. Percent Distributions of Respondent-Assessed Health Status Among Persons 18 Years of Age and Over, by Selected Characteristics, 2001—*Continued*

(Percent.)

Characteristic	Current health status among persons 18 years of age and over [1]						
	Total	Excellent or very good	Standard error for excellent or very good	Good	Standard error for good	Fair or poor	Standard error for fair or poor
Health Insurance Coverage [10]							
Under age 65 years							
Private	100.0	72.9	0.40	21.1	0.37	6.0	0.20
Medicaid	100.0	38.2	1.39	27.5	1.25	34.3	1.36
Other	100.0	42.3	1.96	25.1	1.56	32.6	1.89
Uninsured	100.0	62.7	0.84	26.7	0.76	10.6	0.48
Age 65 years and over							
Private	100.0	42.7	0.97	35.8	0.86	21.5	0.79
Medicaid and Medicare	100.0	18.8	1.96	24.1	2.17	57.1	2.60
Medicare only	100.0	38.8	1.43	34.0	1.35	27.2	1.34
Other	100.0	34.9	3.34	31.1	3.35	34.0	3.30
Uninsured	100.0	28.8	6.65	25.9	6.85	45.3	7.37
Marital Status							
Married	100.0	65.3	0.45	24.2	0.39	10.5	0.29
Widowed	100.0	39.1	0.97	32.6	0.96	28.2	0.94
Divorced or separated	100.0	54.4	0.85	27.4	0.72	18.1	0.64
Never married	100.0	71.8	0.70	21.0	0.63	7.3	0.36
Living with a partner	100.0	64.8	1.33	24.9	1.18	10.3	0.79
Place of Residence [11]							
Large MSA	100.0	66.1	0.48	23.4	0.41	10.5	0.29
Small MSA	100.0	63.5	0.70	24.5	0.55	12.0	0.43
Not in MSA	100.0	58.5	0.84	26.9	0.65	14.6	0.51
Region							
Northeast	100.0	64.2	0.73	25.9	0.64	9.9	0.42
Midwest	100.0	64.7	0.72	24.1	0.57	11.2	0.41
South	100.0	61.8	0.65	24.4	0.51	13.8	0.42
West	100.0	65.3	0.78	23.6	0.66	11.1	0.47
Sex and Age							
Male							
18–44 years	100.0	76.2	0.59	18.9	0.54	4.8	0.29
45–64 years	100.0	58.5	0.88	26.6	0.75	14.9	0.59
65 years and over	100.0	38.7	1.17	34.7	1.08	26.6	1.06
Female							
18–44 years	100.0	72.6	0.55	21.0	0.50	6.4	0.29
45–64 years	100.0	56.5	0.81	27.6	0.73	15.9	0.55
65 years and over	100.0	40.7	0.96	34.2	0.86	25.1	0.83
Hispanic or Latino Origin, Race, and Age							
Hispanic or Latino, male							
18–44 years	100.0	70.3	1.37	23.6	1.25	6.1	0.67
45–64 years	100.0	51.0	2.62	30.7	2.27	18.3	1.94
65 years and over	100.0	34.0	3.97	29.5	3.49	36.5	3.77
Hispanic or Latina, Female							
18–44 years	100.0	68.2	1.29	25.3	1.18	6.5	0.62
45–64 years	100.0	41.9	2.22	34.7	2.09	23.4	1.83
65 years and over	100.0	28.5	2.88	29.1	2.78	42.3	3.32
Not Hispanic or Latino							
White, single race, male							
18–44 years	100.0	78.5	0.70	17.1	0.62	4.4	0.35
45–64 years	100.0	61.9	1.00	25.3	0.86	12.8	0.64
65 years and over	100.0	40.2	1.30	35.1	1.21	24.7	1.14
White, single race, female							
18–44 years	100.0	75.7	0.68	18.9	0.61	5.4	0.35
45–64 years	100.0	60.4	0.96	26.2	0.86	13.4	0.63
65 years and over	100.0	43.1	1.05	34.9	0.98	22.0	0.89
Black or African American, single race, male							
18–44 years	100.0	71.2	1.80	22.8	1.71	6.0	0.84
45–64 years	100.0	40.6	2.38	31.8	2.18	27.6	2.18
65 years and over	100.0	23.4	3.36	33.2	4.06	43.4	4.25
Black or African American, single race, female							
18–44 years	100.0	62.6	1.51	27.1	1.37	10.3	0.90
45–64 years	100.0	40.6	2.16	31.4	1.95	28.0	1.92
65 years and over	100.0	27.3	2.51	31.0	2.63	41.7	2.73

[1]The data in this table are based on a question in the survey that asked respondents, "Would you say [subject name's] health in general was excellent, very good, good, fair, or poor?" This information was obtained during a part of the interview that allowed proxy responses, such that a knowledgeable adult family member could respond on behalf of adults not taking part in the interview (however, the sample in this table is based on the reported health status for the Sample Adult only). "Excellent" and "very good "are combined in this table, as are "fair" and "poor."

[10]Classification of health insurance coverage is based on hierarchy of mutually exclusive categories. Persons with more than one type of health insurance were assigned to the first appropriate category in the hierarchy. Persons under the age of 65 and those 65 years and over were classified separately due to the prominence of Medicare coverage in the older population. The category "private" includes persons who had any type of private coverage either alone or in combination with other coverage. For example, for persons age 65 years and over, "private" includes persons with only private or private in combination in Medicare. The category "uninsured" includes persons who had coverage as well as those who had only Indian Health Service coverage or had only a private plan that paid for one type of service such as accidents or dental care.

[11]MSA is a metropolitan statistical area. Large MSA's have a population of 1,000,000 or more; small MSAs have a population size of less than 1,000,000. "Not in MSA" consists of persons not living in an MSA.

TABLE III-5. PERCENT OF ADULTS AGED 18 YEARS AND OVER, WHO ENGAGED IN REGULAR LEISURE-TIME PHYSICAL ACTIVITY, BY AGE, SEX, RACE, AND HISPANIC ORIGIN, 1997–2003

Activities qualify as "regular physical activity" if persons engage in moderate activity of 30 minutes or more, 5 times a week; or vigorous activity of 20 minutes or more, 3 times a week. In the first half of 2003, 33.2 percent of adults engaged in such physical activity, a little more than the percentages for recent years. Regular physical activity was practiced by 45.8 percent of young persons between 18 and 24 years, but declined successively for older age groups, reaching 27.4 percent for persons between 65 and 74 years of age, and 17.2 percent for those above 74. Among the elderly even small amounts of physical activity may improve cardiovascular functioning.

Men were more likely to be physically active than women, especially in the 18 to 24 year age group, where 51.9 percent of men were active, as compared with 39.7 percent of women. In terms of race and ethnicity, Whites were the most active group with 35.9 percent participating on average, physically active Blacks averaged 26.5 percent, and Hispanic participation was 24.7 percent.

Table III-5. Percent of Adults Aged 18 Years and Over who Engaged in Regular Leisure-Time Physical Activity, by Age, Sex, Race, and Hispanic Origin, 1997–2003

(Percent.)

Characteristic	Percent
Total	
2003 (January–June)	33.2
2002	31.9
2001	32.1
2000	31.9
1999	30.4
1998	32.7
1997	32.1
JANUARY 2003–JUNE 2003	
Age and Sex	
18–24 years	
Total	45.8
Men	51.9
Women	39.7
25–64 years	
Total	33.3
Men	35.4
Women	31.3
65–74 years	
Total	27.4
Men	31.2
Women	24.4
75 years and over	
Total	17.2
Men	23.4
Women	13.2
Age-adjusted[1]	
Total	33.1
Men	36.3
Women	30.3
Race and Hispanic Origin	
Age-sex-adjusted[2]	
Hispanic or Latino[3]	24.7
White, single race	35.9
Black, single race	26.5
Age-adjusted[1]	
Hispanic or Latino[3]	24.8
White, single race	35.9
Black, single race	26.2

Note: Regular leisure-time physical activity is defined as engaging in light-moderate leisure-time physical activity for greater than or equal to 30 minutes at a frequency greater than or equal to five times per week, or engaging in vigorous leisure-time physical activity for greater than or equal to 20 minutes at a frequency greater than or equal to three times per week. The analyses excluded persons with unknown physical activity participation.

[1]Estimates for this Healthy People 2010 leading health indicator are age-adjusted to the year 2000 projected U.S. standard population using five age groups: 18–24 years, 25–44 years 45–64 years, 65–74 years, and 75 years and over.

[2]Estimates are age-sex-adjusted to the year 2000 projected U.S. standard populaton using five age agroups: 18–24 years, 25–34 years, 35–44 years, 45–64 years, and 65 years and over.

[3]Persons of Hispanic origin may be of any race.

Figure III-2. High School Students Not Engaging in Recommended Amounts of Physical Activity by Grade and Sex, 2001

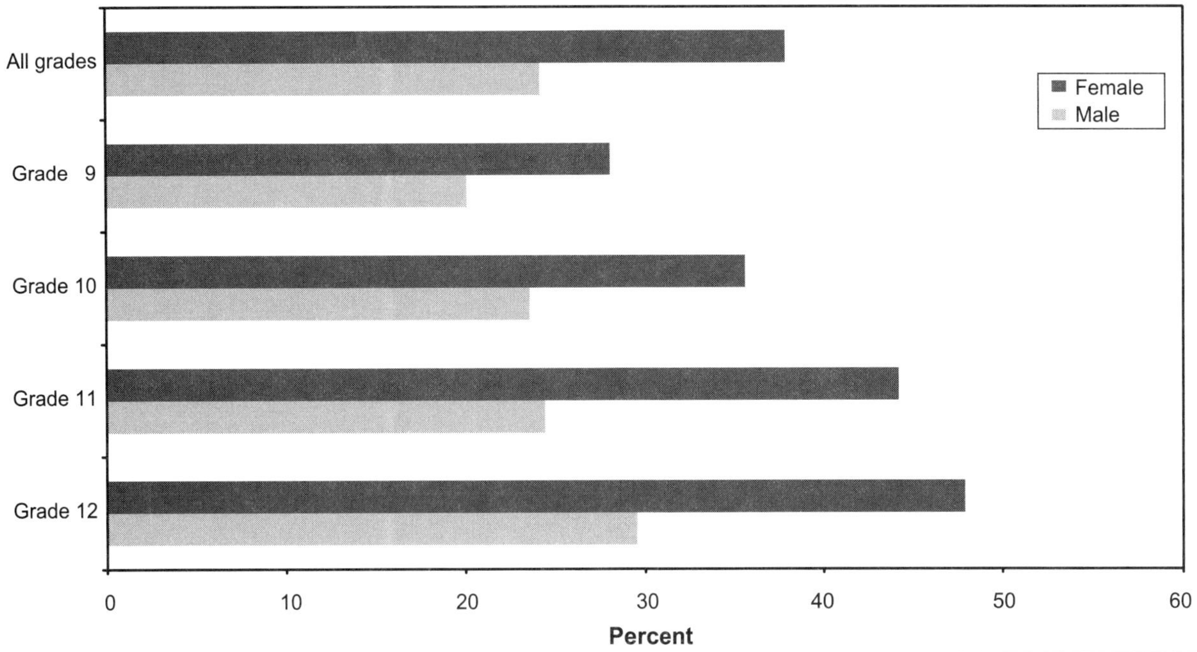

TABLE III-6. HIGH SCHOOL STUDENTS NOT ENGAGING IN RECOMMENDED AMOUNTS OF PHYSICAL ACTIVITY, NEITHER MODERATE OR VIGOROUS, BY GRADE AND SEX, 2001

In 2001, 38 percent of female high school students and 24 percent of male students fell short of even moderate amounts of recommended levels of physical activity. The percentages of students who did not meet the recommended physical activity increased in successive grades, starting with 24.3 percent in grade 9 and reaching 38.9 percent by grade 12. These percentages remained fairly stable between 1999 and 2001.

Table III-6. High School Students Not Engaging in Recommended Amounts of Physical Activity, Neither Moderate or Vigorous, by Grade and Sex, 2001

(Percent.)

Grade	All students		Male students		Female students	
	Percent	Standard error	Percent	Standard error	Percent	Standard error
Grade 9	24.3	1.4	20.1	1.6	28.1	1.9
Grade 10	29.6	0.9	23.6	1.1	35.6	1.7
Grade 11	34.4	1.2	24.4	1.3	44.2	1.6
Grade 12	38.9	1.4	29.5	2.1	47.9	1.2
All grades	31.2	0.7	24.2	0.8	37.9	1.2

Note: The recommended amount of physical activity for high school students is at least 30 minutes of moderate activity (does not cause sweating or hard breathing) on five or more of the past seven days; or at least 20 minutes of vigorous activity (causes sweating and hard breathing) on three or more of the past seven days. The recommended amounts of physical activity for high school students are based on the Healthy People 2010 objectives 22-6 and 22-7 (moderate and vigorous activity in adolescents).

TABLE III-7. OVERWEIGHT, OBESITY, AND HEALTHY WEIGHT AMONG PERSONS 20 YEARS AND OVER, ACCORDING TO SEX, AGE, RACE AND HISPANIC ORIGIN, SELECTED YEARS, 1960–2000

Adults are shown in three partly overlapping categories: healthy weight with BMI over 18.5, but less than 25; overweight with BMI of 25 or above; and obesity with BMI of 30 or above. BMI is a measure that adjusts body weight for height; weight in kilograms is divided by height in meters squared.

For the population of 20 to 74 years of age, 34 percent (age-adjusted) were of healthy weight in 1999–2000, down from 51 percent in 1960–1962. Non-Hispanic White females and non-Hispanic Black males had greater than average percentages with healthy weight, while non-Hispanic Black females had only a 22.3 percent participants in the healthy group. The largest participation in the healthy weight group occurred between age 20 and 34, and the smallest was reported between age 65 and 74.

In 1999–2000, 64.5 percent of the adult population (age-adjusted) was classified as overweight and almost half of these, 31 percent, were reported to be obese; the percentage of obese persons had become about twice as large as it was in 1976–1980, and even larger compared with 1960–1962. Especially high percentages were reported for non-Hispanic Black women (pregnant women excluded) and, to a lesser extent, for Mexican women. (The table does not provide global figures for Hispanics.)

Table III-7. Overweight, Obesity, and Healthy Weight Among Persons 20 Years and Over, According to Sex, Age, Race, and Hispanic Origin, Selected Years, 1960–2000

(Percent.)

Sex, age, race, and Hispanic origin	1999–2000	1999–2000 standard error [1]	1988–1994	1988–1994 standard error [1]	1976–1980 [2]	1971–1974	1960–1962
PERCENT OF POPULATION OVERWEIGHT [3]							
20–74 Years, Age-Adjusted [4]							
Both sexes [5, 6]	64.5	1.5	56.0	0.9	47.4	47.7	44.8
Male	67.0	1.5	61.0	1.0	52.9	54.7	49.5
Female [5]	62.0	2.0	51.2	1.1	42.0	41.1	40.2
Not Hispanic or Latino							
White only, male	67.3	2.0	61.6	1.2	53.8	. . .	. . .
White only, female [5]	57.2	2.7	47.2	1.4	38.7	. . .	. . .
Black or African American only, male	60.3	2.3	58.2	1.2	51.3	. . .	. . .
Black or African American only, female [5]	77.7	1.9	68.5	1.4	62.6	. . .	. . .
Mexican male [7]	74.4	2.8	69.4	1.1	61.6	. . .	. . .
Mexican female [5, 7]	71.8	2.5	69.6	1.7	61.7	. . .	. . .
20 Years and Over, Age-Adjusted [4]							
Both sexes [5, 6]	64.5	1.4	56.0	0.8	. . .	. . .	. . .
Male	67.2	1.4	60.9	1.0	. . .	. . .	. . .
Female [5]	61.9	1.9	51.4	1.0	. . .	. . .	. . .
Not Hispanic or Latino							
White only, male	67.4	1.8	61.6	1.2	. . .	. . .	. . .
White only, female [5]	57.3	2.5	47.5	1.3	. . .	. . .	. . .
Black or African American only, male	60.7	1.9	57.8	1.2	. . .	. . .	. . .
Black or African American only, female [5]	77.3	2.0	68.2	1.3	. . .	. . .	. . .
Mexican male [7]	74.7	2.6	68.9	1.1	. . .	. . .	. . .
Mexican female [5, 7]	71.9	2.4	68.9	1.6	. . .	. . .	. . .
20 Years and Over, Crude							
Both sexes [5, 6]	64.1	1.4	54.9	0.8	. . .	. . .	. . .
Male	66.5	1.4	59.4	1.0	. . .	. . .	. . .
Female [5]	61.8	1.9	50.7	1.0	. . .	. . .	. . .
Not Hispanic or Latino							
White only, male	67.3	1.8	60.6	1.2	. . .	. . .	. . .
White only, female [5]	57.8	2.6	47.4	1.2	. . .	. . .	. . .
Black or African American only, male	58.9	2.3	56.7	1.2	. . .	. . .	. . .
Black or African American only, female [5]	76.5	2.2	66.0	1.4	. . .	. . .	. . .
Mexican male [7]	72.0	3.7	63.9	1.5	. . .	. . .	. . .
Mexican female [5, 7]	69.4	3.1	65.9	1.4	. . .	. . .	. . .
Male							
20–34 years	58.0	2.5	47.5	1.4	41.2	42.8	42.7
35–44 years	67.6	2.7	65.5	1.7	57.2	63.2	53.5
45–54 years	71.3	3.5	66.1	2.1	60.2	59.7	53.9
55–64 years	72.5	3.5	70.5	2.1	60.2	58.5	52.2
65–74 years	77.2	3.6	68.5	2.1	54.2	54.6	47.8
75 years and over	66.4	3.5	56.5	2.0	. . .	. . .	. . .
Female [5]							
20–34 years	51.5	2.8	37.0	1.4	27.9	25.8	21.2
35–44 years	63.6	4.1	49.6	2.4	40.7	40.5	37.2
45–54 years	64.7	3.7	60.3	2.5	48.7	49.0	49.3
55–64 years	73.1	3.7	66.3	1.6	53.7	54.5	59.9
65–74 years	70.1	4.0	60.3	1.8	59.5	55.9	60.9
75 years and over	59.6	3.7	52.3	1.5	. . .	. . .	. . .

[1] Standard errors of estimates for 1988–1994 and 1999–2000 are shown, and 1999–2000 estimates are based on a smaller sample size than estimates for earlier time periods and therefore are subject to greater sampling error.

[2] Data for Mexicans are for 1982–1984.

[3] Body mass index (BMI) greater than or equal to 25.

[4] Age-adjusted to the 2000 standard population using five age groups except for the 1999–2000 estimates, which are age-adjusted using three age groups. However, use of three rather than five age groups had virtually no effect on age-adjusted estimates.

[5] Excludes pregnant women.

[6] Includes persons of all races and Hispanic origins, not just those shown separately.

[7] Persons of Mexican origin may be of any race.

. . . = Not available.

Table III-7. Overweight, Obesity, and Healthy Weight Among Persons 20 Years and Over, According to Sex, Age, Race, and Hispanic Origin, Selected Years, 1960–2000—*Continued*

(Percent.)

Sex, age, race, and Hispanic origin	1999–2000	1999–2000 standard error[1]	1988–1994	1988–1994 standard error[1]	1976–1980[2]	1971–1974	1960–1962
PERCENT OF POPULATION OBESE[8]							
20–74 Years, Age-Adjusted[4]							
Both sexes[5, 6]	30.9	1.6	23.3	0.7	15.1	14.6	13.3
Male	27.7	1.7	20.6	0.7	12.8	12.2	10.7
Female[5]	34.0	2.0	26.0	1.0	17.1	16.8	15.7
Not Hispanic or Latino							
White only, male	27.4	1.9	20.7	0.9	12.4	. . .	. . .
White only, female[5]	30.4	2.3	23.3	1.2	15.4	. . .	. . .
Black or African American only, male	28.9	2.4	21.3	1.0	16.5	. . .	. . .
Black or African American only, female[5]	50.4	2.8	39.1	1.4	31.0	. . .	. . .
Mexican male[7]	29.4	2.5	24.4	1.1	15.7	. . .	. . .
Mexican female[5, 7]	40.1	3.8	36.1	1.4	26.6	. . .	. . .
20 Years and Over, Age-Adjusted[4]							
Both sexes[5, 6]	30.5	1.4	22.9	0.7	. . .	. . .	. . .
Male	27.5	1.6	20.2	0.7	. . .	. . .	. . .
Female[5]	33.4	1.8	25.5	0.9	. . .	. . .	. . .
Not Hispanic or Latino							
White only, male	27.3	1.8	20.3	0.8	. . .	. . .	. . .
White only, female[5]	30.1	2.1	22.9	1.1	. . .	. . .	. . .
Black or African American only, male	28.1	2.3	20.9	1.0	. . .	. . .	. . .
Black or African American only, female[5]	49.7	2.8	38.3	1.4	. . .	. . .	. . .
Mexican male[7]	28.9	2.3	23.8	1.0	. . .	. . .	. . .
Mexican female[5, 7]	39.7	3.6	35.2	1.4	. . .	. . .	. . .
20 Years and Over, Crude							
Both sexes[5, 6]	30.3	1.4	22.3	0.6	. . .	. . .	. . .
Male	27.1	1.6	19.5	0.7	. . .	. . .	. . .
Female[5]	33.3	1.8	25.0	0.9	. . .	. . .	. . .
Not Hispanic or Latino							
White only, male	27.2	1.8	19.9	0.8	. . .	. . .	. . .
White only, female[5]	30.3	2.1	22.7	1.1	. . .	. . .	. . .
Black or African American only, male	28.1	2.6	20.7	1.0	. . .	. . .	. . .
Black or African American only, female[5]	49.4	2.8	36.7	1.4	. . .	. . .	. . .
Mexican male[7]	29.3	2.8	20.6	1.2	. . .	. . .	. . .
Mexican female[5, 7]	37.9	4.1	33.3	1.3	. . .	. . .	. . .
Male							
20–34 years	24.1	1.9	14.1	1.0	8.9	9.7	9.2
35–44 years	25.2	3.0	21.5	1.2	13.5	13.5	12.1
45–54 years	30.1	4.4	23.2	1.7	16.7	13.7	12.5
55–64 years	32.9	4.1	27.2	2.2	14.1	14.1	9.2
65–74 years	33.4	3.9	24.1	1.8	13.2	10.9	10.4
75 years and over	20.4	2.8	13.2	2.1	. . .	. . .	. . .
Female[5]							
20–34 years	25.8	2.6	18.5	1.1	11.0	9.7	7.2
35–44 years	33.9	3.0	25.5	2.1	17.8	17.7	14.7
45–54 years	38.1	4.1	32.4	1.9	19.6	18.9	20.3
55–64 years	43.1	4.2	33.7	1.8	22.9	24.1	24.4
65–74 years	38.8	3.8	26.9	1.5	21.5	22.0	23.2
75 years and over	25.1	4.0	19.2	1.3	. . .	. . .	. . .

[1]Standard errors of estimates for 1988–1994 and 1999–2000 are shown, and 1999–2000 estimates are based on a smaller sample size than estimates for earlier time periods and therefore are subject to greater sampling error.
[2]Data for Mexicans are for 1982–1984.
[4]Age-adjusted to the 2000 standard population using five age groups except for the 1999–2000 estimates, which are age-adjusted using three age groups. However, use of three rather than five age groups had virtually no effect on age-adjusted estimates.
[5]Excludes pregnant women.
[6]Includes persons of all races and Hispanic origins, not just those shown separately.
[7]Persons of Mexican origin may be of any race.
[8]Body mass index (BMI) greater than or equal to 30.
. . . = Not available.

Table III-7. Overweight, Obesity, and Healthy Weight Among Persons 20 Years and Over, According to Sex, Age, Race, and Hispanic Origin, Selected Years, 1960–2000—*Continued*

(Percent.)

Sex, age, race, and Hispanic origin	1999–2000	1999–2000 standard error [1]	1988–1994	1988–1994 standard error [1]	1976–1980 [2]	1971–1974	1960–1962
PERCENT OF POPULATION WITH A HEALTHY WEIGHT [9]							
20–74 Years, Age-Adjusted [4]							
Both sexes [5, 6]	33.6	1.4	41.7	0.9	49.6	48.8	51.2
Male	31.8	1.5	37.9	1.0	45.4	43.0	48.3
Female [5]	35.3	2.0	45.3	1.1	53.7	54.3	54.1
Not Hispanic or Latino							
White only, male	31.3	2.0	37.4	1.2	45.3	...	...
White only, female [5]	39.1	2.6	49.2	1.4	56.7	...	...
Black or African American only, male	38.1	2.2	40.0	1.2	46.6	...	...
Black or African American only, female [5]	22.0	1.9	28.9	1.2	35.0	...	...
Mexican male [7]	25.0	2.7	29.8	1.1	37.1	...	...
Mexican female [5, 7]	26.9	2.6	29.0	1.7	36.4	...	...
20 Years and Over, Age-Adjusted [4]							
Both sexes [5, 6]	33.5	1.3	41.6	0.8	...	...	...
Male	31.7	1.4	37.9	1.0	...	...	...
Female [5]	35.3	1.9	45.0	1.0	...	...	...
Not Hispanic or Latino							
White only, male	31.3	1.8	37.3	1.1	...	...	...
White only, female [5]	39.0	2.5	48.7	1.3	...	...	...
Black or African American only, male	37.4	1.9	40.1	1.2	...	...	...
Black or African American only, female [5]	22.3	2.0	29.2	1.2	...	...	...
Mexican male [7]	24.8	2.5	30.2	1.0	...	...	...
Mexican female [5, 7]	26.6	2.4	29.7	1.6	...	...	...
20 Years and Over, Crude							
Both sexes [5, 6]	33.9	1.3	42.6	0.8	...	...	...
Male	32.3	1.4	39.4	1.0	...	...	...
Female [5]	35.4	1.9	45.7	1.0	...	...	...
Not Hispanic or Latino							
White only, male	31.4	1.8	38.2	1.2	...	...	...
White only, female [5]	38.6	2.5	48.8	1.2	...	...	...
Black or African American only, male	39.3	2.2	41.5	1.2	...	...	...
Black or African American only, female [5]	23.1	2.1	31.2	1.3	...	...	...
Mexican male [7]	27.4	3.5	35.2	1.5	...	...	...
Mexican female [5, 7]	29.1	3.1	32.4	1.5	...	...	...
Male							
20–34 years	39.8	2.5	51.1	1.5	57.1	54.7	55.3
35–44 years	31.4	2.7	33.4	1.7	41.3	35.2	45.2
45–54 years	28.7	3.5	33.6	2.0	38.7	38.5	44.8
55–64 years	26.5	3.2	28.6	2.1	38.7	38.3	44.9
65–74 years	21.5	3.3	30.1	2.2	42.3	42.1	46.2
75 years and over	32.7	3.4	40.9	1.9	...	...	...
Female [5]							
20–34 years	43.9	2.5	57.9	1.3	65.0	65.8	67.6
35–44 years	33.6	4.0	47.1	2.5	55.6	56.7	58.4
45–54 years	33.3	3.8	37.2	2.3	48.7	49.3	47.6
55–64 years	26.9	3.7	31.5	1.5	43.5	41.1	38.1
65–74 years	27.7	3.9	37.0	2.0	37.8	40.6	36.4
75 years and over	37.4	3.7	43.0	1.6	...	...	...

Note: Percents do not sum to 100 because the percent of persons with BMI less than 18.5 is not shown and the percent of persons with obesity is a subset of the percent that is overweight.

[1]Standard errors of estimates for 1988–1994 and 1999–2000 are shown, and 1999–2000 estimates are based on a smaller sample size than estimates for earlier time periods and therefore are subject to greater sampling error.
[2]Data for Mexicans are for 1982–1984.
[4]Age-adjusted to the 2000 standard population using five age groups except for the 1999–2000 estimates, which are age-adjusted using three age groups. However, use of three rather than five age groups had virtually no effect on age-adjusted estimates.
[5]Excludes pregnant women.
[6]Includes persons of all races and Hispanic origins, not just those shown separately.
[7]Persons of Mexican origin may be of any race.
[9]BMI of 18.5 to less than 25 kilograms/meter squared.
. . . = Not available.

Figure III-3. Healthy Weight, Overweight, and Obesity Rates Among Persons 20 Years of Age and Over, 1960–2000

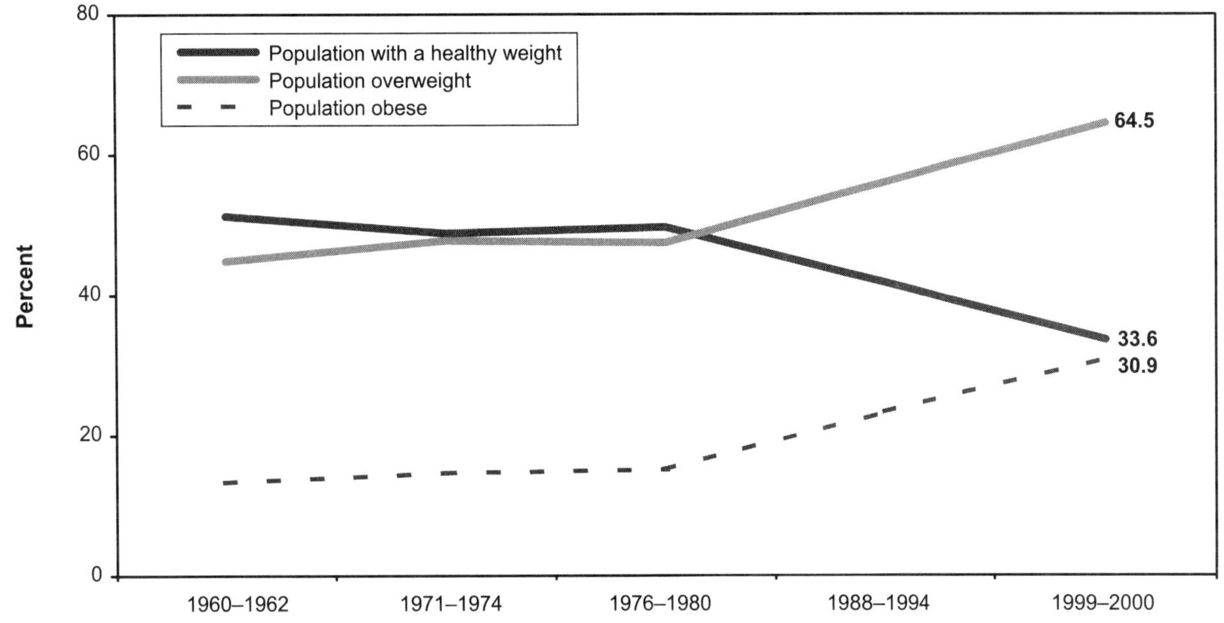

Figure III-4. Overweight Children and Adolescents 6–19 Years of Age, 1971–2000

TABLE III-8. OVERWEIGHT CHILDREN AND ADOLESCENTS 6–19 YEARS OF AGE, ACCORDING TO SEX, AGE, RACE, AND HISPANIC ORIGIN, SELECTED YEARS 1963–2000

In this table overweight is defined in terms of the body mass index (BMI) computed on the 2000 CDC Growth Charts. The cut off points are the sex and age-specific 95th percentiles of BMI on these charts, and overweight is defined to be at or above these points.

In the 6 to 11 year age group, 15.3 percent of children were classified as overweight in 1999–2000, and in the 12 to 19 year age group 15.5 percent fell into this classification. This represented big increases over the last three and a half decades. Mexican boys and non-Hispanic Black girls were more frequently overweight than the average.

Table III-8. Overweight Children and Adolescents 6–19 Years of Age, According to Sex, Age, Race, and Hispanic Origin, Selected Years, 1963–2000

(Percent.)

Sex, age, race, and Hispanic origin	1999–2000	1999–2000 standard error [1]	1988–1994	1988–1994 standard error	1976–1980 [2]	1971–1974	1963–1966 1966–1970 [3]
PERCENT OF POPULATION OVERWEIGHT							
6–11 Years of Age							
Both sexes [4]	15.3	1.7	11.3	1.0	6.5	4.0	4.2
Boys	16.0	2.3	11.6	1.3	6.6	4.3	4.0
Not Hispanic or Latino							
White only	*11.9	3.0	10.7	2.0	6.1	...	...
Black or African American only	17.6	2.8	12.3	1.4	6.8	...	...
Mexican [5]	27.3	3.1	17.5	2.4	13.3	...	...
Girls [6]	14.5	2.4	11.0	1.4	6.4	3.6	4.5
Not Hispanic or Latino							
White only	*12.0	3.6	*9.8	2.0	5.2	...	...
Black or African American only	22.1	3.3	17.0	1.6	11.2	...	...
Mexican [5]	19.6	3.1	15.3	2.5	9.8	...	...
12–19 Years of Age							
Both sexes	15.5	1.2	10.5	0.9	5.0	6.1	4.6
Boys	15.5	1.6	11.3	1.3	4.8	6.1	4.5
Not Hispanic or Latino							
White only	13.0	2.5	11.6	1.9	3.8	...	...
Black or African American only	20.5	2.6	10.7	1.4	6.1	...	...
Mexican [5]	27.5	3.0	14.1	1.8	7.7	...	...
Girls [6]	15.5	1.6	9.7	1.1	5.3	6.2	4.7
Not Hispanic or Latino							
White only	12.2	2.1	8.9	1.7	4.6	...	...
Black or African American only	25.7	2.6	16.3	2.1	10.7	...	...
Mexican [5]	19.4	2.8	*13.4	3.1	8.8	...	...

Note: Overweight is defined as body mass index (BMI) at or above the sex and age-specific 95th percentile BMI cutoff points from the 2000 CDC Growth Charts.

[1]Standard errors of estimates for 1999–2000 are shown, and 1999–2000 estimates are based on a smaller sample size than estimates for earlier time periods and therefore are subject to greater sampling error.
[2]Data for Mexicans are for 1982–1984.
[3]Data for 1963–1965 are for children 6–11 years of age; data for 1966–1970 are for adolescents 12–17 years of age, not 12–19 years.
[4]Includes persons of all races and Hispanic origins, not just those shown separately.
[5]Persons of Mexican origin may be of any race.
[6]Excludes pregnant women.
. . . = Not available.
* = Figure does not meet standards of reliability or precision.

TABLE III-9. LIFE EXPECTANCIES AND EXPECTED YEARS WITH BODY MASS INDEX LESS THAN 25, AND YEARS WITH BODY MASS INDEX LESS THAN 30 FOR SELECTED AGES, 1994–1996

This table shows the average life expectancy remaining that will be spent with a healthy weight of less (BMI < 25) or with a weight less than the obesity level (BMI < 30). The trends by age in percent of life remaining spent at or below a certain BMI show that acquiring overweight tends to increase up to age 45 and decrease thereafter. At age 45, 82.4 percent of total life remaining will be spent below the obesity level, but by age 80 this measure is 91.5 percent. Subtracting these percentages from 100 provides estimates of remaining life of the overweight or obese population. These data present an interesting way of summarizing the prevalence of overweight and obesity as a personal and a public health burden.

Table III-9. Life Expectancies and Expected Years with Body Mass Index Less than 25, and Years with Body Mass Index Less than 30 for Selected Ages, 1994–1996

(Percent.)

Age and sex	Expected years of life	Expected years with BMI[1] less than 25	Years with BMI less than 25 as a percentage of life expectancy	Expected years with BMI less than 30	Years with BMI less than 30 as a percentage of life expectancy
Both Sexes					
20 years	56.9	27.9	49.0	47.6	83.7
30 years	47.5	22.0	46.4	39.3	82.7
45 years	33.8	15.2	44.9	27.8	82.4
65 years	17.5	8.8	50.4	15.0	85.8
70 years	14.1	7.5	53.5	12.3	87.5
75 years	11.1	6.4	57.9	9.9	89.6
80 years	8.3	5.3	63.6	7.6	91.5
Male					
20 years	54.0	21.6	40.0	45.4	84.0
30 years	44.9	16.7	37.1	37.3	83.1
45 years	31.5	11.6	36.8	26.2	83.1
65 years	15.9	7.1	44.9	14.0	88.0
70 years	12.8	6.2	48.8	11.5	90.0
75 years	10.1	5.5	54.2	9.3	92.2
80 years	7.7	4.6	60.3	7.3	94.2
Female					
20 years	59.8	34.2	57.2	49.8	83.4
30 years	50.1	27.4	54.6	41.2	82.4
45 years	35.9	18.6	51.8	29.3	81.8
65 years	18.8	10.2	54.3	15.8	84.2
70 years	15.1	8.6	56.6	13.0	85.9
75 years	11.7	7.1	60.2	10.3	88.0
80 years	8.7	5.7	65.3	7.8	90.1
Gender Difference[2]					
20 years	+5.7	+12.6	X	+4.5	X
30 years	+5.2	+10.7	X	+4.0	X
45 years	+4.3	+7.0	X	+3.1	X
65 years	+2.9	+3.1	X	+1.9	X
70 years	+2.3	+2.3	X	+1.5	X
75 years	+1.7	+1.6	X	+1.1	X
80 years	+1.0	+1.0	X	+0.5	X

[1]BMI refers to body mass index.
[2]Calculated by subtracting expected years for males from expected years for females.
X = Not applicable.

SELECTED CHRONIC DISEASES

TABLE III-10. RATE OF SELECTED CIRCULATORY DISEASES AMONG PERSONS 18 YEARS OF AGE AND OVER, BY SELECTED CHARACTERISTICS, 2001

Nearly 12 percent of adults had been told by a medical professional at some time that they had heart disease, and for 6 percent it was specified that they had coronary heart disease. More than 2 percent of adults had ever been told that they suffered a stroke. A large portion of adults, almost 21 percent, had been told on two or more visits that they had hypertension.

Such findings are considerably more frequent for persons aged 65 and above than for younger men or women. Heart diseases are substantially more frequent for older men than for older women, but older women were more likely to have been told that the had hypertension than men. In regard to poverty status, the not poor reported fewer cases of circulatory diseases than the poor and the near poor.

In a comparison of Whites, Blacks, Asians and American Indians, Asians reported the lowest incidence of circulatory diseases. Hispanics reported a lower frequency of heart diseases than non-Hispanic Whites and Blacks. However, after having been told on two or more visits that they had hypertension, non-Hispanic Whites, 65 years and over, had the lowest frequency (47.5 percent), followed by Hispanics (51.9 percent), and non-Hispanic Blacks (66.2 percent). Among women, Hispanic seniors had a high frequency of hypertension, and a much lower frequency among men.

Table III-10. Rate of Selected Circulatory Diseases Among Persons 18 Years of Age and Over, by Selected Characteristics, 2001

(Percent.)

Characteristic	Selected circulatory diseases [1]							
	Heart disease [2]							
	All types	Standard error for all types	Coronary [3]	Standard error for coronary	Hyper-tension [4]	Standard error for hyper-tension	Stroke	Standard error for stroke
Total [5]	11.5	0.21	6.3	0.16	20.5	0.27	2.4	0.09
Sex								
Male	11.7	0.30	7.4	0.25	19.6	0.39	2.3	0.14
Female	11.4	0.28	5.2	0.19	21.4	0.34	2.5	0.12
Age								
18–44 years	4.6	0.20	1.0	0.09	7.0	0.23	0.4	0.06
45–64 years	13.1	0.39	7.6	0.30	28.7	0.51	2.3	0.16
65–74 years	26.8	0.88	17.7	0.79	47.1	1.03	6.9	0.53
75 years and over	37.1	0.99	25.2	0.92	52.8	1.03	11.4	0.69
Race								
1 race [6]	11.5	0.21	6.2	0.16	20.5	0.27	2.4	0.09
White	12.2	0.24	6.6	0.18	20.1	0.29	2.4	0.10
Black or African American	9.6	0.50	5.2	0.38	26.7	0.84	2.9	0.28
American Indian or Alaska Native	12.6	3.21	8.6	2.57	25.0	3.39	*1.1	0.57
Asian	5.4	0.91	2.9	0.70	13.5	1.42	2.2	0.63
Native Hawaiian or other Pacific Islander	*1.6	1.58	*	-	*14.5	7.14	*6.3	6.06
2 or more races [7]	14.9	2.10	8.3	1.72	21.9	2.44	*2.4	0.82
Black or African American, white	*6.9	4.01	*	-	*16.8	7.08	*	-
American Indian or Alaska Native, white	17.5	3.18	12.1	2.81	23.3	3.59	*1.4	0.75
Hispanic or Latino Origin and Race								
Hispanic or Latino [8]	6.1	0.39	3.5	0.30	14.5	0.59	1.8	0.23
Mexican or Mexican American	5.0	0.43	2.9	0.34	12.9	0.70	1.6	0.28
Not Hispanic or Latino	12.2	0.23	6.6	0.17	21.3	0.29	2.5	0.10
White, single race	12.9	0.26	6.9	0.19	20.7	0.31	2.4	0.11
Black or African American, single race	9.6	0.49	5.3	0.38	26.9	0.86	2.9	0.29
Poverty Status [9]								
Poor	14.7	0.76	8.1	0.58	21.7	0.95	3.5	0.35
Near poor	14.2	0.55	8.5	0.45	26.0	0.74	3.9	0.30
Not poor	9.8	0.28	4.7	0.19	18.3	0.33	1.7	0.11
Region								
Northeast	11.7	0.46	6.4	0.38	19.8	0.57	1.9	0.18
Midwest	12.1	0.46	6.2	0.33	20.9	0.52	2.4	0.19
South	11.6	0.35	6.5	0.27	22.3	0.48	2.5	0.15
West	10.7	0.48	5.7	0.30	17.7	0.57	2.5	0.20
Sex and Age								
Male								
18–44 years	3.6	0.26	1.1	0.14	7.0	0.35	0.3	0.07
45–64 years	14.1	0.58	9.4	0.48	29.2	0.73	2.3	0.23
65 years and over	37.5	1.14	27.1	1.11	47.4	1.19	9.9	0.73
Female								
18–44 years	5.6	0.28	0.9	0.12	7.1	0.30	0.5	0.09
45–64 years	12.2	0.54	5.9	0.38	28.3	0.68	2.3	0.21
65 years and over	27.0	0.82	16.8	0.69	51.4	0.94	8.3	0.51

[1]In separate questions, respondents were asked if they have ever been told by a doctor of other health professional that they had: hypertension (high blood pressure), coronary heart disease, angina (or angina pectoris), heart attack (or mycardial infarction), and any other heart condition or disease not already mentioned or a stroke. A person may be represented in more than one column.
[2]Heart disease includes coronary heart disease, angina pectoris, heart attack, or any other heart condition or disease.
[3]Coronary heart disease includes coronary heart disease, angina pectoris, or heart attack.
[4]Persons had to have been told on two or more different visits that they had hypertension or high blood pressure to be classified as hypertensive.
[5]Total includes other races not shown separately and and persons with unknown poverty status.
[6]The category "1 race" refers to people who indicated only a single race group. Persons who indicated a single race other than the groups shown are included in the total for "1 race" but not shown separately due to small sample sizes. Therefore the frequencies for the category "1 race" will will be greater than the sum of the frequencies for the specific groups shown separately.
[7]The category "2 or more races" refers to all persons who indicated more than one race group. Only two combinations of race groups are shown due to small sample sizes for other combinations. Therefore, the frequencies for the category "2 or more races" will be greater than the sum of the frequencies for the specific combinations shown separately.
[8]Persons of Hispanic or Latino origin may be of any race or any combination of races. Similarly, the category "Not Hispanic or Latino" refers to all persons who are not of Hispanic or Latino origin, regardless of race.
[9]Poverty status is based on family income and family size using the U.S. Census Bureau's poverty thresholds for the previous calendar year. "Poor" persons are defined as below the poverty threshold. "Near poor" persons have incomes of 100 percent to less than 200 percent of the poverty threshold. "Not poor" persons have incomes that are 200 percent of the poverty threshold or greater.
* = Figure does not meet standards of reliability or precision.
- = Quantity zero.

Table III-10. Rate of Selected Circulatory Diseases Among Persons 18 Years of Age and Over, by Selected Characteristics, 2001—*Continued*

(Percent.)

Characteristic	Selected circulatory diseases [1]							
	Heart disease [2]							
	All types	Standard error for all types	Coronary [3]	Standard error for coronary	Hyper-tension [4]	Standard error for hyper-tension	Stroke	Standard error for stroke
Hispanic or Latino Origin, Race, and Age								
Hispanic or Latino [8]								
18–44 years	2.8	0.33	1.1	0.19	5.2	0.39	*0.4	0.11
45–64 years	8.9	0.95	5.9	0.81	26.6	1.49	2.3	0.52
65 years and over	22.9	2.17	15.5	1.92	51.9	2.60	10.5	1.75
Not Hispanic or Latino								
White, single race								
18–44 years	5.2	0.27	1.0	0.12	6.9	0.28	0.4	0.07
45–64 years	13.5	0.46	7.7	0.35	26.8	0.58	1.9	0.17
65 years and over	32.5	0.75	21.7	0.69	47.5	0.79	8.6	0.46
Black or African American, single race								
18–44 years	3.8	0.43	1.0	0.24	10.6	0.72	0.6	0.17
45–64 years	14.3	1.10	8.4	0.87	45.6	1.49	4.8	0.65
65 years and over	27.8	2.04	19.6	1.83	66.2	2.25	10.3	1.48
Hispanic or Latino Origin, Race, Age and Sex								
Hispanic or Latino, male [8]								
18–44 years	1.9	0.38	1.0	0.29	4.6	0.54	*0.3	0.17
45–64 years	10.4	1.64	7.8	1.46	24.7	2.12	3.5	0.96
65 years and over	28.2	4.02	20.1	3.40	40.6	3.75	12.4	3.08
Hispanic or Latina, Female [8]								
18–44 years	3.6	0.55	1.1	0.24	5.8	0.57	*0.4	0.14
45–64 years	7.5	1.02	4.2	0.76	28.3	2.04	*1.3	0.46
65 years and over	19.0	2.44	12.1	2.13	60.2	3.28	9.0	2.03
Not Hispanic or Latino								
White, single race, male								
18–44 years	4.1	0.34	1.1	0.18	7.0	0.40	0.3	0.09
45–64 years	14.8	0.67	9.8	0.57	28.3	0.83	1.8	0.24
65 years and over	39.0	1.26	28.3	1.23	46.3	1.30	9.6	0.79
White, single race, female								
18–44 years	6.4	0.37	0.9	0.16	6.7	0.38	0.6	0.11
45–64 years	12.3	0.65	5.7	0.44	25.3	0.77	2.0	0.25
65 years and over	27.6	0.91	16.7	0.75	48.5	1.04	7.8	0.55
Black or African American, single race, male								
18–44 years	3.6	0.76	*1.1	0.39	10.2	1.23	*0.5	0.25
45–64 years	12.5	1.48	8.0	1.19	40.0	2.32	4.9	1.09
65 years and over	30.3	3.69	20.6	3.35	61.9	3.77	9.6	2.33
Black or African American, single race, female								
18–44 years	4.0	0.50	1.0	0.27	10.8	0.83	*0.7	0.22
45–64 years	15.8	1.52	8.7	1.21	50.1	1.92	4.7	0.75
65 years and over	26.2	2.35	19.0	2.14	69.0	2.58	10.8	1.87
Poverty Status and Age [9]								
Poor								
18–44 years	7.2	0.68	2.5	0.50	7.6	0.72	1.1	0.26
45–64 years	23.7	1.87	14.8	1.45	38.1	2.01	6.4	0.99
65 years and over	32.6	2.07	21.8	1.86	56.8	2.47	9.8	1.31
Near poor								
18–44 years	5.2	0.52	1.4	0.25	8.9	0.67	0.7	0.19
45–64 years	16.9	1.25	10.9	1.11	37.2	1.87	3.9	0.65
65 years and over	33.5	1.48	23.6	1.38	57.0	1.61	11.8	1.02
Not poor								
18–44 years	4.5	0.26	0.8	0.10	6.8	0.30	0.3	0.07
45–64 years	11.6	0.51	6.2	0.36	27.7	0.65	1.7	0.19
65 years and over	30.4	1.08	19.8	0.97	46.3	1.24	7.9	0.66

[1]In separate questions, respondents were asked if they have ever been told by a doctor of other health professional that they had: hypertension (high blood pressure), coronary heart disease, angina (or angina pectoris), heart attack (or mycardial infarction), and any other heart condition or disease not already mentioned or a stroke. A person may be represented in more than one column.

[2]Heart disease includes coronary heart disease, angina pectoris, heart attack, or any other heart condition or disease.

[3]Coronary heart disease includes coronary heart disease, angina pectoris, or heart attack.

[4]Persons had to have been told on two or more different visits that they had hypertension or high blood pressure to be classified as hypertensive.

[8]Persons of Hispanic or Latino origin may be of any race or any combination of races. Similarly, the category "Not Hispanic or Latino" refers to all persons who are not of Hispanic or Latino origin, regardless of race.

[9]Poverty status is based on family income and family size using the U.S. Census Bureau's poverty thresholds for the previous calendar year. "Poor" persons are defined as below the poverty threshold. "Near poor" persons have incomes of 100 percent to less than 200 percent of the poverty threshold. "Not poor" persons have incomes that are 200 percent of the poverty threshold or greater.

* = Figure does not meet standards of reliability or precision.

- = Quantity zero.

Figure III-5. Percent of Selected Circulatory Diseases Among Persons 18 Years of Age and Over by Race and Hispanic Origin, 2001

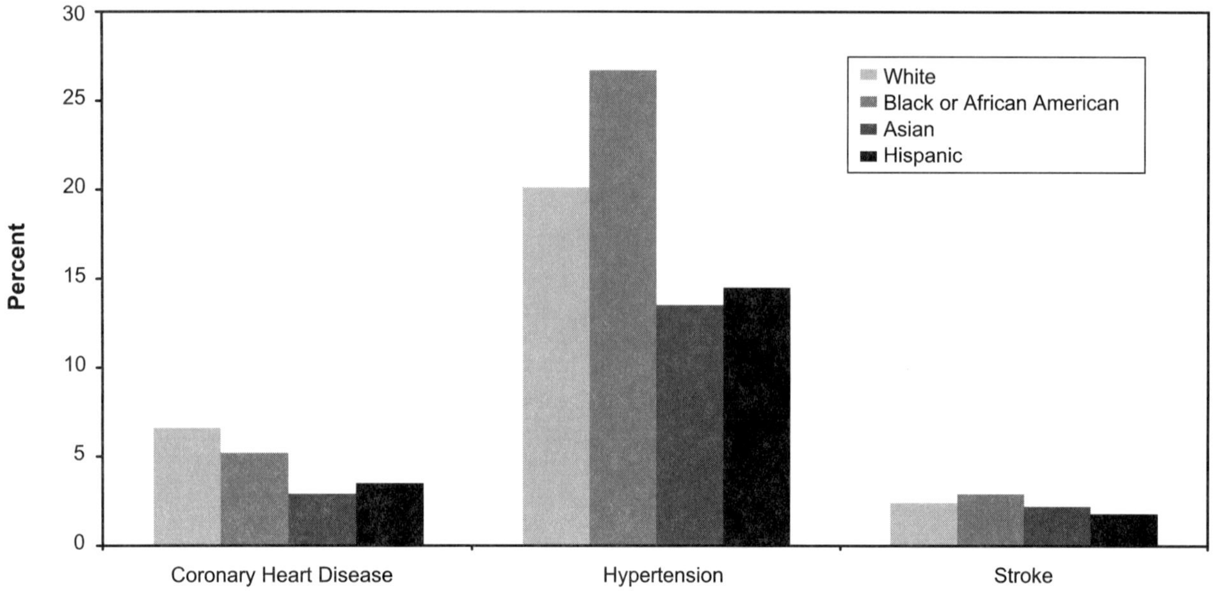

TABLE III-11. HYPERTENSION AMONG PERSONS 20 YEARS OF AGE AND OVER, ACCORDING TO SEX, AGE, RACE, AND HISPANIC ORIGIN, 1988–1994 AND 1999–2000

In 1999–2000 on the basis of age-adjusted figures, 25.6 percent of persons from age 20 to 74 had hypertension. The classification of hypertension is based on either an elevated blood pressure (systolic pressure of at least 140 mmHg or diastolic pressure of at least 90 mmHg) or taking anti-hypertensive medication.

The data shown here are somewhat different from those shown in the previous table. That table included 18 and 19 year old persons who generally do not have elevated blood pressure. Also the previous table was based on household interviews and recollections of doctors' comments, whereas this table is predominantly based on actual examinations of the respondents.

The frequency of hypertension had increased since 1988–1994, perhaps with the general tendency to gain weight. In the earlier period, 21.7 percent of the 20 to 74-year population (or 3.9 percentage points less) had hypertension. Among races, non-Hispanic Blacks were particularly afflicted with hypertension; in 1999–2000, 34.8 percent of Black males had hypertension, and among Black females this percentage increased to 38.6. The frequency of hypertension tends to increase with age. Thus, the frequency of hypertension was below average for persons in their 30s and early 40s, but it rose to 59.9 percent and 68.8 percent for males aged 65–74, and 75 and over, respectively. For females these figures were 70.3 percent and 84.1 percent for the 65–74 year age group, and the 75-year and over group, respectively.

Table III-11. Hypertension Among Persons 20 Years of Age and Over, According to Sex, Age, Race, and Hispanic Origin, 1988–1994 and 1999–2000

(Percent.)

Sex, age, race, and Hispanic origin	1999–2000	Standard error [1]	1988–1994	Standard error [1]
20–74 Years, Age-Adjusted [2]				
Both sexes [3, 4]	25.6	1.3	21.7	0.6
Male	25.9	1.7	23.4	0.9
Female [3]	25.2	1.4	20.0	0.6
Not Hispanic or Latino				
White only, male	24.8	1.9	22.6	1.1
White only, female [3]	22.6	1.7	18.4	0.7
Black or African American only, male	34.8	2.4	34.3	1.1
Black or African American only, female [3]	38.6	2.1	34.9	0.9
Mexican male [5]	24.2	2.1	23.4	1.1
Mexican female [3, 5]	23.8	1.8	20.9	1.0
20 Years and Over, Age-Adjusted [2]				
Both Sexes [3, 4]	29.8	1.2	25.5	0.6
Male	29.2	1.6	26.4	0.9
Female [3]	29.9	1.4	24.4	0.6
Not Hispanic or Latino				
White only, male	28.2	1.7	25.6	1.1
White only, female [3]	27.6	1.6	22.9	0.7
Black or African American only, male	38.6	2.2	37.5	1.2
Black or African American only, female [3]	42.4	2.0	38.2	0.9
Mexican male [5]	28.0	2.1	26.9	1.1
Mexican female [3, 5]	28.0	1.7	25.0	0.9
20 Years and Over, Crude				
Both sexes [3, 4]	28.9	1.3	24.1	0.8
Male	27.4	1.5	23.8	1.0
Female [3]	30.3	1.5	24.4	0.8
Not Hispanic or Latino				
White only, male	28.1	1.6	24.3	1.2
White only, female [3]	30.1	1.9	24.6	1.1
Black or African American only, male	32.7	2.6	31.1	1.3
Black or African American only, female [3]	38.1	3.0	32.3	1.2
Mexican male [5]	19.3	2.1	16.4	1.1
Mexican female [3, 5]	19.9	2.1	15.9	0.7
Male				
20–34 years	*9.8	2.1	7.1	0.8
35–44 years	17.1	2.8	17.1	1.9
45–54 years	32.3	3.0	29.2	2.1
55–64 years	44.1	3.9	40.6	2.3
65–74 years	59.9	4.0	54.4	2.8
75 years and over	68.8	3.4	60.4	2.2
Female [3]				
20–34 years	*	*	2.9	0.6
35–44 years	16.0	2.5	11.2	1.1
45–54 years	30.5	3.4	23.9	2.0
55–64 years	53.0	3.1	42.5	2.3
65–74 years	70.3	3.5	56.1	1.7
75 years and over	84.1	2.7	73.5	1.8

[1]Standards errors of estimates for 1999–2000 are shown, and 1999–2000 estimates are based on a smaller sample size than estimates for earlier time periods and therefore are subject to greater sampling error.
[2]Age-adjusted to the 2000 standard population using five age groups except for the 1999–2000 estimates, which are age-adjusted using three age groups. However, use of three rather than five age groups had virtually no effect on age-adjusted estimates.
[3]Excludes pregnant women.
[4]Includes persons of all races and Hispanic origins, not just those shown separately.
[5]Persons of Mexican origin may be of any race.
* = Figure does not meet standards of reliability or precision.

TABLE III-12. SERUM CHOLESTEROL LEVELS AMONG PERSONS 20 YEARS OF AGE AND OVER, ACCORDING TO SEX, AGE, RACE, AND HISPANIC ORIGIN, SELECTED YEARS, 1960–2000

In 1999–2000, on the basis of age-adjusted figures, 18.0 percent of persons from age 20 to 74 had high cholesterol levels. These are defined as equal or above 240mg/dL (or 6.20mmo/L). The data represented a steady reduction since 1960–1963, when high cholesterol was found in 33.3 percent of the 20–74 age group. In recent data non-Hispanic Whites had slightly higher cholesterol percentages than the average for the nation.

The mean serum cholesterol level for the population above 20 years of age was 204mg/dL. The highest level for males occurred between 45 and 54 years of age, and for females it occurred between 65 and 74 years.

Table III-12. Serum Cholesterol Levels Among Persons 20 Years of Age and Over, According to Sex, Age, Race and Hispanic Origin, Selected Years, 1960–2000

(Percent.)

Sex, age, race, and Hispanic origin	1999–2000	1999–2000 standard error [1]	1988–1994	1988–1994 standard error [1]	1976–1980 [2]	1971–1974	1960–1962
PERCENT OF POPULATION WITH HIGH SERUM CHOLESTEROL							
20–74 Years, Age-Adjusted [3]							
Both sexes [4]	18.0	0.9	19.7	0.6	27.8	28.6	33.3
Male	17.7	1.1	18.8	0.8	26.4	27.9	30.6
Female	18.2	1.3	20.5	0.8	28.8	29.1	35.6
Not Hispanic or Latino							
White only, male	18.3	1.4	18.7	0.9	26.4	. . .	. . .
White only, female	19.3	1.7	20.7	1.0	29.6	. . .	. . .
Black or African American only, male	10.7	2.1	16.4	1.0	25.5	. . .	. . .
Black or African American only, female	16.5	2.5	19.9	0.8	26.3	. . .	. . .
Mexican male [5]	17.8	1.9	18.7	1.5	20.3	. . .	. . .
Mexican female [5]	13.1	1.3	17.7	1.2	20.5	. . .	. . .
20 Years and Over, Age-Adjusted [3]							
Both sexes [4]	18.3	0.8	20.8	0.6	. . .	. . .	. . .
Male	17.2	1.1	19.0	0.7	. . .	. . .	. . .
Female	19.1	1.2	22.0	0.8	. . .	. . .	. . .
Not Hispanic or Latino							
White only, male	17.8	1.3	18.8	0.8	. . .	. . .	. . .
White only, female	19.9	1.6	22.2	1.0	. . .	. . .	. . .
Black or African American only, male	10.6	1.9	16.9	0.9	. . .	. . .	. . .
Black or African American only, female	17.7	2.4	21.4	0.9	. . .	. . .	. . .
Mexican male [5]	17.8	1.9	18.5	1.6	. . .	. . .	. . .
Mexican female [5]	13.9	1.3	18.7	1.3	. . .	. . .	. . .
20 Years and Over, Crude							
Both sexes [4]	17.8	0.9	19.6	0.6	. . .	. . .	. . .
Male	16.7	1.1	17.7	0.7	. . .	. . .	. . .
Female	18.7	1.2	21.3	0.9	. . .	. . .	. . .
Not Hispanic or Latino							
White only, male	17.7	1.4	18.0	0.8	. . .	. . .	. . .
White only, female	20.2	1.6	22.5	1.1	. . .	. . .	. . .
Black or African American only, male	*10.1	2.1	14.7	1.0	. . .	. . .	. . .
Black or African American only, female	15.9	2.2	18.2	0.9	. . .	. . .	. . .
Mexican male [5]	14.8	2.0	15.4	1.3	. . .	. . .	. . .
Mexican female [5]	11.3	1.4	14.3	1.1	. . .	. . .	. . .
Male							
20–34 years	11.0	1.5	8.2	0.9	11.9	12.4	15.1
35–44 years	21.1	3.3	19.4	1.6	27.9	31.8	33.9
45–54 years	22.9	3.4	26.6	2.3	36.9	37.5	39.2
55–64 years	16.5	2.4	28.0	2.1	36.8	36.2	41.6
65–74 years	19.2	2.8	21.9	2.2	31.7	34.7	38.0
75 years and over	*10.1	2.1	20.4	1.8	. . .	. . .	. . .
Female							
20–34 years	9.3	1.4	7.3	1.0	9.8	10.9	12.4
35–44 years	12.8	2.5	12.3	1.3	20.7	19.3	23.1
45–54 years	23.7	4.2	26.7	2.1	40.5	38.7	46.9
55–64 years	26.2	3.1	40.9	1.9	52.9	53.1	70.1
65–74 years	37.4	4.0	41.3	2.4	51.6	57.7	68.5
75 years and over	27.6	2.5	38.2	2.2	. . .	. . .	. . .

[1]Standards errors of estimates for 1999–2000 are shown, and 1999–2000 estimates are based on a smaller sample size than estimates for earlier time periods and therefore are subject to greater sampling error.
[2]Data for Mexicans are for 1982–1984.
[3]Age-adjusted to the 2000 standard population using five age groups except for the 1999–2000 estimates, which are age-adjusted using three age groups. However, use of three rather than five age groups had virtually no effect on age-adjusted estimates.
[4]Includes persons of all races and Hispanic origins, not just those shown separately.
[5]Persons of Mexican origin may be of any race.
* = Figure does not meet standards of reliability or precision.
. . . = Not available.

Table III-12. Serum Cholesterol Levels Among Persons 20 Years of Age and Over, According to Sex, Age, Race and Hispanic Origin, Selected Years, 1960–2000—*Continued*

(Percent.)

Sex, age, race, and Hispanic origin	1999–2000	1999–2000 standard error [1]	1988–1994	1988–1994 standard error [1]	1976–1980 [2]	1971–1974	1960–1962
MEAN SERUM CHOLESTEROL LEVEL, MG/DL							
20–74 Years, Age-Adjusted [3]							
Both sexes [4]	204.0	1.0	205.0	0.8	215.0	216.0	222.0
Male	204.0	1.2	204.0	0.9	213.0	216.0	220.0
Female	203.0	1.4	205.0	0.8	216.0	217.0	224.0
Not Hispanic or Latino							
White only, male	204.0	1.6	204.0	1.0	213.0	. . .	. . .
White only, female	206.0	1.7	206.0	1.1	216.0	. . .	. . .
Black or African American only, male	192.0	2.5	201.0	1.3	211.0	. . .	. . .
Black or African American only, female	200.0	2.8	204.0	0.6	216.0	. . .	. . .
Mexican male [5]	207.0	1.7	206.0	1.6	209.0	. . .	. . .
Mexican female [5]	198.0	2.0	204.0	1.3	209.0	. . .	. . .
20 Years and Over, Age-Adjusted [3]							
Both sexes [4]	204.0	0.9	206.0	0.7	. . .	. . .	. . .
Male	203.0	1.1	204.0	0.9	. . .	. . .	. . .
Female	205.0	1.3	207.0	0.8	. . .	. . .	. . .
Not Hispanic or Latino							
White only, male	204.0	1.5	205.0	1.0	. . .	. . .	. . .
White only, female	207.0	1.6	208.0	1.1	. . .	. . .	. . .
Black or African American only, male	192.0	2.3	202.0	1.3	. . .	. . .	. . .
Black or African American only, female	201.0	2.7	207.0	0.7	. . .	. . .	. . .
Mexican male [5]	207.0	1.7	206.0	1.5	. . .	. . .	. . .
Mexican female [5]	199.0	1.9	206.0	1.3	. . .	. . .	. . .
20 Years and Over, Crude							
Both sexes [4]	203.0	1.0	204.0	0.8	. . .	. . .	. . .
Male	202.0	1.1	202.0	0.9	. . .	. . .	. . .
Female	204.0	1.5	206.0	0.9	. . .	. . .	. . .
Not Hispanic or Latino							
White only, male	203.0	1.5	203.0	1.0	. . .	. . .	. . .
White only, female	207.0	1.6	208.0	1.3	. . .	. . .	. . .
Black or African American only, male	190.0	2.5	198.0	1.3	. . .	. . .	. . .
Black or African American only, female	198.0	2.8	201.0	0.7	. . .	. . .	. . .
Mexican male [5]	203.0	1.8	199.0	1.6	. . .	. . .	. . .
Mexican female [5]	195.0	2.0	198.0	1.5	. . .	. . .	. . .
Male							
20–34 years	189.0	2.0	186.0	1.2	192.0	194.0	198.0
35–44 years	205.0	2.8	206.0	1.6	217.0	221.0	227.0
45–54 years	215.0	2.9	216.0	1.8	227.0	229.0	231.0
55–64 years	210.0	2.5	216.0	2.2	229.0	229.0	233.0
65–74 years	210.0	2.6	212.0	1.9	221.0	226.0	230.0
75 years and over	194.0	2.4	205.0	1.9	. . .	. . .	. . .
Female							
20–34 years	186.0	2.2	184.0	1.3	189.0	191.0	194.0
35–44 years	197.0	2.1	195.0	1.4	207.0	207.0	214.0
45–54 years	213.0	3.1	217.0	2.3	232.0	232.0	237.0
55–64 years	223.0	2.8	235.0	1.6	249.0	245.0	262.0
65–74 years	229.0	3.2	233.0	1.9	246.0	250.0	266.0
75 years and over	219.0	2.5	229.0	2.0	. . .	. . .	. . .

[1]Standards errors of estimates for 1999–2000 are shown, and 1999–2000 estimates are based on a smaller sample size than estimates for earlier time periods and therefore are subject to greater sampling error.
[2]Data for Mexicans are for 1982–1984.
[3]Age-adjusted to the 2000 standard population using five age groups except for the 1999–2000 estimates, which are age-adjusted using three age groups. However, use of three rather than five age groups had virtually no effect on age-adjusted estimates.
[4]Includes persons of all races and Hispanic origins, not just those shown separately.
[5]Persons of Mexican origin may be of any race.
. . . = Not available.

TABLE III-13. LIFE EXPECTANCIES AND EXPECTED YEARS WITHOUT CHRONIC ARTHRITIS AND CHRONIC HEART DISEASES FOR SELECTED AGES, BY SEX, 1994–1996

Estimates based on three years, with 1995 as the midpoint, suggest that a person aged 65 years could expect to live 17.5 more years. Only 8.8 years, just over one half, were expected to be free of chronic arthritis for the average person. Even though females tend to live longer, the expectation for males was that their shorter life would nevertheless include more years free of chronic arthritis. This finding held for every age level.

Females, on the other hand, were more likely than males to be free of chronic heart diseases. At age 65, for example, females were expected to live 18.8 more years, with 13.6 years, on average, free of chronic heart diseases. Males at that age were expected to live 15.9 additional years, but only 10.2 of these years would be without chronic heart disease. Thus of their life expectancy at age 65, females would spend 72.3 percent of the time free of chronic heart disease, while males, also at age 65, would have only 64.5 percent of the remaining life span free of chronic heart disease.

Table III-13. Life Expectancies and Expected Years without Chronic Arthritis and Chronic Heart Diseases for Selected Ages, by Sex, 1994–1996

(Percent.)

Age and sex	Expected years of life	Expected years without chronic arthritis	Years without chronic arthritis as a percentage of life expectancy	Expected years without chronic heart disease	Years without chronic heart diseases as a percentage of life expectancy
Both Sexes					
30 years	47.5	34.7	73.0	40.0	84.1
45 years	33.8	21.6	63.9	26.7	79.0
65 years	17.5	8.8	50.2	12.1	69.1
70 years	14.1	6.8	48.2	9.4	66.5
75 years	11.1	5.1	46.1	7.1	64.2
80 years	8.3	3.8	45.9	5.3	64.1
Male					
30 years	44.9	35.4	78.8	37.1	82.8
45 years	31.5	22.5	71.3	24.0	76.1
65 years	15.9	9.2	58.1	10.2	64.5
70 years	12.8	7.1	55.6	7.7	60.6
75 years	10.1	5.7	56.2	5.8	57.6
80 years	7.7	4.2	54.0	4.5	58.8
Female					
30 years	50.1	34.0	68.0	42.7	85.3
45 years	35.9	20.8	57.9	29.2	81.3
65 years	18.8	8.4	44.8	13.6	72.3
70 years	15.1	6.6	43.6	10.6	70.4
75 years	11.7	4.7	40.1	8.0	68.1
80 years	8.7	3.6	41.6	5.8	67.0
Gender Difference [1]					
30 years	+5.2	-1.3	X	+5.6	X
45 years	+4.3	-1.7	X	+5.2	X
65 years	+2.9	-0.8	X	+3.3	X
70 years	+2.3	-0.5	X	+2.9	X
75 years	+1.7	-0.9	X	+2.2	X
80 years	+1.0	-0.6	X	+1.3	X

[1]Calculated by subtracting expected years for males from expected years for females.
X = Not applicable.

TABLE III-14. RATE OF CANCER AMONG PERSONS 18 YEARS OF AGE AND OVER, BY SELECTED CHARACTERISTICS, 2001

Overall, 7 percent of adults, 18 years of age and over, had ever been told by a doctor or other health professional that they had some form of cancer. Non-Hispanic Whites reported a high frequency of 8.5 percent, whereas Hispanics and non-Hispanic Blacks reported much lower frequencies of only 2.3 percent and 2.8 percent, respectively. There is likely to be a survey-response bias in the large difference between the White and the Black and Hispanic responses. The racial and ethnic distribution in the next table is considered more reliable.

The reported cancer percentages increased with age, reaching 22.1 percent for non-Hispanic Whites, 65 years of age and over. In this age group males had a greater reported frequency of cancer than females, but below 65 years of age females reported slightly more cancer than males. The frequency of breast cancer and prostate cancer increased with age, but cervical cancer appears to have reached its highest frequency with persons 45–64 years of age.

Table III-14. Rate of Cancer Among Persons 18 Years of Age and Over, by Selected Characteristics, 2001

(Percent.)

Characteristic	Any cancer	Standard error for any cancer	Selected type of cancer [1]: Breast cancer	Standard error for breast cancer	Cervical cancer	Standard error for cervical cancer	Prostate cancer	Standard error for prostate cancer
Total [2]	6.9	0.17	1.1	0.07	1.1	0.09	1.5	0.11
Sex								
Male	6.2	0.23	*0.0	0.02	*	-	1.5	0.11
Female	7.5	0.23	2.1	0.13	1.1	0.09	*	-
Age								
18–44 years	2.2	0.14	0.2	0.04	1.1	0.12	*	-
45–64 years	8.3	0.32	1.6	0.15	1.3	0.17	0.7	0.13
65–74 years	18.1	0.76	2.9	0.32	1.2	0.31	7.7	0.87
75 years and over	21.7	0.86	3.8	0.41	0.4	0.13	11.1	1.10
Race								
1 race [3]	6.9	0.17	1.1	0.07	1.1	0.09	1.5	0.11
White	7.9	0.19	1.2	0.08	1.3	0.11	1.7	0.13
Black or African American	2.9	0.29	0.6	0.15	0.4	0.11	1.1	0.24
American Indian or Alaska Native	2.6	1.20	*0.2	0.23	*	-	*1.3	0.93
Asian	0.9	0.32	*0.2	0.15	*0.3	0.34	*0.3	0.29
Native Hawaiian or other Pacific Islander	6.3	6.06	*	-	*	-	*	-
2 or more races [4]	5.8	1.58	*0.8	0.43	*1.3	0.79	*0.7	0.64
Black or African American, White	1.3	1.34	1.3	1.34	*	-	*	-
American Indian or Alaska Native, White	8.1	2.73	0.4	0.35	1.9	1.40	*1.1	1.05
Hispanic or Latino Origin and Race								
Hispanic or Latino [5]	2.3	0.28	0.8	0.20	0.5	0.13	*0.4	0.18
Mexican or Mexican American	1.7	0.28	*0.4	0.16	*0.2	0.11	*0.2	0.10
Not Hispanic or Latino	7.4	0.18	1.1	0.07	1.2	0.10	1.7	0.12
White, single race	8.5	0.21	1.3	0.08	1.3	0.12	1.8	0.14
Black or African American, single race	2.8	0.28	0.5	0.11	0.4	0.11	1.1	0.25
Poverty Status [6]								
Poor	6.3	0.51	1.1	0.27	1.5	0.24	1.5	0.41
Near poor	6.8	0.42	1.3	0.18	1.5	0.23	2.1	0.38
Not poor	6.5	0.21	1.0	0.09	1.2	0.13	1.2	0.13
Region								
Northeast	6.5	0.33	1.1	0.15	1.0	0.19	1.5	0.24
Midwest	6.8	0.34	1.2	0.15	1.0	0.14	1.5	0.23
South	7.3	0.30	1.2	0.12	1.2	0.17	1.6	0.18
West	6.6	0.35	0.9	0.12	1.2	0.21	1.5	0.23
Sex and Age								
Male								
18–44 years	1.2	0.15	*0.0	0.01	X	X	*	-
45–64 years	7.1	0.44	*0.1	0.04	X	X	0.7	0.13
65 years and over	23.2	0.99	*0.1	0.07	X	X	9.1	0.68
Female								
18–44 years	3.1	0.21	0.3	0.07	1.1	0.12	X	X
45–64 years	9.3	0.45	3.0	0.29	1.3	0.17	X	X
65 years and over	17.1	0.71	5.7	0.44	0.8	0.17	X	X
Hispanic or Latino Origin, Race, and Age								
Hispanic or Latino [5]								
18–44 years	1.0	0.21	*0.3	0.17	*0.4	0.15	*	-
45–64 years	3.6	0.75	*1.3	0.54	*0.7	0.35	*0.1	0.08
65 years and over	8.5	1.77	*2.8	1.15	*0.5	0.35	*4.6	2.24
Not Hispanic or Latino								
White, single race								
18–44 years	2.7	0.19	0.1	0.04	1.4	0.17	*	-
45–64 years	9.6	0.38	1.8	0.18	1.5	0.21	0.8	0.16
65 years and over	22.1	0.65	3.7	0.29	0.9	0.21	9.7	0.78
Black or African American, single race								
18–44 years	1.2	0.22	0.2	0.09	0.2	0.10	*	-
45–64 years	4.0	0.69	0.8	0.27	0.9	0.33	*0.9	0.33
65 years and over	8.4	1.33	1.8	0.53	0.3	0.33	8.2	2.18

[1]Respondents were asked if they have ever been told by a doctor or other health care professional that they had a cancer or a malignancy of any kind. They were then asked to name the kind of cancer they had. A person may be represented in more than one column.

[2]Total includes other races not shown separately and persons with unknown poverty status.

[3]The category "1 race" refers to people who indicated only a single race group. Persons who indicated a single race other than the groups shown are included in the total for "1 race" but not shown separately due to small sample sizes. Therefore the frequencies for the category "1 race" will will be greater than the sum of the frequencies for the specific groups shown separately.

[4]The category two or more races refers to all persons who indicated more than one race group. Only two combinations of race groups are shown due to small sample sizes for other combinations. Therefore, the frequencies for the category "2 or more races" will be greater than the sum of the frequencies for the specific combinations shown separately.

[5]Persons of Hispanic or Latino origin may be of any race or any combination of races. Similarly, the category "Not Hispanic or Latino" refers to all persons who are not of Hispanic or Latino origin, regardless of race.

[6]Poverty status is based on family income and family size using the U.S. Census Bureau's poverty thresholds for the previous calendar year. "Poor" persons are defined as below the poverty threshold. "Near poor" persons have incomes of 100 percent to less than 200 percent of the poverty threshold. "Not poor" persons have incomes that are 200 percent of the poverty threshold or greater.

X = Not applicable.

* = Figure does not meet standards of reliability or precision.

- = Quantity zero.

Table III-14. Rate of Cancer Among Persons 18 Years of Age and Over, by Selected Characteristics, 2001—*Continued*

(Percent.)

Characteristic	Selected type of cancer [1]							
	Any cancer	Standard error for any cancer	Breast cancer	Standard error for breast cancer	Cervical cancer	Standard error for cervical cancer	Prostate cancer	Standard error for prostate cancer
Hispanic or Latino Origin, Race, Age, and Sex								
Hispanic or Latino, male [5]								
18–44 years	*0.2	0.11	*	-	*	-	*	-
45–64 years	*1.7	0.56	*	-	*	-	*0.1	0.08
65 years and over	*7.6	2.48	*0.3	0.33	*	-	*4.6	2.24
Hispanic or Latina, female [5]								
18–44 years	1.7	0.40	*0.6	0.34	*0.4	0.15	*	-
45–64 years	5.3	1.32	*2.5	1.02	*0.7	0.35	*	-
65 years and over	9.1	2.26	*4.5	1.95	*0.5	0.35	*	-
Not Hispanic or Latino								
White, single race, male								
18–44 years	1.6	0.20	*	-	*	-	*	-
45–64 years	8.4	0.53	*0.1	0.05	*	-	0.8	0.16
65 years and over	25.9	1.14	*0.1	0.08	*	-	9.7	0.78
White, single race, female								
18–44 years	3.9	0.29	0.3	0.07	1.4	0.17	*	-
45–64 years	10.8	0.55	3.4	0.35	1.5	0.21	*	-
65 years and over	19.3	0.81	6.3	0.50	0.9	0.21	*	-
Black or African American, single race, male								
18–44 years	*0.6	0.28	*	-	*	-	*	-
45–64 years	*3.4	1.04	*	-	*	-	*0.9	0.33
65 years and over	12.5	2.65	*	-	*	-	8.2	2.18
Black or African American, single race, female								
18–44 years	1.7	0.34	*0.3	0.16	*0.2	0.10	*	-
45–64 years	4.5	0.82	*1.4	0.48	*0.9	0.33	*	-
65 years and over	5.8	1.20	3.0	0.87	*0.3	0.33	*	-
Poverty Status and Age [6]								
Poor								
18–44 years	2.7	0.40	0.1	0.07	1.5	0.32	*	-
45–64 years	9.3	1.46	2.4	1.05	2.4	0.61	*0.4	0.28
65 years and over	17.0	1.82	3.4	0.71	*0.5	0.38	12.7	3.56
Near poor								
18–44 years	2.5	0.36	0.4	0.18	1.8	0.36	*	-
45–64 years	7.6	1.01	1.4	0.40	*1.3	0.41	*0.9	0.52
65 years and over	16.7	1.19	3.4	0.57	*1.0	0.36	9.8	1.80
Not poor								
18–44 years	2.3	0.18	0.1	0.04	1.0	0.17	*	-
45–64 years	8.4	0.41	1.5	0.19	1.4	0.25	0.7	0.17
65 years and over	21.6	1.00	3.5	0.44	*1.1	0.40	9.4	1.04

[1]Respondents were asked if they have ever been told by a doctor or other health care professional that they had a cancer or a malignancy of any kind. They were then asked to name the kind of cancer they had. A person may be represented in more than one column.

[5]Persons of Hispanic or Latino origin may be of any race or any combination of races. Similarly, the category "Not Hispanic or Latino" refers to all persons who are not of Hispanic or Latino origin, regardless of race.

[6]Poverty status is based on family income and family size using the U.S. Census Bureau's poverty thresholds for the previous calendar year. "Poor" persons are defined as below the poverty threshold. "Near poor" persons have incomes of 100 percent to less than 200 percent of the poverty threshold. "Not poor" persons have incomes that are 200 percent of the poverty threshold or greater.

* = Figure does not meet standards of reliability or precision.

- = Quantity zero.

TABLE III-15. AGE–ADJUSTED CANCER INCIDENCE RATES FOR SELECTED CANCER SITES, ACCORDING TO SEX, RACE, AND HISPANIC ORIGIN, SELECTED YEARS, 1990–2000

In 2000, the U.S. rate of new cancer cases per 100,000 of population was estimated to be 461.6, based on a large sample covering 12 geographic areas. The rate for men was one-third higher than that for women. For men, however, new cancer cases have been declining slowly since 1990 at a trend rate of 1.4 percent per year. The trend rate for women has not declined, partly due to increases in breast cancer cases. From 1999 to 2000 alone, however, new cancer cases for women did decline by 3 percent.

For men, prostate cancer has had the highest rate of new cases, followed by lung and colon cancer. For women, breast cancer leads in new cases, again followed by lung and colon cancer; the rate for the latter two cancers is lower for women than for men. In general, the lowest rates of new cancer cases have occurred among American Indians and Eskimo natives. Asians and Pacific Islanders and Hispanics also have notably lower rates than non-Hispanic Whites. Black men, on the other hand, have the highest rates in the groups shown here.

Table III-15. Age-Adjusted[1] Cancer Incidence Rates for Selected Cancer Sites, According to Sex, Race and Hispanic Origin, Selected Years, 1990–2000

(Number of new cases per 100,000 population.)

Site, sex, race, and Hispanic origin	1990–2000 EAPC[2]	2000	1999	1998	1997	1996	1995	1994	1990
All Sites									
All persons	-0.5	461.6	474.5	476.9	476.9	471.9	470.9	475.8	477.2
White	-0.5	469.7	482.6	485.1	483.6	478.3	476.8	480.8	484.4
Black or African American	-0.5	506.2	522.1	521.6	531.8	528.0	532.6	540.7	515.7
American Indian or Alaska Native	-2.1	197.0	243.4	244.8	269.5	250.3	271.1	261.1	262.9
Asian or Pacific Islander	-0.2	339.2	343.1	341.0	348.1	336.8	340.5	347.2	337.2
Hispanic or Latino[3]	-0.1	342.1	356.6	360.4	348.1	353.9	358.6	358.4	342.2
White, not Hispanic or Latino	-0.4	480.1	493.2	492.3	494.5	489.3	486.1	489.1	492.2
Male	-1.4	544.6	557.3	556.4	561.3	559.6	561.7	578.2	584.3
White	-1.5	544.7	557.5	557.1	560.1	560.4	560.1	576.9	590.6
Black or African American	-1.0	677.3	693.2	696.7	713.0	704.9	727.1	741.5	689.1
American Indian or Alaska Native	-3.5	193.0	277.4	252.3	311.9	269.6	322.8	305.5	311.1
Asian or Pacific Islander	-0.6	394.5	394.0	383.8	399.5	387.9	397.4	412.4	388.5
Hispanic or Latino[3]	-0.5	406.8	422.3	424.3	416.4	427.6	436.8	441.0	404.0
White, not Hispanic or Latino	-1.5	550.9	563.2	558.8	566.4	567.4	566.1	582.0	599.0
Female	0.1	406.2	420.2	426.0	422.3	415.3	412.3	409.4	414.0
White	0.2	421.0	434.7	440.7	435.9	426.8	424.8	420.1	423.9
Black or African American	-0.1	391.5	409.4	407.9	412.0	411.7	402.5	409.4	406.5
American Indian or Alaska Native	-0.7	205.9	222.3	241.4	241.3	240.0	237.8	229.6	230.7
Asian or Pacific Islander	0.4	301.1	308.7	312.4	312.0	300.3	299.0	298.3	297.1
Hispanic or Latino[3]	0.2	304.9	317.7	322.0	306.6	310.0	311.3	309.3	308.9
White, not Hispanic or Latino	0.4	433.2	448.4	451.4	449.6	440.0	435.8	430.0	430.7
Lung and Brochus									
Male	-2.2	75.3	79.2	82.9	82.6	84.3	87.0	87.4	95.4
White	-2.2	73.8	77.8	81.9	80.9	82.8	85.2	86.8	94.6
Black or African American	-2.2	108.1	117.6	122.8	125.8	128.7	136.7	122.7	134.8
Asian or Pacific Islander	-0.6	63.1	62.1	61.2	62.6	61.2	60.4	60.6	64.4
Hispanic or Latino[3]	-2.4	43.5	43.0	49.9	46.6	47.9	52.1	49.9	58.6
White, not Hispanic or Latino	-2.1	74.8	79.3	82.8	81.7	84.3	85.6	87.5	95.2
Female	0.2	47.2	49.8	50.8	50.3	50.2	49.3	49.1	47.4
White	0.4	49.2	51.8	53.0	52.9	52.4	51.7	50.7	48.7
Black or African American	0.4	53.6	58.1	57.1	50.7	54.2	50.4	55.8	53.4
Asian or Pacific Islander	0.1	27.4	28.8	28.4	29.7	27.9	27.8	29.3	28.4
Hispanic or Latino[3]	-0.7	22.6	23.6	25.6	25.2	25.4	24.2	22.5	24.8
White, not Hispanic or Latino	0.6	50.9	54.3	54.7	55.3	54.6	54.0	52.6	49.8
Colon and Rectum									
Male	-1.3	61.6	63.4	65.6	66.2	64.5	63.0	65.5	72.4
White	-1.4	61.0	63.4	65.4	66.0	64.8	62.5	65.5	73.1
Black or African American	-0.5	71.4	72.9	76.1	73.7	67.5	73.0	72.7	73.1
Asian or Pacific Islander	0.6	57.6	54.5	57.6	59.7	56.8	58.3	58.1	61.4
Hispanic or Latino[3]	0.6	48.1	49.5	51.3	50.1	50.5	45.4	46.8	45.6
White, not Hispanic or Latino	-1.4	61.9	64.8	65.8	65.6	65.1	63.0	66.2	74.4
Female	-0.7	45.3	46.6	48.5	47.1	45.9	45.9	46.5	50.3
White	-0.7	44.8	45.8	48.1	47.0	45.6	45.5	45.9	49.8
Black or African American	-0.4	56.5	57.0	55.9	57.5	53.8	54.8	56.9	61.1
Asian or Pacific Islander	-0.5	37.4	40.6	41.0	35.6	39.4	38.6	39.0	38.3
Hispanic or Latino[3]	0.4	32.7	33.9	33.5	31.9	32.2	32.7	33.4	32.4
White, not Hispanic or Latino	-0.6	46.1	46.9	49.4	48.3	46.8	46.0	47.0	50.8
Prostate									
Male	-2.1	172.0	176.5	166.7	169.9	165.0	165.3	176.9	166.6
White	-2.5	165.8	170.5	160.3	164.6	160.2	159.7	170.1	168.0
Black or African American	-0.5	275.9	274.6	273.0	269.5	266.5	271.2	298.3	218.8
American Indian or Alaska Native	-7.8	24.2	52.5	50.6	71.4	75.4	66.0	59.1	84.0
Asian or Pacific Islander	-1.0	106.6	106.0	94.2	97.7	94.4	104.0	112.6	89.5
Hispanic or Latino[3]	0.0	137.2	140.4	138.1	136.2	133.4	137.4	147.8	114.8
White, not Hispanic or Latino	-2.7	165.0	169.3	157.1	164.7	160.5	159.9	169.1	169.6
Breast									
Female	0.6	132.9	137.0	138.1	135.2	131.7	130.5	128.5	129.3
White	0.7	139.1	143.3	144.1	140.8	136.7	136.0	134.1	134.3
Black or African American	0.4	119.8	122.1	121.4	123.1	122.2	122.3	121.1	116.7
American Indian or Alaska Native	-0.7	52.2	52.0	55.7	57.1	75.8	63.6	59.3	46.8
Asian or Pacific Islander	1.9	94.7	99.7	101.1	99.5	90.5	87.5	80.9	87.1
Hispanic or Latino[3]	0.9	92.0	91.2	91.0	84.7	89.8	88.6	85.9	84.5
White, not Hispanic or Latino	0.9	146.3	151.7	151.4	148.2	144.0	142.0	140.6	138.6
Cervix Uteri									
Female	-2.4	8.7	9.2	9.8	9.8	10.6	9.9	10.5	11.9
White	-2.2	8.7	9.0	9.3	9.2	9.9	9.2	9.9	11.2
Black or African American	-3.1	10.6	12.7	12.4	13.0	13.6	14.3	13.5	16.2
Asian or Pacific Islander	-3.4	8.2	8.3	10.9	11.1	12.9	11.0	14.4	12.1
Hispanic or Latino[3]	-2.7	16.9	17.1	15.7	16.0	18.3	18.0	20.0	21.1
White, not Hispanic or Latino	-2.4	6.7	7.4	7.9	7.8	8.3	7.7	8.2	9.5

[1]Age-adjusted by 5-year age groups to the year 2000 U.S. standard population
[2]Estimated annual percent change (EAPC) has been calculated by fitting a linear regression model to the natural logarithm of the yearly rates from 1990 to 1999.
[3]Persons of Hispanic origin may be of any race.

Table III-15. Age-Adjusted[1] Cancer Incidence Rates for Selected Cancer Sites, According to Sex, Race and Hispanic Origin, Selected Years, 1990–2000—*Continued*

(Number of new cases per 100,000 population.)

Site, sex, race, and Hispanic origin	1990–2000 EAPC[2]	2000	1999	1998	1997	1996	1995	1994	1990
Corpus Uteri									
Female	-0.1	23.5	24.5	24.9	25.3	24.4	24.9	24.4	24.7
White	-0.1	25.2	26.0	26.5	26.9	25.8	26.3	25.8	26.4
Black or African American	0.5	16.6	17.8	18.2	17.9	19.1	17.9	17.9	17.0
Asian or Pacific Islander	1.9	16.5	17.5	17.1	17.8	16.6	17.9	15.8	13.2
Hispanic or Latino[3]	0.2	15.1	16.6	18.1	17.1	15.8	16.7	15.7	16.2
White, not Hispanic or Latino	0.0	26.2	26.8	27.3	27.8	27.0	27.4	27.2	27.0
Ovary									
Female	-0.8	16.3	16.9	16.8	16.9	16.8	17.3	16.9	17.9
White	-0.7	17.5	18.1	17.9	18.0	18.1	18.4	17.8	18.9
Black or African American	-0.4	12.0	12.2	12.6	12.0	10.9	12.6	13.9	12.9
Asian or Pacific Islander	-0.3	11.7	12.6	12.1	13.2	11.8	12.4	11.7	12.8
Hispanic or Latino[3]	-0.5	13.5	13.3	15.5	13.5	14.5	13.9	14.0	14.4
White, not Hispanic or Latino	-0.8	17.9	18.7	17.9	18.4	18.5	18.8	18.4	19.6
Oral Cavity and Pharynx									
Male	-2.1	15.5	15.2	16.4	16.9	17.4	16.9	17.7	19.1
White	-2.1	15.3	15.0	16.2	16.7	17.0	16.8	17.2	18.7
Black or African American	-2.6	19.0	19.3	21.5	19.8	23.3	22.4	24.0	26.1
Asian or Pacific Islander	-1.2	13.3	11.0	13.1	14.7	14.3	11.9	14.2	15.2
Hispanic or Latino[3]	-2.3	8.4	9.9	10.1	10.3	11.1	12.8	11.4	11.0
White, not Hispanic or Latino	-2.0	15.9	15.8	16.6	17.3	17.4	16.7	17.8	19.2
Female	-1.6	6.1	6.2	6.6	6.9	6.9	7.0	6.8	7.3
White	-1.7	6.1	6.1	6.7	6.8	6.8	7.0	6.8	7.4
Black or African American	-1.4	5.3	5.9	6.5	7.1	7.3	6.7	7.1	6.3
Asian or Pacific Islander	-0.7	5.9	6.4	4.5	6.5	5.7	5.2	5.6	6.1
Hispanic or Latino[3]	-0.1	3.7	4.5	3.4	3.8	3.7	3.8	4.7	3.6
White, not Hispanic or Latino	-1.7	6.3	6.2	7.2	7.2	7.3	7.3	7.0	7.7
Stomach									
Male	-1.8	12.5	12.8	12.8	13.4	13.8	13.6	14.3	14.7
White	-1.9	10.5	11.1	11.0	11.3	11.9	12.0	12.4	12.9
Black or African American	-2.1	18.5	16.7	20.4	21.9	22.5	18.5	23.3	21.9
Asian or Pacific Islander	-2.3	22.7	22.7	21.2	24.7	23.8	24.0	24.5	26.9
Hispanic or Latino[3]	-1.5	15.8	19.4	19.1	19.0	17.2	19.4	22.6	19.7
White, not Hispanic or Latino	-2.3	9.7	9.7	9.7	10.1	11.1	10.9	11.2	12.0
Female	-0.9	6.0	6.5	6.4	6.2	6.1	6.2	6.2	6.7
White	-1.4	4.9	5.4	5.2	4.9	5.1	5.2	5.0	5.7
Black or African American	-0.4	8.3	10.1	10.9	11.0	9.3	9.9	10.1	9.9
Asian or Pacific Islander	-2.4	13.1	12.3	12.8	12.3	13.7	13.1	15.1	15.5
Hispanic or Latino[3]	-1.0	10.0	9.3	10.7	10.1	10.0	11.1	8.6	10.8
White, not Hispanic or Latino	-2.0	4.0	4.7	4.3	4.1	4.3	4.4	4.4	5.0
Pancreas									
Male	-0.4	12.3	12.4	12.9	12.9	12.6	12.7	13.0	13.1
White	-0.3	12.0	12.2	12.9	12.5	12.3	12.4	12.4	12.7
Black or African American	-0.7	17.7	18.1	17.1	18.2	19.1	18.9	19.1	19.7
Asian or Pacific Islander	-1.3	10.7	9.1	10.3	12.2	10.6	10.5	14.2	11.2
Hispanic or Latino[3]	-0.2	11.0	9.3	9.7	11.9	11.2	12.4	10.0	10.9
White, not Hispanic or Latino	0.0	12.0	12.6	13.0	12.5	12.1	12.1	12.6	12.4
Female	-0.5	9.4	9.5	10.0	10.1	10.0	10.0	9.9	10.1
White	-0.5	9.2	9.3	9.8	9.6	9.7	9.6	9.7	9.8
Black or African American	-1.3	12.3	13.3	13.8	16.9	15.1	15.7	15.1	13.0
Asian or Pacific Islander	0.4	9.1	8.5	8.6	8.4	7.9	8.1	6.9	10.1
Hispanic or Latino[3]	-0.6	8.5	9.8	9.6	9.6	9.0	8.7	9.6	9.7
White, not Hispanic or Latino	-0.5	9.1	8.9	9.7	9.3	9.7	9.6	9.4	9.6
Urinary Bladder									
Male	-0.3	36.1	36.1	36.6	35.7	35.6	35.2	36.0	37.2
White	-0.3	39.7	39.6	40.3	39.3	39.1	38.6	39.5	40.6
Black or African American	0.0	19.5	22.1	20.3	20.8	19.3	19.5	19.4	19.8
Asian or Pacific Islander	1.1	17.3	17.0	16.5	15.4	16.1	16.6	16.2	15.7
Hispanic or Latino[3]	-1.0	19.3	18.0	18.1	17.7	18.3	18.4	20.9	21.6
White, not Hispanic or Latino	-0.3	40.6	41.0	41.4	40.7	40.6	39.7	40.2	41.7
Female	-0.4	8.9	9.3	9.0	9.3	9.0	9.3	9.1	9.5
White	-0.2	9.7	9.9	9.7	9.9	9.8	10.0	9.8	9.9
Black or African American	-0.6	7.7	8.5	6.6	8.1	7.2	7.4	7.1	8.6
Asian or Pacific Islander	-0.3	4.3	4.2	4.8	5.3	3.9	4.6	3.9	5.3
Hispanic or Latino[3]	-1.1	5.3	4.2	4.5	5.1	5.3	5.2	5.3	5.3
White, not Hispanic or Latino	-0.1	9.9	10.4	10.2	10.5	10.0	10.4	9.9	10.1

[1]Age-adjusted by 5-year age groups to the year 2000 U.S. standard population
[2]Estimated annual percent change (EAPC) has been calculated by fitting a linear regression model to the natural logarithm of the yearly rates from 1990 to 1999.
[3]Persons of Hispanic origin may be of any race.

Table III-15. Age-Adjusted[1] Cancer Incidence Rates for Selected Cancer Sites, According to Sex, Race and Hispanic Origin, Selected Years, 1990–2000—*Continued*

(Number of new cases per 100,000 population.)

Site, sex, race, and Hispanic origin	1990–2000 EAPC[2]	2000	1999	1998	1997	1996	1995	1994	1990
Non-Hodgkins Lymphoma									
Male	0.0	22.6	23.7	22.7	23.8	24.6	25.0	24.7	22.7
White	0.0	23.9	24.7	23.8	24.6	25.7	26.1	25.7	23.7
Black or African American	-0.2	16.4	17.8	17.0	22.9	18.9	21.3	20.3	17.8
Asian or Pacific Islander	0.4	16.1	19.1	15.6	16.4	16.8	16.5	18.2	16.5
Hispanic or Latino[3]	0.2	19.0	17.2	19.6	17.8	21.5	21.5	18.6	17.0
White, not Hispanic or Latino	-0.2	24.0	25.7	24.5	25.0	26.4	26.9	26.6	24.5
Female	0.9	15.3	15.8	16.0	15.9	15.1	15.1	15.3	14.5
White	0.9	16.0	16.8	16.8	16.6	15.8	15.8	16.1	15.4
Black or African American	2.0	11.5	10.6	12.4	11.9	11.4	10.0	8.7	10.4
Asian or Pacific Islander	1.4	11.4	11.2	11.3	11.1	9.5	11.9	12.4	9.2
Hispanic or Latino[3]	0.5	12.4	13.3	13.2	14.0	13.5	12.4	13.7	13.2
White, not Hispanic or Latino	0.9	16.1	17.0	17.1	16.8	15.9	16.1	16.2	15.4
Leukemia									
Male	-1.1	14.6	15.5	16.3	16.4	16.2	17.3	16.3	17.0
White	-1.1	15.4	16.3	17.2	17.6	17.0	18.6	17.3	17.8
Black or African American	-1.1	12.5	12.7	13.2	13.3	13.5	12.7	11.6	15.4
Asian or Pacific Islander	0.3	9.5	10.7	9.9	9.0	11.2	9.9	10.3	8.6
Hispanic or Latino[3]	0.0	11.8	10.7	11.6	11.8	12.2	15.3	10.4	11.4
White, not Hispanic or Latino	-1.1	15.2	16.4	17.3	17.7	17.0	18.9	17.6	17.7
Female	-0.9	9.2	8.7	9.5	9.5	9.7	9.9	9.7	9.7
White	-0.7	9.7	9.2	10.1	10.2	10.2	10.5	10.1	10.1
Black or African American	-1.4	8.6	7.3	7.4	7.7	8.1	8.0	7.7	8.3
Asian or Pacific Islander	-0.5	5.9	6.3	6.7	5.7	6.6	6.4	7.0	6.0
Hispanic or Latino[3]	-0.6	7.2	7.4	8.0	8.1	7.1	8.2	8.2	8.3
White, not Hispanic or Latino	-0.9	9.2	9.0	9.8	10.1	10.1	10.3	9.8	10.0

[1]Age-adjusted by 5-year age groups to the year 2000 U.S. standard population
[2]Estimated annual percent change (EAPC) has been calculated by fitting a linear regression model to the natural logarithm of the yearly rates from 1990 to 1999.
[3]Persons of Hispanic origin may be of any race.

TABLE III-16. FIVE-YEAR RELATIVE CANCER SURVIVAL RATES FOR SELECTED CANCER SITES, ACCORDING TO RACE AND SEX, 1974–1998

For the period 1992–1998, White male cancer patients achieved a 63.5 percent five-year survival rate, while White women attained a 64.2 percent survival rate. For Black males, the five-year survival rates were 54.0 percent and for Black women this rate was 51.0 percent. In all of these cases, the survival rates had risen substantially since 1974–1979. They had improved particularly for prostate cancer, and this led to a greater percent of improvement for men than for women.

The highest survival rates in men were for prostate and bladder cancer and for women they were for melanoma of the skin and breast cancer. The lowest five-year survival rates were for pancreas and esophagus in men, and for pancreas and lung cancer in women.

The observed survival rate is adjusted to factor out expected normal deaths during the five-year period. Without such adjustment survival rates for young patients would be overstated relative to older patients with higher general mortality rates. To make the adjustment the observed survival rate is divided by the expected survival rate of the same group in the general population. Thus if 50 percent of the patient group survived and if the survival rate of that group in the general population for the same five years was 0.9, the adjusted cancer survival rate would be 55.6.

Table III-16. Five-Year Relative Cancer Survival Rates for Selected Cancer Sites, According to Race and Sex, 1974–1998

(Percent of patients.)

Sex and site of cancer	White						Black or African American					
	1992–1998	1989–1991	1986–1988	1983–1985	1980–1982	1974–1979	1992–1998	1989–1991	1986–1988	1983–1985	1980–1982	1974–1979
BOTH SEXES												
All Sites	63.8	60.3	56.7	53.9	52.1	50.9	52.6	46.2	42.6	39.8	39.7	39.2
Oral cavity and pharynx	58.8	55.4	55.2	55.3	55.6	54.9	34.9	32.6	34.7	35.2	31.0	36.5
Esophagus	14.7	11.8	10.7	9.3	7.3	5.5	8.3	8.9	7.2	6.2	5.4	3.3
Stomach	20.9	18.4	19.1	16.2	16.5	15.2	20.0	24.8	19.3	18.9	19.4	15.7
Colon	62.7	63.1	61.6	58.4	55.7	51.9	52.8	53.9	52.9	49.3	49.3	47.3
Rectum	62.4	60.5	59.1	55.9	53.1	49.8	52.7	54.3	51.1	43.9	37.9	40.4
Pancreas	4.3	4.1	3.1	2.9	2.8	2.4	3.9	3.8	6.2	5.0	4.5	3.2
Lung and bronchus	15.0	14.3	13.5	13.8	13.5	13.1	12.3	10.7	11.9	11.3	12.1	11.3
Urinary bladder	82.3	82.1	80.7	78.2	78.9	74.9	64.5	61.9	62.3	59.6	58.3	51.5
Non-Hodgkin's lymphoma	56.1	51.9	52.9	54.4	51.9	48.2	46.1	43.7	50.2	45.1	50.2	50.3
Leukemia	47.3	45.8	44.2	42.0	39.5	36.6	38.4	34.1	38.0	33.6	32.9	30.8
MALE												
All Sites	63.5	57.7	51.8	48.5	46.7	43.4	54.0	43.3	37.7	34.5	34.3	32.1
Oral cavity and pharynx	57.8	52.1	52.2	54.4	54.6	54.4	29.5	27.9	29.3	29.9	26.5	31.2
Esophagus	14.7	11.8	11.2	7.8	6.5	5.2	9.0	8.0	7.0	5.0	4.6	2.3
Stomach	19.5	15.1	16.2	14.6	15.6	13.8	19.2	22.3	15.3	18.5	18.5	15.3
Colon	63.0	63.7	62.5	58.9	56.0	51.0	53.2	53.8	52.6	48.3	46.7	45.4
Rectum	61.6	60.2	58.8	55.2	51.7	48.9	51.6	56.0	46.2	43.1	36.1	36.7
Pancreas	4.3	3.8	2.9	2.7	2.5	2.6	4.2	3.3	6.4	4.4	3.2	2.4
Lung and bronchus	13.3	12.7	12.0	12.1	12.2	11.6	10.8	9.5	11.9	10.3	10.9	10.0
Prostate gland	97.8	91.8	82.7	76.3	74.5	70.3	92.6	80.6	69.3	63.9	64.7	60.8
Urinary bladder	84.4	84.4	82.3	79.5	79.9	75.9	69.4	65.5	67.4	64.8	62.8	58.9
Non-Hodgkin's lymphoma	52.7	47.7	50.1	53.4	50.9	47.2	41.4	38.7	46.6	43.6	47.2	44.7
Leukemia	48.4	46.4	45.5	41.6	39.6	35.6	37.9	29.2	36.7	32.5	30.1	30.7
FEMALE												
All Sites	64.2	62.9	61.5	58.8	57.1	57.4	51.0	49.3	47.9	45.4	46.0	46.9
Colon	62.4	62.5	60.7	58.0	55.4	52.6	52.5	53.9	53.1	50.1	51.3	48.7
Rectum	63.5	60.8	59.5	56.7	54.7	50.7	53.8	52.6	56.0	44.7	40.2	43.8
Pancreas	4.4	4.4	3.4	3.2	3.0	2.3	3.7	4.3	6.0	5.5	5.8	4.1
Lung and bronchus	17.0	16.6	15.8	17.0	16.2	16.7	14.7	13.0	11.7	14.1	15.5	15.5
Melanoma of skin	91.8	91.5	91.3	89.4	88.3	85.8	61.9	94.0	(1)	70.1	(1)	69.9
Breast	87.6	86.1	83.9	79.2	77.1	75.4	72.5	71.1	69.2	63.4	65.7	63.1
Cervix uteri	72.1	72.3	71.7	70.6	68.2	69.7	59.9	62.6	55.5	60.4	61.3	63.0
Corpus uteri	86.0	85.6	84.4	84.5	82.8	87.7	60.5	57.5	57.1	54.2	55.1	59.4
Ovary	52.5	49.6	41.9	40.2	38.7	37.2	52.5	41.6	38.6	41.6	39.1	40.5
Non-Hodgkin's lymphoma	60.2	57.1	56.2	55.5	52.9	49.3	53.8	50.8	54.8	46.9	53.7	57.5

[1]Data for population groups with fewer than 25 cases are not shown.

Figure III-6. Five-Year Relative Cancer Survival Rates, According to Race and Sex, 1974–1998

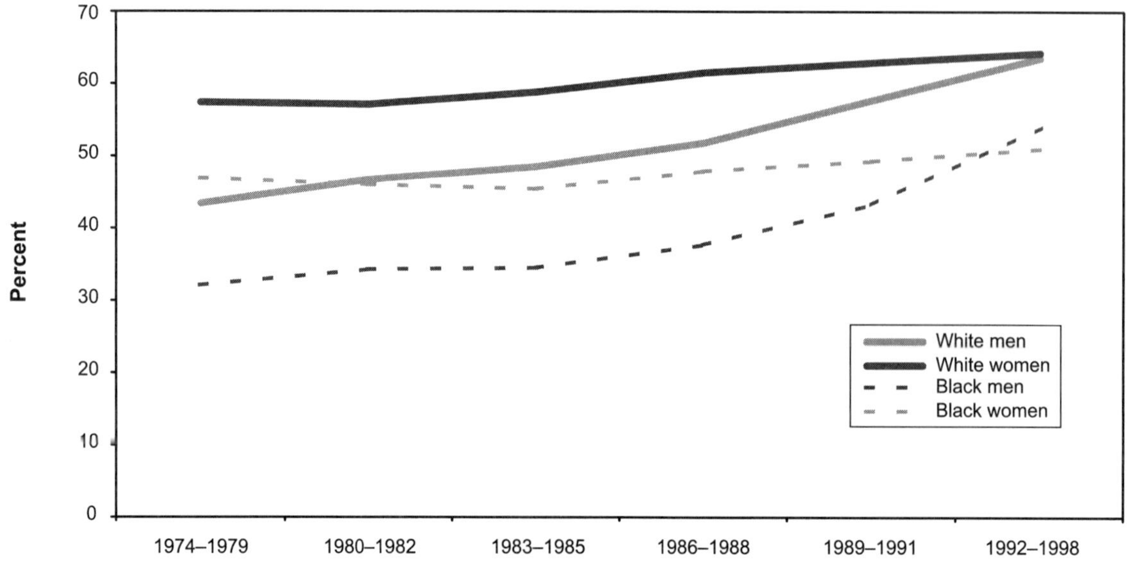

TABLE III-17. ACQUIRED IMMUNODEFICIENCY SYNDROME (AIDS) CASES, ACCORDING TO AGE AT DIAGNOSIS, SEX, DETAILED RACE, AND HISPANIC ORIGIN, SELECTED YEARS, 1985–2002

From the beginning of the epidemic, just prior to 1985 until the end of 2002, a total of 832,112 cases of AIDS have been reported to the Centers for Disease Control and Prevention. In 2002, alone, 42,745 cases were reported, a level roughly similar to those prevailing since 1999. AIDS cases were some 60 percent higher in the mid-90s compared with last year. Since then prevention with measures have reduced infection rates in certain at-risk groups and effective medicines have reduced or delayed the onset of AIDS in HIV-infected people.

All AIDS cases reported in 2002 amounted to 15.0 per 100,000. Black males above 13 years of age had the highest rate of 111.9 per 100,000, and the rate for Hispanic men also was high at 39.3. The rate for Black women per 100,000 was 50.0 and that for Hispanic women was 11.8. For non-Hispanic Whites, the rate was 13.9 for men and 2.3 for women. Overall, the 2002 rate for men above 13 years of age was 28.0, and that for women was 9.2.

Table III-17. Acquired Immunodeficiency Syndrome (AIDS) Cases, According to Age at Diagnosis, Sex, Detailed Race, and Hispanic Origin, Selected Years, 1985–2002

(Percent, number.)

Age at diagnosis, sex, race, and Hispanic orgin	All years [1] (percent distribution) [2]	Cases per 100,000 population (2002) [3]	Number, by year of report							
			All years [3]	2002	2001	2000	1999	1995	1990	1985
All races	X	15.0	831 112	42 745	41 450	40 282	44 580	70 412	41 448	8 159
Male										
All males 13 years and over	100.0	28.0	676 609	31 644	30 663	30 135	34 013	56 689	36 179	7 504
Not Hispanic or Latino										
White	47.7	13.9	322 920	11 221	11 054	11 314	12 691	26 028	20 825	4 746
Black or African American	35.4	111.9	239 650	14 310	13 764	13 082	14 830	20 833	10 239	1 710
American Indian or Alaskan Native	0.3	17.3	2 203	155	149	136	135	197	81	7
Asian or Pacific Islander	0.8	8.2	5 666	381	348	291	295	489	264	49
Hispanic or Latino [4]	15.6	39.3	105 628	5 543	5 318	5 275	6 043	9 111	4 743	992
13–19 years	0.4	1.4	2 632	199	184	145	131	223	107	27
20–29 years	15.4	17.5	104 174	3 433	3 291	3 327	3 972	8 387	6 921	1 501
30–39 years	44.7	56.2	302 420	12 101	12 082	12 543	14 410	25 684	16 668	3 588
40–49 years	27.8	49.5	188 299	10 658	10 261	9 648	10 836	16 151	8 828	1 634
50–59 years	8.7	24.9	58 745	3 959	3 633	3 387	3 479	4 692	2 645	597
60 years and over	3.0	6.5	20 339	1 294	1 212	1 085	1 185	1 552	1 010	157
Female										
All females 13 and over	100.0	9.2	145 696	10 951	10 617	9 958	10 312	12 978	4 544	524
Not Hispanic or Latino										
White	22.0	2.3	32 000	1 930	1 993	1 859	1 896	3 042	1 228	143
Black or African American	61.2	50.0	89 130	7 339	6 963	6 489	6 711	7 586	2 557	280
American Indian or Alaskan Native	0.3	4.5	509	42	42	70	40	38	9	2
Asian or Pacific Islander	0.6	1.4	832	68	67	74	61	73	20	1
Hispanic or Latino [4]	15.9	11.8	23 145	1 561	1 543	1 462	1 599	2 236	726	98
13–19 years	1.4	1.5	1 995	203	171	170	166	157	67	5
20–29 years	20.6	9.6	29 996	1 819	1 717	1 750	1 886	2 676	1 117	178
30–39 years	43.6	18.7	63 504	3 991	4 145	3 973	4 234	5 934	2 087	232
40–49 years	24.1	15.3	35 168	3 377	3 147	2 857	2 789	3 059	780	45
50–59 years	7.0	6.9	10 243	1 150	999	867	916	818	274	26
60 years and over	3.3	1.5	4 790	411	438	341	321	334	219	38
Children										
All children, under 13 years	100.0	0.3	8 807	150	170	189	255	745	725	131
Not Hispanic or Latino										
White	18.2	0.1	1 601	23	30	32	30	117	157	26
Black or African American	61.6	1.2	5 422	99	111	122	171	483	390	87
American Indian or Alaskan Native	0.4	-	31	-	-	1	2	2	5	-
Asian or Pacific Islander	0.7	0.2	58	4	3	3	2	5	4	-
Hispanic or Latino [4]	19.1	0.2	1 685	24	26	30	49	135	169	18
Under 1 year	36.9	1.1	3 249	46	47	61	87	258	298	54
1–12 years	63.1	0.2	5 558	104	123	128	168	487	427	77

[1] Includes cases reported to the Centers for Disease Control and Prevention prior to 1985 and through December 31, 2002.
[2] Percents may not sum to 100 percent due to rounding.
[3] Computed using estimates of July 1, 2001, U.S. resident population by age, sex, race, and Hispanic origin, prepared for the National Center for Health Statistics (NCHS) under a collaboration arrangement with the U.S. Census Bureau.
[4] Persons of Hispanic origin may be of any race.
- = Quantity zero.
X = Not applicable.

Figure III-7. Number of Acquired Immunodeficiency Syndrome (AIDS) Cases by Sex, 1985–2002 (by Year of Report)

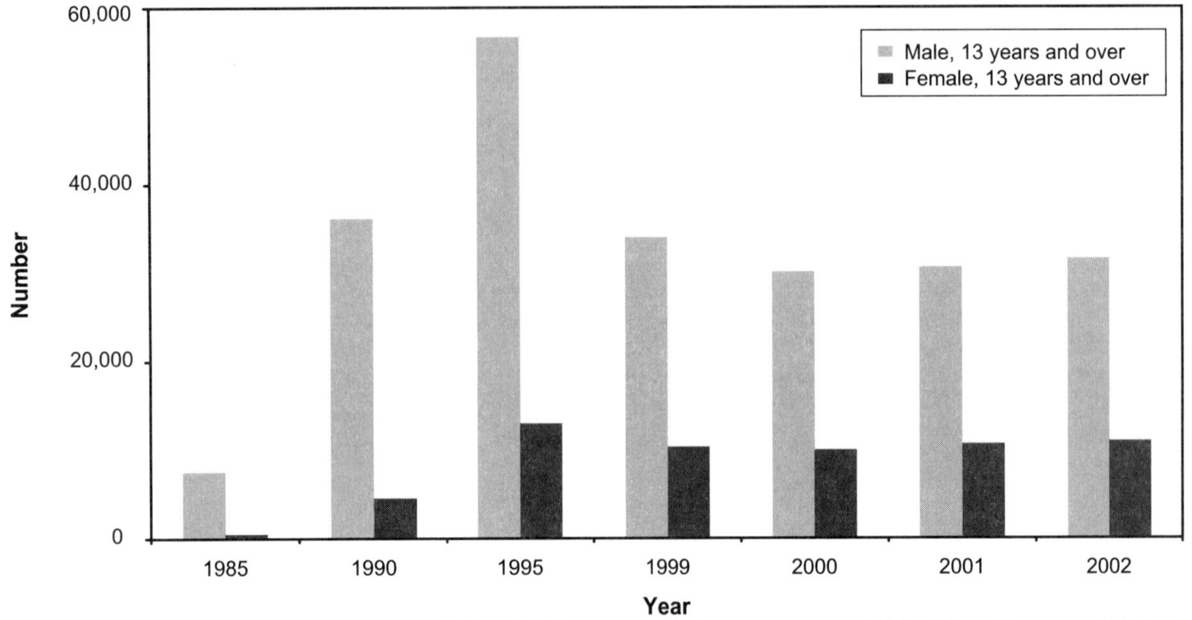

TABLE III-18. RATE OF SELECTED RESPIRATORY DISEASES AMONG PERSONS 18 YEARS OF AGE AND OVER, BY SELECTED CHARACTERISTICS, 2001

Overall, 1.5 percent of adults had ever been told by a medical professional that they had emphysema, and 11 percent had ever been told that they had asthma. Ten percent of adults had been told in the last 12 months that they had hay fever, 17 percent had been told they had sinusitis, and 6 percent had been told that they had chronic bronchitis. Women were told more frequently than men that they had these respiratory diseases, except in the case of emphysema.

Of the major population groups, Asians were afflicted least frequently by respiratory diseases, closely followed by Hispanics. Non-Hispanic Whites reported respiratory diagnoses more frequently than the average for the total population, while non-Hispanic Blacks were closer to the national average. For emphysema, the frequency increased substantially with age for all population groups, but for the other respiratory diseases the highest frequency was often reported in the 45 to 64 year age bracket. For asthma the highest frequency sometimes occurred in the 18–44 year age bracket. To interpret these results, note that the question shifted from whether the respondent had ever been told by a doctor or health professional that they had emphysema or asthma, to whether in the past 12 months they had been told by a doctor or health professional that they had hay fever, sinusitis, or bronchitis. A separate question was asked for each disease.

Table III-18. Rate of Selected Respiratory Diseases Among Persons 18 Years of Age and Over, by Selected Characteristics, 2001

(Percent.)

Characteristic	Selected respiratory disease[1]									
	Emphysema	Standard error for emphysema	Asthma	Standard error for asthma	Hay fever	Standard error for hay fever	Sinusitis	Standard error for sinusitis	Chronic bronchitis	Standard error for chronic brochitis
Total[2]	1.5	0.07	10.9	0.21	10.0	0.20	17.4	0.27	5.5	0.15
Sex										
Male	1.7	0.12	9.4	0.29	8.9	0.27	12.8	0.33	3.8	0.19
Female	1.2	0.09	12.3	0.29	11.1	0.28	21.7	0.38	7.1	0.23
Age										
18–44 years	0.2	0.03	11.8	0.29	10.0	0.27	15.9	0.35	4.5	0.19
45–64 years	1.8	0.14	10.4	0.35	11.6	0.38	21.3	0.48	6.5	0.28
65–74 years	4.7	0.40	9.4	0.55	7.5	0.54	17.2	0.77	6.7	0.48
75 years and over	5.6	0.51	8.0	0.62	6.8	0.54	12.8	0.72	6.9	0.63
Race										
1 race[3]	1.4	0.07	10.8	0.21	9.9	0.20	17.3	0.27	5.5	0.15
White	1.6	0.08	10.9	0.23	10.1	0.22	17.8	0.30	5.7	0.17
Black or African American	0.7	0.13	11.1	0.56	8.8	0.54	17.5	0.72	5.3	0.40
American Indian or Alaska Native	*1.1	0.71	12.2	2.62	13.3	2.57	15.5	2.72	*5.4	1.87
Asian	*0.4	0.22	6.7	0.98	9.9	1.11	10.7	1.20	2.3	0.57
Native Hawaiian or other Pacific Islander	*3.9	2.81	*16.9	6.46	*3.7	2.74	*11.1	4.94	*2.3	2.32
2 or more races[4]	*3.0	1.00	18.6	2.08	16.8	2.15	25.7	2.46	7.1	1.17
Black or African American, White	*	0.00	31.5	8.61	*21.3	7.83	28.6	7.82	*7.3	4.16
American Indian or Alaska Native, White	*5.4	1.87	18.3	3.05	16.5	3.17	27.9	3.87	7.2	1.76
Hispanic or Latino Origin and Race										
Hispanic or Latino[5]	0.6	0.14	8.5	0.46	8.4	0.48	11.1	0.52	3.1	0.29
Mexican or Mexican American	*0.4	0.16	5.4	0.45	6.2	0.54	9.2	0.59	2.3	0.29
Not Hispanic or Latino	1.6	0.08	11.2	0.23	10.2	0.22	18.2	0.29	5.8	0.16
White, single race	1.7	0.09	11.3	0.26	10.4	0.24	18.6	0.33	6.1	0.19
Black or African American, single race	0.7	0.13	10.9	0.56	8.7	0.54	17.5	0.73	5.2	0.39
Poverty Status[6]										
Poor	2.4	0.27	15.2	0.71	8.6	0.55	16.0	0.71	7.9	0.53
Near poor	2.9	0.31	11.9	0.55	8.7	0.46	16.5	0.61	7.1	0.43
Not poor	1.0	0.08	10.7	0.28	11.3	0.29	18.8	0.37	5.1	0.20
Region										
Northeast	1.3	0.16	11.7	0.52	10.8	0.49	16.6	0.57	4.8	0.30
Midwest	1.7	0.16	10.8	0.38	8.5	0.37	16.2	0.51	5.2	0.31
South	1.6	0.13	10.6	0.35	9.5	0.34	20.7	0.49	6.4	0.27
West	1.1	0.12	10.6	0.44	12.2	0.45	13.7	0.50	4.8	0.29
Sex and Age										
Male										
18–44 years	0.2	0.05	10.4	0.42	8.9	0.36	11.3	0.44	3.3	0.25
45–64 years	2.0	0.23	8.1	0.48	10.1	0.52	15.9	0.63	3.8	0.35
65 years and over	7.0	0.60	8.1	0.62	6.0	0.57	11.6	0.74	5.8	0.59
Female										
18–44 years	0.2	0.04	13.1	0.41	11.1	0.39	20.3	0.51	5.8	0.28
45–64 years	1.6	0.18	12.6	0.52	12.9	0.52	26.4	0.69	9.1	0.45
65 years and over	3.8	0.35	9.2	0.55	8.1	0.54	17.9	0.73	7.5	0.53
Hispanic or Latino Origin, Race, and Age										
Hispanic or Latino[5]										
18–44 years	*0.1	0.06	8.4	0.58	7.2	0.52	9.1	0.56	2.4	0.32
45–64 years	*0.8	0.26	8.8	0.83	10.6	1.00	17.1	1.29	4.0	0.56
65 years and over	*3.6	1.27	8.8	1.44	11.1	1.96	10.0	1.51	5.9	1.38
Not Hispanic or Latino										
White, single race										
18–44 years	0.2	0.05	12.7	0.39	10.7	0.35	17.7	0.45	5.2	0.25
45–64 years	1.9	0.17	10.8	0.42	11.9	0.45	21.8	0.58	6.8	0.34
65 years and over	5.5	0.37	8.6	0.45	6.6	0.41	15.5	0.59	7.1	0.46
Black or African American, single race										
18–44 years	*0.1	0.04	11.4	0.77	9.0	0.77	14.9	0.88	3.9	0.46
45–64 years	1.2	0.32	10.2	0.93	8.8	0.82	23.2	1.28	7.5	0.87
65 years and over	3.0	0.85	9.7	1.28	7.3	1.22	17.2	1.70	6.1	1.10

[1]In two separate questions, respondents were asked if they have ever been told by a doctor or other health care professional that they had emphysema or asthma. Respondents were asked in three separate questions if they had been told by a doctor or other health professional in the past 12 months that they had hay fever, sinusitis, or brochitis. A person may be represented in more than one column.

[2]Total includes other races not shown separately and persons with unknown poverty status.

[3]The category "1 race" refers to people who indicated only a single race group. Persons who indicated a single race other than the groups shown are included in the total for "1 race" but not shown separately due to small sample sizes. Therefore the frequencies for the category "1 race" will will be greater than the sum of the frequencies for the specific groups shown separately.

[4]The category two or more races refers to all persons who indicated more than one race group. Only two combinations of race groups are shown due to small sample sizes for other combinations. Therefore, the frequencies for the category "2 or more races" will be greater than the sum of the frequencies for the specific combinations shown separately.

[5]Persons of Hispanic or Latino origin may be of any race or any combination of races. Similarly, the category "Not Hispanic or Latino" refers to all persons who are not of Hispanic or Latino origin, regardless of race.

[6]Poverty status is based on family income and family size using the U.S. Census Bureau's poverty thresholds for the previous calendar year. "Poor" persons are defined as below the poverty threshold. "Near poor" persons have incomes of 100 percent to less than 200 percent of the poverty threshold. "Not poor" persons have incomes that are 200 percent of the poverty threshold or greater.

* = Figure does not meet standards of reliability or precision.

Table III-18. Rate of Selected Respiratory Diseases Among Persons 18 Years of Age and Over, by Selected Characteristics, 2001—*Continued*

(Percent.)

Characteristic	Selected respiratory disease [1]									
	Emphysema	Standard error for emphysema	Asthma	Standard error for asthma	Hay fever	Standard error for hay fever	Sinusitis	Standard error for sinusitis	Chronic bronchitis	Standard error for chronic brochitis
65 years and over	*1.7	0.78	7.6	1.46	15.2	2.57	12.4	2.26	6.8	1.79
Not Hispanic or Latino										
White, single race, male										
18–44 years	0.3	0.07	11.1	0.54	9.5	0.45	12.3	0.57	3.8	0.33
45–64 years	2.1	0.27	8.9	0.59	10.5	0.63	16.2	0.73	4.0	0.42
65 years and over	7.2	0.66	7.8	0.64	5.9	0.60	12.0	0.83	6.0	0.66
White, single race, female										
18–44 years	*0.2	0.05	14.1	0.54	11.9	0.50	22.9	0.66	6.6	0.37
45–64 years	1.8	0.22	12.6	0.61	13.3	0.63	27.1	0.84	9.5	0.54
65 years and over	4.1	0.41	9.2	0.62	7.1	0.56	18.1	0.82	7.8	0.61
Black or African American, single race, male										
18–44 years	*0.1	0.08	10.2	1.11	8.9	1.16	11.0	1.18	2.5	0.71
45–64 years	*1.5	0.60	5.6	1.02	7.2	1.13	16.1	1.69	3.0	0.82
65 years and over	*5.4	1.89	10.0	2.42	*5.2	2.09	12.4	2.37	*5.8	2.09
Black or African American, single race, female										
18–44 years	*0.0	0.04	12.4	1.01	9.0	0.94	18.1	1.26	5.1	0.59
45–64 years	*0.9	0.32	13.9	1.48	10.0	1.18	28.9	1.86	11.1	1.39
65 years and over	*1.5	0.65	9.5	1.40	8.7	1.49	20.3	2.39	6.3	1.19
Poverty Status and Age [6]										
Poor										
18–44 years	*0.4	0.12	15.8	0.95	8.2	0.75	14.0	0.92	6.3	0.63
45–64 years	4.9	0.70	15.1	1.41	10.3	1.14	21.7	1.57	11.1	1.13
65 years and over	7.4	1.24	12.4	1.68	7.7	1.04	15.6	1.51	9.8	1.51
Near poor										
18–44 years	*0.2	0.10	12.8	0.78	8.5	0.61	14.3	0.85	5.7	0.58
45–64 years	4.6	0.83	11.8	1.12	10.5	1.09	22.3	1.46	9.5	1.09
65 years and over	7.8	0.91	10.0	0.96	7.3	0.83	16.2	1.07	8.4	0.85
Not poor										
18–44 years	0.2	0.04	11.3	0.39	11.3	0.38	17.6	0.47	4.4	0.25
45–64 years	1.1	0.15	10.3	0.47	12.5	0.52	21.6	0.63	5.9	0.37
65 years and over	4.4	0.50	8.8	0.66	7.9	0.66	16.3	0.93	6.4	0.68

[1]In two separate questions, respondents were asked if they have ever been told by a doctor or other health care professional that they had emphysema or asthma. Respondents were asked in three separate questions if they had been told by a doctor or other health professional in the past 12 months that they had hay fever, sinusitis, or brochitis. A person may be represented in more than one column.

[6]Poverty status is based on family income and family size using the U.S. Census Bureau's poverty thresholds for the previous calendar year. "Poor" persons are defined as below the poverty threshold. "Near poor" persons have incomes of 100 percent to less than 200 percent of the poverty threshold. "Not poor" persons have incomes that are 200 percent of the poverty threshold or greater.

* = Figure does not meet standards of reliability or precision.

TABLE III-19. RATE OF SELECTED DISEASES AND CONDITIONS AMONG PERSONS 18 YEARS OF AGE AND OVER, BY SELECTED CHARACTERISTICS, 2001

Overall, 6 percent of the adult population in the United States had ever been told by a doctor or health professional that they had diabetes, and 9 percent that they had an ulcer; 2 percent had been told by a doctor or health professional in the past 12 months that they had kidney disease, and 1 percent that they had liver disease; and 20 percent had experienced symptoms in the past 12 months that were consistent with arthritis (that is, pain or stiffness in or around a joint).

The frequency of arthritic symptoms, ulcers, and kidney disease increases with age; the greater longevity of women partly explains the notably higher prevalence of arthritis for females. The frequency of diabetes also increases with age up to the 65–74 year bracket, but it shows a decline for persons 75 years and over. Diabetes has a 23 percent to 24 percent response frequency among Hispanics and non-Hispanic Blacks in the age group of 65 years and over. By comparison the response rate for non-Hispanic Whites in this age group is 14 percent, with males showing a higher frequency than females. For ulcers, non-Hispanic Whites reported a greater frequency than either Hispanics or non-Hispanic Blacks.

Adults in families classified as not poor reported lower frequencies than the poor and the near poor for each of the five diseases considered in this table.

Table III-19. Rate of Selected Diseases and Conditions Among Persons 18 Years of Age and Over, by Selected Characteristics, 2001

(Percent.)

Characteristic	Selected diseases and conditions									
	Diabetes [1]	Standard error for diabetes	Ulcer [1]	Standard error for ulcers	Kidney stones [2]	Standard error for kiney stones	Liver disease [2]	Standard error for liver disease	Arthritic symptoms [3]	Standard error for arthritic symptoms
Total [4]	6.4	0.16	9.3	0.20	1.6	0.08	1.3	0.07	20.3	0.28
Sex										
Male	6.6	0.25	9.0	0.28	1.4	0.11	1.4	0.11	18.4	0.39
Female	6.3	0.21	9.6	0.25	1.8	0.12	1.2	0.09	22.0	0.35
Age										
18–44 years	2.0	0.12	6.5	0.22	0.9	0.09	0.9	0.08	12.0	0.31
45–64 years	9.4	0.35	11.6	0.36	1.8	0.15	2.1	0.16	27.0	0.53
65–74 years	17.0	0.77	13.6	0.65	3.0	0.35	1.5	0.23	32.2	0.95
75 years and over	13.8	0.72	14.9	0.74	4.1	0.43	1.1	0.21	37.7	1.02
Race										
1 race [5]	6.4	0.16	9.3	0.20	1.6	0.08	1.3	0.07	20.1	0.28
White	6.1	0.18	9.8	0.22	1.6	0.09	1.4	0.08	21.4	0.31
Black or African American	9.0	0.50	7.2	0.42	1.9	0.22	0.9	0.14	16.5	0.72
American Indian or Alaska Native	13.6	2.80	13.0	2.32	*2.0	0.80	*2.2	1.10	24.9	3.78
Asian	3.9	0.69	5.0	0.80	*0.8	0.28	*1.3	0.42	8.6	1.13
Native Hawaiian or other Pacific Islander	*7.9	3.77	*4.3	3.99	*	-	*0.7	0.76	*14.9	5.62
2 or more races [6]	9.3	1.90	11.8	1.91	*2.2	0.80	2.3	0.67	30.9	2.79
Black or African American, White	*4.1	2.93	*6.2	3.63	*2.5	2.52	*2.7	2.71	*22.2	7.91
American Indian or Alaska Native, White	7.8	2.25	17.3	3.19	*0.7	0.51	*2.2	0.95	37.5	4.35
Hispanic or Latino Origin and Race										
Hispanic or Latino [7]	6.9	0.41	6.1	0.36	2.3	0.27	1.5	0.19	12.4	0.54
Mexican or Mexican American	7.2	0.51	6.2	0.44	2.0	0.34	1.2	0.21	10.5	0.65
Not Hispanic or Latino	6.4	0.17	9.7	0.21	1.5	0.08	1.3	0.08	21.2	0.31
White, single race	6.0	0.19	10.2	0.24	1.5	0.09	1.4	0.09	22.4	0.34
Black or African American, single race	9.1	0.50	7.3	0.43	2.0	0.23	0.8	0.14	16.5	0.73
Poverty Status [8]										
Poor	8.1	0.59	12.2	0.65	3.1	0.33	2.9	0.40	23.8	0.94
Near poor	8.9	0.46	11.2	0.52	3.1	0.27	1.9	0.24	23.1	0.72
Not poor	5.3	0.21	8.6	0.25	1.1	0.09	1.1	0.09	19.4	0.36
Region										
Northeast	5.6	0.33	7.4	0.37	1.3	0.16	1.1	0.14	18.7	0.61
Midwest	6.2	0.29	10.1	0.41	1.3	0.14	1.2	0.13	23.3	0.61
South	7.3	0.30	9.7	0.32	1.9	0.15	1.3	0.12	19.1	0.46
West	5.8	0.32	9.3	0.48	1.8	0.20	1.8	0.18	20.1	0.61
Sex and Age										
Male										
18–44 years	1.9	0.19	5.7	0.31	0.5	0.08	1.0	0.13	11.7	0.44
45–64 years	10.1	0.51	11.4	0.53	2.1	0.24	2.3	0.23	24.4	0.74
65 years and over	17.4	0.93	16.5	0.83	3.3	0.41	1.4	0.26	30.7	1.13
Female										
18–44 years	2.1	0.16	7.3	0.31	1.3	0.16	0.7	0.09	12.3	0.40
45–64 years	8.8	0.47	11.7	0.49	1.5	0.17	2.0	0.22	29.5	0.70
65 years and over	14.2	0.66	12.5	0.60	3.7	0.36	1.3	0.21	37.7	0.88

[1] In separate questions respondents were asked if they have ever been told by a doctor or other health care professional that they had an ulcer (including a stomach, duodenal, or peptic ulcer) and diabetes (or sugar diabetes). Female respondents were instructed to exclude pregnancy-related diabetes. Responses from persons who said they had "Borderline" diabetes were treated as with respect to diabetes. A person may be represented in more than one column.

[2] In separate questions respondents were asked if they had been told in the last 12 months by a doctor or other health care professional that they had weak or failing kidneys (excluding kidney stones, bladder infection, or incontinence) or any kind of liver condition.

[3] The information in this table regarding arthritic symptoms was obtained from the following questions: "During the past 12 months, have you had pain, aching, stiffness, or swelling in or around the joint?" and, if yes, "Were these symptoms present on most days for at least a month?" Respondents who indicated that the symptoms were present on most days for at least a month were classified in this table as having arthritic symptoms.

[4] Total includes other races not shown separately and persons with unknown poverty status.

[5] The category "1 race" refers to people who indicated only a single race group. Persons who indicated a single race other than the groups shown are included in the total for "1 race" but not shown separately due to small sample sizes. Therefore the frequencies for the category "1 race" will be greater than the sum of the frequencies for the specific groups shown separately.

[6] The category two or more races refers to all persons who indicated more than one race group. Only two combinations of race groups are shown due to small sample sizes for other combinations. Therefore, the frequencies for the category "2 or more races" will be greater than the sum of the frequencies for the specific combinations shown separately.

[7] Persons of Hispanic or Latino origin may be of any race or any combination of races. Similarly, the category "Not Hispanic or Latino" refers to all persons who are not of Hispanic or Latino origin, regardless of race.

[8] Poverty status is based on family income and family size using the U.S. Census Bureau's poverty thresholds for the previous calendar year. "Poor" persons are defined as below the poverty threshold. "Near poor" persons have incomes of 100 percent to less than 200 percent of the poverty threshold. "Not poor" persons have incomes that are 200 percent of the poverty threshold or greater.

* = Figure does not meet standards of reliability or precision.

- = Quantity zero.

Table III-19. Rate of Selected Diseases and Conditions Among Persons 18 Years of Age and Over, by Selected Characteristics, 2001—*Continued*

(Percent.)

Characteristic	Selected diseases and conditions									
	Diabetes [1]	Standard error for diabetes	Ulcer [1]	Standard error for ulcers	Kidney stones [2]	Standard error for kiney stones	Liver disease [2]	Standard error for liver disease	Arthritic symptoms [3]	Standard error for arthritic symptoms
Hispanic or Latino Origin, Race, and Age										
Hispanic or Latino [7]										
18–44 years	1.8	0.22	4.8	0.38	1.6	0.29	1.1	0.20	6.2	0.43
45–64 years	14.6	1.23	7.8	0.93	2.8	0.53	2.5	0.57	23.2	1.49
65 years and over	24.4	2.19	11.8	1.65	5.6	1.25	*1.2	0.40	30.4	2.31
Not Hispanic or Latino										
White, single race										
18–44 years	1.8	0.15	7.2	0.30	0.8	0.11	0.9	0.10	13.9	0.40
45–64 years	7.9	0.40	12.1	0.43	1.5	0.17	2.2	0.19	28.0	0.63
65 years and over	14.0	0.59	14.8	0.56	3.3	0.30	1.4	0.18	35.5	0.79
Black or African American, single race										
18–44 years	3.5	0.43	4.8	0.49	0.9	0.20	*0.3	0.13	9.3	0.75
45–64 years	15.5	1.18	11.1	0.92	3.4	0.52	1.7	0.37	24.5	1.42
65 years and over	23.2	1.87	10.7	1.22	4.0	0.95	*1.2	0.49	34.7	2.10
Hispanic or Latino Origin, Race, and Age and Sex										
Hispanic or Latino, male [7]										
18–44 years	1.7	0.30	4.1	0.55	0.9	0.25	1.1	0.25	5.2	0.59
45–64 years	16.8	2.00	6.4	1.20	4.0	0.92	*2.3	0.72	17.1	1.84
65 years and over	24.0	3.26	12.9	2.59	*5.6	1.86	*1.4	0.66	27.5	3.48
Hispanic or Latina, female [7]										
18–44 years	2.0	0.32	5.4	0.55	2.4	0.52	1.2	0.30	7.2	0.61
45–64 years	12.5	1.51	9.1	1.38	*1.7	0.55	*2.6	0.88	28.7	2.26
65 years and over	24.7	2.86	11.0	2.10	5.7	1.69	*1.1	0.52	32.6	3.04
Not Hispanic or Latino										
White, single race, male										
18–44 years	1.8	0.24	6.5	0.42	0.4	0.09	1.0	0.18	13.8	0.56
45–64 years	8.5	0.55	12.0	0.63	1.8	0.27	2.2	0.27	25.9	0.86
65 years and over	16.9	1.04	17.2	0.94	3.0	0.44	1.3	0.29	31.2	1.25
White, single race, female										
18–44 years	1.8	0.19	8.0	0.40	1.2	0.19	0.7	0.11	14.0	0.55
45–64 years	7.3	0.54	12.2	0.57	1.3	0.19	2.1	0.27	30.0	0.83
65 years and over	11.9	0.69	13.1	0.68	3.5	0.40	1.4	0.24	38.7	1.00
Black or African American, single race, male										
18–44 years	3.0	0.62	3.4	0.61	*0.9	0.29	*0.3	0.20	8.8	1.23
45–64 years	14.3	1.71	10.5	1.31	4.0	0.92	2.5	0.70	18.7	1.94
65 years and over	19.4	3.36	11.1	2.34	*4.3	1.87	*2.5	1.17	34.6	4.01
Black or African American, single race, female										
18–44 years	3.8	0.55	6.1	0.70	1.0	0.28	0.4	0.17	9.7	0.88
45–64 years	16.4	1.59	11.5	1.25	2.9	0.65	1.1	0.33	29.2	1.82
65 years and over	25.7	2.24	10.4	1.42	*3.8	1.15	*0.3	0.21	34.8	2.39
Poverty Status and Age [8]										
Poor										
18–44 years	2.3	0.36	8.6	0.78	2.0	0.36	2.0	0.41	14.7	0.96
45–64 years	16.1	1.74	20.9	1.54	5.4	0.88	5.5	1.19	37.5	2.20
65 years and over	20.7	1.80	14.2	1.46	4.3	0.74	2.5	0.66	41.9	2.13
Near poor										
18–44 years	2.2	0.34	7.0	0.59	2.1	0.33	1.2	0.28	14.0	0.83
45–64 years	17.6	1.38	16.3	1.29	4.1	0.71	3.8	0.66	31.6	1.68
65 years and over	17.2	1.17	16.5	1.09	4.6	0.64	1.7	0.42	37.2	1.56
Not poor										
18–44 years	2.0	0.16	6.5	0.30	0.6	0.10	0.7	0.10	11.9	0.39
45–64 years	7.9	0.42	10.2	0.45	1.0	0.14	1.7	0.19	26.4	0.68
65 years and over	14.2	0.92	14.3	0.84	3.4	0.46	1.4	0.28	34.9	1.14

[1]In separate questions respondents were asked if they have ever been told by a doctor or other health care professional that they had an ulcer (including a stomach, duodenal, or peptic ulcer) and diabetes (or sugar diabetes). Female respondents were instructed to exclude pregnancy-related diabetes. Responses from persons who said they had "Borderline" diabetes were treated as with respect to diabetes. A person may be represented in more than one column.

[2]In separate questions respondents were asked if they had been told in the last 12 months by a doctor or other health care professional that they had weak or failing kidneys (excluding kidney stones, bladder infection, or incontinence) or any kind of liver condition.

[3]The information in this table regarding arthritic symptoms was obtained from the following questions: "During the past 12 months, have you had pain, aching, stiffness, or swelling in or around the joint?" and, if yes, "Were these symptoms present on most days for at least a month?" Respondents who indicated that the symptoms were present on most days for at least a month were classified in this table as having arthritic symptoms.

[7]Persons of Hispanic or Latino origin may be of any race or any combination of races. Similarly, the category "Not Hispanic or Latino" refers to all persons who are not of Hispanic or Latino origin, regardless of race.

[8]Poverty status is based on family income and family size using the U.S. Census Bureau's poverty thresholds for the previous calendar year. "Poor" persons are defined as below the poverty threshold. "Near poor" persons have incomes of 100 percent to less than 200 percent of the poverty threshold. "Not poor" persons have incomes that are 200 percent of the poverty threshold or greater.

* = Figure does not meet standards of reliability or precision.

Figure III-8. Percent of Selected Diseases and Conditions Among Persons Age 18 Years and Over by Sex, 2001

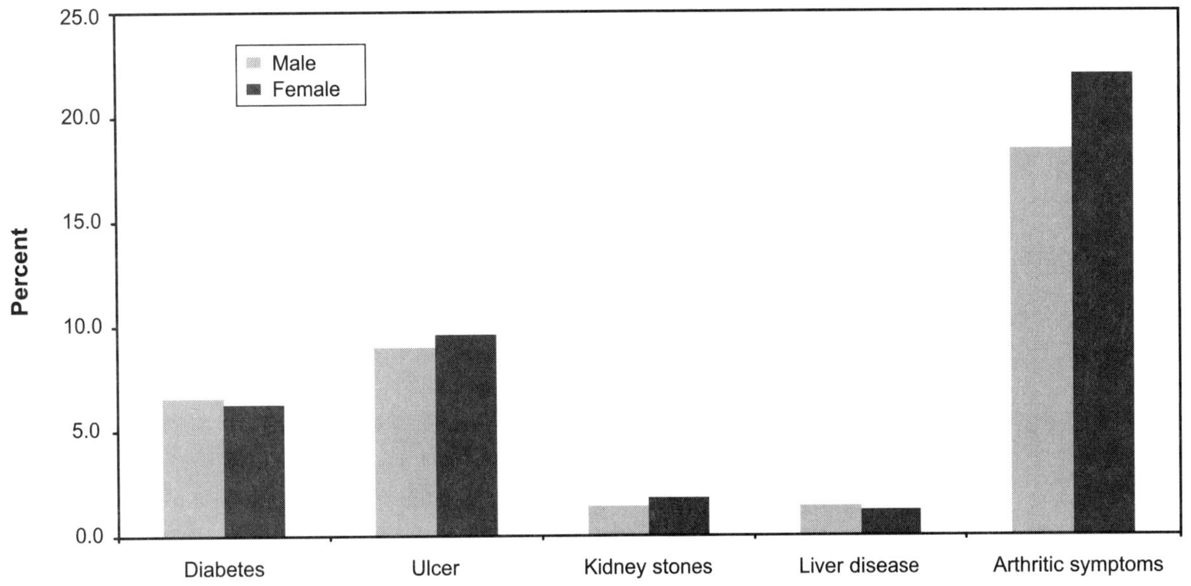

TABLE III-20. RATE OF MIGRAINES AND PAIN IN THE NECK, LOWER BACK, FACE, OR JAW AMONG PERSONS 18 YEARS OF AGE AND OVER, BY SELECTED CHARACTERISTICS, 2001

During the 3 months prior to the interview, 17 percent of the adult population in the United States had experienced a migraine or severe headache, 17 percent had experienced pain in the neck area, 31 percent had experienced pain in the lower back, and 5 percent had experienced pain in the face or jaw area. Respondents were instructed to report pain only if it had lasted a whole day or more; a respondent may be represented in more than one category.

Women were twice as likely to have experienced a recent migraine headache as men (22 percent versus 11 percent), and they also reported other pains more frequently than men. Pain in the lower back showed the least lopsided results, with women reporting a frequency of 33 percent and men 29 percent.

In contrast to many diseases, the frequency of migraine headaches was inversely related to age. As age advances, this frequency steadily declines from 20 percent in the age group of 18 to 44 years to 5 percent for those aged 75 years and over. This inverse age pattern was reported by non-Hispanic White as well as Black respondents, both male and female. The age pattern for Hispanics was slightly different as their greatest frequency of headaches occurred in the 45–64 year age bracket rather than before that.

For all adults, pain in the lower back occurred much more evenly throughout adult life than migraine headaches, but it was most frequent in the 45–64 year age bracket. Non-Hispanic Whites reported higher frequencies of lower back pain, compared with Hispanics and non-Hispanic Blacks. Asians reported the smallest frequency of 20 percent.

Table III-20. Rate of Migraines and Pain in the Neck, Lower Back, Face, or Jaw Among Persons 18 Years of Age and Over, by Selected Characteristics, 2001

(Percent.)

Characteristic	Migraines or severe headaches [1]	Standard error for migraines or severe headaches	Pain in the neck [2]	Standard error for pains in the neck	Pain in the lower back [3]	Standard error for pains in the lower back	Pain in face or jaw [4]	Standard error for pains in face or jaw
Total [5]	16.7	0.26	16.8	0.26	31.1	0.35	5.3	0.15
Sex								
Male	10.6	0.31	13.8	0.35	29.4	0.48	3.4	0.18
Female	22.3	0.40	19.5	0.35	32.7	0.43	7.0	0.22
Age								
18–44 years	20.0	0.38	14.6	0.33	28.4	0.45	5.5	0.21
45–64 years	16.1	0.42	21.0	0.48	35.3	0.58	5.9	0.26
65–74 years	8.3	0.55	15.4	0.74	31.2	0.92	3.8	0.35
75 years and over	5.2	0.47	16.2	0.78	33.0	0.97	3.1	0.37
Race								
1 race [6]	16.6	0.27	16.7	0.26	31.0	0.35	5.3	0.15
White	16.5	0.30	17.4	0.29	32.1	0.39	5.5	0.17
Black or African American	17.0	0.69	13.1	0.59	26.5	0.86	4.1	0.34
American Indian or Alaska Native	15.1	2.98	21.0	3.28	35.2	3.71	*5.9	1.82
Asian	13.7	1.35	10.4	1.18	20.4	1.50	3.7	0.83
Native Hawaiian or other Pacific Islander	*22.4	9.28	*10.4	4.89	24.0	6.82	*5.9	4.20
2 or more races [7]	25.5	2.38	22.8	2.53	38.8	2.80	8.2	1.36
Black or African American, White	*16.3	7.11	*20.2	7.46	33.8	8.65	*9.8	4.67
American Indian or Alaska Native, White	27.0	3.41	24.0	3.74	39.3	4.26	8.0	1.97
Hispanic or Latino Origin and Race								
Hispanic or Latino [8]	18.1	0.61	15.3	0.59	27.7	0.74	4.4	0.31
Mexican or Mexican American	16.1	0.73	12.8	0.67	25.0	0.89	3.8	0.37
Not Hispanic or Latino	16.5	0.29	16.9	0.28	31.5	0.38	5.4	0.16
White, single race	16.5	0.32	17.7	0.32	32.7	0.42	5.7	0.19
Black or African American, single race	16.8	0.68	13.0	0.60	26.3	0.87	4.1	0.34
Poverty Status [9]								
Poor	23.6	0.89	20.1	0.89	36.1	1.12	8.1	0.50
Near poor	19.0	0.70	18.7	0.64	35.0	0.87	6.4	0.42
Not poor	16.3	0.34	16.6	0.34	31.2	0.44	5.0	0.19
Region								
Northeast	14.9	0.55	16.8	0.57	30.9	0.76	4.8	0.30
Midwest	17.5	0.52	17.2	0.50	32.4	0.63	5.3	0.32
South	17.3	0.46	15.9	0.44	29.4	0.64	5.2	0.25
West	16.2	0.60	17.8	0.57	32.8	0.74	5.9	0.35
Sex and Age								
Male								
18–44 years	12.5	0.45	12.1	0.45	26.6	0.62	3.7	0.25
45–64 years	10.1	0.51	17.1	0.65	34.5	0.85	3.5	0.31
65 years and over	4.2	0.44	12.8	0.78	29.0	1.05	2.3	0.34
Female								
18–44 years	27.2	0.56	17.0	0.47	30.2	0.59	7.4	0.32
45–64 years	21.7	0.66	24.7	0.69	36.0	0.75	8.0	0.42
65 years and over	8.9	0.53	18.0	0.72	34.2	0.88	4.4	0.38
Hispanic or Latino Origin, Race, and Age								
Hispanic or Latino [8]								
18–44 years	17.8	0.72	11.5	0.61	24.4	0.86	3.4	0.33
45–64 years	20.6	1.39	22.3	1.37	33.6	1.55	6.7	0.82
65 years and over	13.5	1.71	25.3	2.13	36.6	2.63	5.7	1.12
Not Hispanic or Latino								
White, single race								
18–44 years	20.8	0.48	16.2	0.43	30.7	0.58	6.4	0.28
45–64 years	15.7	0.51	21.3	0.56	35.9	0.68	5.9	0.30
65 years and over	6.2	0.40	15.3	0.59	32.3	0.78	3.1	0.29
Black or African American, single race								
18–44 years	19.0	0.97	10.0	0.73	22.7	1.06	3.3	0.42
45–64 years	16.0	1.12	18.6	1.27	32.9	1.58	5.3	0.70
65 years and over	7.5	1.10	15.4	1.58	29.0	2.03	5.1	0.88

[1]Respondents were asked, "During the past 3 months, did you have a severe headache or migraine?" Respondents were instructed to report pain that had lasted a whole day or more and, conversely, not to report fleeting or minor aches or pains. A person may be represented in more than one column.

[2]Respondents were asked, "During the past 3 months, did you have neck pain?" Respondents were instructed to report pain that had lasted a whole day or more and, conversely, not to report fleeting or minor aches or pains. A person may be represented in more than one column.

[3]Respondents were asked, "During the past 3 months, did you have low back pain?" Respondents were instructed to report pain that had lasted a whole day or more and, conversely, not to report fleeting or minor aches or pains. A person may be represented in more than one column.

[4]Respondents were asked, "During the past 3 months, did you have facial ache or pain in the jaw muscles or the joint in front of the ear?" Respondents were instructed to report pain that had lasted a whole day or more and, conversely, not to report fleeting or minor aches or pains. A person may be represented in more than one column.

[5]Total includes other races not shown separately and persons with unknown poverty status.

[6]The category "1 race" refers to people who indicated only a single race group. Persons who indicated a single race other than the groups shown are included in the total for "1 race" but not shown separately due to small sample sizes. Therefore the frequencies for the category "1 race" will will be greater than the sum of the frequencies for the specific groups shown separately.

[7]The category "2 or more races" refers to all persons who indicated more than one race group. Only two combinations of race groups are shown due to small sample sizes for other combinations. Therefore, the frequencies for the category "2 or more races" will be greater than the sum of the frequencies for the specific combinations shown separately.

[8]Persons of Hispanic or Latino origin may be of any race or any combination of races. Similarly, the category "Not Hispanic or Latino" refers to all persons who are not of Hispanic or Latino origin, regardless of race.

[9]Poverty status is based on family income and family size using the U.S. Census Bureau's poverty thresholds for the previous calendar year. "Poor" persons are defined as below the poverty threshold. "Near poor" persons have incomes of 100 percent to less than 200 percent of the poverty threshold. "Not poor" persons have incomes that are 200 percent of the poverty threshold or greater.

* = Figure does not meet standards of reliability or precision.

Table III-20. Rate of Migraines and Pain in the Neck, Lower Back, Face, or Jaw Among Persons 18 Years of Age and Over, by Selected Characteristics, 2001—*Continued*

(Percent.)

Characteristic	Migraines or severe headaches [1]	Standard error for migraines or severe headaches	Pain in the neck [2]	Standard error for pains in the neck	Pain in the lower back [3]	Standard error for pains in the lower back	Pain in face or jaw [4]	Standard error for pains in face or jaw
Hispanic or Latino Origin, Race, Age, and Sex								
Hispanic or Latino, male [8]								
18–44 years	11.1	0.87	8.6	0.79	21.7	1.16	1.7	0.32
45–64 years	12.7	1.71	17.5	1.86	29.6	2.28	4.0	0.97
65 years and over	9.1	2.30	19.5	3.39	34.7	4.20	*3.9	1.55
Hispanic or Latina, female [8]								
18–44 years	24.6	1.11	14.5	0.93	27.1	1.23	5.2	0.56
45–64 years	27.9	2.08	26.8	2.03	37.3	2.11	9.1	1.28
65 years and over	16.7	2.41	29.5	2.98	38.1	3.21	7.1	1.58
Not Hispanic or Latino								
White, single race, male								
18–44 years	12.8	0.56	13.5	0.57	29.4	0.81	4.2	0.33
45–64 years	9.6	0.60	17.3	0.77	35.5	0.99	3.3	0.36
65 years and over	3.7	0.46	12.3	0.84	29.2	1.14	2.1	0.37
White, single race, female								
18–44 years	28.6	0.73	18.9	0.62	32.0	0.77	8.6	0.42
45–64 years	21.4	0.78	25.1	0.82	36.4	0.88	8.4	0.50
65 years and over	8.2	0.58	17.6	0.80	34.7	0.99	3.9	0.40
Black or African American, single race, male								
18–44 years	13.3	1.32	8.6	1.10	19.1	1.55	2.3	0.60
45–64 years	11.2	1.60	13.5	1.70	30.4	2.25	4.0	0.82
65 years and over	*3.5	1.10	14.4	2.77	26.7	3.58	*2.5	1.03
Black or African American, single race, female								
18–44 years	23.7	1.32	11.2	0.94	25.8	1.30	4.2	0.57
45–64 years	19.9	1.70	22.6	1.76	34.9	2.04	6.3	1.06
65 years and over	10.0	1.60	16.0	1.90	30.5	2.41	6.8	1.33
Poverty Status and Age [9]								
Poor								
18–44 years	25.3	1.17	16.1	1.03	31.4	1.38	7.7	0.66
45–64 years	26.2	1.95	30.7	2.02	47.7	2.12	10.4	1.17
65 years and over	11.8	1.37	20.5	1.70	37.9	2.25	6.3	1.00
Near poor								
18–44 years	21.7	1.00	15.7	0.82	31.6	1.15	6.6	0.63
45–64 years	21.1	1.39	25.9	1.55	42.5	1.70	7.7	0.96
65 years and over	10.3	0.98	18.9	1.35	36.0	1.63	4.6	0.61
Not poor								
18–44 years	19.4	0.50	14.9	0.44	29.2	0.58	5.3	0.26
45–64 years	14.8	0.53	19.6	0.61	34.3	0.71	5.1	0.30
65 years and over	6.2	0.59	15.6	0.89	31.9	1.15	3.0	0.42

[1]Respondents were asked, "During the past 3 months, did you have a severe headache or migraine?" Respondents were instructed to report pain that had lasted a whole day or more and, conversely, not to report fleeting or minor aches or pains. A person may be represented in more than one column.
[2]Respondents were asked, "During the past 3 months, did you have neck pain?" Respondents were instructed to report pain that had lasted a whole day or more and, conversely, not to report fleeting or minor aches or pains. A person may be represented in more than one column.
[3]Respondents were asked, "During the past 3 months, did you have low back pain?" Respondents were instructed to report pain that had lasted a whole day or more and, conversely, not to report fleeting or minor aches or pains. A person may be represented in more than one column.
[4]Respondents were asked, "During the past 3 months, did you have facial ache or pain in the jaw muscles or the joint in front of the ear?" Respondents were instructed to report pain that had lasted a whole day or more and, conversely, not to report fleeting or minor aches or pains. A person may be represented in more than one column.
[8]Persons of Hispanic or Latino origin may be of any race or any combination of races. Similarly, the category "Not Hispanic or Latino" refers to all persons who are not of Hispanic or Latino origin, regardless of race.
[9]Poverty status is based on family income and family size using the U.S. Census Bureau's poverty thresholds for the previous calendar year. "Poor" persons are defined as below the poverty threshold. "Near poor" persons have incomes of 100 percent to less than 200 percent of the poverty threshold. "Not poor" persons have incomes that are 200 percent of the poverty threshold or greater.
* = Figure does not meet standards of reliability or precision.

TABLE III-21. RATE OF HEARING TROUBLE, VISION TROUBLE, AND ABSENCE OF TEETH AMONG PERSONS 18 YEARS OF AGE AND OVER, BY SELECTED CHARACTERISTICS, 2001

At the time of the interview, 17 percent of the U.S. adult population experienced some hearing difficulty (without a hearing aid), and men were somewhat more likely to have this trouble than women. Some 10 percent of the population experienced vision trouble, defined as trouble seeing, even with glasses or contact lenses. Nine percent of the adult population had lost all of their natural teeth.

All of these afflictions tend to increase with advancing age, such that for persons 75 years of age and over, 47 percent reported some degree of hearing impairment, 23 percent reported vision trouble, and 31 percent reported the loss of their teeth.

Non-Hispanic Whites and American Indians were more likely to have hearing impairment than Hispanics and non-Hispanic Blacks. In regard to vision trouble, non Hispanic Blacks joined non-Hispanic Whites and American Indians in reporting relatively high frequencies. Asians stood out as having relatively few of the impairments considered here.

Table III-21. Rate of Hearing Trouble, Vision Trouble, and Absence of Teeth Among Persons 18 Years of Age and Over, by Selected Characteristics, 2001

(Percent.)

Characteristic	Persons 18 years of age and over with selected sensory problems											
	Hearing [1]						Vision [1]				Teeth [1]	
	Good	Standard error for good	A little trouble	Standard error for a little trouble	A lot of trouble or deaf	Standard error for a lot of trouble or deaf	No vision trouble	Standard error for no vision trouble	Vision trouble	Standard error for vision trouble	Absence of natural teeth	Standard error for absense of natural teeth
Total [2]	82.6	0.28	14.0	0.25	3.4	0.13	90.0	0.22	10.0	0.22	8.5	0.18
Sex												
Male	79.2	0.42	16.2	0.37	4.6	0.21	91.4	0.28	8.6	0.28	7.7	0.24
Female	85.8	0.32	11.9	0.29	2.4	0.13	88.7	0.30	11.3	0.30	9.1	0.25
Age												
18–44 years	91.4	0.28	7.6	0.26	1.0	0.09	93.7	0.24	6.3	0.24	2.2	0.14
45–64 years	79.5	0.47	17.3	0.44	3.2	0.21	87.7	0.38	12.3	0.38	9.3	0.35
65–74 years	65.2	1.01	26.7	0.92	8.1	0.62	85.8	0.71	14.2	0.71	24.7	0.90
75 years and over	52.5	1.10	31.1	1.04	16.3	0.82	77.5	0.90	22.5	0.90	31.2	1.07
Race												
1 race [3]	82.7	0.28	13.9	0.25	3.4	0.13	90.1	0.22	9.9	0.22	8.4	0.18
White	80.9	0.32	15.3	0.28	3.9	0.15	90.1	0.25	9.9	0.25	8.8	0.21
Black or African American	91.2	0.52	7.3	0.47	1.5	0.20	89.4	0.55	10.6	0.55	8.1	0.45
American Indian or Alaska Native	78.9	3.29	16.9	2.92	*4.2	1.64	88.6	2.48	11.4	2.48	6.0	1.91
Asian	91.8	1.05	7.5	1.01	*0.7	0.30	94.0	0.89	6.0	0.89	4.6	0.74
Native Hawaiian or other Pacific Islander	92.5	3.53	*3.2	1.80	*4.3	3.13	87.7	6.65	*12.3	6.65	*	-
2 or more races [4]	76.3	2.67	19.6	2.50	4.1	1.08	82.6	2.41	17.4	2.41	11.3	1.75
Black or African American, White	89.6	4.72	*7.8	4.02	*2.6	2.59	94.7	3.67	*5.3	3.67	*9.2	5.49
American Indian or Alaska Native, White	68.4	3.89	26.5	3.78	5.1	1.39	79.7	3.57	20.3	3.57	15.8	2.86
Hispanic or Latino Origin and Race												
Hispanic or Latino [5]	91.5	0.50	6.8	0.45	1.7	0.23	91.4	0.44	8.6	0.44	5.7	0.41
Mexican or Mexican American	91.9	0.57	6.4	0.51	1.7	0.30	91.7	0.55	8.3	0.55	4.3	0.44
Not Hispanic or Latino	81.5	0.31	14.8	0.27	3.6	0.14	89.8	0.24	10.2	0.24	8.8	0.19
White, single race	79.7	0.34	16.2	0.31	4.1	0.16	89.8	0.27	10.2	0.27	9.1	0.22
Black or African American, single race	91.3	0.52	7.2	0.47	1.5	0.20	89.3	0.55	10.7	0.55	8.1	0.46
Poverty Status [6]												
Poor	83.8	0.84	12.6	0.71	3.6	0.40	85.1	0.75	14.9	0.75	12.5	0.70
Near poor	80.3	0.69	14.9	0.61	4.8	0.34	85.8	0.63	14.2	0.63	14.0	0.57
Not poor	82.9	0.36	14.2	0.32	2.9	0.15	91.5	0.27	8.5	0.27	5.9	0.20
Region												
Northeast	85.1	0.61	11.8	0.54	3.0	0.26	91.4	0.43	8.6	0.43	7.8	0.38
Midwest	79.3	0.60	16.7	0.54	3.9	0.26	89.8	0.46	10.2	0.46	8.9	0.36
South	83.8	0.45	12.9	0.38	3.3	0.22	89.5	0.40	10.5	0.40	9.4	0.34
West	82.0	0.65	14.4	0.59	3.5	0.27	89.9	0.45	10.1	0.45	6.7	0.33
Sex and Age												
Male												
18–44 years	89.9	0.43	8.9	0.40	1.2	0.13	95.1	0.28	4.9	0.28	2.0	0.18
45–64 years	73.4	0.76	21.7	0.70	4.9	0.38	88.9	0.53	11.1	0.53	9.4	0.50
65 years and over	50.6	1.19	32.3	1.09	17.1	0.95	82.4	0.92	17.6	0.92	26.1	0.98
Female												
18–44 years	92.8	0.34	6.3	0.31	0.9	0.11	92.4	0.35	7.6	0.35	2.4	0.20
45–64 years	85.2	0.54	13.2	0.52	1.5	0.18	86.6	0.55	13.4	0.55	9.2	0.44
65 years and over	66.0	0.90	26.0	0.82	8.0	0.53	81.7	0.71	18.3	0.71	28.8	0.94

[1]Respondents were asked, "Which statement best describes your hearing without a hearing aid: good, a little trouble, a lot of trouble, deaf?" For the purposes of this table, "a lot of trouble" and "deaf" are combined into one category. Regarding their vision, respondents were asked, "Do you have any trouble seeing, even when wearing glasses or contact lenses?" Lastly, respondents were asked, in one question, "Have you lost all of your upper and lower natural (permanent) teeth?" A person may be represented in more than one column.
[2]Total includes other races not shown separately
[3]The category "1 race" refers to people who indicated only a single race group. Persons who indicated a single race other than the groups shown are included in the total for "1 race" but not shown separately due to small sample sizes. Therefore the frequencies for the category "1 race" will will be greater than the sum of the frequencies for the specific groups shown separately.
[4]The category "2 or more races" refers to all persons who indicated more than one race group. Only two combinations of race groups are shown due to small sample sizes for other combinations. Therefore, the frequencies for the category "2 or more races" will be greater than the sum of the frequencies for the specific combinations shown separately.
[5]Persons of Hispanic or Latino origin may be of any race or any combination of races. Similarly, the category "Not Hispanic or Latino" refers to all persons who are not of Hispanic or Latino origin, regardless of race.
[6]Poverty status is based on family income and family size using the U.S. Census Bureau's poverty thresholds for the previous calendar year. "Poor" persons are defined as below the poverty threshold. "Near poor" persons have incomes of 100 percent to less than 200 percent of the poverty threshold. "Not poor" persons have incomes that are 200 percent of the poverty threshold or greater.
* = Figure does not meet standards of reliability or precision.
- = Quantity zero.

Table III-21. Rate of Hearing Trouble, Vision Trouble, and Absence of Teeth Among Persons 18 Years of Age and Over, by Selected Characteristics, 2001—*Continued*

(Percent.)

Characteristic	Persons 18 years of age and over with selected sensory problems											
	Hearing [1]						Vision [1]				Teeth [1]	
	Good	Standard error for good	A little trouble	Standard error for a little trouble	A lot of trouble or deaf	Standard error for a lot of trouble or deaf	No vision trouble	Standard error for no vision trouble	Vision trouble	Standard error for vision trouble	Absence of natural teeth	Standard error for absense of natural teeth
Hispanic or Latino Origin, Race, and Age												
Hispanic or Latino [5]												
18–44 years	95.6	0.42	3.8	0.40	0.6	0.14	94.9	0.40	5.1	0.40	1.8	0.28
45–64 years	87.3	1.24	10.8	1.15	1.9	0.48	85.4	1.21	14.6	1.21	6.1	0.86
65 years and over	72.1	2.27	18.0	1.79	9.9	1.73	81.4	2.08	18.6	2.08	32.7	2.38
Not Hispanic or Latino												
White, single race												
18–44 years	89.8	0.38	9.0	0.35	1.2	0.12	93.5	0.30	6.5	0.30	2.4	0.18
45–64 years	77.3	0.55	19.1	0.53	3.6	0.25	88.4	0.44	11.6	0.44	9.6	0.40
65 years and over	56.5	0.84	30.7	0.77	12.8	0.60	82.4	0.65	17.6	0.65	26.4	0.76
Black or African American, single race												
18–44 years	95.4	0.51	4.1	0.48	0.5	0.13	93.1	0.61	6.9	0.61	2.0	0.30
45–64 years	88.8	0.99	9.7	0.91	1.5	0.37	85.5	1.09	14.5	1.09	10.1	0.87
65 years and over	76.0	2.07	17.7	1.75	6.4	1.28	79.0	1.88	21.0	1.88	35.3	2.21
Hispanic or Latino Origin, Race, Age, and Sex												
Hispanic or Latino, male [5]												
18–44 years	94.9	0.62	4.7	0.60	*0.4	0.17	95.9	0.53	4.1	0.53	1.9	0.42
45–64 years	85.7	1.93	11.7	1.73	*2.6	0.91	86.8	1.67	13.2	1.67	7.3	1.48
65 years and over	64.5	4.00	23.6	3.42	12.0	3.02	78.3	3.57	21.7	3.57	27.9	3.86
Hispanic or Latina, female [5]												
18–44 years	96.3	0.54	2.9	0.49	*0.7	0.22	93.9	0.59	6.1	0.59	1.8	0.37
45–64 years	88.8	1.55	10.0	1.50	*1.2	0.41	84.1	1.68	15.9	1.68	4.9	0.90
65 years and over	77.7	2.65	13.9	1.93	8.4	2.01	83.7	2.25	16.3	2.25	36.2	3.11
Not Hispanic or Latino												
White, single race, male												
18–44 years	87.6	0.58	10.9	0.54	1.5	0.19	94.9	0.37	5.1	0.37	2.1	0.24
45–64 years	70.2	0.90	24.1	0.86	5.6	0.46	89.4	0.59	10.6	0.59	9.9	0.58
65 years and over	47.1	1.29	34.2	1.23	18.7	1.09	82.9	0.98	17.1	0.98	25.0	1.08
White, single race, female												
18–44 years	91.9	0.46	7.1	0.42	1.0	0.15	92.1	0.45	7.9	0.45	2.6	0.25
45–64 years	84.0	0.64	14.4	0.62	1.6	0.21	87.4	0.65	12.6	0.65	9.4	0.53
65 years and over	63.6	1.02	28.1	0.94	8.3	0.61	82.0	0.81	18.0	0.81	27.5	1.03
Black or African American, single race, male												
18–44 years	97.2	0.61	2.5	0.58	*0.3	0.14	94.6	0.88	5.4	0.88	1.6	0.46
45–64 years	87.0	1.70	11.2	1.60	*1.9	0.60	88.3	1.39	11.7	1.39	8.2	1.22
65 years and over	73.8	3.39	19.3	3.26	6.9	2.02	79.9	3.19	20.1	3.19	35.1	3.67
Black or African American, single race, female												
18–44 years	93.9	0.77	5.4	0.71	*0.7	0.21	91.8	0.82	8.2	0.82	2.3	0.45
45–64 years	90.3	1.22	8.5	1.10	*1.2	0.45	83.2	1.53	16.8	1.53	11.6	1.24
65 years and over	77.4	2.28	16.6	1.86	6.0	1.65	78.3	2.22	21.7	2.22	35.4	2.59
Poverty Status and Age [6]												
Poor												
18–44 years	91.4	0.74	7.5	0.71	1.1	0.28	89.8	0.87	10.2	0.87	3.6	0.53
45–64 years	76.7	1.83	17.8	1.67	5.5	0.97	78.6	1.56	21.4	1.56	17.4	1.53
65 years and over	61.9	2.34	26.8	2.06	11.3	1.56	74.9	1.93	25.1	1.93	43.1	2.13
Near poor												
18–44 years	89.8	0.73	8.9	0.68	1.3	0.27	92.0	0.67	8.0	0.67	2.6	0.38
45–64 years	77.0	1.50	17.8	1.45	5.3	0.83	78.3	1.50	21.7	1.50	16.8	1.41
65 years and over	60.2	1.49	27.0	1.34	12.8	1.02	77.9	1.30	22.1	1.30	38.9	1.59
Not poor												
18–44 years	90.8	0.38	8.2	0.36	1.0	0.11	94.6	0.29	5.4	0.29	1.9	0.18
45–64 years	78.4	0.63	18.7	0.58	2.9	0.25	89.2	0.48	10.8	0.48	7.4	0.41
65 years and over	57.9	1.20	30.2	1.12	11.9	0.85	83.7	0.93	16.3	0.93	20.8	0.94

[1]Respondents were asked, "Which statement best describes your hearing without a hearing aid: good, a little trouble, a lot of trouble, deaf?" For the purposes of this table, "a lot of trouble" and "deaf" are combined into one category. Regarding their vision, respondents were asked, "Do you have any trouble seeing, even when wearing glasses or contact lenses?" Lastly, respondents were asked, in one question, "Have you lost all of your upper and lower natural (permanent) teeth?" A person may be represented in more than one column.

[5]Persons of Hispanic or Latino origin may be of any race or any combination of races. Similarly, the category "Not Hispanic or Latino" refers to all persons who are not of Hispanic or Latino origin, regardless of race.

[6]Poverty status is based on family income and family size using the U.S. Census Bureau's poverty thresholds for the previous calendar year. "Poor" persons are defined as below the poverty threshold. "Near poor" persons have incomes of 100 percent to less than 200 percent of the poverty threshold. "Not poor" persons have incomes that are 200 percent of the poverty threshold or greater.

* = Figure does not meet standards of reliability or precision.

TABLE III-22. RATE OF FEELINGS OF SADNESS, HOPELESSNESS, WORTHLESSNESS OR THAT EVERYTHING IS AN EFFORT AMONG PERSONS 18 YEARS OF AGE AND OVER, BY SELECTED CHARACTERISTICS, 2001

Respondents gave answers to four separate questions on frequency of sadness, hopelessness, worthlessness and manageability of life ("everything is an effort").

More than 13 percent of all adults felt so sad that nothing could cheer them up, either during all, most, or some of the time in the 30 days prior to the interview. Similarly, 7 percent felt hopeless, 5 percent felt worthless, and 15 percent felt like everything was an effort during all, most, or some of the time during the same period. Women were more likely than men to report the feelings covered by each of these four questions.

Non-Hispanic Whites and Asians had the feelings under consideration less frequently than non-Hispanic Blacks. Hispanics are somewhat more likely to report feelings of sadness, hopelessness, and worthlessness than non-Hispanic Whites. Adults in poor families were much more likely (generally at least two times more likely) to report depressive feelings than respondents from families that were not poor. Respondents covered by Medicaid insurance, which insures persons with low incomes, also reported high percentages of depressive or pesssimistic feelings.

While age was not a major factor in distinguishing response frequencies, marital status was important: respondents who were divorced, separated, or widowed had a higher frequency of negative feelings than those married, never married, or living with a partner. Married persons had the lowest frequencies among these categories.

Table III-22. Rate of Feelings of Sadness, Hopelessness, Worthlessness, or that Everything is an Effort Among Persons 18 Years of Age and Over, by Selected Characteristics, 2001

(Percent.)

Characteristic	Selected mental health characterisitics															
	Sadness[1]				Hopelessness[1]				Worthlessness[1]				Everything is an effort[1]			
	All or most of the time	Standard error for all or most of the time	Some of the time	Standard error for some of the time	All or most of the time	Standard error for all or most of the time	Some of the time	Standard error for som eof the time	All or most of the time	Standard error for all or most of the time	Some of the time	Standard error for some of the time	All or most of the time	Standard error for all or most of the time	Some of the time	Standard error for some of the time
Total[2]	3.4	0.12	9.8	0.19	2.2	0.09	4.5	0.14	1.9	0.09	3.5	0.12	5.8	0.17	9.2	0.20
Sex																
Male	2.7	0.15	7.8	0.27	1.7	0.12	4.0	0.19	1.7	0.13	3.0	0.17	5.2	0.23	8.2	0.28
Female	4.1	0.17	11.6	0.27	2.6	0.14	5.1	0.19	2.2	0.13	4.1	0.17	6.3	0.23	10.2	0.27
Age																
18–44 years	3.2	0.15	9.1	0.26	2.1	0.13	4.6	0.19	1.7	0.12	3.3	0.16	5.9	0.23	9.7	0.29
45–64 years	3.8	0.22	10.4	0.35	2.4	0.17	4.8	0.25	2.4	0.17	4.0	0.23	5.9	0.28	9.0	0.33
65–74 years	2.9	0.32	10.3	0.61	1.9	0.25	3.9	0.38	1.9	0.26	3.0	0.34	4.5	0.39	7.9	0.54
75 years and over	4.0	0.41	12.2	0.70	1.9	0.30	3.7	0.38	1.8	0.30	3.8	0.41	5.5	0.49	8.8	0.63
Race																
1 race[3]	3.4	0.12	9.8	0.19	2.2	0.09	4.5	0.14	1.9	0.09	3.5	0.12	5.7	0.17	9.2	0.19
White	3.1	0.13	9.2	0.21	2.1	0.10	4.4	0.15	1.8	0.10	3.4	0.13	5.2	0.18	8.9	0.21
Black or African American	4.8	0.36	12.8	0.60	2.4	0.26	5.2	0.38	2.2	0.26	4.1	0.34	8.5	0.51	10.7	0.55
American Indian or Alaska Native	10.4	2.68	16.7	3.85	*5.1	1.93	10.2	2.79	*8.0	2.48	*4.5	1.95	9.5	2.58	15.5	2.84
Asian	2.9	0.64	11.0	1.24	*1.6	0.48	5.2	0.82	1.1	0.32	2.4	0.58	5.5	0.90	8.5	1.02
Native Hawaiian or other Pacific Islander	*	-	*9.6	4.86	*3.2	3.21	*1.2	1.20	*	-	*4.4	3.43	*	-	*9.8	5.39
2 or more races[4]	4.3	0.94	14.2	1.92	3.6	0.89	5.1	1.12	3.3	0.85	5.1	1.32	8.8	1.40	16.0	2.26
Black or African American, white	*6.3	4.31	*7.1	4.16	*	-	*11.0	5.30	*0.7	0.67	*9.0	4.99	*10.6	4.99	*22.2	8.02
American Indian or Alaska Native, white	*4.2	1.40	17.9	3.17	*4.6	1.53	*4.3	1.54	*4.3	1.48	*6.2	2.16	10.2	2.36	12.9	3.07
Hispanic or Latino Origin and Race																
Hispanic or Latino[5]	4.8	0.36	10.6	0.51	3.7	0.32	5.1	0.36	2.2	0.22	3.6	0.30	6.0	0.41	7.6	0.44
Mexican or Mexican American	4.5	0.46	9.9	0.62	3.6	0.41	5.0	0.45	2.2	0.29	3.7	0.38	6.6	0.56	7.0	0.54
Not Hispanic or Latino	3.3	0.12	9.7	0.20	2.0	0.10	4.5	0.15	1.9	0.09	3.5	0.13	5.7	0.18	9.4	0.21
White, single race	3.0	0.13	9.1	0.23	1.9	0.11	4.3	0.16	1.8	0.10	3.4	0.14	5.3	0.19	9.2	0.23
Black or African American, single race	4.8	0.36	12.8	0.60	2.4	0.26	5.2	0.39	2.2	0.26	4.1	0.34	8.4	0.51	10.7	0.55
Education[6]																
Less than a high school diploma	7.1	0.37	14.9	0.57	4.6	0.34	7.6	0.43	4.0	0.31	6.1	0.36	9.4	0.48	11.2	0.49
High school diploma or GED[7]	3.6	0.23	10.9	0.38	2.0	0.17	4.6	0.26	2.0	0.17	3.7	0.23	6.0	0.30	9.2	0.38
Some college	3.1	0.23	9.3	0.37	1.9	0.17	4.2	0.25	1.9	0.17	3.4	0.24	5.7	0.30	9.4	0.38
Bachelor's degree or higher	1.4	0.15	6.2	0.32	0.9	0.12	2.8	0.21	0.7	0.10	2.0	0.19	2.4	0.21	7.2	0.35
Family Income[8]																
Less than $20,000	7.3	0.34	15.3	0.49	4.9	0.27	8.2	0.39	4.5	0.28	6.1	0.27	10.7	0.42	12.7	0.45
$20,000 or more	2.5	0.12	8.4	0.21	1.5	0.09	3.7	0.14	1.3	0.08	3.0	0.13	4.7	0.17	8.5	0.22
$20,000–$34,999	4.1	0.33	12.1	0.49	2.5	0.26	6.1	0.37	2.0	0.23	4.9	0.34	7.2	0.44	11.0	0.48
$35,000–$54,999	2.8	0.25	8.9	0.45	1.7	0.20	4.0	0.30	1.4	0.20	3.3	0.29	5.1	0.37	9.2	0.45
$55,000–$74,999	2.3	0.29	7.2	0.52	1.2	0.22	3.2	0.36	0.9	0.18	2.5	0.34	4.4	0.41	8.6	0.58
$75,000 or more	1.2	0.18	6.0	0.39	0.8	0.16	2.8	0.27	0.6	0.12	2.1	0.25	3.1	0.30	7.2	0.41
Poverty Status[9]																
Poor	7.8	0.53	16.3	0.84	5.9	0.46	9.6	0.67	5.2	0.45	7.1	0.49	12.2	0.65	13.9	0.80
Near poor	5.7	0.40	13.4	0.55	3.2	0.28	6.6	0.42	2.9	0.29	5.2	0.38	9.0	0.51	11.9	0.53
Not poor	2.3	0.13	8.1	0.24	1.4	0.11	3.7	0.17	1.1	0.09	3.0	0.15	4.5	0.20	8.5	0.25

[1]In four separate questions, respondents were asked how often in the past 30 days they felt: so sad that nothing could cheer them up, hopeless, worthless, or that everything was an effort. Respondents could choose from among five response categories: "All of the time," "Most of the time," "Some of the time," "A little of the time," or "None of the time." For the purposes of this table, "All" and "Most" are combined, and "Some" is shown separately.
[2]Total includes other races not shown separately and persons with unknown poverty status.
[3]The category "1 race" refers to people who indicated only a single race group. Persons who indicated a single race other than the groups shown are included in the total for "1 race" but not shown separately due to small sample sizes. Therefore the frequencies for the category "1 race" will will be greater than the sum of the frequencies for the specific groups shown separately.
[4]The category "2 or more races" refers to all persons who indicated more than one race group. Only two combinations of race groups are shown due to small sample sizes for other combinations. Therefore, the frequencies for the category "2 or more races" will be greater than the sum of the frequencies for the specific combinations shown separately.
[5]Persons of Hispanic or Latino origin may be of any race or any combination of races. Similarly, the category "Not Hispanic or Latino" refers to all persons who are not of Hispanic or Latino origin, regardless of race.
[6]Education is shown only for persons aged 25 years and over.
[7]GED is General Educational Development high school equivalency diploma .
[8]The categories "less than $20,000" and "$20,000 or more" include both persons reporting dollar amounts and persons reporting only that their incomes were within one of these two categories. The indented categories include only those persons who reported dollar amounts.
[9]Poverty status is based on family income and family size using the U.S. Census Bureau's poverty thresholds for the previous calendar year. "Poor" persons are defined as below the poverty threshold. "Near poor" persons have incomes of 100 percent to less than 200 percent of the poverty threshold. "Not poor" persons have incomes that are 200 percent of the poverty threshold or greater.
* = Figure does not meet standards of reliability or precision.
- = Quantity zero.

Table III-22. Rate of Feelings of Sadness, Hopelessness, Worthlessness, or that Everything is an Effort Among Persons 18 Years of Age and Over, by Selected Characteristics, 2001—*Continued*

(Percent.)

Characteristic	Selected mental health characterisitics															
	Sadness [1]				Hopelessness [1]				Worthlessness [1]				Everything is an effort [1]			
	All or most of the time	Standard error for all or most of the time	Some of the time	Standard error for some of the time	All or most of the time	Standard error for all or most of the time	Some of the time	Standard error for som eof the time	All or most of the time	Standard error for all or most of the time	Some of the time	Standard error for some of the time	All or most of the time	Standard error for all or most of the time	Some of the time	Standard error for some of the time
Health Insurance Coverage [10]																
Under age 65 years																
Private	2.2	0.12	7.9	0.23	1.3	0.09	3.5	0.16	1.1	0.08	2.6	0.13	4.3	0.18	8.4	0.24
Medicaid	11.6	0.86	22.3	1.18	9.0	0.77	12.9	0.99	8.6	0.80	10.1	0.86	18.1	1.22	17.4	1.13
Other	7.0	0.95	13.4	1.32	5.6	0.89	8.9	1.21	5.0	0.87	7.1	0.99	10.5	1.21	13.5	1.40
Uninsured	5.4	0.35	12.1	0.55	3.5	0.29	6.4	0.38	2.8	0.27	4.9	0.36	7.9	0.45	10.5	0.56
Age 65 years and over																
Private	2.8	0.29	10.5	0.58	1.4	0.21	3.0	0.31	1.4	0.22	3.0	0.31	4.3	0.37	8.2	0.52
Medicaid and Medicare	9.1	1.56	16.0	1.81	7.3	1.41	10.3	1.60	4.5	1.10	8.1	1.50	11.0	1.60	13.9	1.82
Medicare only	3.6	0.52	10.8	0.89	1.9	0.40	3.8	0.53	2.1	0.41	3.1	0.49	4.6	0.56	7.6	0.74
Other	*2.3	0.85	13.3	2.56	*2.5	0.92	*5.6	1.84	*3.7	1.21	*4.2	1.63	8.3	1.89	*6.2	1.89
Uninsured	*0.8	0.80	*22.5	6.82	*3.8	3.09	*6.1	3.84	*3.8	3.09	*2.0	1.98	*0.8	0.80	*8.4	4.70
Marital Status																
Married	2.5	0.14	7.9	0.24	1.5	0.11	3.6	0.17	1.4	0.11	2.8	0.14	4.6	0.20	8.1	0.25
Widowed	5.6	0.48	16.9	0.76	2.9	0.36	5.6	0.47	2.4	0.30	4.7	0.45	6.2	0.49	11.1	0.62
Divorced or separated	6.5	0.41	15.1	0.56	4.3	0.34	7.9	0.41	3.9	0.34	6.0	0.38	8.9	0.48	11.3	0.52
Never married	3.7	0.28	10.0	0.45	2.4	0.22	5.0	0.34	1.9	0.19	3.9	0.30	6.6	0.38	10.4	0.50
Living with a partner	4.0	0.48	10.9	0.85	3.4	0.48	5.2	0.58	2.8	0.50	3.9	0.50	8.1	0.77	10.6	0.84
Place of Residence [11]																
Large MSA	3.4	0.17	10.1	0.29	2.1	0.13	4.2	0.19	1.8	0.12	3.4	0.18	5.3	0.22	8.6	0.28
Small MSA	3.5	0.20	9.1	0.32	2.3	0.17	4.5	0.23	2.0	0.16	3.5	0.19	6.1	0.30	9.6	0.36
Not in MSA	3.5	0.26	10.3	0.44	2.2	0.21	5.3	0.35	2.2	0.21	3.9	0.26	6.2	0.43	10.1	0.43
Region																
Northeast	3.8	0.26	11.0	0.47	2.1	0.19	4.0	0.30	2.0	0.19	3.2	0.26	4.8	0.35	8.7	0.46
Midwest	3.0	0.23	8.9	0.35	1.9	0.16	4.5	0.28	1.9	0.18	3.4	0.24	6.0	0.33	9.5	0.37
South	3.6	0.19	10.1	0.33	2.3	0.17	4.7	0.24	2.0	0.15	3.5	0.19	6.2	0.29	9.2	0.35
West	3.4	0.26	9.2	0.42	2.4	0.21	4.7	0.29	1.8	0.19	4.1	0.27	5.6	0.39	9.6	0.41
Sex and Age																
Male																
18–44 years	2.3	0.19	7.6	0.36	1.6	0.16	4.1	0.26	1.4	0.17	2.7	0.22	5.3	0.32	8.5	0.40
45–64 years	3.3	0.30	7.8	0.43	2.0	0.23	4.4	0.35	2.0	0.23	3.4	0.32	5.4	0.40	8.1	0.49
65 years and over	2.9	0.36	8.7	0.69	1.7	0.27	2.9	0.37	1.8	0.28	3.1	0.40	4.3	0.46	6.8	0.59
Female																
18–44 years	4.1	0.23	10.5	0.37	2.7	0.19	5.2	0.28	2.0	0.17	3.9	0.23	6.5	0.33	10.8	0.39
45–64 years	4.3	0.31	12.8	0.53	2.9	0.26	5.1	0.35	2.6	0.25	4.6	0.33	6.4	0.37	9.7	0.46
65 years and over	3.8	0.36	12.9	0.62	2.1	0.27	4.5	0.39	2.0	0.26	3.6	0.36	5.4	0.41	9.4	0.55
Hispanic or Latino Origin, Race, Age, and Sex																
Hispanic or Latino, male [5]																
18–44 years	2.9	0.54	8.0	0.89	2.4	0.45	3.6	0.60	1.3	0.33	2.7	0.47	4.2	0.58	5.2	0.61
45–64 years	5.0	0.93	8.7	1.26	*2.8	0.97	5.1	1.22	2.1	0.58	*2.1	0.68	5.2	1.18	5.9	1.14
65 years and over	6.6	1.87	11.9	2.57	*3.8	1.28	7.8	2.14	*4.0	1.55	*4.1	1.52	6.5	1.93	8.0	2.03
Hispanic or Latina, female [5]																
18–44 years	5.0	0.59	11.3	0.83	3.7	0.51	5.7	0.59	1.9	0.34	4.2	0.55	6.6	0.72	9.8	0.84
45–64 years	8.7	1.24	14.9	1.58	6.7	1.13	5.8	0.92	3.8	0.75	3.9	0.82	9.1	1.38	8.5	1.13
65 years and over	5.6	1.30	16.5	2.33	7.6	1.65	8.1	1.87	4.5	1.35	8.0	1.89	8.6	1.68	10.6	2.21
Not Hispanic or Latino																
White, single race, male																
18–44 years	2.1	0.24	6.6	0.40	1.5	0.21	4.0	0.33	1.5	0.22	2.8	0.28	5.1	0.41	8.8	0.49
45–64 years	2.9	0.33	7.2	0.49	1.7	0.24	4.4	0.42	1.8	0.25	3.5	0.38	5.0	0.45	8.0	0.56
65 years and over	2.7	0.39	8.5	0.73	1.6	0.28	2.5	0.39	1.6	0.29	3.1	0.44	4.3	0.50	6.8	0.65
White, single race, female																
18–44 years	3.5	0.29	9.5	0.46	2.4	0.24	4.8	0.35	2.0	0.22	3.6	0.29	5.7	0.39	10.7	0.50
45–64 years	3.6	0.33	12.1	0.61	2.3	0.27	4.8	0.39	2.4	0.28	4.5	0.38	5.8	0.40	9.5	0.54
65 years and over	3.5	0.38	12.2	0.68	1.7	0.28	4.0	0.41	1.7	0.27	3.3	0.38	5.0	0.45	9.2	0.61
Black or African American, single race, male																
18–44 years	2.5	0.57	10.7	1.25	*1.0	0.33	3.9	0.71	*1.2	0.43	3.1	0.66	7.8	1.05	9.3	1.15
45–64 years	5.4	1.01	11.4	1.59	3.0	0.79	4.0	0.97	*2.5	0.78	3.8	0.77	8.5	1.56	10.7	1.72
65 years and over	*2.9	1.56	8.4	1.82	*2.4	1.56	*1.4	0.72	*2.9	1.57	*0.3	0.30	*3.0	1.49	*4.8	1.47
Black or African American, single race, female																
18–44 years	6.5	0.67	13.2	0.97	3.0	0.46	6.3	0.69	2.6	0.47	4.8	0.64	9.9	0.90	12.0	0.94
45–64 years	5.7	1.02	17.1	1.58	3.2	0.71	6.7	1.02	2.8	0.75	6.2	1.00	8.6	1.12	12.3	1.33
65 years and over	4.4	1.11	15.0	1.79	*2.1	0.91	5.8	1.20	*2.0	0.86	3.4	0.79	6.6	1.26	10.0	1.46

[1]In four separate questions, respondents were asked how often in the past 30 days they felt: so sad that nothing could cheer them up, hopeless, worthless, or that everything was an effort. Respondents could choose from among five response categories: "All of the time," "Most of the time," "Some of the time," "A little of the time," or "None of the time." For the purposes of this table, "All" and "Most" are combined, and "Some" is shown separately.

[5]Persons of Hispanic or Latino origin may be of any race or any combination of races. Similarly, the category "Not Hispanic or Latino" refers to all persons who are not of Hispanic or Latino origin, regardless of race.

[10]Classification of health insurance coverage is based on hierarchy of mutually exclusive categories. Persons with more than one type of health insurance were assigned to the first appropriate category in the hierarchy. Persons under the age of 65 and those 65 years and over were classified separately due to the prominence of Medicare coverage in the older population. The category "private" includes persons who had any type of private coverage either alone or in combination with other coverage. For example, for persons age 65 years and over, "private" includes persons with only private or private in combination in Medicare. The category "uninsured" includes persons who had coverage as well as those who had only Indian health service coverage or had only a private plan that paid for one type of service such as accidents or dental care.

[11]MSA is a metropolitan statistical area. Large MSA's have a population of 1,000,000 or more; small MSAs have a population size of less than 1,000,000. "Not in MSA" consists of persons not living in an MSA.

* = Figure does not meet standards of reliability or precision.

DISABILITIES

TABLE III-23. LIMITATION OF ACTIVITY CAUSED BY CHRONIC CONDITIONS, ACCORDING TO SELECTED CHARACTERISTICS, SELECTED YEARS, 1997–2001

In 2001, as many as 12.1 percent of interview respondents reported limitations of activity. In the age group of 18 to 44 years, this percentage was 6.1 percent, but it rose to 34.5 percent for persons 65 years and over. Among races, American Indians had the highest rates of limitations.

The term, limitation of activity, refers to a long-term reduction in a person's performance of the usual type of activities associated with his or her age group, due to physical or mental chronic condition. For persons 65 years or over, additional detail is presented for more stringent criteria of limitations of activities: those pertaining to activities of daily living (ADL) and those pertaining to instrumental activities of daily living (IADL). Examples of ADL are help needed for bathing, dressing, or eating. IADL focuses on persons living independently and includes help needed for preparing meals, shopping and using a telephone. Persons are considered to have IADL limitation if the limitation is chronic. The 2001 survey shows 6.4 percent of persons 65 years of age and over with ADL limitations, and 12.6 percent with IADL limitations.

Figure III-9. Selected Chronic Health Conditions Causing Limitation of Activity Among Working-Age Adults by Age, 1999–2001

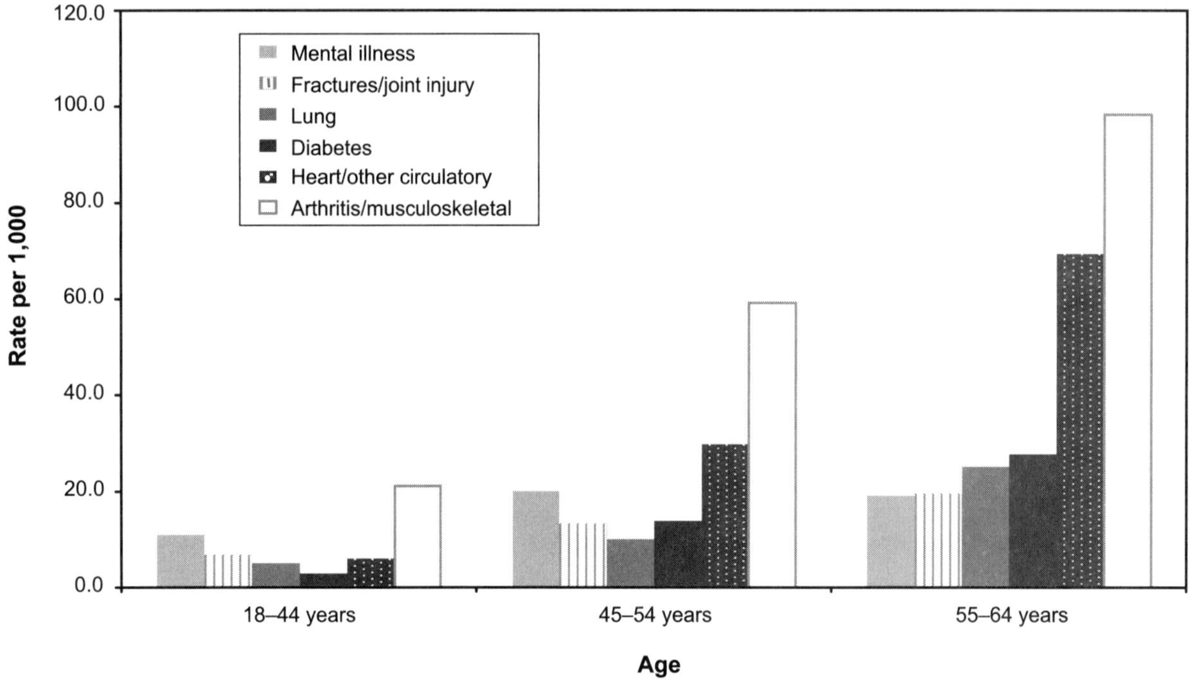

Table III-23. Limitation of Activity Caused by Chronic Conditions, According to Selected Characteristics, Selected Years, 1997–2001

(Percent.)

Characteristic	Persons with any activity limitation [1]				Persons with ADL limitation [2]				Persons with IADL limitation [2]			
	2001	2000	1999	1997	2001	2000	1999	1997	2001	2000	1999	1997
All Ages												
Total [3]	12.1	11.7	12.2	13.3	X	X	X	X	X	X	X	X
Age												
Under 18 years	6.7	6.0	6.0	6.6	X	X	X	X	X	X	X	X
Under 5 years	3.3	3.2	3.1	3.5	X	X	X	X	X	X	X	X
5–17 years	8.0	7.0	7.0	7.8	X	X	X	X	X	X	X	X
18–44 years	6.1	5.8	6.3	7.0	X	X	X	X	X	X	X	X
18–24 years	4.6	3.6	4.4	5.1	X	X	X	X	X	X	X	X
25–44 years	6.6	6.5	6.9	7.6	X	X	X	X	X	X	X	X
45–54 years	13.1	12.4	13.1	14.2	X	X	X	X	X	X	X	X
55–64 years	20.7	19.7	21.1	22.2	X	X	X	X	X	X	X	X
65 years and over	34.5	34.7	35.6	38.7	6.4	6.3	6.3	6.7	12.6	12.7	12.4	13.7
65–74 years	26.0	26.1	27.5	30.0	3.4	3.3	3.1	3.4	6.7	6.6	6.2	6.9
75 years and over	44.7	45.1	45.6	50.2	9.6	9.5	9.9	10.4	18.9	19.3	19.1	21.2
Sex												
Male	12.2	11.7	12.1	13.1	6.1	5.1	4.9	5.2	9.6	9.2	8.4	9.1
Female	11.9	11.5	12.2	13.4	6.6	7.0	7.2	7.7	14.6	15.1	15.1	16.9
Race [4]												
White only	11.8	11.5	12.0	13.1	5.7	5.8	5.8	6.3	11.8	12.1	11.6	13.1
Black or African American only	15.6	14.3	15.3	17.1	11.7	10.2	12.0	11.7	18.7	19.2	20.9	21.3
American Indian and Alaska Native only	18.9	20.1	18.8	23.1	*	*	*	*	*	*	*25.2	*
Asian only	6.7	6.6	6.8	7.5	*9.1	*7.4	*	*	15.8	*10.1	*9.1	*9.1
Native Hawaiian and Other Pacific	*	*	*	...	*	*	*	...	*	*	*	...
Two or more races	19.8	19.8	20.3	...	*	*	*	...	*16.1	*	*	...
Black or African American; White	14.9	*20.3	14.9	...	...	...	...	...	...	...	...	...
American Indian and Alaska Native; White	22.0	25.3	26.0	...	...	...	...	...	...	...	...	...
Hispanic Origin and Race [4]												
Hispanic or Latino	10.6	10.3	10.4	12.8	11.2	8.6	8.6	10.8	17.0	13.4	14.1	16.3
Mexican	10.3	10.4	9.6	12.5	10.6	9.4	8.9	11.4	17.0	16.3	15.6	18.8
Not Hispanic or Latino	12.4	11.9	12.4	13.5	6.1	6.1	6.2	6.5	12.3	12.6	12.3	13.6
White only	12.1	11.7	12.2	13.2	5.5	5.7	5.7	6.1	11.6	12.1	11.5	13.0
Black or African American only	15.5	14.3	15.2	17.0	11.8	10.1	12.0	11.7	18.7	19.1	21.0	21.2
Poverty Status [5]												
Poor	24.1	23.2	24.6	26.8	12.7	9.6	10.1	13.0	24.8	20.2	22.3	26.9
Near poor	18.8	17.5	19.1	19.0	7.4	7.1	6.7	7.5	15.0	15.3	15.1	16.3
Nonpoor	9.9	9.5	10.0	10.5	5.0	5.2	5.5	5.3	9.7	9.4	9.7	10.1
Hispanic Origin and Race and Poverty Status [4]												
Hispanic or Latino												
Poor	17.2	16.9	16.4	19.7	14.1	12.3	*8.4	15.5	23.5	17.6	17.9	25.8
Near poor	11.2	11.4	11.4	13.1	10.5	*7.6	*8.6	11.3	14.6	14.1	14.1	16.5
Nonpoor	7.9	7.4	7.9	9.8	*8.4	*6.5	*8.1	*	*11.7	*7.9	*10.4	*9.7
Not Hispanic or Latino												
White only												
Poor	26.4	25.6	27.6	29.5	11.2	8.9	8.8	12.7	25.2	20.0	21.2	27.2
Near poor	20.8	19.7	21.5	20.7	6.1	6.4	5.7	6.7	14.0	14.8	14.3	15.8
Nonpoor	10.2	9.8	10.2	10.7	4.4	4.8	5.3	5.0	9.0	9.2	9.2	10.0
Black or African American												
Poor	27.0	24.7	28.0	29.4	18.2	9.9	13.9	12.9	29.1	21.8	27.9	27.4
Near poor	22.1	18.3	19.1	20.0	13.4	11.2	15.2	12.0	19.3	20.5	23.5	21.4
Nonpoor	10.3	9.7	10.1	10.7	*12.0	*11.9	*8.9	*10.6	18.3	15.5	18.4	*13.0
Geographic Region												
Northeast	11.1	10.6	11.3	13.0	6.6	5.7	5.8	6.1	11.3	11.5	11.2	12.2
Midwest	13.4	12.3	12.9	13.1	4.9	5.6	5.4	5.8	12.5	13.2	12.3	13.1
South	12.3	11.7	12.6	13.9	7.5	7.4	7.1	8.2	13.3	13.1	13.2	15.8
West	11.5	12.1	11.7	13.0	6.0	5.7	6.7	5.9	12.6	12.6	12.3	12.4
Location of Residence												
Within MSA [6]	11.3	10.9	11.4	12.7	6.1	6.4	6.3	6.6	12.2	12.6	12.1	13.5
Outside MSA [6]	15.3	14.6	15.1	15.5	7.3	6.0	6.4	7.2	13.7	13.2	13.4	14.4

[1] Limitation of activity is assessed by asking respondents a series of questions about limitations in their ability to perform activities usual for their age group because of a physical, mental, or emotional problem.

[2] These estimates are for noninstitutionalized persons and are age-adjusted to the year 2000 standard population using two age groups: 65–74 years and 75 years and over. ADL is activities of daily living and IADL is instrumental activities of daily living. Respondents were asked about needing the help of another person with personal care (ADL) and routine needs such as chores and shopping (IADL) because of a physical, mental, or emotional problems. The first four columns show percentages for the indicated population groups. The last eight columns show percentages for the elderly noninstitutionalized populations only.

[3] Includes all other races not shown separately and unknown poverty status.

[4] The race groups, White, Black, American Indian and Alaska Native (AI/AN), Asian, Native Hawaiian and Other Pacific Islander, and two or more races, include persons of Hispanic and non-Hispanic origin. Persons of Hispanic origin may be of any race.

[5] Poor persons are defined as below the poverty threshold. Near poor persons have incomes of 100 percent to less than 200 percent of the poverty threshold. Nonpoor persons have incomes of 200 percent or greater than the poverty threshold.

[6] MSA = metropolitan statistical area.

* = Figure does not meet standards of reliability or precision.

. . . = Not available.

X = Not applicable.

TABLE III-24. LIFE EXPECTANCIES AND EXPECTED YEARS WITHOUT LIMITATION IN ACTIVITY FOR SELECTED AGES, BY SEX AND TYPE OF ACTIVITY, 1995

Respondents were classified as those who were not limited in performing their major activity and those who were not limited in any way. As an example, at age 45, life expectancy was 33.8 years; of these, 30.3 years, on average, were expected to be without limitation of major activities; and 23.6 years were expected to be without any limitation of activities. As a percent of remaining life expectancy, limitations become more important with advancing age. Only 60 percent of the remaining life of 70 year old persons were expected to by free of any limitations. Males, at that age, had slightly better expectations of 62 percent, versus 59 percent for females.

Table III-24. Life Expectancies and Expected Years without Limitation in Activity for Selected Ages, by Sex and Type of Activity, 1995

(Percent.)

Age and sex	Expected years of life	Expected years without major activity limitation	Years without major activity limitation as a percentage of life expectancy	Expected years without any activity limitation	Years without any activity limitation as a percentage of life expectancy
Both Sexes					
Birth	75.8	71.6	94.4	62.5	82.5
20 years	56.9	52.8	92.7	44.7	78.5
30 years	47.5	43.5	91.6	35.9	75.5
45 years	33.8	30.3	89.5	23.6	69.8
65 years	17.5	15.6	89.1	10.7	61.3
70 years	14.1	12.9	91.3	8.5	60.4
75 years	11.1	9.8	88.7	6.1	55.5
80 years	8.3	7.1	85.1	4.1	49.7
Male					
Birth	72.8	68.5	94.1	60.5	83.2
20 years	54.0	49.8	92.2	43.0	79.6
30 years	44.9	40.8	90.9	34.4	76.8
45 years	31.5	27.9	88.6	22.5	71.2
65 years	15.9	14.1	88.6	9.9	62.4
70 years	12.8	11.8	92.6	8.0	62.4
75 years	10.1	9.1	90.2	5.9	58.9
80 years	7.7	6.8	88.1	4.3	55.4
Female					
Birth	78.8	74.7	94.7	64.6	81.9
20 years	59.8	55.7	93.1	46.3	77.5
30 years	50.1	46.1	92.2	37.3	74.4
45 years	35.9	32.4	90.3	24.6	68.6
65 years	18.8	16.8	89.5	11.4	60.6
70 years	15.1	13.7	90.5	8.9	59.2
75 years	11.7	10.3	87.8	6.3	53.6
80 years	8.7	7.2	83.4	4.0	46.6
Gender Difference [1]					
Birth	+6.1	+6.2	X	+4.0	X
20 years	+5.7	+5.9	X	+3.3	X
30 years	+5.2	+5.4	X	+2.8	X
45 years	+4.3	+4.5	X	+2.2	X
65 years	+2.9	+2.8	X	+1.5	X
70 years	+2.3	+1.8	X	+1.0	X
75 years	+1.7	+1.2	X	+0.4	X
80 years	+1.0	+0.4	X	-0.2	X

[1]Calculated by subtracting expected years for males from expected years for females.
X = Not applicable.

TABLE III-25. PERCENT OF ADULTS AGED 65 YEARS AND OVER WHO NEED HELP WITH PERSONAL CARE FROM OTHER PERSONS, 1997–2003

In early 2003, 6.4 percent of adults aged 65 and over needed personal care assistance from other persons. This was about the same as in recent years. Personal care needs—or activities of daily living (ADL)—include such things as eating, bathing, dressing, and getting around in one's home.

The need for personal care assistance is moderate at about 3 percent for the 65 to 74-year age group, increases to more than 6 percent for persons 75 to 84 years old, and increases sharply to 22 percent at age 85 and over. Personal care assistance needs are somewhat less, percentage wise, for non-Hispanic Whites than for Blacks or Hispanics.

Table III-25. Percent of Adults Aged 65 Years and Over who Need Help with Personal Care from Other Persons, 1997–2003

(Percent.)

Characteristic	Percent
Total	
2003 (January-June)	6.4
2002	6.2
2001	6.5
2000	6.4
1999	6.4
1998	6.3
1997	6.6
JANUARY 2003–JUNE 2003	
Age and Sex	
65–74	
Both sexes	3.3
Men	3.1
Women	3.5
75–84	
Both sexes	6.6
Men	5.5
Women	7.3
85 Years and Over	
Both sexes	22.4
Men	15.4
Women	26.2
Race and Hispanic Origin	
Hispanic or Latino [1]	7.8
White, single race	6.1
Black, single race	9.0

Note: Personal care need or activities of daily living (ADL) include such things as eating, bathing, dressing or getting around inside the person's home. The analyses excluded persons with unknown information on personal care needs.

[1]Persons of Hispanic origin may be of any race.

TABLE III-26. LIFE EXPECTANCIES AND EXPECTED YEARS WITHOUT FUNCTIONAL DEPENDENCY FOR SELECTED AGES, 1995

The table shows estimated years of remaining life without need for personal care help (for eating, bathing etc.); this is shown under the heading of ADL, that is, Activities of Daily Living. Freedom from need for help for other routine needs, including shopping, using the telephone, and preparing one's food (classified as IADL, or Instrumental Activities of Daily Living) is also measured. The effects of both limitations are shown on a combined basis; that is, the table shows remaining years of life without need for assistance for either ADL or IADL.

For example, at age 70 persons were expected to live 14.1 years on average. Of these, 12.9 years (91 percent) were expected to be lived without need for help in ADL, and 11.2 years (79 percent) would pass without need for help in either ADL or IADL. While limitations become more prevalent at higher age level, even 80 year old males were expected to have 6.8 more years (88 percent) without need for help with ADL and 5.7 more years (74 percent) without need for help with either ADL or IADL, out of a total life expectancy of 7.7 years. For 80 year old females, the total life expectancy is 8.7 years, years passed without need for help with ADL are 7.2 (83 percent) and years spent without either need for help with ADL or IADL are estimated to be 5.4 (63 percent). The findings suggest that women spend a lower percentage of remaining life without need for assistance, and the gender difference in the number of fully independent years remaining diminishes as age increases.

Table III-26. Life Expectancies and Expected Years without Functional Dependency for Selected Ages, 1995

(Percent.)

Age and sex	Expected years of life	Expected years without needing help in ADL [1]	Years without ADL help as a percentage of life expectancy	Expected years without needing help in ADL or IADL [2]	Years without ADL or IADL help as a percentage of life expectancy
Both Sexes					
Birth	71.5	70.1	98.0	67.8	94.7
20 years	56.9	55.5	97.5	53.2	93.4
30 years	47.5	46.1	97.1	43.8	92.3
45 years	33.8	32.4	96.0	30.3	89.7
65 years	17.5	16.2	92.8	14.5	82.7
70 years	14.1	12.9	91.3	11.2	79.3
75 years	11.1	9.8	88.7	8.2	73.9
80 years	8.3	7.1	85.1	5.5	66.8
Male					
Birth	68.5	67.3	98.2	65.8	96.0
20 years	54.0	52.9	97.9	51.3	95.1
30 years	44.9	43.7	97.5	42.2	94.1
45 years	31.5	30.4	96.5	29.0	92.1
65 years	15.9	14.9	93.8	13.8	86.5
70 years	12.8	11.8	92.6	10.7	83.9
75 years	10.1	9.1	90.2	8.0	79.6
80 years	7.7	6.8	88.1	5.7	74.4
Female					
Birth	74.5	72.8	97.7	69.7	93.6
20 years	59.8	58.1	97.2	55.0	92.1
30 years	50.1	48.4	96.7	45.4	90.7
45 years	35.9	34.3	95.6	31.5	87.8
65 years	18.8	17.3	92.2	15.1	80.1
70 years	15.1	13.7	90.5	11.5	76.3
75 years	11.7	10.3	87.8	8.3	70.6
80 years	8.7	7.2	83.4	5.4	62.8
Gender Difference [3]					
Birth	+6.1	+5.5	X	+3.9	X
20 years	+5.7	+5.2	X	+3.7	X
30 years	+5.2	+4.7	X	+3.2	X
45 years	+4.3	+3.9	X	+2.5	X
65 years	+2.9	+2.4	X	+1.3	X
70 years	+2.3	+1.8	X	+0.8	X
75 years	+1.7	+1.2	X	+0.3	X
80 years	+1.0	+0.4	X	-0.3	X

[1]ADL refers to activities of daily living.
[2]IADL refers to instrumental activities of daily living
[3]Calculated by subtracting expected years for males from expected years for females.
X = Category not applicable.

TABLE III-27. SELECTED CHRONIC HEALTH CONDITIONS CAUSING LIMITATION OF ACTIVITY AMONG WORKING AGE ADULTS BY AGE, 1999–2001

This table focuses on six major chronic conditions that cause a limitation of activity of persons of working age, that is, from 18 to 64 years. The limitations include ADL and IADL, previously defined in Table 26, as well as having a job outside the home, walking, remembering, and other activities. Arthritis and other musculoskeletal conditions were the most frequent types of afflictions for all age groups, causing limitations for almost 10 percent of persons from 55 to 64 years. Heart and circulatory problems also were prominent, especially for the older age group. Fractures and joint injury and mental illness, each, caused limitations of 2 percent among the older age group. Mental illness is the only condition presented that did not have steadily increased impact with age, but rather plateaued after age 45–54.

Persons who reported more than one chronic health condition as the cause of their activity limitation were counted in each category.The data are shown per 1,000 of population.

Table III-27. Selected Chronic Health Conditions Causing Limitation of Activity Among Working Age Adults by Age, 1999–2001

(Percent.)

Chronic health condition	18–44 years		45–54 years		55–64 years	
	Rate	Standard error	Rate	Standard error	Rate	Standard error
Mental illness	10.9	0.4	20.1	0.8	19.1	1.0
Fractures/joint injury	6.8	0.3	13.3	0.6	19.6	1.0
Lung	5.0	0.2	10.0	0.6	25.1	1.2
Diabetes	2.9	0.2	13.8	0.7	27.7	1.2
Heart/other circulatory	6.0	0.3	29.8	1.0	69.3	2.1
Arthritis/other muscoskeletal	21.1	0.6	59.2	1.4	98.3	2.4

Note: Data are for the civilian noninstitutionalized population. Selected chronic health conditions include the four leading casuses of activity limitation among adults in each age category. Conditions refer to response categories in the National Health Interview Survey; some conditions include several response categories. "Mental illness" includes depression, anxiety, or emotional problem, and other mental conditions. "Heart/other circulatory" includes heart problem, stroke problem, hyptertension or high blood pressure, and other circulatory system conditions. "Arthritis/other musculoskeletal" includes arthritis, rheumatism, back or neck problem, and other muscoskeletal system conditions. Persons who reported more than one chronic health condition as the casuse of their activity limitation were counted in each reported category.

TABLE III-28. OCCUPATIONAL INJURIES WITH LOST WORKDAYS IN THE PRIVATE SECTOR, ACCORDING TO INDUSTRY, SELECTED YEARS, 1980–2001

The number of injuries with lost workdays was 2.4 million in 2001, representing 2.6 percent of employees (on a full time work basis). The injury percentages were highest in transportation, communications and public utilities (4.2 percent), construction (3.9 percent), manufacturing (3.6 percent), and agriculture (3.6 percent). Injuries were least frequent in finance, insurance and real estate (0.7 percent).

The frequency of injuries has declined considerably since 1980 and 1990, when the overall injury incidence was 3.9 percent in each of those years. Partly this reflected the shift to a more service oriented economy, but the decline in injuries also took place within the goods producing industries. In manufacturing the frequency of injuries declined from 5.3 percent in 1990 to 3.6 percent in 2001; in construction the corresponding decline was from 6.6 percent to 3.9 percent; and the injury rate in mining fell from 4.9 percent to 2.3 percent. These data are based on surveys of business establishments.

Table III-28. Occupational Injuries with Lost Workdays in the Private Sector, According to Industry, Selected Years, 1980–2001

(Rate per 100 full-time employees, number in thousands.)

Industry	2001	2000	1999	1998	1997	1995	1990	1985	1980
Injuries with Lost Workdays (Percent) [1]									
Total private sector [2]	2.6	2.8	2.8	2.9	3.1	3.4	3.9	3.6	3.9
Agriculture, fishing, and forestry [2]	3.6	3.5	3.3	3.8	3.8	4.2	5.7	5.6	5.6
Mining	2.3	3.0	2.5	2.7	3.2	3.8	4.9	4.7	6.4
Construction	3.9	4.0	4.1	4.0	4.4	4.8	6.6	6.8	6.5
Manufacturing	3.6	4.0	4.0	4.2	4.3	4.6	5.3	4.4	5.2
Transportation, communication, and public utilities	4.2	4.1	4.3	4.2	5.0	5.0	5.4	4.9	5.4
Wholesale trade	2.5	3.0	3.2	3.2	3.3	3.5	3.6	3.5	3.8
Retail trade	2.4	2.4	2.5	2.6	2.7	2.9	3.4	3.1	2.9
Finance, insurance, and real estate	0.7	0.7	0.7	0.6	0.8	0.9	1.1	0.9	0.8
Services	2.1	2.2	2.1	2.3	2.5	2.7	2.7	2.5	2.3
Number of Injuries with Lost Workdays (Thousands)									
Total private sector [2]	2 409.4	2 587.0	2 575.9	2 612.0	2 646.3	2 767.6	2 987.3	2 484.7	2 491.0
Agriculture, fishing, and forestry [2]	53.5	52.4	47.5	53.8	49.0	51.7	57.2	45.2	39.3
Mining	13.9	17.0	14.2	16.9	19.5	22.8	35.6	43.9	66.2
Construction	237.6	246.1	240.2	217.0	216.8	217.9	296.3	272.8	242.6
Manufacturing	614.5	727.7	744.6	782.6	782.9	838.1	975.0	825.1	1 009.5
Transportation, communication, and public utilities	276.4	274.6	274.3	261.3	293.0	289.2	293.3	243.5	263.0
Wholesale trade	177.1	201.7	210.6	211.1	203.9	214.7	211.5	188.4	191.1
Retail trade	411.0	420.6	420.7	434.7	433.9	459.6	483.9	399.9	330.2
Finance, insurance, and real estate	45.0	45.9	45.8	39.6	49.5	52.2	63.7	45.5	38.1
Services	580.5	601.1	578.0	594.9	597.8	621.4	570.8	420.6	311.1

[1] Incidence rates calculated as (N/EH) * 200,000 where N = total number of injuries with lost workdays in a calender year, EH = total hours worked by all full-time and part-time employees in a calendar year, and 200,000 = base for 100 full-time employees working 40 hours per week, 50 weeks in a year.
[2] Excludes farms with fewer than 11 employees.

USE OF ADDICTIVE SUBSTANCES

TABLE III-29. USE OF SELECTED SUBSTANCES IN THE PAST MONTH BY PERSONS 12 YEARS OF AGE AND OVER, ACCORDING TO AGE, SEX, RACE, AND HISPANIC ORIGIN, 1999–2002

Based on in-person interviews, 6.2 percent of persons 12 years of age and over used marijuana at least once during the previous month in 2002. The most frequent use, 17.3 percent, was for persons in the 18 to 25 year age group. Another high percentage was for 16 and 17 year olds. The percentages rapidly tapered off at earlier and later ages. Marijuana is classified as a category of illicit drugs. Separate data are shown for the nonmedical use of prescription-type psychotherapeutic drugs, which are consumed by 2.6 percent of the population.

Turning to alcohol, 51 percent of the population of 12 years and over had at least one drink of alcohol during the month prior to the interview. The highest percentage of 61.4 was among persons 26 to 34 years old. Binge alcohol use was as high as 40.9 percent in the 18 to 25 year age group. Males indulged in "binges" about twice as often as females. The definition of a binge is drinking at least five drinks on the same occasion during at least one day in the past 30 days. If the respondent had a "binge" on five or more days during the month, he or she is additionally reported in the category of "heavy alcohol use." Among the population above 11 years, 6.7 percent were heavy alcohol users in 2002, and among specific races American Indians, Native Hawaiians and Other Pacific Islanders, were the most frequent users, at a level of around 8.5 percent for each group.

Cigarette use, which was steady in 2000 and 2001 at 24.9 percent for the population of 12 years and above, increased to 26.0 percent in 2002. Respondents who smoked one cigarette or more in the past 30 days were counted as cigarette users. Smoking was most frequent in the 18 to 25-year age group, where 40.8 percent used cigarettes. Overall, men had a 5 percentage point lead over women in cigarette use in 2002, but for those between 12 and 17 years, 13.6 percent of females used cigarettes and as compared with 12.3 percent of males.

Table III-29. Use of Selected Substances in the Past Month by Persons 12 Years of Age and Over, According to Age, Sex, Race, and Hispanic Origin, 1999–2002

(Percent.)

Characteristic	Any illicit drug [1]				Marijuana				Nonmedical use of any psychotherapeutic drug [2]			
	2002	2001	2000	1999	2002	2001	2000	1999	2002	2001	2000	1999
12 years and over	8.3	7.1	6.3	6.3	6.2	5.4	4.8	4.7	2.6	2.1	1.7	1.8
Age												
12–13 years	4.2	3.8	3.0	3.9	1.4	1.5	1.1	1.5	1.7	1.8	1.6	1.8
14–15 years	11.2	10.9	9.8	9.8	7.6	7.6	6.9	6.9	4.0	3.5	3.0	3.4
16–17 years	19.8	17.8	16.4	15.4	15.7	14.9	13.7	13.2	6.2	4.4	4.3	3.4
18–25 years	20.2	18.8	15.9	16.4	17.3	16.0	13.6	14.2	5.4	4.8	3.6	3.7
26–34 years	10.5	8.8	7.8	6.8	7.7	6.8	5.9	5.4	3.6	2.4	2.1	1.5
35 years and over	4.6	3.5	3.3	3.4	3.1	2.4	2.3	2.2	1.6	1.3	1.0	1.3
Sex												
Male	10.3	8.7	7.7	8.1	8.1	7.0	6.2	6.5	2.7	2.2	1.8	1.9
Female	6.4	5.5	5.0	4.6	4.4	3.8	3.5	3.1	2.6	2.0	1.7	1.7
Age and Sex												
12–17 years	11.6	10.8	9.7	9.8	8.2	8.0	7.2	7.2	4.0	3.2	3.0	2.9
Male	12.3	11.4	9.8	10.1	9.1	8.9	7.7	7.8	3.6	2.7	2.7	2.6
Female	10.9	10.2	9.5	9.4	7.2	7.1	6.6	6.7	4.3	3.8	3.3	3.1
Hispanic Origin and Race [3]												
Not Hispanic or Latino												
White only	8.5	7.2	6.4	6.2	6.5	5.6	4.9	4.7	2.8	2.3	1.8	1.9
Black or African American only	9.7	7.4	6.4	7.5	7.4	5.6	5.2	5.9	2.0	1.6	1.2	1.4
American Indian and Alaska Native only	10.1	9.9	12.6	10.4	6.7	8.0	10.1	6.9	3.2	2.3	3.9	3.5
Native Hawaiian and Other Pacific Islander only	7.9	7.5	6.2	*	4.4	7.1	2.5	*	3.8	1.1	3.5	0.3
Asian only	3.5	2.8	2.7	3.2	1.8	1.7	1.4	2.3	0.7	0.8	1.1	0.9
2 or more races	11.4	12.6	14.8	10.3	9.0	9.6	12.5	8.5	3.5	5.3	2.3	2.7
Hispanic or Latino	7.2	6.4	5.3	6.1	4.3	4.2	3.6	4.2	2.9	1.9	1.7	1.7

Characteristic	Alcohol use				Binge alcohol use [4]				Heavy alcohol use [5]			
	2002	2001	2000	1999	2002	2001	2000	1999	2002	2001	2000	1999
12 years and over	51.0	48.3	46.6	46.4	22.9	20.5	20.6	20.2	6.7	5.7	5.6	5.7
Age												
12–13 years	4.3	4.4	4.6	4.4	1.8	1.9	2.0	1.8	0.3	0.2	0.2	0.2
14–15 years	16.6	16.6	15.7	15.4	9.2	9.2	9.3	9.0	1.9	1.7	1.8	1.6
16–17 years	32.6	30.8	29.1	29.6	21.4	20.8	20.3	19.3	5.6	5.7	6.0	5.4
18–25 years	60.5	58.8	56.8	57.2	40.9	38.7	37.8	37.9	14.9	13.6	12.8	13.3
26–34 years	61.4	59.9	58.3	57.4	33.1	30.1	30.3	29.3	9.0	7.8	7.6	7.5
35 years and over	52.1	48.7	46.8	46.6	18.6	16.2	16.4	16.0	5.2	4.2	4.1	4.2
Sex												
Male	57.4	54.8	53.6	53.2	31.2	28.2	28.3	28.1	10.8	9.2	8.7	9.2
Female	44.9	42.3	40.2	40.2	15.1	13.4	13.5	12.9	3.0	2.6	2.7	2.4
Age and Sex												
12–17 years	17.6	17.3	16.4	16.5	10.7	10.6	10.4	10.1	2.5	2.5	2.6	2.4
Male	17.4	17.2	16.2	16.7	11.4	11.2	11.2	11.3	3.1	3.1	3.2	3.0
Female	17.9	17.3	16.5	16.3	9.9	9.9	9.6	8.9	1.9	1.9	2.0	1.8
Hispanic Origin and Race [3]												
Not Hispanic or Latino												
White only	55.0	52.7	50.7	50.3	23.4	21.5	21.2	21.1	7.5	6.4	6.2	6.2
Black or African American only	39.9	35.1	33.7	34.3	21.0	16.8	17.7	16.3	4.4	4.1	4.0	3.6
American Indian and Alaska Native only	44.7	35.0	35.1	33.9	27.9	21.8	26.2	20.0	8.7	7.1	7.2	5.8
Native Hawaiian and Other Pacific Islander only	*	*	*	*	25.2	17.0	*	*	8.3	4.0	*	*
Asian only	37.1	31.9	28.0	30.7	12.4	10.1	11.6	10.8	2.6	1.5	1.4	2.5
2 or more races	49.9	43.2	41.6	41.4	19.8	19.4	17.5	20.2	7.5	6.7	5.2	7.7
Hispanic or Latino	42.8	39.5	39.8	38.6	24.8	21.3	22.7	21.7	5.9	4.4	4.4	5.4

Note: Because of methodological differences among the National Survey on Drug Use & Health (formerly called NHSDA), Monitoring the Future Study (MTF), and Youth Risk Behavior Survey (YRBS), rates of substance use measured by these surveys are not directly comparable.

[1] Any illicit drug includes marijuana/hashish, cocaine (including crack), heroin, hallucinogens (including LSD and PCP), inhalants, or any prescription-type psychotherapeutic drug used nonmedically.
[2] Psychotherapeutic drugs include prescription-type pain relievers, tranquilizers, stimulants, or sedatives; does not include over-the-counter drugs.
[3] Persons of Hispanic origin may be of any race.
[4] Binge alcohol use is defined as drinking five or more drinks on the same occasion on at least 1 day in the past 30 days. Occasion is defined as at the same time or within a couple of hours of each other.
[5] Heavy alcohol use is defined as drinking five or more drinks on the same occasion on each of 5 or more days in the past 30 days; all heavy alcohol users are also 'binge' alcohol users.
* = Figure does not meet standards of reliability or precision.

Table III-29. Use of Selected Substances in the Past Month by Persons 12 Years of Age and Over, According to Age, Sex, Race, and Hispanic Origin, 1999–2002—*Continued*

(Percent.)

Characteristic	Any tobacco [6]				Cigarettes				Cigars			
	2002	2001	2000	1999	2002	2001	2000	1999	2002	2001	2000	1999
12 years and over	30.4	29.5	29.3	30.2	26.0	24.9	24.9	25.8	5.4	5.4	4.8	5.5
Age												
12–13 years	3.8	3.9	4.0	4.8	3.2	3.2	3.4	4.1	0.7	1.0	0.8	1.1
14–15 years	13.4	13.4	14.2	16.4	11.2	11.4	12.1	14.3	3.8	3.7	4.3	5.1
16–17 years	29.0	28.1	28.9	30.3	24.9	24.4	25.2	26.1	9.3	8.1	8.6	9.7
18–25 years	45.3	43.9	42.9	44.6	40.8	39.1	38.3	39.7	11.0	10.4	10.4	11.5
26–34 years	38.2	36.2	35.6	38.2	32.7	30.5	29.7	31.5	5.3	5.9	5.7	7.0
35 years and over	27.9	27.3	27.3	27.4	23.4	22.7	22.9	23.3	4.1	4.4	3.4	3.9
Sex												
Male	37.0	35.6	35.2	36.6	28.7	27.1	26.9	28.3	9.4	9.4	8.4	9.5
Female	24.3	23.8	23.9	24.3	23.4	23.0	23.1	23.4	1.7	1.6	1.5	1.7
Age and Sex												
12–17 years	15.2	15.1	15.6	17.3	13.0	13.0	13.4	14.9	4.5	4.3	4.5	5.4
Male	16.0	15.8	16.3	18.7	12.3	12.4	12.8	14.8	6.2	5.8	6.4	7.7
Female	14.4	14.4	14.8	15.8	13.6	13.6	14.1	15.0	2.7	2.6	2.5	2.9
Hispanic Origin and Race [3]												
Not Hispanic or Latino												
White only	32.0	31.3	31.0	31.9	26.9	26.1	25.9	27.0	5.5	5.6	5.0	5.8
Black or African American only	28.8	27.7	26.7	26.6	25.3	23.9	23.3	22.5	6.8	5.8	5.1	5.9
American Indian and Alaska Native only	44.3	44.9	55.0	43.1	37.1	38.0	42.3	36.0	5.2	8.6	10.7	6.3
Native Hawaiian and Other Pacific Islander only	28.8	28.5	*	*	*	27.7	*	*	4.1	2.8	3.1	4.4
Asian only	18.6	13.6	17.9	18.7	17.7	12.9	16.5	16.7	1.1	1.4	1.6	1.9
2 or more races	38.1	34.1	38.9	34.0	35.0	31.1	32.3	29.8	5.5	5.3	5.1	5.3
Hispanic or Latino	25.2	22.9	22.2	24.7	23.0	20.9	20.7	22.6	5.0	4.2	3.5	4.3

Note: Because of methodological differences among the National Survey on Drug Use & Health (formerly called NHSDA), Monitoring the Future Study (MTF), and Youth Risk Behavior Survey (YRBS), rates of substance use measured by these surveys are not directly comparable.

[3]Persons of Hispanic origin may be of any race.
[6]Any tobacco product includes cigarettes, smokeless tobacco (i.e., chewing tobacco or snuff), cigars, or pipe tobacco.
* = Figure does not meet standards of reliability or precision.

TABLE III-30. PERCENT OF HIGH SCHOOL SENIORS, TENTH GRADERS, AND EIGHTH GRADERS WHO USE SELECTED SUBSTANCES, ACCORDING TO SEX AND RACE, SELECTED YEARS, 1980–2002

This survey presents separate data for the use by seniors, 10th graders, and 8th graders of cigarettes, marijuana, cocaine, inhalants, ecstasy, and alcohol. Widespread use, at least once during the past month, was in a range from 7 percent to 31 percent for cigarettes and marijuana. Use was even more prevalent for alcohol in a range between 15 percent and 53 percent, depending on grade, sex, and race. The other substances were used much less frequently. Inhalants, used by 3.8 percent of 8th graders, were in the lead, but this represented a reduction since 1995 when inhalants had their top attraction. Cocaine had its greatest use of 2.8 percent by White, male seniors.

More specifically, 26.7 percent of seniors in 2002 had used at least one cigarette in the preceding month and 48.6 percent had had an alcoholic drink. These percentages are not greatly different from those that apply to the entire population of 12 years and above, but they do show a declining trend since 1980. Binge drinking (as defined in Table 29) was indulged in by 28.6 percent of seniors, but only by 12.4 percent of 8th graders. This compares with an average of 22.9 percent for the entire population of 12 years and above. Marijuana was used by 21.5 percent of seniors, representing a fairly steady fraction since 1995. More males used marijuana than females.

These data are based on a survey of drug abuse by Monitoring the Future (MTF)—conducted by the University of Michigan—and are not fully comparable in scope with the National Household Survey on Drug Abuse—sponsored by the Substance Abuse and Mental Health Services Administration—which provides the data shown in Table 29.

Table III-30. Percent of High School Seniors, Tenth Graders, and Eighth Graders who Use Selected Substances, According to Sex and Race, Selected Years, 1980–2002

(Percent using substance in the past month.)

Substance, sex, race, and grade in school	Percent using in the last month								
	2002	2001	2000	1999	1998	1995	1991	1990	1980
Cigarettes									
All seniors	26.7	29.5	31.4	34.6	35.1	33.5	28.3	29.4	30.5
Male	27.4	29.7	32.8	35.4	36.3	34.5	29.0	29.1	26.8
Female	25.5	28.7	29.7	33.5	33.3	32.0	27.5	29.2	33.4
White	30.9	34.1	36.6	39.1	41.0	37.3	31.8	32.5	31.0
Black or African American	11.3	12.9	13.6	14.9	14.9	15.0	9.4	12.0	25.2
All tenth graders	17.7	21.3	23.9	25.7	27.6	27.9	20.8	...	...
Male	16.7	20.9	23.8	25.2	26.2	27.7	20.8	...	...
Female	18.6	21.5	23.6	25.8	29.1	27.9	20.7	...	...
White	20.8	24.0	27.3	29.1	32.4	31.2	23.9	...	...
Black or African American	9.1	10.9	11.3	11.0	13.8	12.2	6.4	...	...
All eighth graders	10.7	12.2	14.6	17.5	19.1	19.1	14.3	...	...
Male	11.0	12.2	14.3	16.7	18.0	18.8	15.5	...	...
Female	10.4	12.0	14.7	17.7	19.8	19.0	13.1	...	...
White	11.1	12.8	16.4	19.0	21.1	21.7	15.0	...	...
Black or African American	7.3	8.0	8.4	10.7	10.8	8.2	5.3	...	...
Marijuana									
All seniors	21.5	22.4	21.6	23.1	22.8	21.2	13.8	14.0	33.7
Male	25.3	25.6	24.7	26.3	26.5	24.6	16.1	16.1	37.8
Female	17.4	19.1	18.3	19.7	18.8	17.2	11.2	11.5	29.1
White	22.8	23.9	22.0	23.4	24.2	21.5	15.0	15.6	34.2
Black or African American	16.4	16.5	17.5	20.4	18.3	17.8	6.5	5.2	26.5
All tenth graders	17.8	19.8	19.7	19.4	18.7	17.2	8.7	...	...
Male	19.3	22.7	23.3	21.8	20.3	19.2	10.1	...	...
Female	16.4	16.8	16.2	17.0	17.2	15.0	7.3	...	...
White	19.1	20.4	20.1	20.2	19.5	17.7	9.4	...	...
Black or African American	14.4	16.5	17.0	14.7	14.5	15.1	3.8	...	...
All eighth graders	8.3	9.2	9.1	9.7	9.7	9.1	3.2	...	...
Male	9.5	11.0	10.2	10.5	10.3	9.8	3.8	...	...
Female	7.1	7.3	7.8	8.8	8.8	8.2	2.6	...	...
White	7.9	8.6	8.3	8.5	8.9	9.0	3.0	...	...
Black or African American	7.1	7.7	8.5	10.0	9.4	7.0	2.1	...	...
Cocaine									
All seniors	2.3	2.1	2.1	2.6	2.4	1.8	1.4	1.9	5.2
Male	2.7	2.5	2.7	3.3	3.0	2.2	1.7	2.3	6.0
Female	1.8	1.6	1.6	1.8	1.7	1.3	0.9	1.3	4.3
White	2.8	2.3	2.2	2.8	2.7	1.7	1.3	1.8	5.4
Black or African American	0.2	0.6	1.0	0.5	0.4	0.4	0.8	0.5	2.0
All tenth graders	1.6	1.3	1.8	1.9	2.1	1.7	0.7	...	...
Male	1.8	1.5	2.1	2.2	2.4	1.8	0.7	...	...
Female	1.4	1.2	1.4	1.6	1.8	1.5	0.6	...	...
White	1.7	1.2	1.7	1.9	2.0	1.7	0.6	...	...
Black or African American	0.4	0.3	0.4	0.3	0.8	0.4	0.2	...	...
All eighth graders	1.1	1.2	1.2	1.3	1.4	1.2	0.5	...	...
Male	1.1	1.1	1.3	1.4	1.5	1.1	0.7	...	...
Female	1.1	1.2	1.1	1.2	1.2	1.2	0.4	...	...
White	1.0	1.1	1.1	1.1	1.0	1.0	0.4	...	...
Black or African American	0.5	0.4	0.5	0.3	0.6	0.4	0.4	...	...
Inhalants									
All seniors	1.5	1.7	2.2	2.0	2.3	3.2	2.4	2.7	1.4
Male	2.2	2.3	2.9	2.5	2.9	3.9	3.3	3.5	1.8
Female	0.8	1.1	1.7	1.5	1.7	2.5	1.6	2.0	1.0
White	1.3	1.8	2.1	2.1	2.6	3.7	2.4	3.0	1.4
Black or African American	1.2	1.3	2.1	0.4	1.0	1.1	1.5	1.5	1.0
All tenth graders	2.4	2.5	2.6	2.6	2.9	3.5	2.7	...	...
Male	2.3	2.5	3.0	2.9	3.2	3.8	2.9	...	...
Female	2.4	2.4	2.2	2.2	2.6	3.2	2.6	...	...
White	2.6	2.5	2.8	2.9	3.3	3.9	2.9	...	...
Black or African American	1.5	0.9	1.5	0.8	1.1	1.2	2.0	...	...
All eighth graders	3.8	4.0	4.5	5.0	4.8	6.1	4.4	...	...
Male	3.5	3.6	4.1	4.6	4.8	5.6	4.1	...	...
Female	3.9	4.3	4.8	5.3	4.7	6.6	4.7	...	...
White	3.9	4.1	4.5	5.6	5.3	7.0	4.5	...	...
Black or African American	2.7	2.6	2.3	2.3	2.2	2.3	2.3	...	...
MDMA (Ecstasy)									
All seniors	2.4	2.8	3.6	2.5	1.5	...	...	...	...
Male	2.6	3.7	4.1	2.6	2.3	...	...	...	...
Female	2.1	2.0	3.1	2.5	0.8	...	...	...	...
White	2.5	2.8	3.9	2.7	1.8	...	...	...	...
Black or African American	0.5	0.9	1.9	0.0	0.2	...	...	...	...
All tenth graders	1.8	2.6	2.6	1.8	1.3	...	...	...	...
Male	1.6	3.5	2.5	1.7	1.4	...	...	...	...
Female	1.8	1.6	2.5	1.9	1.1	...	...	...	...
White	2.3	2.6	2.5	2.1	1.3	...	...	...	...
Black or African American	0.5	1.0	1.8	0.3	0.7	...	...	...	...
All eighth graders	1.4	1.8	1.4	0.8	0.9	...	...	...	...
Male	1.5	1.9	1.6	0.9	1.0	...	...	...	...
Female	1.3	1.8	1.2	0.7	0.7	...	...	...	...
White	1.0	2.0	1.4	0.9	0.9	...	...	...	...
Black or African American	0.6	1.1	0.8	0.4	0.4	...	...	...	...

. . . = Not available.

Table III-30. Percent of High School Seniors, Tenth Graders, and Eighth Graders who Use Selected Substances, According to Sex and Race, Selected Years, 1980–2002—*Continued*

(Percent using substance in the past month.)

Substance, sex, race, and grade in school	Percent using in the last month								
	2002	2001	2000	1999	1998	1995	1991	1990	1980
Alcohol [1]									
All seniors	48.6	49.8	50.0	51.0	52.0	51.3	54.0	57.1	72.0
Male	52.3	54.7	54.0	55.3	57.6	55.7	58.4	61.3	77.4
Female	45.1	45.1	46.1	46.8	46.9	47.0	49.0	52.3	66.8
White	52.7	55.3	55.3	54.9	57.6	54.8	57.7	62.2	75.8
Black or African American	30.7	29.6	29.3	30.8	33.6	37.4	34.4	32.9	47.7
All tenth graders	35.4	39.0	41.0	40.0	38.8	38.8	42.8	. . .	. . .
Male	35.3	41.1	43.3	42.3	40.0	39.7	45.5	. . .	. . .
Female	35.7	36.8	38.6	38.1	37.7	37.8	40.3	. . .	. . .
White	39.0	41.0	44.3	43.4	42.5	41.3	45.7	. . .	. . .
Black or African American	23.2	26.0	24.7	24.6	24.2	24.9	30.2	. . .	. . .
All eighth graders	19.6	21.5	22.4	24.0	23.0	24.6	25.1	. . .	. . .
Male	19.1	22.3	22.5	24.8	24.0	25.0	26.3	. . .	. . .
Female	20.0	20.6	22.0	23.3	21.9	24.0	23.8	. . .	. . .
White	20.4	22.5	23.9	25.6	24.0	25.4	26.0	. . .	. . .
Black or African American	14.7	14.9	15.1	16.8	15.4	17.3	17.8	. . .	. . .
Binge Drinking [2]									
All seniors	28.6	29.7	30.0	30.8	31.5	29.8	29.8	32.2	41.2
Male	34.2	36.0	36.7	38.1	39.2	36.9	37.8	39.1	52.1
Female	23.0	23.7	23.5	23.6	24.0	23.0	21.2	24.4	30.5
White	32.9	34.5	34.4	34.8	36.6	32.9	32.9	36.2	44.6
Black or African American	10.4	12.6	11.0	11.9	12.7	15.5	11.8	11.6	17.0
All tenth graders	22.4	24.9	26.2	25.6	24.4	24.0	22.9	. . .	. . .
Male	23.8	28.6	29.8	29.7	26.7	26.4	26.4	. . .	. . .
Female	21.0	21.4	22.5	21.8	22.2	21.5	19.5	. . .	. . .
White	24.6	26.4	28.5	27.7	26.8	25.7	24.4	. . .	. . .
Black or African American	12.4	12.3	12.9	12.9	12.5	12.3	14.4	. . .	. . .
All eighth graders	12.4	13.2	14.1	15.2	13.7	14.5	12.9	. . .	. . .
Male	12.5	13.7	14.4	16.4	14.4	15.1	14.3	. . .	. . .
Female	12.1	12.4	13.6	13.9	12.7	13.9	11.4	. . .	. . .
White	12.3	13.1	14.6	15.2	13.5	14.5	12.6	. . .	. . .
Black or African American	9.9	8.8	9.3	10.8	9.1	10.0	9.9	. . .	. . .

[1]In 1993 the alcohol question was changed to indicate that a "drink" meant "more than a few sips." 1993 data, available electronically, are based on a half sample.
[2]Five or more alcoholic drinks in a row at least once in the prior 2-week period.
. . . = Not available.

Figure III-10. Percent of High School Seniors Using Selected Substances in the Past Month

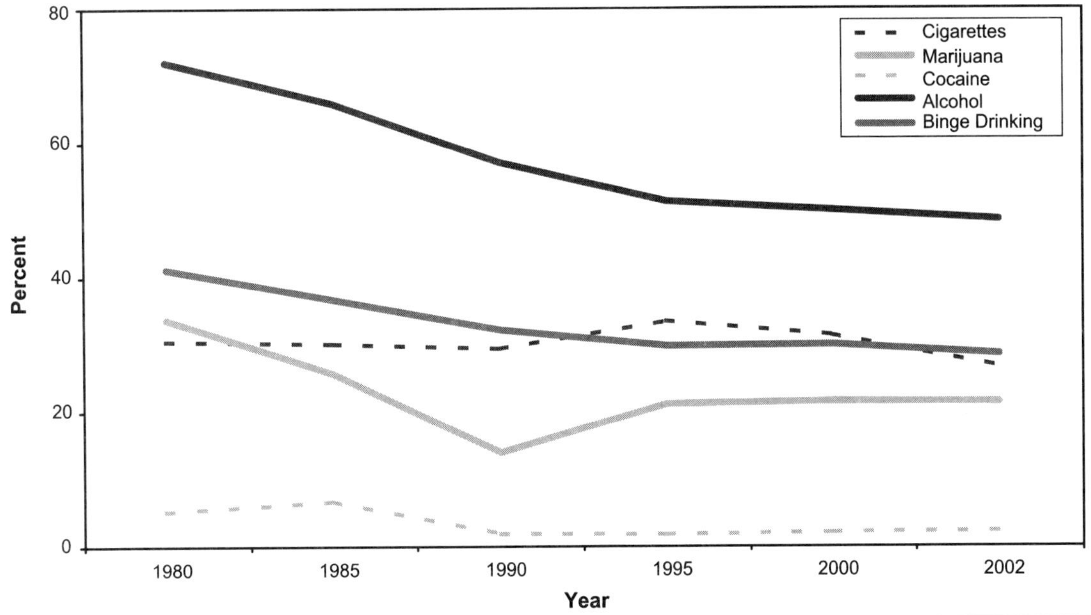

TABLE III-31. CURRENT CIGARETTE SMOKING BY PERSONS 18 YEARS OF AGE AND OVER, ACCORDING TO SEX, RACE, AND AGE, SELECTED YEARS, 1965–2001

One of the big successes in health promotion has been the pronounced decline in smoking during recent decades. In 1965, 41.9 percent of persons 18 years of age and over were current cigarette smokers, but in 2001 this percentage had shrunk to 22.7. In a large interview survey, persons who have smoked 100 cigarettes in their entire life and who now smoke every day or some days are defined as current smokers. Age-adjusted, current smokers included 24.7 percent of all males, but only 17.9 percent of Black females.

Among males in 2001, smoking was most frequent in the 18 to 24-year age group, then remained fairly steady until age 64, and thereafter dropped sharply. Smoking of females by age group also showed the sharp drop at advanced age, but the highest frequency of smoking among women occurred in the 35–44 year age bracket.

Table III-31. Current Cigarette Smoking by Persons[1] 18 Years of Age and Over, According to Sex, Race, and Age, Selected Years, 1965–2001

(Percent.)

Sex, race, and age	2001[2]	2000[2]	1999[2]	1998[2]	1997[2]	1995	1990	1985	1983	1979	1974	1965
18 Years and Over, Age-Adjusted[3]												
All persons	22.7	23.1	23.3	24.0	24.6	24.6	25.3	29.9	31.9	33.3	37.0	41.9
Male	24.7	25.2	25.2	25.9	27.1	26.5	28.0	32.2	34.8	37.0	42.8	51.2
Female	20.8	21.1	21.6	22.1	22.2	22.7	22.9	27.9	29.4	30.1	32.2	33.7
White male[4]	24.9	25.5	25.0	26.0	26.8	26.2	27.6	31.3	34.2	36.4	41.7	50.4
Black or African American male[4]	27.6	25.7	28.4	29.0	32.4	29.4	32.8	40.2	41.7	43.9	53.6	58.8
White female[4]	22.1	22.0	22.5	23.0	22.8	23.4	23.5	27.9	29.6	30.3	32.0	33.9
Black or African American female[4]	17.9	20.7	20.5	21.1	22.5	23.5	20.8	30.9	31.3	30.5	35.6	31.8
18 Years and Over, Crude												
All persons	22.8	23.3	23.5	24.1	24.7	24.7	25.5	30.1	32.1	33.5	37.1	42.4
Male	25.2	25.7	25.7	26.4	27.6	27.0	28.4	32.6	35.1	37.5	43.1	51.9
Female	20.7	21.0	21.5	22.0	22.1	22.6	22.8	27.9	29.5	29.9	32.1	33.9
White male[4]	25.1	25.8	25.3	26.3	27.2	26.6	28.0	31.7	34.5	36.8	41.9	51.1
Black or African American male[4]	27.6	26.1	28.6	29.0	32.2	28.5	32.5	39.9	40.6	44.1	54.3	60.4
White female[4]	21.7	21.6	22.1	22.6	22.5	23.1	23.4	27.7	29.4	30.1	31.7	34.0
Black or African American female[4]	18.0	20.8	20.6	21.1	22.5	23.5	21.2	31.0	32.2	31.1	36.4	33.7
All Males												
18–24	30.4	28.5	29.5	31.3	31.7	27.8	26.6	28.0	32.9	35.0	42.1	54.1
25–34 years	27.2	29.0	29.1	28.5	30.3	29.5	31.6	38.2	38.8	43.9	50.5	60.7
35–44 years	27.4	30.2	30.0	30.2	32.1	31.5	34.5	37.6	41.0	41.8	51.0	58.2
45–64 years	26.4	26.4	25.8	27.7	27.6	27.1	29.3	33.4	35.9	39.3	42.6	51.9
65 years and over	11.5	10.2	10.5	10.4	12.8	14.9	14.6	19.6	22.0	20.9	24.8	28.5
White Male[4]												
18–24	32.5	30.9	30.5	34.1	34.0	28.4	27.4	28.4	32.5	34.3	40.8	53.0
25–34 years	29.0	29.9	30.8	29.2	30.4	29.9	31.6	37.3	38.6	43.6	49.5	60.1
35–44 years	27.8	30.6	29.5	29.6	32.1	31.2	33.5	36.6	40.8	41.3	50.1	57.3
45–64 years	25.1	25.8	24.5	27.0	26.5	26.3	28.7	32.1	35.0	38.3	41.2	51.3
65 years and over	10.7	9.8	10.0	10.0	11.5	14.1	13.7	18.9	20.6	20.5	24.3	27.7
Black or African American Male[4]												
18–24	21.6	20.8	23.6	19.7	23.5	*14.6	21.3	27.2	34.2	40.2	54.9	62.8
25–34 years	23.8	23.3	22.7	25.2	31.6	25.1	33.8	45.6	39.9	47.5	58.5	68.4
35–44 years	29.9	30.8	34.8	36.1	33.9	36.3	42.0	45.0	45.5	48.6	61.5	67.3
45–64 years	34.3	32.2	35.7	37.3	39.4	33.9	36.7	46.1	44.8	50.0	57.8	57.9
65 years and over	21.1	14.2	17.3	16.3	26.0	28.5	21.5	27.7	38.9	26.2	29.7	36.4
All Females												
18–24	23.4	25.1	26.3	24.5	25.7	21.8	22.5	30.4	35.5	33.8	34.1	38.1
25–34 years	23.0	22.5	23.5	24.6	24.8	26.4	28.2	32.0	32.6	33.7	38.8	43.7
35–44 years	25.7	26.2	26.5	26.4	27.2	27.1	24.8	31.5	33.8	37.0	39.8	43.7
45–64 years	21.4	21.6	21.0	22.5	21.5	24.0	24.8	29.9	31.0	30.7	33.4	32.0
65 years and over	9.2	9.3	10.7	11.2	11.5	11.5	11.5	13.5	13.1	13.2	12.0	9.6
White Female[4]												
18–24	27.2	28.7	29.6	28.1	29.4	24.9	25.4	31.8	36.5	34.5	34.0	38.4
25–34 years	25.5	25.1	25.5	26.9	26.1	27.3	28.5	32.0	32.2	34.1	38.6	43.4
35–44 years	27.0	26.6	26.9	26.6	27.5	27.0	25.0	31.0	34.8	37.2	39.3	43.9
45–64 years	21.6	21.4	21.2	22.5	20.9	24.3	25.4	29.7	30.6	30.6	33.0	32.7
65 years and over	9.4	9.1	10.5	11.2	11.7	11.7	11.5	13.3	13.2	13.8	12.3	9.8
Black or African American Female[4]												
18–24	10.0	14.2	14.8	*8.1	11.5	*8.8	10.0	23.7	32.0	31.8	35.6	37.1
25–34 years	16.8	15.5	18.2	21.5	22.5	26.7	29.1	36.2	38.0	35.2	42.2	47.8
35–44 years	24.0	30.2	28.8	30.0	30.1	31.9	25.5	40.2	32.7	37.7	46.4	42.8
45–64 years	22.6	25.6	22.3	25.4	28.4	27.5	22.6	33.4	36.3	34.2	38.9	25.7
65 years and over	9.3	10.2	13.5	11.5	10.7	13.3	11.1	14.5	*13.1	*8.5	*8.9	7.1

[1]Beginning in 1993 current cigarette smokers reported ever smoking 100 cigarettes in their lifetime and smoking now every day or some days.
[2]Data starting in 1997 are not strictly comparable with data for earlier years due to the 1997 questionnaire redesign.
[3]Estimates are age-adjusted to the year 2000 standard population using five age groups: 18-24 years, 25-34 years, 35-44 years, 45-64 years, 65 years and over.
[4]The race groups, White, Black, American Indian and Alaska Native (AI/AN), Asian, Native Hawaiian and Other Pacific Islander, and two or more races, include persons of Hispanic and non-Hispanic origin. Persons of Hispanic origin may be of any race.
* = Figure does not meet standards of reliability or precision.

TABLE III-32. AGE-ADJUSTED PREVALENCE OF CURRENT SMOKING BY PERSONS 25 YEARS OF AGE AND OVER, ACCORDING TO SEX, RACE, AND EDUCATION, SELECTED YEARS, 1974–2001

In general the frequency of smoking declined sharply the higher the level of education. Without a high school degree, 30.9 percent of all persons 25 years and over were current smokers in 2001, but among persons with a bachelor's degree only 10.8 percent were smokers. Other levels of education fell in between these categories. Among Black males the educational contrast was even greater: for those without a high school degree, 37.9 percent were smokers, and for those with a college degree 11.3 percent smoked.

The rankings according to educational achievement have been the same since 1974, but the percent of decline in smoking has differed. For instance, 12.8 percent of the population without a high school degree "dropped" smoking between 1974 and 2001, from 43.7 percent to 30.9 percent (these are not the same people in many cases), while a larger percentage of the population with college degrees—16.4 percent—discontinued cigarette smoking.

Table III-32. Age-Adjusted[1] Prevalence of Current Smoking by Persons[2] 25 Years of Age and Over, According to Sex, Race, and Education, Selected Years, 1974–2001

(Percent.)

Sex, race, and education	2001[3]	2000[3]	1999[3]	1998[3]	1997[3]	1995	1990	1985	1983	1979	1974
All Persons[4]	22.1	22.6	22.7	23.4	24.0	24.5	25.4	30.0	31.6	33.1	36.9
No high school diploma or GED	30.9	31.9	32.2	34.4	33.5	35.6	36.7	40.8	40.7	40.7	43.7
High school diploma or GED	28.2	29.2	28.0	28.9	29.9	29.1	29.1	32.0	33.5	33.6	36.2
Some college, no bachelor's degree	22.3	21.7	23.3	23.5	23.7	22.6	23.4	29.5	30.3	33.2	35.9
Bachelor's degree or higher	10.8	10.9	11.1	10.9	11.4	13.6	13.9	18.5	20.5	22.6	27.2
All Males[4]	23.9	24.8	24.5	25.1	26.4	26.4	28.2	32.8	35.1	37.3	42.9
No high school diploma or GED	34.7	36.4	36.2	37.5	39.1	39.7	42.0	45.7	47.1	47.6	52.3
High school diploma or GED	30.3	32.1	30.4	32.0	32.2	32.7	33.1	35.5	37.4	38.9	42.4
Some college, no bachelor's degree	24.4	23.3	24.8	25.4	25.5	23.7	25.9	32.9	33.3	36.5	41.8
Bachelor's degree or higher	11.2	11.6	11.8	11.0	12.5	13.8	14.5	19.6	21.7	22.7	28.3
White Males[4,5]	23.8	24.7	24.2	24.8	25.8	25.9	27.6	31.7	34.4	36.7	41.9
No high school diploma or GED	35.4	38.6	36.3	37.4	38.5	38.7	41.8	45.0	47.7	47.6	51.5
High school diploma or GED	30.5	32.5	30.5	32.2	31.8	32.9	32.9	34.8	37.0	38.5	42.0
Some college, no bachelor's degree	24.6	23.6	24.7	25.2	25.6	23.3	25.4	32.2	32.9	36.4	41.6
Bachelor's degree or higher	11.2	11.3	11.8	10.9	12.0	13.4	14.4	19.1	21.0	22.5	27.8
Black or African American Males[4,5]	28.4	26.5	29.1	30.4	33.8	31.6	34.5	42.1	42.8	44.4	53.4
No high school diploma or GED	37.9	38.3	43.8	42.9	44.6	41.9	41.6	50.5	46.0	49.7	58.1
High school diploma or GED	33.4	29.1	32.5	32.8	39.0	36.6	37.4	41.8	47.7	48.6	*50.7
Some college, no bachelor's degree	24.2	20.0	23.4	28.4	27.0	26.4	28.1	41.8	44.9	39.2	*45.3
Bachelor's degree or higher	11.3	14.7	11.3	*15.3	14.5	*17.3	*20.8	*32.0	*31.7	*36.8	*41.4
All Females[4]	20.4	20.6	20.9	21.7	21.7	22.9	22.9	27.5	28.5	29.5	32.0
No high school diploma or GED	27.2	27.3	28.2	31.3	28.2	31.7	31.8	36.5	35.2	34.8	36.6
High school diploma or GED	26.5	26.7	25.9	26.2	27.9	26.4	26.1	29.5	30.7	29.8	32.2
Some college, no bachelor's degree	20.5	20.4	21.9	21.8	22.0	21.6	21.0	26.3	27.3	30.0	30.1
Bachelor's degree or higher	10.5	10.1	10.4	10.7	10.3	13.3	13.3	17.1	18.9	22.5	25.9
White Females[4,5]	21.4	21.1	21.4	22.3	21.9	23.1	23.3	27.3	28.6	29.7	31.7
No high school diploma or GED	29.6	28.6	29.5	33.0	29.7	32.4	33.4	36.7	35.6	35.8	36.8
High school diploma or GED	28.4	27.9	27.2	27.1	28.3	26.8	26.5	29.4	30.8	29.9	31.9
Some college, no bachelor's degree	21.3	21.1	22.3	22.2	22.1	22.2	21.2	26.7	27.8	30.7	30.4
Bachelor's degree or higher	10.9	10.2	10.5	11.5	10.5	13.5	13.4	16.5	18.7	21.9	25.5
Black or African American Females[4,5]	19.1	21.6	21.4	23.0	24.1	25.7	22.4	32.0	31.2	30.3	35.6
No high school diploma or GED	26.3	31.2	30.1	32.8	27.1	32.3	26.3	39.4	36.5	31.6	36.1
High school diploma or GED	21.3	25.4	22.4	24.3	29.1	27.8	24.1	32.1	34.6	32.6	40.9
Some college, no bachelor's degree	17.4	20.4	22.3	21.7	24.3	20.8	22.7	23.9	*27.1	*28.9	32.3
Bachelor's degree or higher	11.6	10.8	13.4	9.0	12.5	17.3	17.0	26.6	*36.8	*43.3	*36.3

[1]Estimates are age-adjusted to the year 2000 standard population using five age groups: 18–24 years, 25–34 years, 35–44 years, 45–64 years, 65 years and over.
[2]Beginning in 1993 current cigarette smokers reported ever smoking 100 cigarettes in their lifetime and smoking now on every day or some days.
[3]Data starting in 1997 are not strictly comparable with data for earlier years due to the 1997 questionnaire redesign.
[4]Includes unknown education. Education quantities shown apply to 1997 and subsequent years. GED stands for General Educational Development high school equivalency diploma. In 1974–1995 the following categories based on the number of years of school completed were used: less than 12 years, 12 years, 13–15 years, 16 years or more.
[5]The race groups, White, Black, American Indian and Alaska Native (AI/AN), Asian, Native Hawaiian and Other Pacific Islander, and two or more races, include persons of Hispanic and non-Hispanic origin. Persons of Hispanic origin may be of any race.
* = Figure does not meet standards of reliability or precision.

TABLE III-33. CURRENT CIGARETTE SMOKING BY ADULTS, ACCORDING TO SEX, RACE, HISPANIC ORIGIN, AGE, AND EDUCATION, AVERAGE ANNUAL, SELECTED YEARS, 1990–2001

This table, with its details on race, indicates that American Indians have been the most frequent smokers and Asians the ones who smoke least. In between, Black men, but not Black women, tend to smoke more than Whites, and Hispanics tend to smoke less than non-Hispanics. It may also be noted that persons who reported belonging to more than one race, a classification innovation that began with the 2000 Census, are indicated to have an especially high frequency of smoking.

By age group, the data confirm the large drop off in smoking by persons 65 years and older that applies both to men and women, and to Hispanics as well as White and Black non-Hispanics. The decline in smoking as educational attainment increases is pronounced for most groups, but is weaker for Hispanics.

Table III-33. Current Cigarette Smoking by Adults[1], According to Sex, Race, Hispanic Origin, Age, and Education, Average Annual, Selected Years, 1990–2001

(Percent.)

Characteristic	Male			Female		
	1999–2001 [2]	1995–1998 [2]	1990–1992	1999–2001 [2]	1995–1998 [2]	1990–1992
18 Years and Over, Age-Adjusted [3]						
All persons [4]	25.1	26.5	27.9	21.2	22.1	23.7
Race [5]						
White only	25.2	26.4	27.4	22.2	22.9	24.3
Black or African American only	27.2	30.7	33.9	19.7	21.8	23.1
American Indian and Alaska Native only	30.4	40.5	34.2	34.7	28.9	36.7
Asian only	20.3	18.1	24.8	6.7	11.0	6.3
Native Hawaiian and Other Pacific Islander	*	. . .	. . .	*	. . .	. . .
Two or more races	34.5	. . .	. . .	30.8	. . .	. . .
American Indian and Alaska Native; White	38.7	. . .	. . .	39.0	. . .	. . .
Hispanic Origin and Race [5]						
Hispanic or Latino	22.3	24.4	25.7	12.1	13.7	15.8
Mexican	22.0	24.5	26.2	10.6	12.0	14.8
Not Hispanic or Latino	25.5	26.9	28.1	22.3	23.1	24.4
White only	25.6	26.9	27.7	23.5	24.1	25.2
Black or African American only	27.3	30.7	33.9	19.7	21.9	23.2
18 Years and Over, Crude						
All persons	25.5	27.0	28.4	21.0	22.0	23.6
Race [5]						
White only	25.4	26.8	27.8	21.8	22.6	24.1
Black or African American only	27.4	30.6	33.2	19.8	21.8	23.3
American Indian and Alaska Native only	32.0	39.2	35.5	36.9	31.2	37.3
Asian only	21.3	20.0	24.9	6.9	11.2	6.3
Native Hawaiian and Other Pacific Islander	*	. . .	. . .	*	. . .	. . .
2 or more races	35.9	. . .	. . .	31.6	. . .	. . .
American Indian and Alaska Native; White	41.0	. . .	. . .	40.3	. . .	. . .
Hispanic Origin, Race and Age [5]						
Hispanic or Latino	23.2	25.5	26.5	12.5	13.8	16.6
Mexican	22.8	25.2	27.1	10.9	11.6	15.0
Not Hispanic or Latino	25.8	27.2	28.5	22.0	22.9	24.2
White only	25.6	27.0	28.0	22.8	23.5	24.8
Black or African American only	27.5	30.6	33.3	19.8	21.9	23.3
18–24 years						
Hispanic or Latino	22.8	26.5	19.3	12.9	12.0	12.8
Not Hispanic or Latino						
White only	32.7	35.5	28.9	30.8	31.6	28.7
Black or African American only	21.9	21.3	17.7	13.0	9.8	10.8
25–34 years						
Hispanic or Latino	23.4	25.9	29.9	12.5	12.6	19.2
Not Hispanic or Latino						
White only	30.8	30.5	32.7	27.4	28.5	30.9
Black or African American only	23.3	28.5	34.6	16.9	22.0	29.2
35–44 years						
Hispanic or Latino	25.4	26.2	32.1	14.1	17.6	19.9
Not Hispanic or Latino						
White only	29.6	31.5	32.3	28.3	28.1	27.3
Black or African American only	32.0	34.7	44.1	27.5	30.3	31.3
45–64 years						
Hispanic or Latino	24.7	26.8	26.6	13.5	14.7	17.1
Not Hispanic or Latino						
White only	25.2	26.8	28.4	22.1	22.3	26.1
Black or African American only	34.0	38.8	38.0	23.6	26.9	26.1
65 years and over						
Hispanic or Latino	12.5	14.7	16.1	5.9	9.4	6.6
Not Hispanic or Latino						
White only	10.0	10.6	14.2	9.9	11.6	12.3
Black or African American only	17.6	20.9	25.2	11.0	11.2	10.7
Education, Hispanic Origin and Race [5, 6]						
25 years and over, age-adjusted [7]						
No high school diploma or GED						
Hispanic or Latino	24.2	27.6	30.2	12.1	13.3	15.8
Not Hispanic or Latino						
White only	43.5	43.9	46.1	39.3	40.7	40.4
Black or African American only	40.0	44.6	45.4	29.4	30.0	31.3
High school diploma or GED						
Hispanic or Latino	24.2	26.7	29.6	12.5	16.4	18.4
Not Hispanic or Latino						
White only	31.9	32.8	32.9	29.2	28.8	28.4
Black or African American only	31.5	35.7	38.2	23.0	26.6	25.4
Some college or more						
Hispanic or Latino	17.2	16.6	20.4	11.1	13.5	14.3
Not Hispanic or Latino						
White only	17.6	18.3	19.3	16.7	17.2	18.1
Black or African American only	19.2	23.3	25.6	17.0	18.9	22.8

[1]Beginning in 1993 current cigarette smokers reported ever smoking 100 cigarettes in their lifetime and smoking now on every day or some days.
[2]Data starting in 1997 are not strictly comparable with data for earlier years due to the 1997 questionnaire redesign.
[3]Estimates are age-adjusted to the year 2000 standard population using five age groups: 18–24 years, 25–34 years, 35–44 years, 45–64 years, 65 years and over.
[4]Includes all other races not shown separately.
[5]The race groups, White, Black, American Indian and Alaska Native (AI/AN), Asian, Native Hawaiian and Other Pacific Islander, and two or more races, include persons of Hispanic and non-Hispanic origin. Persons of Hispanic origin may be of any race.
[6]Education categories shown are for 1997 and subsequent years. GED stands for General Educational Development high school equivalency diploma. In years prior to 1997 the following categories based on number of years of school completed were used: less than 12 years, 12 years, 13 years or more.
[7]Estimates are age adjusted to the year 2000 standard population using four age groups, 25–34 years, 35–44 years, 45–64 years, 65 years and over.
. . . = Not available.
* = Figure does not meet standards of reliability or precision.

TABLE III-34. ALCOHOL CONSUMPTION BY PERSONS 18 YEARS OF AGE AND OVER, ACCORDING TO SELECTED CHARACTERISTICS, SELECTED YEARS, 1997–2001

Respondents, 18 years of age and over, were divided into several categories: Lifetime abstainers (22.6 percent in 2001—age-adjusted); former drinkers—infrequent or regular in the past, but without a drink for a year (14.9 percent in 2001); current infrequent drinkers—had fewer than 12 drinks in the past year (12.8 percent in 2001): and current regular drinkers—those who had at least 12 drinks in the past year (48.7 percent in 2001). A fractional percentage of current drinkers was not allocated in the table.

The table classifies drinkers in a number of additional ways. One panel classifies current drinkers according to the frequency with which they have consumed at least five drinks in a single day. The majority of current drinkers—68 percent—never had five drinks in one day during the preceding year. The next category, 17 percent, had five drinks on 1 to 11 days during the year. The heavier drinkers, representing 15 percent, drank five or more drinks on 12 or more days of the year.

Drinking, like smoking, became much less frequent at age 65 and above, and there was a gradual reduction starting with 45 years. Also, since 1997 (the first year in the table) drinking has been declining over time.

TABLE III-35. COCAINE-RELATED EMERGENCY DEPARTMENT EPISODES, ACCORDING TO AGE, SEX, RACE, AND HISPANIC ORIGIN, SELECTED YEARS, 1990–2001

In 2001, there were almost 200,000 cocaine-related episodes in hospital emergency departments, which is equivalent to 78 cases for 100,000 of population. These cases may represent drug overdose, unexpected drug reactions, chronic abuse, detoxification, or other drug related reasons. These episodes have been increasing since 1990. Occurrences are about twice as frequent for men as for women and they peak in the 26–34 year age bracket. Blacks have had a proportionately high incidence of such episodes, especially so for men and women age 35 and over.

Figure III-11. Cocaine-Related Emergency Department Episodes by Age, 1990–2001

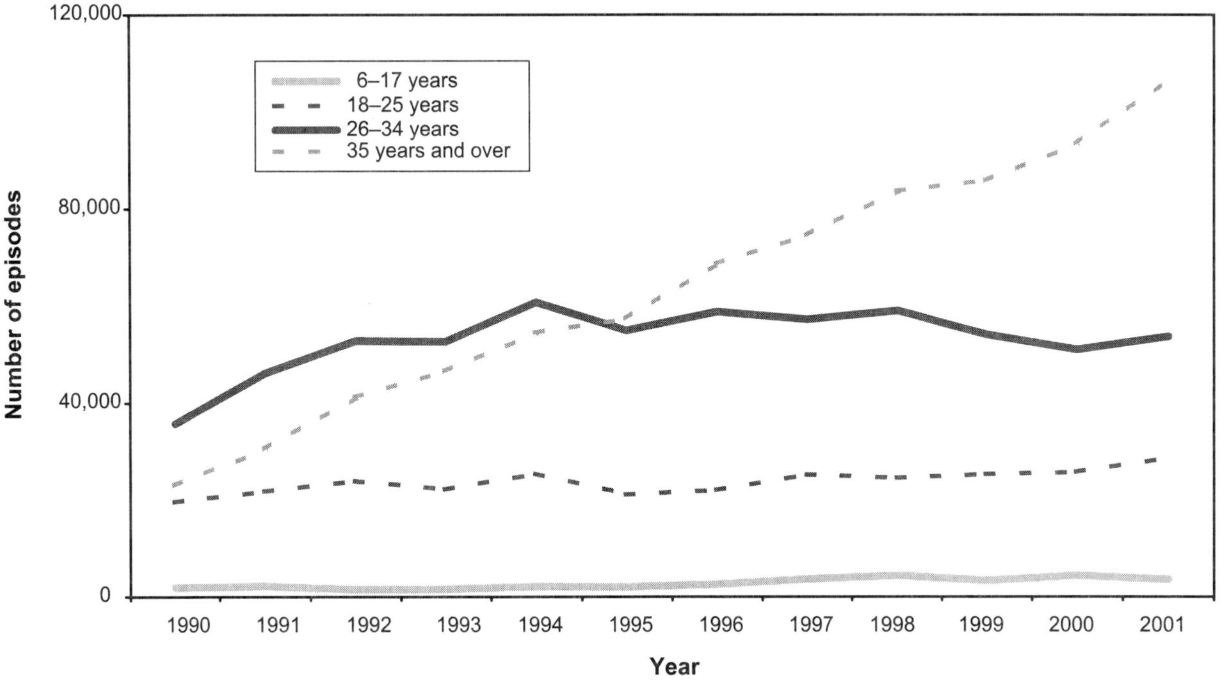

Table III-34. Alcohol Consumption by Persons 18 Years of Age and Over, According to Selected Characteristics, Selected Years, 1997–2001

(Percent.)

Characteristic	Both sexes			Male			Female		
	2001	2000	1997	2001	2000	1997	2001	2000	1997
DRINKING STATUS PERCENT DISTRIBUTION [1]									
18 Years and Over, Age-Adjusted [2]									
All	100.0	100.0	100.0	100.0	100.0	100.0	100.0	100.0	100.0
Lifetime abstainer	22.6	24.1	21.2	15.2	17.5	14.0	29.2	29.9	27.6
Former drinker	14.9	14.4	15.7	16.0	14.9	16.2	14.0	14.2	15.3
Infrequent	8.5	8.2	9.0	7.7	7.0	7.7	9.1	9.2	10.1
Regular	6.4	6.3	6.7	8.3	7.8	8.5	4.8	5.0	5.2
Current drinker	62.5	61.5	63.1	68.8	67.7	69.8	56.8	55.8	57.0
Infrequent	12.8	14.7	15.0	9.2	11.1	11.7	16.2	18.2	18.1
Regular	48.7	46.7	48.1	58.4	56.6	58.1	39.7	37.7	38.9
18 Years and Over, Crude									
All	100.0	100.0	100.0	100.0	100.0	100.0	100.0	100.0	100.0
Lifetime abstainer	22.5	24.0	21.1	15.2	17.5	14.0	29.3	30.1	27.7
Former drinker	14.8	14.3	15.5	15.5	14.3	15.6	14.1	14.3	15.4
Infrequent	8.4	8.1	8.9	7.5	6.8	7.5	9.2	9.3	10.1
Regular	6.4	6.2	6.6	8.0	7.5	8.1	4.9	5.0	5.2
Current drinker	62.7	61.7	63.4	69.3	68.2	70.5	56.6	55.6	57.0
Infrequent	12.9	14.7	15.0	9.3	11.1	11.7	16.2	18.1	18.1
Regular	48.8	46.9	48.4	58.9	57.2	58.8	39.5	37.5	38.8
PERCENT CURRENT DRINKERS AMONG ALL PERSONS									
All Persons									
18–44 years	69.0	67.4	69.4	75.0	73.0	74.8	63.2	61.9	64.2
18–24 years	63.6	59.4	62.2	69.6	64.0	66.7	57.7	54.9	57.7
25–44 years	70.8	69.9	71.6	76.8	76.0	77.2	65.0	64.1	66.1
45–64 years	62.5	62.0	63.3	67.8	68.1	70.8	57.5	56.3	56.2
45–54 years	65.6	65.1	67.1	70.1	70.3	73.8	61.2	60.1	60.7
55–64 years	57.6	57.2	57.3	64.2	64.6	65.8	51.6	50.6	49.4
65 years and over	42.0	42.1	43.4	50.9	50.0	52.7	35.5	36.2	36.6
65–74 years	45.8	46.9	48.6	55.2	52.6	56.7	38.2	42.2	42.0
75 years and over	37.6	36.3	36.6	45.1	46.6	46.7	32.6	29.7	30.2
Race [3]									
White only	65.8	64.6	66.0	71.0	69.7	71.8	61.0	59.9	60.7
Black or African American only	46.6	46.8	47.8	56.9	56.2	56.9	38.6	39.4	40.9
American Indian and Alaska Native only	51.5	54.2	53.9	62.8	62.6	66.1	38.6	46.9	45.2
Asian only	44.7	43.0	45.8	59.7	55.9	60.1	30.1	29.3	31.6
Native Hawaiian and Other Pacific Islander	*	*	. . .	*	*	. . .	*	*	. . .
Two or more races	68.5	61.6	. . .	69.9	70.5	. . .	67.1	52.7	. . .
Hispanic Origin and Race [3]									
Hispanic or Latino	49.8	52.1	53.4	61.2	63.7	64.6	39.0	41.2	42.1
Mexican	49.8	50.6	53.0	63.0	64.4	66.9	36.7	36.8	38.9
Not Hispanic or Latino	49.8	52.1	64.1	61.2	63.7	70.2	39.0	41.2	58.7
White only	67.7	66.0	67.5	72.0	70.4	72.7	63.7	61.9	62.9
Black or African American only	46.5	46.8	47.8	57.0	56.4	57.1	38.4	39.3	40.7
Geographic Region									
Northeast	68.4	68.0	68.7	73.8	73.2	74.4	63.7	63.7	63.8
Midwest	67.1	65.6	66.8	71.7	70.7	73.0	63.0	61.1	61.1
South	55.5	54.3	56.2	63.2	62.1	63.9	48.5	47.1	49.2
West	64.0	62.8	64.9	70.5	68.4	71.5	57.6	57.2	58.9
Location of Residence									
Within MSA [4]	64.1	63.0	64.7	70.4	69.0	71.0	58.3	57.5	59.1
Outside MSA [4]	56.9	56.0	57.4	62.8	62.6	65.7	51.6	50.3	49.5
PERCENT DISTRIBUTION OF CURRENT DRINKERS									
Level of Alcohol Consumption in Past Year [5]									
18 years and over, age-adjusted [2]									
All drinking levels	100.0	100.0	100.0	100.0	100.0	100.0	100.0	100.0	100.0
Light	68.8	70.6	69.6	59.1	60.4	59.5	79.6	82.0	81.0
Moderate	23.4	22.2	22.5	32.6	32.0	31.8	13.0	11.5	12.0
Heavier	7.9	7.1	7.9	8.3	7.7	8.7	7.3	6.5	7.0
18 Years and Over, Crude									
All drinking levels	100.0	100.0	100.0	100.0	100.0	100.0	100.0	100.0	100.0
Light	69.0	70.8	69.8	59.2	60.5	59.6	79.9	82.3	81.4
Moderate	23.2	22.1	22.3	32.4	31.8	31.7	12.9	11.3	11.7
Heavier	7.8	7.1	7.9	8.4	7.7	8.8	7.2	6.4	6.9

[1]Drinking status categories are based on self-reported responses to questions about alcohol consumption. Lifetime abstainers had fewer than 12 drinks in their lifetime. Former drinkers had at least 12 drinks in their lifetime and none in the past year. Former infrequent drinkers are former drinkers who had fewer than 12 drinks in any one year. Former regular drinkers are former drinkers who had at least 12 drinks in any one year. Current drinkers had 12 drinks in their lifetime and at least one drink in the past year. Current infrequent drinkers are current drinkers who had fewer than 12 drinks in the past year. Current regular drinkers are current drinkers who had at least 12 drinks in the past year.
[2]Estimates are age-adjusted to the year 2000 standard population using four age groups: 18–24 years, 25–44 years, 45–64 years, and 65 years and over.
[3]The race groups, White, Black, American Indian and Alaska Native (AI/AN), Asian, Native Hawaiian and Other Pacific Islander, and two or more races, include persons of Hispanic and non-Hispanic origin. Persons of Hispanic origin may be of any race.
[4]MSA = metropolitan statistical area.
[5]Level of alcohol consumption categories are based on self-reported responses to questions about average alcohol consumption and defined as follows: light drinkers: three drinks or fewer per week; moderate drinkers: more than three drinks and up to 14 drinks per week for men and more than three drinks and up to seven drinks per week for women; heavier drinkers: more than 14 drinks per week for men and more than seven drinks per week for women.
. . . = Not available.
* = Figure does not meet standards of reliability or precision.

Table III-34. Alcohol Consumption by Persons 18 Years of Age and Over, According to Selected Characteristics, Selected Years, 1997–2001—*Continued*

(Percent.)

Characteristic	Both sexes			Male			Female		
	2001	2000	1997	2001	2000	1997	2001	2000	1997
NUMBER OF DAYS IN PAST YEAR WITH 5 OR MORE DRINKS									
18 Years and Over, Crude									
All current drinkers	100.0	100.0	100.0	100.0	100.0	100.0	100.0	100.0	100.0
No days	67.6	68.2	65.9	57.2	56.9	54.7	79.0	80.6	78.6
At least 1 day	32.4	31.8	34.1	42.8	43.1	45.3	21.0	19.4	21.4
1–11 days	17.1	17.4	18.5	19.9	21.2	22.0	14.0	13.3	14.6
12 or more days	15.3	14.4	15.6	22.8	21.9	23.4	7.0	6.1	6.8
CURRENT DRINKERS WHO DRANK 5 OR MORE DRINKS ON AT LEAST ONE DAY									
All Persons									
18 years and over, age adjusted	30.6	30.3	32.4	40.5	41.2	43.3	19.9	18.5	20.2
18 years and over, crude	32.4	31.8	34.1	42.7	43.1	45.3	20.9	19.4	21.4
18–44 years	41.4	40.4	42.4	52.2	52.3	54.6	29.2	27.0	28.7
18–24 years	51.8	52.2	51.6	60.1	60.8	61.5	41.7	42.4	40.2
25–44 years	38.3	37.1	40.0	49.7	49.9	52.8	25.7	22.8	25.7
45–64 years	23.8	23.5	25.3	34.7	35.0	36.1	11.9	10.8	12.9
45–54 years	26.4	25.5	28.5	38.0	37.9	40.1	14.0	12.0	15.3
55–64 years	18.9	19.9	19.6	28.7	29.9	28.9	8.0	8.7	8.3
65 years and over	8.2	9.1	11.2	13.1	14.8	17.8	3.1	3.4	4.4
65–74 years	10.2	11.3	13.9	15.5	18.3	21.6	4.0	4.2	5.5
75 years and over	5.3	5.8	6.7	9.1	9.5	11.0	*	*	*2.5
Race [3]									
White only	31.8	31.0	33.3	42.0	41.8	44.4	20.7	19.2	20.9
Black or African American only	20.6	23.9	23.6	27.5	34.1	31.7	12.5	12.7	14.9
American Indian and Alaska Native only	34.6	45.0	54.5	35.9	47.0	70.5	*27.9	34.7	38.4
Asian only	23.2	20.3	25.5	28.9	25.4	30.7	*12.9	10.8	16.6
Native Hawaiian and Other Pacific Islander	*	*	. . .	*	*	. . .	*	*	. . .
Two or more races	41.6	44.0	. . .	59.9	53.1	. . .	24.3	31.5	. . .
Hispanic Origin and Race [3]									
Hispanic or Latino	32.2	31.9	36.8	41.9	43.0	46.3	17.9	16.0	22.3
Mexican	35.6	37.5	39.0	45.9	49.0	50.1	17.7	17.9	20.3
Not Hispanic or Latino	32.2	31.9	31.9	41.9	43.0	42.7	17.9	16.0	20.0
White only	32.0	31.2	33.2	42.2	42.1	44.5	21.3	19.7	21.0
Black or African American only	20.5	23.8	23.4	27.4	33.8	31.7	12.5	12.7	14.4
Geographic Area									
Northeast	30.0	28.9	31.3	41.4	39.7	43.1	18.8	18.1	18.9
Midwest	33.6	33.3	33.8	44.4	44.7	44.7	22.4	21.5	21.6
South	27.9	27.4	30.9	36.1	37.1	40.5	18.0	15.6	19.2
West	31.7	32.0	33.4	41.4	43.1	44.6	20.1	18.8	20.8
Location of Residence									
Within MSA [4]	30.1	30.0	31.6	39.9	40.4	42.4	19.1	18.5	19.8
Outside MSA [4]	32.8	31.3	34.8	42.2	43.1	45.7	22.5	18.5	21.2

[3]The race groups, White, Black, American Indian and Alaska Native (AI/AN), Asian, Native Hawaiian and Other Pacific Islander, and two or more races, include persons of Hispanic and non-Hispanic origin. Persons of Hispanic origin may be of any race.
[4]MSA = metropolitan statistical area.
. . . = Not available.
* = Figure does not meet standards of reliability or precision.

Table III-35. Cocaine-Related Emergency Department Episodes, According to Age, Sex, Race, and Hispanic Origin, Selected Years, 1990–2001

(Number, rate per 100,000 population.)

Age, sex, race, and Hispanic origin	2001	2000	1999	1998	1997	1996	1995	1991	1990
NUMBER									
All Races, Both Sexes [1]									
All ages [2]	193 034	174 881	168 751	172 011	161 083	152 420	135 711	101 189	80 355
6–17 years	3 514	4 402	3 299	4 362	3 642	2 595	2 051	2 210	1 877
18–25 years	28 666	25 753	25 264	24 507	25 218	22 060	21 110	21 766	19 614
26–34 years	53 693	51 007	54 058	59 008	57 143	58 729	54 881	46 137	35 639
35 years and over	106 810	93 357	85 869	83 730	74 600	68 717	57 341	30 582	23 054
Male									
Not Hispanic or Latino									
White									
All ages [2]	43 387	36 508	35 378	32 767	32 778	28 644	25 634	19 385	15 512
6–17 years	935	897	666	1 302	898	604	493	486	527
18–25 years	9 726	7 294	7 367	6 069	6 644	4 967	5 459	5 284	3 810
26–34 years	12 282	11 143	11 421	11 302	11 697	11 405	10 426	8 777	6 724
35 years and over	20 424	17 148	15 893	14 075	13 464	11 645	9 226	4 747	4 432
Black or African American									
All ages [2]	53 282	49 612	49 944	55 562	54 257	51 685	48 872	36 597	27 745
6–17 years	91	305	404	236	388	348	304	244	241
18–25 years	3 756	3 836	4 066	4 153	4 725	3 886	4 735	5 743	5 104
26–34 years	11 924	11 608	13 433	17 578	18 052	18 558	18 756	16 232	12 160
35 years and over	37 437	33 758	31 978	33 511	30 850	28 741	25 016	14 110	10 202
Hispanic or Latino [3]									
All ages [2]	18 293	16 774	15 111	14 844	11 540	12 575	7 886	6 571	4 821
6–17 years	485	612	899	725	402	431	181	201	144
18–25 years	4 108	4 268	4 027	3 871	3 467	3 725	1 892	1 831	1 774
26–34 years	6 080	5 510	4 582	4 694	3 575	4 342	2 901	2 723	1 758
35 years and over	7 615	6 375	5 540	5 536	4 077	4 054	2 907	1 801	1 125
Female									
Not Hispanic or Latino									
White									
All ages [2]	27 365	22 419	20 884	19 687	17 593	15 593	13 566	9 541	8 331
6–17 years	838	1 208	837	1 125	1 021	542	495	529	486
18–25 years	5 675	4 259	4 348	4 368	3 742	3 344	2 962	2 765	2 663
26–34 years	8 936	7 471	8 022	6 621	6 771	6 540	5 976	4 427	3 636
35 years and over	11 801	9 414	7 667	7 504	6 043	5 155	4 126	1 808	1 539
Black or African American									
All ages [2]	26 257	25 480	27 625	28 361	27 298	25 713	24 138	19 149	14 833
6–17 years	175	99	125	80	100	89	153	210	177
18–25 years	1 824	1 947	2 012	2 245	3 407	2 803	3 307	3 892	3 820
26–34 years	6 927	7 962	9 994	11 312	11 004	11 082	10 831	9 481	7 418
35 years and over	17 305	15 453	15 473	14 687	12 752	11 712	9 822	5 512	3 369
Hispanic or Latino [3]									
All ages [2]	6 491	6 598	5 224	6 238	5 063	5 042	3 515	2 356	1 719
6–17 years	550	901	146	625	675	250	128	183	64
18–25 years	1 112	1 699	1 167	1 505	1 287	1 296	901	616	634
26–34 years	2 409	1 967	2 091	2 278	1 698	2 116	1 280	1 044	663
35 years and over	2 419	2 029	1 811	1 821	1 402	1 378	1 203	513	357
RATE									
Both Sexes									
6 years and over, age-adjusted [4]	77.6	70.8	69.2	70.7	66.4	63.0	56.2	41.0	. . .
6 years and over, crude	76.1	70.7	69.4	71.5	67.7	64.8	58.3	45.2	. . .
6–11 years	*	*	*	*	*	*	*	*	. . .
12–17 years	14.5	18.8	14.0	18.8	16.0	11.5	9.3	10.6	. . .
18–25 years	85.5	88.9	89.5	88.2	91.8	80.1	76.2	76.9	. . .
26–34 years	176.4	154.6	161.9	173.1	164.5	166.7	153.7	120.5	. . .
35 years and over	76.2	67.7	63.7	63.2	57.4	54.0	46.0	26.5	. . .
Male									
6 years and over, age-adjusted [4]	104.5	95.7	93.4	96.4	91.2	87.1	77.5	56.2	. . .
6 years and over, crude [2]	101.8	94.8	93.0	96.7	92.2	88.7	79.9	61.6	. . .
6–11 years	*	*	*	*	*	*	*	*	. . .
12–17 years	14.3	16.7	17.4	20.7	15.3	12.9	10.5	9.5	. . .
18–25 years	112.8	118.5	120.5	115.2	116.1	101.9	98.1	102.7	. . .
26–34 years	220.8	193.8	195.5	219.7	211.3	212.7	196.2	152.8	. . .
35 years and over	108.0	97.1	92.0	92.2	85.6	81.2	69.2	40.7	. . .
Female									
6 years and over, age-adjusted [4]	51.4	46.3	46.0	46.1	42.6	40.0	35.5	26.5	. . .
6 years and over, crude [2]	50.4	46.4	46.4	46.7	43.5	41.3	37.0	29.1	. . .
6–11 years	*	*	*	*	*	*	*	*	. . .
12–17 years	14.5	20.9	10.2	16.7	16.6	10.0	7.8	11.0	. . .
18–25 years	55.6	58.2	57.7	61.7	66.4	57.5	54.1	53.0	. . .
26–34 years	130.0	112.9	127.3	125.0	117.0	118.9	108.6	86.1	. . .
35 years and over	46.6	40.1	37.9	36.6	31.3	29.2	24.8	13.6	. . .

[1] Includes other races and unknown race, Hispanic origin, and/or sex.
[2] Includes unknown age.
[3] Persons of Hispanic origin may be of any race.
[4] Age-adjusted to the year 2000 standard population using five age groups.
[5] Includes unknown age and sex.
. . . = Not available.
* = Figure does not meet standards of reliability or precision.

AMBULATORY CARE

TABLE III-36. PERCENT DISTRIBUTION OF HEALTH CARE VISITS TO DOCTOR'S OFFICES, EMERGENCY DEPARTMENTS, AND HOME VISITS WITHIN THE LAST 12 MONTHS, ACCORDING TO SELECTED CHARACTERISTICS, SELECTED YEARS, 1997–2001

The number of health care visits by patients during a 12-month period ending in 2001 differed widely, but the most frequent count was one to three visits. Only for persons 75 years old and above, the most frequent count increased to the range four to nine visits. At the same time, 16.5 percent of all persons did not have any appointments during the year, but for persons aged 75 and above, this figure dropped to 5.8 percent.

The frequency of health care visits was not substantially different between Blacks and Whites, but Asians and Hispanics tended to have a lower number of visits.

Table III-36. Percent Distribution of Health Care Visits to Doctor's Offices, Emergency Departments, and Home Visits within the Last 12 Months, According to Selected Characteristics, Selected Years, 1997–2001

(Percent distribution.)

Characteristic	Number of health care visits											
	None			1–3 visits			4–9 visits			10 or more visits		
	2001	1999	1997	2001	1999	1997	2001	1999	1997	2001	1999	1997
All persons [1, 2]	16.5	17.5	16.5	45.8	45.8	46.2	24.4	23.3	23.6	13.3	13.4	13.7
Age												
Under 18 years	11.6	12.4	11.8	54.6	54.4	54.1	26.1	25.0	25.2	7.6	8.2	8.9
Under 6 years	5.5	5.9	5.0	45.8	45.9	44.9	37.9	36.8	37.0	10.8	11.3	13.0
6–17 years	14.6	15.5	15.3	58.9	58.5	58.7	20.5	19.4	19.3	6.1	6.7	6.8
18–44 years	23.3	24.2	21.7	46.1	45.8	46.7	18.9	17.8	19.0	11.8	12.3	12.6
18–24 years	25.4	24.8	22.0	44.7	46.1	46.8	19.5	17.8	20.0	10.5	11.4	11.2
25–44 years	22.6	24.0	21.6	46.5	45.7	46.7	18.7	17.8	18.7	12.2	12.6	13.0
45–64 years	15.6	16.9	16.9	42.9	42.4	42.9	25.7	25.0	24.7	15.9	15.7	15.5
45–54 years	17.1	18.4	17.9	44.9	43.2	43.9	23.6	22.8	23.4	14.4	15.7	14.8
55–64 years	13.3	14.7	15.3	39.6	41.1	41.3	28.9	28.4	26.7	18.2	15.8	16.7
65 years and over	7.1	7.9	8.9	32.3	34.3	34.7	35.6	34.1	32.5	25.0	23.7	23.8
65–74 years	8.1	8.6	9.8	35.8	36.9	36.9	33.5	33.2	31.6	22.6	21.3	21.6
75 years and over	5.8	7.2	7.7	28.2	31.1	31.8	38.1	35.1	33.8	27.9	26.6	26.6
Sex [2]												
Male	21.3	23.1	21.3	46.5	45.5	47.1	21.6	20.6	20.6	10.7	10.8	11.0
Female	11.9	12.0	11.8	45.1	46.1	45.4	27.1	25.9	26.5	15.9	15.9	16.3
Race [2, 3]												
White only	15.9	16.9	16.0	45.7	45.7	46.1	24.8	23.8	23.9	13.5	13.6	14.0
Black or African American only	16.4	18.4	16.8	46.4	46.2	46.1	24.0	21.9	23.2	13.2	13.5	13.9
American Indian and Alaska Native only	*21.4	20.6	17.1	36.4	34.3	38.0	25.4	27.8	24.2	16.9	17.2	20.7
Asian only	20.8	23.1	22.8	48.3	47.3	49.1	22.3	19.4	19.7	8.6	10.2	8.3
Native Hawaiian and Other Pacific Islander only	*	*	. . .	*	*	. . .	*	*	. . .	*	*	. . .
Two or more races	18.0	15.2	. . .	41.2	40.8	. . .	23.5	22.2	. . .	17.3	21.8	. . .
Hispanic Origin and Race [2, 3]												
Hispanic or Latino	27.0	26.2	24.9	40.2	44.3	42.3	20.7	19.2	20.3	12.0	10.3	12.5
Mexican	31.4	30.2	28.9	39.2	43.0	40.8	19.6	18.2	18.5	9.8	8.7	11.8
Not Hispanic or Latino	15.0	16.2	15.4	46.5	46.0	46.7	25.0	23.9	24.0	13.5	13.9	13.9
White only	14.3	15.5	14.7	46.4	46.0	46.6	25.4	24.5	24.4	13.9	14.1	14.3
Black or African American only	16.4	18.4	16.9	46.4	46.2	46.1	24.0	21.9	23.1	13.1	13.5	13.8
Respondent–Assessed Health Status [2]												
Fair or poor	9.0	9.8	7.8	22.1	25.9	23.3	27.7	24.3	29.0	41.3	40.1	39.9
Good to excellent	17.3	18.1	17.2	48.0	47.7	48.4	24.3	23.2	23.3	10.5	11.0	11.1
Poverty Status [2, 4]												
Poor	21.7	21.5	20.3	37.2	39.2	37.1	23.4	21.3	22.7	17.7	18.0	19.9
Near poor	20.4	22.2	19.9	41.4	41.6	42.8	22.9	21.5	21.8	15.3	14.7	15.5
Nonpoor	14.0	14.9	14.0	47.4	47.0	48.0	25.8	25.0	25.0	12.8	13.1	13.0
Hispanic Origin, Race and Poverty Status [2, 3, 4]												
Hispanic or Latino												
Poor	34.3	31.2	30.6	32.7	38.2	33.8	18.1	18.7	20.0	14.9	11.8	15.6
Near poor	28.9	30.2	29.1	39.3	42.1	39.0	20.2	17.5	20.9	11.6	10.1	11.0
Nonpoor	19.9	21.0	18.7	44.6	46.8	48.6	24.7	21.9	20.3	10.8	10.2	12.3
Not Hispanic or Latino												
White only												
Poor	16.2	17.2	16.3	38.7	38.9	37.7	26.4	23.3	24.0	18.8	20.7	22.1
Near poor	17.1	19.8	17.1	41.3	40.8	43.7	24.1	23.3	22.3	17.6	16.1	17.0
Nonpoor	13.1	14.0	13.2	47.5	46.9	47.6	26.1	25.5	25.7	13.3	13.6	13.4
Not Hispanic or Latino												
Black or African American only												
Poor	17.3	18.0	17.8	38.1	39.9	37.4	24.0	23.1	23.3	20.5	19.0	21.5
Near poor	18.1	19.9	18.9	44.9	44.0	43.0	23.4	20.5	23.4	13.6	15.6	14.7
Nonpoor	14.6	16.3	15.6	47.4	48.2	50.5	26.6	23.7	23.3	11.4	11.8	10.6

[1]Includes all other races not shown separately, unknown poverty status, and unknown health insurance status.
[2]Estimates are age-adjusted to the year 2000 standard population using six age groups: Under 18 years, 18–44 years, 45–54 years, 55–64 years, 65–74 years, and 75 years and over.
[3]The race groups, White, Black, American Indian and Alaska Native (AI/AN), Asian, Native Hawaiian and Other Pacific Islander, and two or more races, include persons of Hispanic and non-Hispanic origin. Persons of Hispanic origin may be of any race.
[4]Poor persons are defined as below the poverty threshold. Near poor persons have incomes of 100 percent to less than 200 percent of poverty threshold. Nonpoor persons have incomes of 200 percent or greater than the poverty threshold.
[5]Health insurance categories are mutually exclusive. Persons who reported both Medicaid and private coverage are classified as having private coverage. Persons 65 years of age and over who reported Medicare HMO (health maintenance organization) and some other type of health insurance coverage are classified as having Medicare HMO.
[6]MSA = metropolitan statistical area.
. . . = Not available.
* = Figure does not meet standards of reliability or precision.

The effect of poverty on the number of health care visits is not always in one direction, but for the highest category, 10 visits or more, poor respondents showed a higher frequency of care than others. This could reflect more ailments and/or more insurance coverage under Medicaid.

Table III-36. Percent Distribution of Health Care Visits to Doctor's Offices, Emergency Departments, and Home Visits within the Last 12 Months, According to Selected Characteristics, Selected Years, 1997–2001—*Continued*

(Percent distribution.)

Characteristic	Number of health care visits											
	None			1–3 visits			4–9 visits			10 or more visits		
	2001	1999	1997	2001	1999	1997	2001	1999	1997	2001	1999	1997
Health Insurance Status [4,5]												
Under 65 years of age												
Insured	14.1	15.4	14.3	49.1	48.6	49.0	24.2	23.2	23.6	12.6	12.7	13.1
Private	14.4	15.9	14.7	50.6	49.9	50.6	24.0	22.9	23.1	11.0	11.3	11.6
Medicaid	10.4	10.7	9.8	35.4	35.6	35.5	26.3	26.0	26.5	27.8	27.6	28.2
Uninsured	37.5	37.3	33.7	41.4	41.6	42.8	14.6	13.2	15.3	6.5	7.9	8.2
65 years of age and over												
Medicare HMO	5.0	5.7	8.9	30.0	34.2	35.8	41.1	34.6	33.1	23.9	25.5	22.3
Private	5.5	6.7	7.3	34.6	34.9	35.9	35.2	34.9	34.0	24.8	23.5	22.7
Medicaid	6.1	*7.3	9.3	18.7	21.4	19.2	31.6	34.8	27.9	43.5	36.5	43.7
Medicare fee-for-service only	14.1	14.0	15.5	30.5	35.8	34.0	34.2	31.0	28.1	21.2	19.2	22.4
Poverty Status and Health Insurance Status [4,5]												
Under 65 years of age												
Poor												
Insured	14.0	14.6	13.7	41.1	41.4	38.8	24.9	23.2	24.5	20.0	20.7	22.9
Uninsured	43.2	39.8	36.7	34.6	39.3	38.8	15.3	12.6	14.9	6.9	8.3	9.5
Near poor												
Insured	15.8	17.0	15.6	44.7	44.9	45.5	22.7	22.6	22.3	16.8	15.5	16.6
Uninsured	35.3	38.0	34.5	40.9	40.2	41.8	16.6	13.4	15.6	7.2	8.4	8.1
Nonpoor												
Insured	13.6	14.7	13.4	49.8	49.1	50.3	25.0	24.2	24.2	11.6	12.0	12.1
Uninsured	31.9	32.9	29.1	46.0	43.7	45.4	15.5	14.6	17.0	6.6	8.8	8.4
Geographic Region												
Northeast	11.8	12.8	13.2	47.2	46.4	45.9	26.6	25.6	26.0	14.3	15.2	14.9
Midwest	14.9	16.2	15.9	47.2	46.7	47.7	24.0	23.8	22.8	13.9	13.3	13.6
South	17.7	18.9	17.2	45.2	45.5	46.1	24.4	22.5	23.3	12.8	13.2	13.5
West	20.5	20.9	19.1	44.1	44.8	44.8	22.8	21.9	22.8	12.7	12.4	13.3
Location of Residence												
Within MSA [6]	16.4	17.4	16.2	45.7	45.9	46.4	24.6	23.4	23.7	13.2	13.2	13.7
Outside MSA [6]	16.7	17.7	17.3	46.1	45.1	45.4	23.6	22.9	23.3	13.6	14.4	13.9

[4]Poor persons are defined as below the poverty threshold. Near poor persons have incomes of 100 percent to less than 200 percent of poverty threshold. Nonpoor persons have incomes of 200 percent or greater than the poverty threshold.
[5]Health insurance categories are mutually exclusive. Persons who reported both Medicaid and private coverage are classified as having private coverage. Persons 65 years of age and over who reported Medicare HMO (health maintenance organization) and some other type of health insurance coverage are classified as having Medicare HMO.
[6]MSA = metropolitan statistical area.
* = Figure does not meet standards of reliability or precision.

TABLE III-37. VACCINATIONS OF CHILDREN 19–35 MONTHS OF AGE FOR SELECTED DISEASES, ACCORDING TO RACE, HISPANIC ORIGIN, POVERTY STATUS, AND RESIDENCE IN METROPOLITAN STATISTICAL AREAS (MSA), 1995–2002

According to a 2002 sample survey, 78 percent of all children 19 to 35 months of age (including 71 percent of Black children) received a combined series of shots (4:3:1:3), which consisted of 4 or more doses of diphtheria and tetanus toxoids and pertussis vaccine (DTP) or some variants of these items, three or more doses of any polio virus vaccine, one or more doses of a measles-containing vaccine (MCV), and 3 or more doses of Haemophilus influenzae type b vaccine (Hib).

Children, who did not receive the combined series (4:3:1:3), frequently received vaccinations in other combinations. For example, 90 percent of all children received three doses or more of polio shots (including 87 percent of Black children), and 92 percent were immunized for measles, mumps and rubella (including 90 percent of Black children).

The table also shows differences between children raised above and below the poverty line. Children below the poverty line received vaccinations for the combined series by about 10 percentage points less, when compared with children above the poverty line.

Table III-37. Vaccinations of Children 19–35 Months of Age for Selected Diseases, According to Race, Hispanic Origin, Poverty Status, and Residence In Metropolitan Statistical Area (MSA), 1995–2002

(Percent.)

Vaccination and year	Race and Hispanic origin								Poverty status		Location of residence		
	Not Hispanic or Latino										Inside MSA		
	All	White	Black or African American	American Indian or Alaskan Native	Asian [1]	Native Hawaiian or Pacific Islander [1]	2 or more races	Hispanic or Latino [2]	Below poverty	At or above poverty	Central city	Remaining areas	Outside MSA
PERCENT													
Combined Series (4:3:1:3) [3]													
2002	78	80	71	*	83	*	74	76	72	79	75	80	77
2001	77	79	71	76	77	...	...	77	72	79	75	78	79
2000	76	79	71	69	75	...	...	73	71	78	73	78	79
1999	78	81	74	75	77	...	...	75	73	81	77	79	80
1998	79	82	73	78	79	...	...	75	74	82	77	81	81
1997	76	79	73	73	71	...	...	73	71	79	74	78	77
1996	76	79	74	82	78	...	...	71	69	79	74	78	77
1995	74	76	70	69	76	...	...	68	67	77	72	75	75
DTP/DT/DTaP (4 or more doses)													
2002	82	84	76	*	88	*	78	79	75	84	79	84	80
2001	82	84	76	77	84	...	...	83	77	84	81	83	82
2000	82	84	76	75	85	...	...	79	76	84	80	83	83
1999	83	86	79	80	87	...	...	80	79	85	82	84	83
1998	84	87	77	83	89	...	...	81	80	86	82	85	85
1997	82	84	77	80	80	...	...	78	76	84	80	83	81
1996	81	83	79	85	85	...	...	77	74	84	79	83	81
1995	78	80	74	71	84	...	...	75	71	81	77	79	78
Polio (3 or more doses)													
2002	90	91	87	*	92	95	87	90	88	91	89	91	90
2001	89	90	85	88	90	...	...	91	87	90	88	90	91
2000	90	91	87	90	93	...	...	88	87	90	88	90	91
1999	90	90	87	88	90	...	...	89	87	91	89	90	90
1998	91	92	88	85	93	...	...	89	90	92	89	91	93
1997	91	92	89	90	89	...	...	90	89	92	90	91	92
1996	91	92	90	90	90	...	...	89	88	92	89	92	92
1995	88	89	84	86	90	...	...	87	85	89	87	88	89
Measels, Mumps, and Rubella													
2002	92	93	90	84	95	94	89	91	90	92	90	93	90
2001	91	92	89	94	90	...	...	92	89	92	91	92	91
2000	91	92	88	87	90	...	...	90	89	91	90	91	91
1999	92	92	90	92	93	...	...	90	90	92	91	92	90
1998	92	93	89	91	92	...	...	91	90	93	92	92	93
1997	90	91	89	92	90	...	...	88	86	92	90	91	91
1996	91	91	90	89	93	...	...	88	87	92	90	91	91
1995	90	91	87	88	95	...	...	88	86	91	90	90	89
Hib (3 or more doses)													
2002	93	94	92	*	95	93	90	92	90	94	92	94	93
2001	93	94	90	91	92	...	...	93	90	94	91	94	93
2000	93	95	93	90	92	...	...	91	90	95	92	94	95
1999	94	95	92	91	90	...	...	92	91	95	92	95	93
1998	93	95	90	90	92	...	...	92	91	95	92	94	94
1997	93	94	91	86	89	...	...	90	90	94	91	93	94
1996	91	93	89	91	92	...	...	89	87	93	90	93	92
1995	91	93	88	93	90	...	...	89	88	93	91	92	92
Hepatitis B (3 or more doses)													
2002	90	91	88	*	94	94	84	90	88	90	89	91	90
2001	89	90	85	86	90	...	...	90	87	90	88	90	89
2000	90	91	89	91	91	...	...	88	87	91	89	90	92
1999	88	89	87	*	88	...	...	87	87	89	87	89	88
1998	87	88	84	82	89	...	...	86	85	88	85	88	87
1997	84	85	82	83	88	...	...	81	81	85	82	85	85
1996	82	82	82	79	85	...	...	81	78	83	81	83	81
1995	68	68	66	52	80	...	...	70	65	69	69	71	59
Varicella [4]													
2002	81	79	83	71	87	*	79	82	79	81	81	83	75
2001	76	75	75	69	82	...	...	80	74	77	78	78	68
2000	68	66	67	62	77	...	...	70	64	69	69	70	60
1999	58	56	58	*	64	...	...	61	55	58	59	61	47
1998	43	42	42	28	53	...	...	47	41	44	45	45	34
1997	26	28	21	20	36	...	...	22	17	29	26	29	17
PCV (3 or more doses) [5]													
2002	41	44	34	33	55	*	38	37	33	43	41	45	32

[1]Prior to the data year 2002 the category "Asian" included Native Hawaiian and Other Pacific Islander.
[2]Persons of Hispanic origin may be of any race.
[3]The 4:3:1:3 combined series consists of four or more doses of diphtheria and tetanus toxoids and pertussis vaccine (DTP), diphtheria and tetanus toxoids (DT), or diphtheria and tetanus toxoids and acellular pertussis vaccine (DTaP), three or more doses of any poliovirus vaccine, one or more doses of a measles-containing vaccine (MCV), and three or more doses of Haemophilus influenzae type b vaccine (Hib).
[4]Recommended in 1996. Data collection for varicella began in July 1996.
[5]Pneumococcal conjugate vaccine. Recommended in 2000. Data collection for PCV began in July 2001.
. . . = Not available.
* = Figure does not meet standards of reliability or precision.

Table III-37. Vaccinations of Children 19–35 Months of Age for Selected Diseases, According to Race, Hispanic Origin, Poverty Status and Residence In Metropolitan Statistical Area (MSA), 1995–2002—*Continued*

(Percent.)

Vaccination and year	Not Hispanic or Latino				Hispanic or Latino [1]	
	White		Black or African American			
	Below poverty	At or below poverty	Below poverty	At or above poverty	Below poverty	At or above poverty
Combined Series (4:3:1:3) [2]						
2002	69	78	70	73	63	72
2001	68	80	69	79	68	73
2000	72	79	71	77	70	77
1999	77	83	72	74	73	79
1998	76	82	72	77	73	78
1997	73	80	69	72	70	74
1996	71	80	69	74	73	79
1995	72	81	68	72	75	76

Note: Final estimates from the National Immunization Survey include an adjustment for children with missing immunization provider data. Poverty status is based on family income and family size using Census Bureau poverty thresholds.

[1] Persons of Hispanic origin may be of any race.
[2] The 4:3:1:3 combined series consists of four or more doses of diphtheria and tetanus toxoids and pertussis vaccine (DTP), diphtheria and tetanus toxoids (DT), or diphtheria and tetanus toxoids and acellular pertussis vaccine (DTaP), three or more doses of any poliovirus vaccine, one or more doses of a measles-containing vaccine (MCV), and three or more doses of Haemophilus influenzae type b vaccine (Hib).

Figure III-12. Percent of Children 19–35 Months Vaccinated for Selected Diseases, According to Race and Hispanic Origin, 2002

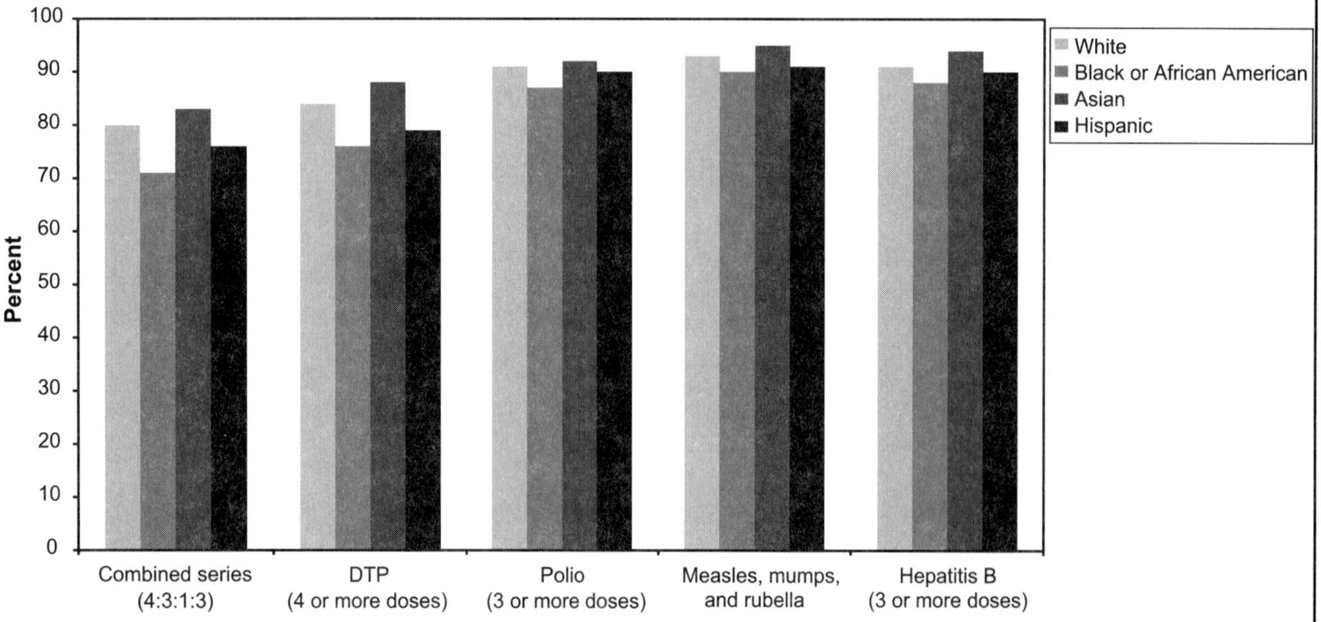

TABLE III-38. VACCINATION COVERAGE AMONG CHILDREN 19–35 MONTHS OF AGE, ACCORDING TO GEOGRAPHIC DIVISION, STATE, AND SELECTED URBAN AREAS, 1995–2002

This table reports the frequency of vaccinations for the combined series (4:3:1:3). For the United States in 2002, this frequency was 78 percent, up from 74 percent in 1995. Among the states and regions, New England had the highest 2002 coverage in a range between 83 percent and 89 percent for the component states. States with low coverage, on the other hand, were Oklahoma and New Mexico (both 67 percent), Louisiana (69 percent), and Texas and Montana (both 71 percent).

Table III-38. Vaccination Coverage Among Children 19–35 Months of Age, According to Geographic Division, State, and Selected Urban Areas, 1995–2002

(Percent.)

Geographic division and state	2002	2001	2000	1999	1998	1997	1996	1995
PERCENT OF CHILDREN 19–35 MONTHS OF AGE WITH 4:3:1:3 SERIES[1]								
United States	78	77	76	78	79	76	76	74
New England								
Connecticut	86	84	85	86	90	86	88	86
Maine	83	82	83	83	86	87	86	88
Massachusetts	89	81	85	85	87	88	87	81
New Hampshire	87	84	83	85	82	85	83	89
Rhode Island	86	84	82	87	86	82	84	83
Vermont	87	88	83	91	86	87	87	87
Middle Atlantic								
New Jersey	80	76	76	81	82	76	75	70
New York	81	81	75	81	85	75	80	74
Pennsylvania	77	82	78	86	83	79	79	77
East North Central								
Illinois	80	76	75	77	78	74	75	78
Indiana	78	74	76	74	78	72	70	74
Michigan	84	74	75	74	78	75	75	68
Ohio	77	75	72	78	78	72	78	71
Wisconsin	82	83	80	85	78	81	77	74
West North Central								
Iowa	80	79	83	83	82	77	81	83
Kansas	73	76	76	79	82	84	72	70
Minnesota	79	79	86	85	82	78	84	75
Missouri	77	78	78	75	85	79	75	75
Nebraska	79	80	79	82	76	74	78	71
North Dakota	79	83	81	80	79	80	80	79
South Dakota	81	79	78	82	74	77	81	79
South Atlantic								
Delaware	81	79	75	78	79	80	81	68
District of Columbia	72	74	71	78	71	71	76	69
Florida	77	77	74	80	79	74	79	74
Georgia	82	80	81	82	80	78	81	77
Maryland	81	78	78	79	77	81	79	77
North Carolina	87	85	87	82	83	80	78	80
South Carolina	80	81	80	81	88	81	85	78
Virginia	77	78	74	80	80	72	76	69
West Virginia	79	81	76	81	82	81	71	71
East South Central								
Alabama	80	83	81	78	82	87	74	73
Kentucky	74	79	81	88	82	78	76	81
Mississippi	78	84	81	82	84	80	81	79
Tennessee	80	84	81	78	82	79	79	74
West South Central								
Arkansas	74	74	72	77	73	80	70	73
Louisiana	69	69	75	77	78	77	79	77
Oklahoma	67	76	71	73	75	70	72	74
Texas	71	74	69	72	74	74	71	71
Mountain								
Arizona	70	73	72	72	76	71	70	69
Colorado	64	75	74	76	76	74	80	75
Idaho	73	74	74	69	76	71	65	66
Montana	71	82	77	83	82	75	75	71
Nevada	78	72	74	73	76	70	67	67
New Mexico	67	71	68	73	71	73	78	74
Utah	79	74	77	80	76	69	65	65
Wyoming	77	81	79	83	80	75	77	71
Pacific								
Alaska	78	74	77	80	81	75	72	74
California	76	75	75	75	76	74	74	70
Hawaii	81	73	75	82	79	77	80	75
Oregon	75	73	79	72	76	72	70	71
Washington	73	76	77	75	81	79	78	76

[1]The 4:3:1:3 combined series consists of four or more doses of diphtheria and tetanus toxoids and pertussis vaccine (DTP), diphtheria and tetanus toxoids (DT), or diphtheria and tetanus toxoids and acellular pertussis vaccine (DTaP), three or more doses of any poliovirus vaccine, one or more doses of a measles-containing vaccine (MCV), and three or more doses of Haemophilus influenzae type b vaccine (Hib).

Among urban areas, Franklin County (Columbus), Ohio, and Santa Clara County in California had a high, 84 percent rate, of vaccination coverage. The immunization rate was low in Newark, New Jersey (60 percent), in New Orleans, Louisiana (63 percent), and in Houston, Texas (64 percent).

Table III-38. Vaccination Coverage Among Children 19–35 Months of Age, According to Geographic Division, State, and Selected Urban Areas, 1995–2002—*Continued*

(Percent.)

Geographic division and state	2002	2001	2000	1999	1998	1997	1996	1995
New England								
Boston, Massachusetts	80	85	79	84	89	86	85	85
Middle Atlantic								
New York City, New York	81	76	68	78	81	72	78	72
Newark, New Jersey	60	64	63	67	64	68	64	67
Philadelphia, Pennsylvania	74	74	74	81	80	81	74	67
East North Central								
Chicago, Illinois	72	69	65	71	64	66	72	70
Cuyahoga County (Cleveland), Ohio	74	73	73	74	75	70	79	72
Detroit, Michigan	66	63	59	66	70	60	60	54
Franklin County (Columbus), Ohio	84	78	77	78	78	73	80	75
Marion County (Indianapolis), Indiana	75	72	69	79	78	80	72	77
Milwaukee County (Milwaukee), Wisconsin	70	70	69	74	73	72	70	69
South Atlantic								
Baltimore, Maryland	75	72	70	72	81	84	80	*
Dade County (Miami), Florida	73	78	78	84	75	75	79	78
District of Columbia	72	74	71	78	71	71	76	67
Duval County (Jacksonville), Florida	77	76	79	78	79	69	76	69
Fulton/DeKalb Counties (Atlanta), Georgia	79	75	80	83	71	74	76	*
East South Central								
Davidson County (Nashville), Tennessee	80	82	73	73	80	76	80	72
Shelby County (Memphis), Tennessee	73	74	77	75	71	70	70	69
Jefferson County (Birmingham), Alabama	82	87	79	85	85	83	76	86
West South Central								
Bexar County (San Antonio), Texas	76	73	68	70	79	79	74	76
Dallas County (Dallas), Texas	76	67	67	72	71	75	68	70
El Paso County (El Paso), Texas	77	69	70	73	78	63	61	72
Houston, Texas	64	69	65	63	61	62	62	64
Orleans Parish (New Orleans), Louisiana	63	68	70	72	79	69	72	78
Mountain								
Maricopa County (Phoenix), Arizona	73	72	71	71	77	70	72	67
Pacific								
King County (Seattle), Washington	77	72	75	77	86	81	82	84
Los Angeles County (Los Angeles), California	77	73	77	76	76	72	75	68
San Diego County (San Diego), California	78	80	76	75	77	76	74	72
Santa Clara County (Santa Clara), California	84	77	76	82	84	69	80	76

Note: Urban areas were chosen because they were high risk for under vaccination. Final estimates from the National Immunization Survey include an adjustment for children with missing immunization provider data.

* = Figure does not meet standards of reliability or precision.

TABLE III-39. NO HEALTH CARE VISITS TO AN OFFICE OR CLINIC WITHIN THE PAST 12 MONTHS AMONG CHILDREN UNDER 18 YEARS OF AGE, ACCORDING TO SELECTED CHARACTERISTICS, ANNUAL AVERAGE, 1997–1998 AND 2000–2001

Household interviews on a sample basis indicated that in 2000 and 2001, 12.7 percent of all children under 18 did not receive medical attention during the past 12 months. This percentage was roughly the same in the 1997–1998 interview period. When medical visits of children under 6 years of age were compared with those between 6 and 17 years of age, a much higher rate of visits was found for the younger age group. No visits occurred for 6.6 percent of the younger group, but 15.8 percent for older children.

The percentage of no visits was high for Hispanic children as well as for poor households. For poor Hispanics it was especially high. Insurance status was associated with considerable differences in the utilization of medical care. On average 30.6 percent of uninsured children had no visits in the 2000–2001 survey period, as compared with 10.5 percent of children with insurance. When lack of insurance was combined with poverty status, the percentages of no visits increased further: in the under 6-year age group, 23.8 percent had no visits, and in the 6-to 17-year age bracket 44.4 percent of the children had no medical visits.

Table III-39. No Health Care Visits to an Office or Clinic within the Past 12 Months Among Children Under 18 Years of Age, According to Selected Characteristics, Annual Average, 1997–1998 and 2000–2001

(Percent.)

Characteristic	Under 18 years of age		Under 6 years of age		6–17 years of age	
	2000–2001	1997–1998	2000–2001	1997–1998	2000–2001	1997–1998
PERCENT OF CHILDREN WITHOUT A HEALTH CARE VISIT [1]						
All children [2]	12.7	12.8	6.6	5.7	15.8	16.3
Race [3]						
White only	12.0	12.2	6.6	5.5	14.6	15.5
Black or African American only	15.1	14.3	6.3	6.5	19.0	18.1
American Indian and Alaska Native only	22.4	13.8	*	*	*25.0	*17.6
Asian only	15.2	16.3	*8.7	*5.6	18.9	22.1
Native Hawaiian and Other Pacific Islander only	*	. . .	*	. . .	*	. . .
Two or more races	8.5	. . .	*	. . .	13.6	. . .
Hispanic Origin and Race [3]						
Hispanic or Latino	19.7	19.3	10.0	9.7	25.5	25.3
Not Hispanic or Latino	11.4	11.6	5.8	4.8	14.0	14.9
White only	10.3	10.7	5.5	4.3	12.6	13.7
Black or African American only	14.8	14.5	6.5	6.5	18.5	18.3
Poverty Status [4]						
Poor	18.5	17.7	10.4	8.1	22.9	23.8
Near poor	17.1	16.1	8.8	7.0	21.5	20.8
Nonpoor	9.5	9.7	4.0	3.9	12.1	12.5
Hispanic Origin, Race and Poverty Status [3, 4]						
Hispanic or Latino						
Poor	23.2	23.4	11.5	11.8	30.4	31.2
Near poor	21.8	20.1	12.9	9.3	27.4	27.3
Nonpoor	13.6	12.9	*4.8	7.1	18.4	16.2
Not Hispanic or Latino						
White only						
Poor	16.1	13.7	*11.0	*5.4	18.9	19.3
Near poor	14.1	14.0	6.9	6.0	17.9	18.1
Nonpoor	8.6	9.1	3.8	3.4	10.7	11.6
Black or African American only						
Poor	16.1	16.1	*8.2	*7.5	19.8	21.0
Near poor	18.1	16.6	*7.8	*7.0	23.0	21.1
Nonpoor	12.4	12.2	*4.1	*4.0	15.9	15.6
Health Insurance Status [5]						
Insured	10.5	10.4	5.2	4.5	13.2	13.4
Private	10.3	10.4	4.7	4.3	12.9	13.1
Medicaid	11.0	10.1	6.3	5.0	14.2	14.4
Uninsured	30.6	28.8	18.8	14.6	35.7	34.9
Poverty Status and Health Insurance Status [4]						
Poor						
Insured	13.6	13.2	7.5	5.7	17.2	18.4
Uninsured	37.8	34.5	23.8	19.6	44.4	40.6
Near poor						
Insured	13.7	12.5	5.9	4.4	18.0	16.9
Uninsured	31.5	27.9	22.9	16.9	35.5	32.7
Nonpoor						
Insured	8.9	8.9	3.6	3.6	11.4	11.4
Uninsured	21.9	23.0	*13.2	*8.8	25.2	28.6
Geographic Region						
Northeast	6.5	7.0	4.6	3.1	7.3	8.9
Midwest	10.5	12.2	5.3	5.9	12.9	15.3
South	15.1	14.3	7.4	5.6	18.9	18.5
West	16.8	16.3	8.2	7.9	21.3	20.7
Location of Residence						
Within MSA [6]	12.3	12.3	6.5	5.4	15.2	15.9
Outside MSA [6]	14.6	14.6	6.8	6.9	18.1	17.9

[1]Respondents were asked how many times a doctor or other health care professional was seen in the past 12 months at a doctor's office, clinic, or some other place. Excluded are visits to emergency rooms, hospitalizations, home visits, and telephone calls. Beginning in 2000, dental visits were also excluded.
[2]Includes all other races not shown separately, unknown poverty status, and unknown health insurance status.
[3]The race groups, White, Black, American Indian and Alaska Native (AI/AN), Asian, Native Hawaiian and Other Pacific Islander, and two or more races, include persons of Hispanic and non-Hispanic origin. Persons of Hispanic origin may be of any race.
[4]Poor persons are defined as below the poverty threshold. Near poor persons have incomes of 100 percent to less than 200 percent of poverty threshold. Nonpoor persons have incomes of 200 percent or greater than the poverty threshold.
[5]Health insurance categories are mutually exclusive. Persons who reported both Medicaid and private coverage are classified as having private coverage. Persons 65 years of age and over who reported Medicare HMO (health maintenance organization) and some other type of health insurance coverage are classified as having Medicare HMO.
[6]MSA = metropolitan statistical area.
* = Figure does not meet standards of reliability or precision.
. . . = Not available.

Figure III-13. Percent of Children Under the Age of 18 with No Health Care Visits to an Office or Clinic within the Past 12 Months, According to Poverty Status, 2000–2001

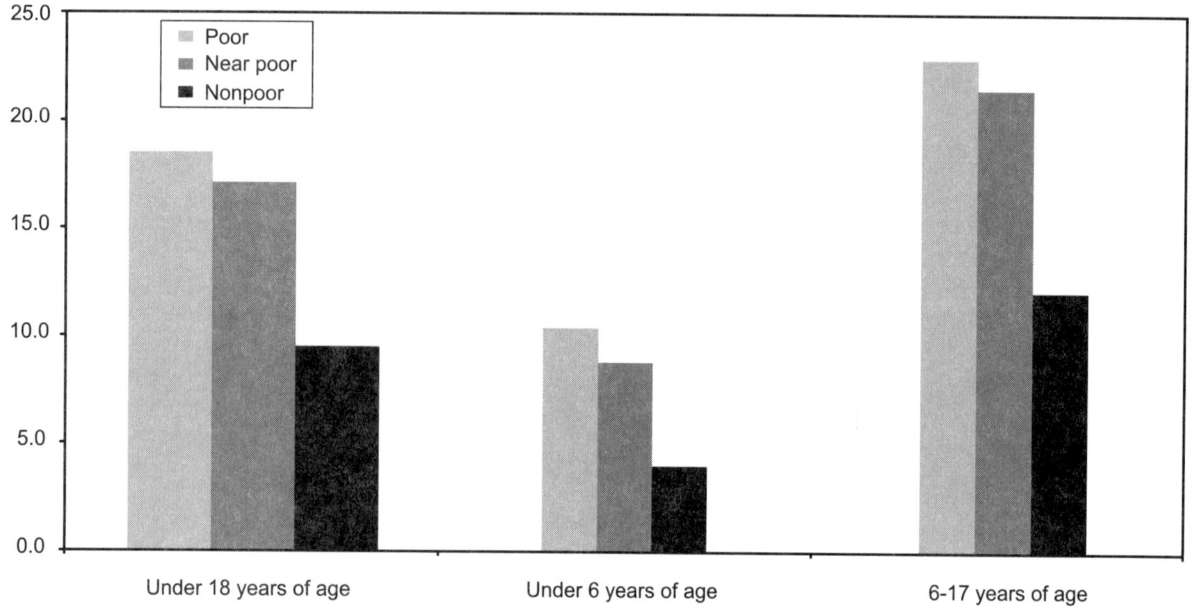

TABLE III-40. NO USUAL SOURCE OF HEALTH CARE AMONG CHILDREN UNDER 18 YEARS OF AGE, ACCORDING TO SELECTED CHARACTERISTICS, ANNUAL AVERAGE, SELECTED YEARS, 1993–2001

In this household interview survey the respondent was asked whether there is a place where (the child) usually goes when sick or when you need advice about his (or her) health. Persons who reported the emergency department as their usual source of care were defined as having no usual source of care. The question was phrased slightly differently prior to 1997.

Among children under 6 years of age, 4.4 percent were reported to have no usual source of care, and for children from 6 to 17 years, 7.3 percent were in this category. These figures were smaller than in 1993. Hispanics had a greater than average frequency of no usual source of health care, and the frequency for Blacks also was greater but only for the older children, age 6 to 17.

Medical insurance had a big impact: for insured children under 6 years, only 2.5 percent lacked a usual source of care, while for those from 6 to 17 years of age, only 4.0 percent were lacking. For uninsured children these percentages were 20.7 and 32.5, respectively. Respondents that were both uninsured and poor reported no usual place of care for 29.0 percent of children under 6 years, and 42.9 percent of youngsters in the 7–17 year bracket.

Table III-40. No Usual Source of Health Care Among Children Under 18 Years of Age, According to Selected Characteristics, Annual Average, Selected Years, 1993–2001

(Percent.)

Characteristic	Percent of children without a usual source of health care [1]								
	Under 18 years of age			Under 6 years of age			6–17 years of age		
	2000–2001 [2]	1997–1998 [2]	1993–1994	2000–2001 [2]	1997–1998 [2]	1993–1994	2000–2001 [2]	1997–1998 [2]	1993–1994
PERCENT OF CHILDREN WITHOUT A USUAL SOURCE OF HEALTH CARE									
All children [3]	6.4	6.7	7.7	4.4	4.5	5.2	7.3	7.8	9.0
Race [4]									
White only	5.6	5.8	7.0	4.2	4.1	4.7	6.3	6.7	8.3
Black or African American only	7.4	8.9	10.3	4.0	5.6	7.6	9.0	10.4	11.9
American Indian and Alaska Native only	*6.7	*10.8	*9.3	*	*	*	*	*	*8.7
Asian only	9.9	10.7	9.7	*7.5	*	*3.4	11.3	14.4	13.5
Native Hawaiian and Other Pacific Islander only	*	. . .	. . .	*	. . .	. . .	*	. . .	. . .
Two or more races	6.0	. . .	. . .	*	. . .	. . .	*8.0	. . .	. . .
Hispanic Origin and Race [4]									
Hispanic or Latino	14.1	13.2	14.3	9.3	7.6	9.3	17.0	16.7	17.7
Not Hispanic or Latino	4.8	5.6	6.7	3.2	3.8	4.4	5.6	6.4	7.8
White only	3.9	4.5	5.7	2.8	3.4	3.7	4.4	5.0	6.7
Black or African American only	7.4	8.8	10.2	4.0	5.4	7.7	9.0	10.4	11.6
Poverty Status [5]									
Poor	12.2	12.4	13.9	8.2	8.2	9.4	14.4	15.0	16.8
Near poor	9.4	10.1	9.8	7.1	6.5	6.7	10.7	12.0	11.6
Nonpoor	3.3	3.5	3.7	2.2	2.0	1.8	3.8	4.2	4.6
Hispanic Origin, Race, and Poverty Status [4,5]									
Hispanic or Latino									
Poor	19.2	17.0	19.6	12.8	8.4	12.7	23.3	22.7	24.8
Near poor	15.6	16.0	15.3	11.4	10.2	9.9	18.1	19.9	18.9
Nonpoor	6.2	5.7	5.0	*4.2	*3.1	*2.7	7.3	7.3	6.5
Not Hispanic or Latino									
White only									
Poor	9.6	11.4	10.2	*	10.7	6.5	11.0	12.0	12.7
Near poor	6.5	6.6	8.7	*5.4	4.5	6.3	7.1	7.7	10.1
Nonpoor	2.6	2.8	3.4	1.7	1.6	1.6	3.0	3.3	4.2
Black or African American only									
Poor	8.7	9.1	13.7	*4.7	*5.4	10.9	10.5	11.2	15.5
Near poor	8.3	12.5	9.1	*5.0	*7.2	*6.0	9.8	15.0	10.8
Nonpoor	4.1	6.4	4.6	*	*4.0	*	4.8	7.4	5.8
Health Insurance Status [6]									
Insured	3.5	3.6	5.0	2.5	2.6	3.3	4.0	4.2	5.9
Private	2.9	3.1	3.8	1.9	2.0	1.9	3.4	3.6	4.6
Medicaid	5.1	5.5	8.9	3.8	4.1	6.4	6.0	6.7	11.3
Uninsured	28.9	27.8	23.5	20.7	19.0	18.0	32.5	31.6	26.0
Poverty Status and Health Insurance Status [5]									
Poor									
Insured	5.5	6.2	9.1	3.5	4.8	6.0	6.7	7.2	11.5
Uninsured	38.4	35.5	29.4	29.0	26.5	25.0	42.9	39.2	31.5
Near poor									
Insured	5.3	5.1	6.0	4.3	3.4	4.0	5.9	6.1	7.2
Uninsured	27.0	27.1	22.9	20.4	18.4	18.0	30.1	30.9	25.3
Nonpoor									
Insured	2.5	2.5	2.9	1.7	1.6	1.5	2.8	3.0	3.6
Uninsured	19.7	19.7	14.5	*13.2	*10.6	6.4	22.2	23.4	18.1
Geographic Region									
Northeast	2.3	3.1	4.1	2.0	*2.5	2.9	2.5	3.5	4.8
Midwest	4.8	4.6	5.2	3.6	4.0	4.1	5.4	4.9	5.9
South	7.6	8.4	10.9	4.9	5.3	7.3	9.0	9.9	12.7
West	9.4	9.8	8.6	6.2	5.5	5.3	11.1	12.0	10.6
Location of Residence									
Within MSA [7]	6.4	6.8	7.7	4.6	4.4	5.0	7.3	8.0	9.2
Outside MSA [7]	6.3	6.4	7.8	3.5	4.7	6.0	7.6	7.2	8.7

[1]Persons who report the emergency department as the place of their usual source of care are defined as having no usual source of care.
[2]Data starting in 1997 are not strictly comparable with data for earlier years due to the 1997 questionnaire redesign.
[3]Includes all other races not shown separately, unknown poverty status, and unknown health insurance status.
[4]The race groups, White, Black, American Indian and Alaska Native (AI/AN), Asian, Native Hawaiian and Other Pacific Islander, and two or more races, include persons of Hispanic and non-Hispanic origin. Persons of Hispanic origin may be of any race.
[5]Poor persons are defined as below the poverty threshold. Near poor persons have incomes of 100 percent to less than 200 percent of poverty threshold. Nonpoor persons have incomes of 200 percent or greater than the poverty threshold.
[6]Health insurance categories are mutually exclusive. Persons who reported both Medicaid and private coverage are classified as having private coverage. Persons 65 years of age and over who reported Medicare HMO (health maintenance organization) and some other type of health insurance coverage are classified as having Medicare HMO.
[7]MSA = metropolitan statistical area.
* = Figure does not meet standards of reliability or precision.
. . . = Not available.

TABLE III-41. EMERGENCY DEPARTMENT VISITS WITHIN THE PAST 12 MONTHS AMONG CHILDREN UNDER 18 YEARS OF AGE, ACCORDING TO SELECTED CHARACTERISTICS, SELECTED YEARS, 1997–2001

In 2001, 20.7 percent of children under 18 years of age had one or more visits to an emergency department of a hospital. For children under 6 years, 24.9 percent paid at least one visit to emergency, while those aged 6 to 17 years had a visitation frequency of 18.6 percent. Blacks made visits somewhat more frequently than Whites, especially for the younger age group.

Poor households made visits to emergency departments

Table III-41. Emergency Department Visits within the Past 12 Months Among Children Under 18 Years of Age, According to Selected Characteristics, Selected Years, 1997–2001

(Percent.)

Characteristic	Under 18 years of age			Under 6 years of age			6–17 years of age		
	2001	1999	1997	2001	1999	1997	2001	1999	1997
PERCENT OF CHILDREN WITH 1 OR MORE DEPARTMENT VISITS									
All children [1]	20.7	17.9	19.9	24.9	23.3	24.3	18.6	15.3	17.7
Race [2]									
White only	20.2	17.1	19.4	24.2	21.9	22.6	18.2	14.8	17.8
Black or African American only	22.6	22.5	24.0	29.2	32.3	33.1	19.7	18.2	19.4
American Indian and Alaska Native only	27.0	33.3	*24.1	*	*29.5	*24.3	*26.5	*36.2	*24.0
Asian only	11.4	9.4	12.6	*8.6	*13.4	20.8	*13.0	*7.4	8.6
Native Hawaiian and Other Pacific Islander only	*	*	...	*	*	...	*	*	...
Two or more races	31.1	23.3	...	33.7	28.7	...	29.2	*19.7	...
Hispanic Origin and Race [2]									
Hispanic or Latino	19.4	15.9	21.1	26.1	21.4	25.7	15.4	12.6	18.1
Not Hispanic or Latino	20.9	18.3	19.7	24.7	23.8	24.0	19.2	15.7	17.6
White only	20.6	17.4	19.2	24.3	22.1	22.2	18.9	15.3	17.7
Black or African American only	22.6	22.5	23.6	28.8	32.5	32.7	19.8	18.2	19.2
Poverty Status [3]									
Poor	25.2	24.4	25.4	28.1	31.6	29.9	23.5	20.6	22.5
Near poor	22.1	22.2	22.6	28.5	30.4	28.8	18.6	17.8	19.4
Nonpoor	19.5	15.4	17.4	23.5	19.0	21.0	17.7	13.8	15.8
Hispanic Origin and Race and Poverty Status [2, 3]									
Hispanic or Latino									
Poor	21.7	16.4	22.0	26.3	21.0	24.8	18.7	13.0	20.1
Near poor	17.4	15.2	20.8	24.8	21.7	28.9	13.1	11.6	15.6
Nonpoor	21.0	17.2	20.3	28.4	23.0	22.7	16.8	14.3	18.9
Not Hispanic or Latino									
White only									
Poor	28.2	26.3	26.3	30.0	34.9	28.0	27.2	22.4	25.1
Near poor	23.0	24.5	23.0	28.7	33.0	26.5	19.8	20.1	21.2
Nonpoor	19.4	15.1	17.4	23.1	18.1	20.6	17.7	13.8	15.9
Black or African American only									
Poor	26.2	29.8	29.8	31.9	42.5	40.9	23.3	23.4	22.8
Near poor	26.0	23.5	23.6	35.8	33.7	33.6	21.2	18.7	19.1
Nonpoor	20.2	18.8	17.8	27.0	27.1	23.8	17.3	15.6	15.5
Health Insurance Status [4]									
Insured	21.0	18.1	19.8	25.1	23.1	24.4	18.9	15.7	17.5
Private	18.6	15.4	17.5	22.0	18.9	20.9	17.1	13.9	15.9
Medicaid	28.5	28.8	28.2	32.4	35.2	33.0	25.9	24.2	24.1
Uninsured	17.4	16.4	20.2	22.1	25.5	23.0	15.5	12.7	18.9
Poverty Status and Health Insurance Status [3]									
Poor									
Insured	27.6	26.9	26.6	30.3	32.8	31.4	26.0	23.4	23.2
Uninsured	15.0	15.8	20.9	*16.7	25.8	20.9	14.2	11.9	20.9
Near poor									
Insured	22.6	23.3	22.7	28.8	31.2	29.2	19.0	18.7	19.2
Uninsured	19.3	18.3	22.2	*25.8	26.6	27.3	16.7	14.8	20.1
Nonpoor									
Insured	19.3	15.4	17.3	23.3	18.7	20.8	17.5	13.9	15.7
Uninsured	19.8	16.1	18.8	24.3	25.8	23.7	18.3	12.0	16.7
Geographic Region									
Northeast	20.1	17.1	18.5	23.7	20.3	20.7	18.6	15.5	17.4
Midwest	21.2	18.4	19.5	25.6	24.1	26.0	19.2	15.8	16.4
South	22.2	19.2	21.8	27.0	25.7	25.6	19.8	16.1	19.9
West	18.1	15.9	18.5	22.1	21.4	23.5	15.9	13.1	15.9
Location of Residence									
Within MSA [5]	19.8	16.7	19.7	24.1	22.0	23.9	17.6	14.0	17.4
Outside MSA [5]	24.3	22.4	20.8	28.5	29.1	26.2	22.4	19.7	18.6
PERCENT OF CHILDREN WITH 2 OR MORE DEPARTMENT VISITS									
All children [1]	6.8	5.5	7.1	9.5	8.7	9.6	5.5	4.0	5.8
Race [2]									
White only	6.4	4.7	6.6	8.9	7.3	8.4	5.2	3.4	5.7
Black or African American only	8.5	9.1	9.6	14.3	15.8	14.9	5.9	6.1	6.9
American Indian and Alaska Native only	*13.5	*	*	*	*	*	*	*	*
Asian only	*4.9	*	*5.7	*	*	*12.9	*	*	*
Native Hawaiian and Other Pacific Islander only	*	*	...	*	*	...	*	*	...
Two or more races	8.8	10.5	...	*9.3	*15.7	...	*8.4	*	...

[1]Includes all other races not shown separately, unknown poverty status, and unknown health insurance status.

[2]The race groups, White, Black, American Indian and Alaska Native (AI/AN), Asian, Native Hawaiian and Other Pacific Islander, and two or more races, include persons of Hispanic and non-Hispanic origin. Persons of Hispanic origin may be of any race.

[3]Poor persons are defined as below the poverty threshold. Near poor persons have incomes of 100 percent to less than 200 percent of poverty threshold. Nonpoor persons have incomes of 200 percent or greater than the poverty threshold.

[4]Health insurance categories are mutually exclusive. Persons who reported both Medicaid and private coverage are classified as having private coverage. Persons 65 years of age and over who reported Medicare HMO (health maintenance organization) and some other type of health insurance coverage are classified as having Medicare HMO.

[5]MSA = metropolitan statistical area.

* = Figure does not meet standards of reliability or precision.

. . . = Not available.

more frequently than the nonpoor, in part reflecting the lack of attachment to a physician by some of the poor. Also, many poor households have Medicaid insurance, and the data show that this group made particularly frequent use of emergency departments: 32.4 percent of this group made at least one visit for children under 6 years of age, and 25.9 percent for children between 6 and 17 years of age.

Visits to emergency departments are more frequent by residents outside of a metropolitan area (MSA) than by those within a metropolitan area.

Table III-41. Emergency Department Visits within the Past 12 Months Among Children Under 18 Years of Age, According to Selected Characteristics, Selected Years, 1997–2001—*Continued*

(Percent.)

Characteristic	Under 18 years of age			Under 6 years of age			6–17 years of age		
	2001	1999	1997	2001	1999	1997	2001	1999	1997
Hispanic Origin and Race [2]									
Hispanic or Latino	6.4	5.2	8.9	8.7	7.9	11.8	5.1	3.6	7.0
Not Hispanic or Latino	6.9	5.5	6.8	9.7	8.8	9.2	5.6	4.0	5.7
White only	6.5	4.7	6.2	9.3	7.4	7.8	5.2	3.4	5.5
Black or African American only	8.4	9.1	9.3	14.0	15.9	14.6	5.9	6.1	6.8
Poverty Status [3]									
Poor	10.5	10.5	11.2	12.9	15.5	14.4	9.2	7.7	9.1
Near poor	8.5	7.6	8.6	13.1	12.4	12.7	6.0	5.0	6.4
Nonpoor	5.6	3.9	5.2	8.0	6.1	6.7	4.4	3.0	4.6
Hispanic Origin and Race and Poverty Status [2, 3]									
Hispanic or Latino									
Poor	7.8	5.7	10.6	7.9	8.1	13.9	7.7	*	8.4
Near poor	5.7	6.0	8.1	8.8	9.9	12.2	*3.9	*	*5.4
Nonpoor	6.0	5.5	7.4	9.8	9.2	8.2	*3.8	*3.6	7.0
Not Hispanic or Latino									
White only									
Poor	12.5	10.8	11.0	*16.0	18.5	12.4	10.7	*7.3	10.1
Near poor	9.0	7.7	8.4	14.7	13.0	11.8	5.9	*5.0	6.6
Nonpoor	5.3	3.5	5.0	7.6	5.2	6.0	4.3	2.8	4.5
Black or African American only									
Poor	11.6	14.6	12.9	17.3	22.2	19.6	*8.7	10.9	*8.7
Near poor	10.0	9.5	9.5	*16.5	*15.7	*14.0	*6.8	*	*7.5
Nonpoor	6.7	6.3	5.1	12.1	*13.1	*8.1	*4.3	*3.7	*4.0
Health Insurance Status [4]									
Insured	6.9	5.6	7.0	9.6	8.6	9.6	5.6	4.1	5.7
Private	5.4	3.8	5.2	7.4	5.7	6.8	4.5	3.0	4.5
Medicaid	11.7	12.6	13.1	14.6	17.1	16.2	9.8	9.4	10.4
Uninsured	5.6	4.9	7.7	8.3	9.0	9.8	4.5	*3.2	6.8
Poverty Status and Health Insurance Status [3]									
Poor									
Insured	11.8	12.1	12.0	14.0	16.9	15.4	10.5	9.2	9.6
Uninsured	*5.5	*4.8	8.0	*	*	*8.7	*4.5	*	*7.7
Near poor									
Insured	8.9	8.3	8.6	14.1	13.4	12.7	6.0	5.4	6.4
Uninsured	*6.5	*5.1	8.3	*	*	*12.2	*	*	6.8
Nonpoor									
Insured	5.5	3.9	5.1	7.8	5.9	6.4	4.4	3.0	4.5
Uninsured	*5.5	*4.5	7.1	*	*	*11.8	*	*	*5.0
Geographic Region									
Northeast	6.1	4.9	6.2	6.8	6.5	7.6	5.8	4.0	5.4
Midwest	7.1	5.8	6.6	11.3	9.8	10.4	5.1	4.0	4.8
South	7.8	6.1	8.0	11.3	9.8	10.1	6.0	4.3	6.9
West	5.6	4.7	7.1	7.0	7.6	10.0	4.8	3.3	5.6
Location of Residence									
Within MSA [5]	6.8	5.0	7.2	9.2	8.0	9.6	5.5	3.4	5.9
Outside MSA [5]	7.1	7.4	6.8	10.8	11.3	9.7	5.5	5.8	5.6

[1]Includes all other races not shown separately, unknown poverty status, and unknown health insurance status.

[2]The race groups, White, Black, American Indian and Alaska Native (AI/AN), Asian, Native Hawaiian and Other Pacific Islander, and two or more races, include persons of Hispanic and non-Hispanic origin. Persons of Hispanic origin may be of any race.

[3]Poor persons are defined as below the poverty threshold. Near poor persons have incomes of 100 percent to less than 200 percent of poverty threshold. Nonpoor persons have incomes of 200 percent or greater than the poverty threshold.

[4]Health insurance categories are mutually exclusive. Persons who reported both Medicaid and private coverage are classified as having private coverage. Persons 65 years of age and over who reported Medicare HMO (health maintenance organization) and some other type of health insurance coverage are classified as having Medicare HMO.

[5]MSA = metropolitan statistical area.

* = Figure does not meet standards of reliability or precision.

. . . = Not available.

TABLE III-42. NO USUAL SOURCE OF HEALTH CARE AMONG ADULTS 18–64 YEARS OF AGE, ACCORDING TO SELECTED CHARACTERISTICS, ANNUAL AVERAGE, 1993–2001

About 16.5 percent of adults between age 18 and 64 indicated that they had no usual place to go to in order to find medical care in 2001. If they responded that they go to an emergency department, this answer was counted as an absence of a usual place. Non-Hispanic Blacks had a somewhat higher frequency than non-Hispanic Whites of lacking a usual health care connection, but a much bigger lack of 30.8 percent occurred among Hispanics. Moreover, 43.7 percent of Hispanics below the poverty line lacked a usual place of health care.

Insurance status was an important factor in these data. About 46.3 percent of persons without health insurance lacked a usual place for health services, and for the uninsured poor this percentage was 51.7. For those with health insurance only 9.7 percent were without a place for medical service. On a regional basis, the West was more lacking in having a usual place for health services than the other regions of the country: 20 percent in the West were without such a place, but only 12.1 percent in the Northeast. The other regions are between these figures.

TABLE III-43. EMERGENCY DEPARTMENT VISITS WITHIN THE PAST 12 MONTHS AMONG ADULTS 18 YEARS OF AGE AND OVER, ACCORDING TO SELECTED CHARACTERISTICS, SELECTED YEARS, 1997–2001

In 2001, 19.7 percent of adults, aged 18 years and over, made one or more visits to an emergency department connected with a hospital, and 6.4 percent made two or more visits. The fewest visits were made by Asians, Hispanics, and non-Hispanic Whites, in that order. Blacks made substantially more visits than Whites, and visits were most frequent for American Indians. By age, the most frequent visitors were between 18 and 24 years of age as well those 75 years and over. Persons between those age groups made fewer visits.

Insurance status did not cause a very large difference in the utilization of emergency departments, except for persons covered by Medicaid. As many as 39.7 percent of persons between 18 and 64 years of age, who were insured by Medicaid, went to an emergency department one or more times in 2001, and 21.7 percent went two times or more. By comparison, persons in this age bracket, who had private insurance, made one or more visits with a frequency of 17.2 per cent, and two or more visits with a frequency of 4.7 percent. For persons 65 years and over with Medicaid, the percentage was 36.0 for one or more visits and 18.7 for two or more visits.

Table III-42. No Usual Source of Health Care Among Adults 18–64 Years of Age, According to Selected Characteristics, Annual Average, 1993–2001

(Percent.)

Characteristic	2000–2001 [1]	1999–2000 [1]	1997–1998 [1]	1995–1996	1993–1994
PERCENT OF ADULTS WITHOUT A USUAL SOURCE OF HEALTH CARE [2]					
All adults 18–64 years old [3, 4]	16.5	17.7	17.5	16.6	18.5
Age					
18–44 years	20.3	21.5	21.1	19.6	21.7
18–24 years	26.1	26.9	27.0	22.6	26.6
25–44 years	18.4	19.7	19.3	18.8	20.3
45–64 years	9.9	10.9	11.2	11.3	12.8
45–54 years	10.9	12.0	12.6	12.2	14.1
55–64 years	8.3	9.2	9.0	9.8	11.1
Sex [4]					
Male	21.8	23.8	23.2	21.0	23.3
Female	11.4	11.7	11.9	12.5	13.9
Race [4,5]					
White only	15.8	16.8	16.9	16.3	18.2
Black or African American only	16.8	18.7	18.7	17.6	19.2
American Indian and Alaska Native only	15.9	18.7	20.7	15.9	19.1
Asian only	18.5	21.4	21.1	20.7	24.0
Native Hawaiian and Other Pacific Islander only	*	*	. . .	. . .	. . .
Two or more races	21.1	20.4	. . .	. . .	. . .
American Indian and Alakan Native; White	24.0	26.5	. . .	. . .	. . .
Hispanic Origin and Race [4,5]					
Hispanic or Latino	30.8	30.4	28.6	26.2	28.8
Mexican	34.6	33.7	33.4	28.1	30.5
Not Hispanic or Latino	14.6	16.0	16.1	15.5	17.5
White only	13.9	15.2	15.4	15.0	17.0
Black or African American only	16.7	18.7	18.6	17.4	18.9
Poverty Status [4,6]					
Poor	27.4	27.4	28.0	24.9	28.2
Near poor	25.3	26.5	25.3	22.3	24.6
Nonpoor	12.4	13.8	13.8	13.5	14.8
Hispanic Origin and Race and Poverty Status [4,5,6]					
Hispanic or Latino					
Poor	43.7	39.7	41.0	32.6	38.0
Near poor	36.0	37.4	32.6	31.6	35.7
Nonpoor	19.3	20.5	18.7	18.2	18.3
Not Hispanic or Latino					
White only					
Poor	21.5	22.7	24.6	22.8	27.1
Near poor	22.5	23.5	22.6	20.3	22.7
Nonpoor	11.6	12.9	13.3	13.0	14.4
Black or African American only					
Poor	20.7	22.5	22.9	21.1	23.8
Near poor	20.7	23.6	25.5	21.2	21.6
Nonpoor	12.7	14.4	14.0	13.6	14.6
Health Insurance Status [4,7]					
Insured	9.7	11.0	11.4	11.4	13.3
Private	9.7	11.2	11.5	11.3	13.1
Medicaid	9.6	9.5	10.0	12.5	15.2
Uninsured	46.3	47.3	45.3	40.9	41.5
Poverty Status and Health Insurance Status [4,6]					
Poor					
Insured	11.6	12.4	14.0	13.6	16.8
Uninsured	51.7	49.8	50.3	42.1	45.7
Near poor					
Insured	12.4	13.1	14.4	13.1	15.3
Uninsured	47.8	49.3	46.3	41.5	42.9
Nonpoor					
Insured	9.0	10.5	10.9	10.8	12.3
Uninsured	41.4	44.2	41.3	39.4	37.0
Geographic Region					
Northeast	12.1	12.9	13.2	13.3	14.5
Midwest	15.2	16.8	14.9	14.5	15.8
South	17.8	19.6	20.5	18.4	21.6
West	20.0	19.8	19.8	19.5	20.5
Location of Residence					
Within MSA [8]	16.7	17.8	17.6	16.9	18.8
Outside MSA [8]	15.8	16.9	17.1	15.4	17.4

[1]Data starting in 1997 are not strictly comparable with data for earlier years due to the 1997 questionnaire redesign.
[2]Persons who report the emergency department as the place of their usual source of care are defined as having no usual source of care.
[3]Includes all other races not shown separately, unknown poverty status, and unknown health insurance status.
[4]Estimates are for persons 18-64 years of age and are age adjusted to the year 2000 standard population using three age groups: 18–44 years, 45–54 years, and 55–64 years of age.
[5]The race groups, White, Black, American Indian and Alaska Native (AI/AN), Asian, Native Hawaiian and Other Pacific Islander, and two or more races, include persons of Hispanic and non-Hispanic origin. Persons of Hispanic origin may be of any race.
[6]Poor persons are defined as below the poverty threshold. Near poor persons have incomes of 100 percent to less than 200 percent of poverty threshold. Nonpoor persons have incomes of 200 percent or greater than the poverty threshold.
[7]Health insurance categories are mutually exclusive. Persons who reported both Medicaid and private coverage are classified as having private coverage. Persons 65 years of age and over who reported Medicare HMO (health maintenance organization) and some other type of health insurance coverage are classified as having Medicare HMO.
[8]MSA = metropolitan statistical area.
* = Figure does not meet standards of reliability or precision.
. . . = Not available.

Table III-43. Emergency Department Visits within the Past 12 Months Among Adults 18 Years of Age and Over, According to Selected Characteristics, Selected Years, 1997–2001

(Percent.)

Characteristic	2001	2000	1999	1997
PERCENT OF ADULTS WITH 1 OR MORE EMERGENCY DEPARTMENT VISITS				
All adults 18 years and over [1]	19.7	20.2	17.2	19.6
Age				
18–44 years	19.8	20.6	17.7	20.7
18–24 years	24.0	25.9	21.7	26.3
25–44 years	18.4	18.9	16.5	19.0
45–64 years	18.0	17.6	14.6	16.2
45–54 years	17.7	17.9	14.3	15.7
55–64 years	18.5	17.0	15.1	16.9
65 years and over	22.3	23.7	19.9	22.0
65–74 years	19.7	21.6	17.3	20.3
75 years and over	25.4	26.2	23.1	24.3
Sex				
Male	18.9	18.8	16.1	19.1
Female	20.5	21.6	18.2	20.2
Race [2]				
White only	19.1	19.4	16.6	19.0
Black or African American only	25.2	26.5	22.2	25.9
American Indian and Alaska Native only	33.9	30.5	29.2	24.8
Asian only	12.7	13.6	9.7	11.6
Native Hawaiian and Other Pacific Islander only	*	*	*	...
2 or more races	25.5	32.9	24.4	...
American Indian and Alakan Native; White	25.4	33.9	26.0	...
Hispanic Origin and Race [2]				
Hispanic or Latino	18.4	18.4	15.3	19.2
Mexican American	15.6	17.4	14.4	17.8
Not Hispanic or Latino	20.0	20.6	17.5	19.7
White only	19.4	19.8	16.9	19.1
Black or African American only	25.3	26.5	22.2	25.9
Poverty Status [3]				
Poor	27.5	30.2	27.6	29.2
Near poor	26.2	25.1	21.7	24.9
Nonpoor	18.2	18.6	15.4	17.5
Hispanic Origin, Race, and Poverty Status [2, 3]				
Hispanic or Latino				
Poor	19.9	24.4	17.1	22.9
Near poor	20.1	19.4	15.9	19.2
Nonpoor	17.6	17.1	14.5	17.9
Not Hispanic or Latino				
White only				
Poor	29.6	30.6	29.4	30.8
Near poor	27.9	26.8	22.2	25.5
Nonpoor	18.0	18.2	15.5	17.2
Black or African American only				
Poor	32.1	38.0	33.5	35.5
Near poor	28.8	29.9	27.8	30.8
Nonpoor	22.3	24.1	18.4	20.7
Health Insurance Status [4]				
18–64 years of age				
Insured	19.2	19.5	16.1	18.8
Private	17.2	17.6	14.5	16.9
Medicaid	39.7	42.3	35.4	37.6
Uninsured	18.9	19.6	18.3	20.0
65 years and over				
Medicare HMO	23.6	24.4	20.1	20.2
Private	21.0	23.3	19.3	21.3
Medicaid	36.0	35.9	30.0	35.2
Medicare fee-for-service only	21.5	20.1	19.2	22.0

[1]Includes all other races not shown separately, unknown poverty status, and unknown health insurance status.

[2]The race groups, White, Black, American Indian and Alaska Native (AI/AN), Asian, Native Hawaiian and Other Pacific Islander, and two or more races, include persons of Hispanic and non-Hispanic origin. Persons of Hispanic origin may be of any race.

[3]Poor persons are defined as below the poverty threshold. Near poor persons have incomes of 100 percent to less than 200 percent of poverty threshold. Nonpoor persons have incomes of 200 percent or greater than the poverty threshold.

[4]Health insurance categories are mutually exclusive. Persons who reported both Medicaid and private coverage are classified as having private coverage. Persons 65 years of age and over who reported Medicare HMO (health maintenance organization) and some other type of health insurance coverage are classified as having Medicare HMO.

[5]MSA = metropolitan statistical area.

* = Figure does not meet standards of reliability or precision.

. . . = Not available.

Table III-43. Emergency Department Visits within the Past 12 Months Among Adults 18 Years of Age and Over, According to Selected Characteristics, Selected Years, 1997–2001—*Continued*

(Percent.)

Characteristic	2001	2000	1999	1997
Poverty Status and Health Insurance Status [3]				
18–64 years of age				
Poor				
Insured	30.8	33.6	29.8	32.1
Unsured	20.0	26.0	22.7	24.4
Near poor				
Insured	28.0	27.3	23.1	26.6
Uninsured	23.6	20.1	18.6	21.3
Nonpoor				
Insured	17.2	17.6	14.7	16.6
Uninsured	17.8	19.2	16.3	19.0
Geographic Region				
Northeast	19.8	20.0	16.9	19.5
Midwest	19.6	20.1	17.2	19.3
South	20.9	21.3	17.7	20.9
West	17.6	18.7	16.4	17.7
Location of Residence				
Within MSA[5]	19.4	19.6	16.6	19.1
Outside MSA[5]	21.3	22.5	19.5	21.5
PERCENT OF ADULTS WITH 2 OR MORE EMERGENCY DEPARTMENT VISITS				
All adults 18 years of age and over [1]	6.4	6.9	5.2	6.7
Age				
18–44 years	6.5	7.0	5.6	6.8
18–24 years	8.7	8.9	7.3	9.1
25–44 years	5.8	6.4	5.0	6.2
45–64 years	5.6	5.6	4.3	5.6
45–54 years	5.5	5.8	4.3	5.5
55–64 years	5.9	5.3	4.3	5.7
65 years and over	7.5	8.6	5.6	8.1
65–74 years	7.1	7.4	4.7	7.1
75 years and over	8.0	10.1	6.7	9.3
Sex				
Male	5.7	5.8	4.3	5.9
Female	7.2	8.0	6.0	7.5
Race [2]				
White only	6.1	6.4	4.7	6.2
Black or African American only	9.4	10.7	8.8	11.1
American Indian and Alaska Native only	15.5	*12.8	*11.7	13.1
Asian only	*2.6	*3.8	*	*2.9
2 or more races	8.8	11.4	11.4	. . .
American Indian and Alakan Native; White	*6.1	*9.2	*13.9	. . .
Hispanic Origin and Race [2]				
Hispanic or Latino	7.0	7.1	4.5	7.4
Mexican American	5.6	7.1	4.1	6.4
Not Hispanic or Latino	6.4	6.9	5.3	6.7
White only	6.1	6.4	4.8	6.2
Black or African American only	9.4	10.7	8.8	11.0
Poverty Status [3]				
Poor	13.1	14.3	11.7	13.7
Near poor	10.4	10.6	8.0	10.0
Nonpoor	5.1	5.3	4.1	5.0
Hispanic Origin, Race, and Poverty Status [2, 3]				
Hispanic or Latino				
Poor	10.1	11.3	6.6	10.2
Near poor	7.7	7.6	5.0	8.4
Nonpoor	5.3	6.1	3.8	5.5
Not Hispanic or Latino				
White only				
Poor	13.7	14.3	11.7	14.1
Near poor	11.0	11.5	7.6	9.8
Nonpoor	5.1	5.0	4.1	4.8
Black or African American only				
Poor	15.7	19.0	16.8	17.9
Near poor	12.8	13.1	13.0	12.9
Nonpoor	7.0	8.4	5.7	7.8

[1]Includes all other races not shown separately, unknown poverty status, and unknown health insurance status.
[2]The race groups, White, Black, American Indian and Alaska Native (AI/AN), Asian, Native Hawaiian and Other Pacific Islander, and two or more races, include persons of Hispanic and non-Hispanic origin. Persons of Hispanic origin may be of any race.
[3]Poor persons are defined as below the poverty threshold. Near poor persons have incomes of 100 percent to less than 200 percent of poverty threshold. Nonpoor persons have incomes of 200 percent or greater than the poverty threshold.
[5]MSA = metropolitan statistical area.
* = Figure does not meet standards of reliability or precision.
. . . = Not available.

Table III-43. Emergency Department Visits within the Past 12 Months Among Adults 18 Years of Age and Over, According to Selected Characteristics, Selected Years, 1997–2001—*Continued*

(Percent.)

Characteristic	2001	2000	1999	1997
Health Insurance Status [4]				
18–64 years of age				
Insured	6.2	6.4	4.7	6.1
Private	4.7	5.1	3.7	4.7
Medicaid	21.7	21.0	17.4	19.7
Uninsured	6.6	7.0	7.0	7.5
65 years and over				
Medicare HMO	8.8	8.5	5.7	6.7
Private	6.4	7.9	5.3	6.9
Medicaid	18.7	18.3	12.8	20.2
Medicare fee-for-service only	6.9	7.3	4.4	9.4
Poverty Status and Health Insurance Status [3]				
18–64 years of age				
Poor				
Insured	15.5	17.4	13.3	15.9
Unsured	8.5	10.6	10.3	10.0
Near poor				
Insured	11.7	11.6	8.7	10.3
Uninsured	9.0	7.7	7.5	9.1
Nonpoor				
Insured	4.6	4.9	3.7	4.5
Uninsured	5.0	6.4	6.5	5.4
Geographic Region				
Northeast	6.1	6.2	5.1	6.9
Midwest	6.0	6.9	5.1	6.2
South	7.3	7.6	5.7	7.3
West	5.6	6.3	4.5	6.0
Location of Residence				
Within MSA[5]	6.3	6.6	4.9	6.4
Outside MSA[5]	7.0	7.8	6.4	7.8

[3]Poor persons are defined as below the poverty threshold. Near poor persons have incomes of 100 percent to less than 200 percent of poverty threshold. Nonpoor persons have incomes of 200 percent or greater than the poverty threshold.
[4]Health insurance categories are mutually exclusive. Persons who reported both Medicaid and private coverage are classified as having private coverage. Persons 65 years of age and over who reported Medicare HMO (health maintenance organization) and some other type of health insurance coverage are classified as having Medicare HMO.
[5]MSA = metropolitan statistical area.

TABLE III-44. DENTAL VISITS IN THE PAST YEAR ACCORDING TO SELECTED CHARACTERISTICS, SELECTED YEARS, 1997–2001

In 2001, 64.6 percent of persons between the age of 18 and 64 had a dental visit during the past year. For persons between 2 to 17 years of age, a visit took place by 73.3 percent. And for those 65 years old or above, 56.3 percent had a dental appointment. A survey showed that 28 percent to 30 percent of persons in the older age group had lost their natural teeth and that this considerably reduced their dental visits. Dental visits occurred for about 70 percent of the seniors with natural teeth.

As already mentioned, 64.6 percent of persons of all races aged 18–64 had a dental visit. This compares with 57.1 percent for non-Hispanic Blacks, and 49.2 percent for Hispanics. For persons classified as poor visits occurred at a rate of 45.8 percent, and for poor Hispanics, the rate was down to 35.1 percent.

Table III-44. Dental Visits in the Past Year According to Selected Characteristics, Selected Years, 1997–2001

(Percent.)

Characteristic	2 years of age and over			2–17 years of age			18–64 years of age			65 years of age and over		
	2001	1999	1997	2001	1999	1997	2001	1999	1997	2001	1999	1997
PERCENT OF PERSONS WITH A DENTAL VISIT IN THE PAST YEAR												
Total [1]	65.6	65.2	64.9	73.3	72.6	72.7	64.6	64.6	64.1	56.3	55.0	54.8
Sex												
Male	62.6	62.5	62.6	72.7	72.3	72.3	60.2	60.4	60.4	56.1	54.7	55.4
Female	68.5	67.8	67.2	73.9	72.8	73.0	68.9	68.5	67.7	56.5	55.2	54.4
Race [2]												
White only	67.4	67.2	66.5	74.9	74.5	74.0	66.6	66.6	65.7	58.5	56.8	56.8
Black or African American only	56.9	56.2	56.5	68.0	67.6	68.8	57.2	55.8	57.0	37.5	39.7	35.4
American Indian and Alaska Native only	53.9	56.2	51.5	72.9	58.2	66.8	47.7	55.2	49.9	*50.7	*50.6	*
Asian only	64.9	63.6	61.8	74.4	69.6	69.9	64.3	63.1	60.3	53.4	53.2	53.9
Native Hawaiian and Other Pacific Islander only	*	*	. . .	*	*	. . .	*	*	. . .	*	*	. . .
Two or more races	56.3	58.6	. . .	69.3	73.0	. . .	57.1	57.8	. . .	*34.5	*35.1	. . .
Black or African American; White	52.7	63.7	. . .	57.6	68.7	. . .	55.5	58.8	. . .	*	*	. . .
American Indian and Alakan Native; White	58.7	55.8	. . .	79.2	70.3	. . .	53.6	53.5	. . .	39.0	*	*39.0
Hispanic Origin and Race [2]												
Hispanic or Latino	51.2	52.3	52.9	60.5	59.3	61.0	49.2	50.6	50.8	42.6	44.0	47.8
Not Hispanic or Latino	67.5	66.9	66.4	75.8	74.9	74.7	66.7	66.3	65.7	57.2	55.6	55.2
White only	69.6	68.9	68.2	77.8	77.0	76.4	68.7	68.3	67.5	59.4	57.3	57.2
Black or African American only	56.9	56.1	56.5	68.1	67.7	68.8	57.1	55.7	56.9	37.6	39.6	35.3
Poverty Status [3]												
Poor	47.0	46.2	47.2	61.0	57.8	62.0	45.8	46.0	46.4	30.6	31.9	30.3
Near poor	49.7	48.5	48.9	63.2	61.6	61.6	46.9	46.1	46.4	40.0	38.9	39.6
Nonpoor	72.0	72.0	72.3	79.3	79.9	79.7	70.5	70.8	71.1	67.0	64.4	66.3
Hispanic Origin, Race and Poverty Status [2, 3]												
Hispanic or Latino												
Poor	38.7	41.5	41.9	54.2	49.6	56.8	35.1	39.7	39.0	29.8	32.1	33.0
Near poor	43.2	43.8	46.2	59.2	54.0	54.1	39.8	41.0	42.6	30.6	34.8	49.2
Nonpoor	64.4	63.8	65.1	71.1	72.0	74.8	62.8	62.0	62.5	60.8	58.9	56.5
Not Hispanic or Latino												
White only												
Poor	51.5	49.8	49.9	62.9	62.6	63.3	52.1	50.6	50.3	32.3	31.9	31.1
Near poor	52.9	50.2	51.0	64.4	63.2	64.8	51.0	48.0	48.2	42.1	39.6	41.2
Nonpoor	73.7	73.6	73.6	81.7	81.8	80.7	72.0	72.4	72.5	68.6	65.4	67.6
Black or African American only												
Poor	45.1	44.9	46.7	63.2	61.0	66.7	42.9	42.1	44.5	25.2	33.5	26.2
Near poor	47.3	47.6	44.9	64.2	66.3	60.1	44.9	45.2	44.7	33.4	30.9	23.6
Nonpoor	64.1	64.2	65.4	72.4	72.7	75.5	65.4	64.7	66.2	46.1	51.5	48.9
Geographic Region												
Northeast	72.2	70.9	69.6	79.6	78.5	77.5	72.2	71.5	69.6	59.6	54.3	55.5
Midwest	68.4	68.1	68.3	77.4	76.8	76.4	68.0	67.6	67.4	55.0	54.3	57.6
South	60.2	60.6	60.0	68.8	68.0	68.0	58.7	59.4	59.4	52.0	52.4	49.0
West	65.7	64.7	64.9	70.7	69.9	71.5	64.4	63.3	62.9	62.6	61.9	61.9
Location of Residence												
Within MSA [4]	67.0	67.1	66.5	73.9	73.1	73.6	66.0	66.8	65.7	59.1	58.1	57.6
Outside MSA [4]	60.3	58.3	59.1	70.7	70.7	69.3	59.1	56.2	58.0	47.2	45.0	46.1

[1]Includes all other races not shown separately, unknown poverty status, and unknown health insurance status.

[2]The race groups, White, Black, American Indian and Alaska Native (AI/AN), Asian, Native Hawaiian and Other Pacific Islander, and two or more races, include persons of Hispanic and non-Hispanic origin. Persons of Hispanic origin may be of any race.

[3]Poor persons are defined as below the poverty threshold. Near poor persons have incomes of 100 percent to less than 200 percent of poverty threshold. Nonpoor persons have incomes of 200 percent or greater than the poverty threshold.

[4]MSA = metropolitan statistical area.

* = Figure does not meet standards of reliability or precision.

. . . = Not available.

TABLE III-45. UNTREATED DENTAL CARIES (CAVITIES), ACCORDING TO AGE, SEX, RACE AND HISPANIC ORIGIN, AND POVERTY STATUS, SELECTED YEARS 1971–1994

Persons between 18 and 64 years of age in the 1988–1994 survey period who had untreated dental caries, amounted to 28.2 percent of all respondents, considerably less than the 48.4 percent reported for the period of 1971–1974. In the more recent survey, these percentages were smaller than average for children and also somewhat smaller for persons aged 65 and above.

Untreated caries were much more frequent among the minorities reported in the survey, namely Blacks and Mexicans. The incidence of untreated caries was especially large among the poor, with the age group of 18 to 64 years reporting 42.3 percent for non-Hispanic Whites, 59.0 percent for non-Hispanic Blacks, and 52.4 percent for persons of Mexican origin.

TABLE III-46. USE OF MAMMOGRAPHY FOR WOMEN 40 YEARS OF AGE AND OVER, ACCORDING TO SELECTED CHARACTERISTICS, SELECTED YEARS, 1987–2000

This table shows the percent of women, 40 years and over, who had a mammogram during the past two years. For 2000, this percentage was 70.3, showing consecutive increases since 1987, when this percentage was 28.7. Women in the 50–64 age bracket were the highest users of mammogram tests with a frequency of 78.6 percent.

Coverage for minorities, age 40 and over, was smaller than for Whites, although the non-Hispanic Blacks had a relatively high usage of 67.9 percent. About 55.2 percent of all women in this age group, classified as poor, received a mammogram during a two-year period ending in 2000.

The frequency of tests was influenced by educational attainment: using the high frequency age group of 50 to 64 as example, 84.1 percent of women with some college or more had the procedure, 76.6 percent of women with a high school degree (or equivalent), and 66.3 percent of women without a high school degree (or equivalent).

Table III-45. Untreated Dental Caries (Cavities), According to Age, Sex, Race, Hispanic Origin, and Poverty Status, Selected Years, 1971–1994

(Percent.)

Sex, race, Hispanic origin, and poverty status	2–5 years			6–17 years of age			18–64 years of age			65–74 years		
	1988–1994	1982–1984	1971–1974	1988–1994	1982–1984	1971–1974	1988–1994	1982–1984	1971–1974	1988–1994	1982–1984	1971–1974
PERCENT OF PESONS WITH UNTREATED DENTAL CARIES												
Total [1]	18.7	...	24.4	23.1	...	55.0	28.2	...	48.4	25.4	...	29.7
Sex												
Male	19.2	...	26.1	22.6	...	54.8	31.2	...	48.4	29.9	...	30.2
Female	18.1	...	22.7	23.7	...	55.2	25.3	...	48.5	21.5	...	28.3
Race and Hispanic Origin [2]												
Not Hispanic or Latino												
White only	14.4	...	23.7	18.9	...	52.3	23.6	...	45.2	22.7	...	28.1
Black or African American only	25.1	...	28.2	33.0	...	70.9	47.9	...	68.1	46.7	...	41.5
Mexican	34.9	23.1	...	37.2	42.8	...	39.9	45.4	...	43.8	44.3	...
Poverty Status [3]												
Poor	28.8	...	30.7	36.3	...	70.4	47.3	...	63.6	46.7	...	34.3
Near poor	24.3	...	29.8	29.2	...	60.2	42.7	...	56.3	39.3	...	35.6
Nonpoor	9.7	...	17.5	14.5	...	46.3	19.5	...	43.1	19.4	...	26.2
Race, Hispanic Origin, and Poverty Status [2, 3]												
Not Hispanic or Latino												
White only												
Poor	25.4	...	31.9	32.5	...	68.1	42.3	...	58.4	39.0	...	33.3
Near poor and nonpoor	12.4	...	22.1	16.7	...	50.3	21.6	...	44.3	22.7	...	28.0
Black or African American only												
Poor	27.5	...	29.0	35.6	...	73.4	59.0	...	73.1	50.1	...	39.8
Near poor and nonpoor	23.0	...	26.5	31.2	...	67.4	43.4	...	65.8	43.6	...	41.1
Mexican												
Poor	38.5	22.6	...	45.8	46.4	...	52.4	56.3	...	55.5	54.4	...
Near poor and nonpoor	30.5	22.0	...	27.6	39.3	...	31.5	41.0	...	35.6	30.8	...

Note: Excludes persons without teeth, most of whom are 65 years and over. In this age group they are estimated to account for 46, 37, and 33 percent, respectively in the three annual averages shown above.

[1]Includes all other races not shown separately, unknown poverty status, and unknown health insurance status.
[2]In 1971–1974, data are for White persons and Black persons. Persons of Mexican origin may be of any race.
[3]Poverty status is based on family income and family size. Poor persons are defined as below the poverty threshold. Near poor persons have incomes of 100 percent to less than 200 percent of the poverty threshold. Nonpoor persons have incomes of 200 percent or greater than the poverty threshold.
. . . = Not available.

Table III-46. Use of Mammography for Women 40 Years of Age and Over, According to Selected Characteristics, Selected Years, 1987–2000

(Percent.)

Sex, race, Hispanic origin, and poverty status	2000	1999	1998	1994	1993	1991	1990	1987
PERCENT OF WOMEN HAVING A MAMMOGRAM WITHIN THE PAST TWO YEARS								
40 years and over, age adjusted [1]	70.3	70.3	67.0	61.0	59.7	54.7	51.7	29.0
40 years and over, crude [1]	70.3	70.3	66.9	60.9	59.7	54.6	51.4	28.7
Age								
40-49 years	64.2	67.2	63.4	61.3	59.9	55.6	55.1	31.9
50-64 years	78.6	76.5	73.7	66.5	65.1	60.3	56.0	31.7
65 years and over	68.0	66.8	63.8	55.0	54.2	48.1	43.4	22.8
65-74 years	74.0	73.9	69.4	63.0	64.2	55.7	48.7	26.6
75 years and over	61.3	58.9	57.2	44.6	41.0	37.8	35.8	17.3
Race [2]								
White only	71.4	70.6	67.4	60.6	60.0	55.6	52.2	29.6
Black or African American only	67.8	71.0	66.0	64.3	59.1	48.0	46.4	24.0
American Indian and Alaska Native only	47.3	63.0	45.2	65.8	49.8	54.5	43.2	*
Asian only	53.3	58.3	60.2	55.8	55.1	45.9	46.0	*
Native Hawaiian and Other Pacific Islander only	*	*	. . .	. . .	. . .	. . .	. . .	. . .
Two or more races	69.2	70.2	. . .	. . .	. . .	. . .	. . .	. . .
Hispanic Origin and Race [2]								
40 years and over, crude								
Hispanic or Latino	61.4	65.7	60.2	51.9	50.9	49.2	45.2	18.3
Not Hispanic or Latino	71.0	70.7	67.5	61.5	60.3	54.9	51.8	29.4
White only	72.1	71.1	68.0	61.3	60.6	56.0	52.7	30.3
Black or African American only	67.9	71.0	66.0	64.4	59.2	47.7	46.0	23.8
Age, Hispanic Origin and Race [2]								
40–49 years								
Hispanic or Latino	54.2	61.6	55.2	47.5	52.6	44.0	45.1	*15.3
Not Hispanic or Latino								
White only	67.1	68.3	64.4	62.0	61.6	58.1	57.0	34.3
Black or African American only	60.9	69.2	65.0	67.2	55.6	48.0	48.4	27.8
50–64 years								
Hispanic or Latino	66.4	69.7	67.2	60.1	59.2	61.7	47.5	23.0
Not Hispanic or Latino								
White only	80.5	77.9	75.3	67.5	66.2	61.5	58.1	33.6
Black or African American only	77.7	75.0	71.2	63.6	65.5	52.4	48.4	26.4
65 years and older								
Hispanic or Latino	68.2	67.2	59.0	48.0	35.7	40.9	41.1	*
Not Hispanic or Latino								
White only	68.3	66.8	64.3	54.9	54.7	49.1	43.8	24.0
Black or African American only	65.5	68.1	60.6	61.0	56.3	41.6	39.7	14.1
Age and Poverty Status [3]								
40 years and over, crude								
Poor	55.2	56.9	50.5	44.2	41.1	35.2	30.8	16.4
Near poor or nonpoor	72.2	71.5	69.3	63.4	61.8	57.5	54.1	31.3
40–49 years								
Poor	47.2	52.5	44.9	43.0	36.1	33.0	32.2	23.0
Near poor or nonpoor	65.9	68.7	65.0	63.4	62.1	58.1	57.0	33.4
50–64 years								
Poor	62.7	61.1	53.5	46.2	47.3	37.3	29.9	15.1
Near poor or nonpoor	80.6	77.4	76.7	68.8	66.8	63.0	58.5	34.3
65 years and over								
Poor	55.4	57.3	52.3	43.9	40.4	35.2	30.8	13.6
Near poor or nonpoor	70.0	67.8	66.2	57.7	56.4	51.1	46.2	25.5
Age and Education [4]								
40 years and over, crude								
No high school diploma or GED	57.7	56.7	54.5	48.2	46.4	40.0	36.4	17.8
High school diploma or GED	69.6	69.2	66.7	61.0	59.0	55.8	52.7	31.3
Some college or more	76.1	77.3	72.8	69.7	69.5	65.2	62.8	37.7
40–49 years of age								
No high school diploma or GED	46.9	48.8	47.3	50.4	43.6	40.8	38.5	15.1
High school diploma or GED	59.0	60.8	59.1	55.8	56.6	52.0	53.1	32.6
Some college or more	70.5	74.4	68.3	68.7	66.1	63.7	62.3	39.2
50–64 years of age								
No high school diploma or GED	66.3	62.3	58.8	51.6	51.4	43.6	41.0	21.2
High school diploma or GED	76.6	77.2	73.3	67.8	62.4	60.8	56.5	33.8
Some college or more	84.1	81.2	79.8	74.7	78.5	72.7	68.0	40.5
65 years and over								
No high school diploma or GED	57.5	56.6	54.7	45.6	44.2	37.7	33.0	16.5
High school diploma or GED	72.0	68.4	66.8	59.1	57.4	54.0	47.5	25.9
Some college or more	74.1	77.1	71.3	64.3	64.8	57.9	56.7	32.3

[1] Includes all other races not shown separately, unknown poverty status, and unknown education.
[2] The race groups White, Black, American Indian and Alaskan Native (AI/AN), Asian, Native Hawaiian and Other pacific Islander, and 2 or more races, include persons of Hispanic and Non-Hispanic origin. Persons of Hispanic origin may be of any race.
[3] Poor persons are defined as below the poverty threshold. Near poor persons have incomes of 100 percent to less than 200 percent of the poverty threshold. Nonpoor persons have incomes of 200 percent or greater than the poverty threshold.
[4] Education quantities shown are for 1998 and subsequent years. GED stands for General Educational Development high school equivalency diploma. In years prior to 1998 the following categories based on number of years of school completed were used: Less than 12 years, 12 years, 13 years or more.
* = Figure does not meet standards of reliability or precision.
. . . = Not available.

TABLE III-47. USE OF PAP SMEARS FOR WOMEN 18 YEARS OF AGE AND OVER, ACCORDING TO SELECTED CHARACTERISTICS, SELECTED YEARS 1987–2000

A high percentage of women, 81.3 percent, had Pap smear tests in 2000 or within the past three years. This represented an increase over 1987, when this rate was 74.4 percent. Women between 18 and 50 years of age were the most frequent users of Pap smears. Among population groups, non-Hispanic Blacks had Pap smears most frequently, followed by non-Hispanic Whites, and then Hispanics. Persons classified as poor took the test less frequently, especially those aged 65 and over. Low and moderate levels of educational attainment also were associated with a reduced frequency of Pap smears.

Table III-47. Use of Pap Smears for Women 18 Years of Age and Over According to Selected Characteristics, Selected Years, 1987–2000

(Percent.)

Characteristic	2000	1999	1998	1994	1993	1987
PERCENT OF WOMEN HAVING A PAP SMEAR WITHIN THE PAST THREE YEARS						
18 years and over, age-adjusted [1]	81.4	80.8	79.2	76.6	77.5	74.1
18 years and over, crude [1]	81.3	80.8	79.1	76.6	77.6	74.4
Age						
18–49 years	85.2	86.5	84.4	82.4	84.2	82.6
50–64 years	83.7	80.2	79.9	75.6	74.4	68.2
65 years and over	64.6	61.0	59.8	57.6	58.0	50.8
65–74 years	71.6	70.0	67.0	64.9	64.7	57.9
75 years and over	56.8	50.8	51.2	47.3	47.8	40.4
Race [2]						
18 years and over, crude						
White only	81.4	80.6	78.9	76.3	77.2	74.1
Black or African American only	85.1	85.7	84.2	83.0	82.6	80.7
American Indian and Alaska Native only	76.9	92.2	74.6	73.3	78.2	85.4
Asian only	66.3	64.4	68.5	67.2	69.6	51.9
Native Hawaiian and Other Pacific Islander only	*	*	. . .	. . .	. . .	. . .
Two or more races	80.2	86.9	. . .	. . .	. . .	. . .
Hispanic Origin and Race [2]						
18 years and over, crude						
Hispanic or Latino	76.9	76.3	75.2	74.3	77.2	67.6
Not Hispanic or Latino	81.8	81.3	79.6	76.9	77.6	74.9
White only	81.9	81.0	79.3	76.5	77.2	74.7
Black or African American only	85.2	86.0	84.2	83.3	82.7	80.9
Age, Hispanic Origin and Race [2]						
18–49 years						
Hispanic or Latino	78.5	77.5	77.0	80.1	81.1	74.2
Not Hispanic or Latino						
White only	86.7	88.1	85.5	82.6	84.7	83.3
Black or African American only	88.5	90.3	88.9	89.2	87.8	89.3
50–64 years						
Hispanic or Latino	75.9	78.4	76.6	67.6	72.3	50.7
Not Hispanic or Latino						
White only	85.0	80.4	80.4	76.1	74.8	70.1
Black or African American only	83.9	82.9	81.6	78.5	77.9	70.9
65 years and older						
Hispanic or Latino	66.9	63.7	59.8	44.1	57.3	41.7
Not Hispanic or Latino						
White only	64.3	60.5	59.7	58.4	57.5	51.8
Black or African American only	67.3	64.5	61.7	60.9	61.5	44.8
Age and Poverty Status [3]						
18 years and over, crude						
Poor	72.1	73.9	69.4	68.1	69.4	64.2
Near poor or nonpoor	83.5	82.9	81.7	78.2	78.8	77.0
18–49 years						
Poor	76.5	79.8	76.8	78.0	76.1	76.6
Near poor or nonpoor	86.8	88.1	86.1	83.3	85.7	84.0
50–64 years						
Poor	73.0	72.1	63.6	57.9	62.2	50.5
Near poor or nonpoor	84.4	80.9	82.7	77.6	75.5	70.8
65 years and over						
Poor	53.7	50.7	47.1	44.3	47.2	33.2
Near poor or nonpoor	67.4	62.6	62.2	60.8	59.6	55.8
Age and Education [4]						
25 years and over, crude						
No high school diploma or GED	70.0	66.1	65.0	60.7	61.8	57.1
High school diploma or GED	79.9	79.3	77.4	75.8	78.1	76.4
Some college or more	88.1	87.8	86.9	85.1	84.3	84.0
25–49 years of age						
No high school diploma or GED	80.1	77.6	76.9	71.8	73.3	73.8
High school diploma or GED	85.4	86.3	83.7	80.9	84.8	84.1
Some college or more	91.0	92.3	90.6	88.9	89.2	89.4
50–64 years of age						
No high school diploma or GED	73.3	71.9	66.8	65.7	63.4	55.0
High school diploma or GED	81.4	79.2	80.2	75.7	75.2	70.9
Some college or more	88.8	83.8	84.8	81.9	80.6	78.7
65 years and over						
No high school diploma or GED	56.7	51.8	52.4	48.0	50.8	44.0
High school diploma or GED	67.0	63.7	60.7	61.4	61.9	55.4
Some college or more	69.8	68.8	67.9	66.9	62.9	59.4

[1] Includes all other races not shown separately, unknown poverty status, and unknown education.
[2] The race groups, White, Black, American Indian and Alaska Native (AI/AN), Asian, Native Hawaiian and Other Pacific Islander, and two or more races, include persons of Hispanic and non-Hispanic origin. Persons of Hispanic origin may be of any race.
[3] Poor persons are defined as below the poverty threshold. Near poor persons have incomes of 100 percent to less than 200 percent of poverty threshold. Nonpoor persons have incomes of 200 percent or greater than the poverty threshold.
[4] Education categories shown are for 1998 and subsequent years. GED stands for General Educational Development high school equivalency diploma. In years prior to 1998 the following categories based on number of years of school completed were used: less than 12 years, 12 years, 13 years or more.
* = Figure does not meet standards of reliability or precision.
. . . = Not available.

TABLE III-48. AMBULATORY CARE VISITS TO PHYSICIAN OFFICES AND HOSPITAL OUTPATIENT AND EMERGENCY DEPARTMENTS BY SELECTED CHARACTERISTICS, SELECTED YEARS, 1995–2001

Total visits to doctors and outpatient facilities in 2001 were estimated to have slightly exceeded 1 billion. This astoundingly large figure is the result of 3.83 visits during the year by an average resident of the United States.

Visits to physicians' offices averaged 314 per 100 persons, those to hospital outpatient departments occurred 30 times per 100 persons, and those to emergency departments took place 38 times per 100.

Females made about one-third more visits to all three places than males. For males, the rate of visits is lowest, at 207 per 100, between 18 and 44 years of age. Thereafter it rises steadily with advancing age, reaching an average

Table III-48. Ambulatory Care Visits to Physician Offices and Hospital Outpatient and Emergency Departments by Selected Characteristics, Selected Years, 1995–2001

(Number in thousands, rate per 100 people.)

Age, sex, and race	All places					Physician offices				
	2001	2000	1999	1998	1995	2001	2000	1999	1998	1995
NUMBER OF VISITS (Thousands)										
Total	1 071 692	1 014 848	944 122	1 005 078	860 859	880 487	823 542	756 734	829 280	697 082
Under 18 years	216 202	212 165	183 072	213 486	194 644	168 663	163 459	135 627	168 520	150 351
18–44 years	317 826	315 774	300 051	328 475	285 184	244 288	243 011	227 005	260 379	219 065
45–64 years	279 955	255 894	240 688	237 700	188 320	239 106	216 783	201 911	203 296	159 531
45–54 years	154 518	142 233	130 824	132 146	104 891	130 795	119 474	108 597	112 316	88 266
55–64 years	125 437	113 661	109 864	105 555	83 429	108 310	97 309	93 315	90 979	71 264
65 years and over	257 709	231 014	220 311	225 416	192 712	228 430	200 289	192 190	197 085	168 135
65–74 years	126 828	116 505	106 066	115 526	102 605	112 978	102 447	92 642	102 306	90 544
75 years and over	130 881	114 510	114 245	109 890	90 106	115 452	97 842	99 548	94 779	77 591
NUMBER OF VISITS PER 100 PERSONS										
Total, age-adjusted	385	374	352	377	334	316	304	283	312	271
Total, crude	383	370	347	373	329	314	300	279	308	266
Under 18 years	298	293	254	297	275	233	226	188	235	213
18–44 years	289	291	277	303	264	222	224	209	240	203
45–64 years	437	422	410	419	364	373	358	344	358	309
45–54 years	397	385	368	384	339	336	323	305	327	286
55–64 years	499	481	477	473	401	431	412	405	407	343
65 years and over	764	706	679	697	612	678	612	592	609	534
65–74 years	702	656	596	643	560	625	577	521	569	494
75 years and over	837	766	779	764	683	739	654	679	659	588
Age										
Male, age-adjusted	337	325	309	321	290	275	261	246	261	232
Male, crude	325	314	297	310	277	264	251	235	251	220
Under 18 years	302	302	255	303	273	235	231	189	239	209
18–44 years	207	203	206	201	190	152	148	150	149	139
45–54 years	327	316	300	302	275	273	260	247	251	229
55–64 years	429	428	427	435	351	371	367	361	379	300
65–74 years	670	614	580	608	508	598	539	510	538	445
75 years and over	852	771	758	739	711	758	670	663	640	616
Female, age-adjusted	431	420	393	431	377	356	345	317	360	309
Female, crude	437	424	396	433	378	362	348	320	362	310
Under 18 years	295	285	252	291	277	231	221	187	231	217
18–44 years	370	377	345	401	336	291	298	267	328	265
45–54 years	464	451	432	462	400	397	384	361	399	339
55–64 years	563	529	522	506	446	485	453	445	433	382
65–74 years	728	692	610	672	603	648	609	530	595	534
75 years and over	828	763	792	780	666	726	645	689	671	571
Race and Age [1]										
White, age-adjusted	402	380	356	376	339	338	315	292	316	282
White, crude	407	381	357	376	338	343	316	293	317	281
Under 18 years	322	306	258	292	295	260	243	197	235	237
18–44 years	305	301	284	305	267	242	239	222	248	211
45–54 years	410	386	368	380	334	354	330	312	328	286
55–64 years	513	480	474	462	397	451	416	410	406	345
65–74 years	722	641	597	639	557	651	568	526	572	496
75 years and over	858	764	781	768	689	764	658	687	669	598
Black or African American, age-adjusted	327	353	355	400	309	212	239	239	281	204
Black or African American, crude	302	324	322	373	281	189	214	211	259	178
Under 18 years	201	264	237	315	193	102	167	144	217	100
18–44 years	266	257	267	317	260	154	149	155	207	158
45–54 years	397	383	398	426	387	286	269	277	310	281
55–64 years	476	495	543	561	414	341	373	404	411	294
65–74 years	567	656	611	660	553	426	512	485	511	429
75 years and over	632	745	780	725	534	473	568	608	537	395

[1]Beginning in 1999, the instruction for the race item on the patient record form was changed so that more than one race could be recorded. In previous years, only one racial category could be checked.

of 852 visits per 100 males age 75 and over. The pattern is somewhat different for females, because their frequency of visits increases during their main reproductive years between 18 and 44 years of age. Thus females, per 100, average 295 visits when they are under 18 years of age, 370 between age 18 and 44, and 464 when their age is between 45 and 54.

The frequency of visits to all places by Blacks was about 25 percent less than the rate for Whites. However, Blacks had a higher rate of visits than Whites to hospital outpatient and emergency departments.

Table III-48. Ambulatory Care Visits to Physician Offices and Hospital Outpatient and Emergency Departments by Selected Characteristics, Selected Years, 1995–2001—*Continued*

(Number in thousands, rate per 100 people.)

Age, sex, and race	Hospital outpatient departments					Hospital emergency rooms				
	2001	2000	1999	1998	1995	2001	2000	1999	1998	1995
NUMBER OF VISITS (Thousands)										
Total	83 715	83 289	84 623	75 412	67 232	107 490	108 017	102 765	100 385	96 545
Under 18 years	21 299	21 076	21 758	18 551	17 636	26 239	27 630	25 688	26 415	26 657
18–44 years	27 430	26 947	29 514	26 032	24 299	46 109	45 816	43 532	42 064	41 820
45–64 years	21 590	20 772	20 891	17 980	14 811	19 260	18 339	17 886	16 425	13 978
45–54 years	12 016	11 558	11 541	9 859	8 029	11 707	11 201	10 686	9 970	8 595
55–64 years	9 574	9 214	9 350	8 120	6 782	7 552	7 138	7 200	6 455	5 383
65 years and over	13 396	14 494	12 461	12 849	10 486	15 883	16 232	15 659	15 482	14 090
65–74 years	7 299	7 515	6 969	6 869	6 004	6 551	6 543	6 455	6 350	6 057
75 years and over	6 097	6 979	5 493	5 979	4 482	9 332	9 690	9 205	9 132	8 033
NUMBER OF VISITS PER 100 PERSONS										
Total, age-adjusted	30	31	31	28	26	39	40	38	37	37
Total, crude	30	30	31	28	26	38	39	38	37	37
Under 18 years	29	29	30	26	25	36	38	36	37	38
18–44 years	25	25	27	24	22	42	42	40	39	39
45–64 years	34	34	36	32	29	30	30	31	29	27
45–54 years	31	31	32	29	26	30	30	30	29	28
55–64 years	38	39	41	36	33	30	30	31	29	26
65 years and over	40	44	38	40	33	47	50	48	48	45
65–74 years	40	42	39	38	33	36	37	36	35	33
75 years and over	39	47	37	42	34	60	65	63	63	61
Age										
Male, age-adjusted	25	26	26	23	21	37	38	37	36	37
Male, crude	24	25	25	23	21	37	38	37	36	36
Under 18 years	29	29	29	26	25	38	41	37	39	40
18–44 years	16	17	18	16	14	39	38	38	37	37
45–54 years	26	26	25	23	20	29	30	29	28	26
55–64 years	30	32	37	28	26	28	30	30	28	25
65–74 years	35	38	35	35	29	38	36	35	35	34
75 years and over	37	42	34	42	34	57	59	61	57	61
Female, age-adjusted	35	35	37	33	31	40	41	39	38	37
Female, crude	35	35	37	33	31	40	41	39	38	37
Under 18 years	29	29	31	26	25	34	35	34	35	35
18–44 years	34	33	36	32	31	45	46	42	41	40
45–54 years	36	36	40	34	32	31	31	31	30	29
55–64 years	45	45	44	44	38	32	31	32	30	26
65–74 years	45	46	43	41	36	35	37	37	35	32
75 years and over	40	49	39	41	34	61	69	64	67	61
Race and Age [1]										
White, age-adjusted	28	28	28	25	23	36	37	35	35	34
White, crude	28	28	28	25	23	36	37	35	35	34
Under 18 years	28	27	28	23	23	34	36	33	34	35
18–44 years	23	23	25	21	20	40	39	37	36	36
45–54 years	28	28	28	25	23	28	28	27	27	25
55–64 years	34	36	36	30	28	28	28	28	26	24
65–74 years	37	38	36	33	29	33	35	34	33	32
75 years and over	36	44	34	38	31	58	63	61	61	60
Black or African American, age-adjusted	51	51	54	55	48	65	62	62	63	58
Black or African American, crude	49	48	51	52	45	64	62	60	62	58
Under 18 years	44	40	42	43	39	55	57	51	55	53
18–44 years	41	40	45	44	38	71	68	68	67	64
45–54 years	56	61	66	63	55	56	53	55	54	51
55–64 years	79	70	83	90	73	56	52	57	59	47
65–74 years	70	85	89	88	*77	71	50	50	84	47
75 years and over	67	85	*79	85	66	91	92	93	103	73

[1]Beginning in 1999, the instruction for the race item on the patient record form was changed so that more than one race could be recorded. In previous years, only one racial category could be checked.

* = Figure does not meet standards of reliability or precision.

TABLE III-49. INJURY-RELATED VISITS TO HOSPITAL EMERGENCY DEPARTMENTS BY SEX, AGE, AND INTENT AND MECHANISM OF INJURY, ANNUAL AVERAGE, SELECTED YEARS, 1995–2001

Almost 40 million injury related visits to hospital emergency departments took place, taking an average for 2000 and 2001. There were 14.4 visits, for every 100 persons, 16.1 by males, and 12.7 by females.

The age group that was most-injury prone consisted of young men between 18 and 24 years of age; they made 24.2 visits to emergency departments for each representative 100. These were made up of 17.6 visits for unintentional injuries, 2.3 visits for intentional ones, and some for additional injuries of an undetermined nature or resulting from adverse effects of medical treatment. The largest categories among the unintentional injuries were motor vehicle traffic (3.8 per 100), and “struck by or against objects or persons” (3.3 per 100).

Females 18–24 years in age, incurred 17.5 visits per 100 to emergency departments, substantially fewer than their male contemporaries. In the area of injuries from motor vehicle traffic, however, females had 3.7 visits per 100, only slightly less than the 3.8 figure for males.

Females 18–24 years in age, incurred 17.5 visits per 100 to emergency departments, substantially fewer than their male contemporaries. In the area of injuries from motor vehicle traffic, however, females had 3.7 visits per 100, only slightly less than the 3.8 figure for males.

Table III-49. Injury-Related Visits to Hospital Emergency Departments by Sex, Age, Intent and Mechanism of Injury, Annual Average, Selected Years, 1995–2001

(Number, rate per 10,000 persons.)

Sex, age, and intent and mechanism of injury [1]	Injury-related visits in thousands				Injury-related visits per 10,000 persons			
	2000–2001	1999–2000	1997–1998	1995–1996	2000–2001	1999–2000	1997–1998	1995–1996
BOTH SEXES								
All Ages [2]	39 918	39 029	36 111	36 081	1 441.9	1 428.1	1 344.2	1 360.9
MALE								
All Ages [2]	21 852	21 286	19 838	20 030	1 606.2	1 585.3	1 500.0	1 530.7
Under 18 Years [2]	6 517	6 364	6 057	6 238	1 759.4	1 722.2	1 651.9	1 720.2
Unintentional injuries [3]	5 387	5 457	5 192	5 478	1 454.5	1 476.7	1 416.1	1 510.5
Falls	1 428	1 303	1 241	1 402	385.4	352.6	338.4	386.5
Struck by or against objects or persons	1 428	1 378	1 468	1 011	385.4	372.8	400.5	278.9
Motor vehicle traffic	435	432	405	453	117.5	116.9	110.6	125.0
Cut or pierce	403	455	482	493	108.9	123.2	131.5	136.0
Intentional injuries	255	242	229	290	68.8	65.6	62.4	80.0
18–24 Years [2]	3 225	3 096	2 805	2 980	2 418.6	2 361.6	2 222.5	2 396.9
Unintentional injuries [3]	2 347	2 416	2 169	2 423	1 760.3	1 842.7	1 718.8	1 948.7
Falls	281	307	255	299	210.5	233.9	202.1	240.8
Struck by or against objects or persons	443	405	456	387	332.3	308.6	361.7	311.0
Motor vehicle traffic	511	469	404	347	383.4	357.5	320.2	279.4
Cut or pierce	366	394	310	304	274.3	300.5	245.7	244.8
Intentional injuries	309	322	269	335	231.8	245.9	213.1	269.2
25–44 Years [2]	7 287	7 251	6 788	7 245	1 798.3	1 796.9	1 660.4	1 767.4
Unintentional injuries [3]	5 234	5 528	5 122	5 757	1 291.7	1 370.0	1 252.8	1 404.3
Falls	867	850	779	817	213.9	210.8	190.5	199.4
Struck by or against objects or persons	774	781	849	619	191.0	193.6	207.8	151.0
Motor vehicle traffic	793	848	831	912	195.7	210.1	203.2	222.6
Cut or pierce	777	764	741	860	191.8	189.4	181.1	209.8
Intentional injuries	458	511	526	701	113.0	126.5	128.8	171.0
45–64 Years [2]	3 218	2 972	2 755	2 240	1 066.9	1 030.9	1 020.4	883.4
Unintentional injuries [3]	2 314	2 325	2 108	1 845	767.0	806.7	781.0	727.6
Falls	542	582	512	445	179.7	202.0	189.5	175.6
Struck by or against objects or persons	286	232	202	186	94.6	80.6	74.8	73.3
Motor vehicle traffic	323	316	312	244	107.2	109.6	115.6	96.3
Cut or pierce	332	294	289	203	110.1	101.9	107.2	79.9
Intentional injuries	121	99	107	86	40.0	34.3	39.7	33.8
65 Years and over [2]	1 605	1 603	1 434	1 327	1 142.8	1 158.7	1 056.6	1 000.7
Unintentional injuries [3]	1 206	1 207	1 109	1 009	858.8	872.1	817.2	760.6
Falls	544	579	492	505	387.5	418.1	362.3	380.9
Struck by or against objects or persons	112	112	84	*39	79.8	*80.7	*61.9	*29.4
Motor vehicle traffic	144	114	124	99	102.4	*82.5	91.7	74.7
Cut or pierce	117	102	117	*81	83.3	74.0	86.4	*61.1
Intentional injuries	14	10	19	*	*	*	*	*

[1]Intent and mechanism of injury are based on the first-listed external cause of injury code. Intentional injuries include suicide attempts and assaults.
[2]Includes all injury-related visits not shown separately in table including those with undetermined intent (less than 0.5 percent in 2000–2001), insufficient or no information to code cause of injury (about 16.9 percent in 2000–2001), and resulting from adverse effects of medical treatment (about 3.3 percent in 2000–2001).
[3]Includes unintentional injury-related visits with mechanism of injury not shown in table.
* = Figure does not meet standards of reliability or precision.

Table III-49. Injury-Related Visits to Hospital Emergency Departments by Sex, Age, Intent and Mechanism of Injury, Annual Average, Selected Years, 1995–2001—*Continued*

(Number, rate per 10,000 persons.)

Sex, age, and intent and mechanism of injury [1]	Injury-related visits in thousands				Injury-related visits per 10,000 persons			
	2000–2001	1999–2000	1997–1998	1995–1996	2000–2001	1999–2000	1997–1998	1995–1996
FEMALE								
All Ages [2]	18 066	17 743	16 273	16 051	1 271.4	1 267.4	1 183.1	1 186.4
Under 18 Years [2]	4 269	4 443	4 100	4 372	1 207.0	1 259.0	1 172.0	1 263.9
Unintentional injuries [3]	3 408	3 722	3 395	3 760	963.7	1 054.7	970.5	1 087.0
Falls	966	1 025	821	1 040	273.2	290.6	234.6	300.7
Struck by or against objects or persons	670	728	704	477	189.5	206.4	201.4	137.9
Motor vehicle traffic	437	430	403	447	123.6	122.0	115.4	129.3
Cut or pierce	208	232	265	253	58.9	65.7	75.9	73.0
Intentional injuries	166	149	178	220	46.9	42.3	50.9	63.6
18–24 Years [2]	2 338	2 219	2 025	1 900	1 749.3	1 688.1	1 606.2	1 523.4
Unintentional injuries [3]	1 657	1 579	1 494	1 430	1 239.5	1 200.9	1 185.3	1 146.7
Falls	296	234	222	268	221.6	178.0	176.2	214.5
Struck by or against objects or persons	181	170	180	134	135.1	129.6	143.1	107.4
Motor vehicle traffic	500	469	473	373	373.9	357.1	374.9	298.8
Cut or pierce	162	156	121	131	120.9	118.3	96.0	105.3
Intentional injuries	192	219	227	239	143.3	166.8	179.9	191.7
25–44 Years [2]	5 645	5 584	5 050	5 098	1 345.3	1 332.7	1 194.2	1 205.8
Unintentional injuries [3]	3 834	3 976	3 720	3 877	913.9	948.9	879.9	916.8
Falls	903	947	830	817	215.2	225.9	196.2	193.3
Struck by or against objects or persons	358	382	447	380	85.3	91.3	105.7	89.8
Motor vehicle traffic	849	788	821	872	202.3	188.0	194.3	206.2
Cut or pierce	373	434	378	338	89.0	103.5	89.3	79.8
Intentional injuries	406	425	400	422	96.8	101.5	94.7	99.8
45–64 Years [2]	3 098	2 933	2 649	2 369	963.9	952.9	919.1	873.7
Unintentional injuries [3]	2 155	2 180	1 980	1 857	670.5	708.2	686.8	685.2
Falls	755	749	659	600	234.8	243.5	228.5	221.5
Struck by or against objects or persons	186	192	224	160	57.8	62.3	77.6	58.8
Motor vehicle traffic	345	324	331	343	107.2	105.2	114.7	126.5
Cut or pierce	183	175	192	127	56.8	56.8	66.6	46.9
Intentional injuries	86	125	88	64	26.8	40.5	30.4	23.5
65 Years and Over [2]	2 716	2 564	2 449	2 313	1 417.4	1 367.8	1 314.2	1 256.1
Unintentional injuries [3]	2 081	2 013	2 009	1 931	1 086.2	1 073.8	1 078.0	1 049.0
Falls	1 326	1 219	1 239	1 230	691.9	650.4	664.7	667.9
Struck by or against objects or persons	118	103	146	82	61.5	54.8	78.2	44.8
Motor vehicle traffic	130	132	163	169	67.8	70.6	87.5	91.6
Cut or pierce	65	72	68	42	*34.0	*38.3	*36.7	*22.7
Intentional injuries	22	20	31	*	*	*	*	*

[1]Intent and mechanism of injury are based on the first-listed external cause of injury code. Intentional injuries include suicide attempts and assaults.
[2]Includes all injury-related visits not shown separately in table including those with undetermined intent (less than 0.5 percent in 2000–2001), insufficient or no information to code cause of injury (about 16.9 percent in 2000–2001), and resulting from adverse effects of medical treatment (about 3.3 percent in 2000–2001).
[3]Includes unintentional injury-related visits with mechanism of injury not shown in table.
* = Figure does not meet standards of reliability or precision.

TABLE III-50. AMBULATORY CARE VISITS TO PRIMARY CARE AND SPECIALIST PHYSICIANS, ACCORDING TO SELECTED CHARACTERISTICS AND TYPE OF PHYSICIAN, SELECTED YEARS, 1980–2001

In 2001, 51.3 percent of doctor visits were to primary care physicians, that is, general and family practice doctors, internal medicine practitioners, or pediatricians. This percentage has declined gradually from 56.6 percent in 1980, reflecting a decline in the share of general and family practice appointments.

The remainder of the visits, 48.7 percent, were to specialists, including obstetricians and gynecologists (together 7.9 percent). Generally more than one-half of visits by adults were to specialists. Children frequently visit pediatricians, who are primary care physicians, and the age group under 18 years, shows only a 20 percent frequency of seeing a specialist.

In 2001—but not in earlier years—Blacks, in the 18–44 year age group, are shown to have a substantially greater preference for specialists when compared with White patients.

Table III-50. Ambulatory Care Visits to Primary Care and Specialist Physicians, According to Selected Characteristics and Type of Physician, Selected Years, 1980–2001

(Percent.)

Age, sex, and race	Type of primary care physician															
	All primary care				General and family practice				Internal medicine				Pediatrics			
	2001	2000	1990	1980	2001	2000	1990	1980	2001	2000	1990	1980	2001	2000	1990	1980
PERCENT OF ALL PHYSICIAN OFFICE VISITS																
All Persons	51.3	51.1	54.9	56.6	23.9	24.1	29.9	33.5	15.3	15.3	13.8	12.1	12.1	11.7	11.2	10.9
Under 18 years	80.2	78.6	78.3	76.6	16.6	19.9	26.5	26.1	*	*	2.9	2.0	62.1	57.3	48.9	48.5
18–44 years	42.3	41.7	44.3	43.6	28.2	28.2	31.9	34.3	13.2	12.7	11.8	8.6	*0.9	*0.9	0.7	0.7
45–64 years	43.7	46.7	50.9	56.0	25.5	26.4	32.1	36.3	18.3	20.1	18.6	19.5	0.0	*	*	*
45–54 years	43.2	46.7	49.4	54.6	25.8	27.8	32.0	37.4	17.3	18.7	17.1	17.1	0.0	*	*	*
55–64 years	44.4	46.7	52.4	57.3	25.0	24.7	32.1	35.4	19.4	21.7	20.0	21.8	0.0	*	*	*
65 years and over	47.4	45.0	51.5	60.3	22.9	20.2	28.1	37.5	24.5	24.5	23.3	22.7	*	*	*	*
65–74 years	46.8	44.6	51.2	59.5	25.0	19.7	28.1	37.4	21.7	24.5	23.0	22.1	*	*	*	*
75 years and over	48.1	45.4	51.8	61.3	20.9	20.8	28.0	37.6	27.2	24.5	23.7	23.5	*	*	*	*
SEX AND AGE																
Male																
Under 18 years	80.0	77.4	77.9	77.1	16.1	18.3	24.1	25.6	*	*	3.0	2.0	62.5	58.0	50.7	49.4
18–44 years	52.2	50.4	51.7	50.5	35.2	34.2	35.9	38.0	16.2	14.4	15.0	11.5	*0.9	*1.7	*	*
45–64 years	46.6	48.9	50.5	55.0	26.7	28.7	31.0	34.4	19.9	19.8	19.2	20.5	0.0	*	*	*
65 years and over	47.3	43.1	51.1	57.9	22.3	19.3	27.7	35.6	25.0	23.8	23.3	22.3	0.0	*	*	*
Female																
Under 18 years	80.5	79.9	78.8	76.0	17.1	21.7	29.1	26.6	*	*	2.8	2.0	61.7	56.5	46.9	47.4
18–44 years	37.2	37.6	41.0	40.4	24.7	25.3	30.0	32.5	11.6	11.9	10.3	7.3	*0.9	*	*	*
45–64 years	41.8	45.2	51.1	56.7	24.7	24.9	32.8	37.7	17.2	20.2	18.2	18.9	0.0	*	*	*
65 years and over	47.5	46.3	51.7	61.8	23.3	20.9	28.3	38.7	24.1	25.0	23.3	22.9	*	*	*	*
RACE AND AGE [1]																
White																
Under 18 years	79.7	77.3	78.2	76.5	16.8	21.2	27.1	26.4	*	*	2.3	2.0	61.3	54.7	48.8	48.2
18–44 years	43.2	41.0	43.2	43.8	28.8	29.2	31.9	34.5	13.5	11.0	10.6	8.6	*0.9	*0.8	*	*
45–64 years	44.2	44.6	49.4	55.4	26.2	27.3	31.5	36.0	17.9	17.1	17.6	19.2	0.0	*	*	*
65 years and over	47.5	43.6	50.7	60.0	23.0	20.3	27.5	36.6	24.4	23.0	23.1	23.3	*	*	*	*
Black or African American																
Under 18 years	71.2	86.4	82.1	77.1	*14.4	*	20.2	23.7	*	*	*	*	56.3	75.0	52.1	51.2
18–44 years	31.2	44.3	50.4	41.4	19.8	22.0	31.9	31.7	*11.0	20.9	18.1	9.0	*	*	*	*
45–64 years	42.3	59.4	58.2	61.3	18.0	23.3	31.2	38.6	*24.3	35.9	26.9	22.6	0.0	*	*	*
65 years and over	49.7	52.1	57.8	63.3	*21.9	*18.5	28.9	49.0	*27.7	33.4	28.7	14.2	0.0	*	*	*

[1]Beginning in 1999, the instruction for the race item on the Patient Record Form was changed so that more than one race could be recorded. In previous years, only one racial category could be checked.

* = Figure does not meet standards of reliability or precision.

Table III-50. Ambulatory Care Visits to Primary Care and Specialist Physicians, According to Selected Characteristics and Type of Physician, Selected Years, 1980–2001—*Continued*

(Percent.)

Age, sex, and race	Type of specialist physician											
	All specialists				Obstetrics and gynecology				All other specialists			
	2001	2000	1990	1980	2001	2000	1990	1980	2001	2000	1990	1980
PERCENT OF ALL PHYSICIAN OFFICE VISITS												
All Persons	48.7	48.9	45.1	43.4	7.9	7.9	9.0	9.6	40.8	40.9	36.1	33.8
Under 18 years	19.8	21.4	21.7	23.4	*	*1.1	1.2	1.3	18.7	20.3	20.5	22.2
18–44 years	57.7	58.3	55.7	56.4	19.3	20.7	21.5	21.7	38.4	37.5	34.1	34.7
45–64 years	56.3	53.3	49.1	44.0	6.7	4.6	4.8	4.2	49.6	48.8	44.3	39.8
45–54 years	56.8	53.3	50.6	45.4	8.3	5.6	6.5	5.6	48.5	47.7	44.2	39.8
55–64 years	55.6	53.3	47.6	42.7	4.8	3.3	3.2	2.9	50.8	50.1	44.4	39.8
65 years and over	52.6	55.0	48.5	39.7	*2.1	1.5	1.2	1.4	50.5	53.5	47.3	38.4
65–74 years	53.2	55.4	48.8	40.5	*2.7	2.1	1.6	1.7	50.5	53.4	47.2	38.8
75 years and over	51.9	54.6	48.2	38.7	*1.5	*1.0	*0.7	1.0	50.4	53.6	47.5	37.7
SEX AND AGE												
Male												
Under 18 years	20.0	22.6	22.1	22.9	X	X	X	X	20.0	22.3	21.9	22.7
18–44 years	47.8	49.6	48.3	49.5	X	X	X	X	47.7	48.5	48.2	49.2
45–64 years	53.4	51.1	49.5	45.0	X	X	X	X	53.4	50.6	49.4	44.4
65 years and over	52.7	56.9	48.9	42.1	X	X	X	X	52.7	56.9	48.8	41.8
Female												
Under 18 years	19.5	20.1	21.2	24.0	*	2.1	2.3	2.5	17.3	18.0	18.9	21.5
18–44 years	62.8	62.4	59.0	59.6	29.1	30.2	31.4	31.7	33.7	32.2	27.6	27.9
45–64 years	58.2	54.8	48.9	43.3	11.2	7.3	7.9	6.7	47.0	47.5	40.9	36.6
65 years and over	52.5	53.7	48.3	38.2	*3.6	2.6	1.9	2.1	48.9	51.1	46.4	36.1
RACE AND AGE [1]												
White												
Under 18 years	20.3	22.7	21.8	23.5	*1.1	*1.2	1.0	1.1	19.2	21.5	20.8	22.4
18–44 years	56.8	59.0	56.8	56.2	19.2	20.8	21.8	21.0	37.6	38.2	35.0	35.2
45–64 years	55.8	55.4	50.6	44.6	6.3	4.8	4.9	4.1	49.5	50.6	45.7	40.4
65 years and over	52.5	56.4	49.3	40.0	*2.0	1.5	1.3	1.4	50.5	54.9	48.1	38.6
Black or African American												
Under 18 years	*28.8	*13.6	17.9	22.9	*	*	*3.4	2.8	*26.3	*12.7	14.5	20.1
18–44 years	68.8	55.7	49.6	58.6	*23.9	20.7	18.6	27.1	44.9	35.0	31.0	31.5
45–64 years	57.7	40.6	41.8	38.7	*	*2.4	4.0	4.8	48.5	38.3	37.9	33.9
65 years and over	50.3	47.9	42.2	36.7	*	*	*	*	45.9	47.0	41.3	35.4

[1]Beginning in 1999, the instruction for the race item on the Patient Record Form was changed so that more than one race could be recorded. In previous years, only one racial category could be checked.
X = Category not applicable.
* = Figure does not meet standards of reliability or precision.

Figure III-14. Ambulatory Care Visits to Primary Care and Specialist Physician According to Type of Physician, 2001

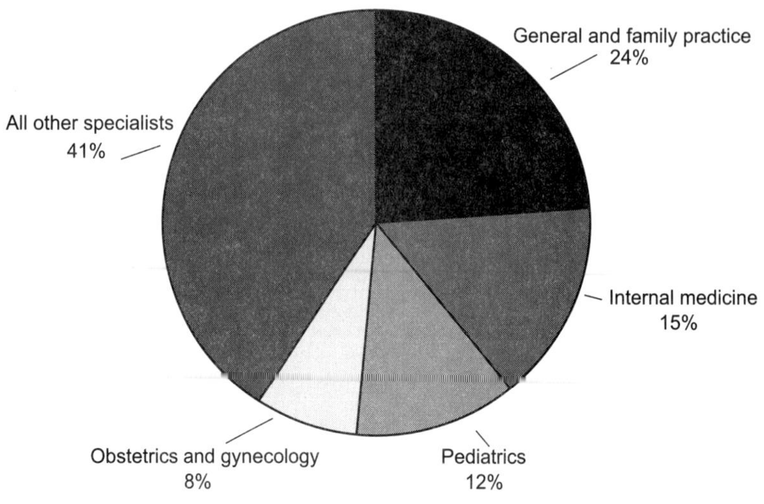

TABLE III-51. SUBSTANCE ABUSE CLIENTS IN SPECIALTY TREATMENT UNITS, ACCORDING TO SUBSTANCE ABUSED, GEOGRAPHIC DIVISION AND STATE, SELECTED YEARS, 1998–2002

A one-day census of treatment providers furnished information on clients for substance abuse treatment. The survey covers inpatients as well as outpatients who have been seen on a scheduled appointment basis. In 2002, such clients together amounted to 488 per 100,000 of the U.S. population. Of these 235 were both alcohol and drug abusers, 103 were alcohol abusers only, and 149 were drug abusers only.

There was considerable variation in the frequency of clients by state. The District of Columbia had the highest rate of clients, 1,309 per 100,000; drug abuse clients only (605 per 100,000) were especially frequent in this jurisdiction. For all treatment clients, this was followed by New York (925), Colorado (909), and Oregon (816). The lowest rate of clients was reported in Arkansas (175 per 100,000), Tennessee (201) and Texas (224).

Table III-51. Substance Abuse Clients in Specialty Treatment Units, According to Substance Abused, Geographic Division, and State, Selected Years, 1998–2002

(Rate per 100,000 population.)

Geographic division and state	All clients			Clients with both alcoholism and drug abuse			Alcoholism clients only			Drug abuse clients only		
	2002	2000	1998	2002	2000	1998	2002	2000	1998	2002	2000	1998
United States	488.0	434.9	461.9	235.4	211.5	228.6	103.1	98.0	109.8	149.4	125.4	123.5
New England	648.9	617.6	703.7	291.6	289.0	365.5	126.3	118.5	160.8	231.0	210.1	177.4
Maine	613.0	462.0	809.2	309.3	241.7	406.2	165.9	152.8	290.2	137.9	67.5	112.7
New Hampshire	303.1	324.1	338.4	141.3	180.7	174.6	118.2	95.8	132.5	43.6	47.6	31.3
Vermont	456.2	536.3	507.0	261.3	275.0	278.2	124.0	198.6	166.4	70.9	62.7	62.4
Massachusetts	682.9	671.4	824.4	328.2	331.1	461.2	129.4	126.4	171.8	225.4	213.8	191.4
Rhode Island	689.2	704.2	768.0	212.2	257.6	355.4	143.1	107.8	155.0	334.0	338.8	257.5
Connecticut	754.0	674.5	585.5	303.2	280.0	258.4	104.9	86.2	101.0	345.9	308.3	226.1
Middle Atlantic	650.0	556.7	554.3	319.9	272.9	259.7	82.8	79.1	86.7	247.3	204.7	207.9
New York	925.5	779.0	773.0	456.7	376.7	330.2	115.8	102.6	114.2	353.1	299.7	328.6
New Jersey	461.5	349.2	367.1	189.0	150.3	178.6	47.8	48.2	56.3	224.7	150.7	132.2
Pennsylvania	374.3	369.5	357.0	207.8	202.4	209.7	58.1	65.3	66.4	108.4	101.8	80.9
East North Central	475.8	421.0	473.8	219.9	194.1	224.5	127.5	121.4	141.1	128.4	105.5	108.3
Ohio	408.2	409.9	452.5	233.0	226.6	253.9	96.0	105.2	119.7	79.2	78.2	79.0
Indiana	548.9	313.9	339.7	277.0	159.3	148.8	148.8	90.4	116.4	123.1	64.2	74.5
Illinois	453.4	419.8	466.5	184.4	179.0	230.2	104.9	104.3	113.3	164.1	136.4	122.9
Michigan	547.4	541.6	614.6	224.8	217.5	249.3	154.9	171.5	198.8	167.7	152.7	166.5
Wisconsin	456.9	351.6	431.8	195.9	157.4	189.0	171.6	139.2	172.3	89.3	55.0	70.5
West North Central	344.9	322.2	355.0	199.3	179.9	199.9	79.4	79.3	97.3	66.3	63.0	57.8
Minnesota	229.6	205.9	264.0	128.7	100.2	140.4	54.3	54.7	67.1	46.6	51.0	56.5
Iowa	334.6	229.0	300.9	189.7	117.1	150.5	88.4	76.1	107.9	56.5	35.8	42.4
Missouri	404.4	378.9	387.0	237.3	216.9	249.2	75.6	73.1	73.7	91.5	88.9	64.1
North Dakota	330.9	251.6	549.3	176.3	120.4	258.7	116.0	114.4	224.0	38.6	16.8	66.6
South Dakota	407.8	290.6	445.9	216.1	148.4	201.9	149.3	115.9	211.2	42.4	26.4	32.8
Nebraska	374.2	323.1	396.2	238.5	162.5	220.2	76.5	96.6	122.4	59.3	64.0	53.6
Kansas	408.9	544.7	411.6	235.0	351.7	231.0	95.3	111.1	109.1	78.6	81.9	71.6
South Atlantic	427.0	406.1	382.0	196.4	199.9	191.7	88.6	99.1	94.2	142.0	107.1	96.0
Delaware	623.2	589.8	603.5	473.3	371.3	306.3	60.5	122.5	127.5	89.4	96.0	169.6
Maryland	808.1	696.6	557.6	327.5	293.3	256.0	153.2	133.0	117.2	327.4	270.3	184.3
District of Columbia	1 309.5	1 395.4	1 447.8	569.2	616.3	879.7	135.3	195.8	199.6	605.0	583.2	368.5
Virginia	391.3	388.1	364.5	185.6	205.9	184.9	98.4	103.1	102.3	107.2	79.2	77.3
West Virginia	304.6	317.4	295.8	157.8	123.2	103.5	89.0	143.2	142.0	57.9	51.0	50.3
North Carolina	407.1	471.8	402.0	203.1	237.6	214.6	96.6	140.6	115.5	107.4	93.6	71.9
South Carolina	339.0	414.1	306.6	162.6	191.5	116.3	102.5	138.0	112.6	73.9	84.6	77.6
Georgia	276.9	199.3	252.0	135.6	100.5	115.5	67.6	47.5	65.4	73.6	51.3	71.1
Florida	398.4	352.3	365.1	171.3	185.0	199.1	65.2	71.8	70.2	162.0	95.4	95.8
East South Central	301.5	307.8	327.8	141.1	136.7	151.8	70.0	79.9	89.1	90.4	91.1	86.9
Kentucky	541.6	534.6	442.4	247.5	236.8	199.1	162.1	178.1	161.4	132.1	119.6	81.9
Tennessee	201.2	195.5	280.0	94.0	77.8	110.9	33.2	42.4	71.4	74.1	75.3	97.7
Alabama	254.1	233.0	244.8	110.4	81.7	117.1	43.2	37.7	47.4	100.5	113.6	80.3
Mississippi	237.6	328.7	391.3	134.3	199.0	221.7	58.1	80.0	86.7	45.2	49.7	83.0
West South Central	244.3	268.8	329.3	129.9	150.0	186.1	37.6	42.2	63.8	76.8	76.6	79.5
Arkansas	175.5	141.2	326.1	95.6	73.7	190.6	28.7	22.2	66.6	51.3	45.4	68.9
Louisiana	327.6	311.2	472.6	160.6	157.8	268.8	55.2	42.3	88.0	111.8	111.1	115.8
Oklahoma	309.6	260.9	315.2	179.6	143.2	125.4	65.7	70.6	96.7	64.3	47.0	93.2
Texas	224.2	278.0	299.7	119.5	159.9	177.3	30.3	39.9	52.1	74.4	78.2	70.3
Mountain	630.2	596.8	584.8	291.0	261.2	278.1	178.2	194.5	175.1	161.0	141.2	131.7
Montana	318.0	244.1	321.9	190.7	108.2	172.8	83.6	96.8	107.8	43.7	39.1	41.3
Idaho	359.7	257.2	278.4	223.0	169.5	178.6	82.5	55.9	58.4	54.3	31.9	41.3
Wyoming	450.6	523.9	406.2	253.7	317.7	193.0	138.8	154.5	160.2	58.1	51.7	53.0
Colorado	909.4	843.4	722.4	406.5	333.5	326.7	350.4	351.1	267.3	152.5	158.7	128.4
New Mexico	678.7	669.7	714.0	308.6	281.7	296.6	183.7	215.3	275.3	186.4	172.7	142.1
Arizona	668.3	661.0	532.6	276.1	277.4	236.5	148.2	201.8	137.9	244.0	181.9	158.2
Utah	490.4	409.6	704.7	286.9	206.4	351.7	88.2	79.4	145.4	115.4	123.8	207.5
Nevada	428.3	464.5	548.4	170.8	229.3	322.2	95.5	90.1	116.7	162.0	145.1	109.5
Pacific	628.1	466.4	525.1	309.6	229.8	256.8	145.0	98.2	120.1	173.5	138.4	148.2
Washington	745.9	676.4	674.8	429.4	391.7	398.4	189.2	169.1	182.7	127.3	115.6	93.7
Oregon	815.9	759.7	653.4	461.9	431.4	347.8	160.6	149.4	138.5	193.4	178.9	167.0
California	597.3	401.7	491.4	275.3	180.7	223.7	136.6	79.0	106.2	185.4	142.0	161.5
Alaska	564.1	536.1	583.0	317.7	250.2	287.8	198.9	228.7	252.2	47.5	57.2	43.0
Hawaii	346.6	254.3	299.6	169.1	128.1	169.1	72.1	38.3	64.6	105.4	87.9	66.0

TABLE III-52. ADDITIONS TO MENTAL HEALTH ORGANIZATIONS, ACCORDING TO TYPE OF SERVICE AND ORGANIZATION, SELECTED YEARS, 1986–1998

By "addition" is meant a new admission, a readmission, a return from long-term leave, or a transfer from another service. This table indicates the volume of care by state and local, Department of Veterans' Affairs, and private organizations. Around the clock treatment and less than 24-hour care are shown separately.

In 1998, there were 2.3 million additions to around the clock treatment and 4.2 million additions to less than 24-hour care. This represented a rate of 873 per 100,000 (or somewhat less than 1 percent) for the former treatment and a rate of 1,571 per 100,000 for the latter treatment. The largest providers of 24-hour treatment were non-federal general hospitals, but in the case of less than overnight care the top providers were "all other organizations," consisting of freestanding psychiatric outpatient clinics, partial care organizations, and multi-service mental health organizations.

Table III-52. Additions to Mental Health Organizations, According to Type of Service and Organization, Selected Years, 1986–1998

(Number, Rate per 100,000)

Service and organization	Additions [1] in thousands					Additions per 100,000 civilian population				
	1998 [2]	1994 [2]	1992	1990	1986	1998 [2]	1994 [2]	1992	1990	1986
24–Hour Hospital and Residential Treatment [3]										
All organizations	2 349	2 267	2 092	2 035	1 819	872.9	874.6	830.1	833.7	759.9
State and county mental hospitals	206	238	275	276	333	76.8	92.0	109.3	113.2	139.1
Private psychiatric hospitals	489	485	470	407	235	181.6	187.1	186.4	166.5	98.0
Non-federal general hospital psychiatric services	1 154	1 067	951	960	849	428.9	411.5	377.4	393.2	354.8
Department of Veterans' Affairs medical centers [4]	147	173	181	198	180	54.7	66.9	71.6	81.2	75.1
Residential treatment centers for emotionally disturbed children	50	47	36	42	25	18.4	18.0	14.4	17.0	10.2
All other organizations [5]	303	257	179	153	198	112.5	99.0	70.9	62.6	82.7
Less Than 24–Hour Care [6]										
All organizations	4 228	3 516	3 164	3 298	2 955	1 571.2	1 356.8	1 255.2	1 352.4	1 233.4
State and county mental hospitals	42	42	50	48	68	15.7	16.1	19.7	19.8	28.4
Private psychiatric hospitals	237	214	206	163	132	87.9	82.4	81.8	66.9	55.2
Non-federal general hospital psychiatric services	663	498	480	659	533	246.3	192.0	190.2	270.0	222.4
Department of Veterans' Affairs medical centers [4]	149	132	159	184	133	55.3	51.1	63.1	75.3	55.3
Residential treatment centers for emotionally disturbed children	155	167	121	100	67	57.6	64.6	48.0	40.8	28.1
All other organizations [5]	2 982	2 464	2 149	2 145	2 022	1 108.4	950.7	852.4	879.6	844.0

[1]Admissions plus transfers from another service.
[2]Beginning in 1994, data for supportive residential clients (moderately staffed housing arrangements such as supervised apartments, group homes, and halfway houses) are included in the totals and all other organizations.This change affects the comparability of trend data prior to 1994 with data for 1994 and later years.
[3]These data exclude mental health care provided in non-psychiatric units of hospitals such as general medical units.
[4]Includes Department of Veterans' Affairs (VA) neuropsychiatric hospitals,VA general hospital psychiatric services, and VA psychiatric outpatient clinics.
[5]Includes freestanding psychiatric outpatient clinics, partial care organizations, and multiservice mental health organizations.
[6]Formerly reported as partial care and outpatient treatment, the survey format was changed in 1994 and the reporting of these services was combined due to similarities in the care provided. These data exclude office-based mental health care (psychiatrists, psychologists, licensed clinical social workers, and psychiatric nurses).

TABLE III-53. HOME HEALTH CARE PATIENTS, ACCORDING TO AGE, SEX, AND DIAGNOSIS, SELECTED YEARS, 1992–2000

Home health care is provided by home health care agencies to individuals and families in their place of residence for promoting health or minimizing the effects of disability and illness including terminal illness. Almost 1.4 million patients were reported in 2000 at the time of a survey, a little more than in 1992, but below the levels of intervening years. Almost 30 percent of the patients were under 65 years of age, but, in terms of rates per 10,000 of population, the largest client group was 85 years and over. Female clients were much more frequent than males. Care for diseases of the circulatory system ranked as most frequent, and accounted for 23.6 percent of diagnosed illnesses.

Table III-53. Home Health Care Patients, According to Age, Sex, and Diagnosis, Selected Years, 1992–2000

(Number, rate per 10,000, percent.)

Age, sex, and diagnosis	2000	1998	1996	1994	1992
NUMBER OF CURRENT PATIENTS					
Total Home Health Care Patients	1 355 290	1 881 768	2 427 483	1 889 327	1 232 200
CURRENT PATIENTS PER 10,000 POPULATION					
Total	48.7	69.6	90.6	71.8	47.8
Age					
Under 65 years, crude	16.4	25.0	27.8	21.0	12.6
65 years and over, crude	277.0	375.7	526.3	424.9	295.4
65 years and over, age-adjusted	276.5	381.0	546.6	449.6	315.8
65–74 years	130.2	202.0	240.1	209.1	151.7
75–84 years	347.6	470.3	753.6	542.2	398.3
85 years and over	694.1	885.4	1 253.4	1 206.1	775.9
Sex					
Male, total	35.1	47.9	60.9	47.8	32.6
Under 65 years, crude	15.6	22.9	22.1	17.8	10.9
65 years and over, crude	199.6	255.2	386.4	303.1	219.2
65 years and over, age-adjusted	216.4	277.6	438.3	350.0	255.8
65–74 years	100.7	159.7	187.0	169.9	121.8
75–84 years	270.0	321.4	598.7	427.5	322.0
85 years and over	553.9	653.0	1 044.3	893.1	635.2
Female, total	61.8	90.4	118.9	94.7	62.4
Under 65 years, crude	17.2	27.0	33.6	24.2	14.3
65 years and over, crude	332.6	460.4	623.9	508.9	347.4
65 years and over, age-adjusted	315.5	445.8	615.0	506.6	351.5
65–74 years	154.6	236.3	283.2	240.6	175.3
75–84 years	400.4	568.8	854.0	614.5	445.3
85 years and over	754.9	981.7	1 337.0	1 327.6	830.7
PERCENT DISTRIBUTION					
Age [1]					
Under 65 years	29.5	31.3	27.0	25.7	23.1
65 years and over	70.5	68.7	73.0	74.3	76.9
65–74 years	17.3	19.7	18.4	20.6	22.6
75–84 years	31.3	29.9	35.3	31.2	33.9
85 years and over	21.9	19.1	19.4	22.4	20.4
Sex					
Male	35.2	33.6	32.9	32.5	33.2
Female	64.8	66.4	67.1	67.5	66.8
Primary Admission Diagnosis [2]					
Malignant neoplasms	4.9	3.8	4.8	5.7	5.7
Diabetes	7.8	6.1	8.5	8.1	7.7
Diseases of the nervous system and sense organs	6.1	7.6	5.8	8.0	6.3
Diseases of the circulatory system	23.6	23.6	25.6	27.2	25.9
Diseases of heart	10.9	12.3	10.9	14.3	12.6
Cerebrovascular diseases	7.3	5.1	7.8	6.1	5.8
Diseases of the respiratory system	6.8	7.9	7.7	6.1	6.6
Decubitus ulcers	1.9	1.2	1.0	1.1	1.9
Diseases of the musculoskeletal system and connective tissue	9.8	8.3	8.8	8.3	9.4
Osteoarthritis	3.5	2.7	3.2	2.8	2.5
Fractures, all sites	4.1	4.0	3.3	3.7	3.8
Fracture of neck of femur (hip)	1.5	1.1	1.3	1.7	1.4
Other	34.9	37.5	34.6	31.8	32.7

[1]Denominator excludes persons with unknown age.
[2]Denominator excludes persons with unknown diagnosis.

TABLE III-54. HOSPICE PATIENTS, ACCORDING TO AGE, SEX, AND DIAGNOSIS, SELECTED YEARS 1992–2000

Hospice services are provided in the home or on an inpatient basis. In 2000, just prior to a survey of hospice agencies, there were a little over 100,000 patients of hospices. This was about twice as many as in 1992, and it represented a rate of 3.8 per 10,000 of population. Rates of attendance increase with age and are somewhat higher for women than for men. About 52 percent of the patients had cancer and 13 percent had heart diseases.

Table III-54. Hospice Patients, According to Age, Sex, and Diagnosis, Selected Years, 1992–2000

(Number, number per 10,000, percent.)

Age, sex, and diagnosis	2000	1998	1996	1994	1992
NUMBER OF CURRENT PATIENTS					
Total Hospice Patients	105 496	79 837	59 363	60 783	52 100
CURRENT PATIENTS PER 10,000 POPULATION					
Total	3.8	3.0	2.2	2.3	2.0
Age					
Male					
Under 65 years, crude	0.8	0.7	0.5	0.8	0.5
65 years and over, crude	24.9	18.2	13.9	12.9	13.1
65 years and over, age-adjusted	24.9	18.4	14.4	13.6	13.7
65–74 years	10.1	9.9	7.8	7.3	7.8
75–84 years	31.9	22.0	16.9	16.9	19.2
85 years and over	67.3	44.7	34.7	30.6	23.4
Sex					
Male, total	3.3	2.6	2.0	2.1	1.9
Under 65 years, crude	0.8	0.7	0.5	0.9	0.5
65 years and over, crude	24.8	18.5	14.8	12.5	13.9
65 years and over, age-adjusted	26.9	20.3	16.1	14.4	16.0
65–74 years	13.0	10.2	10.4	7.0	6.3
75–84 years	32.6	25.2	18.5	18.2	25.8
85 years and over	69.9	49.2	33.9	34.8	28.8
Female, total	4.3	3.3	2.4	2.5	2.1
Under 65 years, crude	0.9	0.8	0.6	0.7	0.4
65 years and over, crude	25.0	18.0	13.2	13.2	12.6
65 years and over, age-adjusted	23.3	17.3	12.9	13.2	12.6
65–74 years	7.6	9.6	5.8	7.5	8.9
75–84 years	31.5	19.9	15.9	16.1	15.1
85 years and over	66.2	42.9	35.0	29.0	21.4
PERCENT DISTRIBUTION					
Age [1]					
Under 65 years	18.6	21.6	21.3	30.1	19.5
65 years and over	81.4	78.4	78.7	69.9	80.5
65–74 years	17.2	22.7	24.5	22.2	27.3
75–84 years	37.0	32.9	32.4	30.1	38.6
85 years and over	27.3	22.7	21.9	17.6	14.6
Sex					
Male	42.6	42.7	44.9	44.7	46.1
Female	57.4	57.3	55.1	55.3	53.9
Primary Admission Diagnosis [2]					
Malignant neoplasms	51.9	55.5	58.3	57.2	65.7
Large intestine and rectum	4.9	6.4	4.0	8.0	9.0
Treachea, bronchus, and lung	12.3	13.0	15.8	12.5	21.1
Breast	4.8	4.9	6.2	4.8	3.9
Prostate	7.7	6.1	6.6	5.9	6.0
Diseases of the heart	12.8	9.7	8.3	9.3	10.2
Diseases of the respiratory system	6.5	10.6	7.3	6.6	4.3
Other	28.8	24.3	26.1	27.0	19.8

[1]Denominator excludes persons with unknown age.
[2]Denominator excludes persons with unknown diagnosis.

Figure III-15. Total Number of Hospice Patients, 1992–2000

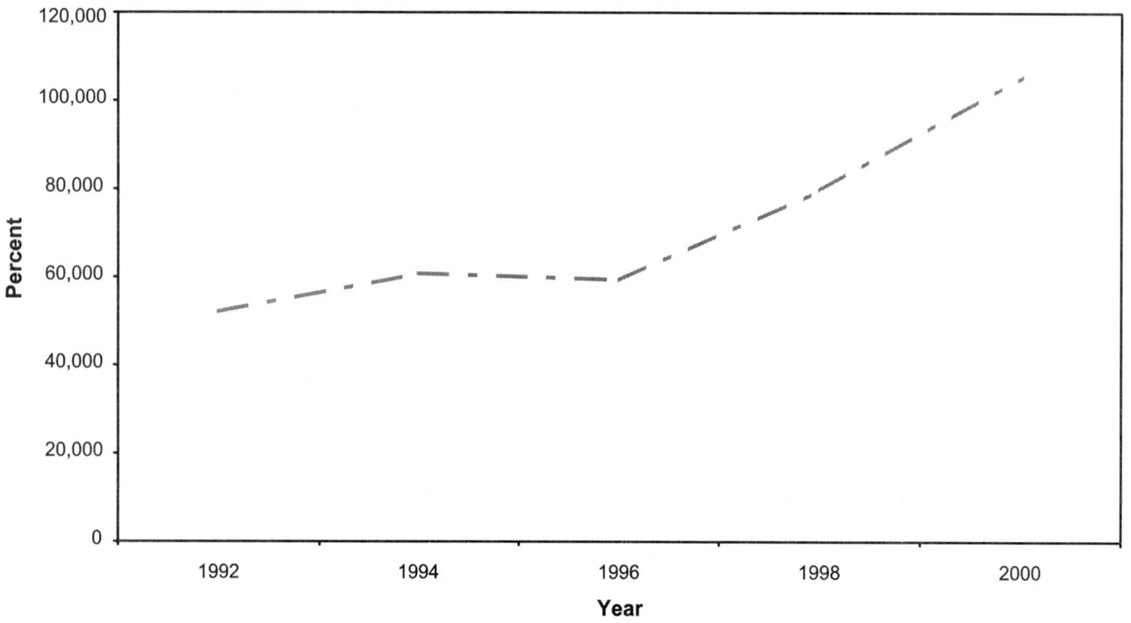

Table III-55. Discharges, Days of Care, and Average Length of Stay in Short-Stay Hospitals, According to Selected Characteristics, Selected Years, 1997–2001

(Number per 1,000 population, number.)

Characteristic	Number per 1,000 population						Number of days		
	Discharges			Days of care			Average length of stay		
	2001	1999	1997	2001	1999	1997	2001	1999	1997
TOTAL [1]	122.0	119.7	124.3	554.2	555.1	601.2	4.5	4.6	4.8
Age									
Under 18 years	78.6	76.3	90.8	312.4	302.6	319.0	4.0	4.0	3.5
Under 6 years	184.6	183.2	203.5	674.0	664.8	632.6	3.7	3.6	3.1
6–17 years	27.0	24.3	34.0	136.5	*126.5	163.1	5.1	*5.2	4.8
18–44 years	94.8	95.8	96.8	348.9	352.8	358.8	3.7	3.7	3.7
45–64 years	125.2	125.6	124.9	616.0	592.5	631.1	4.9	4.7	5.1
45–54 years	103.7	110.1	99.2	465.9	473.9	527.5	4.5	4.3	5.3
55–64 years	159.2	149.6	164.8	853.5	775.5	792.4	5.4	5.2	4.8
65 years and over	286.6	269.7	274.4	1 560.6	1 620.5	1 852.5	5.4	6.0	6.8
65–74 years	231.2	229.8	249.1	1 239.0	1 386.4	1 595.2	5.4	6.0	6.4
75 years and over	351.9	318.5	307.3	1 941.2	1 907.6	2 188.4	5.5	6.0	7.1
UNDER 65 YEARS OF AGE									
Total [1]	97.8	97.6	102.2	406.6	398.9	416.4	4.2	4.1	4.1
Sex									
Male	76.1	77.9	79.1	372.9	374.0	374.9	4.9	4.8	4.7
Female	119.0	116.7	124.7	439.7	422.8	456.6	3.7	3.6	3.7
Race [2]									
White only	93.2	94.7	100.8	369.4	368.7	385.8	4.0	3.9	3.8
Black or African American only	130.3	122.8	126.3	657.2	638.3	688.6	5.0	5.2	5.5
American Indian and Alaska Native only	169.2	128.3	111.9	*767.6	*570.0	*494.3	*4.5	*4.4	*4.4
Asian only	68.0	78.4	61.7	228.7	*249.5	*268.6	3.4	*3.2	*4.4
Native Hawaiian and Other Pacific Islander only	*	*	. . .	*	*	. . .	*	*	. . .
Two or more races	139.5	139.1	. . .	586.2	688.8	. . .	*4.2	*5.0	. . .
Hispanic Origin and Race [2]									
Hispanic or Latino	101.5	90.1	109.9	406.3	389.8	416.7	4.0	4.3	3.8
Not Hispanic or Latino	97.5	98.7	101.2	407.9	401.5	415.4	4.2	4.1	4.1
White only	92.6	95.4	99.6	368.2	368.4	382.7	4.0	3.9	3.8
Black or African American only	129.4	122.7	125.7	656.0	625.5	692.6	5.1	5.1	5.5
Poverty Status [3]									
Poor	167.9	174.0	196.8	857.7	992.9	971.0	5.1	5.7	4.9
Near poor	136.2	150.3	125.5	646.5	671.4	553.7	4.7	4.5	4.4
Nonpoor	86.5	82.1	85.6	316.7	291.9	312.1	3.7	3.6	3.6
Hispanic Origin, Race and Poverty Status [2, 3]									
Hispanic or Latino									
Poor	167.8	122.3	163.9	723.9	555.2	625.1	4.3	4.5	3.8
Near poor	93.5	97.0	93.9	362.0	*434.9	421.4	3.9	*4.5	4.5
Nonpoor	79.8	79.2	95.4	288.1	295.4	297.9	3.6	3.7	3.1
Not Hispanic or Latino									
White only									
Poor	145.7	181.2	222.2	812.3	1 042.5	1 053.4	5.6	5.8	4.7
Near poor	144.3	166.8	132.8	725.4	683.8	539.1	5.0	4.1	4.1
Nonpoor	86.6	81.5	85.7	305.4	289.7	306.8	3.5	3.6	3.6
Black or African American only									
Poor	230.4	223.6	195.9	1 243.3	1 424.2	1 260.0	5.4	6.4	6.4
Near poor	163.7	154.1	142.0	732.2	*920.3	819.2	4.5	*6.0	5.8
Nonpoor	96.8	86.7	92.3	489.8	332.7	389.0	5.1	3.8	4.2
Health Insurance Status [4]									
Insured	104.3	101.6	108.1	433.2	416.8	442.5	4.2	4.1	4.1
Private	84.4	80.2	85.6	311.8	287.8	310.2	3.7	3.6	3.6
Medicaid	296.2	332.5	311.6	1 495.1	1 695.5	1 575.3	5.0	5.1	5.1
Uninsured	64.2	75.7	75.3	270.9	304.2	296.3	4.2	4.0	3.9
Poverty Status and Health Insurance Status [3]									
Poor									
Insured	210.9	217.1	243.9	1 123.1	1 294.5	1 272.5	5.3	6.0	5.2
Unsured	90.1	101.5	110.0	366.5	476.9	459.4	4.1	4.7	4.2
Near poor									
Insured	164.7	184.3	149.2	788.6	837.5	663.8	4.8	4.5	4.4
Uninsured	72.9	75.2	73.4	*333.1	295.5	302.1	*4.6	3.9	4.1
Nonpoor									
Insured	88.9	84.1	88.1	327.4	299.7	316.0	3.7	3.6	3.6
Uninsured	54.5	58.8	59.8	166.4	197.6	*253.5	3.1	3.4	*4.2

[1] Includes all other races not shown separately, unknown poverty status, and unknown health insurance status.
[2] The race groups, White, Black, American Indian and Alaska Native (AI/AN), Asian, Native Hawaiian and Other Pacific Islander, and two or more races, include persons of Hispanic and non-Hispanic origin. Persons of Hispanic origin may be of any race.
[3] Poor persons are defined as below the poverty threshold. Near poor persons have incomes of 100 percent to less than 200 percent of poverty threshold. Nonpoor persons have incomes of 200 percent or greater than the poverty threshold.
[4] Health insurance categories are mutually exclusive. Persons who reported both Medicaid and private coverage are classified as having private coverage. Persons 65 years of age and over who reported Medicare HMO (health maintenance organization) and some other type of health insurance coverage are classified as having Medicare HMO.
[5] MSA = metropolitan statistical area.
* = Figure does not meet standards of reliability or precision.
. . . = Not available.

INPATIENT CARE

TABLE III-55. DISCHARGES, DAYS OF CARE, AND AVERAGE LENGTH OF STAY IN SHORT-STAY HOSPITALS, ACCORDING TO SELECTED CHARACTERISTICS, SELECTED YEARS, 1997–2001

"Discharges" occur when inpatient hospitalizations are completed, and thus they give a measure of the number of hospital stays. In 2001, there were 122 discharges per each 1000 of the population. On average patients had stayed 4.5 days, so that hospital days per 1,000 population were 554.

Poor persons had longer average hospital stays than the nonpoor and, by race, non-Hispanic Blacks had longer stays than non-Hispanic Whites and than Hispanics. Medicaid patient had both a higher rate of discharge and a longer length of stay, so that they spent almost five times as many days in the hospital (per 1,000) as privately insured persons. The table also gives separate statistics for persons 65 years of age and above. One item of interest is that Medicare patients, whose provider is an HMO, had fewer days in the hospital than fee-for-service members of Medicare.

Table III-55. Discharges, Days of Care, and Average Length of Stay in Short-Stay Hospitals, According to Selected Characteristics, Selected Years, 1997–2001—*Continued*

(Number per 1,000 population, number.)

Characteristic	Number per 1,000 population						Number of days		
	Discharges			Days of care			Average length of stay		
	2001	1999	1997	2001	1999	1997	2001	1999	1997
Geographic region									
Northeast	87.5	85.6	96.0	403.9	381.6	455.4	4.6	4.5	4.7
Midwest	100.5	99.6	108.7	400.0	359.9	384.4	4.0	3.6	3.5
South	110.6	112.8	111.8	456.3	463.9	466.1	4.1	4.1	4.2
West	82.5	80.0	82.9	332.4	348.3	327.2	4.0	4.4	3.9
Location of Residence									
Within MSA [5]	96.1	94.0	99.3	400.1	383.0	411.8	4.2	4.1	4.1
Outside MSA [5]	104.7	111.9	113.2	432.7	459.4	435.9	4.1	4.1	3.8
65 YEARS OF AGE AND OVER									
Total [1]	288.8	272.1	276.9	1 574.3	1 635.3	1 878.4	5.5	6.0	6.8
Sex									
Male	304.4	280.2	291.6	1 746.2	1 551.7	2 077.4	5.7	5.5	7.1
Female	277.1	264.0	265.2	1 457.2	1 676.5	1 727.4	5.3	6.4	6.5
Hispanic Origin and Race [2]									
Hispanic or Latino	304.4	289.8	312.7	1 568.9	1 882.8	2 512.1	5.2	6.5	8.0
Not Hispanic or Latino	287.8	271.2	274.6	1 574.6	1 618.0	1 846.3	5.5	6.0	6.7
White only	287.7	271.4	274.8	1 525.4	1 586.4	1 808.2	5.3	5.8	6.6
Black or African American only	336.7	300.7	290.8	2 311.8	2 064.8	2 423.5	6.9	6.9	8.3
Poverty status [3]									
Poor	416.1	394.7	357.4	2 220.6	2 169.0	2 690.9	5.3	5.5	7.5
Near poor	310.7	328.5	329.6	1 849.7	1 954.8	2 498.3	6.0	6.0	7.6
Nonpoor	266.7	247.3	256.6	1 345.9	1 510.4	1 680.3	5.0	6.1	6.5
Health Insurance Status [4]									
Medicare HMO	252.8	241.9	217.8	1 305.8	1 396.0	1 355.3	5.2	5.8	6.2
Private	286.3	270.9	271.9	1 485.9	1 592.8	1 756.1	5.2	5.9	6.5
Medicaid	484.0	455.0	539.7	2 663.2	3 286.7	3 810.6	5.5	7.2	7.1
Medicare fee-for-service only	264.2	266.0	252.9	1 717.7	1 565.3	1 906.6	6.5	5.9	7.5
Geographic Region									
Northeast	272.0	288.0	265.0	1 467.7	1 873.4	1 828.5	5.4	6.5	6.9
Midwest	280.0	244.0	285.2	1 517.9	1 475.3	1 971.1	5.4	6.0	6.9
South	312.3	298.1	298.1	1 757.4	1 783.8	2 140.2	5.6	6.0	7.2
West	275.7	238.5	237.2	1 434.7	1 284.6	1 299.2	5.2	5.4	5.5
Location of Residence									
Within MSA [5]	286.8	265.3	271.3	1 584.1	1 653.3	1 875.9	5.5	6.2	6.9
Outside MSA [5]	295.8	295.3	295.1	1 540.8	1 574.8	1 893.6	5.2	5.3	6.4

[1]Includes all other races not shown separately, unknown poverty status, and unknown health insurance status.

[2]The race groups, White, Black, American Indian and Alaska Native (AI/AN), Asian, Native Hawaiian and Other Pacific Islander, and two or more races, include persons of Hispanic and non-Hispanic origin. Persons of Hispanic origin may be of any race.

[3]Poor persons are defined as below the poverty threshold. Near poor persons have incomes of 100 percent to less than 200 percent of poverty threshold. Nonpoor persons have incomes of 200 percent or greater than the poverty threshold.

[4]Health insurance categories are mutually exclusive. Persons who reported both Medicaid and private coverage are classified as having private coverage. Persons 65 years of age and over who reported Medicare HMO (health maintenance organization) and some other type of health insurance coverage are classified as having Medicare HMO.

[5]MSA = metropolitan statistical area.

* = Figure does not meet standards of reliability or precision.

. . . = Not available.

TABLE III-56. DISCHARGES, DAYS OF CARE, AND AVERAGE LENGTH OF STAY IN NON–FEDERAL SHORT–STAY HOSPITALS, FOR DISCHARGES WITH THE DIAGNOSIS OF HUMAN IMMUNODEFICIENCY VIRUS (HIV) AND FOR ALL DISCHARGES, SELECTED YEARS, 1986–2001

The data in this table are based on a sample of hospital records and thus are slightly different in magnitude from those shown in the previous table on discharges, which were based on household interviews.

Hospital discharges in 2001 totaled 32.7 million, and discharges of patients diagnosed with HIV amounted to 185,000. For persons between 20 and 49 years of age, there were 93,000 male HIV discharges and 55,000 female discharges. The highest level shown for HIV discharges occurred in 1995, amounting to 249,000.

The average length of hospital stay for HIV was 7.8 days, whereas for all patients it was 4.9 days. Hospital care in total, per 1,000 of population, was received during the year for 561.9 days, while HIV care was received for 5.06 days for each 1,000 of the total population.

Table III-56. Discharges, Days of Care, and Average Length of Stay in Non-Federal Short-Stay Hospitals for Discharges with the Diagnosis of Human Immunodeficiency Virus (HIV) and for all Discharges, Selected Years, 1986–2001

(Number, rate per 1,000 population.)

Type of discharge, sex, and age	2001 [1]	2000 [2]	1999	1998	1997	1995	1990	1987 [3]	1986 [3]
Discharges (Thousands)									
HIV discharges	185	173	180	189	178	249	146	67	44
Male, 20–49 years	93	88	101	113	107	162	102	51	35
Female, 20–49 years	55	48	52	51	46	55	27	*	*
All discharges	32 653	31 706	32 132	31 827	30 914	30 722	30 788	33 387	34 256
Male, 20–49 years	3 333	3 195	3 149	3 154	3 116	3 360	3 649	4 075	4 300
Female, 20–49 years	7 679	7 350	7 396	7 639	7 322	7 593	8 228	8 980	9 027
Discharges (Per 1,000 Population)									
HIV discharges	0.65	0.62	0.65	0.69	0.66	0.94	0.58	0.28	0.18
Male, 20–49 years	1.51	1.43	1.68	1.88	1.77	2.72	1.79	0.96	0.67
Female, 20–49 years	0.89	0.77	0.85	0.84	0.76	0.91	0.47	*	*
All discharges	115.1	112.9	116.6	116.5	114.3	115.7	122.3	138.8	143.7
Male, 20–49 years	53.9	52.1	52.3	52.6	51.8	56.5	64.2	76.8	82.2
Female, 20–49 years	123.4	118.8	121.0	125.2	120.8	125.9	142.2	163.6	166.7
Days of Care (Thousands)									
HIV discharges	1 435	1 257	1 310	1 503	1 448	2 326	2 188	936	714
Male, 20–49 years	713	723	669	892	855	1 408	1 645	724	573
Female, 20–49 years	454.0	299.0	384.0	365.0	364.0	559.0	341.0	*	*
All discharges	159 365	155 857	160 128	160 914	157 458	164 627	197 422	214 942	218 496
Male, 20–49 years	16 435	15 665	15 278	16 085	15 529	17 984	22 539	26 295	26 488
Female, 20–49 years	26 502	24 883	25 415	25 976	24 955	26 596	34 473	39 356	40 620
Days of Care (Per 1,000 Population)									
HIV discharges	5.06	4.47	4.75	5.50	5.35	8.76	8.69	3.89	2.99
Male, 20–49 years	11.55	11.79	11.11	14.86	14.22	23.70	28.96	13.64	10.95
Female, 20–49 years	7.29	4.83	6.28	5.98	6.00	9.27	5.90	*	*
All discharges	561.9	554.8	581.1	589.2	582.3	620.2	784.0	893.6	916.5
Male, 20–49 years	266.0	255.4	253.8	268.0	258.3	302.7	396.8	495.2	506.4
Female, 20–49 years	425.9	402.1	415.7	425.8	411.7	441.0	595.7	717.1	750.2
Average Length of Stay (Days)									
HIV discharges	7.8	7.3	7.3	8.0	8.1	9.3	14.9	14.1	16.4
Male, 20–49 years	7.7	8.2	6.6	8.0	8.0	8.7	16.2	14.1	16.4
Female, 20–49 years	8.2	6.3	7.4	7.1	7.9	10.2	12.6	*	*
All discharges	4.9	4.9	5.0	5.1	5.1	5.4	6.4	6.4	6.4
Male, 20–49 years	4.9	4.9	4.9	5.1	5.0	5.4	6.2	6.5	6.2
Female, 20–49 years	3.5	3.4	3.4	3.4	3.4	3.5	4.2	4.4	4.5

Note: Excludes newborn infants.

[1]Rates for 2001 were computed using 2000-based postcensal estimates.
[2]Rates for 2000 were computed using Census 2000 counts.
[3]Comparisons of data from 1986 and 1987 with data from later years should be made with caution as estimates of change may reflect improvements in the design rather than true changes in hospital use.
* = Figure does not meet standards of reliability or precision.

TABLE III-57. RATES OF DISCHARGES AND DAYS OF CARE IN NON–FEDERAL SHORT–STAY HOSPITALS, ACCORDING TO SEX, AGE, AND SELECTED FIRST–LISTED DIAGNOSIS, 1990, 2000, AND 2001

This table—based on hospital records—focuses on the first-listed diagnosis of hospital patients, who are subdivided by age group. The table also shows that days of care for patients have substantially declined from 1990 to 2000 and 2001, even though the number of patients has edged down only a little.

In 2001, for hospitalized children under 18 years of age, the leading diagnosis was pneumonia. For patients between 18 and 44, mental illness and injuries and poisoning both were near or at the top.

Table III-57. Rates of Discharges and Days of Care in Non-Federal Short-Stay Hospitals, According to Sex, Age, and Selected First-Listed Diagnosis, 1990, 2000 and 2001

(Number per 1,000 population.)

Sex, age, and first–listed diagnosis	Discharges			Days of care		
	2001 [1]	2000 [2]	1990	2001 [1]	2000 [2]	1990
BOTH SEXES [3]	115.3	113.3	125.2	563.2	557.8	818.9
MALE						
All Ages [3]	100.2	99.1	113.0	535.7	536.0	805.8
Under 18 Years [3]	43.8	40.9	46.3	200.7	195.6	233.6
Pneumonia	6.2	5.4	5.3	21.2	17.3	22.6
Asthma	2.0	3.5	3.3	4.8	7.4	9.3
Injuries and poisoning	5.2	5.0	6.8	*21.3	21.4	30.1
Fracture, all sites	1.7	1.8	2.2	6.0	7.2	9.3
18–44 Years [3]	46.2	45.0	57.9	225.9	217.5	351.7
Alcohol and drug [4]	4.0	4.0	3.7	*20.4	19.1	33.1
Serious mental illness [5]	*6.1	*5.3	3.4	*47.7	*43.6	47.1
Diseases of heart	2.7	2.7	3.0	10.0	9.4	16.3
Intervertebral disc disorders	1.3	1.5	2.6	3.0	3.2	10.7
Injuries and poisoning	7.4	7.3	13.1	35.7	33.2	65.7
Fracture, all sites	2.7	2.5	4.0	12.7	12.8	22.7
45–64 Years [3]	112.2	112.7	140.3	578.0	570.5	943.4
Malignant neoplasms	6.3	6.2	10.6	48.5	42.1	99.1
Trachea, bronchus, lung	0.9	0.9	2.7	*8.0	5.2	19.1
Diabetes	3.4	3.7	2.9	18.7	22.5	21.2
Alcohol and drug [4]	3.7	3.5	3.5	17.1	15.8	29.7
Serious mental illness [5]	4.7	*4.0	2.5	*43.9	*34.6	34.8
Diseases of heart	24.5	26.4	31.7	98.4	101.5	185.0
Ischemic heart disease	15.7	17.7	22.6	57.2	63.8	128.2
Acute myocardial infarction	5.5	5.9	7.4	25.3	27.8	55.8
Congestive heart failure	3.5	3.3	3.0	17.4	17.2	19.7
Cerebrovascular diseases	3.3	3.8	4.1	18.5	19.8	40.7
Pneumonia	3.4	3.4	3.5	18.4	20.5	27.4
Injuries and poisoning	9.8	8.8	11.6	50.0	49.8	82.6
Fracture, all sites	2.7	2.5	3.3	13.3	16.2	24.2
65–74 Years [3]	260.9	264.9	287.8	1 473.0	1 489.6	2 251.5
Malignant neoplasms	18.0	17.6	27.9	136.2	121.2	277.6
Large intestine and rectum	2.8	3.0	3.0	24.0	27.3	34.2
Trachea, bronchus, lung	2.8	2.8	6.4	23.0	19.2	55.7
Prostate	3.4	3.7	5.1	*15.4	14.0	33.1
Diabetes	4.9	4.7	4.4	25.1	29.0	39.8
Serious mental illness [5]	*2.9	*3.4	2.5	*	39.9	43.8
Diseases of heart	68.2	70.6	69.4	315.0	331.9	487.2
Ischemic heart disease	38.6	39.6	42.0	169.7	171.1	285.2
Acute myocardial infarction	12.1	12.5	14.0	69.5	66.5	122.4
Congestive heart failure	12.9	13.4	11.4	74.7	76.8	90.2
Cerebrovascular diseases	13.1	13.2	13.8	62.1	59.0	114.8
Pneumonia	13.3	12.8	11.4	75.1	82.0	107.8
Hyperplasia of prostate	4.2	5.4	14.4	10.1	15.0	65.0
Osteoarthritis	7.8	9.6	5.0	35.1	46.7	44.9
Injuries and poisoning	17.3	17.9	17.6	127.5	105.7	139.0
Fracture, all sites	5.0	4.7	4.5	38.5	29.9	45.9
Fracture of neck of femur (hip)	2.0	*2.0	1.5	*15.6	*15.9	*18.1
75 Years and Over [3]	471.9	467.8	478.5	2 801.1	2 890.5	4 231.6
Malignant neoplasms	24.8	21.9	41.0	171.7	165.4	408.3
Large intestine and rectum	4.3	4.2	5.4	38.8	44.1	80.7
Trachea, bronchus, lung	4.1	3.0	5.4	28.8	18.3	53.4
Prostate	*4.3	3.2	9.7	*18.0	*19.4	65.6
Diabetes	6.1	6.5	4.6	39.6	43.2	51.2
Serious mental illness [5]	3.1	2.9	*2.6	*28.3	*32.7	*40.5
Diseases of heart	115.6	113.3	106.2	584.8	601.4	855.7
Ischemic heart disease	49.5	53.0	49.1	246.4	276.3	398.1
Acute myocardial infarction	21.1	23.0	23.1	136.3	136.6	227.5
Congestive heart failure	33.5	30.6	31.0	171.9	175.6	242.3
Cerebrovascular diseases	24.9	30.3	30.2	129.4	171.5	298.3
Pneumonia	39.5	37.2	38.6	254.5	233.5	393.6
Hyperplasia of prostate	7.2	6.8	17.9	*25.2	21.6	109.2
Osteoarthritis	8.2	6.2	5.8	36.4	28.8	60.7
Injuries and poisoning	31.6	33.6	31.2	206.2	257.9	341.3
Fracture, all sites	13.7	14.4	13.7	99.4	*119.3	145.1
Fracture of neck of femur (hip)	7.9	8.5	8.5	56.6	63.3	97.8

[1]Rates for 2001 were computed using 2000-based postcensal estimates.
[2]Rates for 2000 were computed using Census 2000 counts.
[3]Includes discharges with first-listed diagnoses not shown in table.
[4]Includes abuse, dependence, and withdrawal. These estimates are for non-federal short-stay hospitals and do not include alcohol and drug discharges from other types of facilities or programs such as the Department of Veterans' Affairs or day treatment programs.
[5]These estimates are for non-federal short-stay hospitals and do not include serious mental illness discharges from other types of facilities or programs such as the Department of Veterans' Affairs or long-term hospitals.
* = Figure does not meet standards of reliability or precision.

Those between 45 and 64 years of age predominantly were hospitalized for heart diseases, followed by injuries and poisoning, and cancer. Hospitalizations increased markedly for the age group between 65 and 74, but the ranking of diagnoses is similar to the previous age group, both for males and females. Relative to the previous year, there is a slight uptick of injuries in 2001 that could reflect the terrorist attacks.

Hospital patients aged 75 years and over, were in the hospital most frequently for heart diseases, and for men this was followed by pneumonia, stroke, and cancer. For women pneumonia was less frequent. Hospital discharges for cancer declined considerably between 1990 and 2001, partly reflecting declines in prostate and breast cancer hospitalizations.

Table III-57. Rates of Discharges and Days of Care in Non-Federal Short-Stay Hospitals, According to Sex, Age, and Selected First-Listed Diagnosis, 1990, 2000 and 2001—*Continued*

(Number per 1,000 population.)

Sex, age, and first–listed diagnosis	Discharges			Days of care		
	2001 [1]	2000 [2]	1990	2001 [1]	2000 [2]	1990
FEMALE						
All Ages, [3]	130.9	127.8	139.0	592.9	581.2	840.5
Under 18 years [3]	43.0	39.6	46.4	184.3	161.5	218.7
Pneumonia	5.0	4.8	4.0	17.1	17.2	17.4
Asthma	1.3	2.4	2.2	2.8	5.5	6.8
Injuries and poisoning	3.4	3.1	4.3	12.1	*12.0	16.7
Fracture, all sites	0.8	0.9	1.3	2.5	2.3	6.4
18–44 Years [3]	129.1	124.9	146.8	421.5	401.3	582.0
Delivery	66.3	64.6	69.9	167.0	160.3	195.0
Alcohol and drug [4]	2.0	*2.1	1.6	*10.4	*10.8	14.1
Serious mental illness [5]	6.1	*5.4	3.7	45.4	*41.1	54.3
Diseases of heart	1.8	1.7	1.3	6.3	6.3	7.2
Intervertebral disc disorders	1.0	1.0	1.5	2.7	2.4	7.3
Injuries and poisoning	4.8	4.3	6.7	18.7	18.1	36.6
Fracture, all sites	1.1	1.0	1.6	4.8	4.5	10.7
45–64 Years [3]	111.9	110.2	131.0	551.6	533.7	886.5
Malignant neoplasms	6.1	6.1	12.7	37.6	34.7	107.4
Trachea, bronchus, lung	0.8	0.5	1.7	5.2	3.4	14.8
Breast	1.2	1.3	2.8	3.0	2.6	12.1
Diabetes	2.8	2.9	2.9	12.8	15.0	25.8
Alcohol and drug [4]	1.5	1.5	1.0	*7.0	*7.1	8.0
Serious mental illness [5]	5.6	4.6	4.0	58.1	42.7	60.5
Diseases of heart	15.0	14.6	16.6	64.5	59.5	101.1
Ischemic heart disease	8.2	7.8	9.9	32.5	29.5	57.4
Acute myocardial infarction	2.6	2.0	2.8	14.6	10.0	21.6
Congestive heart failure	2.8	2.9	2.1	14.3	13.6	15.8
Cerebrovascular diseases	3.1	3.5	3.0	14.9	19.5	32.1
Pneumonia	3.4	3.6	3.4	18.4	20.8	26.5
Injuries and poisoning	8.5	7.7	9.4	45.9	41.2	63.3
Fracture, all sites	2.5	2.7	3.1	11.5	13.3	25.0
65–74 Years [3]	252.3	246.1	241.1	1 431.5	1 397.1	1 959.3
Malignant neoplasms	14.7	14.1	20.9	106.7	101.0	189.8
Large intestine and rectum	2.1	1.7	2.4	19.0	15.2	34.9
Trachea, bronchus, lung	2.3	2.4	2.6	16.5	*17.5	26.9
Breast	2.2	2.8	3.9	5.4	*	17.6
Diabetes	5.9	4.6	5.8	36.4	26.1	46.8
Serious mental illness [5]	4.7	4.0	3.9	60.1	46.3	62.8
Diseases of heart	48.8	52.1	45.1	237.9	256.0	316.9
Ischemic heart disease	22.1	23.3	24.4	108.0	113.9	153.8
Acute myocardial infarction	8.0	8.0	7.5	59.6	52.8	58.1
Congestive heart failure	11.1	12.7	9.2	63.1	68.4	81.8
Cerebrovascular diseases	11.7	12.3	11.3	61.8	59.4	96.0
Pneumonia	11.5	11.7	8.7	74.7	73.5	81.8
Osteoarthritis	10.0	9.3	6.9	46.6	43.6	68.9
Injuries and poisoning	18.9	18.3	17.8	117.9	109.9	166.2
Fracture, all sites	7.9	7.7	8.4	41.9	43.8	97.3
Fracture of neck of femur (hip)	3.4	3.2	3.6	20.3	21.1	59.6
75 Years and Over [3]	455.2	459.0	409.6	2 694.5	2 832.3	3 887.1
Malignant neoplasms	18.0	17.6	22.1	136.2	125.7	257.3
Large intestine and rectum	3.2	3.4	4.6	29.6	28.4	69.8
Trachea, bronchus, lung	1.8	1.9	2.1	13.8	14.0	20.6
Breast	2.6	2.5	3.9	9.5	8.9	22.0
Diabetes	6.3	6.3	4.6	32.8	34.0	55.3
Serious mental illness [5]	4.2	4.7	4.2	45.7	49.3	78.4
Diseases of heart	94.4	99.1	84.6	476.7	523.7	672.8
Ischemic heart disease	34.0	35.5	33.7	160.0	185.6	253.2
Acute myocardial infarction	16.9	16.5	13.1	101.6	110.7	125.9
Congestive heart failure	31.4	32.2	28.0	175.2	181.8	236.6
Cerebrovascular diseases	28.2	27.6	29.6	146.9	156.9	302.0
Pneumonia	26.7	30.5	23.9	180.8	209.8	260.1
Osteoarthritis	10.0	8.7	5.3	44.0	40.4	54.1
Injuries and poisoning	47.1	44.7	46.3	284.9	275.6	489.2
Fracture, all sites	30.1	30.0	31.5	179.9	190.1	352.7
Fracture of neck of femur (hip)	17.8	17.9	18.8	113.8	125.3	236.3

Note: Excludes newborn infants.

[1]Rates for 2001 were computed using 2000-based postcensal estimates.
[2]Rates for 2000 were computed using Census 2000 counts.
[3]Includes discharges with first-listed diagnoses not shown in table.
[4]Includes abuse, dependence, and withdrawal. These estimates are for non-federal short-stay hospitals and do not include alcohol and drug discharges from other types of facilities or programs such as the Department of Veterans' Affairs or day treatment programs.
[5]These estimates are for non-federal short-stay hospitals and do not include serious mental illness discharges from other types of facilities or programs such as the Department of Veterans' Affairs or long-term hospitals.
* = Figure does not meet standards of reliability or precision.

TABLE III-58. NUMBER OF DISCHARGES AND AVERAGE LENGTH OF STAY IN NON–FEDERAL SHORT–STAY HOSPITALS, ACCORDING TO SEX, AGE, AND SELECTED FIRST–LISTED DIAGNOSIS, 1990, 2000, AND 2001

The previous table provided information in terms of rates per 1000, while this table shows the number discharges (in thousands) and the days of care (in number days). For males, the average length of hospitalization was 5.3 days in 2001, down considerably from 7.1 in 1990. The average length of hospitalization for females was 4.5 days. For serious mental illness and cancer, hospital stays tend to be longer than average. For men fractures also resulted relatively long hospital stays, averaging 7.8 and 7.3 days at age groups 65–74 and 75 and over, respectively. Women incurred shorter hospitalizations for fractures, averaging 5.3 and 6.0 days, respectively for those two age groups.

Table III-58. Number of Discharges and Average Length of Stay in Non-Federal Short-Stay Hospitals, According to Sex, Age, and Selected First-Listed Diagnosis, 1990, 2000, and 2001

(Number.)

Sex, age, and first–listed diagnosis	Discharges[1] (in thousands)			Average length of stay (in days)		
	2001	2000	1990	2001	2000	1990
BOTH SEXES	32 653	31 706	30 788	4.9	4.9	6.5
MALE						
All ages	12 852	12 514	12 280	5.3	5.4	7.1
Under 18 Years	1 629	1 515	1 572	4.6	4.8	5.0
Pneumonia	229	199	178	3.4	3.2	4.3
Asthma	76	129	111	2.3	2.1	2.8
Injuries and poisoning	192	185	232	*4.1	4.3	4.4
Fracture, all sites	63	68	76	3.5	3.9	4.2
18–44 Years	2 573	2 498	3 120	4.9	4.8	6.1
Alcohol and drug[2]	225	224	201	*5.1	4.7	8.9
Serious mental illness[3]	*338	*296	184	*7.9	*8.2	13.8
Diseases of heart	150	148	163	3.7	3.5	5.4
Intervertebral disc disorders	74	81	138	2.2	2.2	4.2
Injuries and poisoning	412	408	704	4.8	4.5	5.0
Fracture, all sites	148	141	217	4.8	5.0	5.6
45–64 Years	3 519	3 424	3 115	5.2	5.1	6.7
Malignant neoplasms	199	188	235	7.7	6.8	9.4
Trachea, bronchus, lung	28	26	60	*8.8	6.0	7.1
Diabetes	107	114	65	5.5	6.0	7.3
Alcohol and drug[2]	115	106	77	4.7	4.5	8.5
Serious mental illness[3]	148	*120	56	*9.3	*8.8	13.7
Diseases of heart	768	802	704	4.0	3.8	5.8
Ischemic heart disease	492	539	502	3.6	3.6	5.7
Acute myocardial infarction	172	178	165	4.6	4.7	7.5
Congestive heart failure	110	101	66	5.0	5.2	6.7
Cerebrovascular diseases	102	116	91	5.7	5.2	10.0
Pneumonia	108	104	77	5.4	6.0	7.9
Injuries and poisoning	307	266	257	5.1	5.7	7.2
Fracture, all sites	85	77	74	4.9	6.4	7.2
65–74 Years	2 165	2 199	2 268	5.6	5.6	7.8
Malignant neoplasms	149	146	220	7.6	6.9	9.9
Large intestine and rectum	23	24	24	8.7	9.2	11.4
Trachea, bronchus, lung	23	23	50	8.4	6.8	8.7
Prostate	28	31	40	*4.6	3.8	6.5
Diabetes	41	39	34	5.1	6.2	9.1
Serious mental illness[3]	*24	*28	20	*	*11.7	17.4
Diseases of heart	566	586	547	4.6	4.7	7.0
Ischemic heart disease	320	329	331	4.4	4.3	6.8
Acute myocardial infarction	100	104	110	5.8	5.3	8.8
Congestive heart failure	107	112	90	5.8	5.7	7.9
Cerebrovascular diseases	109	109	108	4.7	4.5	8.3
Pneumonia	111	106	90	5.6	6.4	9.5
Hyperplasia of prostate	35	45	113	2.4	2.8	4.5
Osteoarthritis	65	80	39	4.5	4.9	9.0
Injuries and poisoning	143	149	139	7.4	5.9	7.9
Fracture, all sites	41	39	36	7.8	6.4	10.2
Fracture of neck of femur (hip)	17	*17	12	*7.8	*7.9	*11.8
75 Years and Over	2 966	2 878	2 203	5.9	6.2	8.8
Malignant neoplasms	156	135	189	6.9	7.6	10.0
Large intestine and rectum	27	26	25	9.1	10.6	15.0
Trachea, bronchus, lung	26	18	25	7.1	6.1	10.0
Prostate	*27	20	45	*4.2	*6.1	6.8
Diabetes	38	40	21	6.5	6.6	11.0
Serious mental illness[3]	19	18	*12	*9.1	*11.2	*15.5
Diseases of heart	727	697	489	5.1	5.3	8.1
Ischemic heart disease	311	326	226	5.0	5.2	8.1
Acute myocardial infarction	133	141	106	6.4	5.9	9.9
Congestive heart failure	210	188	143	5.1	5.7	7.8
Cerebrovascular diseases	157	186	139	5.2	5.7	9.9
Pneumonia	248	229	178	6.4	6.3	10.2
Hyperplasia of prostate	45	42	82	*3.5	3.2	6.1
Osteoarthritis	52	38	27	4.4	4.6	10.5
Injuries and poisoning	199	207	144	6.5	7.7	10.9
Fracture, all sites	86	89	63	7.3	*8.3	10.6
Fracture of neck of femur (hip)	50	52	39	7.1	7.5	11.5

[1]Includes discharges with first-listed diagnoses not shown in table.

[2]Includes abuse, dependence, and withdrawal. These estimates are for non-federal short-stay hospitals and do not include alcohol and drug discharges from other types of facilities or programs such as the Department of Veterans' Affairs or day treatment programs.

[3]These estimates are for non-federal short-stay hospitals and do not include serious mental illness discharges from other types of facilities or programs such as the Department of Veterans' Affairs or long-term hospitals.

* = Figure does not meet standards of reliability or precision.

Table III-58. Number of Discharges and Average Length of Stay in Non-Federal Short-Stay Hospitals, According to Sex, Age, and Selected First-Listed Diagnosis, 1990, 2000, and 2001—*Continued*

(Number.)

Sex, age, and first–listed diagnosis	Discharges [1] (in thousands)			Average length of stay (in days)		
	2001	2000	1990	2001	2000	1990
FEMALE						
All Ages	19 801	19 192	18 508	4.5	4.5	6.0
Under 18 years	1 523	1 397	1 500	4.3	4.1	4.7
Pneumonia	176	168	129	3.5	3.6	4.4
Asthma	45	85	71	2.2	2.3	3.1
Injuries and poisoning	120	111	138	3.6	*3.8	3.9
Fracture, all sites	30	32	42	2.9	2.5	5.0
18–44 Years	7 178	6 941	8 018	3.3	3.2	4.0
Delivery	3 685	3 588	3 815	2.5	2.5	2.8
Alcohol and drug [2]	112	*116	85	*5.2	*5.2	9.1
Serious mental illness [3]	338	*300	200	7.5	*7.6	14.8
Diseases of heart	99	95	73	3.5	3.7	5.4
Intervertebral disc disorders	55	58	84	2.7	2.3	4.7
Injuries and poisoning	267	237	366	3.9	4.2	5.5
Fracture, all sites	62	57	85	4.3	4.4	6.9
45–64 Years	3 705	3 534	3 129	4.9	4.8	6.8
Malignant neoplasms	203	195	303	6.1	5.7	8.5
Trachea, bronchus, lung	28	17	41	6.2	6.4	8.6
Breast	40	40	67	2.5	2.1	4.3
Diabetes	94	93	70	4.5	5.2	8.9
Alcohol and drug [2]	51	47	23	4.5	4.8	8.2
Serious mental illness [3]	184	146	95	10.4	9.4	15.2
Diseases of heart	497	470	397	4.3	4.1	6.1
Ischemic heart disease	272	251	237	4.0	3.8	5.8
Acute myocardial infarction	86	64	68	5.6	5.0	7.6
Congestive heart failure	91	94	51	5.2	4.6	7.4
Cerebrovascular diseases	102	113	72	4.9	5.5	10.7
Pneumonia	113	117	80	5.4	5.7	7.9
Injuries and poisoning	280	248	225	5.4	5.3	6.7
Fracture, all sites	83	87	75	4.6	4.9	7.9
65–74 Years	2 527	2 479	2 421	5.7	5.7	8.1
Malignant neoplasms	147	142	210	7.3	7.2	9.1
Large intestine and rectum	21	17	24	8.9	9.0	14.5
Trachea, bronchus, lung	23	25	26	7.2	*7.1	10.2
Breast	22	29	40	2.4	*	4.5
Diabetes	59	47	59	6.1	5.6	8.0
Serious mental illness [3]	47	40	39	12.8	11.7	16.3
Diseases of heart	489	525	453	4.9	4.9	7.0
Ischemic heart disease	222	235	245	4.9	4.9	6.3
Acute myocardial infarction	80	81	75	7.4	6.6	7.8
Congestive heart failure	111	128	92	5.7	5.4	8.9
Cerebrovascular diseases	118	124	114	5.3	4.8	8.5
Pneumonia	115	117	87	6.5	6.3	9.4
Osteoarthritis	100	94	69	4.7	4.7	10.0
Injuries and poisoning	189	185	179	6.2	6.0	9.3
Fracture, all sites	79	77	85	5.3	5.7	11.5
Fracture of neck of femur (hip)	34	32	36	5.9	6.7	*16.7
75 Years and Over	4 867	4 840	3 440	5.9	6.2	9.5
Malignant neoplasms	192	186	185	7.6	7.1	11.7
Large intestine and rectum	34	36	39	9.3	8.4	15.1
Trachea, bronchus, lung	19	20	18	7.7	7.3	9.9
Breast	28	27	33	*3.7	*3.5	5.7
Diabetes	67	67	39	5.2	5.4	11.9
Serious mental illness [3]	45	49	35	10.9	10.5	18.7
Diseases of heart	1 010	1 045	711	5.0	5.3	8.0
Ischemic heart disease	363	375	283	4.7	5.2	7.5
Acute myocardial infarction	181	174	110	6.0	6.7	9.6
Congestive heart failure	336	339	235	5.6	5.6	8.5
Cerebrovascular diseases	301	292	249	5.2	5.7	10.2
Pneumonia	286	322	201	6.8	6.9	10.9
Osteoarthritis	107	91	45	4.4	4.7	10.2
Injuries and poisoning	503	472	389	6.1	6.2	10.6
Fracture, all sites	322	316	265	6.0	6.3	11.2
Fracture of neck of femur (hip)	190	189	158	6.4	7.0	12.5

Note: Excludes newborn infants.

[1]Includes discharges with first-listed diagnoses not shown in table.
[2]Includes abuse, dependence, and withdrawal. These estimates are for non-federal short-stay hospitals and do not include alcohol and drug discharges from other types of facilities or programs such as the Department of Veterans' Affairs or day treatment programs.
[3]These estimates are for non-federal short-stay hospitals and do not include serious mental illness discharges from other types of facilities or programs such as the Department of Veterans' Affairs or long-term hospitals.
* = Figure does not meet standards of reliability or precision.

TABLE III-59. AMBULATORY AND INPATIENT PROCEDURES, ACCORDING TO PLACE, SEX, AGE, AND TYPE OF PROCEDURE, SELECTED YEARS 1994–2001

This table shows the frequency of different medical procedures. Between 1994 and 1996, information is provided for both inpatient and outpatient procedures. For those years, the National Survey of Ambulatory Surgery was available and it showed that in 1996 ambulatory procedures amounted to about 45 percent of the combined number of outpatient and inpatient procedures. Since then this particular information has been available only for the number of inpatient procedures.

Inpatient procedures in 2001 totaled almost 145 per 1,000 of population. The frequency of inpatient procedures

Table III-59. Ambulatory and Inpatient Procedures, According to Place, Sex, Age, and Type of Procedure, Selected Years, 1994–2001

(Number of procedures per 1,000 population.)

Sex, age, and procedure category	Ambulatory [1]			Inpatient [2]					
	1996	1995	1994	2001 [3]	2000 [4]	1999	1996	1995	1994
PROCEDURES PER 1,000 POPULATION									
BOTH SEXES [5]	120.1	113.6	107.9	144.9	142.8	151.5	153.3	152.7	157.9
MALE									
All Ages [5]	115.0	107.9	102.5	124.9	124.1	133.4	136.1	135.1	139.1
Under 18 Years [5]	44.7	43.0	45.6	39.9	34.7	37.6	36.3	37.4	36.0
Myringotomy with insertion of tube	8.2	8.5	9.1	*0.2	0.3	*0.2	0.4	0.3	0.4
Tonsillectomy, with or without adenoidectomy	3.6	4.1	3.6	0.3	*0.2	*0.2	0.4	0.4	0.4
Reduction of fracture	1.2	1.0	1.0	1.4	1.3	1.3	1.5	1.2	1.6
18–44 Years [5]	63.7	61.3	58.4	50.0	49.2	52.5	58.1	59.5	62.7
Cardiac catheterization	0.6	*0.3	0.5	1.1	1.1	1.2	1.2	1.1	1.1
Endoscopy of small or large intestine with or without biopsy	5.8	5.2	5.2	1.7	1.5	1.5	1.5	1.8	2.0
Cholecystectomy	0.5	*0.3	*0.2	0.5	0.4	0.5	0.4	0.5	0.5
Reduction of fracture	1.4	1.0	1.0	2.0	2.1	2.2	2.4	2.5	2.3
Arthroscopy of the knee	3.4	4.0	3.5	*	*	*	*0.2	0.3	0.5
Excision or destruction of intervertebral disc	*0.3	*	*	1.2	1.3	1.4	1.4	1.5	1.7
Angiocardiography with contrast material	0.6	0.4	0.7	1.5	1.6	1.7	1.8	1.8	1.8
45–64 Years [5]	155.9	146.8	132.7	161.5	165.8	177.9	185.9	180.7	189.0
Coronary angioplasty	*	*	*	5.7	5.7	7.2	6.4	5.6	5.6
Coronary artery bypass graft [6]	-	-	-	5.1	5.6	6.2	7.2	7.6	6.7
Cardiac catheterization	5.4	3.8	3.3	10.0	11.4	12.6	12.7	11.7	11.7
Endoscopy of small or large intestine with or without biopsy	21.8	21.1	20.2	6.1	6.0	5.8	6.4	6.5	7.2
Cholecystectomy	1.1	*0.7	*0.5	1.5	1.6	1.7	2.1	1.8	2.1
Prostatectomy	*	*	*	1.8	1.6	1.7	1.9	2.2	2.5
Reduction of fracture	0.8	*0.8	*0.6	2.0	1.9	1.9	2.0	2.3	2.3
Arthroscopy of the knee	4.4	4.5	3.7	*	*	*	*	*	*0.3
Excision or destruction of intervertebral disc	*	*	*	2.1	2.2	2.4	2.6	2.2	2.6
Angiocardiography with contrast material	6.4	5.1	4.6	11.9	14.7	15.7	17.7	15.8	16.0
65–74 Years [5]	316.4	282.5	271.6	376.3	388.6	417.1	417.7	419.5	426.4
Coronary angioplasty	*	*	*	12.0	12.2	12.6	11.7	9.4	10.0
Extraction of lens	36.9	33.4	31.6	*	*	*	*	*	*
Insertion of prosthetic lens (pseudophakos)	29.6	26.0	25.8	*	*	*	*	*	*
Coronary artery bypass graft [6]	-	-	-	14.2	13.5	15.9	19.2	18.3	15.4
Cardiac catheterization	10.2	7.1	5.7	22.9	22.4	24.1	23.1	23.6	22.3
Pacemaker insertion or replacement	*	*	*	4.2	4.4	5.0	5.7	4.8	5.6
Carotid endarterectomy	*	-	*	2.7	3.0	3.1	3.9	4.2	3.3
Endoscopy of small or large intestine with or without biopsy	40.2	42.8	42.6	16.8	16.6	17.1	16.6	16.5	18.4
Cholecystectomy	*	*	*	3.6	3.8	3.7	4.0	4.4	4.5
Prostatectomy	*1.5	*	*	7.4	8.6	9.1	10.3	12.3	14.2
Reduction of fracture	*	*	*	3.0	2.8	2.8	2.4	2.5	2.8
Total hip replacement	-	-	-	1.9	2.7	2.5	2.3	2.5	1.7
Angiocardiography with contrast material	13.5	9.3	9.0	27.6	28.0	30.3	29.7	30.5	31.0
75 Years and Over [5]	378.7	355.1	339.2	564.6	548.1	593.3	578.9	567.2	584.1
Coronary angioplasty	*	-	*	12.0	10.3	10.9	7.4	8.2	6.5
Extraction of lens	71.6	71.5	61.7	*	*	*	*	*	*
Insertion of prosthetic lens (pseudophakos)	55.1	53.6	47.9	*	*	*	*	*	*
Coronary artery bypass graft [6]	-	-	-	11.4	12.4	11.5	11.6	12.5	10.7
Cardiac catheterization	7.1	4.7	*3.8	22.5	21.5	21.4	19.7	19.2	18.1
Pacemaker insertion or replacement	*	*	*	20.0	14.6	13.9	16.4	15.4	15.4
Carotid endarterectomy	-	-	*	4.9	4.3	4.2	4.6	4.6	3.6
Endoscopy of small or large intestine with or without biopsy	48.9	43.4	43.2	38.8	33.0	33.6	35.2	36.4	35.9
Cholecystectomy	*	*	*	5.9	4.6	4.5	5.8	5.5	6.2
Prostatectomy	*2.2	*2.3	*2.1	10.7	10.2	10.8	12.3	15.3	16.1
Reduction of fracture	*	*	*	6.4	6.6	6.3	6.6	6.4	6.4
Total hip replacement	-	-	-	2.3	1.8	2.5	2.2	2.1	2.2
Angiocardiography with contrast material	10.3	5.5	*3.8	25.5	27.0	25.5	25.9	24.0	24.2

[1]Data are from the National Survey of Ambulatory Surgery (conducted from 1994–1996) and exclude ambulatory surgery procedures for patients who became inpatients.
[2]Inpatient data are from the National Hospital Discharge Survey and exclude newborn infants.
[3]Rates for 2001 were computed using 2000-based postcensal estimates.
[4]Rates for 2000 were computed using Census 2000 counts.
[5]Includes procedures not listed in table.
[6]Data in the main body of the table are for all-listed coronary artery bypass grafts. Often, more than one coronary bypass procedure is performed during a single operation.
* = Figure does not meet standards of reliability or precision.
- Quantity zero.

increased substantially with age, and for patients aged 65 and over they averaged between 330 and 560 per 1,000 of population, depending on the specific age and sex group.

The most common procedures were endoscopy, angiocardiography, and cardiac catheterization, and also reduction of fracture for women 75 years of age and over.

Table III-59. Ambulatory and Inpatient Procedures, According to Place, Sex, Age, and Type of Procedure, Selected Years, 1994–2001—*Continued*

(Number of procedures per 1,000 population.)

Sex, age, and procedure category	Ambulatory [1]			Inpatient [2]					
	1996	1995	1994	2001 [3]	2000 [4]	1999	1996	1995	1994
FEMALE									
All Ages [5]	126.1	120.1	114.2	167.0	163.4	171.5	173.0	172.7	179.2
Under 18 Years [5]	34.8	34.3	35.3	36.7	34.6	38.0	38.4	39.0	40.3
Myringotomy with insertion of tube	5.6	6.3	6.7	*0.2	*0.2	*	*0.3	0.3	0.4
Tonsillectomy, with or without adenoidectomy	4.7	4.4	4.8	*0.1	*0.2	*0.2	0.3	0.4	0.5
Reduction of fracture	0.8	0.7	*0.5	0.7	0.7	0.5	0.5	0.7	0.8
18–44 Years [5]	102.2	96.9	94.9	189.5	184.0	185.8	195.0	195.4	203.6
Cardiac catheterization	*0.3	*0.2	*	0.5	0.6	0.6	0.4	0.5	0.5
Endoscopy of small or large intestine with or without biopsy	8.4	7.4	6.6	1.8	1.7	1.7	1.9	1.9	2.1
Cholecystectomy	2.2	1.9	1.4	2.2	1.9	1.8	2.0	2.4	2.2
Bilateral destruction or occlusion of fallopian tubes	5.7	6.4	5.6	5.8	5.7	5.9	6.2	5.9	6.6
Hysterectomy	*0.2	*0.3	*	5.9	5.7	5.7	5.6	5.9	5.4
Cesarean section [7]	-	-	-	17.0	15.0	14.7	14.6	13.8	15.0
Repair of current obstetrical laceration	*	*	*	20.4	19.6	19.4	18.3	16.8	15.7
Reduction of fracture	0.5	0.5	0.4	1.0	0.9	1.1	1.1	1.1	1.1
Arthroscopy of the knee	2.0	2.0	1.8	*	*	*	*0.1	*0.1	0.2
Excision or destruction of intervertebral disc	*	*	*	0.9	0.9	1.0	1.0	0.9	1.2
Lumpectomy	2.2	2.0	2.5	*	*	*	*	*0.1	*0.1
Mastectomy	*	*	*	0.3	0.2	0.2	0.2	0.2	0.3
45–64 Years [5]	173.2	165.9	155.5	151.7	147.7	160.9	162.2	162.6	173.4
Coronary angioplasty	*	*	*	2.1	2.1	2.6	2.0	2.0	2.1
Coronary artery bypass graft	-	-	-	1.8	1.4	1.8	2.0	1.7	2.0
Cardiac catheterization	2.4	2.0	2.2	5.8	5.4	6.1	6.0	5.4	6.0
Endoscopy of small or large intestine with or without biopsy	22.9	24.3	22.1	5.4	5.4	5.9	5.6	6.2	6.5
Cholecystectomy	3.3	2.3	1.8	2.5	2.5	2.9	3.4	3.5	3.7
Hysterectomy	*	*	*	7.8	7.9	7.9	7.9	7.1	7.2
Reduction of fracture	0.8	*0.7	*0.7	2.0	2.1	2.0	2.3	2.2	2.2
Arthroscopy of the knee	3.5	3.4	2.8	*	*	*	*	*	*
Excision or destruction of intervertebral disc	*	*	*	1.9	1.9	2.1	1.8	1.6	2.0
Lumpectomy	4.6	5.0	4.9	*0.2	0.4	*0.3	0.4	0.4	0.5
Mastectomy	*0.4	*	*	1.1	1.1	1.1	1.3	1.5	1.6
Angiocardiography with contrast material	3.3	2.7	3.0	7.5	7.6	8.5	8.4	8.1	8.5
65–74 Years [5]	291.9	272.5	254.6	329.8	318.8	340.8	334.4	326.1	328.3
Coronary angioplasty	*	*	*	5.8	6.3	6.3	5.7	4.6	4.9
Extraction of lens	47.8	48.3	41.8	*	*	*	*	*	*
Insertion of prosthetic lens (pseudophakos)	35.9	35.7	33.5	*	*	*	*	*	*
Coronary artery bypass graft [6]	-	*	-	4.7	5.1	6.5	6.7	6.1	5.1
Cardiac catheterization	5.4	3.6	3.3	13.1	13.5	15.0	14.7	12.4	12.6
Pacemaker insertion or replacement	*	*	*	4.3	3.8	5.6	3.7	3.9	4.3
Carotid endarterectomy	-	-	-	2.0	2.0	2.6	2.2	2.3	1.7
Endoscopy of small or large intestine with or without biopsy	45.5	41.0	39.0	18.0	18.1	18.2	14.7	18.2	16.1
Cholecystectomy	2.3	*1.6	*1.3	4.2	4.6	4.7	4.7	4.6	5.1
Hysterectomy	*	*	*	4.0	3.7	3.6	3.7	4.3	4.7
Reduction of fracture	*	*	*	5.2	4.9	4.4	5.0	4.4	4.7
Total hip replacement	*	-	-	2.8	2.5	3.3	2.9	2.7	2.6
Lumpectomy	4.9	4.7	4.4	*	*0.7	*	*0.6	*	*
Mastectomy	*	*	*	2.0	2.3	2.1	2.3	2.3	2.8
Angiocardiography with contrast material	6.9	5.0	4.8	17.9	17.5	19.9	20.3	17.5	18.0
75 Years and Over [5]	319.3	304.6	274.3	455.7	467.4	503.0	468.4	468.0	476.8
Coronary angioplasty	*	*	*	5.9	6.0	6.0	4.3	4.2	4.0
Extraction of lens	82.6	82.0	70.6	*	*	*	*	*	*
Insertion of prosthetic lens (pseudophakos)	61.4	61.3	54.5	*	*	*	*	*	*
Coronary artery bypass graft [6]	-	-	-	3.8	4.4	5.8	4.6	4.1	3.4
Cardiac catheterization	3.5	*1.8	*1.5	13.9	13.5	14.0	11.6	11.3	10.4
Pacemaker insertion or replacement	*1.1	*	*	10.8	10.3	10.4	11.3	10.1	11.5
Carotid endarterectomy	-	*	-	2.3	1.9	2.1	2.3	2.0	2.0
Endoscopy of small or large intestine with or without biopsy	38.6	39.3	34.5	35.6	35.8	38.2	33.2	35.0	36.0
Cholecystectomy	*1.2	*	*	4.9	4.1	5.3	5.0	5.5	4.3
Hysterectomy	*	*	-	2.1	2.3	2.3	2.7	2.4	2.4
Reduction of fracture	*	*	*	14.4	14.4	16.0	16.9	14.6	13.9
Total hip replacement	*	*	-	3.9	2.9	3.3	3.5	3.3	3.2
Lumpectomy	2.9	2.5	2.7	*	*0.6	*	*	*0.7	*
Mastectomy	*	*	*	2.0	2.0	2.0	1.8	2.6	2.4
Angiocardiography with contrast material	5.5	*2.1	2.3	18.8	17.4	17.8	15.6	15.6	14.5

[1]Data are from the National Survey of Ambulatory Surgery (conducted from 1994–1996) and exclude ambulatory surgery procedures for patients who became inpatients.
[2]Inpatient data are from the National Hospital Discharge Survey and exclude newborn infants.
[3]Rates for 2001 were computed using 2000-based postcensal estimates.
[4]Rates for 2000 were computed using Census 2000 counts.
[5]Includes procedures not listed in table.
[6]Data in the main body of the table are for all-listed coronary artery bypass grafts. Often, more than one coronary bypass procedure is performed during a single operation.
[7]Cesarean sections accounted for 22.0 percent of deliveries in 1994, 20.8 percent in 1995, 21.8 percent in 1996, 21.5 percent in 1997, 22.5 percent in 1998, 22.1 percent in 1999, 22.9 percent in 2000, and 25.3 percent in 2001.
* = Figure does not meet standards of reliability or precision.
- Quantity zero.

TABLE III-60. HOSPITAL ADMISSIONS, AVERAGE LENGTH OF STAY, AND OUTPATIENT VISITS, ACCORDING TO TYPE OF OWNERSHIP AND SIZE OF HOSPITAL, AND PERCENT OF OUTPATIENT SURGERY, SELECTED YEARS, 1975–2001

Community hospitals provide by far the greatest number of hospital admissions. These are short-stay hospitals (that is, average length of stay is less than 30 days); they include general hospitals open to the public as well as some special hospitals, such as those for obstetrics and orthopedics. They exclude hospitals attached to institutions, such as prisons and college infirmaries, and they exclude facilities for the mentally retarded as well as facilities for alcoholism and chemical dependency. Some 80 percent of admissions at community hospitals are to nonprofit facilities, while the remainder goes to state and local units and to for-profit hospitals. A hospital may have as few as 6 beds, but many provide 500 beds or more. Federal hospitals are listed separately and are operated by the federal government.

Table III-60. Hospital Admissions, Average Length of Stay, and Outpatient Visits, According to Type of Ownership and Size of Hospital, and Percent of Outpatient Surgery, Selected Years, 1975–2001

(Number, percent.)

Type of ownership and size of hospital	2001	2000	1999	1995	1990	1980	1975
Admissions (Thousands)							
All hospitals	35 644	34 891	34 181	33 282	33 774	38 892	36 157
Federal	1 001	1 034	1 072	1 559	1 759	2 044	1 913
Non-federal [1]	34 644	33 946	33 109	31 723	32 015	36 848	34 243
Community [2]	33 814	33 089	32 359	30 945	31 181	36 143	33 435
Nonprofit	27 983	24 453	23 871	22 557	22 878	25 566	23 722
For profit	4 197	4 141	3 905	3 428	3 066	3 165	2 646
State-local government	4 634	4 496	4 583	4 961	5 236	7 413	7 067
6–24 beds	140	141	145	124	95	159	174
25–49 beds	1 030	995	959	944	870	1 254	1 431
50–99 beds	2 422	2 355	2 317	2 299	2 474	3 700	3 675
100–199 beds	6 778	6 735	6 684	6 288	5 833	7 162	7 017
200–299 beds	6 630	6 702	6 389	6 495	6 333	6 596	6 174
300–399 beds	5 328	5 135	5 419	4 693	5 091	5 358	4 739
400–499 beds	3 779	3 617	3 045	3 413	3 644	4 401	3 689
500 beds or more	7 706	7 410	7 400	6 690	6 840	7 513	6 537
Average length of Stay (Days)							
All hospitals	6.7	6.8	7.0	7.8	9.1	9.9	11.4
Federal	13.2	12.8	14.0	13.1	14.9	16.8	20.3
Non-federal [1]	6.6	6.6	6.8	7.5	8.8	9.6	10.9
Community [2]	5.7	5.8	5.9	6.5	7.2	7.6	7.7
Nonprofit	5.6	5.7	5.8	6.4	7.3	7.7	7.8
For profit	5.4	5.4	5.5	5.8	6.4	6.5	6.6
State-local government	6.7	6.7	6.9	7.4	7.7	7.3	7.6
6–24 beds	4.0	4.2	4.5	5.5	5.4	5.3	5.6
25–49 beds	5.0	5.1	5.2	5.7	6.1	5.8	6.0
50–99 beds	6.4	6.4	6.7	7.0	7.2	6.7	6.8
100–199 beds	5.7	5.7	5.9	6.4	7.1	7.0	7.1
200–299 beds	5.6	5.7	5.7	6.2	6.9	7.4	7.5
300–399 beds	5.4	5.5	5.6	6.1	7.0	7.6	7.8
400–499 beds	5.6	5.6	5.9	6.3	7.3	7.9	8.1
500 beds or more	6.1	6.2	6.3	7.1	8.1	8.7	9.1
Outpatient Visits (Thousands) [3]							
All hospitals	612 276	592 673	573 461	483 195	368 184	262 951	254 844
Federal	64 035	63 402	70 060	59 934	58 527	50 566	51 957
Non-federal [1]	548 242	531 972	503 401	423 261	309 657	212 385	202 887
Community [2]	538 480	521 405	495 346	414 345	301 329	202 310	190 672
Nonprofit	404 901	393 168	370 784	303 851	221 073	142 156	131 435
For profit	44 706	43 378	39 896	31 940	20 110	9 696	7 713
State-local government	88 873	84 858	84 667	78 554	60 146	50 459	51 525
6–24 beds	4 556	4 555	4 650	3 644	1 471	1 155	915
25–49 beds	27 941	27 007	23 870	19 465	10 812	6 227	5 855
50–99 beds	51 331	49 385	46 156	38 597	27 582	17 976	16 303
100–199 beds	114 921	114 183	110 336	91 312	58 940	36 453	35 156
200–299 beds	99 596	99 248	90 878	84 080	60 561	36 073	32 772
300–399 beds	75 242	73 444	75 849	54 277	43 699	30 495	29 169
400–499 beds	59 580	52 205	43 867	44 284	33 394	25 501	22 127
500 beds or more	105 314	101 378	99 741	78 685	64 870	48 430	48 375
Outpatient Surgery, Percent of Total Surgeries [4]							
Community hospitals [2]	63	63	62	58	50	16	. . .

[1]The category of non-federal hospitals comprises psychiatric, tuberculosis and other respiratory diseases hospitals, and long term and short term general and other special hospitals

[2]Community hospitals are non-federal short-term general and special hospitals whose facilities and services are available to the public.

[3]Outpatient visits include visits to the emergency department, outpatient department, referred visits (pharmacy, EKG, radiology), and outpatient surgery.

[4]The American Hospital Association and this survey count a surgical episode involving more than one surgical procedure as one surgical operation, in contrast to the National Hospital Discharge Survey and the National Survey of Ambulatory Surgery which count up to four and six procedures that are performed in a single surgical episode as separate surgical operations.

. . . = Not available.

Admissions at all hospitals were estimated at 35.6 million in 2001. Outpatient visits at these same hospitals totaled 612.3 million, also in 2001. At community hospitals, surgeries for outpatients totaled 63 percent of all community hospital surgeries. Outpatient surgeries have become much more important than they were in 1980.

TABLE III-61. NURSING HOME RESIDENTS 65 YEARS OF AGE AND OVER, ACCORDING TO AGE, SEX, AND RACE, SELECTED YEARS, 1973–1999

In 1999, there were about 1.5 million residents in nursing homes, some 500,000 between 75 and 84 years of age, and 750,000 aged 85 years and over. Women were clients more frequently than men; per 1000 of population 65 years and over, 55 women were in nursing homes, but only 27 men. Because these figures are rates, they adjust for the fact that the female population is substantially larger in the top age brackets than the male population. When a head count is used the dominance of female clients increases: their number is 1.092 million, and the male number is 378,000.

The rate of Black clients in nursing homes is greater than the rate for Whites. In 1999, there were 51.1 Blacks (per 1,000), 65 years and over, in nursing homes, in comparison with 42.1 Whites.

Table III-61. Nursing Home Residents 65 Years of Age and Over, According to Age, Sex, and Race, Selected Years, 1973–1999

(Number, rate per 1,000 population.)

Age, sex, and race	Residents				Residents per 1,000 population			
	1999	1995	1985	1973–1974	1999	1995	1985	1973–1974
All Persons								
65 years and over, age-adjusted [1]	...	...	...	...	43.3	45.9	54.0	58.5
65 years and over, crude	1 469 500	1 422 600	1 318 300	961 500	42.9	42.4	46.2	44.7
65–74 years	194 800	190 200	212 100	163 100	10.8	10.1	12.5	12.3
75–84 years	517 600	511 900	509 000	384 900	43.0	45.9	57.7	57.7
85 years and over	757 100	720 400	597 300	413 600	182.5	198.6	220.3	257.3
Male								
65 years and over, age-adjusted [1]	...	...	...	...	30.6	32.8	38.8	42.5
65 years and over, crude	377 800	356 800	334 400	265 700	26.5	26.1	29.0	30.0
65–74 years	84 100	79 300	80 600	65 100	10.3	9.5	10.8	11.3
75–84 years	149 500	144 300	141 300	102 300	30.8	33.3	43.0	39.9
85 years and over	144 200	133 100	112 600	98 300	116.5	130.8	145.7	182.7
Female								
65 years and over, age-adjusted [1]	...	...	...	...	49.8	52.3	61.5	67.5
65 years and over, crude	1 091 700	1 065 800	983 900	695 800	54.6	53.7	57.9	54.9
65–74 years	110 700	110 900	131 500	98 000	11.2	10.6	13.8	13.1
75–84 years	368 100	367 600	367 700	282 600	51.2	53.9	66.4	68.9
85 years and over	612 900	587 300	484 700	315 300	210.5	224.9	250.1	294.9
White [2]								
65 years and over, age-adjusted [1]	...	...	...	...	41.9	45.4	55.5	61.2
65 years and over, crude	1 279 600	1 271 200	1 227 400	920 600	42.1	42.3	47.7	46.9
65–74 years	157 200	154 400	187 800	150 100	10.0	9.3	12.3	12.5
75–84 years	440 600	453 800	473 600	369 700	40.5	44.9	59.1	60.3
85 years and over	681 700	663 000	566 000	400 800	181.8	200.7	228.7	270.8
Black or African American [2]								
65 years and over, age-adjusted [1]	...	...	...	...	55.6	50.4	41.5	28.2
65 years and over, crude	145 900	122 900	82 000	37 700	51.1	45.2	35.0	22.0
65–74 years	30 300	29 700	22 500	12 200	18.2	18.4	15.4	11.1
75–84 years	58 700	47 300	30 600	13 400	66.5	57.2	45.3	26.7
85 years and over	56 900	45 800	29 000	12 100	183.1	167.1	141.5	105.7

Note: Excludes residents in personal care or domiciliary care homes.

[1] Age-adjusted by the direct method to the year 2000 population standard using the following three age groups: 65–74 years, 75–84 years, and 85 years and over.

[2] Beginning in 1999, the instruction for the race item on the Current Resident Questionnaire was changed so that more than one race could be recorded. In previous years, only one racial category could be checked. Estimates in this table are for single race residents.

. . . = Not available.

TABLE III-62. NURSING HOME RESIDENTS 65 YEARS OF AGE AND OVER, ACCORDING TO SELECTED FUNCTIONAL STATUS AND AGE, SEX, AND RACE, 1985, 1995, AND 1999

This table shows the percent of nursing home residents who are functionally dependent in regard to mobility or eating, or who are incontinent. They are considered dependent if they require assistance or special equipment, whether for mobility or eating. In 1999, there were 80.4 percent of residents, age 65 and over, who were dependent in regard to mobility; 65.7 percent were incontinent; a lower percentage of 47.4 was dependent in regard to eating; and 36.9 were dependent in all three functions. Comparing White and Black residents, Black percentages of impairment were higher than those of Whites for each of the functional categories and for each separate age group.

Table III-62. Nursing Home Residents 65 Years of Age and Over, According to Selected Functional Status and Age, Sex, and Race, 1985, 1995, and 1999

(Percent.)

Characteristic	Functional status [1]											
	Dependent mobility			Incontinent			Dependent eating			Dependent mobility, eating, and incontinent		
	1999	1995	1985	1999	1995	1985	1999	1995	1985	1999	1995	1985
All Persons												
65 years and over, age-adjusted [2]	80	79	76	66	64	55	47	45	41	37	37	33
65 years and over, crude	80	79	75	66	64	55	47	45	41	37	37	32
65-74 years	74	73	61	59	62	43	43	44	34	32	36	26
75-84 years	78	77	71	64	63	55	47	45	39	35	35	31
85 years and over	84	82	83	69	65	58	49	45	44	39	38	36
Male												
65 years and over, age-adjusted [2]	77	77	71	67	64	54	45	42	36	35	34	28
65 years and over, crude	76	76	68	66	64	52	45	43	35	35	35	27
65-74 years	71	71	56	60	63	39	45	44	33	35	37	24
75-84 years	77	77	66	69	65	54	45	44	33	35	36	26
85 years and over	78	78	79	67	63	58	46	40	39	35	33	31
Female												
65 years and over, age-adjusted [2]	82	80	77	65	64	55	48	46	42	37	37	34
65 years and over, crude	82	80	77	66	64	55	48	46	42	38	37	34
65-74 years	76	75	65	58	61	45	42	44	34	29	35	27
75-84 years	78	77	72	62	62	55	47	46	42	36	35	33
85 years and over	85	83	84	69	66	58	50	46	45	40	39	37
White [3]												
65 years and over, age-adjusted [2]	80	79	75	65	63	55	46	44	40	36	36	32
65 years and over, crude	80	79	74	65	63	54	46	44	40	36	36	32
65-74 years	73	71	60	57	60	42	41	42	33	29	34	25
75-84 years	78	76	70	64	62	54	46	45	39	35	35	30
85 years and over	84	82	83	68	65	58	48	44	44	38	37	36
Black or African American [3]												
65 years and over, age-adjusted [2]	82	83	83	72	69	61	56	52	49	47	44	38
65 years and over, crude	82	82	81	71	69	60	55	52	48	46	44	38
65-74 years	79	80	71	65	68	49	53	51	43	43	43	34
75-84 years	80	78	83	68	69	70	50	50	48	41	42	41
85 years and over	85	88	87	77	70	58	61	54	52	52	46	38

Note: Excludes residents in personal care or domicilary care homes.

[1] Nursing home residents who are dependent in mobility and eating require the assistance of a person or special equipment. Nursing home residents who are incontinent have difficulty in controlling bowels and/ or bladder or have an ostomy or indwelling catheter.

[2] Age-adjusted by the direct method to the 1995 National Nursing Home Survey population using the following three age groups: 65–74 years, 75–84 years, and 85 years and over.

[3] Beginning in 1999, the instruction for the race item on the Current Resident Questionnaire was changed so that more than one race could be recorded. In previous years, only one racial category could be checked. Estimates in this table are for single race residents.

CHAPTER IV: MARRIAGE AND DIVORCE

This chapter begins with data on marriage and divorce rates between 2001 and 2003. It then shows available earlier data on the timing and duration of marriages and on divorces. The percentages of persons that were never married are shown by age group, sex, race, and Hispanic ethnicity. The marriage experience of cohorts from earlier decades is set forth starting with those born between 1925 and 1939 and ending with births between 1955 and 1964.

TABLE IV-1. PROVISIONAL MARRIAGES AND DIVORCES, SEPTEMBER 2003

The marriage rate has fallen somewhat between 2001 and 2003, from an annual rate of 8.2 to 7.6 per 1,000 total population. Reported divorce rates appear relatively stable at or near 4.0 per 1,000 total population annually. Events to nonresidents are included in marriage and divorce figures (but not in birth and death data). Divorce figures include reported annulments. Marriages are not reported from Oklahoma, while divorces are not reported from California, Hawaii, Indiana, Louisiana, and Oklahoma. The 12-month divorce rates shown in Table 1 are based solely on the combined divorces and population of the reporting states and the District of Columbia. Table 3 shows monthly divorce counts by state, but NCHS does not publish aggregate 1-month, cumulative counts and rates, or 12-month counts for divorce because they would be based on less than the full complement of states.

TABLE IV-2. PROVISIONAL NUMBER AND RATE OF MARRIAGES, BY MONTH, JANUARY 2002–SEPTEMBER 2003

This table shows monthly aggregate marriage counts and rates for the United States. The marriage rate listed is an annual rate that would have occurred had the observed monthly rate persisted for 12 months. There is a noticeable increase in marriage counts and rates during the warmer months.

Table IV-1. Provisional Marriages and Divorces, September 2003

(Number, rate per 1,000 total population.)

Item	September				January–September				12 months ending with September				
	Number		Annual rate		Number		Annual rate		Number		Annual rate		
	2003	2002	2003	2002	2003	2002	2003	2002	2003	2002	2003	2002	2001
Marriages [1]	217 000	216 000	9.0	9.1	1 663 000	1 718 000	7.6	8.0	2 200 000	2 267 000	7.6	7.9	8.2
Divorces [2]	. . .	. . .	. . .	. . .	. . .	. . .	. . .	. . .	. . .	. . .	3.0	4.0	4.0

[1]Marriage rates may be underestimated due to incomplete reporting in Oklahoma.
[2]Divorce rates exclude data for California, Hawaii, Indiana, Louisiana, and Oklahoma. Populations for these rates also exclude these states.
. . . = Not available.

Table IV-2. Provisional Number and Rate of Marriages, by Month, January 2002–September 2003

(Number, rate per 1,000 total population.)

Year and month	Number	Annual rate per 1,000 population [1]
2002		
January	113 000	4.6
February	152 000	6.9
March	156 000	6.4
April	172 000	7.3
May	211 000	8.6
June	227 000	9.6
July	237 000	9.7
August	235 000	9.6
September	216 000	9.1
October	214 000	8.7
November	157 000	6.6
December	165 000	6.7
2003		
January	141 000	5.7
February	146 000	6.6
March	150 000	6.1
April	168 000	7.0
May	178 000	7.2
June	229 000	9.6
July	221 000	8.9
August	214 000	8.7
September	217 000	9.0

[1]Marriage rates may be underestimated due to incomplete reporting in Oklahoma.

TABLE IV-3. PROVISIONAL NUMBER OF MARRIAGES AND DIVORCES FOR EACH STATE, SEPTEMBER 2002 AND SEPTEMBER 2003, CUMULATIVE FIGURES, 2001–2003

This table shows state-specific counts for marriages and divorces for the most recent month reported, September 2003, and for comparison September 2002. In addition, it provides cumulative months January to September 2003, and for comparison, the same nine months in 2001 and 2002. Marriages are not reported from Oklahoma, while divorces are not reported from California, Hawaii, Indiana, Louisiana, and Oklahoma. Certain large states, such as California, New York, and Texas, have falling marriage counts for the nine-month periods across the 3 years, in agreement with the falling rates shown in Table 1. However, this pattern is not universal across states. Note the large number of marriages in Nevada, well out of proportion to its population.

Table IV-3. Provisional Number of Marriages and Divorces for each State, September 2002 and September 2003, Cumulative Figures, 2001–2003

(Number.)

Area	Marriages					Divorces				
	September		January–September			September		January–September		
	2003	2002	2003	2002	2001	2003	2002	2003	2002	2001
Alabama	3 874	3 493	32 263	32 387	32 308	2 151	2 076	17 685	19 381	18 486
Alaska	629	343	4 386	3 124	4 307	362	297	2 661	1 569	2 041
Arizona [1]	2 468	2 615	26 596	27 611	30 599	1 948	1 784	15 684	19 567	17 665
Arkansas	3 142	3 027	27 671	28 874	29 445	1 566	2 220	12 571	12 724	12 709
California	12 935	22 504	158 353	163 524	179 738	. . .	. . .	. . .	. . .	. . .
Colorado	3 133	4 496	28 908	27 819	28 906	1 647	1 448	14 244	15 572	15 735
Connecticut	1 671	1 850	12 510	12 574	14 432	937	897	7 757	7 417	8 123
Delaware	254	827	3 156	3 797	3 657	268	152	2 002	1 892	2 475
District of Columbia	327	316	2 382	2 377	3 082	128	78	493	575	1 331
Florida	12 028	12 234	118 697	124 518	119 300	7 155	7 254	63 285	64 420	64 393
Georgia	5 864	4 599	43 218	43 738	36 251	2 744	2 545	27 114	23 656	19 270
Hawaii	2 686	2 263	20 582	19 129	18 193	. . .	438	. . .	3 288	3 533
Idaho	1 791	1 628	11 489	10 968	11 325	510	653	5 279	5 338	5 241
Illinois	9 230	9 102	62 378	63 027	69 327	2 951	3 021	26 107	27 987	30 173
Indiana	4 122	2 811	33 608	33 164	35 380	. . .	. . .	. . .	. . .	. . .
Iowa	2 010	2 305	15 624	14 859	16 385	729	680	5 864	6 428	7 073
Kansas	1 613	1 666	14 587	14 708	15 550	815	707	6 732	7 778	6 751
Kentucky	3 796	4 470	29 192	28 507	27 141	1 987	1 542	15 956	15 937	15 663
Louisiana	3 066	2 351	29 954	27 605	28 294	. . .	. . .	. . .	. . .	. . .
Maine.	1 642	1 599	8 028	7 895	8 871	563	464	4 313	3 158	4 477
Maryland	3 768	3 906	28 060	28 654	28 254	1 310	1 233	11 806	12 734	10 989
Massachusetts	5 835	4 917	26 791	26 819	29 493	1 790	760	12 088	12 489	11 756
Michigan	7 851	7 558	49 158	49 120	50 070	2 752	3 045	27 877	28 122	29 541
Minnesota	4 744	4 110	25 886	24 339	26 312	1 119	1 092	11 198	11 281	11 583
Mississippi	1 652	1 626	13 698	14 681	14 519	1 334	1 570	10 405	11 043	10 780
Missouri	4 095	4 261	31 997	32 379	32 907	1 808	2 181	16 750	17 419	17 608
Montana	787	698	5 423	5 162	5 325	277	224	2 165	1 795	2 788
Nebraska	1 168	1 315	9 764	10 530	11 254	498	518	4 609	4 801	4 536
Nevada	12 905	12 291	106 931	92 381	116 390	1 213	1 416	11 240	15 977	9 928
New Hampshire	1 288	1 182	7 795	8 191	7 546	400	297	3 441	4 470	4 083
New Jersey	5 493	5 485	40 494	35 405	39 842	2 453	2 300	21 051	22 130	21 891
New Mexico [2,3]	702	1 067	7 628	11 530	11 906	348	652	5 652	5 963	6 807
New York	12 575	13 653	89 370	101 682	113 213	6 026	7 353	48 412	55 094	49 989
North Carolina	11 005	6 096	43 529	43 370	43 327	1 432	3 368	22 124	28 226	28 395
North Dakota	586	559	3 538	3 444	3 136	134	97	1 279	1 328	1 363
Ohio	3 425	8 425	47 155	60 921	62 532	1 832	3 633	29 568	34 500	34 216
Oklahoma	. . .	. . .	. . .	. . .	. . .	. . .	. . .	. . .	. . .	. . .
Oregon	3 870	2 717	18 207	20 197	20 569	1 115	1 549	10 453	12 292	12 073
Pennsylvania	8 076	7 278	56 852	54 201	54 095	3 354	3 020	27 337	27 397	27 467
Rhode Island	1 011	1 072	6 429	6 350	6 707	305	339	2 595	2 539	2 514
South Carolina	3 057	2 869	24 777	26 773	31 322	1 225	913	9 415	9 309	11 130
South Dakota	738	726	5 213	5 375	5 484	226	202	1 871	1 915	1 847
Tennessee	8 208	2 443	57 273	59 954	58 271	2 915	1 620	22 679	24 103	22 595
Texas	18 790	16 011	126 181	150 820	153 493	7 616	7 192	60 287	62 617	63 262
Utah	1 599	2 817	11 324	16 450	17 972	849	899	6 595	6 853	7 320
Vermont	715	762	4 130	4 045	4 287	361	273	2 017	2 033	1 837
Virginia	5 438	5 104	46 131	47 882	48 547	2 364	2 367	21 827	22 960	22 272
Washington	3 046	3 809	30 655	31 088	33 640	1 760	2 083	20 512	20 326	20 272
West Virginia	1 329	1 774	10 561	11 451	10 223	957	848	6 740	6 757	7 096
Wisconsin	4 528	4 661	26 189	26 508	26 999	1 278	1 419	13 444	13 122	13 215
Wyoming	666	511	3 909	3 822	4 097	214	239	2 087	2 059	2 136
Puerto Rico	1 471	1 716	17 906	17 803	20 897	1 213	1 482	13 652	12 939	12 208

[1]Figures for marriages include licenses issued for some counties.
[2]Figures for marriages are marriage licenses issued.
[3]Figures for divorces include estimates for some counties.
. . . = Not available.

TABLE IV. MARITAL HISTORY FOR PEOPLE 15 YEARS AND OVER, BY AGE AND SEX, FALL 1996

Tables 4 through 8 are based on figures projected onto U.S. population in 1996 based on survey of 37,000 households. In 1996, 31 percent of men and 24 percent of women age 15 and over had never married. Among teenagers, 99 percent of males and 97 percent of females were never married. For people age 50 and over, only a small proportion of men and women had never married—5 percent or less. An older average age at first marriage for men is reflected in the difference between the percent never married for men and women during the ages at which most people marry. Among the population 20 to 24 years old, 82 percent of men and 67 percent of women had never married (a difference of 15 percentage points). For 30 to 34 year olds, the difference was 10 percentage points, while for 40 to 49 year olds, the difference was only 3 percentage points.

Table IV-4. Marital History for People 15 Years and Over, by Age and Sex, Fall 1996

(Numbers in thousands, percent.)

Characteristic	Total, 15 years and over		15 to 19 years	20 to 24 years	25 to 29 years	30 to 34 years	35 to 39 years	40 to 49 years	50 to 59 years	60 to 69 years	70 years and over
	Estimate	90-percent confidence interval									
Men											
Total (in thousands)	99 005	98 090–99 920	9 680	8 658	9 445	10 568	11 138	19 381	12 157	9 038	8 940
Percent	100.0	X	100.0	100.0	100.0	100.0	100.0	100.0	100.0	100.0	100.0
Never married	30.6	30.0–31.2	99.0	81.7	48.6	28.7	19.2	12.0	5.3	5.2	4.4
Ever married	69.4	68.8–70.0	1.0	18.3	51.4	71.3	80.8	88.0	94.7	94.8	95.6
Married once	53.8	53.2–54.4	1.0	17.8	48.8	62.2	64.9	63.3	65.5	70.5	73.9
Still married [1]	44.4	43.8–45.0	0.8	16.2	41.8	53.0	54.0	51.6	54.5	59.1	55.5
Married twice	12.6	12.2–13.0	-	0.5	2.4	8.4	13.8	20.0	22.1	18.7	17.0
Still married [1]	10.1	9.7–10.5	-	0.4	2.0	7.1	11.1	16.0	17.5	15.4	12.6
Married 3 or more times	3.1	2.9–3.3	-	-	0.2	0.7	2.1	4.8	7.0	5.6	4.7
Still married [1]	2.3	2.1–2.5	-	-	0.2	0.5	1.8	3.8	5.1	4.4	3.2
Ever divorced	20.4	19.9–20.9	-	1.3	7.3	16.1	24.3	34.1	35.7	26.8	18.0
Currently divorced	8.2	7.8–8.6	-	0.9	5.2	8.1	10.8	13.9	13.1	9.2	4.5
Ever widowed	3.9	3.7–4.1	-	-	0.2	0.2	0.3	1.2	3.5	9.2	25.1
Currently widowed	2.6	2.4–2.8	-	-	0.1	0.2	0.1	0.6	1.9	5.1	18.6
Women											
Total (in thousands)	106 477	105 562–107 392	9 253	8 749	9 725	10 769	11 342	20 029	12 982	10 400	13 228
Percent	100.0	X	100.0	100.0	100.0	100.0	100.0	100.0	100.0	100.0	100.0
Never married	23.6	23.1–24.1	96.6	67.0	35.3	18.7	14.1	8.6	5.0	3.7	4.2
Ever married	76.4	75.9–76.9	3.4	33.0	64.7	81.3	85.9	91.4	95.0	96.3	95.8
Married once	59.9	59.3–60.5	3.4	31.7	58.8	68.5	66.8	65.8	69.4	74.7	76.2
Still married [1]	42.0	41.4–42.6	3.0	27.4	48.9	57.0	53.3	49.2	50.3	48.4	28.1
Married twice	13.4	13.0–13.8	-	1.2	5.5	11.4	16.0	20.6	19.3	17.3	16.0
Still married [1]	9.1	8.7–9.5	-	1.1	4.6	9.6	12.0	15.5	13.2	10.5	6.0
Married 3 or more times	3.1	2.9–3.3	-	-	0.4	1.4	3.1	5.0	6.3	4.3	3.6
Still married [1]	1.8	1.6–2.0	-	-	0.3	1.1	2.3	3.4	3.9	2.4	0.9
Ever divorced	22.2	21.7–22.7	-	3.2	12.2	20.8	27.8	37.0	34.6	26.1	16.7
Currently divorced	10.3	9.9–10.7	-	2.1	7.1	9.5	12.8	17.2	16.6	11.5	5.6
Ever widowed	12.1	11.7–12.5	-	0.2	0.3	0.7	1.8	3.8	10.7	25.5	58.3
Currently widowed	10.5	10.1–10.9	-	0.2	0.2	0.3	1.0	2.6	8.2	21.6	54.5

Note: Figures projected onto U.S. population in 1996 based on survey of 37,000 households from the U.S. Census Bureau Survey of Income and Program Participation.

[1] Includes those currently separated.
X = Not applicable.
- = Quantity zero or rounds to zero.

Figure IV-1. Distribution of Men 15 Years and Over, by Marital Status and Age, Fall 1996

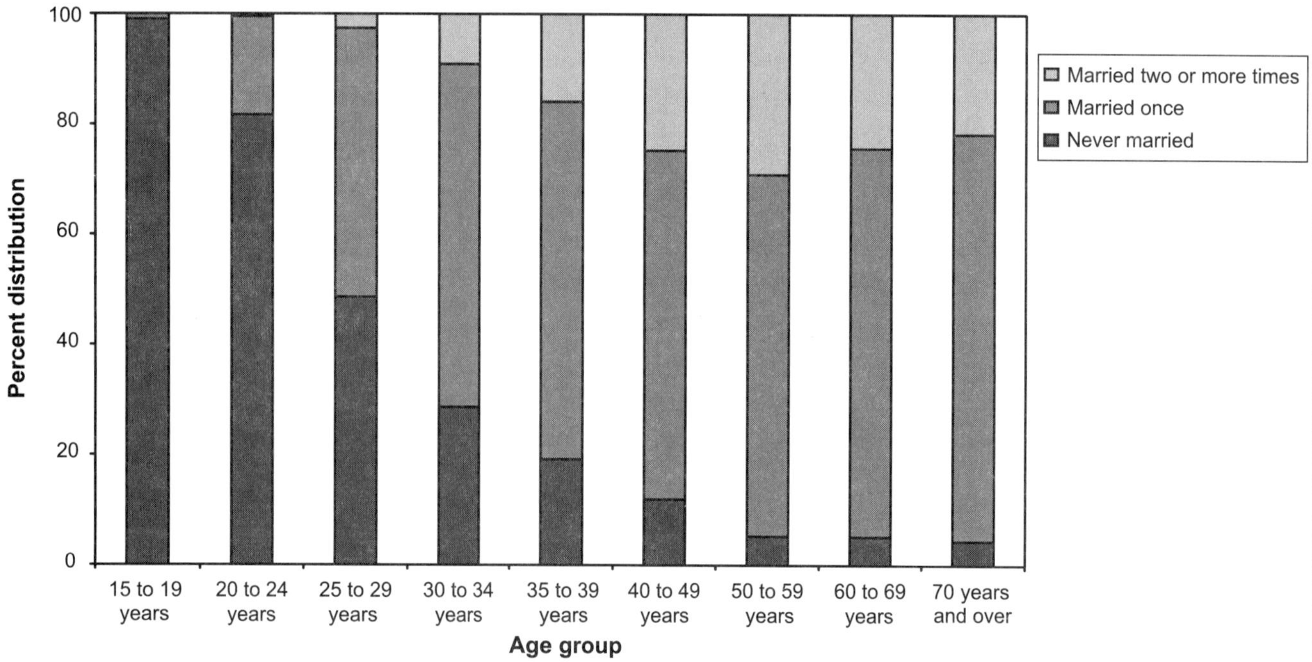

Figure IV-2. Distribution of Women 15 Years and Over, by Marital Status and Age, Fall 1996

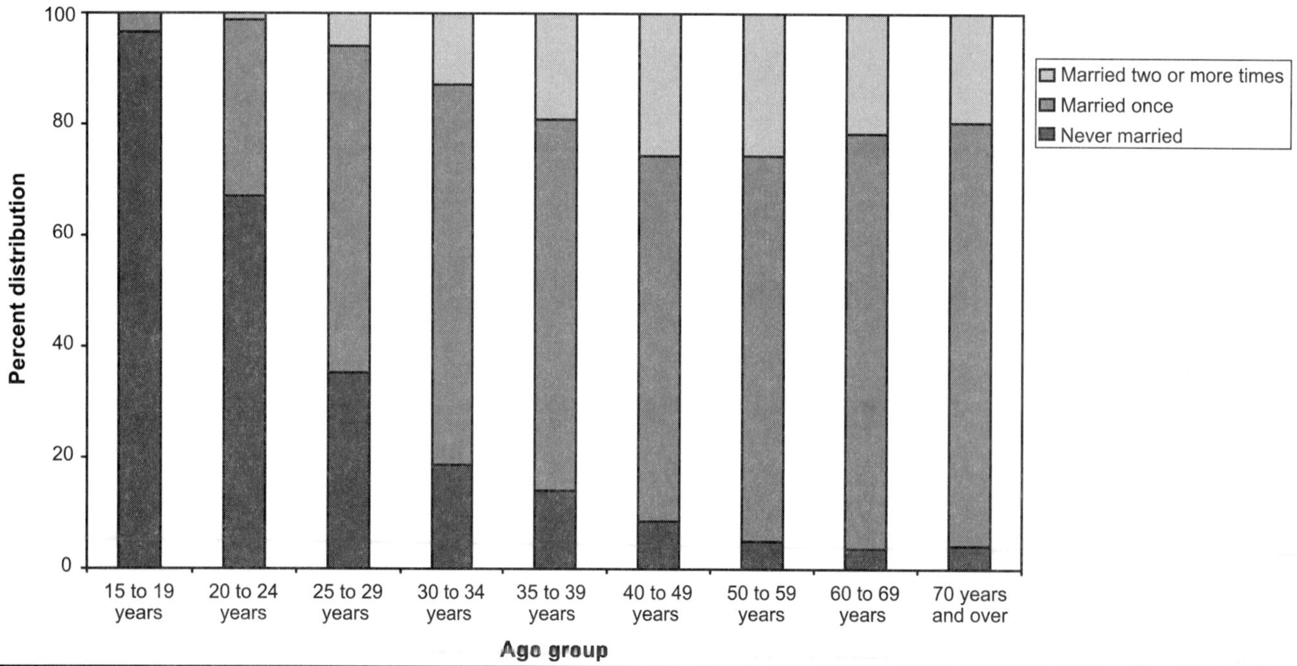

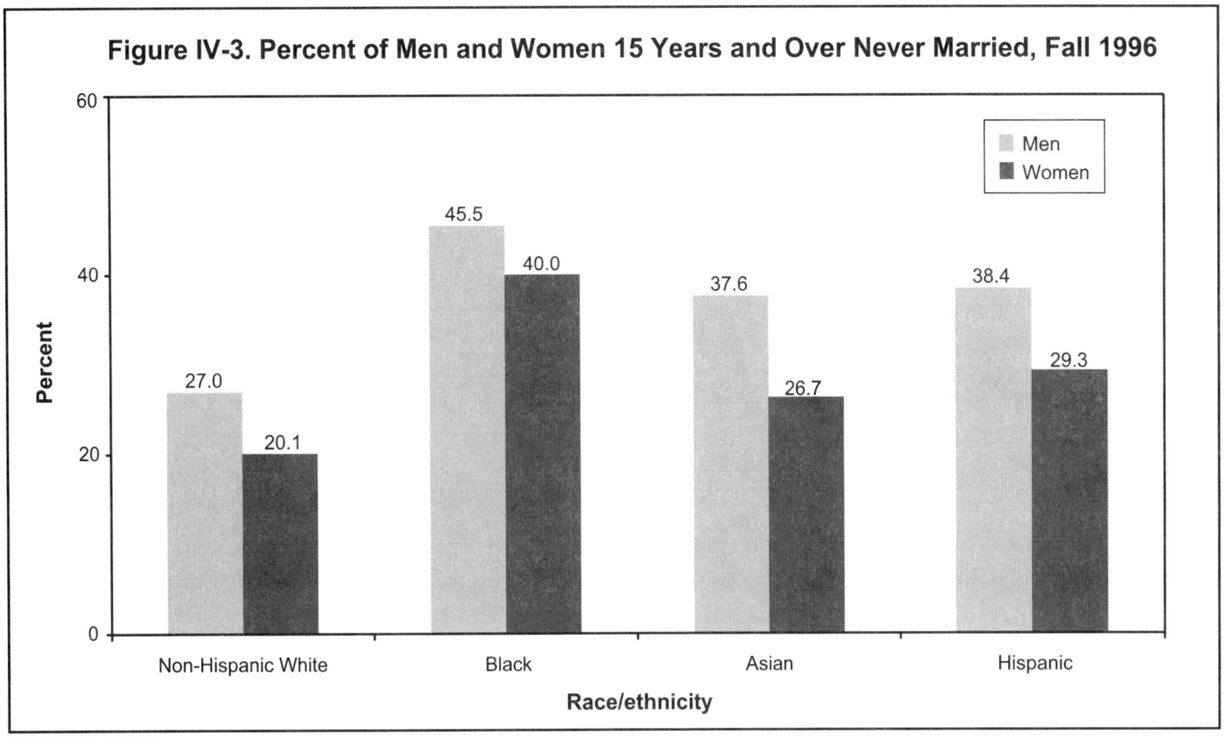

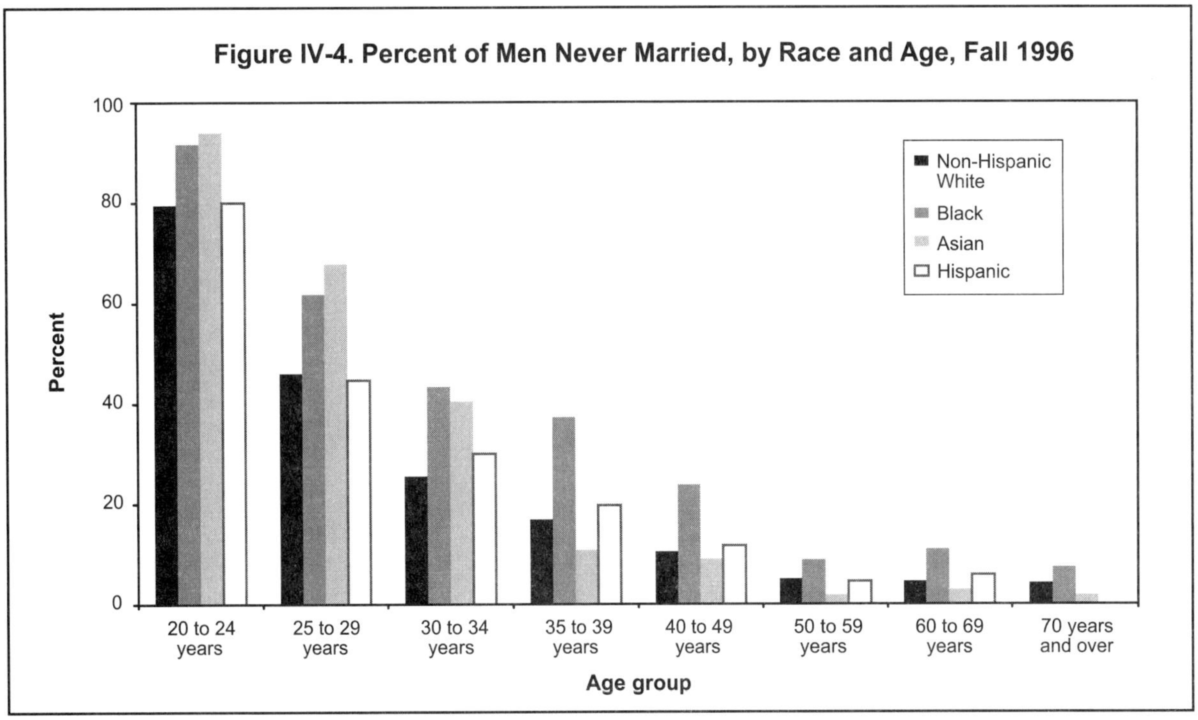

TABLE IV-5. MARITAL HISTORY FOR PEOPLE 15 YEARS AND OVER, BY AGE, SEX, RACE, AND HISPANIC ORIGIN, FALL 1996

Black men and women age 25 to 29 were more likely to be never married than their White counterparts. Among Asians and Pacific Islanders in this age group, a much higher proportion of men than women were never married (68 percent vs. 42 percent). Relatively few Whites and Hispanics, aged 50 to 59 years, were never married, the percentages ranging between 4.2 percent and 6.7 percent, but these shares were higher for Blacks with 8.9 percent of Black men and 10.2 percent of Black women never married in their 50s.

Table IV-5. Marital History for People 15 Years and Over, by Age, Sex, Race, and Hispanic Origin, Fall 1996

(Numbers in thousands, percent.)

Characteristic	Total, 15 years and over	15 to 19 years	20 to 24 years	25 to 29 years	30 to 34 years	35 to 39 years	40 to 49 years	50 to 59 years	60 to 69 years	70 years and over
WHITE NON-HISPANIC										
Men										
Total (in thousands)	74 604	6 518	5 799	6 628	7 425	8 368	14 882	9 818	7 424	7 742
Percent	100.0	100.0	100.0	100.0	100.0	100.0	100.0	100.0	100.0	100.0
Never married	27.0	99.1	79.5	46.0	25.5	16.9	10.5	5.1	4.6	4.3
Ever married	73.0	0.9	20.5	54.0	74.5	83.1	89.5	94.9	95.4	95.7
Married once	55.8	0.9	19.9	51.0	64.1	66.0	62.9	64.4	70.4	75.1
Still married [1]	46.5	0.8	18.0	43.5	54.9	55.1	52.2	54.4	60.3	57.6
Married twice	13.7	-	0.6	2.8	9.7	14.6	21.2	22.8	19.2	16.3
Still married [1]	11.0	-	0.5	2.4	8.1	12.0	17.1	18.0	15.8	12.5
Married 3 or more times	3.5	-	-	0.2	0.7	2.5	5.4	7.8	5.9	4.4
Still married [1]	2.7	-	-	0.2	0.5	2.1	4.3	5.7	4.7	3.0
Ever divorced	22.1	-	1.6	8.8	18.2	25.9	35.6	36.9	27.4	16.9
Currently divorced	8.6	-	1.1	6.2	9.2	11.3	13.8	13.2	9.0	4.1
Ever widowed	4.1	-	-	-	0.3	0.3	1.2	3.3	8.9	24.2
Currently widowed	2.7	-	-	-	0.2	-	0.6	1.8	5.0	17.9
Women										
Total (in thousands)	79 268	6 169	5 839	6 635	7 495	8 257	15 077	10 065	8 353	11 378
Percent	100.0	100.0	100.0	100.0	100.0	100.0	100.0	100.0	100.0	100.0
Never married	20.1	96.6	64.0	31.1	14.3	10.9	6.8	4.2	3.2	3.9
Ever married	79.9	3.4	36.0	68.9	85.7	89.1	93.2	95.8	96.8	96.1
Married once	61.7	3.4	34.6	61.3	70.5	67.8	65.3	68.7	74.6	76.7
Still married [1]	44.3	2.9	30.1	51.3	59.8	55.9	51.0	53.1	51.2	29.4
Married twice	14.6	-	1.3	7.0	13.5	17.4	21.9	20.0	17.7	15.7
Still married [1]	10.1	-	1.2	5.8	11.3	13.3	16.7	14.3	11.2	6.4
Married 3 or more times	3.6	-	-	0.6	1.7	3.8	5.9	7.2	4.4	3.6
Still married [1]	2.2	-	-	0.4	1.4	2.9	4.1	4.4	2.7	0.9
Ever divorced	23.3	0.1	3.8	14.1	23.5	29.5	38.7	35.1	259.1	16.3
Currently divorced	10.3	0.1	2.5	7.7	10.1	12.8	17.1	16.0	10.6	5.4
Ever widowed	13.0	-	0.3	0.4	0.7	1.8	3.4	9.5	24.3	57.6
Currently widowed	11.3	-	0.2	0.2	0.2	1.0	2.1	6.8	20.3	53.6
WHITE										
Men										
Total (in thousands)	83 603	7 621	7 025	7 829	8 737	9 366	16 404	10 603	7 992	8 026
Percent	100.0	100.0	100.0	100.0	100.0	100.0	100.0	100.0	100.0	100.0
Never married	28.2	99.0	79.6	45.8	25.9	17.0	10.6	5.1	4.7	4.2
Ever married	71.8	1.0	20.4	54.2	74.1	83.0	89.4	95.0	95.3	95.8
Married once	55.4	1.0	19.9	51.6	64.7	66.2	63.6	65.0	70.8	75.0
Still married [1]	46.5	0.9	18.0	44.5	55.9	55.6	52.7	54.9	60.6	57.4
Married twice	13.1	-	0.6	2.5	8.8	14.3	20.7	22.5	18.8	16.5
Still married [1]	10.5	-	0.5	2.1	7.4	11.6	16.6	17.8	15.5	12.6
Married 3 or more times	3.3	-	-	0.2	0.7	2.5	5.1	7.5	5.8	4.3
Still married [1]	2.5	-	-	0.2	0.4	2.1	4.1	5.5	4.6	3.0
Ever divorced	21.1	-	1.5	7.9	16.6	25.1	34.9	36.3	27.0	16.9
Currently divorced	8.3	-	1.1	5.6	8.4	10.9	13.8	13.0	8.9	4.1
Ever widowed	3.9	-	-	-	0.3	0.2	1.1	3.4	8.8	24.4
Currently widowed	2.6	-	-	-	0.2	-	0.5	1.9	4.8	18.1
Women										
Total (in thousands)	88 262	7 241	6 925	7 825	8 648	9 258	16 544	11 022	8 958	11 841
Percent	100.0	100.0	100.0	100.0	100.0	100.0	100.0	100.0	100.0	100.0
Never married	20.9	95.9	63.2	30.8	14.8	11.2	6.9	4.2	3.5	4.2
Ever married	79.1	4.1	36.8	69.2	85.2	88.8	93.1	95.8	96.5	95.8
Married once	61.6	4.1	35.4	62.1	70.8	68.2	66.0	69.4	74.9	76.6
Still married [1]	44.4	3.7	30.8	52.1	60.2	56.1	51.2	52.8	50.9	29.1
Married twice	14.1	-	1.3	6.6	12.7	17.0	21.5	19.6	17.4	15.6
Still married [1]	9.7	-	1.2	5.5	10.7	12.9	16.3	13.9	11.0	6.3
Married 3 or more times	3.4	-	-	0.5	1.7	3.5	5.6	6.8	4.3	3.6
Still married [1]	2.1	-	-	0.4	1.4	2.7	3.8	4.2	2.6	0.9
Ever divorced	22.9	0.1	3.6	13.6	22.2	28.6	37.9	34.6	25.7	16.4
Currently divorced	10.1	0.1	2.4	7.6	9.5	12.4	16.9	16.2	10.8	5.5
Ever widowed	12.3	0.1	0.3	0.3	0.7	1.8	3.5	9.6	23.9	57.5
Currently widowed	10.7	0.1	0.2	0.2	0.3	1.0	2.3	7.0	20.0	53.6

Note: Figures projected onto U.S. population in 1996 based on survey of 37,000 households from the U.S. Census Bureau Survey of Income and Program Participation.

[1] Includes those currently separated.
- = Quantity zero or rounds to zero.

Table IV-5. Marital History for People 15 Years and Over, by Age, Sex, Race, and Hispanic Origin, Fall 1996—*Continued*

(Numbers in thousands, percent.)

Characteristic	Total, 15 years and over	15 to 19 years	20 to 24 years	25 to 29 years	30 to 34 years	35 to 39 years	40 to 49 years	50 to 59 years	60 to 69 years	70 years and over
BLACK										
Men										
Total (in thousands)	11 006	1 500	1 146	1 138	1 286	1 297	2 074	1 125	788	653
Percent	100.0	100.0	100.0	100.0	100.0	100.0	100.0	100.0	100.0	100.0
Never married	45.5	98.9	91.7	61.8	43.4	37.3	23.8	8.9	11.0	7.4
Ever married	54.5	1.1	8.3	38.2	56.6	62.7	76.2	91.1	89.0	92.6
Married once	41.9	1.1	8.3	36.2	49.5	50.9	56.6	64.7	64.5	58.0
Still married[1]	28.6	0.7	8.1	28.7	37.0	36.1	37.1	43.5	39.0	31.3
Married twice	8.1	-	-	1.9	6.0	8.3	13.5	17.7	14.8	14.5
Still married[1]	8.1	-	-	1.9	6.0	8.3	13.5	17.7	14.8	14.5
Married 3 or more times	1.7	-	-	-	-	-	2.0	5.0	-	-
Still married[1]	1.1	-	-	-	-	-	1.0	3.0	-	-
Ever divorced	18.5	0.2	-	4.3	14.1	21.1	32.5	36.4	29.2	33.3
Currently divorced	8.8	0.2	-	2.8	7.3	11.4	16.2	16.0	13.4	11.0
Ever widowed	4.0	-	0.2	1.0	0.2	1.1	1.7	5.1	14.3	32.4
Currently widowed	2.8	-	0.2	0.6	-	0.9	0.9	2.8	9.0	25.0
Women										
Total (in thousands)	13 380	1 509	1 361	1 424	1 518	1 531	2 478	1 418	1 057	1 086
Percent	100.0	100.0	100.0	100.0	100.0	100.0	100.0	100.0	100.0	100.0
Never married	40.0	98.8	83.7	57.7	39.4	33.6	20.9	10.2	6.3	5.8
Ever married	60.0	1.2	16.3	42.3	60.6	66.4	79.1	89.8	93.7	94.2
Married once	47.8	1.2	15.5	41.1	54.7	53.8	59.8	67.6	69.9	69.4
Still married[1]	24.1	0.9	12.3	30.4	36.0	29.4	30.1	29.0	25.7	16.7
Married twice	10.6	-	0.8	1.2	5.8	11.5	17.3	18.5	18.5	21.5
Still married[1]	5.6	-	0.6	0.8	4.7	7.1	12.0	9.1	7.7	3.8
Married 3 or more times	1.6	-	-	-	0.1	1.2	2.0	3.7	5.2	3.2
Still married[1]	0.7	-	-	-	0.1	0.6	1.4	1.7	1.6	1.0
Ever divorced	20.6	-	1.2	6.2	17.0	27.0	35.2	37.7	33.3	20.6
Currently divorced	12.4	-	0.6	5.3	11.5	17.5	21.1	2.8	19.4	7.2
Ever widowed	11.9	-	-	0.3	0.4	2.0	6.1	18.8	38.3	66.6
Currently widowed	10.7	-	-	-	0.4	1.3	5.3	16.7	33.1	63.0
ASIAN										
Men										
Total (in thousands)	3 345	379	368	366	402	369	727	335	198	200
Percent	100.0	100.0	100.0	100.0	100.0	100.0	100.0	100.0	100.0	100.0
Never married	37.6	100.0	94.0	67.8	40.5	10.8	9.0	1.8	2.9	1.8
Ever married	62.4	-	6.0	32.2	59.5	89.2	91.0	98.2	97.1	98.2
Married once	55.4	-	6.0	30.5	54.6	82.0	77.3	88.4	85.8	84.7
Still married[1]	50.1	-	4.6	26.5	48.0	77.6	70.2	83.9	83.0	62.8
Married twice	5.2	-	-	0.9	4.0	6.2	9.2	7.6	11.2	7.8
Still married[1]	5.2	-	-	0.9	4.0	6.2	9.2	7.6	11.2	7.8
Married 3 or more times	1.1	-	-	-	0.9	0.9	2.2	1.1	-	4.7
Still married[1]	0.9	-	-	-	0.9	0.9	1.5	1.1	-	4.7
Ever divorced	9.2	-	-	4.6	10.6	10.2	18.9	12.4	5.8	9.5
Currently divorced	3.8	-	-	3.6	5.7	3.1	8.5	4.7	-	1.4
Ever widowed	2.5	-	-	0.9	-	-	2.3	0.9	6.8	23.6
Currently widowed	1.4	-	-	-	-	-	-	0.0	1.0	-
Women										
Total (in thousands)	3 796	368	360	391	473	417	804	439	317	226
Percent	100.0	100.0	100.0	100.0	100.0	100.0	100.0	100.0	100.0	100.0
Never married	26.7	100.0	76.6	41.5	18.4	4.6	5.1	9.9	3.7	2.9
Ever married	73.3	-	23.4	58.5	81.6	95.4	94.9	90.1	96.3	97.1
Married once	65.7	-	23.4	57.1	75.2	86.8	81.8	74.6	90.2	86.4
Still married[1]	52.8	-	21.2	53.1	70.0	79.0	69.3	57.0	58.7	29.9
Married twice	7.0	-	-	1.4	6.4	8.6	11.9	13.8	6.1	7.7
Still married[1]	5.3	-	-	1.4	4.8	7.8	10.1	11.4	2.7	-
Married 3 or more times	0.6	-	-	-	-	-	-	1.0	-	3.0
Still married[1]	0.4	-	-	0.1	0.5	0.7	2.3	2.5	3.3	4.3
Ever divorced	11.4	-	2.2	5.4	9.8	13.7	21.7	20.9	4.7	8.3
Currently divorced	5.5	-	2.2	4.0	4.5	5.8	10.7	8.5	3.6	1.6
Ever widowed	8.2	-	-	-	1.2	0.5	1.7	11.9	30.4	62.9
Currently widowed	7.8	-	-	-	0.8	0.5	1.7	9.5	28.8	62.9

Note: Figures projected onto U.S. population in 1996 based on survey of 37,000 households from the U.S. Census Bureau Survey of Income and Program Participation.

[1] Includes those currently separated.
\- = Quantity zero or rounds to zero.

Table IV-5. Marital History for People 15 Years and Over, by Age, Sex, Race, and Hispanic Origin, Fall 1996—*Continued*

(Numbers in thousands, percent.)

Characteristic	Total, 15 years and over	15 to 19 years	20 to 24 years	25 to 29 years	30 to 34 years	35 to 39 years	40 to 49 years	50 to 59 years	60 to 69 years	70 years and over
HISPANIC										
Men										
Total (in thousands)	9 911	1 248	1 317	1 308	1 449	1 113	1 684	860	625	306
Percent	100.0	100.0	100.0	100.0	100.0	100.0	100.0	100.0	100.0	100.0
Never married	38.4	98.6	80.0	44.7	30.0	19.7	11.7	4.6	5.9	2.1
Ever married	61.6	1.4	20.0	55.3	70.0	80.3	88.3	95.4	94.1	97.9
Married once	52.0	1.4	19.6	54.8	65.3	66.3	69.1	72.5	76.1	70.6
Still married	44.6	1.2	18.1	49.7	58.7	58.2	56.8	58.8	64.0	50.5
Married twice	6.2	-	0.4	0.5	3.4	9.0	12.0	15.4	11.3	15.5
Still married	6.2	-	0.4	0.5	3.4	9.0	12.0	15.4	11.3	15.5
Married 3 or more times	1.5	-	-	-	0.2	1.7	3.1	3.8	4.8	4.5
Still married	1.2	-	-	-	0.2	1.5	2.4	2.6	4.0	2.8
Ever divorced	13.4	0.2	1.1	2.9	8.4	19.0	29.0	29.1	21.1	21.1
Currently divorced	5.8	0.2	0.8	2.4	3.9	8.0	14.0	10.3	6.8	5.4
Ever widowed	2.0	-	-	0.5	-	0.3	0.9	4.2	7.3	29.0
Currently widowed	1.3	-	-	0.5	-	-	0.6	3.1	3.3	21.3
Women										
Total (in thousands)	9 913	1 177	1 212	1 295	1 258	1 113	1 662	1 013	665	521
Percent	100.0	100.0	100.0	100.0	100.0	100.0	100.0	100.0	100.0	100.0
Never married	29.3	91.7	59.8	30.5	19.5	13.3	8.9	6.7	6.9	9.0
Ever married	70.7	8.2	40.2	69.5	80.5	86.7	91.1	93.3	93.1	91.0
Married once	60.3	8.2	38.4	65.8	71.7	71.9	72.4	74.8	77.0	75.7
Still married	44.3	7.7	33.4	55.0	60.7	58.4	51.9	48.0	46.4	21.9
Married twice	9.3	-	1.8	3.7	7.2	13.8	17.0	15.9	14.2	13.2
Still married	6.3	-	1.8	3.4	6.7	9.7	11.3	9.5	9.8	3.9
Married 3 or more times	1.1	-	-	-	1.6	1.0	1.6	2.6	2.0	2.2
Still married	0.8	-	-	-	1.4	0.6	1.1	1.9	1.2	0.7
Ever divorced	16.6	-	3.1	10.8	14.0	20.8	30.2	29.6	23.8	18.5
Currently divorced	8.7	-	1.3	7.4	5.7	9.6	15.9	17.8	13.2	7.8
Ever widowed	6.5	0.3	0.3	0.2	0.9	1.5	4.5	11.7	19.3	55.2
Currently widowed	5.8	0.3	0.3	0.2	0.8	0.6	3.5	10.0	16.8	53.2

Note: Figures projected onto U.S. population in 1996 based on survey of 37,000 households from the U.S. Census Bureau Survey of Income and Program Participation.

[1]Includes those currently separated.
- = Quantity zero or rounds to zero.

Figure IV-5. Percent of Women Never Married, by Race and Age, Fall 1996

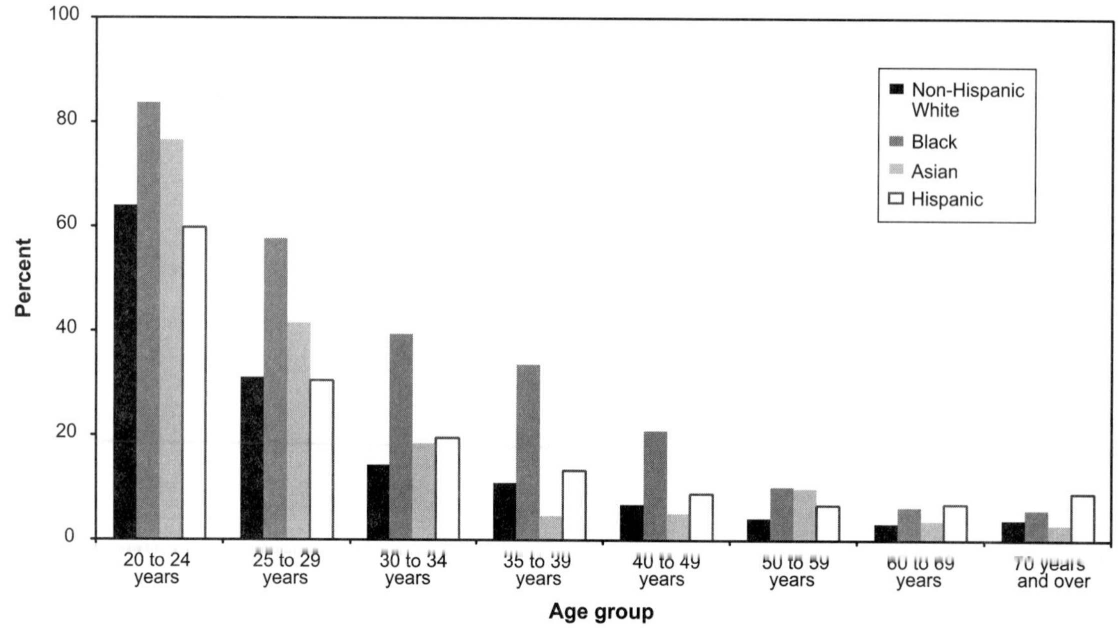

TABLE IV-6. MARITAL HISTORY BY SEX FOR SELECTED BIRTH COHORTS, 1925–1974, FALL 1996

Because people born in a specified period of time ("birth cohort") experience similar historical circumstances as they move through the life cycle, it is useful to compare trends in marital patterns by birth cohort. There have been delays in age of marriage, as well as increases in divorce rates, during the period covered by this table, which starts with the 1925–1934 birth cohort who formed the forefront of adults who parented the post WWII baby boom of 1946–1964. Since this data was collected in 1996, the 1955–1964 birth cohort is the most recent cohort available that had lived long enough to examine delays in marriage (percent married by age 30). The rise in divorce rates also led to higher percentages of persons in more recent cohorts who were able to marry more than once in their lifetimes.

Table IV-6. Marital History by Sex for Selected Birth Cohorts, 1925–1974, Fall 1996

(Numbers in thousands, percent.)

Characteristic	1925 to 1934	1935 to 1944	1945 to 1954	1955 to 1964	1965 to 1974
MEN (in thousands)	8 668	11 446	18 094	22 019	18 666
Percent Ever Married By:					
20 years	20.4	25.5	24.8	18.6	11.3
25 years	67.7	69.9	62.3	49.1	X
30 years	84.7	84.7	77.0	69.2	X
35 years	89.9	88.9	83.9	X	X
40 years	91.9	91.3	87.9	X	X
45 years	93.0	92.9	X	X	X
50 years	93.7	94.1	X	X	X
60 years	94.5	X	X	X	X
Percent Ever Divorced By:					
20 years	0.7	0.5	1.1	1.3	0.8
25 years	3.4	4.9	8.4	8.2	X
30 years	6.9	11.9	18.2	15.8	X
35 years	11.1	20.0	25.7	X	X
40 years	15.1	25.7	31.3	X	X
45 years	18.3	29.5	X	X	X
50 years	21.3	32.3	X	X	X
60 years	24.4	X	X	X	X
Percent Married Two Times or More By:					
25 years	1.7	2.1	3.0	2.8	1.5
30 years	4.2	6.3	9.9	8.4	X
35 years	7.2	12.5	17.0	X	X
40 years	10.6	18.4	22.3	X	X
45 years	14.1	22.6	X	X	X
50 years	17.7	25.6	X	X	X
60 years	22.0	X	X	X	X
WOMEN (in thousands)	10 374	12 157	18 664	22 624	18 955
Percent Ever Married By:					
20 years	50.3	51.9	44.9	33.6	25.2
25 years	83.5	82.8	75.4	63.4	X
30 years	91.5	90.0	84.1	78.8	X
35 years	93.7	92.5	88.5	X	X
40 years	94.7	94.3	90.9	X	X
45 years	95.2	95.0	X	X	X
50 years	95.6	95.3	X	X	X
60 years	96.0	X	X	X	X
Percent Ever Divorced By:					
20 years	2.0	2.0	2.4	3.0	2.0
25 years	5.0	6.9	11.4	12.0	X
30 years	8.6	13.2	21.3	19.9	X
35 years	12.5	20.3	28.6	X	X
40 years	16.0	26.5	33.9	X	X
45 years	19.1	29.6	X	X	X
50 years	21.5	32.2	X	X	X
60 years	23.7	X	X	X	X
Percent Married Two Times or More By:					
25 years	3.2	4.0	5.2	5.5	3.5
30 years	6.1	7.7	11.4	12.3	X
35 years	9.1	12.4	18.0	X	X
40 years	12.1	17.3	22.8	X	X
45 years	14.7	20.9	X	X	X
50 years	17.3	23.6	X	X	X
60 years	20.4	X	X	X	X

Note: Figures projected onto U.S. population in 1996 based on survey of 37,000 households from the U.S. Census Bureau Survey of Income and Program Participation.

X = Cohort had not lived to stated age at the time of the survey.

TABLE IV-7. PERCENT REACHING STATED ANNIVERSARY BY MARRIAGE COHORT AND SEX FOR FIRST AND SECOND MARRIAGES, FALL 1996

The proportion of both men and women reaching various anniversaries declined from marriages begun in the late 1940s to marriages begun in the early 1980s. For example, 90 percent of women who first married in 1945–1949 reached their 10th anniversary, compared with only 73 percent of those who first married in 1980–1984. This reflected the rapid increase in likelihood of divorce over the last half of the twentieth century. The survey suggests that for women, second marriages entered in 1975–1984 were less likely to reach their 10th anniversary than were first marriages during the same time period (although it should be noted that the effect of mortality would be more prominent in second marriages).

TABLE IV-8. MEDIAN AGE AT MARITAL EVENT FOR PEOPLE 15 YEARS AND OVER, BY SEX, RACE, AND HISPANIC ORIGIN, FALL 1996

This table shows the median age for marital events associated with first and second marriages for those people who have experienced those events. In 1996, the median age for first marriage was between 21 and 22 for non-Hispanic White, Hispanic, and Black women. Women's median age at separation from first marriage ranged from 28 to 29 years and their median age at divorce from first marriage was about 29 years. For men, the corresponding ages were roughly two years greater. The median age for widowhood during first marriage was considerably lower for Black women compared with non-Hispanic White and Hispanic women, reflecting higher mortality rates among Black men. Differences indicated in this table may be partly explained by different age distributions in specific population subgroups, in addition to actual variation in age-specific marital behavior.

Table IV-7. Percent Reaching Stated Anniversary by Marriage Cohort and Sex for First and Second Marriages, Fall 1996

(Numbers in thousands, percent.)

Sex and year of marriage	Number of marriages (thousands)	Anniversary [1]							
		5th	10th	15th	20th	25th	30th	35th	40th
FIRST MARRIAGES									
Men									
1945 to 1949	3 634	94.7	90.8	86.5	82.5	79.3	76.0	72.9	70.8
1950 to 1954	4 009	96.8	90.5	85.2	80.4	76.6	73.5	70.6	67.4
1955 to 1959	4 706	94.2	86.4	78.7	73.6	69.6	65.8	62.9	X
1960 to 1964	5 153	94.4	82.2	74.2	67.4	62.7	58.9	X	X
1965 to 1969	6 538	90.5	76.3	66.1	61.0	56.5	X	X	X
1970 to 1974	7 332	87.4	71.8	62.8	57.7	X	X	X	X
1975 to 1979	7 196	86.3	71.2	62.3	X	X	X	X	X
1980 to 1984	7 679	86.5	73.9	X	X	X	X	X	X
1985 to 1989	7 954	87.6	X	X	X	X	X	X	X
Women									
1945 to 1949	5 336	95.0	90.2	86.1	81.2	75.3	70.4	65.2	59.0
1950 to 1954	5 195	94.6	89.2	82.9	77.2	71.9	65.9	61.1	56.1
1955 to 1959	5 293	94.6	87.5	78.7	71.7	66.3	62.3	58.2	X
1960 to 1964	5 657	92.6	82.1	72.0	63.7	58.6	55.1	X	X
1965 to 1969	7 222	90.0	76.1	65.4	59.4	53.8	X	X	X
1970 to 1974	8 188	86.7	70.7	62.0	55.9	X	X	X	X
1975 to 1979	8 006	85.4	69.6	61.4	X	X	X	X	X
1980 to 1984	8 349	86.4	73.0	X	X	X	X	X	X
1985 to 1989	8 347	86.4	X	X	X	X	X	X	X
SECOND MARRIAGES									
Men									
1975 to 1979	2 178	89.3	68.9	X	X	X	X	X	X
1980 to 1984	2 800	87.6	69.0	X	X	X	X	X	X
1985 to 1989	2 669	86.7	X	X	X	X	X	X	X
Women									
1975 to 1979	2 301	82.4	63.8	X	X	X	X	X	X
1980 to 1984	2 937	84.1	64.2	X	X	X	X	X	X
1985 to 1989	3 016	83.0	X	X	X	X	X	X	X

Note: Figures projected onto U.S. population in 1996 based on survey of 37,000 households from the U.S. Census Bureau Survey of Income and Program Participation.

[1] Persons reaching stated anniversary for specified marital order.
X = Not applicable.

Table IV-8. Median Age at Marital Event for People 15 Years and Over, by Sex, Race, and Hispanic Origin, Fall 1996

(Median age in years for those who experienced the specified event.)

Sex and year of marriage	Total	White	White non-Hispanic	Black	Asian and Pacific Islander	Hispanic (of any race)
MEN						
First Marriage						
Age when married	23.5	23.4	23.3	24.2	26.7	23.8
Age when separated [1]	29.3	29.2	29.2	30.3	30.5	29.6
Age when divorced	30.5	30.3	30.3	32.3	31.6	31.0
Age when widowed	59.6	60.3	60.8	54.0	B	49.7
Second Marriage						
Age when married	34.0	33.9	33.9	36.1	36.7	33.9
Age when separated [1]	37.2	37.2	37.2	39.3	B	36.0
Age when divorced	39.3	39.0	39.1	42.5	B	37.7
Age when widowed	62.3	62.8	63.1	B	B	B
WOMEN						
First Marriage						
Age when married	21.3	21.3	21.2	21.8	24.1	21.4
Age when separated [1]	27.8	27.5	27.5	28.8	30.0	27.8
Age when divorced	29.0	28.8	28.7	30.8	31.3	29.1
Age when widowed	57.8	59.3	59.7	50.3	53.8	51.7
Second Marriage						
Age when married	32.0	31.7	31.8	34.9	34.3	31.8
Age when separated [1]	35.2	34.9	34.8	37.2	B	36.3
Age when divorced	37.0	36.8	36.7	40.3	B	39.0
Age when widowed	59.0	59.4	59.5	56.5	B	B

Note: Figures projected onto U.S. population in 1996 based on survey of 37,000 households from the U.S. Census Bureau Survey of Income and Program Participation.

[1]For those who divorced.
B = Base less than 100,000. Median not calculated.

TABLE IV-9. MEDIAN DURATION OF MARRIAGE EVENTS FOR PEOPLE 15 YEARS AND OVER, BY SEX AND RACE AND ETHNICITY, FALL 1996

The median number of years a first marriage lasted that ended in divorce was about 8 years for men and women overall. Half the men and women who remarried after divorce did so within about 3 years or less. The median duration of second marriages that ended in divorce was shorter than for first marriages that ended in divorce.

Table IV-9. Median Duration of Marriage Events for People 15 Years and Over, by Sex, Race, and Hispanic Origin, Fall 1996

(Duration in years.)

Subject	Total	White	White non-Hispanic	Black	Asian and Pacific Islander	Hispanic (of any race)
Duration of First Marriage for those whose First Marriage Ended in Divorce						
Men	7.8	7.8	7.8	8.3	7.8	7.6
Women	7.9	7.8	7.8	8.3	8.0	8.0
Duration Between First Marriage and First Separation for those Who Separated						
Men	6.6	6.6	6.6	6.7	6.1	6.2
Women	6.6	6.6	6.6	6.5	7.0	6.4
Duration Between First Separation and First Divorce for those Who Divorced						
Men	0.8	0.8	0.8	1.0	0.8	0.9
Women	0.8	0.8	0.8	1.1	0.8	0.9
Duration Between First Divorce and Remarriage for those whose First Marriages Ended in Divorce and Who had Remarried						
Men	3.3	3.2	3.2	4.1	3.5	3.4
Women	3.1	3.1	3.0	4.1	3.4	3.3
Duration of Second Marriage for those whose Second Marriage Ended in Divorce						
Men	7.3	7.2	7.2	9.8	B	7.5
Women	6.8	6.6	6.6	8.3	B	B

Note: Figures projected onto U.S. population in 1996 based on survey of 37,000 households from the U.S. Census Bureau Survey of Income and Program Participation.

B = Base less than 100,000. Median not calculated.

TABLE IV-10 NUMBER OF WOMEN 15-44 YEARS OF AGE AND PERCENT DISTRIBUTION, BY PAST COHABITATION AND MARITAL STATUS AND BY AGE AT INTERVIEW, RACE, AND HISPANIC ORIGIN, 1995

Table 10 shows the results from a survey of 10,847 women in National Survey of Family Growth projected onto the total U.S. population of women aged 15–44. Cohabitation is defined as living with a partner but not being married. Past marital status is classified as never married versus ever married, with each group being further split into subgroups of never cohabitated versus ever cohabitated in the past. These four groups are mutually exclusive and cover all respondents, summing to 100 percent. Almost 28 percent of women in this age range have never married or cohabitated. This percentage is considerably larger for younger women and decreases as age increases. About 62 percent of women in this age range have ever been married, and roughly half of these have ever cohabitated in the past. The remaining 10 percent have cohabitated but never married. Non-Hispanic White women are more likely to have experienced both cohabitation and marriage, while non-Hispanic Black women are more likely to have experienced neither cohabitation nor marriage.

TABLE IV-11 NUMBER OF WOMEN 15-44 YEARS OF AGE AND PERCENT DISTRIBUTION, BY CURRENT COHABITATION AND MARITAL STATUS AND BY AGE AT INTERVIEW, RACE, AND HISPANIC ORIGIN, 1995

Table 11 shows the results from a survey of 10,847 women in National Survey of Family Growth projected onto the total U.S. population of women aged 15–44. Cohabitation is defined as living with a partner but not being married. Cohabitation is classified as currently cohabiting or not currently cohabiting. The category not currently cohabiting is further split into never married, formerly married or currently married. Currently cohabiting may include persons formerly or never married, but these were not listed separately. These four groups are mutually exclusive and cover all respondents, summing to 100 percent. Roughly 50 percent of women in this age range are currently married, and seven percent are currently cohabiting. About one third are not cohabiting and have never married. The remaining 10 percent are not cohabiting and are formerly married (separated, divorced, or widowed). The percent cohabiting is larger for those in their 20s and then decreases as age increases. The most striking difference by race and ethnicity are the higher percentage not cohabiting and never married, as well as lower percent currently married, among non-Hispanic Black women.

Table IV-10. Number of Women 15–44 Years of Age and Percent Distribution, by Past Cohabitation and Marital Status and by Age at Interview, Race, and Hispanic Origin, 1995

(Number in thousands, percent.)

Age at interview and race and Hispanic orgin	Number [1] (thousands)	Total	Past cohabitation and marital status			
			Never married		Ever married	
			Never cohabited	Ever cohabited	Never cohabited	Ever cohabited
Total	60 201	100.0	27.5	10.2	31.4	30.9
Age at Interview						
15–19 years	8 961	100.0	88.6	7.0	2.6	1.9
20–24 years	9 041	100.0	45.5	20.2	16.2	18.1
25–29 years	9 693	100.0	20.3	15.4	30.4	33.9
30–34 years	11 065	100.0	10.8	9.3	37.8	42.1
35–39 years	11 211	100.0	7.1	6.4	42.9	43.6
40–44 years	10 230	100.0	5.5	4.1	51.5	38.9
Race/Hispanic Origin						
Hispanic	6 702	100.0	28.2	10.4	35.1	26.3
White non-Hispanic	42 522	100.0	24.7	8.9	32.7	33.7
Black non-Hispanic	8 210	100.0	39.7	17.3	20.2	22.9
Other non-Hispanic [2]	2 767	100.0	33.2	8.3	35.1	23.4

[1]Results from survey of 10,847 women in National Survey of Family Growth projected onto total population.
[2]Includes Asian and Pacific Islander women and American Indian women, not shown separately.

Table IV-11. Number of Women 15–44 Years of Age and Percent Distribution, by Current Cohabitation and Marital Status and by Age at Interview, Race, and Hispanic Origin, 1995

(Number in thousands, percent.)

Age at interview and race and Hispanic origin	Number [1] (thousands)	Total	Current cohabitation and marital status			
			Currently cohabiting [2]	Not cohabiting		
				Never married	Formerly married	Currently married
Total	60 201	100.0	7.0	33.4	10.3	49.3
Age at Interview						
15–19 years	8 961	100.0	4.1	91.5	0.6	3.8
20–24 years	9 041	100.0	11.2	56.1	5.5	27.2
25–29 years	9 693	100.0	9.8	28.9	8.8	52.5
30–34 years	11 065	100.0	7.5	16.2	11.6	64.7
35–39 years	11 211	100.0	5.3	11.9	15.0	67.9
40–44 years	10 230	100.0	4.4	8.8	18.1	68.6
Race/Hispanic Origin						
Hispanic	6 702	100.0	8.2	32.8	11.6	47.4
White non-Hispanic	42 522	100.0	7.0	29.4	9.3	54.3
Black non-Hispanic	8 210	100.0	6.9	52.5	15.5	25.2
Other non-Hispanic [3]	2 767	100.0	4.6	39.1	7.6	48.8

[1] Results from survey of 10,847 women in National Survey of Family Growth projected onto total population.
[2] Defined as living with a partner but not being married. May include persons formerly or never married.
[3] Includes Asian and Pacific Islander women and American Indian women, not shown separately.

TABLE IV-12. PROBABILITY OF FIRST MARRIAGE, BY AGE AND SELECTED CHARACTERISTICS, 1995

This table shows the probability that a woman will marry for the first time according to selected characteristics. Eight percent of women married by their 18th birthday, 25 percent by their 20th birthday, and 76 percent by their 30th birthday. Hispanic and non-Hispanic White women were more likely to marry by age 25 compared with Black and Asian women, although by age 30, Asian women have comparable probabilities to Hispanics and non-Hispanic Whites. Black women are significantly less likely to marry by age 30 compared to other groups. Early marriage was more likely in women without high school graduation or with mothers who had not graduated high school. Residents of the Northeast United States showed considerably more delay in marriage by age 25, although they had somewhat caught up by age 30.

Figure IV-6. Median Age for First Marriage, Fall 1996

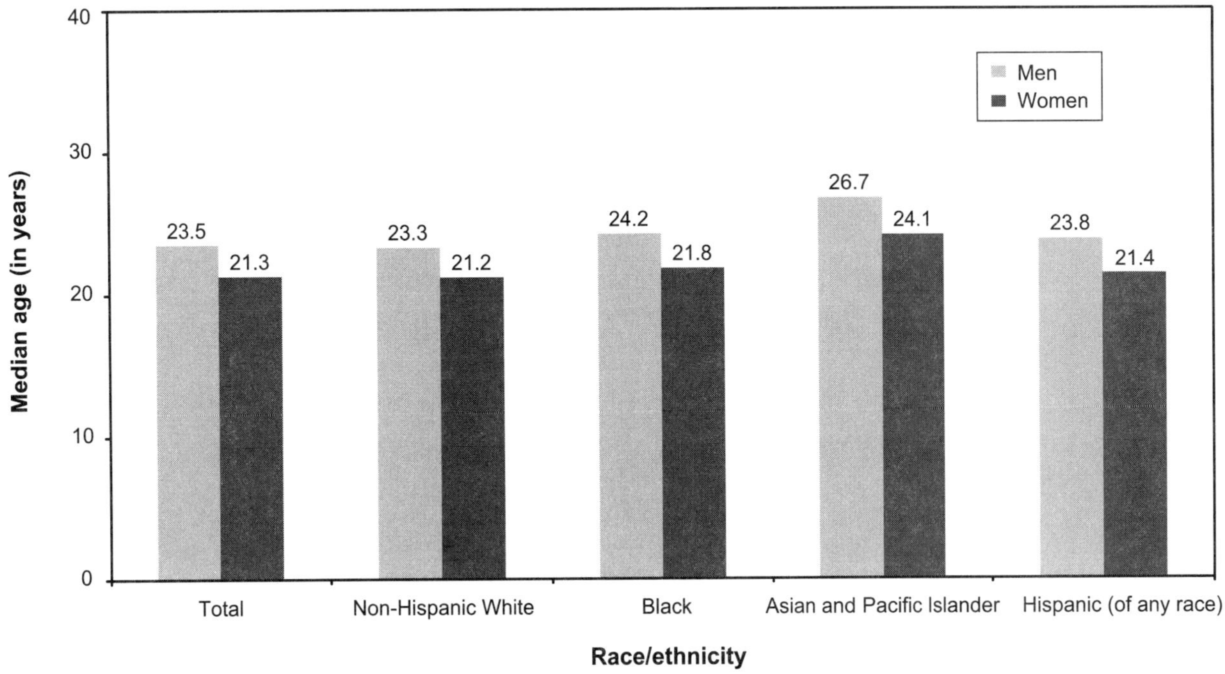

Table IV-12. Probability of First Marriage, by Age and Selected Characteristics, 1995

Characteristic	Probability of marriage by [1]				
	Age 16	Age 18	Age 20	Age 25	Age 30
TOTAL [2]	0.01 (0.001)	0.08 (0.003)	0.25 (0.006)	0.59 (0.007)	0.76 (0.006)
Race/Hispanic Origin					
Hispanic	0.03 (0.005)	0.13 (0.010)	0.29 (0.014)	0.61 (0.015)	0.77 (0.014)
Non-Hispanic White	0.01 (0.002)	0.08 (0.004)	0.26 (0.008)	0.63 (0.008)	0.81 (0.007)
Non-Hispanic Black	0.01 (0.002)	0.05 (0.006)	0.16 (0.010)	0.37 (0.013)	0.52 (0.017)
Non-Hispanic Asian	0.01 (0.006)	0.03 (0.012)	0.13 (0.028)	0.44 (0.042)	0.77 (0.043)
Education [3]					
Less than high school	0.07 (0.008)	0.27 (0.014)	0.45 (0.016)	0.68 (0.018)	0.75 (0.017)
High school	0.01 (0.002)	0.11 (0.006)	0.36 (0.011)	0.69 (0.010)	0.82 (0.009)
More than high school	0.00 (0.001)	0.03 (0.003)	0.13 (0.006)	0.50 (0.009)	0.73 (0.009)
Mother's Education					
Less than high school	0.03 (0.003)	0.16 (0.008)	0.35 (0.010)	0.64 (0.011)	0.77 (0.010)
High school	0.01 (0.002)	0.06 (0.004)	0.24 (0.009)	0.62 (0.010)	0.79 (0.008)
More than high school	0.00 (0.001)	0.03 (0.003)	0.13 (0.007)	0.49 (0.012)	0.72 (0.012)
Family Income					
Less than $25,000	0.02 (0.003)	0.10 (0.006)	0.27 (0.010)	0.53 (0.012)	0.64 (0.014)
$25,000–$49,999	0.01 (0.002)	0.09 (0.005)	0.27 (0.010)	0.63 (0.010)	0.78 (0.009)
$50,000 or more	0.01 (0.001)	0.05 (0.004)	0.20 (0.009)	0.59 (0.012)	0.83 (0.009)
Religious Affiliation					
Any	0.01 (0.001)	0.08 (0.003)	0.25 (0.006)	0.60 (0.007)	0.77 (0.006)
Catholic	0.01 (0.002)	0.06 (0.005)	0.21 (0.009)	0.57 (0.010)	0.78 (0.009)
Fundamentalist	0.03 (0.007)	0.12 (0.013)	0.33 (0.018)	0.69 (0.018)	0.82 (0.018)
Other Protestant	0.01 (0.002)	0.08 (0.005)	0.27 (0.009)	0.60 (0.010)	0.76 (0.009)
Other	0.00 (0.002)	0.06 (0.011)	0.21 (0.019)	0.59 (0.027)	0.76 (0.024)
None	0.02 (0.005)	0.09 (0.010)	0.23 (0.017)	0.53 (0.021)	0.70 (0.018)
Importance of Religion					
Very important	0.02 (0.002)	0.08 (0.004)	0.27 (0.008)	0.62 (0.009)	0.78 (0.008)
Somewhat important	0.01 (0.002)	0.08 (0.005)	0.23 (0.008)	0.58 (0.011)	0.76 (0.010)
Not important	0.00 (0.002)	0.06 (0.008)	0.17 (0.014)	0.46 (0.021)	0.69 (0.020)
Intact Family of Origin Status					
Intact two-parent	0.01 (0.002)	0.07 (0.004)	0.24 (0.007)	0.60 (0.008)	0.79 (0.007)
Otherwise	0.02 (0.002)	0.09 (0.006)	0.25 (0.010)	0.56 (0.012)	0.72 (0.011)
Generalized Anxiety Disorder					
Never	0.01 (0.001)	0.08 (0.003)	0.23 (0.007)	0.58 (0.008)	0.76 (0.007)
Ever	0.01 (0.003)	0.10 (0.007)	0.29 (0.011)	0.63 (0.013)	0.77 (0.012)
Region of Residence					
Northeast	0.00 (0.001)	0.05 (0.004)	0.16 (0.011)	0.50 (0.013)	0.73 (0.012)
Midwest	0.00 (0.001)	0.06 (0.006)	0.24 (0.014)	0.61 (0.016)	0.77 (0.011)
South	0.02 (0.003)	0.12 (0.007)	0.31 (0.011)	0.63 (0.013)	0.78 (0.010)
West	0.01 (0.002)	0.07 (0.005)	0.24 (0.012)	0.59 (0.014)	0.76 (0.015)

Note: Figures projected onto U.S. population in 1995 based on cycle 5 of National Survey of Family Growth, a face-to-face survey of 10,847 women age 15–44 years.

[1]Figures are presented as probability (standard error).
[2]Includes data for American Indian women, not shown separately.
[3]Based on women ages 20 years and over at interview.

Figure IV-7. Percent of Women Currently Married, by Age Group, 1995

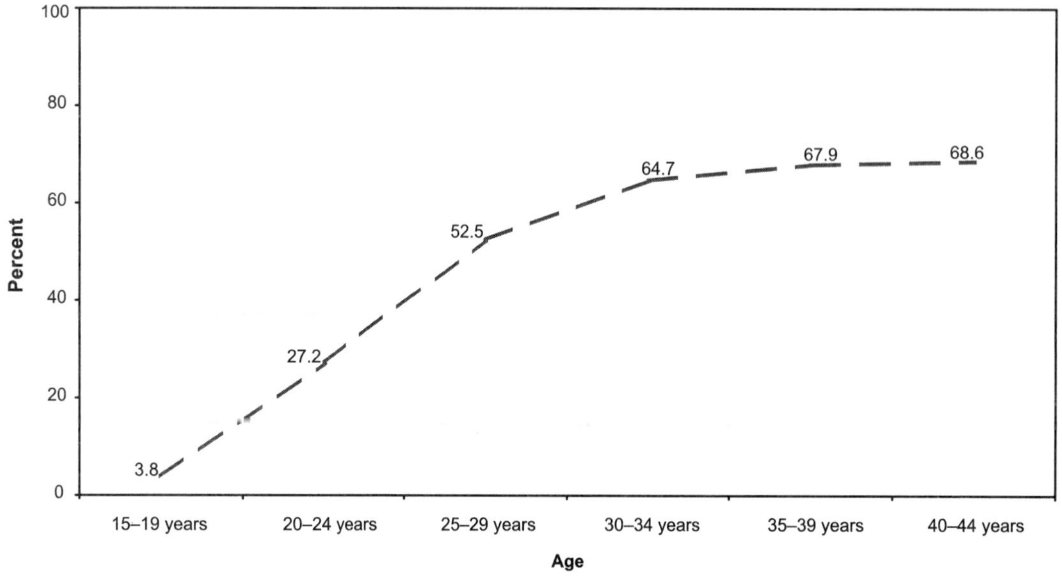

Figure IV-8. Probability of First Marriage by Age 25 Years, by Characteristics of the Community, 1995

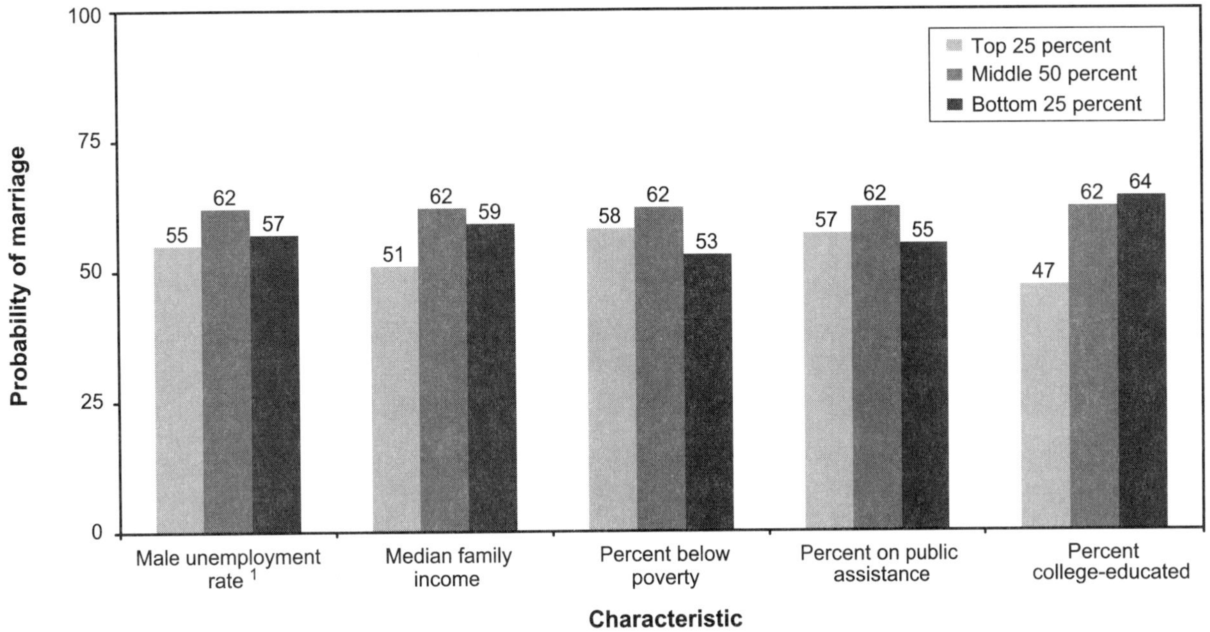

1 Bottom 25 percent denotes least unemployment.

Figure IV-9. Probability of First Marriage by Age 18 Years, by Characteristics of the Community, 1995

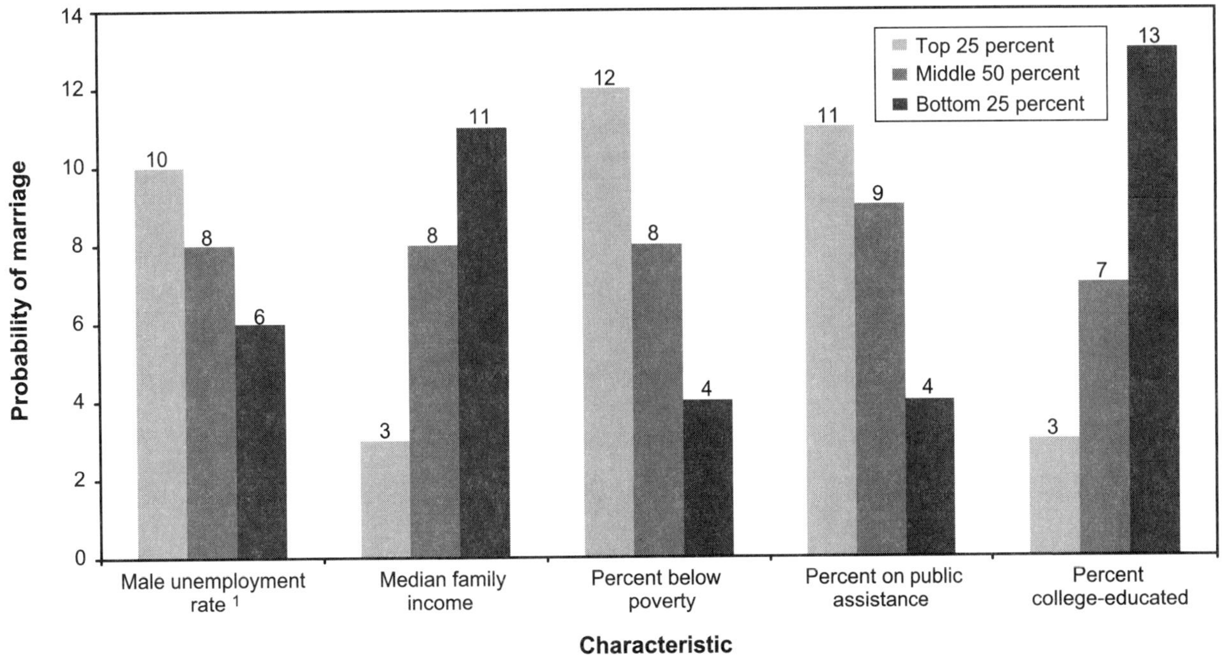

1 Bottom 25 percent denotes least unemployment.

TABLE IV-13. PROBABILITY OF FIRST MARRIAGE, BY AGE AND SELECTED CONTEXTUAL CHARACTERISTICS, 1995

The "contextual characteristics" in this table were derived from respondents' census tracts and county statistics, rather than from asking subjects directly. Areas with relatively high unemployment showed increased early marriage, but a considerably lower probability of marriage by age 30. Living in areas with lower family income was also associated in higher probably of early marriage, but lower probability by age 30. Living outside an urban area (a standard metropolitan statistical area) and in low crime rate areas was associated with higher probabilities of marriage at each age shown.

Table IV-13. Probability of First Marriage, by Age and Selected Contextual Characteristics, 1995

Characteristic	Probability of marriage by [1]				
	Age 16	Age 18	Age 20	Age 25	Age 30
Male Unemployment Rate [2]					
Bottom 25 percent	0.01 (0.002)	0.06 (0.005)	0.20 (0.009)	0.57 (0.013)	0.78 (0.011)
Middle 50 percent	0.02 (0.002)	0.08 (0.005)	0.26 (0.009)	0.62 (0.009)	0.79 (0.008)
Top 25 percent	0.02 (0.003)	0.10 (0.007)	0.28 (0.011)	0.55 (0.013)	0.69 (0.013)
Median Family Income					
Bottom 25 percent	0.02 (0.003)	0.11 (0.006)	0.30 (0.010)	0.59 (0.012)	0.73 (0.011)
Middle 50 percent	0.01 (0.002)	0.08 (0.004)	0.25 (0.009)	0.62 (0.009)	0.79 (0.008)
Top 25 percent	0.00 (0.001)	0.03 (0.004)	0.14 (0.010)	0.51 (0.016)	0.75 (0.013)
Percent Below Poverty [3]					
Bottom 25 percent	0.00 (0.001)	0.04 (0.004)	0.15 (0.009)	0.53 (0.012)	0.77 (0.011)
Middle 50 percent	0.01 (0.002)	0.08 (0.004)	0.26 (0.009)	0.62 (0.010)	0.80 (0.008)
Top 25 percent	0.02 (0.003)	0.12 (0.007)	0.31 (0.010)	0.58 (0.012)	0.70 (0.012)
Percent on Public Assistance					
Bottom 25 percent	0.00 (0.001)	0.04 (0.004)	0.17 (0.009)	0.55 (0.013)	0.78 (0.011)
Middle 50 percent	0.02 (0.002)	0.09 (0.004)	0.26 (0.009)	0.62 (0.011)	0.80 (0.009)
Top 25 percent	0.02 (0.003)	0.11 (0.006)	0.29 (0.011)	0.57 (0.013)	0.68 (0.013)
Percent College-Educated					
Bottom 25 percent	0.03 (0.003)	0.13 (0.007)	0.34 (0.011)	0.64 (0.012)	0.75 (0.011)
Middle 50 percent	0.01 (0.002)	0.07 (0.004)	0.24 (0.008)	0.62 (0.009)	0.80 (0.008)
Top 25 percent	0.00 (0.001)	0.03 (0.004)	0.13 (0.009)	0.47 (0.015)	0.71 (0.015)
Crime Rate					
Bottom 25 percent	0.01 (0.003)	0.10 (0.008)	0.28 (0.013)	0.66 (0.015)	0.83 (0.012)
Middle 50 percent	0.01 (0.002)	0.07 (0.004)	0.24 (0.009)	0.58 (0.010)	0.75 (0.009)
Top 25 percent	0.01 (0.003)	0.07 (0.006)	0.21 (0.011)	0.53 (0.014)	0.72 (0.012)
Percent of Women Never Married					
Bottom 25 percent	0.02 (0.003)	0.12 (0.007)	0.34 (0.012)	0.72 (0.013)	0.87 (0.011)
Middle 50 percent	0.01 (0.002)	0.07 (0.004)	0.23 (0.008)	0.60 (0.010)	0.78 (0.009)
Top 25 percent	0.01 (0.002)	0.05 (0.004)	0.17 (0.009)	0.44 (0.013)	0.62 (0.011)
Metropolitan Status					
SMSA [4], central city	0.01 (0.002)	0.06 (0.005)	0.21 (0.010)	0.49 (0.011)	0.66 (0.012)
SMSA [4] other	0.01 (0.002)	0.07 (0.004)	0.23 (0.008)	0.61 (0.011)	0.79 (0.008)
Not SMSA [4]	0.02 (0.004)	0.12 (0.008)	0.34 (0.016)	0.70 (0.017)	0.84 (0.013)

Note: Figures projected onto U.S. population in 1995 based on cycle 5 of National Survey of Family Growth, a face-to-face survey of 10,847 women age 15–44 years.

[1]Figures are presented as probability (standard error).
[2]Bottom 25 percent denotes least unemployment.
[3]Bottom 25 percent denote least level of poverty.
[4]SMSA is standard metropolitan statistical area.

NOTES AND DEFINITIONS

INTRODUCTION:

This section provides the sources for the data in this book. It also mentions some sources of related interest. The section is followed by a brief glossary.

Table A-1 on worldwide population and life expectancy data was obtained from the U.S. Census Bureau's *Global Population Profile: 2002*. The underlying data are shown in Tables 5 and 12 of Appendix A. The URL is: <http://www.census.gov/ipc/prod/wp02/tabA-05.pdf> or for Table 12 use: <tabA-12.pdf>.

CHAPTER I: BIRTHS

The statistical tables of Chapter I – Births – were obtained from *Births: Final Data for 2002*, by Martin JA, Hamilton BE, Sulton PD, Ventura SJ, Menacker F, and Munson ML; National Vital Statistics Reports; vol 52 no. 10. Hyattsville, Maryland: National Center for Health Statistics. December 2003.

Chapter I contains 50 regular tables and seven tables with a letter attached to the number, which were text tables in the original document. More detailed data are available at the National Center for Health Statistics (NCHS) home page at <http://www.cdc.gov/nchs/datawh/statab/unpubd/natality/natab99.htm>.

The data shown in the tables are based on 100 percent of the birth certificates registered in the states and the District of Columbia. Some additional data are shown for U.S. territories.

Tabulations of births beginning with 1980 data are by race of mother; for earlier years they are by race of child. Race and Hispanic origin are reported independently on the birth certificate. While Hispanics can belong to any race, the majority of women of Hispanic origin are reported as White and the bulk of the remainder are reported as Black. The tables of Chapter I usually show data for these categories: White total, non-Hispanic White, Black total, non-Hispanic Black, and Hispanic. There are also data for American Indians and Asian or Pacific Islanders (API). Some data are presented for Hispanic subgroups and for API subgroups by country of origin.

To calculate birth and fertility rates, the underlying populations had to be calculated. For 2002 these were estimated by using the 2000 Census, and adjusting it to a July 1, 2002 level. This census included an option for persons to report more than one race as appropriate for themselves. Since birth certificates currently collect only one race for the mother, the census data had to be "bridged" to return the data to estimates of single race categories. These procedures are discussed in two publications: Ingram DD et al. "U.S. Census 2000 population with bridged race categories." National Center for Health Statistics. *Vital Health Stat* 2(135), 2003; and Schenker N, Parker JD. "From single race reporting to multiple race reporting: Using imputation methods to bridge the transition." *Statistics in Medicine* 22:1571-87. 2003.

Birth certificates provided nearly complete coverage of most of the requested characteristics. For characteristics with missing entries, the reports that had blanks were subtracted from the total, and the percentages in the tables reflected the distributions of those who were reporting. These "not stated" data generally were quite small; but they exceeded 5 percent for weight gain during pregnancy and for the father's age, race, and Hispanic origin. Table 1 in the technical notes to *Births: Final Data for 2002*, by the National Center for Health Statistics, gives detailed data on the extent of missing information.

CHAPTER II: MORTALITY

Two publications were used for the mortality data:

- *Deaths: Final Data for 2001*, by Arias E, Anderson RN, Hsiang-Ching K, Murphy SL, Kochanek KD. National Vital Statistics Reports; vol 52 no. 3. National Center for Health Statistics, Hyattsville, Maryland. September 2003.

- *Deaths: Preliminary Data for 2002*, by Kochanek KD and Smith BL. National Vital Statistics Reports; vol 52 no. 13. February 2004.

Chapter II contains 35 basic tables for final data for 2001; five supplementary tables from the text of the original document for 2001; and nine tables from the preliminary data for 2002.

The data are based on information from all resident death certificates filed in the 50 states and the District of Columbia. It is believed that more than 99 percent of deaths occurring in this country are registered. Tables that show data by states also include information for Puerto Rico and U.S. territories. Cause-of-death statistics are classified in accordance with the *International Classification of Diseases, Tenth Revision (ICS-10)*.

To obtain death rates, population data were used in the denominator. In contrast to the actual counts of deaths, these were estimates based on the 2000 Census, adjusted to July 1 of the year in question. As discussed in the section for Births, the census data also had to be adjusted to "bridge" entries for multiple-race persons back to the earlier single race categories. Data more detailed than those shown here are available at <www.cdc.gov/nchs/about/major/dvs/mortdata.htm> and on the forthcoming CD-ROM titled, *Vital Statistics of the United States, Mortality 2001*.

All of the data exclude fetal deaths. Many death certificates were multiple-cause coded, but the statistics derived by NCHS are based solely on the underlying cause of death. This is defined as the disease or injury which initiated the train of events leading directly to death. Computer software is used to input the multiple cause codes, and to employ World Health Organization rules to select the underlying cause.

Beginning with data for 2001, the National Center for Health Statistics introduced Codes *U01-*U03 for classifying deaths due to acts of terrorism.

Persons that are classified as members of a minority group in the census are sometimes misclassified as White in death certificates. Persons may also be erroneously reported as White in the census. Estimates of approximate percentages of appropriate corrections in death rates are as follows: White, -1.0 percent; Black, -5.0 percent; American Indian, +20.6 percent; Asian/Pacific Islander, +10.7 percent; Hispanics, +1.6 percent.

Cause of death as shown in the principal tables here can also be classified in alternative frameworks. For instance, injury mortality data are classified by intent, such as, suicide and homicide, in Tables 9 to 17, but they can also be organized by mechanism, such as, firearm or poisoning, as shown in Tables 18 to 24. In addition, the CDC occasionally makes an estimate of the number of deaths from behavioral risk factors or causal agents that cause fatal diseases, such as tobacco, poor diet and physical inactivity, pollutants and asbestos, etc. (See <www.cdc.gov/nccdphp/factsheets/death_causes2000.htm>.)

The Current Population Survey of the U.S. Census Bureau is the source of population estimates by marital status, educational attainment, and regarding the origin by country of Hispanics.

The preliminary 2002 data cover an estimated 96 percent of infant deaths, about 97 percent of deaths to persons aged 1 year and over. Cause of death information was available for an estimated 92 percent to 93 percent of deaths. There is a tendency for preliminary data to slightly underestimate rates for certain causes of death where the cause is pending investigation, such as for homicide, suicide, accidents, and sudden infant death syndrome (SIDS). Rates due to conditions that are highly seasonal, such as influenza and pneumonia or chronic lower respiratory diseases, may also be misestimated.

In general, death rates are on an annual basis per 1,000 or per 100,000 of the estimated population residing in the specified area. Infant mortality rates are per 1,000 or per 100,000 live births.

CHAPTER III: HEALTH

The Health chapter consists of 62 tables. They were obtained from several different publications of the CDC's National Center for Health Statistics (NCHS) <http://www.cdc.gov/nchs>.

The sources are:

A. Freid VM, Pragar K, MacKay AP, Xia H. *Chartbook on Trends in the Health of Americans. Health, United States, 2003.* Hyattsville, MD: National Center for Health Statistics, 2003. <http://www.cdc.gov/nchs/data/hus/hus03.pdf> (as of March 2004). For an interactive lookup of just the trend tables, go to <http://www.cdc.gov/nchs/products/pubs/pubd/hus/trendtables.htm> (as of March 2004).

B. Ni H, Coriaty-Nelson Z, Schiller J, Hao C, Cohen RA, Barnes P. Early release of selected estimates based on data from the January-June 2003 *National Health Interview Survey*. Hyattsville, MD: National Center for Health Statistics, December 2003. <http://www.cdc.gov/nchs/data/nhis/earlyrelease/earlyrelease200312.pdf> (as of March 2004).

C. *Summary Health Statistics for U.S. Adults: National Health Interview Survey, 2001.* Series 10, Number 218. DHHS Publication Number (PHS) 2004-1546. Hyattsville, MD: National Center for Health Statistics, January 2004. <http://www.cdc.gov/nchs/data/series/sr_10/sr10_218.pdf> (as of March 2004).

D. Freid VM, Pragar K, MacKay AP, Xia H. *Chartbook on Trends in the Health of Americans. Health, United States, 2003.* Hyattsville, MD: National Center for Health Statistics, 2003. <http://www.cdc.gov/nchs/data/hus/hus03.pdf> (as of March 2004). For an interactive lookup of just the charts, go to <http://www.cdc.gov/nchs/data/hus/hus03cht.pdf> (as of March 2004).

E. Molla MT, Madans JH, Wagener DK, Crimmins EM. *Summary Measures Of Population Health: Report of Findings on Methodologic and Data Issues.* DHHS Publication Number (PHS) 2004-1258. Hyattsville, MD: National Center for Health Statistics, 2003. <http://www.cdc.gov/nchs/data/misc/pophealth.pdf> (as of March 2004).

F. *Current Population Survey* of the U.S. Census Bureau – this source is used only for Table 1A.

The 62 tables of the Health chapter are listed below, according to the code shown above, and the specific sources of each table are discussed.

MEASURES OF HEALTH

Table 1. (Source A) Trend Table 51. Data are based on interviews and examinations of samples of the civilian noninstitutionalized population or special population subgroups, information from death certificates and birth certificates; and air quality measurements.

Table 1-A. (Source F) Table H101, Health Insurance Coverage Status, based on the Annual Demographic Survey, March Supplement, of the Current Population Survey. <http://ferret.bls.census.gov/macro/032002/health/h01_001.htm>.

Table 2. (Source B) Section 11. Interview respondents are asked to assess their own health and that of family members living in the same household.

Table 3. (Source E) Table 3.1. Based on data from NCHS, especially health surveys, hospital discharges, and nursing home statistics; detailed mortality data; construction of life expectancy tables; and population data.

Table 4. (Source C) Table 21. The Health Interview Survey is conducted annually by face-to-face interviews with an adult who is asked to report information (other than health) for other family members. Interviewers are from the U.S. Census Bureau working on behalf of NCHS. The multistage probability sample is representative of the civilian noninstitutionalized U.S. population. The sample for 2001 consisted of 38,932 households. Some frequencies are underestimated, because nonresponse and unknowns are included in the totals but excluded as an answer. [For further information, see Appendix I of (Source C).]

Table 5. (Source B) Section 7. Sample of the Adult Core component of health questionnaire of the 1997–2003 National Health Interview Surveys.

Table 6. (Source D) Fig. 13. Data are derived from the Youth Risk Behavior Survey, a survey conducted by State Education Departments and sponsored by the CDC.

Table 7. (Source A) Table 68. Data are based on measured height and weight (BMI) of a sample of the civilian noninstitutionalized population.

Table 8. (Source A) Table 69. Data are based on physical examinations of appropriate sample of the civilian noninstitutionalized population.

Table 9. (Source E) Table 3.7. See remarks for Table 3.

SELECTED CHRONIC DISEASES

Table 10. (Source C) Table 2. See remarks for Table 4.

Table 11. (Source A) Table 66. Data are based on physical examinations of a sample of the civilian noninstitutionalized population.

Table 12. (Source A) Table 67. Ibid.

Table 13. (Source E) Table 3.5. See remarks for Table 3.

Table 14. (Source C) Table 6. See remarks for Table 4.

Table 15. (Source A) Table 54. Data are based on the Surveillance, Epidemiology, and End Results (SEER) Program, and its 12 population-based cancer registries.

Table 16. (Source A) Table 55. Same as Table 15, but based on only nine cancer registries.

Table 17. (Source A) Table 53. Data are based on reporting by state health departments.

Table 18. (Source C) Table 4. See remarks for Table 4 above.

Table 19. (Source C) Table 8. Ibid.

Table 20. (Source C) Table 10. Ibid.

Table 21. (Source C) Table 12. Ibid.

Table 22. (Source C) Table 14. Ibid.

DISABILITIES

Table 23. (Source A) Table 56. Data are based on household interviews of a sample of the civilian noninstitutionalized population.

Table 24. (Source E) Table 3.2. See remarks for Table 3 above.

Table 25. (Source B) Section 12. Family Core component of the 1997–2003 National Health Interview Surveys.

Table 26. (Source E) Table 3.4. See remarks for Table 3 above.

Table 27. (Source D) Fig. 19. CDC, NCHS, and particularly the National Health Interview Surveys.

Table 28. (Source A) Table 50. Data are based on employer records from a sample of business establishments.

USE OF ADDICTIVE SUBSTANCES

Table 29. (Source A) Table 62. Data are based on household interviews of a sample of the civilian noninstitutionalized population, 12 years of age and over.

Table 30. (Source A) Table 63. Data are based on a survey of high school and middle school students in the coterminous United States.

Table 31. (Source A) Table 59. Data are based on household interviews of a sample of the civilian noninstitutionalized population.

Table 32. (Source A) Table 60. Ibid.

Table 33. (Source A) Table 61. Ibid.

Table 34. (Source A) Table 65. Ibid.

Table 35. (Source A) Table 64. Data are weighted national estimates based on a sample of emergency departments.

AMBULATORY CARE

Table 36. (Source A) Table 70. Data are based on household interviews of a sample of the civilian noninstitutionalized population.

Table 37. (Source A) Table 71. Data are based on telephone interviews of a sample of the civilian noninstitutionalized population, supplemented by a survey of immunization providers for interview participants.

Table 38. (Source A) Table 72. Ibid.

Table 39. (Source A) Table 73. Data are based on household interviews of a sample of the civilian noninstitutionalized population.

Table 40. (Source A) Table 74. Ibid.

Table 41. (Source A) Table 75. Ibid.

Table 42. (Source A) Table 76. Ibid.

Table 43. (Source A) Table 77. Ibid.

Table 44. (Source A) Table 78. Ibid.

Table 45. (Source A) Table 79. Data are based on dental examinations of a sample of the civilian noninstitutionalized population.

Table 46. (Source A) Table 80. Data are based on household interviews of a sample of the civilian noninstitutionalized population.

Table 47. (Source A) Table 81. Ibid.

Table 48. (Source A) Table 82. Data are based on reporting by a sample of office-based physicians and hospital outpatient and emergency departments.

Table 49. (Source A) Table 83. Data are based on reporting by a sample of hospital emergency departments.

Table 50. (Source A) Table 84. Data are based on reporting by a sample of office-based physicians.

Table 51. (Source A) Table 85. Data are based on a one-day census of treatment providers.

Table 52. (Source A) Table 86. Data are based on inventories of mental health organizations.

Table 53. (Source A) Table 87. Data are based on a survey of current home health care patients.

Table 54. (Source A) Table 88. Data are based on a survey of current hospice patients.

INPATIENT CARE

Table 55. (Source A) Table 89. Data are based on household interviews of a sample of the civilian noninstitutionalized population.

Table 56. (Source A) Table 91. Data are based on a sample of hospital records.

Table 57. (Source A) Table 92. Ibid.

Table 58. (Source A) Table 93. Ibid.

Table 59. (Source A) Table 94. Data are based on a sample of inpatient and ambulatory surgery records.

Table 60. (Source A) Table 95. Data are based on reporting by a census of hospitals.

Table 61. (Source A) Table 96. Data are based on a sample of nursing home residents.

Table 62. (Source A) Table 97. Ibid.

CHAPTER IV: MARRIAGE AND DIVORCE

Information on the total numbers and rates of marriages and divorces at the national and state levels are published in the NCHS National Vital Statistics Reports. The collection of detailed data was suspended beginning in January 1996, due to limitations in the information collected by the states as well as budgetary considerations.

The most recent comprehensive analyses of detailed marriage and divorce data were published in *Advance Report of Final Marriage Statistics, 1989–90* (<http://www.cdc.gov/nchs/products/pubs/pubd/mvsr/supp/44-43/44-43.htm#43_12s>) and *Advance Report of Final Divorce Statistics, 1989–90* (<http://www.cdc.gov/nchs/products/pubs/pubd/mvsr/supp/44-43/44-43.htm#43_9s>).

It should be noted that marriages are not reported adequately from Oklahoma (treated as a nonreporting state), while divorces are not reported from California, Hawaii, Indiana, Louisiana, and Oklahoma. Events to nonresidents are included in marriage and divorce figures (in contrast to birth and death data where nonresidents are excluded). Divorce figures include reported annulments. NCHS ceased publishing divorce counts and rates based on provisional data from the combined 50 states because it is no longer statistically feasible to calculate estimates of monthly divorce counts for nonreporting states. Instead, NCHS publishes only the 12-month divorce rates shown in Table 1, based solely on combined counts and populations for reporting states and the District of Columbia. Aggregate 1-month, cumulative counts and rates, and 12-month counts for divorces have been dropped because of their limited usefulness when based on less than the full complement of states.

Tables 1 through 3 are adapted from:

Sutton PD, Munson ML. *Births, Marriages, Divorces, and Deaths: Provisional Data for September 2003. National Vital Statistics Reports*; volume 52, no. 16. Hyattsville, MD: National Center for Health Statistics, 2004. <http://www.cdc.gov/nchs/data/nvsr/nvsr52/nvsr52_16.pdf>.

Tables 4, and 6 through 9, as well as Figures 1 through 6, derive from:

Kreider RM and Fields JM. *Number, Timing, and Duration of Marriages and Divorces: Fall 1996. Current Population Report,* P70-80. Washington, DC: U.S. Census Bureau, 2001. <http://www.census.gov/prod/2002pubs/p70-80.pdf>.

Table 5 is posted on the U.S. Census Bureau Web site and represents a more detailed presentation of the data from Kreider and Fields (2001):

Census Bureau Internet Table: Marital History for People 15 Years Old and Over by Age, Sex, Race, and Ethnicity: Fall 1996. Release date February 8, 2002. U.S. Census Bureau Survey of Income and Program Participation (SIPP), 1996 Panel Wave 2 Topical Module. <http://www.census.gov/population/socdemo/marital-hist/p70-80/tab01.pdf>.

Tables 10 through 13, as well as Figures 7 through 9, derive from:

Bramlett MD and Mosher WD. *Cohabitation, marriage, divorce, and remarriage in the United States. National Center for Health Statistics*. (PHS) 98-1998. *Vital Health Stat* 23(22), 2002. <http://www.cdc.gov/nchs/data/series/sr_23/sr23_022.pdf>.

GLOSSARY

ADL — see limitations of activity

Age-adjusted death rate — this concept will be illustrated by an example: Suppose a standard population has been selected in a country (in the United States the population of the year 2000 was selected). The standard population consists of 10,000 younger persons and 5,000 older persons. Thus the younger persons have a weight of 2/3 and the older persons have a weight of 1/3. There is a specific population with a larger admixture of elderly people. It consists of 1,000 younger persons, with one death during the year, and 1,000 older persons, with two deaths. The death rate for the specific population is 3 divided by 2,000, or 1.5 per 1,000.

For age-adjustment, the question is what would have been the death rate in the specific population, if the effects of the age distribution were standardized. To get the answer, the weights of the age distribution of the standard population are applied. The one per 1,000 death gets a weight of 2/3 and the two per 1,000 deaths get a weight of 1/3. When combining these two terms, we obtain 2/3 plus 2/3, or 4/3, which is 1.33. Thus the age-specific death rate of 1.5 has been adjusted to 1.33.

Birth rate — the birth rate, generally stated per 1,000, relates the number of live births to the population in a specified group; the population could be females in a certain age group or it could be the total population of the United States. The **fertility rate** shows live births per 1,000 for a specified group of women of childbearing age, defined as women aged 15–44 years. While **birth rates and fertility rates** usually describe events during a single year, **total fertility rates** show the number of children that the representative woman would have over her lifetime if the current fertility rates in each age group were to remain constant.

Birthweight — low birthweight is defined as less than 2,500 grams (5 lb 8 oz) and **very low birthweight** is considered less than 1,500 grams (3 lb 4 oz).

BMI — body mass index is a measure that adjusts body weight for height. It is calculated as weight in kilograms divided by height in meters squared. **Healthy weight** for adults is defined as a BMI of 18.5 to less than 25; **overweight** is greater than or equal to a BMI of 25; **obesity** is considered greater than or equal to a BMI of 30. These guidelines were developed by the Department of Agriculture.

Cause of death — in these data every death is attributed to one underlying condition, based on information from the death certificate and World Health Organization. The underlying cause is the disease or injury that initiated the train of events leading directly to death. Generally more medical information is reported on death certificates than the underlying cause of death, thus requiring analytic approaches.

Fertility rate — see birth rate

Hispanic origin is an ethnic classification. Hispanics can be of any race, but most of them classify themselves as White. Hispanic Blacks rank second.

Hypertension — a person is considered to have hypertension if the systolic blood pressure is 140 mm Hg or more or if the diastolic pressure measures 90 or more, or if the person is taking antihypertensive medication.

IADL — see limitations of activity

Infant mortality — rates of infant deaths, per 1,000, under one year. **Neonatal mortality rate** applies under 28 days, and the **postneonatal rate** measures deaths from 28 days of life to one year.

Life expectancy — the expected life span at birth, or the remaining life span as of a later age, if current mortality rates in each age bracket were to remain unchanged.

Limitations of activity — a reduction in performance of activities usually associated with a person's age group, due to physical or mental chronic conditions. Stringent limitations are presented as **ADL**, activities of daily living, such as help needed for bathing, dressing, or eating; and also as **IADL**, instrumental activities of daily living. The latter includes help needed for preparing meals, shopping, and using a telephone.

Period of gestation — a period of gestation under 37 weeks is classified as **preterm**. If less than 32 weeks of gestation are completed, the term **very preterm** is applied. A period of gestation of 37 to 41 weeks is labeled **term**, and if the pregnancy lasts 42 weeks and over, it is called **postterm**.

Short-stay hospital — these are hospitals that provide general, rather than specialized care and have an average length of stay of less than 30 days.

INDEX

Tables are denoted by boldface page references and figures are denoted by italicized page references.

D

I